ESSENTIALS OF NEUROIMAGING

SECOND EDITION

ESSENTIALS OF NEUROIMAGING

SECOND EDITION

J. ROBERT KIRKWOOD, M.D.

Associate Professor
Department of Radiology
Tufts University School of Medicine
Boston, Massachusetts
Chief of Neuroradiology
Department of Radiology
Baystate Medical Center
Springfield, Massachusetts
President
Radiology and Imaging, Inc.
Springfield, Massachusetts

CHURCHILL
LIVINGSTONE

New York, Edinburgh, London, Madrid, Melbourne, San Francisco, Tokyo

Library of Congress Cataloging-in-Publication Data

Distributed in the United Kingdom by Churchill Livingstone, Robert Stevenson House, 1–3 Baxter's Place, Leith Walk, Edinburgh EH1 3AF, and by associated companies, branches, and representatives throughout the world.

Accurate indications, adverse reactions, and dosage schedules for drugs are provided in this book, but it is possible that they may change. The reader is urged to review the package information data of the manufacturers of the medications mentioned.

The Publishers have made every effort to trace the copyright holders for borrowed material. If they have inadvertently overlooked any, they will be pleased to make the necessary arrangements at the first opportunity.

Acquisitions Editor: *Michael J. Houston*
Production Editor: *Dorothy J. Birch*
Production Supervisor: *Sharon Tuder*
Cover Design: *Jeannette Jacobs*

Printed in the United States of America

First published in 1995   7   6   5   4   3   2   1

Preface to the Second Edition

Essentials in Neuroimaging provides readers with enough information on neuroanatomy, pertinent neurophysiology, and neuroimaging for them to gain a clear understanding of neuroimaging and the knowledge needed to accurately diagnose most neuroradiologic problems. It is most useful to residents in radiology or neurology who are preparing for board examinations, to fellows in neuroradiology, to general radiologists in practice who must deal with neuroradiological problems, to radiologists who need an efficient modern review text as they prepare for the Certificate of Added Qualification (CAQ) examination in neuroradiology, and to neurologists in practice. Many study boxes outline complex materials and help with overall review of the material. The diagnosis or title of figures is in bold face at the beginning of each caption to make identification easy. The book presents more than just the basics, but is less than an exhaustive reference text. Uncommon entities appear, but with their essential elements described briefly. More important common entities command the most emphasis.

This second edition builds upon the first. It is larger than the first because of additional sections on brain and spine anatomy and a new chapter covering the cranial nerves, the orbit, and the temporal bone. It is also heavily illustrated. I have included many new anatomic drawings designed especially for a radiologist. These drawings clarify complex anatomy that is important for image study design and interpretation. Additional new radiologic images illustrate both anatomy and pathology. The extensively revised chapter on congenital disease includes a modern classification for developmental and genetic abnormalities. I have retained the angiography section in Chapter 1 because it provides the concise anatomy of cerebral vessels and a discussion of angiographic principles. This knowledge remains essential for performing and interpreting cerebral angiography and understanding magnetic resonance angiography of the brain and cervical vessels. I hope you are satisfied with this edition.

J. Robert Kirkwood, M.D.

Preface to the First Edition

In the process of teaching neuroimaging to radiology residents, I find it difficult to direct them to one manageable source that can serve as a basic text. To be sure, there are excellent texts on CT scanning, MRI, and angiography; but to study the entire topic from these sources requires considerable expense and effort. The information is frequently too exhaustive and the multiplicity of texts often results in considerable repetition.

This book is intended to provide a good understanding of neuroimaging from a single readable source. It is more than an introduction, as it provides in-depth discussions of the basic principles of neurodiagnosis and gives the information necessary to diagnose accurately almost all processes that involve the nervous system and the spine without being overly lengthy. The book is primarily designed for use by residents in radiology as both a learning and board review text. Fellows in neuroradiology, radiologists in general practice, neurologists, and neurosurgeons may also find the book helpful.

The book is heavily illustrated. The cases are carefully selected to demonstrate specific points necessary for diagnosis and clinical management of the patient. MRI, CT scanning, and angiography are given the most emphasis. Commonly, an entity is compared as it appears on multiple modalities, mostly using images from the same patient.

The most common diseases are given a larger number of illustrations so that the full spectrum of the process can be seen. A plain-film example is used only when it is essential to the disease process. The images are labeled to make important points, and the captions are relatively detailed. However, the diagnosis of each illustration is given in bold print at the beginning of the caption to make identification easy.

The text is weighted according to the importance of the disease process; thus, stroke, glioma, and trauma receive much more detailed discussion than more obscure entities. Many tables and boxes are used to set off important lists of differential diagnoses or components of a disease process. In the text specific diseases are indicated in bold type. These features will allow one to thumb the pages for board review.

A single-author book has the advantage of consistency of style. However, it also presents certain potential hazards. To lessen these pitfalls, I have had the entire text reviewed for accuracy and completeness by other neuroradiologists and neurologists, and for clarity and length by general radiologists and radiology residents. I hope that the reader will find the result satisfactory.

J. Robert Kirkwood, M.D.

Acknowledgments

Again, many people directly or indirectly helped me with this book. My wife, Gale, and our children tolerated many hours away from the family, sequestered in the den ("the mole hole"), often tying up the phone with Medline searching. My sons, Rusty and Tim, helped me establish the computer necessary to create the anatomic drawings. Dr. Eckart Sachsse, Chairman, Department of Radiology, Baystate Medical Center, and Radiology and Imaging, Inc., provided me with encouragement, financial support, and academic time to help complete the task. Kent Spiry and Todd Lajoie of the Media Services Department, Baystate Medical Center, did the tedious precise work of producing the radiographic images added to this edition. Linda Fratini, department secretary, helped immensely with copying, organizing, and mailing. Ann Ruzycka, from Churchill Livingstone, helped with decisions and kept me on sched-ule. Picker International and 3M Corporations gave me technical support and, with their equipment, created the newer high quality images used for the figures.

The residents in radiology, particularly Tom Vaughan, Brian Murphy, Ram Chavali, Greg Blackman, Jeff Bertaldo, Wendy Hanaffee, Chris Petrecca, Steve Lee, Dave Fontaine, Vivian Miller, Alex Boutselis, Alex Raslavicus, Kevin Kon, and Steve Krosnick gave me con-stant encouragement, input about topics for inclusion, and critical review, especially for clarity of writing. Dr. Rick Hicks, my close associate in neuroradiology, not only picked up extra clinical work, but freely contributed facts, cases, and advice. The team of technologists at the Radiology Department helped with their consistent qual-ity of imaging and their efforts in sending images to my computer workstation to be included in the book.

Contents

1
Anatomy and Techniques

THE SKULL

The skull rests on the superior segment of the vertebral column. It consists of the cranium and the facial bones. The cranium is made up of eight generally flat bones with concave inner margins attached by cranial sutures to form a vault (Fig. 1-1). The brain is contained within the vault, surrounded by CSF and the meninges.

The intracranial space is divided by the tentorium, a horizontal fold of dura, into the anterior and posterior compartments (fossae), which communicate through a central hole called the tentorial hiatus, or *incisura*. The anterior supratentorial cranial fossa is divided into two lateral halves by the vertical midline dural falx. The outflow tracts of the brain exit through various foramina throughout the base of the skull.

Tissue covers the cranium and comprises the skin, subcutaneous tissue, the galea aponeurotica and the occipitofrontalis muscle, a subgaleal fascial cleft, and the

---

periosteum of the skull, called the pericranium. Most of the arteries and veins of the scalp course within the subcutaneous tissues. A deep middle temporal artery ascends over the lateral temporal squamosa deep to the galea, often creating a vertical linear groove in the cortex simulating a fracture.

Membranous bones form most of the skull (Fig. 1-2). They consist of an outer table of thick cortical bone, a diploic space of cancellous bone containing marrow and a rich network of diploic veins, and an inner table of cortical bone. The anterior and inferior portion of the occipital bone, the basiocciput, is formed from cartilage. The posterior superior occipital plana develops from membrane and may remain separated from the remainder of the occipital bone by the interparietal sutures.

Sutures join the cranial bones along interdigitated margins. The bones grow at the edges along suture margins. The major growth of the skull occurs during the first 6 years of life, with another small growth during puberty, mostly from development of the cranial sinuses. The sutures close normally between 20 and 26 years of age, but there is considerable individual variation. Premature closure of the cranial sutures during infancy, called craniosynostosis (craniostenosis), results in distortion or restriction of skull growth (see Ch. 12). In the immature skull, nonossified membranous segments, called fontanelles, cover the angles of the parietal bone, where it meets adjacent bones (Fig. 1-3). The largest fontanelles occur at the *bregma* (anterior fontanelle at the junction of the sagittal and coronal suture), and the *lambda* (posterior fontanelle, at the junction of the sagittal and

---

THE BONES OF THE CRANIUM

Frontal (2)

Parietal (2)

Occipital

Temporal (2)

Sphenoid

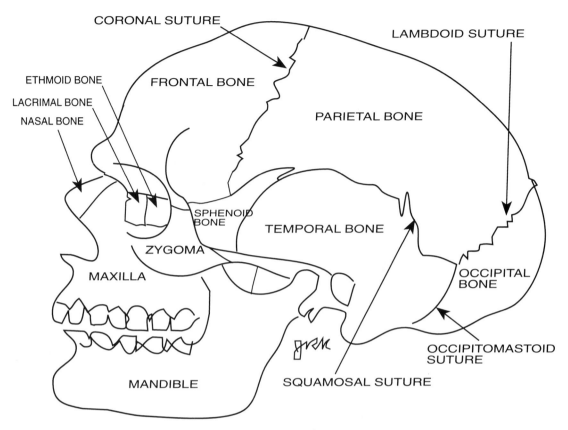

Fig. 1-1 The calvarium. The skull seen from the side, showing the major cranial bones and sutures.

lambdoid sutures). These normally ossify from the adjacent bones, closing by 1 to 3 years of age. However, separate ossification centers may occur within fontanelles, called intersutural (wormian) bones (Fig. 1-4). Many congenital developmental abnormalities are associated with numerous wormian bones of the sutures. The cranial bones may have variable ossification centers or incomplete ossification centers or incomplete ossification resulting in focal thin areas, the most common of which are the "parietal foramina" lateral to the sagittal suture within the parietal bone (Fig. 1-5).

The temporal bone has two major components: a lateral squamosal portion, and a medially directed petrous portion. The inferior squamous portion has an expanded cancellous region, the mastoid sinuses, forming the mastoid tip. The petrous portion contains the cochlea and labyrinthine structures, the internal auditory canal (IAC), the middle ear cavity, and the external auditory canal (Fig. 1-6). The anterior margin of the dural tento-

rium attaches along the superior ridge of the petrous bone. (For more detailed anatomy of the temporal bone, see Ch. 13.)

The sphenoid bone forms the anterior medial base of the skull and has a complex shape. It has a central body, which joins the basal portion of the occipital bone (basiocciput) at the spheno-occipital synchondrosis. This synchondrosis may persist and be visible on radiographs into adult life. The *clivus* is the posterior surface and concavity of the sphenoid and basiocciput, which accommodates the pons. It extends from the superior body of the sphenoid through the basiocciput to the anterior rim of the foramen magnum, a point called the *basion* (Fig. 1-7).

Two lateral vertically oriented greater wings of the sphenoid are convex posteriorly, forming the anterior margin of the middle cranial fossa. An additional pair of laterally directed smaller (lesser) wings lie more superiorly and form the posterior ridge of the floor of

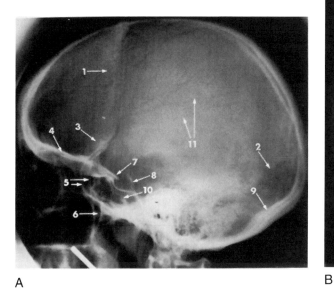

A

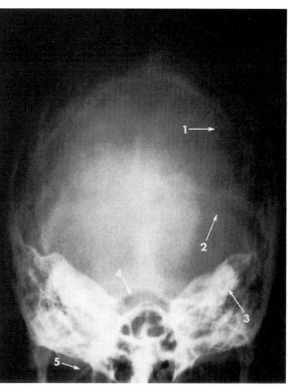

B

Fig. 1-2 (A) Lateral view, normal skull. (1) Coronal suture; (2) lambdoid suture; (3) pterion (note normal focal density); (4) lateral portion of orbital roofs (floor of anterior cranial fossa); (5) greater sphenoid wings (anterior walls of middle cranial fossae); (6) pterygopalatine fossa; (7) anterior clinoid processes; (8) dorsum sellae; (9) internal occipital protuberance (note the shallow groove for the torcular Herophili); (10) carotid sulcus; and (11) middle meningeal artery grooves. Also note (a) the tuberculum sellae seen through the base of the anterior clinoid processes; (b) the planum sphenoidale extending anteriorly from the tuberculum; (c) the clivus cortex sloping inferoposteriorly from the dorsum sellae; (d) the normally prominent nasopharyngeal soft tissue in this young patient; and (e) the normal skull lucency at the frontal and occipital poles and temporal squamosa. **(B) Towne view, normal skull.** (1) Lambdoid suture; (2) groove for transverse sinus; (3) otic capsule within petrous bone; (4) posterior lip of foramen magnum; (5) inferior orbital fissure. The occipital bone extends from lambdoid sutures to the foramen magnum. The transverse sinus grooves indicate the level of tentorial attachment. The vestibule and internal auditory canal are seen within the otic capsule. Mastoid air cells are present lateral to the otic capsules, and are mildly sclerotic on the right side. (From Yock, 1984, with permission.)

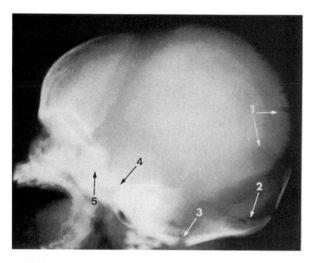

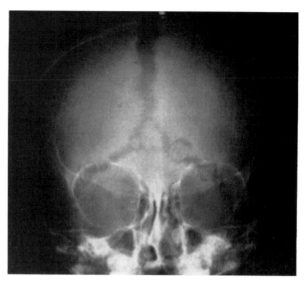

Fig. 1-3 Normal infant skull. (1) Parietal fissures; (2) mendosal suture; (3) innominate synchondrosis; (4) spheno-occipital synchondrosis; (5) intersphenoid synchondrosis. Note the broad, lucent bands at the coronal, lambdoid, and squamosal sutures. (From Yock, 1984, with permission.)

Fig. 1-4 "Split" sutures. In this view, the sagittal and lambdoid sutures are widely separated as the result of increased intracranial pressure. Note the wormian bones. (From Yock, 1984, with permission.)

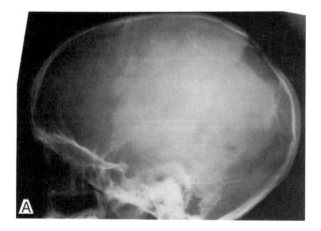

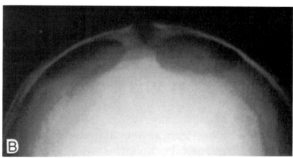

Fig. 1-5 (A & B) Parietal foramina. Symmetric parasagittal posterior parietal bone defects are well defined. Note the superior sagittal sinus groove in Fig. B. (From Yock, 1984, with permission.)

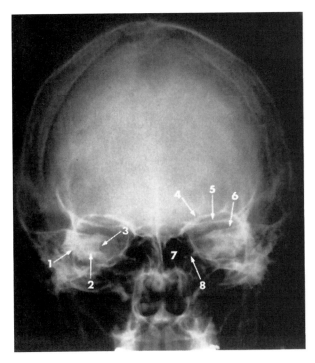

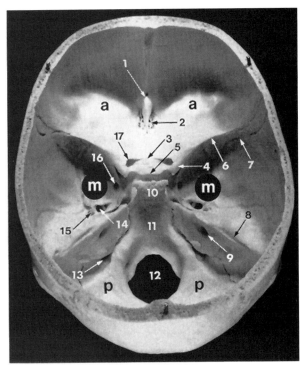

Fig. 1-6 Straight posteroanterior view, normal skull. (1) Vestibule of inner ear; (2) floor of internal auditory canal; (3) posterior wall of internal auditory meatus; (4) orbital roof; (5) sphenoid ridge; (6) superior surface of petrous pyramid (petrous ridge); (7) superimposed ethmoid and sphenoid sinuses; (8) ethmomaxillary plate. The crista falciformis is seen as short, horizontal, linear density within the lateral portion of the internal auditory canal. (From Yock, 1984, with permission.)

Fig. 1-7 Skull base, intracranial aspect. (1) Crista galli; (2) cribriform plate; (3) limbus sphenoidale; (4) anterior clinoid process; (5) tuberculum sellae; (6) sphenoid ridge; (7) groove for middle meningeal artery (with a short bone tunnel lateral to the arrow); (8) petrous ridge, with a groove for the superior petrosal sinus at the site of tentorial attachment; (9) internal auditory meatus; *arrow,* the crescentic posterior cortical edge seen on posteroanterior films and tomograms; (10) dorsum sellae, with the posterior clinoid processes superolaterally; (11) clivus; (12) foramen magnum; (13) jugular foramen (the sigmoid sinus groove approaches the foramen posterolaterally); (14) foramen ovale; (15) foramen spinosum (origin of middle meningeal arterial grooves); (16) foramen rotundum; (17) cranial end of optic canal. The anterior cranial fossa floor slopes inferiorly as it extends medially, so it is deepest in the midline. The petrous pyramids angle about 45 degrees anteriorly in the lateral-to-medial direction. Also note (a) the planum sphenoidale extending from the limbus to the cribriform plate; (b) the chiasmatic sulcus located between the limbus and the tuberculum sellae, and joining the optic canals; (c) the optic strut running between the anterior clinoid process and the tuberculum, forming the floor of the optic canal and separating it from the superior orbital fissure; (d) the superior orbital fissure, partially seen between the anterior clinoid process and the foramen rotundum; (e) the petrous apex (at the numeral 14) adjacent to the clivus; and (f) the jugular tubercles, ridges of bone medial to the jugular foramina at the base of the clivus. a, anterior cranial fossa; m, middle cranial fossa; p, posterior cranial fossa. (From Yock, 1984, with permission.)

the frontal fossa. A gap between the greater and lesser wings forms the superior orbital fissure (Fig. 1-8). The ciliary nerves, the oculomotor nerve (cranial nerve, [CN III]), the trochlear nerve (CN, IV), the ophthalmic division of the trigeminal nerve (CN V), the abducens nerve (CN VI), and the meningeal artery branches pass anteriorly through this fissure to reach the orbit. An orbital vein and the recurrent branch of the lacrimal artery supplying the dura pass through in the opposite direction. The medial portion of the greater wings is pierced by the foramen rotundum, which carries the maxillary division of CN V into the inferior orbital fissure. The inferior orbital fissure is formed by the anterior margin of the greater wing of the sphenoid and the posterior maxillary floor of the orbit.

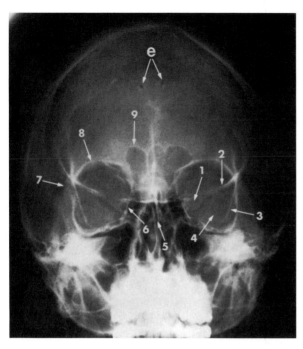

Fig. 1-8 **Normal skull, inclined posteroanterior (Cadwell) view.** (1) Lesser wing of sphenoid bone; (2) sphenoid ridge; (3) innominate or "oblique orbital" line; (4) greater wing of sphenoid bone; (5) floor of sella turcica; (6) lamina papyracea; (7) frontozygomatic synchondrosis; (8) orbital roof; (9) frontal sinus. e: emissary foramina. Also note (a) the superior orbital fissure between the lesser and greater sphenoid wings, (b) the planum sphenoidale, seen as dense, horizontal cortical bone superior to the sellar floor. (From Yock, 1984, with permission.)

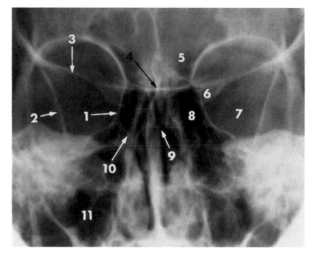

Fig. 1-9 **Normal sinuses, Caldwell view.** (1) Lamina papyracea; (2) oblique orbital (innominate) line; (3) sphenoid ridge; (4) planum sphenoidale; (5) frontal sinus; (6) lesser sphenoid wing; (7) greater sphenoid wing; (8) ethmoid sinus; (9) floor of the sella turcica; (10) ethmomaxillary plate; (11) maxillary sinus. The superior orbital fissures are seen between the sphenoid wings. (From Yock, 1984, with permission.)

Two bilateral vertically oriented pterygoid plates extend inferiorly from the body just lateral to the midline. Their medial and lateral components provide attachments for the pterygoid muscles of the mandible.

The body of the sphenoid forms the central floor of the intracranial fossa. Anteriorly, the bone is hollow forming the paired sphenoid sinus cells. The anterior superior surface is a flat sturdy segment called the planum sphenoidale, which forms the posterior medial floor of the anterior (frontal) intracranial fossa (Fig. 1-9). The cribriform plate of the ethmoid bone lies anterior to this. The posterior portion of the gyrus rectus of the frontal lobe lies on this ledge. The limbus marks the posterior margin of the planum. Behind this a shallow sulcus forms called the chiasmatic (optic) groove. However, the optic chiasm is not contained within this groove; it is more posterior related to the anterior floor of the third ventricle. The posterior margin of the optic groove is called the tuberculum sellae. From this point posteriorly, a deep concavity forms the sella turcica for the pituitary gland (Fig. 1-10), overhung anteriorly by the bilateral anterior clinoid processes. The posterior wall of the sella is formed by the dorsum sellae which is topped by small posterior clinoid processes. The sella is covered with a dural diaphragm (the diaphragma sellae), formed by an anterior extension of the tentorium. This is pierced by the pituitary infundibulum (stalk). In about 25 percent of people, the diaphragm is incomplete, allowing inferior herniation of the arachnoid along with CSF into the sella. This arachnoid sac compresses the pituitary, creating what is known as the *empty sella syndrome.* The floor of the sella is not covered by dura and expands in response to the pulsations of arachnoid, which has herniated into the sella.

The floor of the skull is formed by the occipital, temporal, and sphenoid bones. These bones are pierced by many foramina and canals that allow passage for the brain stem, nerve roots, arteries, and veins. These are shown in Figure 1-7.

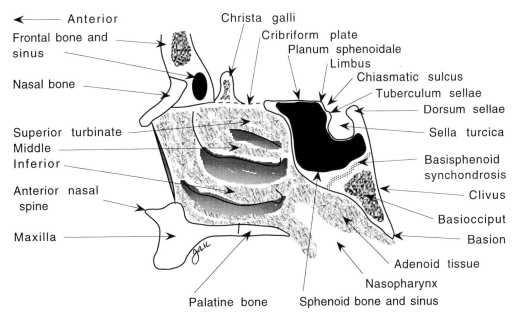

Fig. 1-10 Sphenoid bone, frontal fossa, and nasal cavity. Sagittal view.

Membranes of the Brain

Three tissue membranes surround the brain and spinal cord within the calvarium and spinal canal. They are the *dura mater*, the *arachnoid*, and the *pia mater*. The dura mater is the outermost membrane. This membrane consists of two layers of different cellular matrix. The outer *periosteal layer* is intimately attached to the inner table of the skull and forms the equivalent of the periosteum. The inner *meningeal layer* is made up of looser tissue and is related to the arachnoid on its inner surface. The two layers of the dura are loosely joined. Under certain conditions the two layers may separate to form the subdural space, which becomes filled with blood (subdural hematoma)[SDH]), CSF (subdural hygroma), or pus (subdural empyema) (see Fig. 7-7).

The meningeal layer separates and grows to form the large nearly rigid dural septa within the calvarium. The midline parasagittal septum is called the *falx cerebri* and is within the interhemispheric fissure dividing the brain in half. Posteriorly, the septum extends laterally to form bilateral horizontal septa that attach to the petrous ridges of the temporal bone and the posterior clinoid processes of the sphenoid bone of the skull. Together these horizontal septa form the *tentorium cerebelli*

and create the roof of the posterior fossa. Anteriorly, it continues to form the roof of the sella turcica, the diaphragma sellae. This membrane is pierced by the infundibulum (pituitary stalk) (see Figs. 1-13 to 1-34).

The medial margins of the tentorium outline a large central rounded opening, the *tentorial incisura*, through which passes the mesencephalon. The tentorium slopes upward in the midline. Where it meets the falx, the straight venous sinus is formed as a cavity between its layers. A small parasagittal midline septum, the *falx cerebelli*, divides the superior posterior fossa.

At the foramen magnum, the meningeal layer of the dura mater continues inferiorly to form the dural tube of the spinal canal. The dura tube is not attached to bone within the spinal canal. The tissue external to the dural tube within the spinal canal is called the epidural space (see Fig. 14-2), and it is filled with fat.

The pia mater is the innermost layer of the meninges. It consists of two layers: the inner *intima pia*, and the superficial *epipial layer*. The intima pia is adherent to the underlying nervous tissue, the contours of which it follows closely. The epipial layer blends with the arachnoid trabeculae. The pia forms a strong barrier around the brain and spinal cord. The pia is pierced by exiting cranial and spinal cord. The pia is pierced by exiting cranial and spinal nerves and by blood vessels. Along the spinal cord, the epipial membrane forms a series of lateral bands that extend from the lateral margin of the spinal cord to the inner margin of the dural tube. These are the *dentate ligaments*, which support the cord within the spinal canal.

The *arachnoid* is the membrane between the dura and the pia. It consists of two layers: a flat peripheral membrane lining the inner surface of the dura, and gossimer trabeculae that connect the arachnoid membrane with the epipial layer of the pia. Together, the pia and the arachnoid constitute the *leptomeninges*. The *subarachnoid space* occupies the region between the arachnoid membrane and the pia and is crossed by the trabeculae. Normally, CSF fills this space. The arachnoid and the subarachnoid space extend to surround the olfactory, optic, trigeminal, and vestibulocochlear cranial nerves and the spinal nerves within the segmental dural root sleeves.

The pia and arachnoid membranes invaginate into the brain at sites of blood vessel penetration. This produces a cuff of space about the blood vessel, which is an extension of the subarachnoid space. This perivascular space is called the *Virchow-Robin space*. These spaces vary considerably in size, even at birth, and generally increase with age or after longstanding hypertension (Fig. 1-11).

At certain sites within the calvarium and spinal canal, the subarachnoid space is quite large. The enlargements are called *cisterns*. The cisterns are most common at the base of the brain and take their names from adjacent structures. For example, the suprasellar cistern is just above the sella turcica and contains the optic nerves and pituitary stalk. The quadrigeminal plate cistern is just behind the quadrigeminal plate. The major cisterns of the subarachnoid space are shown in Figure 1-12. The cisterns are important neuroradiologic structures that obliterate with brain swelling, meningeal tumors, or infection.

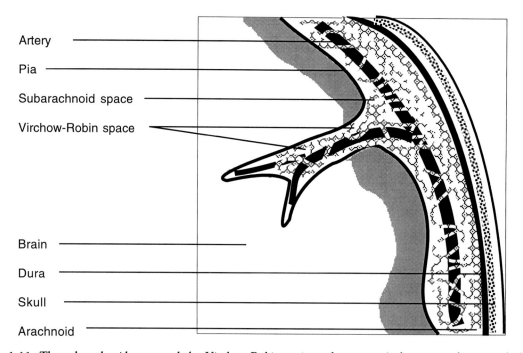

Fig. 1-11 The subarachnoid space and the Virchow-Robin perivascular space. A drawing in the coronal plane illustrates the subarachnoid space, arachnoid, and the extension of the subarachnoid space around arteries penetrating the cerebrum (the Virchow-Robin space). The spaces may continue deeper into the brain than drawn here. This is a magnified view, not to scale, and the Virchow-Robin space should not be confused with a sulcus.

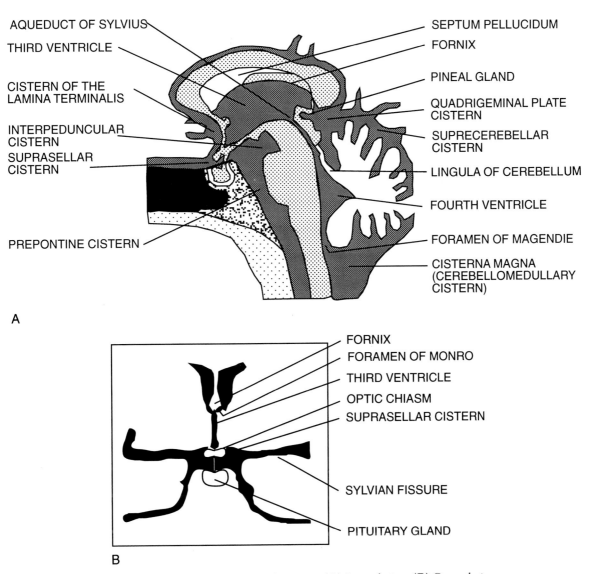

Fig. 1-12 Intracranial subarachnoid cisterns. (A) Sagittal view. **(B)** Coronal view.

The subarachnoid space is continuous with the intraventricular cavities through the foramina of Magendie (medial) and Luschka (lateral) at the inferior fourth ventricle. These foramina drain the fourth ventricle, and so the whole ventricular system, into the cisterna magna.

The *ependyma* is the single cell layer that lines the internal surface of the cerebral ventricles. At specific sites in the ventricular roofs, the pia and ependyma meet and push out together to form the choroid plexuses and the tela choroidia, the membrane that attaches the choroid plexus to the ventricular walls. The choroid plexus is composed of pial blood vessels and specialized lining cells that secrete CSF into the ventricular cavities. The major sites of choroid are the atria of the lateral ventricles, the roof of the third ventricle, and the inferior roof of the

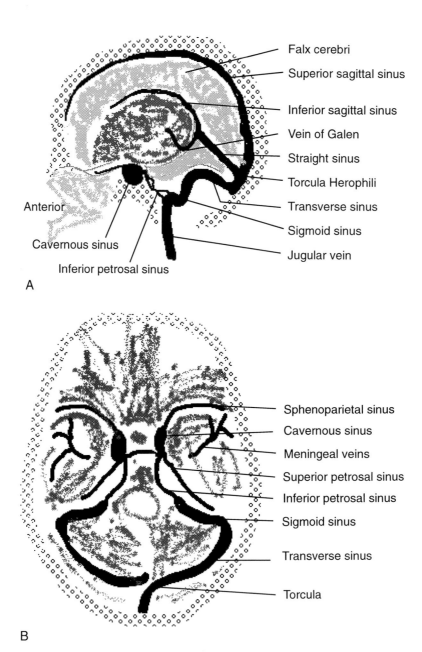

Fig. 1-13 **The dural venous sinuses. (A)** Sagittal view. **(B)** Axial view.

fourth ventricle. The choroid plexuses produce about 70 percent of the total CSF volume. The remaining CSF is produced by the direct flow of intercellular brain fluid into the ventricles. The choroid plexus can both secrete CSF and remove solute from CSF. Calcified psammoma bodies form normally within the choroid, increasing with age.

The CSF acts as a buffer for volume changes of the brain. As the brain volume increases, CSF is displaced from the ventricles, cisterns, and subarachnoid spaces. CSF also clears metabolites and provides a conduit for neuropeptide hormonal communication within the brain. The normal CSF pressure is 100 to 150 mmH$_2$O while recumbent.

The *dural venous sinuses* are blood-filled cavities that form within the dural membrane. They form conduits to receive the drainage from cerebral veins. The cerebral veins penetrate the dura to empty into the sinuses (Fig. 1-13). Emissary veins also extend from the dural sinuses through the calvarium into the extracranial subcutaneous tissue. *Arachnoid villi* protrude through the dura into the sinus cavity and provide passive drainage of CSF from the subarachnoid space into the venous system (Fig. 1-14). Large dural venous lakes may form at the site of arachnoid granulations and produce local impressions in the calvarium.

The arachnoid villi act as pressure sensitive one-way valves with a diameter of about 6 μm. They work on a pressure gradient between the CSF and the venous blood within the sinus. Flow of CSF through the villi into the dural sinus begins when the CSF pressure is 3 to 6 cmH$_2$O greater than the venous pressure within the

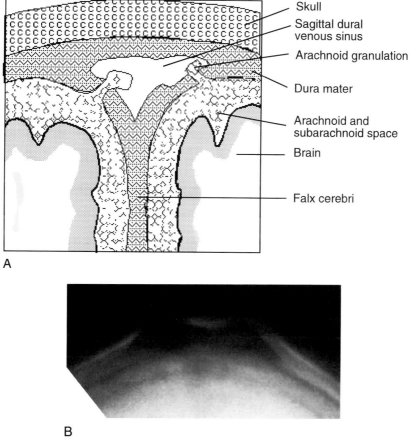

A

B

Fig. 1-14 **(A)** Arachnoid granulation. **(B)** Pacchionian impressions, posteroanterior view. Smooth, well-defined, parasagittal erosions arise from the inner table, causing thinning of the diploic space. (Fig. A modified from Carpenter and Sutin, 1983; Fig. B from Yock, 1984, with permission.)

dural sinus. The greater the pressure gradient, the greater the CSF flow across the villi into the dural sinuses. The villi are closed to reverse flow. Villi are permiable to metabolites and relatively large molecular substances, but they become plugged with debris from subarachnoid hemorrhage. Obstruction of the villi from any cause produces communicating hydrocephalus (see Ch. 11).

The meninges are incompletely visualized with imaging. While the falx and tentorium are seen on axial and coronal images, the pia and arachnoid are not seen directly. With contrast enhancement, the falx and tentorium consistently become bright on CT, but become bright inconsistently with MRI. Tumors, inflammation, and trauma cause increased enhancement of the meninges on both CT and MRI. This is discussed in more detail in the pertinent sections that follow.

The meninges are an important site for intracranial disease. They become involved with primary and metastatic tumors, infections, granulomatous reactions, and bleeding.

DISEASES INVOLVING THE INTRACRANIAL MENINGES

Malignant tumors
    Neoplastic meningitis (metastasis)
    Lymphoma, leukemia
    Melanocytic tumor
    Sarcomas, many types
    Hemangiopericytoma
Benign tumors/cysts
    Meningioma
    Chondroma
    Lipoma
    Arachnoid cyst
    Epidermoid cyst
Inflammation
    Meningitis
        Bacterial, fungal, parasitic
    Granulomatous meningitis
        Sarcoid, Wegener's, rheumatoid
    Tuberculosis
    Siderosis
Subarachnoid hemorrhage

THE NERVOUS SYSTEM

The nervous system is divided into two major categories: the central nervous system (CNS) and the peripheral nervous system (PNS). The CNS consists of the brain and the spinal cord. The PNS consists of the cranial and spinal nerves.

Neurons are the functional cellular units of the nervous system. They are the seat of thought, action, emotion, and memory. Neurons communicate with other neurons and cells by means of their cytoplasmic extensions, called axons. Action potentials created at the origin in the neuron or at terminals of the axon propagate along the axon to a junction with another cell called a synapse. Various neurotransmitters released at axon terminals carry impulses from one cell to another through the narrow synaptic spaces (junctions). Through this means, information is distributed throughout the body. Axons may be long or short, depending upon the location of their ultimate connections.

Synapses alter their sensitivity to action potentials, depending on frequency of use. This permits learning and memory within the nervous system. The patterns of connections and their alterations with time determine responses of the individual. The most important and complex collection of neurons and axons is the CNS.

THE BRAIN

THE DIVISIONS OF THE BRAIN

Cerebral hemispheres (telencephalon)
    Cerebral cortex
    Centrum semiovale
    Basal ganglia
Brain stem
    Diencephalon
        Anterior commissure
        Hypothalamus
        Epithalamus (pineal, habenula)
        Thalamus
        Subthalamus
        Geniculate bodies

(Continued)

Mesencephalon (midbrain)
  Ventral portion
    Crus cerebri
  Dorsal portion (tegmentum)
    Oculomotor nucleus
    Red nucleus
    Reticular formation
    Substantia nigra
  Posterior to aqueduct (tectum)
    Posterior commissure
    Collicular plate
Metencephalon (pons)
  Ventral portion
    Longitudinal tracts
      Corticospinal
      Corticobulbar
      Corticopontine
    Transverse tracts
      Middle cerebellar peduncle
    Pontine nuclei
  Dorsal portion (tegmentum)
    Cranial nerves V to VIII
    Reticular formation
    Tracts
      Spinal thalamic
      Medial lemniscus
      Medial longitudinal fascic-
        ulus
    Locus ceruleus
Myelencephalon (medulla)
  Nuclei
    Cranial nerves IX to XII
    Olivary complex
  Reticular formation
  Decussation of the corticospinal tract
  Inferior cerebellar peduncle
  Corticobulbar tracts
  Area prostrema
  Spinal trigeminal tract
Cerebellum
  Hemispheres
  Dentate nuclei
  Cerebellar peduncles
    Superior (brachium conjunctivum)
    Middle (brachium pontis)
    Inferior (restiform)

The brain is made up of the cerebral hemispheres, the brain stem, and the cerebellum (Fig. 1-15). The cerebral hemispheres—the *telencephalon*—are nearly mirror images of each other separated by the central cleft called the *interhemispheric fissure*. On their surface, they consist of cortical convolutions of a thin layer of gray matter separated by deep clefts called sulci. The outer convex surface of a convolution is called a *gyrus*. The cortical gray matter contains most of the neurons of the brain.

A large body containing the axons of neurons lies beneath the gray matter. This is called the white matter. As a whole, this deep hemispheric white matter constitutes the *centrum semiovale*. The *corona radiata* of the white matter is the bundle of direct corticospinal and cortico-bulbar fibers coursing through the centrum semiovale to enter and transverse the *internal capsule*. Deep in the central portion, the hemispheres are connected by a broad curved band of white matter called the *corpus callosum*. This is the great central commissure forming the depth of the interhemispheric fissure. The hemispheres are also connected by smaller bands of white matter: the anterior and posterior commissure.

Collections of gray matter lie deep within the central portion of each hemisphere. As a whole, this grouping constitutes the basal ganglia (Table 1-1). The thalamus is sometimes included as part of the basal ganglia, but it is a diencephalic (brain stem) structure, and is primarily a sensory way station (Fig. 1-16).

The Cerebral Hemispheres

The cerebral hemispheres are subdivided into regions, or lobes, that are more or less clearly demarcated. They take their names from the adjacent skull bones. The frontal, parietal, occipital, and temporal regions form the superficial lobes separated by large sulci. These lobes also constitute the neocortex. The insula and limbic regions (lobes) are deeper and not lobular in shape. The limbic lobe forms part of the limbic system, the most ancient of the cortical regions in evolution.

THE LOBES OF THE BRAIN

Frontal	Occipital
Parietal	Insula
Temporal	Limbic

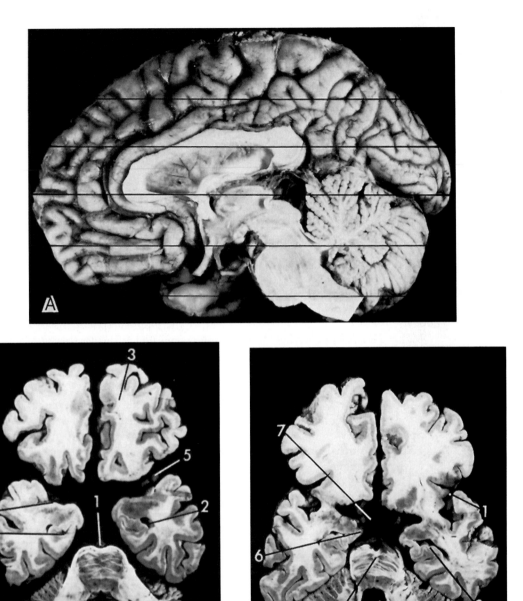

Fig. 1-15 Standard CT slices through the head with corresponding postmortem brain slices cut in approximately the same plane. **(A)** View of brain illustrating levels of slices. **(B)** (1) Basilar artery; (2) temporal horn; (3) frontal lobes; (4) middle cerebral artery; (5) sylvian fissure; (6) temporal lobe; (7) hippocampus; (8) cerebellar peduncle; (9) cerebellar cortex; (10) inferior vermis. **(C)** (1) Sylvian fissure; (2) temporal horn; (3) pons; (4) fourth ventricle; (5) cerebellar cortex; (6) uncus; (7) internal carotid artery. (*Figure continues.*)

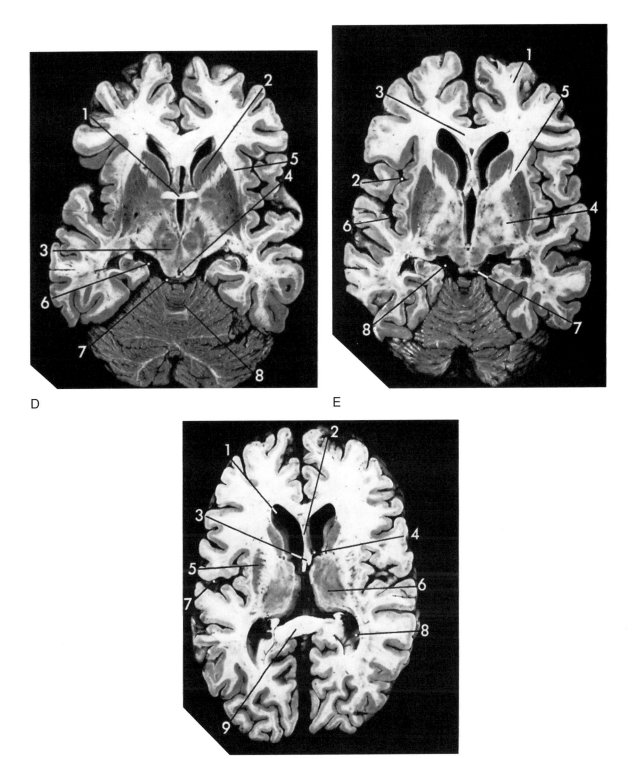

D

E

F

Fig. 1-15 *(Continued).* **(D)** (1) Anterior commissure; (2) head, caudate nucleus; (3) red nucleus; (4) aqueduct; (5) external capsule; (6) choroidal fissure; (7) quadrigeminal plate cistern; (8) superior vermis. **(E)** (1) Frontal lobe white matter; (2) third ventricle; (3) rostrum of corpus callosum; (4) anterior limb of internal capsule; (5) posterior limb of internal capsule; (6) insula; (7) insula; (8) colliculi; (9) quadrigeminal plate cistern. **(F)** (1) Lateral ventricle; (2) septum pellicidum; (3) fornix; (4) fdoramen of Monro; (5) putamen and globus pallidus; (6) thalamus; (7) sylvian fissure; (8) atrium; (9) splenium of corpus callosum. *(Figure continues.)*

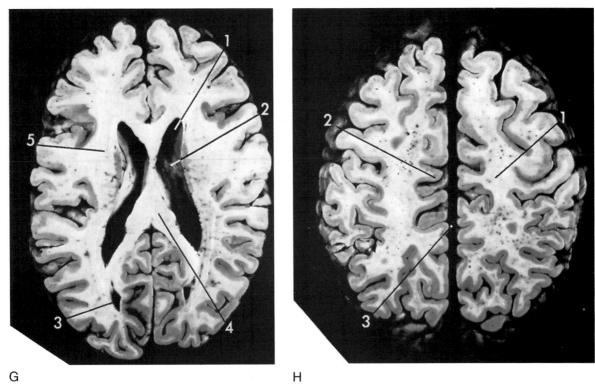

G   H

Fig. 1-15 *(Continued)*. **(G)** (1) Frontal horn; (2) body; (3) occipital horn; (4) corpus callosum; (5) centrum semiovale. **(H)** (1) Centrum semiovale; (2) interhemispheric cortex; (3) interhemispheric fissure. (From Heinz, 1984, with permission.)

FRONTAL LOBE

The frontal lobe is the largest and contains the primary motor area in the precentral gyrus and the anterior wall of the central sulcus. The inferior frontal gyrus over the lower lateral surface contains the motor control of speech called Broca's speech area, normally found within the dominant hemisphere. The olfactory bulb and tract lie just beneath the orbital surface of the frontal lobes (see Fig. 13-1). The frontal gyrus medial to this tract is the gyrus rectus. The cingulate gyrus, actually part of the

limbic system, resides on the medial inferior surface of the frontal lobe, paralleling the corpus callosum and separated from it by the callosal sulcus. The cingulate sulcus is peripheral to the gyrus. The callosal sulcus continues around the splenium of the corpus callosum to become the hippocampal sulcus along the medial temporal lobe (Fig. 1-17).

PARIETAL LOBE

The central sulcus clearly defines the anterior margin of the parietal lobe, but the other margins of this lobe are less clearly defined (Fig. 1-18). The primary somesthetic area is the gyrus immediately posterior to the central sulcus, which contains the major sensory representation in the cortex. It is bounded posteriorly by the postcentral sulcus. This sulcus extends medially to the interhemispheric fissure, where it outlines the posterior margin of

Table 1-1 The Basal Ganglia

Caudate nucleus	Striatum (neostriatum)	
Putamen	} Lentiform nucleus	} Corpus striatum
Globus pallidus		
Amygdala-nuclear complex		

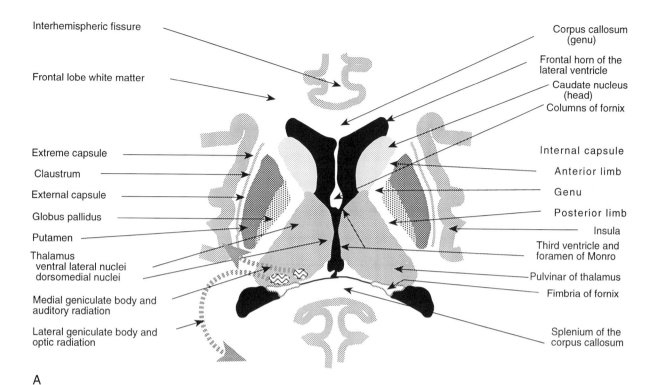

Interhemispheric fissure

Frontal lobe white matter

Extreme capsule

Claustrum

External capsule

Globus pallidus

Putamen

Thalamus
  ventral lateral nuclei
  dorsomedial nuclei

Medial geniculate body and
auditory radiation

Lateral geniculate body and
optic radiation

Corpus callosum
(genu)

Frontal horn of the
lateral ventricle

Caudate nucleus
(head)

Columns of fornix

Internal capsule

Anterior limb

Genu

Posterior limb

Insula

Third ventricle and
foramen of Monro

Pulvinar of thalamus

Fimbria of fornix

Splenium of the
corpus callosum

A

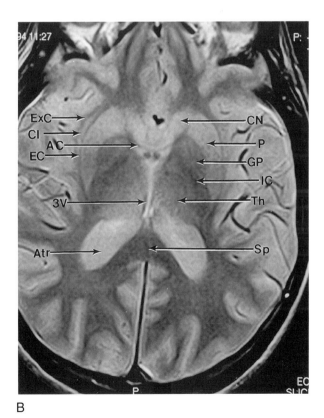

B

Fig. 1-16 Basal ganglia. **(A)** Axial view. **(B)** Normal view. SE30/2722 PD MRI of the region of the low basal ganglia. 3V, third ventricle; AC, anterior commissure; Atr, atrium of the lateral ventricle; Cl, claustrum; CN, caudate nucleus; EC, external capsule; ExC, extreme capsule; GP, globus pallidus; IC, internal capsule; P, putamen; Sp, splenium of the corpus callosum; Th, thalamus.

the paracentral lobule. The cortex posterior to the post-central sulcus is divided by the intraparietal sulcus into superior and inferior parietal lobules. The inferior parietal lobule contains the supramarginal and angular gyri, which are concerned with understanding of speech and language (Wernicke's area). The posterior parietal lobe blends into the lateral occipital lobe (Fig. 1-19).

TEMPORAL LOBE

The temporal lobe is large and occupies the temporal fossa of the skull. It lies ventral to the great lateral sulcus, the sylvian fissure. The lateral surface is separated into the superior, middle, and inferior temporal gyri. The superior temporal gyrus continues posteriorly into the region of the angular gyrus. It is also a lan-

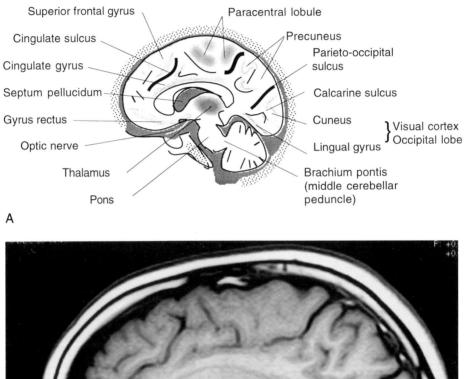

A

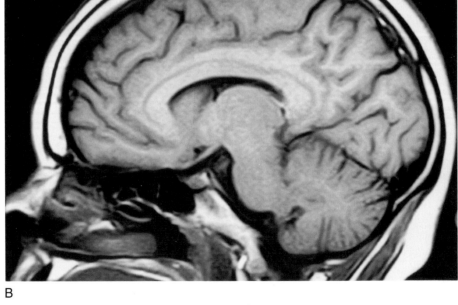

B

Fig. 1-17 **(A)** Midline sagittal view of brain. **(B)** This sagittal T_1 MRI correlates with the drawing in Fig. A.

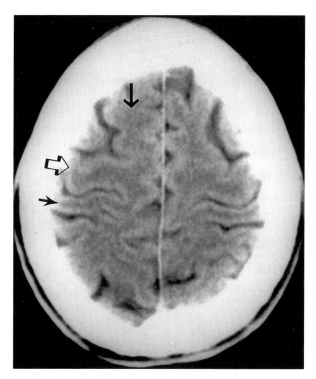

Fig. 1-18 Normal high cortex. An axial T$_1$ MRI identifies the posterior superior frontal gyrus *(larger black arrow)*, the precentral gyrus or motor strip *(open arrow)*, and the postcentral gyrus or sensory strip *(smaller black arrow)*.

OCCIPITAL LOBE

The occipital lobe is relatively small and lies posterior to the parietal and temporal lobes. The margins are poorly defined. It is divided by the calcarine sulcus into the superior cuneus and the inferior lingual gyri (Fig. 1-17). The primary visual cortex is located along both sides of the calcarine sulcus. The occipital lobes lie on the superior surface of the tentorium.

INSULA

The insula is the triangular-shaped cortical region that forms the medial floor of the sylvian fissure. The posterior portion is the *Heschl* region allocated to the temporal lobe (Fig. 1-20). The middle cerebral vessels arborize over the insula before they course laterally around the opercular margins. The anterior insula is sometimes included within the limbic system. The insula is covered by the *opercular* (overhanging) regions of the frontal, parietal, and temporal lobes, which form ridges lining the margin of the lateral sulcus *(sylvian fissure)*.

LIMBIC LOBE

The limbic lobe is deep and central within the cerebral hemispheres. It consists of the *subcallosal, cingulate,* and *parahippocampal gyri,* and the adjacent *hippocampal for-*

guage interpretation and association area. The temporal lobe continues medially to form the posterior portion of the flat *insula* at the depth of the sylvian fissure. The posterior portion of this region has vertical gyri (the *gyri of Heschl*) that constitute the primary auditory cortex (Fig. 1-20). The *parahippocampal gyrus* and the *uncus* form the lateral and anteromedial surfaces of the undersurface of the lobe. The anterior portion of the temporal lobe lies within the temporal fossa. The posterior portion lies on the superior surface of the tentorium. The thin curved *hippocampus* is a strip of deep gray matter along the medial margin of the temporal horn of the lateral ventricle and is part of the *limbic system.* It is thought to have a primary role in memory formation (Fig. 1-21). *Somer's sector* is a portion of hippocampal neurons especially sensitive to the effects of anoxia.

THE LIMBIC SYSTEM

Limbic lobe
    Subcallosal gyrus
    Cingulate gyrus
    Parahippocampal gyrus
    Hippocampal formation
    Dentate gyrus
Amygdaloid complex
Fornix
Mammillary bodies
Septal nuclei
Hypothalamus
Epithalamus
Anterior thalamic nuclei

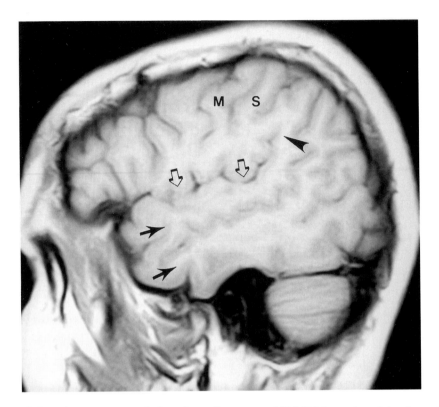

Fig. 1-19 Normal, lateral cortex in sagittal plane. Laterally positioned T₁ MRI shows the sylvian fissure (*open arrows*) and the superior and inferior temporal gyri (*open arrows*) and the superior and inferior temporal gyri (*black arrows*). The angular gyrus is at the posterior sylvian fissure (*arrowhead*). The motor (M) and sensory (S) cortex extends laterally over the convexity.

mation along the medial temporal lobe. The limbic lobe has many intimate functional associations with other regions of the brain, particularly deep adjacent nuclei and their tracts. The fornix provides the major efferent connection of the hippocampus with the diencephalon (the hypothalamus, particularly the mammillary bodies, the olfactory system, and the thalamus). Together with the limbic lobe, this functional group is called the *limbic system* (Fig. 1-22). The limbic system is sometimes referred to as the visceral brain because of its role in the emotional state of an individual and its connection with the neuroendocrine and autonomic systems. It also has a central role in learning and memory.

The Central White Matter

The *corpus callosum* is the large thick band of densely myelinated fibers connecting the white matter of the two hemispheres (Fig. 1-23). The callosal fiber tracts ex-

tending through the centrum semiovale to the frontal and occipital lobes are called the anterior and posterior forceps. The posterior band arching over the posterior horn of the lateral ventricle is the tapetum and lies between the optic radiations and the lateral wall of the posterior horn.

The corpus callosum creates the floor of the interhemispheric fissure and much of the roof of both lateral ventricles. It has four anatomic regions: the rostrum, genu, body, and splenium. The *anterior commissure* is a small band of white matter crossing just anterior to the columns of the fornix (see Figs. 1-15D and 1-25). It connects the olfactory structures and the anterior temporal lobes. The *posterior commissure* is a small band lying immediately rostral to the superior colliculus of the tectum of the mesencephalon, at the point where the posterior third ventricle enters the aqueduct (see Fig. 1-27). It is involved with complex reflexes of the eye to light, but much of the function of its many different fibers is un-

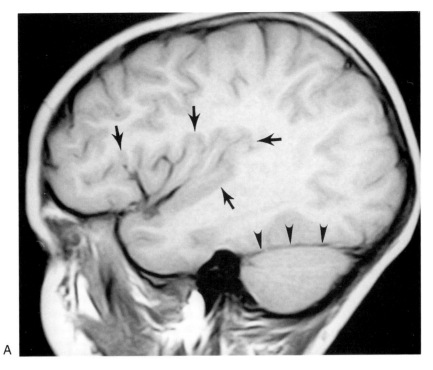

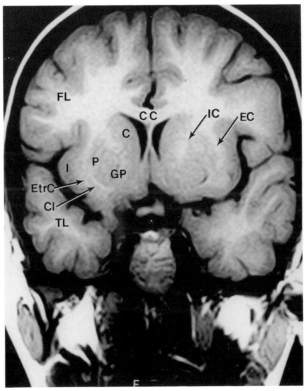

Fig. 1-20 Normal insula and basal ganglia. (A) Sagittal T$_1$ MRI cut through the depth of the sylvian (lateral) fissure shows the flat insular cortex *(arrows)*. Branches of the middle cerebral artery course over the insula. The tentorium is seen between the posterior inferior temporal lobe and the cerebellum *(arrowheads)*. **(B)** Normal coronal T$_1$ nonenhanced MRI through the frontal horns, basal ganglia, and insula. C, caudate nucleus; CC, corpus callosum; Cl, claustrum; EC, external capsule; EtrC, extreme capsule; FL, frontal lobe; GP, globus pallidus; I, insula; IC, internal capsule; P, putamen; TL, temporal lobe.

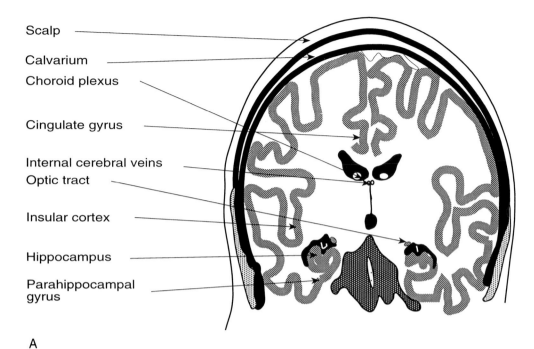

Scalp

Calvarium

Choroid plexus

Cingulate gyrus

Internal cerebral veins

Optic tract

Insular cortex

Hippocampus

Parahippocampal gyrus

A

CG

ICV

SF

CP

TL

PG

FL

FO

I

OT

TH

HF

B

Fig. 1-21 Normal coronal view through hippocampus. (A) Drawing of coronal MRI through hippocampal formation. **(B)** Coronal T_1 Gd-enhanced MRI through hippocampal formation. CG, cingulate gyrus; CP, choroid plexus; FL, frontal lobe; FO, frontal operculum; HF, hippocampal formation; I, insular cortex; ICV, internal cerebral veins; OT, optic tract; PG, parahippocampal gyrus; SF, sylvian fissure; TH, temporal horn; TL, temporal lobe.

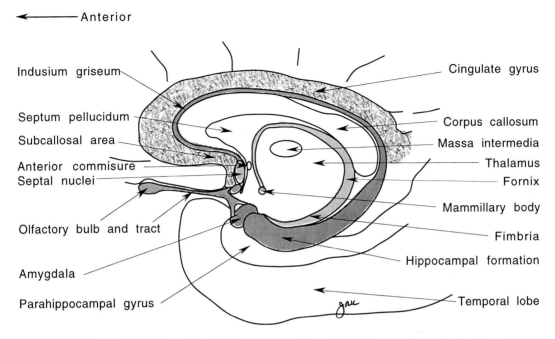

←——————— Anterior

Indusium griseum

Septum pellucidum

Subcallosal area

Anterior commisure

Septal nuclei

Olfactory bulb and tract

Amygdala

Parahippocampal gyrus

Cingulate gyrus

Corpus callosum

Massa intermedia

Thalamus

Fornix

Mammillary body

Fimbria

Hippocampal formation

Temporal lobe

Fig. 1-22 Limbic system. Sagittal view. (Modified from Carpenter and Sutin, 1983, with permission.)

known. The posterior commissure divides the rostral diencephalon from the mesencephalon.

Projection fibers (axons) of cortical neurons extend through the white matter to the brain stem and spinal cord, conducting motor and sensory impulses between the cortex and body. Within the centrum semiovale this fan-like band of radial fibers is called the corona radiata. At the base of the hemispheres the fibers converge to form the dense V-shaped internal capsule (see Fig. 1-16). The internal capsule is segmented into the anterior limb, the posterior limb, and the genu. The internal capsule splits the basal ganglia anteriorly and separates the basal ganglia from the thalamus posteriorly. As in the cerebral cortical regions, there is ordered anatomic representation of the fibers.

The efferent corticospinal fibers continue downward in the brain stem as the *crus cerebri* (cerebral peduncles) of the anterior mesencephalon (see Fig. 1-28). The small perforating arteries at the base of the brain supply the region of the basal ganglia, thalamus, and internal capsules. Infarcts in the distribution of these perforating vessels produce defined clinical stroke syndromes and small cystic regions called lacunes.

White matter association fibers connect various cortical regions within a hemisphere. Short fibers

connecting adjacent gyri course transversely around the depths of the cerebral sulci and are called *arcuate fibers* (U-fibers). They have a different MRI signal intensity from the deeper white matter and can be imaged as a specific structure. Longer association fibers connect lobar regions and are organized into the superior longitudinal fasciculus, the arcuate fasciculus, the uncinate fasciculus, and the cingulum. These association fibers run orthogonal to the corona radiata.

The Basal Ganglia

The basal ganglia are nuclear masses of gray matter deep within the cerebral hemispheres (see Fig. 1-16). The largest is the *putamen*. It is bounded laterally by the *external capsule* and medially by the *globus pallidus*. The globus pallidus is bounded medially by the internal capsule. The anterior commissure crosses along its inferior border.

The *caudate nucleus* is more complex in shape. The head lies anteromedial to the internal capsule and protrudes into the lateral aspect of the frontal horn of

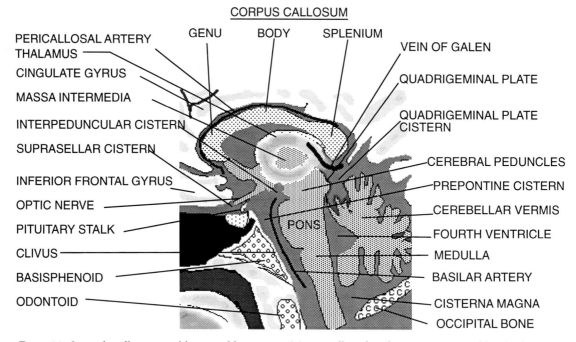

Fig. 1-23 Sagittal midline central brain and brain stem. Major midline deep brain structures and basal subarachnoid cisterns.

the lateral ventricle. The body is a tubular structure extending posteriorly over the dorsomedial border of the thalamus and just beneath the ependyma of the body of the lateral ventricle. The thalamostriate subependymal vein lies in the groove formed by the body of the caudate as it courses backward over the thalamus. The tail of the caudate nucleus arcs inferiorly and forward, following the temporal horn of the lateral ventricle to end in the amygdaloid complex in the anteromedial temporal lobe (see Fig. 1-22).

The basal ganglia have rich communications with the cerebral cortex and substantia nigra. Primarily they regulate motor action and muscle tone. Lesions within the nuclei often create various forms of dyskinesia (abnormal involuntary movements) and changes in muscle tone, particularly rigidity. The dyskinesias (tremor, athetosis, chorea, and ballism) result from the loss of normal inhibitory influence of the basal ganglia on the motor neurons. The *claustrum*, a thin strip of gray matter sandwiched between the external capsule and extreme capsule and sometimes considered part of the basal ganglia, is more closely related to the neocortex of the cerebral hemispheres. Its specific function is unknown, although it has broad cortical connections.

The Ventricles

The brain forms about central cavities called ventricles, which contain CSF (Fig. 1-24). The ependyma, a single specialized cell layer, lines the walls of the ventricles. The largest cavities are the lateral ventricles, the paired mirror-image structures within the major lobes of the cerebral hemispheres. The lateral ventricles are segmented into the frontal horn, the body, the atrium (trigone), the temporal horn (inferior horn), and the occipital horn (posterior horn). The lateral ventricles communicate with the thin midline third ventricle through the paired foramina of Monro. The columns of the *fornix* curve upward from the mammillary bodies adjacent to these foramina (see Fig. 1-22).

Choroid plexus is present within the bodies, atria, and temporal horns, arising at sites of junction of pia and ependyma. The choroid plexus enlarges within the atria, to form a tuft called the glomus. Heavy calcium deposits often form within the glomus. Small branches from the major cerebral arteries, called choroidal arteries, feed the choroid plexus.

The choroid is specialized tissue that produces a large fraction of the CSF within the ventricles. The CSF pro-

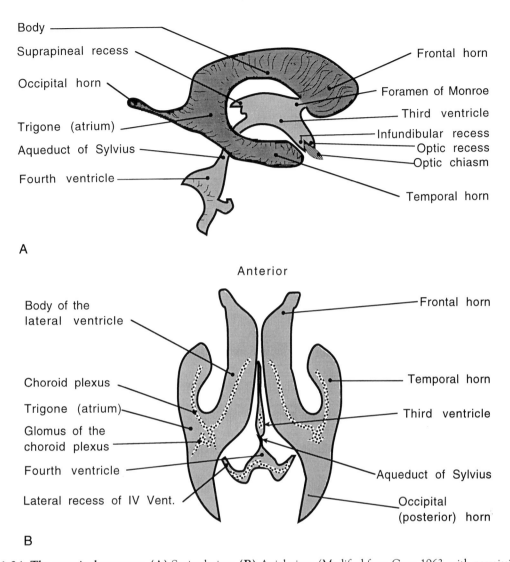

Body
Suprapineal recess
Occipital horn
Trigone (atrium)
Aqueduct of Sylvius
Fourth ventricle
Frontal horn
Foramen of Monroe
Third ventricle
Infundibular recess
Optic recess
Optic chiasm
Temporal horn

A

Anterior

Body of the lateral ventricle
Choroid plexus
Trigone (atrium)
Glomus of the choroid plexus
Fourth ventricle
Lateral recess of IV Vent.
Frontal horn
Temporal horn
Third ventricle
Aqueduct of Sylvius
Occipital (posterior) horn

B

Fig. 1-24 The ventricular system. (A) Sagittal view. **(B)** Axial view. (Modified from Goss, 1963, with permission.)

duced within the ventricle must flow out through the outlets of the fourth ventricle to be absorbed by the arachnoid villi at the sagittal sinus. Obstruction to the outflow or absorption of CSF causes ventricular dilatation termed *hydrocephalus*. The ventricles become smaller in response to brain swelling or larger in response to brain atrophy.

The Brain Stem

The most superior portion of the brain stem is the diencephalon. Its major parts are the hypothalamus, the subthalamus, the thalamus, the epithalamus, and the ge-

niculate bodies. The diencephalon is divided in the parasagittal midline by the third ventricle, except for the midline communication of the bilateral thalamic nuclei through the midline adhesion called the massa intermedia. The size of the massa intermedia varies considerably and is particularly large in some dysgenetic syndromes as the Chiari II malformation (see Ch. 12).

The hypothalamus is a small but important region of the brain, made up of multiple nuclei. It lies along the walls of the anterior inferior third ventricle, which separates the region into two halves, except at the inferior region, where a midline connection forms the floor of the third ventricle. The hypothalamus is anteroinferior to the thalamus. The nuclei of the hypothalamus constitute the main control of the sympathetic and parasympathetic nervous systems. The nuclei connect with the pituitary gland through the infundibulum (pituitary stalk). The tuber cinereum, an inferior bulge just posterior to the infundibulum, is the site of nuclei that communicate chemically with the anterior pituitary gland through the hypophyseal portal system. Paraventricular and supraoptic nuclei provide axons directly into the posterior lobe of the pituitary gland, known as the neurohypophysis (Fig. 1-25), which is an extension of the hypothalamus. Lesions of the hypothalamus may cause diabetes insipidus, hypogonadism and other syndromes of hypopituitarism, precocious puberty, hyperthermia, hyperphagia (obesity), hypophagia, emotional instability (particularly rage reactions), and sleep disorders.

The optic chiasm is contiguous with the hypothalamus just anterior and inferior to the pituitary infundibulum and the tuber cinereum of the hypothalamus. The optic tracts sweep posterolaterally from the chiasm, around the hypothalamus and upper crura cerebri, to reach the lateral geniculate bodies. Most of the fibers of the optic tract enter the lateral geniculate body, but some pass by to enter the superior colliculus. The *lateral geniculate body* lies lateral and inferior to the posterior thalamus, the pulvinar, and lateral to the adjacent medial geniculate body. The geniculate bodies are considered subdivisions of the thalamus. The optic radiation arises from the lateral geniculate body, runs laterally, crossing in front of the atrium of the lateral ventricle and into the temporal and parietal lobe white matter, and then sweeps posteriorly in a broad band to project on the visual cortex of the medial occipital lobe (Fig. 1-26).

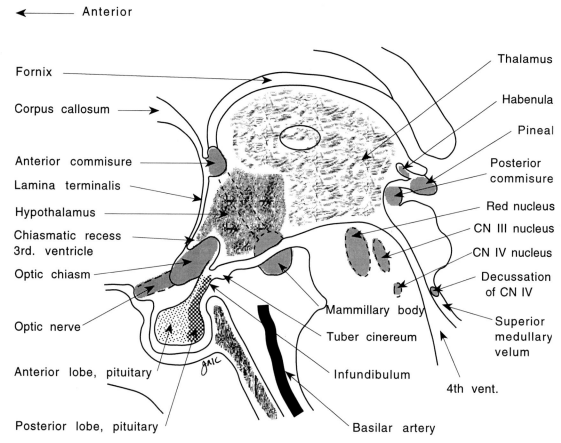

Anterior

Fig. 1-25 Pituitary gland and hypothalamus. Sagittal view. (Modified from Carpenter and Sutin, 1983, with permission.)

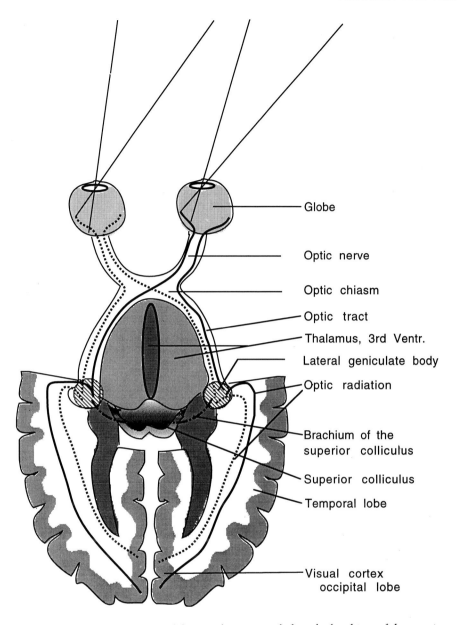

Globe

Optic nerve

Optic chiasm

Optic tract

Thalamus, 3rd Ventr.

Lateral geniculate body

Optic radiation

Brachium of the
superior colliculus

Superior colliculus

Temporal lobe

Visual cortex
occipital lobe

Fig. 1-26 The visual system. Axial drawing of the visual system, including the brachium of the superior colliculus, a connection between the lateral geniculate body and the superior colliculus. (Modified from Netter, 1958, with permission.)

Paired small mammillary bodies protrude from the undersurface of the hypothalamus posterior to the infundibulum (see Fig. 1-25). They have multiple communications with the limbic system through its tracts within the fornix. These bodies are easily seen on the midline

sagittal MRI in the roof of the suprasellar cistern. They atrophy in many cases of Wernicke's encephalopathy.

The thalamus is a large paramedian bilateral collection of nuclei separated by the posterior two-thirds of the third ventricle. It acts as a relay and association station, primarily for sensory function. The nuclear groups are named for their anatomic position. The large posterior nuclear complex is the pulvinar.

The bilateral paired medial and lateral geniculate bodies are part of the posterior thalamic nuclear complex. The medial geniculate body is a relay nucleus for the auditory system, receiving fibers from the inferior colliculus and providing a large radiation to the posterior insula (gyrus of Heschl), the major auditory cortex. The lateral geniculate body is the relay nucleus for the visual pathway.

The epithalamus is composed of the pineal gland, the habenula, and the epithelial roof of the third ventricle. The pineal is a midline unpaired structure related to the posterior margin of the third ventricle. It often calcifies with age and is a midline marker on plain skull radiographs.

The pineal seems to be light sensitive and has a role in circadian rhythms, and possibly depression. Lesions in this region may rarely produce precocious puberty. The habenula is a relay station of the limbic system (Fig. 1-27).

THE MESENCEPHALON

The mesencephalon, or midbrain, is a short segment of the brain stem. It contains the aqueduct of Sylvius, the narrow CSF-containing channel that connects the posterior third ventricle with the upper fourth ventricle. It is surrounded by the periaqueductal gray matter. Glioma, subependymoma, larger pineal tumors, and other masses in the region easily obstruct the thin aqueduct.

The mesencephalon is divided into three segments. The *tectum* is that part posterior to the aqueduct that forms the collicular (quadrigeminal) plate. The *tegmentum* is that portion anterior to the aqueduct containing the reticular formation, ascending white matter tracts, the nuclei of the oculomotor (CN III) and trochlear (CN IV) nerves, and the red nucleus surrounded by fibers of the superior cerebellar peduncle (Fig. 1-28; see also Figs.

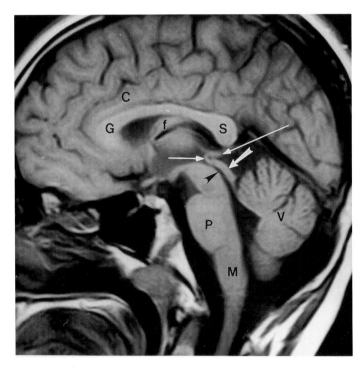

Fig. 1-27 Normal midline sagittal brain. A T_1 sagittal MRI through the midline of the brain shows the aqueduct (*arrowhead*), the posterior commissure (*short thin white arrow*), the pineal gland (*long thin white arrow*), and the collicular plate (*larger white arrow*). The vermis of the cerebellum (V), the pons (P), the medulla (M), the cingulate gyrus (C), the genu of the corpus callosum (G), the splenium of the corpus callosum (S), and the fornix (f) are also seen.

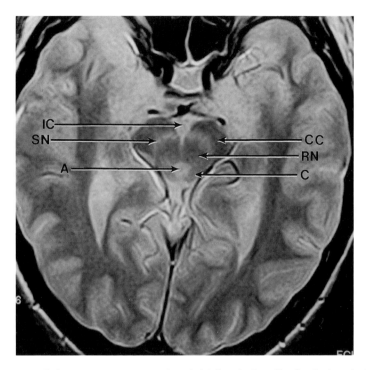

Fig. 1-28 Normal mesencephalon, T₂ MRI. A, aqueduct (of Sylvius); C, colliculus (inferior); CC, crus cerebri; IC, interpeduncular cistern; RN, red nucleus; SN, substantial nigra.

13-32 and 13-33). The large bilateral crura cerebri make up the anterolateral surface and consist of the corticospinal white matter tracts. Between the tegmentum and the crura lies the substantia nigra. The portion of the mesencephalon containing the tegmentum and the crus, but excluding the aqueduct and the tectal plate, is called the *cerebral peduncle* (Fig. 1-29).

The *superior colliculi* are paired bulges of the rostral tectal plate (see Fig. 13-32). They receive topographically organized fibers from the optic tracts and are involved in visual reflexes used to follow or react to visual and auditory stimuli. The pretectal region and the posterior commissure lie just rostral to the tectal plate. Lesions in this region may produce changes in the pupillary light response and to an inability to gaze upward (Parinaud syndrome, or Parinaud's ophthalmoplegia).

The *inferior colliculi* lie at the caudal end of the tectal plate (see Fig. 13-32). These relay nuclei within the auditory system send and receive impulses from the auditory cortex by way of the medial geniculate bodies.

The paired bilateral *red nuclei* lie within the reticular formation of the tegmentum. They are surrounded by the decussating fibers of the superior cerebellar peduncle (brachium conjunctivum) from which they receive afferent fibers (see Fig. 1-15D). The nuclei serve as a relay point for impulses primarily from the precentral cortex to the cerebellum and spinal cord. A lesion of the red nuclei will produce contralateral movement disorders (e.g., ataxia, chorea, tremor) and an ipsilateral oculomotor palsy from involvement of the adjacent CN III, collectively known as Benedikt syndrome.

The *reticular formation* consists of groups of nuclei within the tegmentum of the mesencephalon, extending inferiorly through the pons and into the medulla. It has various zones, the largest being the medial region accounting for two-thirds of the reticular mass. This medial zone is considered the reticular activation system, the effector region for the CNS that arouses and alerts the individual. Some regions of the reticular formation have inhibitory affects. The reticular formation within the medulla is involved with cardiovascular and respiratory control and activation mechanisms. Lesions within the brain stem are likely to produce disturbances of consciousness and coma from injury to the reticular system.

Focal upper brain stem lesions may produce the unusual akinetic mutism (coma vigil), in which consciousness is affected, but eye movements remain intact.

The *substantia nigra* lies in the anterior tegmentum of the mesencephalon, along the posterior margin of the crura cerebri. It is composed of two major layers: the posterior pars compacta, which contains pigmented cells, and the anterior pars reticulata, which contains relatively few neurons. The neurons of the pars compacta contain large amounts of dopamine, which they distribute to the corpus striatum through axonal flow in strionigral tracts. In Parkinson's disease, damage to the pars compacta leads to depletion of dopamine within the substantia nigra and therefore in the corpus striatum as well. MRI is able to visualize alterations within the substantial nigra in some patients with Parkinson's disease.

The oculomotor nerve arises from a nuclear complex within the central mesencephalon anterior to the periaqueductal gray matter at the level of the superior colliculi. This complex innervates the extraocular muscles (except for the lateral rectus and the superior oblique muscles), the levator palpebrae muscle, and the sphincter of the iris (Edinger-Westphal nucleus of the complex through the ciliary nerves). The nerve roots course directly anteriorly through the tegmentum, passing through the red nuclei, to exit medial to the crura cerebri into the interpeduncal cistern. From there the nerve courses anteriorly in the cistern between the posterior cerebral artery (PCA) and the superior cerebellar artery (SCA), travels within the superior portion of the lateral dural wall of the cavernous sinus, and enters the orbit through the superior orbital fissure.

Lesions of the CN III, either from intrinsic disease or external pressure (aneurysm of the posterior communicating artery [PCoA], parasellar tumor), cause oculomotor palsy and diplopia . There is eyelid droop and pupillary dilation (mydriasis), as CN III is the pathway for the pupillary-like reflex. Anisocoria (pupillary inequality) results from lesions of the nerve, the oculomotor nuclear complex and the pretectal fiber system coursing to the nucleus. The ciliary fibers controlling the iris are in the periphery of the third nerve, accounting for pupillary sparing (diplopia without pupillary dilation) when vasculopathy injures the central portion of the nerve as with diabetes mellitus. Acute onset of CN III palsy in an adult requires immediate imaging to define an enlarging aneurysm. A larger lesion within the mesencephalon, such as an infarction, may affect the ipsilateral root of the oculomotor nerve and the crus cerebri, causing an alternating hemiplegia with ipsilateral oculomotor palsy and contralateral hemiplagia (Weber syndrome).

Active pupillary dilation, as from emotional stress, operates through a different neural pathway. The pathway begins within the frontal lobe cortex, passes through the hypothalamus, and descends through the dorsolateral tegmentum of the entire brain stem to reach the C8–T1 segment. It then exits to the cervical sympathetic chain to synapse in the cervical sympathetic ganglion. Fibers from this ganglion ascend around the internal carotid artery to reach the orbit by joining the ciliary fibers of CN III and V. The ciliary nerves innervate the eyelid and the muscles of the iris. Lesions affecting any portion of this pathway will result in Horner syndrome, characterized by slight pupillary constriction, minimal ptosis, impaired ipsilateral facial sweating, and preserved constriction to light. The eyelid will not retract, and the pupil will not enlarge in response to an emotional event, such as fear.

The trochlear nerve (CN IV) originates in a small nucleus centrally in the dorsal tegmentum. The nerve roots decussate within the rostral superior medullary velum (roof) of the upper fourth ventricle to exit the posterior brain stem just caudal to the inferior colliculus (Fig. 1-29). The roots course around the stem within the ambient cistern, pass forward near CN III between the PCA and SCA to enter the midportion of the lateral wall of the cavernous sinus, just inferior to CN III. It innervates the superior oblique muscle. Solitary lesions of the fourth nerve are rare, but there are some reports of lesions due to trauma or hemangiomas associated with the nerve.

The Pons

The pons is the more superior segment of the hindbrain (rhombencephalon) (see Figs. 13-37 and 13-39). It is relatively large and is segmented into a dorsal tegmental portion and a ventral portion. The tegmentum contains the nuclei for CN V, VI, VII, and VIII, as well as the reticular formation and multiple ascending sensory tracts, the largest being the *medial* and *lateral lemnisci*, constituting the major spinothalamic tracts. The ventral portion contains the well-organized transverse fibers of the middle cerebellar peduncle (brachium pontis), the longitudinal fibers of the crura cerebri, and many pontine nuclei nestled within the transverse cerebellar tracts.

The superior medullary velum, the superior mesencephalic roof of the fourth ventricle, continues dorsally to

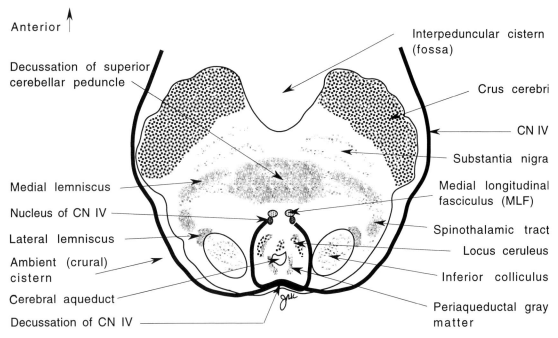

Anterior

Decussation of superior cerebellar peduncle

Interpeduncular cistern (fossa)

Crus cerebri

CN IV

Substantia nigra

Medial lemniscus

Nucleus of CN IV

Lateral lemniscus

Ambient (crural) cistern

Cerebral aqueduct

Decussation of CN IV

Medial longitudinal fasciculus (MLF)

Spinothalamic tract

Locus ceruleus

Inferior colliculus

Periaqueductal gray matter

Fig. 1-29 Inferior mesencephalon, CN IV. Axial view. (Modified from Carpenter and Sutin, 1983, with permission.)

form the upper roof of the main portion of the fourth ventricle. The lingula of the cerebellum lies directly on the dorsal surface of this thin membrane. The moderately large efferent tracts of the cerebellum arise from the dentate and other cerebellar nuclei and form a bundle called the superior cerebellar peduncle (brachium conjunctivum). These paired bundles ascend along the dorsolateral border of the upper fourth ventricle and pons, enter the tegmentum of the pons, decussate within the mesencephalon at the level of the inferior colliculus, and then surround the red nuclei. A few fibers synapse with the red nucleus, but the bulk continues superiorly to synapse in the ventrolateral nucleus of the thalamus.

The *locus ceruleus* nuclear complex lies just anterolateral to the gray matter surrounding the superior floor of the fourth ventricle. This nucleus has broad CNS distribution, contains fibers with norepinephrine, and has a role in cortical and spinal activation and in sleep.

The large CN V enters the pons at its anterolateral middle portion (belly). It is the primary sensory nerve for the face, oral cavity, nasal cavity, and paranasal sinuses. Its motor fibers supply the muscles of the jaw. The three major peripheral divisions—the ophthalmic, maxillary, and mandibular—enter the *trigeminal (Gasserian) gan-*

glion, which lies in the dural pouch on the posterolateral surface of the upper clivus called *Meckel's cave.* From here the nerve forms a large single root, coursing posteriorly through the prepontine cistern, to enter the anterolateral belly of the pons. It passes posteriorly through the anterior fibers to reach the multiple CN V nuclei within the tegmentum of the pons, medulla, and upper spinal cord (see Fig. 13-32).

The connections within the pons are complex. Root fibers for tactile sense and pressure synapse are within the primary sensory nucleus, which lies in the posterolateral tegmentum of the mid-pons. The root fibers for pain and temperature enter the descending tract of CN V, which extends inferiorly to the upper cervical spinal cord. The descending tract runs inferior to the primary nucleus in the posterolateral tegmentum. The tract parallels the long spinal trigeminal nucleus with which it communicates according to somatotopic organization. The motor fibers of CN V originate in the motor nucleus that lies medial to the primary sensory nucleus. The nerve follows the main trigeminal root, passing along the inferior trigeminal ganglion without synapse, and following the mandibular division to the jaw muscles (see Ch. 13 for further details on CN V).

CN VI (abducens nerve) arises from a nucleus in the lateral portion of the medial eminence on the floor of the fourth ventricle. The efferent fibers pass anteriorly through the belly of the pons lateral to the corticospinal tract, to exit at the ventral pontomedullary junction. The nerve takes a long superior route within the prepontine cistern along the clivus and enters the inferior portion of the cavernous sinus. It alone traverses within the cavernous sinus near the internal carotid artery (ICA), and enters the orbit through the superior orbital fissure. It supplies the lateral rectus muscle. A lesion of the nerve results in lateral rectus palsy, medial adduction of the eye, and horizontal diplopia. Because of its long course, the nerve may be affected by remote masses within the posterior fossa, or hydrocephalus. A lesion within the nearby medial longitudinal fasciculus causes inability to adduct the ipsilateral eye with attempts at lateral gaze (internuclear ophthalmoplegia). Lesions of the nuclei and surrounding interconnecting tracts cause a variety of ocular syndromes because of the interruption of the coordination necessary for effective conjugate eye movements.

The facial nerve (CN VII) has both motor and sensory function. The motor fibers originate in the motor nucleus in the ventrolateral tegmentum, anterior to the spinal trigeminal nucleus. Its efferent fibers loop posteriorly around the abducens nucleus (the internal genu) before traversing the pons, to enter the anterior superior IAC. The sensory component of the facial nerve supplies taste perception in the anterior two-thirds of the tongue. The fibers pass posteriorly from the tongue through the lingual nerve into the chorda tympani, across the superior part of the tympanic membrane, ending in the geniculate ganglion. From the ganglion, the sensory fibers continue in the nervus intermedius, which courses along with the facial nerve into the brain stem. The tract ends in the solitary nucleus in the upper portion of the medulla. Sensory fibers from the skin of the external auditory canal and posterior auricular region also travel in this nerve to end in the descending spinal trigeminal tract. Efferent fibers from the superior salivatory nucleus in the dorsal reticular formation course in the nervous intermedius. A branch from this nerve leaves the skull as the superior petrosal nerve to enter the pterygopalatine ganglion and ultimately innervate the lacrimal gland. A branch through the chorda tympani supplies the submandibular salivary glands.

Lesions of the facial nerve produce symptoms that vary with the site of damage. Central lesions in the cerebral cortex produce paralysis of the contralateral inferior face, known as central facial syndrome. Lesions in the peripheral nerve cause complete ipsilateral facial paralysis. Loss of taste, salivation, and tearing results if the lesion is proximal to the nerve branches supplying the appropriate territory. However, in practice, facial nerve lesions do not often follow strict rules, and a muddled clinical picture usually prevails (see in Ch. 13).

The vestibulocochlear nerve (CN VIII) has two distinct parts: the cochlear segment conducting auditory impulses, and the vestibular segment conducting balance and positional information (Fig. 1-30). The auditory nerve begins in the spiral ganglion adjacent to the organ of Corti, the neural transducer of the cochlea. Sound waves are conducted from the tympanic membrane through the oscicles of the middle ear cavity to the footplate of the stapes on the surface of the membrane of the oval window. The wave motion is transmitted to the fluid within the cochlea and transduced into neural impulses by the organ of Corti. The auditory component of the nerve travels from the cochlea in the anterior inferior compartment of the IAC through the cerebellopontine angle cistern, to enter the pons. From here the path is complex. Some fibers connect with the dorsal and ventral cochlear nuclei on the lateral surface of the inferior cerebellar peduncle. Others go to the superior olivary nucleus. From these sites, fibers travel within the lateral lemniscus superiorly either to the inferior colliculi or to the nearby brachium of the inferior colliculi, and then to the medial geniculate body. The fibers are both crossed and uncrossed. From the medial geniculate body, fibers radiate laterally to the posterior insula, the primary auditory cortex of Heschl.

Destruction of the cochlear nerve causes neurosensory deafness. Lesions within the brain stem affecting the lateral lemniscus cause diminution of hearing, but not complete deafness, because of the crossed representation within the tracts. Temporal lobe lesions within the auditory region of the temporal lobe lead to loss of unilateral hearing. Conductive hearing loss refers to loss of the transmission of sound waves through the oscicular chain to the cochlea. This occurs from chronic infection, oscicular dislocation, or otosclerosis.

The vestibular nerve originates from ganglia within the IAC. The nerves form superior and inferior bundles and occupy the superior and inferior portions of the posterior half of the IAC. The vestibular nerve innervates the semicircular canals and the vestibule of the labyrinthine system. It is concerned with head position and bal-

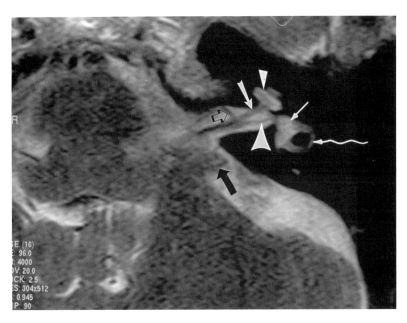

Fig. 1-30 Normal IAC. Axial T₂ MRI shows the left IAC filled with bright CSF. The cochlea lies just anterior to the lateral extent of the canal *(small arrowhead)*. The vestibule *(small white arrow)* and the lateral semicircular canal *(squiggle arrow)* lie more posteriorly. The cochlear division *(large white arrow)* and the vestibular division *(large arrowhead)* of the eight CN fill this level of the canal. The AICA undulates through the canal *(open arrow)*. The flocculus of the cerebellum protrudes into the posterior inferior C-P angle cistern *(black arrow)*.

ance. The nerve exits medially from the meatus of the IAC (porus acousticus) along with the facial nerve, nervus intermedius, and auditory nerve. It enters the pons, divides, and sends fibers to the large elongated mass of vestibular nuclei in the dorsolateral tegmentum in the inferolateral floor of the fourth ventricle. From here major distribution is to the spinal cord, the cerebellum, and the higher centers by way of the medial longitudinal fasciculus and olivary complex.

Lesions within the vestibular pathways result in imbalance, vertigo, nystagmus, and various conjugate eye disorders. A major lesion within the mesencephalon at the intercollicular level will cause decerebrate rigidity because of interruption of inhibitory influence of vestibular nerves on the overall stimulation effect of the brain stem reticular formation.

The Medulla

The medulla is the most caudal portion of the brain stem and forms a transition to the spinal cord (Fig. 1-31; see also Fig. 13-51). Centrally, it contains the reticular formation, which at this level controls respiration and

cardiovascular activity. This is the location of the nuclei of CN IX to XII. As a group these control swallowing and the larynx. The large olivary complex occupies the anterolateral regions just posterior to the corticospinal tracts. These have broad connections with the CNS, particularly the cerebellum and vestibular system. The corticospinal tracts decussate at the lower level of the medulla. The inferior cerebellar peduncle forms a large white matter tract on the posterolateral surface. This carries most of the spinocerebellar afferent fibers to the cerebellum. Many corticobulbar fibers from the cerebral cortex project to the nuclei and reticular formation of lower brain stem. Bilateral interruption of these descending fibers causes pseudobulbar palsy syndrome, characterized by difficulty chewing, swallowing, breathing, and speaking.

Infarction of the dorsolateral portion of the medullary produces a variable constellation of symptoms known as Wallenberg syndrome. It includes ipsilateral Horner syndrome, and contralateral loss of pain and temperature. The ipsilateral face may also be affected from injury to the descending tracts of CN V. Injury to the vestibular nuclei and the inferior cerebellar peduncle causes ver-

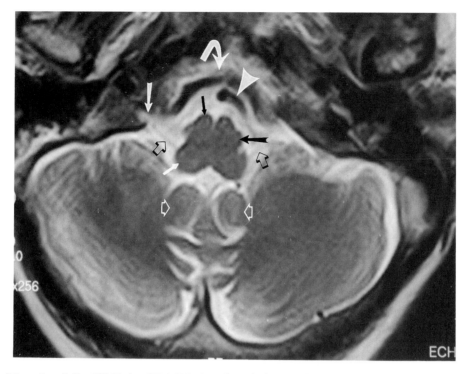

Fig. 1-31 Normal medulla, CN X. Axial T$_2$ MRI taken through the medulla at the level of the inferior olivary nuclear complex *(larger black arrow)*. The cortical spinal tract forms the anterior bump *(smaller black arrow)*. CN X exits posterolaterally *(black open arrows)* and courses to the pars venosa of the jugular foramen *(larger white arrow)*. The inferior cerebellar peduncle lies along the posterolateral medulla *(smaller white arrow)*. The cerebellar tonsils project into the valecula *(white open arrows)*. The left vertebral artery joins the basilar artery *(white arrowhead)* just posterior to the clivus (the posterior surface of the basioccipital bone, *curved white arrow*).

tigo, vomiting, nystagmus, and ipsilateral ataxia. Bilateral vascular or neoplastic lesions may occur in the medulla. This causes the "locked-in syndrome" with quadriplegia and an inability to speak, but preservation of consciousness.

The Cerebellum

The cerebellum is a moderately large structure that occupies the posterior intracranial fossa (Fig. 1-32). It consists of two hemispheres connected by midline cerebellar tissue called the vermis. Cortical gray matter contains specialized cells, the central medulla has myelinated white matter and there are four pairs of deep nuclei. The largest are the dentate nuclei next to the vermis and the lateral walls of the fourth ventricle. The fastigial nucleus is central near the apex of the roof of the

fourth ventricle. The cerebellar hemispheres contain layers of tissue called folia. A large horizontal fissure divides the hemispheres into superior and inferior halves.

The cerebellum is concerned with control of coordination, muscle tone, and balance. It affects the ipsilateral body. The major efferent fiber tracts exit the cerebellum from the dentate nuclei through the superior cerebellar peduncle and red nuclei. The major afferent tracts enter through the inferior cerebellar peduncle (restiform body). Lesions of the cerebellum produce ipsilateral intention tremor, loss of control of fine motor movements, decreased muscle tone, and nystagmus, more prominent when looking toward the side of the lesion. Speech disturbance is also common.

The fourth ventricle is a tent-shaped CSF-containing structure central to the cerebellum. The superior and inferior medullary vela form the roof. The floor is formed

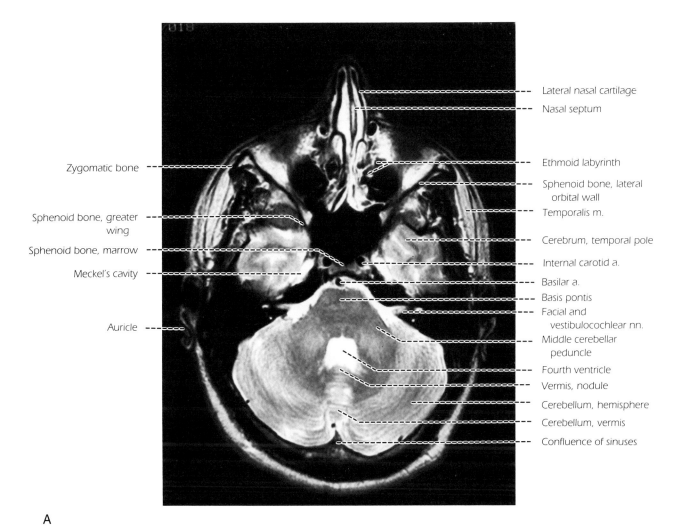

Zygomatic bone

Sphenoid bone, greater wing

Sphenoid bone, marrow

Meckel's cavity

Auricle

Lateral nasal cartilage

Nasal septum

Ethmoid labyrinth

Sphenoid bone, lateral orbital wall

Temporalis m.

Cerebrum, temporal pole

Internal carotid a.

Basilar a.

Basis pontis

Facial and vestibulocochlear nn.

Middle cerebellar peduncle

Fourth ventricle

Vermis, nodule

Cerebellum, hemisphere

Cerebellum, vermis

Confluence of sinuses

A

Fig. 1-32 (A) Normal cerebellum viewed with T_2-weighted MRI. Axial T_2-weighted image through the temporal pole, brain stem, and cerebellum. The portion of the brain stem shown here is the pons. The basis pontis is identified. Ventral to the basis pontis is the basilar artery. The middle cerebellar peduncle connects the pons with the cerebellum. The fourth ventricle is between the cerebellum and pons. The nodule of the vermis is dorsal or superior to the fourth ventricle. The facial and vestibulocochlear nerves (CN VII and VIII) are seen exiting from the pons. Other structures seen in this section include the nasal septum, lateral nasal cartilage, ethmoid labyrinth, temporalis muscle, internal carotid artery, confluence of sinuses, greater wing of the sphenoid bone, zygomatic bone, and bone marrow of the sphenoid bone. (*Figure continues.*)

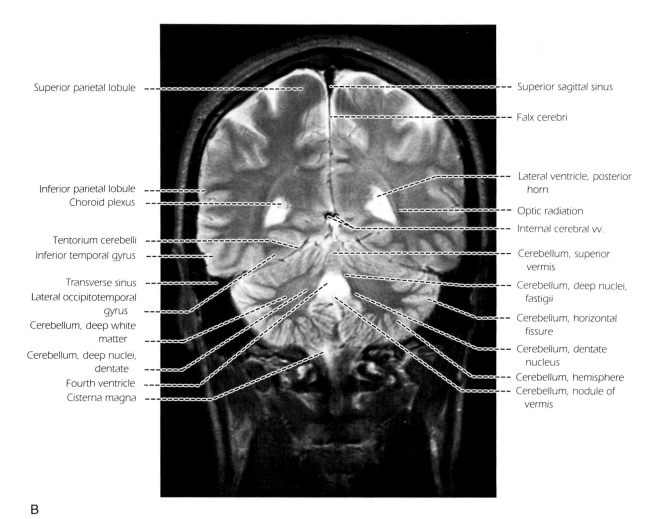

Superior parietal lobule

Inferior parietal lobule
Choroid plexus

Tentorium cerebelli
Inferior temporal gyrus

Transverse sinus
Lateral occipitotemporal
gyrus
Cerebellum, deep white
matter
Cerebellum, deep nuclei,
dentate
Fourth ventricle
Cisterna magna

Superior sagittal sinus

Falx cerebri

Lateral ventricle, posterior
horn

Optic radiation

Internal cerebral vv.

Cerebellum, superior
vermis

Cerebellum, deep nuclei,
fastigii

Cerebellum, horizontal
fissure

Cerebellum, dentate
nucleus

Cerebellum, hemisphere
Cerebellum, nodule of
vermis

B

Fig. 1-32 *(Continued.)* **(B)** Coronal T_2-weighted image through the parietal and temporal lobes. The superior sagittal sinus is located between the two parietal lobes. Within the temporal lobe, the inferior temporal and lateral occipitotemporal gyri are found. The tentorium cerebelli separates the cerebral hemisphere from the cerebellum. Deep within the cerebellum is the corpus medullare, the white matter core of the cerebellum. The fourth ventricle is seen with the bright signal intensity of cerebrospinal fluid within it. The cisterna magna is caudal to the cerebellum. The horizontal fissure of the cerebellum divides it into dorsal and ventral surfaces. The optic radiation is seen as a decreased signal intensity lateral to the trigone of the lateral ventricle. (From Yuh et al., 1994, with permission.)

by the dorsum of the pontine tegmentum. A choroid plexus is present, related to the inferior medullary velum. The fourth ventricle communicates with the third ventricle superiorly through the aqueduct. It communicates with the cisterna magna at the base of the brain through the single midline *foramen of Magendie,* and through the bilateral laterally positioned *foramina of Luschka.*

The Blood-Brain Barrier

Brain cells are extremely sensitive to variations in the chemical environment of the extracellular space. In order to ensure proper function, a system called the blood-brain barrier strictly regulates the transport of substances from the blood into the extracellular space and the CSF. The regulation occurs at two sites. One is at the interface formed by the capillaries and the extracellular space in the brain, and the other is at the interface of the choroid plexus and the CSF. The two are slightly different in function and anatomy.

The blood-brain barrier develops during invasion of the blood vessels into the fetal brain and it is well developed by birth. The barrier is present within nearly all sites of the CNS. The exceptions are a few highly vascular sites thought to have secretory functions responsive to chemical changes in the blood. These sites, which lack a blood-brain barrier, are the pineal, the neurohypophysis, the median eminence (pituitary stalk and adjacent tissue at the floor of the third ventricle), the organum vasculosum of the lamina terminalis (just below the anterior commissure), and the area prostrema (small paired structures at the inferior floor of the fourth ventricle just dorsal and rostral to the obex, or entrance, of the central spinal canal). Lying outside the blood-brain barrier, these structures may show visible contrast enhancement on CT and MRI. The pituitary stalk and the neurohypophysis consistently show contrast enhancement.

Tight junctions between the endothelial cells account for the blood-brain barrier within the capillaries. The tight junctions prevent the passive transfer of smaller molecules into the extracellular space and CSF, molecules that easily transfer into the extracellular space of the somatic tissue. There is little pinocytic transport of molecules across the capillary endothelium. Fat-soluble molecules are the only compounds that easily enter the brain tissue across the endothelium.

Aqueous intravascular radiographic contrast agents are held back by the blood-brain barrier and do not normally enter the extracellular space of the brain from the bloodstream. These agents enter the extracellular space and CSF only in the event of breakdown of the blood-brain barrier. This breakdown can be the result of the effects of a wide variety of influences, including infarction, infection, tumor, and trauma. Neovascularity, such as occurs with malignant tumor, or postsurgical scar, does not have a blood-brain barrier. This permits diffusion of contrast into the abnormal region tissue where it becomes visible by CT or MRI, highlighting the abnormal regions.

Within the choroid plexus, a blood-CSF barrier occurs at the level of the epithelial cells, rather than at the endothelium of the choroidal vessels. The cuboidal cells lining the surface of the choroid are closely held together by tight junctions, and the cells prevent the passage of blood molecules into the CSF. CSF is secreted by the cuboidal cells into the ventricular cavities. Injected contrast agent diffuses from vessels and collects in the interstitium of the choroid but does not diffuse beyond the cuboidal cell layer into the CSF. This accounts for the normally intense enhancement of the choroid plexus on post-contrast CT scans, and the minimal contrast accumulation in the CSF under normal circumstances. CSF contrast enhancement (accumulation of contrast within the CSF) occurs only with diffuse meningeal diseases, such as neoplastic meningitis, or menigiomatosis, and then only rarely. It is not understood why the choroid and dura do not enhance as greatly after gadolinium on MRI scans, as compared with iodinated contrast on CT scans.

CEREBRAL ANGIOGRAPHY

Cerebral angiography is serial radiography of the skull during opacification of the blood vessels with an aqueous contrast agent. Diagnosis of intracranial pathology is then made by (1) observing abnormalities involving the blood vessels themselves (aneurysm, vascular malformation, occlusion, arteritis, tumor neovascularity), or (2) inferring the presence of pathology by the observation of vascular displacements (tumor, hematoma, edema, abscess, hydrocephalus). Because the technique cannot image the brain parenchyma directly and because of its difficulty, danger, and expense, angiography has largely been displaced by CT and MRI. Nevertheless, cerebral angiography remains as the main technique for a number of CNS problems.

The contrast is most effectively delivered by a catheter selectively placed in the carotid and vertebral arteries via the femoral approach. Rarely, catheterization through the axillary artery or direct brachial or common carotid needle placement is an option if the femoral route is not available. Magnification filming is employed when possible, but it is not essential for most diagnostic work. Subtraction films can better demonstrate the vascular anatomy, particularly at the base of the skull. The smallest-diameter catheter possible is used, as it decreases the incidence of thrombus formation on the surface of the catheter. Heparin and Teflon-coated wire guides are generally preferred. Once within the arterial system, the catheter is double flushed with a heparinized saline solution every minute to keep it clear of blood. A continuous high-pressure infusion may also be used. Nonionic contrast agents do not prevent blood clotting, and so they cannot be used as a flush solution or catheter-filling agent. Ionic contrast agents prevent clotting. Meglumine iothalamate or any one of the low-osmolar contrast agents is recommended for contrast. Air must never enter a cerebral catheter. Table 1-2 provides guidelines for contrast flow rate and volume.

Angiography has been associated with significant complications. Stroke is the most common severe complication resulting from embolization of clot or air into a cerebral vessel. This problem may occur despite

INDICATIONS FOR CEREBRAL ANGIOGRAPHY

Common
  Atherosclerosis: extracranial, intracranial stenosis, occlusion
  Thromboembolism: arteries, and dural venous sinuses (diagnosis and thrombolysis)
  Aneurysm: berry, mycotic, traumatic, dissecting
  Vasculitis
  Vascular malformations: cerebral, dural
  Trauma: laceration, pseudoaneurysm, occlusion, intimal injury
Less common
  Definition of vascular supply to tumor
  Evaluation prior to neuroradiologic intervention: embolization, angioplasty, thrombolysis
  Topographic localization of vascular lesions
  Detection of vascular tumor: hemangioblastoma
  Definition of tumor type: glioma, meningioma, metastasis, hemangioblastoma, choroid plexus papilloma (rarely used)
  Determination of brain death

Table 1-2 Rates for Contrast and Filming[a,b]

Vessel Studied	Flow Rate	Volume (ml)	Film (DSA) Rate
Common carotid			
Bifurcation only	7–8 ml/sec	8	2 for 2
With intracranial	8 ml/sec	12	2 for 4, 1 for 6
Internal carotid	5–6/sec	8–10	2 for 4, 1 for 6
Vertebral	4–5/sec	5–8	2 for 5, 1 for 6
External carotid	1–3/sec	4–8	1 for 12

Abbreviation: DSA, digital subtraction angiography.
[a] For evaluation of AVM: contrast may be increased and filming rate increased to greater than 3 per second.
[b] For DSA, contrast may be diluted to approximately 30 percent concentration.

seemingly impeccable techniques (Fig. 1-33). Once an embolism has occurred, little can be done to alter the condition, except to apply steroid therapy and routine support of blood pressure and oxygenation. It is possible that thrombolysis will be used in the future as treatment for documented embolization from angiography. Stroke occurs with about a $1:1,000$ frequency when the procedure is performed by an experienced angiographer. Death from angiography may occur from stroke, cardiac arrest, severe contrast anaphylaxis, undetected femoral hemorrhage, or diffuse atheromatous embolization from a severely diseased aorta into visceral organs. The complications of angiography and preventive measures are given in Tables 1-3 and 1-4 (Fig. 1-34).

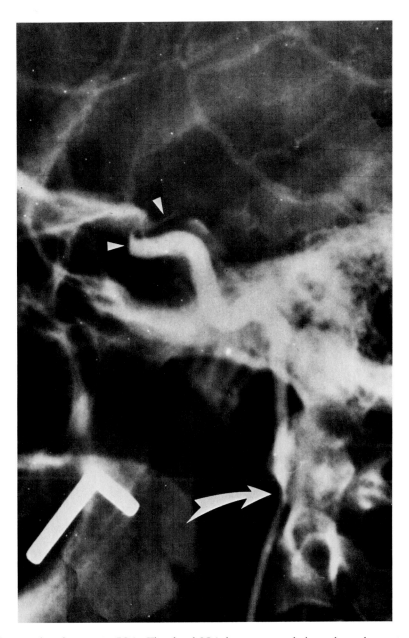

Fig. 1-33 Catheter-induced spasm in ICA. The distal ICA has contracted about the catheter tip *(arrow)* causing complete cessation of flow. A stagnant column of contrast is seen distal to the catheter. Gravity forms a fluid level at the termination of the contrast column and causes a small trickle of contrast to flow into the supraclinoid segment *(arrowheads)*. This problem tends to occur in young women and children.

Table 1-3 Complications of Cerebral Angiography

Complication	Approximate Rate
Death	1 : 5,000+
Stroke	1 : 1,000
Transient cortical blindness	1 : 1,000
TIA	1 : 500
Contrast reaction	1 : 100
Femoral hematoma, pseudoaneurysm, fistula (Fig. 1-34)	1 : 50

Aortic Arch

The most proximal branch of the aortic arch is the innominate artery (Fig. 1-35). It arises as a wide trunk from the superior margin of the anterior aortic arch and, after about 1.5 cm, divides into the right common carotid artery and the right subclavian artery. Normally, the right common carotid artery is the anterior branch. The left common carotid artery normally originates as a separate vessel from the superior margin of the aortic arch just distal to the innominate artery. The left common carotid artery frequently has a partial or complete common origin from the brachiocephalic artery. The left

Table 1-4 Prevention of Complications of Angiography

Preventive Measure	Rationale
Pretreatment with aspirin	Prevent platelet thrombus on wire guide and catheter
Heparin an Teflon-coated wire guides	Prevent thrombus on guide
Pretreatment with steroid	Stabilize endothelium; prevent transient cortical blindness and seizures
Systolic blood pressure < 180 mmHg	Decrease incidence of stroke and hematoma
Preangiography hydration and mannitol	Prevent contrast-induced renal failure in those with renal compromise, diabetes, and myeloma
Low-osmolar contrast	Prevent cardiac decompensation in those with heart disease; decrease chance of adverse reaction in those with prior contrast reaction
Monitor electrocardiogram	Detect and control cardiac arrhythmia
Keep procedure to less than 2 hours in duration	Incidence of complications increases after 2 hours
Keep catheter low in ICA and vertebral arteries	Prevent catheter-induced spasm around catheter (Fig. 1-33)
Avoid injecting costocervical trunk	Prevent quadriplegia from contrast toxicity to cervical spinal cord

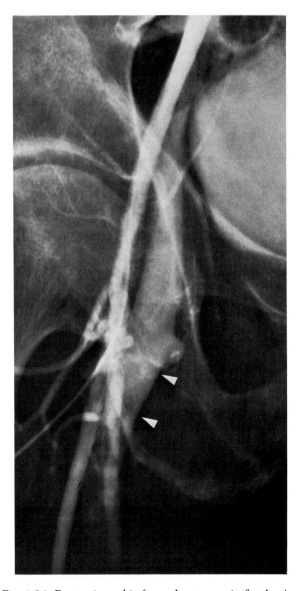

Fig. 1-34 Postangiographic femoral artery–vein fistula. An arteriovenous fistula may form at a puncture site. Here the femoral vein fills rapidly (*arrowheads*) during right femoral arteriography. Usually both artery and vein must be punctured for a fistula to form.

subclavian artery arises constantly just distal to the left common carotid artery. An aberrant right subclavian artery may rarely arise from the posterior aorta distal to the origin of the left subclavian artery.

The right and left vertebral arteries originate from the corresponding subclavian arteries, posteromedially about

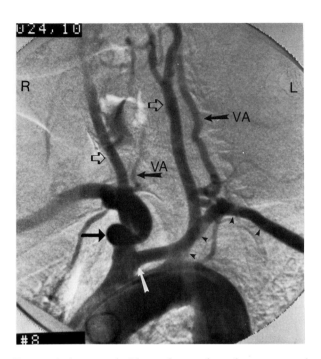

Fig. 1-35 Aortic arch. The study is performed using a pigtail catheter placed as far anterior in the aortic arch as possible. The left common carotid artery has an anomalous origin from the base of the innominate artery *(white arrow)*. The right subclavian artery arises from the posterior wall of the innominate artery *(black arrow)*. The vertebral arteries can be seen (VA, *arrow*) and the left vertebral artery is dominant. Both common carotid arteries *(open arrows)* and the left subclavian arteries are sen *(arrowheads)*.

2 cm distal to the aortic arch. The exact site of origin is variable. Occasionally, the left vertebral artery arises directly from the aorta from a position between the left common carotid and left subclavian arteries. The left vertebral artery is usually dominant. It is preferable to catheterize the dominant vertebral artery for posterior fossa angiography.

Common Carotid Artery

The common carotid artery is that portion of the vessel from the origin at the aorta to the common carotid bifurcation (Fig. 1-35). The bifurcation divides the vessel into the internal and external carotid arteries (Fig. 1-36). The internal carotid artery (ICA) is larger and is almost always posterior. The bifurcation is most commonly at the C4 vertebral level but may be low in the neck.

External Carotid Artery

The external carotid artery (ECA) is the smaller of the two branches arising from the carotid bifurcation. It normally courses anteromedially to the internal carotid artery. It gives rise to numerous important branches (Fig. 1-37).

INTERNAL MAXILLARY ARTERY

The internal maxillary artery is the main trunk of the ECA and begins by coursing superiorly parallel to the ICA. It gives rise to the external occipital, superficial temporal, and middle meningeal arteries. It then goes

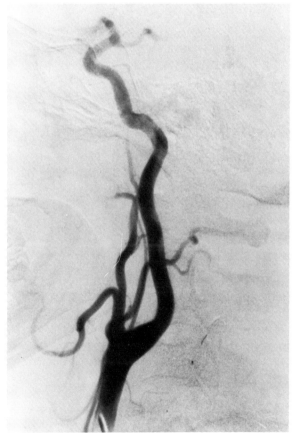

Fig. 1-36 Common carotid bifurcation. The CCA divides into the ECA (smaller anterior branch) and the ICA (larger posterior branch). The origin of the ICA is usually dilated and is called the "bulb." Normally, the vessel margins are perfectly smooth. (From Heinz, 1984, with permission.)

CHECKLIST FOR REVIEWING A CAROTID ANGIOGRAM

1. Venous and arterial midline markers for shift
2. Sylvian point and triangle for position
3. Position of horizontal portion of the middle cerebral artery
4. Position of venous angle
5. Capillary "brain stain" phase for subtle masses
6. Relationship of the cortical veins to the inner table
7. Late arteries and collateral flow
8. Early veins
9. Aneurysm, characteristic locations
10. Extracranial carotid and vertebral arteries
11. DMV, size
12. Vascular lumen (atherosclerosis, vasculitis)
13. Neovascularity

---

deep into the pterygoid fossa, to supply the maxillary sinuses. The distal branches of the internal maxillary artery anastomose with the distal branches of the ophthalmic artery, providing an important collateral arterial pathway to the intracranial ICA.

MIDDLE MENINGEAL ARTERY

The middle meningeal artery arises from the internal maxillary artery and goes superiorly through the foramen spinosum to enter the skull. From there it courses between the dura and the inner table of the skull along the floor of the temporal fossa and the greater sphenoid wing to reach the region of the pterion. Here it normally divides into the anterior and posterior meningeal branches, supplying the dura of the frontal and parietal convexities. The branches are normally accompanied by one or two parallel extradural veins. Together these vessels produce the meningeal vascular channels seen on lateral skull radiography.

The middle meningeal artery is important for the diagnosis of intracranial pathology. This vessel and its

branches enlarge with dural pathology, including meningioma and vascular malformation. Rarely, meningeal vessels contribute collateral circulation to the middle cerebral artery after stroke or to the blood supply of peripheral invasive intracerebral tumors (glioblastoma, gliosarcoma) that reach the cortical surface of the brain. Laceration and hemorrhage from a middle meningeal artery is the most common cause of a post-traumatic epidural hematoma.

EXTERNAL OCCIPITAL ARTERY

The external occipital artery is a posterior branch of the internal maxillary artery, which supplies the deep posterior muscles of the upper neck and the posterior scalp. It has important anastomoses with the deep muscular branches of the ipsilateral vertebral artery that may provide collateral pathways from the vertebral artery. The occipital artery may also supply a meningioma that has invaded the posterior calvarium.

SUPERFICIAL TEMPORAL ARTERY

The superficial temporal artery is the largest branch of the internal maxillary artery. It courses superiorly over the temporalis muscle to supply the anterior scalp. Its importance lies mainly in the diagnosis of temporal arteritis. It may also supply a meningioma that has invaded the calvarium.

ASCENDING PHARYNGEAL ARTERY

The ascending pharyngeal artery arises from the proximal internal maxillary artery and courses superiorly to the base of the skull. This small artery sometimes hypertrophies, to supply meningeal or acoustic tumors adjacent to the petrous pyramid.

SELECTIVE EXTERNAL CAROTID ARTERIOGRAPHY

Selective external carotid arteriography is performed to evaluate meningioma, vasculitis, glomus tumors, juvenile angiofibroma, and abnormalities of the calvarium or scalp. The catheter is rotated away from the internal carotid artery and advanced a small way into the proximal

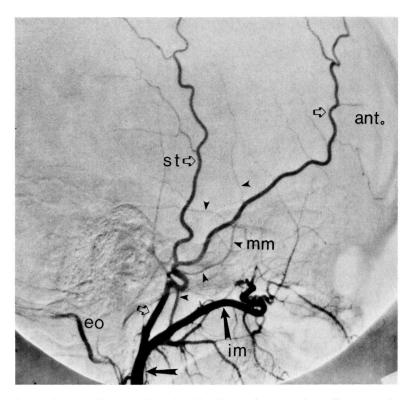

Fig. 1-37 External carotid artery. The major branch of the ECA is the internal maxillary artery (im, *black arrows*). It gives rise to the external occipital artery (eo), superficial temporal artery (st, *open arrows*), and middle meningeal artery (mm, *arrowheads*).

internal maxillary artery just distal to the lingual artery. Placing the catheter too far up the internal maxillary artery will likely result in localized spasm. The usual contrast injection is 2 ml of 60 percent contrast, for a total of 6 to 8 ml, followed by filming for 12 seconds at a 1/second rate. When dural arteriovenous malformations are evaluated, the filming rate is increased to 2 to 3 films/sec. With the standard ionic contrast agents the selective external carotid injection is intensely painful, likened to a hot iron placed on the face. The use of low-osmolar contrast agents significantly reduces the pain of ECA angiography and improves the film quality from reduced patient motion.

Internal Carotid Artery

The ICA courses posterolaterally from the bifurcation within the carotid sheath. It lies medial to the internal jugular vein. The vessel is divided anatomically into seg-

ments. For selective ICA angiography, the small 4 to 5 Fr. catheter is positioned in the vessel at about the C2 level. Contrast is injected at the rate of 6 ml/sec for a total of 8 ml, with intracranial filming of 2 frames/sec for 4 seconds, and 1 frame/sec for 6 seconds. For large aneurysms and vascular malformations, higher frame rates are used during the arterial phase.

CERVICAL SEGMENT

The cervical segment extends from the bifurcation to the entrance into the carotid canal of the petrous bone. Cervical sympathetic nerves course along this segment. Normally, there are no angiographically visible branches along this portion. Insignificant loops may occur (Fig. 1-38).

Atherosclerosis occurs at the proximal portion at the bifurcation, whereas fibromuscular disease, dissection, and traumatic lesions occur in the middle cervical seg-

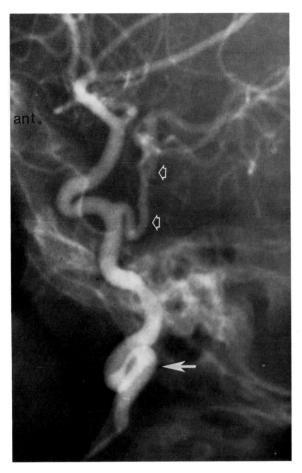

Fig. 1-38 ICA, primitive trigeminal artery. There is an insignificant loop in the cervical portion of the ICA (*arrow*). There is also a primitive trigeminal artery arising from the proximal cavernous portion to connect with the basilar artery (*open arrows*).

horizontally and anteromedially through the bony canal, lies just intracranially adjacent to the foramen lacerum, and then courses superiorly along the lateral clivus to the entrance to the cavernous sinus. This segment gives rise to two small branches. The caroticotympanic branch, which supplies the tympanic cavity, is not seen unless it is enlarged to supply a glomus tympanicum tumor. The pterygoid branch (vidian artery) arises from the horizontal canalicular portion to anastomose with pterygoid branches of the internal maxillary artery. It is sometimes seen angiographically as a collateral vessel faintly filling the petrous portion of the ICA when the proximal portion is occluded (see Fig. 2-38B).

The petrous segment is rarely involved with disease. Spontaneous dissection occurs usually as an extension from dissection in the cervical portion. The upper clival portion is in close relationship with the inferior border of the gasserian ganglion, and a neuroma in this region may anteriorly displace this segment (see Fig. 5-55). A primitive acoustic artery may rarely be seen connecting the petrous portion of the ICA with the basilar artery.

CAVERNOUS SEGMENT

The cavernous segment lies within the cavernous sinus, lateral to the sella and medial to CN III and IV and to branches of CN V and VI. The carotid artery turns sharply anterior as it enters the posterior cavernous sinus and then turns sharply superior to enter the dura just above and medial to the anterior clinoid process. This segment is also referred to as the carotid siphon. A primitive trigeminal artery may arise at the posterior siphon and connect with the basilar artery (Fig. 1-38). The cavernous segment gives rise to two important vessel groups.

ment. The highly vascular carotid body tumor occurs in the crotch of the bifurcation. The primitive hypoglossal artery may rarely be seen coursing through the hypoglossal canal, connecting the cervical portion of the ICA with the basilar artery.

PETROUS SEGMENT

The petrous segment extends from the entrance to the carotid canal at the base of the petrous bone to the entrance to the cavernous sinus. The vessel first courses

Meningohypophyseal Vessels
The meningohypophyseal vessels arise from the proximal curve of the cavernous portion of the ICA. There are usually three branches: (1) the tentorial branch, coursing posteriorly near the free edge of the tentorial hiatus; (2) the dorsal (clival) branch, coursing along the dura of the posterior clinoid process and the clivus; and (3) the inferior hypophyseal branch, which supplies the pituitary gland. There is anastomosis with the opposite side. These vessels, seen normally as small twigs, can hypertrophy to supply meningiomas or dural arteriovenous malforma-

tions of the tentorium, cerebellopontine angle, or clivus, providing important angiographic information (Fig. 1-39).

Cavernous Branches

The cavernous branches are numerous, small, short vessels arising from the cavernous portion of the ICA, supplying the dura of the cavernous sinus and gasserian ganglion. As with any dural branches, they hypertrophy to supply meningiomas or neurinomas of the paracavernous region. They may also enlarge as part of external carotid collateral circulation to the ICA when the proximal ICA is occluded (see Fig. 2-38) or as a dural arteriovenous malformation, emptying directly into the cavernous sinus.

SUPRACLINOID SEGMENT

The supraclinoid (intradural) segment extends from the anterior clinoid to the terminal bifurcation into the middle and anterior cerebral arteries. This segment gives rise to three important vessels (Fig. 1-40).

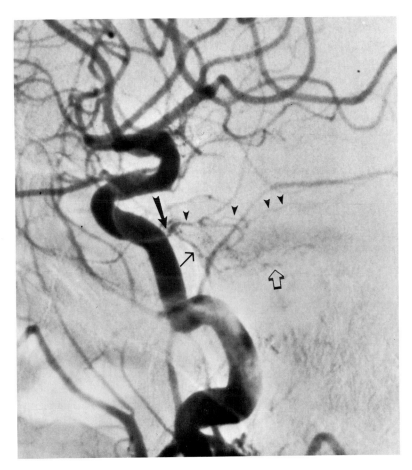

Fig. 1-39 Meningohypophyseal vessels, cerebellopontine angle meningioma. The meningohypophyseal vessels are hypertrophied to supply a meningioma, seen as a blush *(open arrow)*. The trunk *(thick black arrow)* arises from the proximal cavernous portion of the ICA. A tentorial branch courses straight posteriorly *(arrowheads)*. The dorsal branch courses inferiorly *(thin arrow)*.

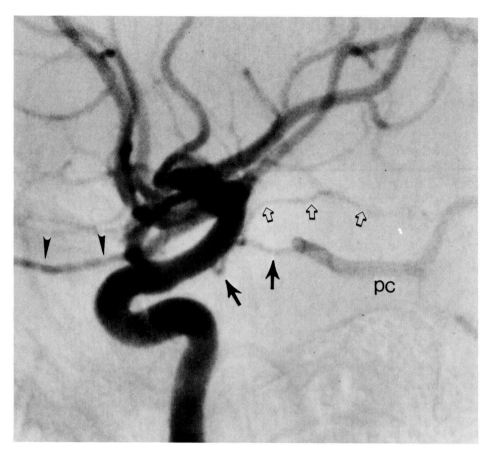

Fig. 1-40 Normal ICA. The supraclinoid segment of the ICA is shown. The ophthalmic artery is the first to arise from this segment, coursing anteriorly into the orbit *(arrowheads)*. The second vessel to arise is the posterior communicating artery *(arrows)*, coursing posteriorly to the ipsilateral posterior cerebral artery (pc). The third vessel to arise is the anterior choroidal artery *(open arrows)*.

Ophthalmic Artery

The ophthalmic artery arises first from the anteromedial aspect of the ICA, usually within the subdural space just as the ICA enters the intradural intracranial space. It courses anteriorly underneath the anterior clinoid process, through the optic canal, and into the orbit. From a lateral inferior position, it curves superiorly over the optic nerve to the anterosuperior and medial portion of the orbit. It exits the orbit anteriorly near the trochlea. It gives off anterior and posterior ethmoidal branches, which supply the ethmoid sinuses and the dura of the anterior falx (anterior falx artery), the cribriform plate, and the planum sphenoidale. These branches hypertrophy to supply meningiomas, which commonly occur

in these regions. The recurrent meningeal artery arising from the proximal ophthalmic artery may give rise to the entire middle meningeal artery. The ocular complex, including the central retinal artery, supplies the globe. It produces the "choroidal blush" outlining the posterior globe. Berry aneurysms occasionally arise from the ICA at the origin of the ophthalmic artery or just beyond. The artery may be involved with temporal arteritis.

Posterior Communicating Artery

The PCoA is the next branch to arise from the supraclinoid ICA. It undulates backward to join the proximal portion of the PCA. This artery forms the lateral segment of the anastomotic circle of Willis. The PCoA may be large, hypoplastic, or atretic. It is an important vessel for potential collateral blood flow to maintain ICA flow when there is proximal ICA stenosis or occlusion. When it is large, it continues without caliber change into the PCA, forming the "fetal type" origin of the PCA from the ICA. In this instance, the connection of the PCA to the basilar artery is lost. This anatomic variation occurs in about 30 percent of cases. When present, this connection allows occipital infarction to occur from carotid vascular disease.

A "funnel-shaped" origin of the PCoA is common. It is termed an infundibulum (see Fig. 2-26). It should not be considered an aneurysm unless the enlargement is more than 3 mm on nonmagnified films, if the margins are more rounded than triangular, and the enlargement does not align with the center of the main trunk of the PCoA. ICA berry aneurysms are common at the origin of the PCA. The PCA is stretched downward by transtentorial herniation of brain tissue and severe hydrocephalus.

Anterior Choroidal Artery

The anterior choroidal artery arises a few millimeters distal to the PCA (Figs. 1-40 and 1-41). This small vessel undulates posteriorly adjacent to the optic tract within the suprasellar cistern. After giving off tiny branches to the optic tracts and geniculate bodies, it dives into the choroidal fissure of the medial temporal lobe. Here it forms a characteristic kink, seen on the lateral angiographic projection. It supplies the choroid plexus, primarily within the temporal horn, as well as portions of the glomus within the atrium. Its posterior curve outlines the inferior portion of the pulvinar of the thalamus.

The anterior choroidal artery enlarges to supply intraventricular tumors, particularly meningioma or choroid plexus papilloma, and it occasionally contributes to vascular malformations of the temporal lobe or thalamus. It becomes stretched and displaced both downward and medially with uncal herniation of the temporal lobe (see Fig. 1-44B).

Anterior Cerebral Artery Complex

The anterior cerebral artery complex is composed of the short anterior cerebral artery (A-1 segment), the very short anterior communicating artery (ACoA), and the

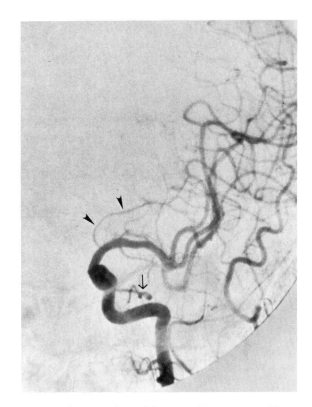

Fig. 1-41 Anterior choroidal artery, AP projection. The anterior choroidal artery is well seen with its characteristic course (*arrowheads*). Note the ophthalmic artery running anteriorly into the orbit (*arrow*). The anterior cerebral artery is atretic.

pericallosal artery and its branches. The anterior cerebral artery is the medial of the two terminal branches of the ICA. It arcs anteromedially from the distal ICA to the midline underneath the inferior portion of the genu of the corpus callosum just anterior to the thin lamina terminalis. At this point, the very short (2-mm) ACoA connects it with the opposite anterior cerebral artery. The anterior cerebral artery may be hypoplastic or atretic (Fig. 1-41). When it is, both pericallosal arteries fill from an angiographic injection of the contralateral ICA. Portions of the pericallosal group may arise from contralateral pericallosal arteries. The anterior cerebral artery gives rise to small lenticulostriate branches and the larger recurrent artery of Heubner, which supply the medial and anterior portion of corpus striatum (caudate nucleus, globus pallidus, and putamen), the anterior inferior internal capsule, and the septum pellucidum.

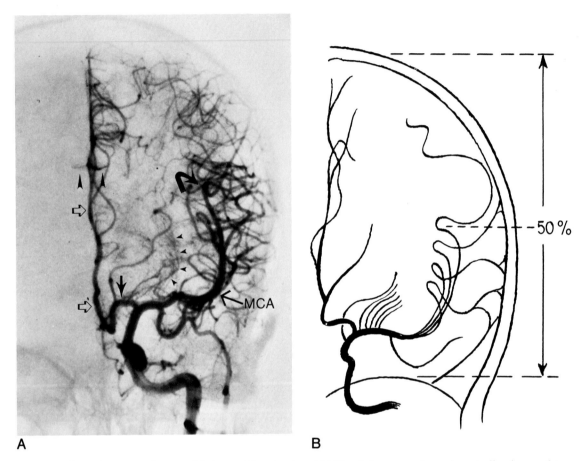

A   B

Fig. 1-42 Carotid angiography, arterial phase, AP projection. (A) The A-1 segment *(arrow)*, pericallosal artery *(open arrows)*, and callosal cistern blush called the "mustache" *(arrowheads)* are shown. The curved arrow points to the sylvian point in the posterior portion of the sylvian fissure. The angiographic sylvian point is usually halfway between the vertex and the ipsilateral roof of the orbit. Note the lenticulostriate arteries *(small arrowheads)*. **(B)** Diagrammatic representation of AP projection. (Fig. B from Heinz, 1984, with permission.)

The pericallosal artery is the principal continuation of the anterior cerebral artery (Fig. 1-42). It undulates within the interhemispheric fissure, curving close to the perimeter of the corpus callosum. Posteriorly, it breaks up into a plexus of small vessels that extend slightly laterally above the corpus callosum in the pericallosal sulcus inferior to the cingulate gyrus. This plexus is seen on the frontal angiographic projection as laterally directed horizontal vessels resembling a mustache. The "mustache" is tilted downward by masses that occur medially within the parietal lobe. This trait may be the only angiographic sign of such a mass.

Other branches of the pericallosal artery are the orbitofrontal branch, supplying the inferior medial frontal lobe, and the variable callosal marginal artery, which generally runs in the sulcus just peripheral to the cingulate gyrus. These vessels continue over the medial convexity of the frontal and parietal lobes to supply the parasagittal convexity for 2 cm lateral to the midline.

The pericallosal arterial complex is the major arterial midline marker. It shifts to correspond with herniation of the brain across the midline through the large opening in the more anterior portion of the relatively rigid midline falx. This shift is termed subfalcial herniation and occurs secondary to unilateral mass lesions within the calvarium. The shape of the shifted pericallosal artery varies with the location of the unilateral mass. Three basic patterns are encountered:

1. *Round shift—frontal masses:* The round shift is recognized as a curve of the pericallosal artery across the midline with the greatest amount of shift in the anterior segment of the complex (Fig. 1-43). It occurs secondary to a mass in the frontal region usually anterior to the coronal suture. The curve is reflecting the mass itself (tumor plus surrounding edema).
2. *Square shift—temporal, large posterior parietal masses:* The square shift results from a mass within the temporal fossa or a large mass in the posterior parietal or occipital region (Fig. 1-44). The square appearance is caused by the forward portion of the brain herniating as a whole.
3. *Posterior shift—all posterior locations:* With the posterior (distal) shift, only the more posterior portion of the brain is displaced across the midline (Fig. 1-45). The anterior portion may remain in the midline. The shift results from a mass in the parietal or occipital region.

The anterior cerebral artery complex may be affected by hydrocephalus, arterial occlusive disease, and masses in the suprasellar, subfrontal, or deep midline regions.

Circle of Willis

The circle of Willis is the anastomotic pentagon of vessels at the base of the brain lying within the suprasellar cistern. It is composed of the ACoA, both anterior cerebral arteries, both PCoAs, and the proximal segment of both PCAs. The anastomotic arch is complete and equal in about 20 percent of cases. Commonly, one or more of the anastomotic segments is absent or hypoplastic (Fig. 1-46).

Middle Cerebral Artery

The MCA is the more lateral of the two terminal branches of the ICA (Figs. 1-42 and 1-47). It curves laterally immediately underneath the anterior perforated substance of the brain (inferior surface of the basal ganglia) and above the temporal lobe. Numerous perforating small end-arteries arise from the horizontal segment of the MCA and are called the lenticulostriate arteries. These supply the basal ganglia. Laterally, it curves around the insula to enter the sylvian fissure. It normally bifurcates into anterior and posterior branches. These branches then undulate posterosuperiorly within the sylvian fissure lying on the surface of the insula. They give off additional branches, which turn sharply laterally, to exit the sylvian fissure between the parietal and temporal opercula (overhanging covering) to distribute over the cortical surface of the hemisphere. The anterior branch gives origin to the orbital frontal, precentral, central, and anterior parietal arteries, which supply the lateral frontal and anterior parietal hemisphere. They anastomose with distal branches of the anterior cerebral complex over the medial hemispheric convexity. The posterior branch gives origin to the posterior parietal, parieto-occipital, and posterior temporal branches, which supply the lateral posterior parietal, occipital, and temporal lobes. They anastomose with distal branches of the posterior cerebral artery over the posterior occipital convexity and the lateral and inferior surface of the temporal lobe.

The *sylvian point* refers to the point at which the most posterosuperior of the arteries on the insula makes its

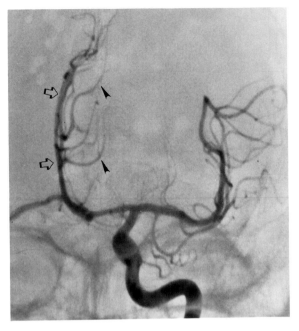

A

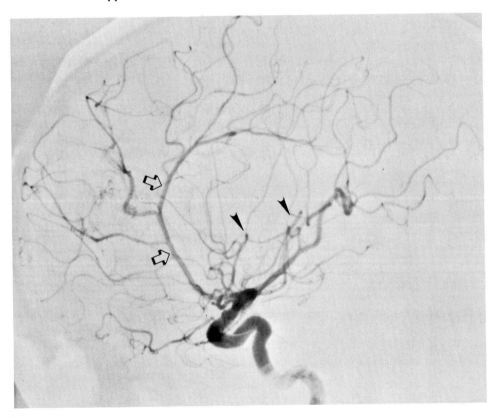

B

Fig. 1-43 "Round" shift, frontal lobe mass. (A) AP arterial angiography shows a rounded contralateral shift of the pericallosal group of vessels *(arrows)*. Note the "frontal polar sign" as the more peripheral vessels return to the midline before the pericallosal artery because of the falx *(arrowheads)*. **(B)** The pericallosal artery is stretched as it is displaced across the midline *(arrows)*. The sylvian vessels are displaced posteriorly and inferiorly *(arrowheads)*. There is wide separation and stretching of all the frontal vessels.

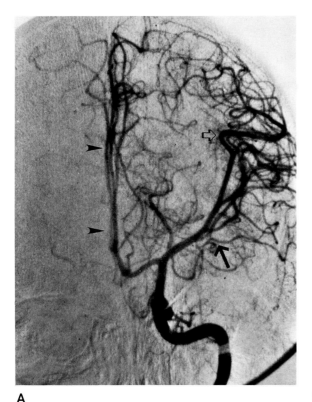
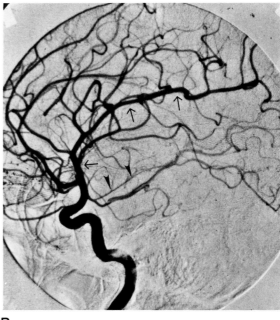

A   B

Fig. 1-44 "Square" shift, temporal lobe mass. (A) AP arteriography shows a slight "square" shift of the pericallosal group of vessels across the midline *(arrowheads)*. Little shift compared with the size of the mass is typical for temporal lobe lesions. The middle cerebral artery is elevated, and the sylvian vessels are displaced medially *(black arrow)*. The sylvian point is elevated *(open arrow)*. **(B)** Lateral arteriography shows marked elevation of the middle cerebral artery and sylvian vessels *(arrows)*. Note the stretching and downward displacement of the anterior choroidal artery, indicating uncal herniation *(arrowheads)*.

initial turn laterally to exit the sylvian fissure. On the frontal view it is most easily recognized as the most posterior and medial loop of the middle cerebral group within the sylvian fissure (see Fig. 1-10). It normally lies at a point halfway between horizontal lines drawn from the external table at the vertex and the orbital roof and is about one-third of the distance from the inner table to the midline. The sylvian point is elevated by masses within or underneath the temporal lobe (Fig. 1-44). It is depressed by masses within, or adjacent to, the parietal lobe (Fig. 1-45A). It is pushed forward by masses in the posterior parietal or occipital lobes (Fig. 1-45B).

The *sylvian triangle* is an upside-down triangle defined angiographically on the lateral projection (Fig. 1-47). The superior border of the triangle is formed by a line drawn connecting the points at which each of the middle cerebral arteries makes its turn deep within the sylvian fissure to course laterally around the parietal operculum. This line roughly parallels the planum sphenoidale and bifurcates the sweep of the pericallosal artery. The apex of the triangle is at the anterior clinoid process. The posterior side of the triangle is defined by the clinoparietal line (a line drawn from a point 2 cm anterior to the lambda to the anterior clinoid process on nonmagnified angiography). The sylvian triangle

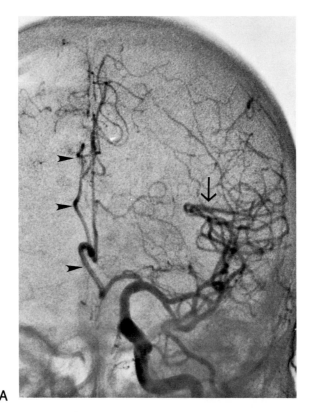

A

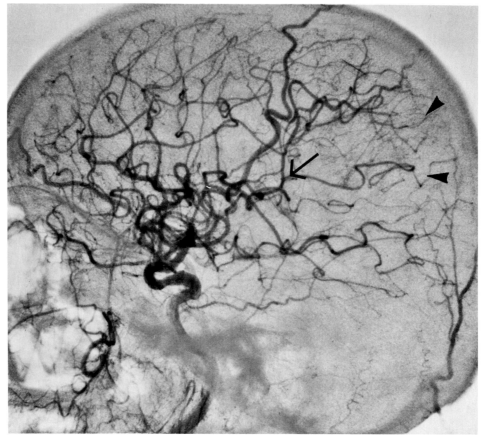

B

52

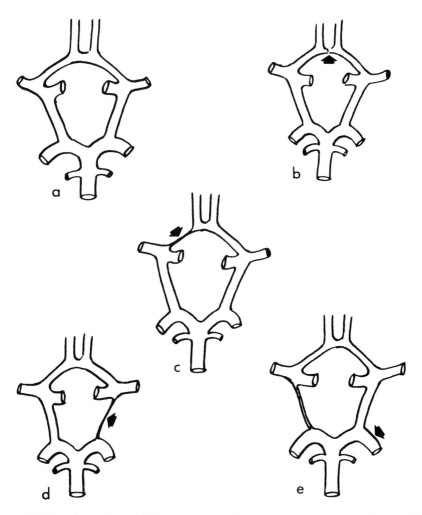

Fig. 1-46 Circle of Willis and variations. (a) The complete circle. A common variation not illustrated is atresia of the proximal segment of the posterior cerebral artery so that the posterior communicating artery becomes the sole supply of its distal segment. This variation is the "fetal type" of communication (see Fig. 1-47B). (From Heinz, 1984, with permission.)

is deformed by masses in the frontal (Fig. 1-43), parietal (Fig. 1-45), occipital, and temporal lobes (Fig. 1-44). The horizontal segment of the MCA is also elevated by masses within the temporal lobe (Fig. 1-44A).

The capillary or intermediate phase of the angiogram represents the flooding of the capillaries with contrast, which occurs at 3 to 6 seconds on normal angiography, best seen with the subtraction technique. It produces a brain stain, which is useful for defining small mass effects

Fig. 1-45 Posterior shift, posterior parietal mass. (A) AP angiography shows the posterior portion of the pericallosal artery shifted more than the anterior portion (*arrowheads*). Note the downward displacement of the sylvian point (*arrow*). **(B)** A large mass is present posteriorly, causing anterior and downward displacement of the sylvian vessels (*arrow*) and stretching of the posterior parietal convexity vessels (*arrowheads*).

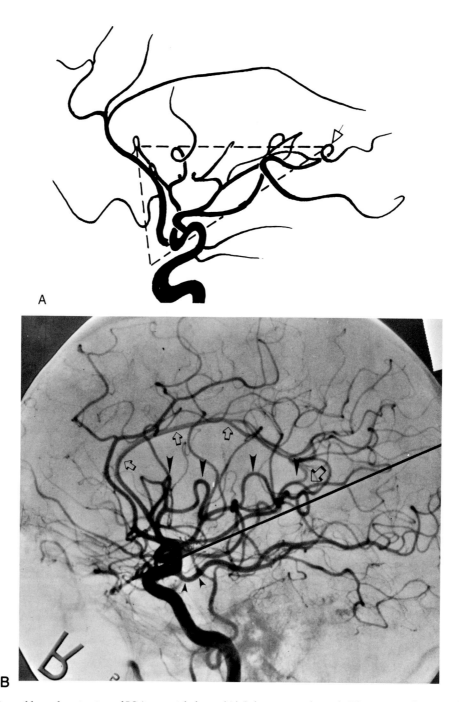

Fig. 1-47 Normal lateral projection of ICA, arterial phase. (A) Sylvian point *(arrow)*. The two vessels arising from the supraclinoid ICA are the proximal posterior communicating artery and the more distal anterior choroidal artery. **(B)** Lateral angiography outlines the normal sylvian vessels. The posteroinferior border of the sylvian triangle is defined by the clinoparietal line paralleling the posterior branch of the MCA within the sylvian fissure. The superior border of the triangle is outlined by the opercular branches *(large arrowheads)*. The sylvian point is indicated *(large open arrow)*. Note the normal sweep of the pericallosal artery *(small open arrows)*. There is a "fetal type" posterior cerebral artery *(small arrowheads)*. Note the normal undulating course of all the cortical arteries. Mass lesions cause straightening as well as displacement of cortical vessels. (Fig. A from Heinz, 1984, with permission.)

or ischemic regions. The normal brain stain is uniform in density throughout the hemisphere (Fig. 1-48). Masses, edema, or infarctions are seen as regions of void within the stain.

Cerebral Veins

The cerebral veins are important for angiographic diagnosis. They are altered by cerebral masses, vascular malformations, extracerebral fluid accumulations, and thrombosis. They can be anatomically classified into superficial and deep veins.

SUPERFICIAL CEREBRAL VEINS

The superficial veins course over the surface of the gyri of the cerebral hemisphere (Fig. 1-49) and drain to the dural sinuses. The veins fill with contrast sequentially in an order determined by their distance from the terminal carotid artery. Thus, the frontal and temporal veins fill first, followed by the parietal veins and finally the occipital veins. Frequently, there is a large vein of Trolard, which drains the parietal lobe superiorly to the sagittal sinus; a vein of Labbe, which drains the posterior temporal lobe into the lateral sinus; and a sylvian vein, which drains the insula and opercular regions into the cavernous sinus. The sylvian veins fill relatively early. Numerous unnamed smaller cortical veins drain the lateral and interhemispheric cortical surfaces to the nearest dural sinus.

The presence of an early-filling vein is a basic angiographic sign and is always an abnormal finding. For a vein to be classified as early filling, it must fill at least 1 second earlier than the veins in its immediate vicinity. It is most commonly seen with a malignant primary or metastatic brain tumor (see Ch. 5) or arteriovenous malformation (see Ch. 4). It may also be seen whenever there is an increase in the regional blood flow that occurs with the luxury perfusion surrounding an infarction (see Fig. 2-24B), an abscess, or recent seizure. In some cases, it may be the only angiographic abnormality demonstrated. When seen alone, it is nonspecific, and the nature of the abnormality then cannot be diagnosed.

The cortical veins are the best markers of the brain surface on angiography. Whereas the cortical arteries course within the depths of the cerebral sulci and so may be at some distance from the inner table of the skull, the

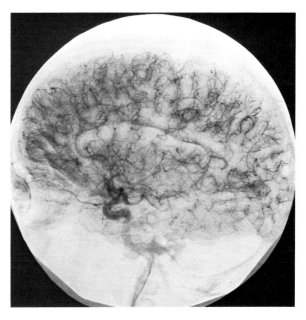

Fig. 1-48 Normal ICA arteriography, capillary phase. The capillary, or "brain stain," phase is intermediate between the arterial and venous phases. It shows a cortical blush for all normally perfused regions of the brain. The perfusion pattern should be homogeneous. Because mass lesions decrease regional blood flow, they can be identified by voids in the brain stain phase. (From Heinz, 1984, with permission.)

veins run on the cortical surface and are normally close to the inner table of the calvarium. They are displaced inwardly from the inner table by extracerebral fluid collections (subdural and epidural hematoma or hygroma), meningioma, or epidural extension of a calvarial metastasis. They traverse the enlarged subarachnoid space associated with brain atrophy and are subject to tears and hemorrhage with trauma.

Although a relatively infrequent occurrence, sagittal sinus thrombosis may cause cerebral infarction. The dural sinuses, especially the sagittal and lateral sinuses, must be specifically observed to determine normal contrast filling. Normally, the posterior portion of the sagittal sinus shows greater filling. However, there are always some contrast "defects" in the opacified sinus because of the entry of unopacified blood from the contralateral side or the posterior fossa. There is also considerable asymmetry of the lateral sinuses, and one side is usually dominant. Reversal of flow in the cortical veins and collateral deep venous drainage must be seen for a reliable diagnosis of sinus thrombosis (see Ch. 2).

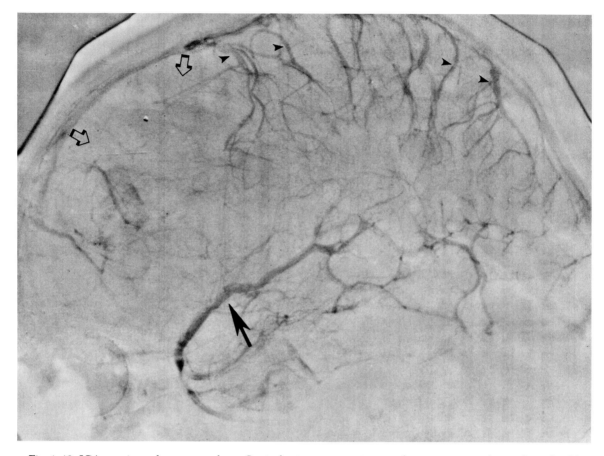

Fig. 1-49 ICA arteriography, venous phase. Cortical veins are emptying into the superior sagittal sinus *(arrowheads)*. A large sylvian vein is seen *(black arrow)*. There is decreased venous filling in the frontal lobe because of a tumor mass *(open arrows)*.

DEEP CEREBRAL VEINS

The deep venous system refers to the paired midline internal cerebral veins, the subependymal veins of the lateral ventricles, and the deep medullary veins of the cerebral white matter. They have importance in the diagnosis of midline brain shifts, hydrocephalus, and cerebral pathology of the white matter.

The paired internal cerebral veins course within the cavum velum interpositum immediately above the roof of the third ventricle. They comprise the most impor-

tant angiographic midline marker and are generally more sensitive to the detection of midline shifts than the pericallosal arteries. The posterior portion of the internal cerebral veins are held close to the midline by their entry into the vein of Galen, but the anterior portion is free to pivot with midline herniations of the brain. Therefore, with hemispheric shifting, the anterior portion pivots away from the side of the mass lesion. This situation can be recognized on the frontal projection of the angiogram (Fig. 1-50). The internal cerebral vein is normally seen as a thick, short band of contrast superimposed on the vein of Galen

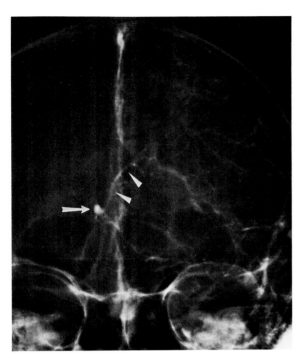

Fig. 1-50 Internal cerebral vein, contralateral shift. The anterior portion of the internal cerebral vein is shifted across the midline (*arrow*). There also is contralateral displacement of the thalamostriate vein (*arrowheads*).

and the straight sinus. With shift, the internal cerebral vein can be seen as a humped, foreshortened channel displaced across the midline away from the mass.

The subependymal veins are numerous, relatively constant veins that run just underneath the ependyma of the lateral ventricles. The most important, the thalamostriate vein (Fig. 1-51A), drains from the lateral superior angle of the body of the lateral ventricle and courses anteromedially and inferiorly in the groove between the tail of the caudate nucleus and the thalamus, to meet the internal cerebral vein at the foramen of Monro. Usually the anterior septal midline vein enters at this point as well, which is called the deep venous angle (Fig. 1-51B). Frequently, the thalamostriate vein is not present, and its draining function is replaced by a direct lateral ventricular vein. This vein also drains from the lateral superior angle but then courses directly

medially along the lateral wall and then the floor of the body of the lateral ventricle to enter the internal cerebral vein at some point posterior to the foramen of Monro. Its point of entry is called the false venous angle (Fig. 1-51C). It is important to recognize this variant so that displacement of the venous angle is not mistakenly diagnosed.

Because they course along the walls of the lateral ventricles, the subependymal veins are excellent markers of ventricular size. The deep veins form what is referred to as the cast of the ventricular cavities. As seen on the frontal projection, the thalamostriate vein outlines the lateral margin of the frontal horn. With ventricular enlargement, the thalamostriate vein is stretched and bowed laterally (Fig. 1-52). The width across the top of a lateral ventricle is normally less than 20 mm when measured on the nonmagnified frontal view. The outline of the ventricle can also be seen on the lateral projection.

The deep medullary veins of the white matter course along the parallel fibers of the corona radiata and drain deep into the subependymal veins at the lateral superior angle of the ventricular system. On high-quality films, they may be seen normally as a series of short, faint, brush-like veins perpendicular to the ventricle fading off into the paraventricular white matter (Fig. 1-51). They enlarge and become much more visible in the presence of any process that results in an increase in the blood flow through the white matter. Prominent deep medullary veins most commonly indicate a glioblastoma. They may be less commonly seen with deep arteriovenous malformations, cortical venous or venous sinus thrombosis (collateral flow), veno-occlusive abnormalities, lymphoma, progressive multifocal leukoencephalopathy, deep infection, and fulminant multiple sclerosis. A checklist for reviewing a carotid angiogram is shown below.

VERTEBRAL ANGIOGRAPHY

Vertebral angiography defines the vessels that supply all of the posterior fossa, the posterior thalamic nuclei, the choroid plexus of the atria of the lateral ventricles, and the occipital and posterior inferior temporal lobes, unless the PCA arises directly from the carotid artery.

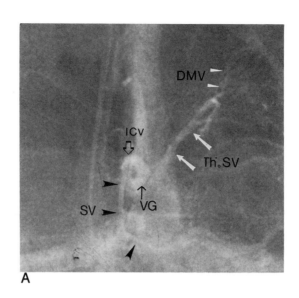

A

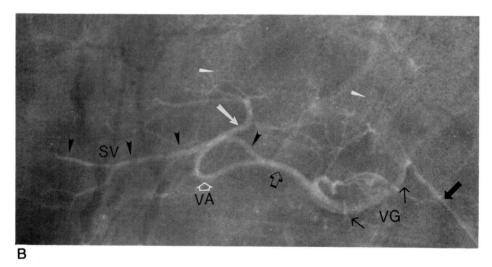

B

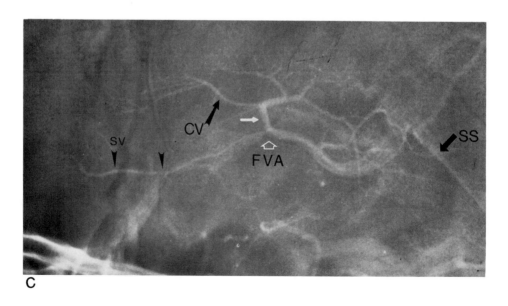

C

58

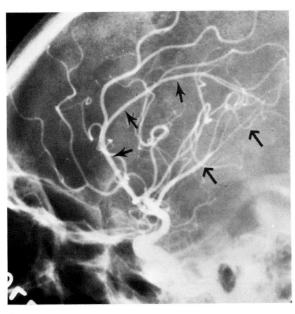

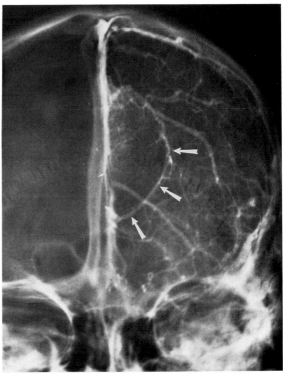

A

B

Fig. 1-52 Hydrocephalus. (A) Lateral arterial phase carotid angiography shows stretching of the pericallosal vessels (*sharp arrows*) and elevation of the sylvian vessels from temporal horn enlargement (*blunt arrows*). **(B)** AP venous phase shows lateral displacement and stretching of the thalamostriate vein (*arrows*), representing frontal horn enlargement.

Vertebral Artery

Throughout its course along the cervical spine, the vertebral artery passes through the transverse foramina of the upper six vertebral segments. It gives off numerous segmental muscular branches. These branches anastomose with the cervical muscular branches of the ipsilat- eral ECA, providing the potential for collateral circulation. If the proximal vertebral artery is occluded, blood flows from the external carotid system through the muscular branches to fill the distal vertebral artery (see Fig. 2-42). If the common carotid artery is occluded, blood flows from the vertebral into the ECA and then to the ICA via the bifurcation (see Fig. 2-40B).

Fig. 1-51 Normal deep veins. (A) AP view obliqued slightly to the right shows the midline septal vein (*black arrowheads*), internal cerebral vein (*open arrow*), and vein of Galen (*thin black arrow*). The tiny deep medullary veins (*white arrowheads*) are seen draining into the thalamostriate vein (*white arrows*), which outlines the lateral margin of the posterior frontal horn. **(B)** Lateral projection. The deep medullary veins (*white arrowheads*) drain into the thalamostriate vein (*white arrow*), which courses anteroinferiorly to meet the internal cerebral vein (*black open arrow*) at the venous angle (*white open arrow*). The septal vein courses along the septum pellucidum (*arrowheads*). In this patient, the septal vein enters the internal cerebral vein posterior to the venous angle. The vein of Galen (*thin black arrows*) is a short, large vein in the midline draining into the straight sinus (*black arrow*). **(C)** False venous angle. Here the thalamostriate vein is atretic, and there is a direct lateral vein (*white arrow*) emptying into the internal cerebral vein at the false venous angle (*white open arrow*). Note the anterior caudate vein (*thin black arrow*) and the septal vein (*arrowheads*).

After characteristically zigzagging through the transverse process of the atlas, it curves posteriorly to the atlanto-occipital joint, to pierce the dura and pass superiorly through the foramen magnum. The vertebral artery often becomes smaller as it enters the intradural space. The vertebral arteries give off anterior and posterior meningeal branches, which supply the dura of the foramen magnum, the falx cerebelli, and the occipital dura.

Basilar Artery

The large basilar artery is formed from the junction of the two vertebral arteries. It continues superiorly more or less in the midline posterior to the clivus and close to the anterior surface of the pons. It tends to curve away from the side of the dominant vertebral artery; but it may meander, especially if it becomes ectatic from atherosclerosis. At a point slightly above the dorsum sellae, it divides into the bilateral PCAs. Along its course, the basilar artery gives off numerous small pontine perforating arteries, analogous to the perforating arteries from the MCA to the basal ganglia.

The basilar artery may be significantly diseased with atherosclerosis, resulting in the numerous symptoms of vertebrobasilar insufficiency. Severe atherosclerosis may result in a giant aneurysm of the entire basilar artery (see Ch. 3). Berry aneurysms occur at the distal tip of the basilar artery and at the origins of the posterior inferior cerebellar arteries (PICA) and the anterior inferior cerebellar arteries (AICA). Masses within the cerebellum and pons displace the basilar artery forward against the clivus (see Fig. 1-55). Masses arising from the clivus (i.e., meningioma) or basisphenoid bone (i.e., chordoma, metastasis) displace the basilar artery backward (see Ch. 5).

Posteroinferior Cerebellar Arteries

The important bilateral PICAs arise from each distal vertebral artery as it lies anterolateral to the upper medulla (Fig. 1-53). The level of origin and the course of

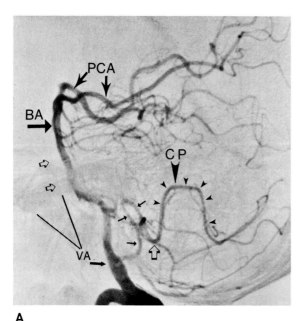
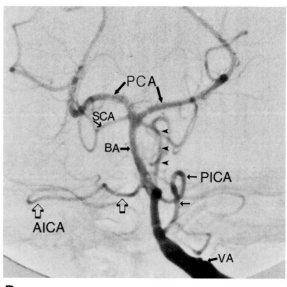

Fig. 1-53 Normal vertebral arteriography, arterial phase. (A) Lateral arterial projection shows the basilar artery (BA) separated from the posterior margin of the clivus *(small open arrows)*. The lateral medullary segment of the PICA *(small arrows)* arises from the distal vertebral artery (VA). The caudal loop of the posterior medullary segment is indicated *(large open arrow)*. The choroidal loop is well formed in this patient *(arrowheads)*, and the choroidal point is indicated (CP). **(B)** Towne projection, showing the lateral medullary segment *(small arrows)* and the choroidal loop *(arrowheads)* of the PICA. The choroidal loop is in its normal position just lateral to the midline. The AICA is well seen on the opposite side *(open arrows)*.

the PICA are variable, but certain generalities obtain. The PICA begins by making a short superior curve to reach the lateral surface of the medulla and then courses downward along the lateral surface of the medulla ending in an upward curve. These portions are called the anterior and the lateral medullary segments. The inferior curve is usually referred to as the caudal loop. At the caudal loop, the artery normally divides into medial and lateral branches.

The medial branch continues superiorly and medially, to run along the anterior surface of the cerebellar tonsil and then on the inferior surface of the posterior medullary velum (inferior roof) of the fourth ventricle, to reach the apex of the triangle of the roof of the fourth ventricle (the fastigium). It then loops backward and downward over the posterior surface of the tonsil and continues posteriorly near the midline along the undersurface of the vermis. The relatively broad superior curve of the medial branch is called the choroidal loop. At its apex, it usually gives off small variable branches to the choroid plexus of the fourth ventricle. This area is called the choroidal point and is an important marker of the position and size of the fourth ventricle.

The choroidal point is generally considered to lie at the apex of the curve of the choroidal loop. Numerous systems have been designed to measure its position. The preferred system is the F-T line because it is proportional and is not subject to the vagaries of different magnification (Fig. 1-54). This line is drawn from the anterior margin of the foramen magnum to the torcula. The choroidal point should lie close to the junction of the anterior and middle thirds of the line. There are numerous variations to the appearance of the choroidal loop, and at times the position of the choroidal point is an estimate.

The PICA is the arterial marker for the position and size of the fourth ventricle. Masses within the posterior compartment of the posterior fossa (cerebellar hemispheres and vermis) displace the fourth ventricle and, thus, the choroidal loop anteriorly and inferiorly (Fig. 1-55). Masses in the anterior compartment (pons, clivus) displace the choroidal loop posteriorly. Masses within the fourth ventricle separate and broaden the two PICA loops.

The PICA is the most important midline marker within the posterior fossa. On the anteroposterior projection, the choroidal loop and the inferior vermian branch are displaced contralaterally by cerebellar hemispheric masses (Fig. 1-56).

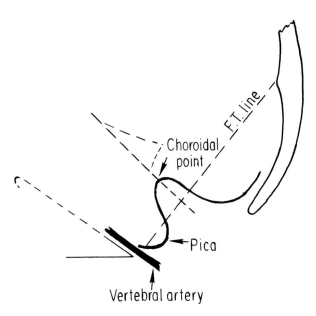

Fig. 1-54 Lateral vertebral arteriogram and choroidal point. The lateral vertebral angiogram illustrates the (PICA) and the relationship of the choroidal point to the fourth ventricle. The choroidal point should fall at the juncture of the anterior and middle thirds of a line (F.T. line) drawn between the anterior lip of the foramen magnum (F) and the torcular herophili (T). (From Heinz, 1984, with permission.)

Anteroinferior Cerebellar Arteries

The paired AICAs arise from the lower portion of the basilar artery. They course laterally around the pons, to meet the exiting CN VII and CN VIII. The artery follows the nerves into the internal auditory canal, makes a loop, and continues laterally over the anterior cerebellar hemisphere (Fig. 1-53B; see also Fig. 1-59B). The size of the AICA is usually reciprocal with the size of the ipsilateral PICA. At times the PICA may be absent on one side, and the AICA supplies the distribution of both the AICA and PICA (Fig. 1-57). The AICA lies within the cerebellopontine angle cistern and is displaced by large tumors in this region, such as acoustic neurinoma (see Ch. 5) and meningioma. Berry aneurysms may rarely occur at its origin.

Superior Cerebellar Arteries

The paired superior cerebellar arteries arise from the distal basilar artery. They pass around the upper pons or mesencephalon within the ambient cisterns just below

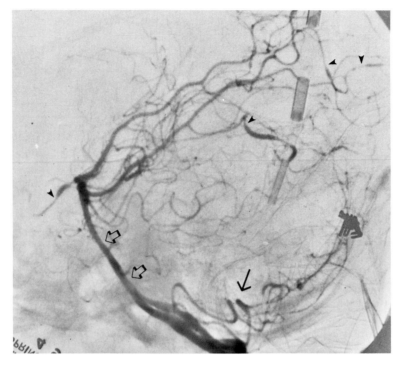

Fig. 1-55 Posterior compartment hemispheric mass. The lateral vertebral study shows anterior and inferior displacement of the choroidal loop of the PICA *(arrow)*. The overall mass causes stretching of hemispheric branches throughout and anterior displacement of the basilar artery against the clivus *(open arrows)*. Scattered vasospasm is also seen *(arrowheads)*, the etiology of which was never determined.

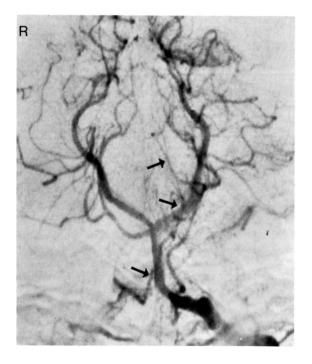

Fig. 1-56 Right hemispheric cerebellar mass. A large right hemispheric mass is causing contralateral displacement of both PICAs *(arrows)*.

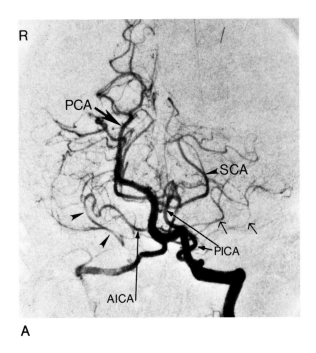

A

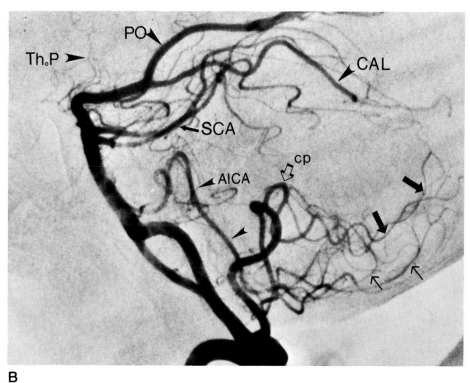

B

Fig. 1-57 **Normal vertebral angiography, AICA. (A)** Towne arterial phase shows the AICA on the right continuing inferiorly to supply the region of the PICA *(arrowheads)*. The left superior cerebellar artery (SCA) is well seen because the proximal segment of the left PCA is atretic and does not fill from the vertebral injection. Hemispheric branches of the left PICA are seen *(small arrows)*. **(B)** Lateral projection shows the inferior continuation of the AICA to supply the region of the PICA on its side *(arrowheads)*. The choroidal point of the PICA is well seen *(open arrow)*, although the choroidal loop is poorly formed. Hemispheric branches *(small arrows)* and the inferior vermian branches *(large arrows)* are well seen. PO, parietal occipital branch of the PCA; CAL, calcarine branch of the PCA; Th.P, thalamoperforate branches.

the tentorium and then divide into medial and lateral branches. The medial branches reach the midline in the quadrigeminal plate cistern and then continue over the superior surface of the vermis. These arteries outline the superior margin of the vermis and cerebellar hemispheres (Fig. 1-58). They are most useful in the diagnosis of herniation, either upward of the vermis due to a cerebellar mass or downward due to transtentorial herniation of the cerebrum. The lateral branches spreading out over the top of the cerebellar hemispheres are stretched by intracerebellar masses.

Posterior Cerebral Arteries

The large paired PCAs arise from the terminal bifurcation of the basilar artery (Fig. 1-53). Each circles the mesencephalon just above the level of CN III and the tentorium. Early in its course, it is joined by the PCoA. About 30 percent of the time, the PCoA is the dominant supplier of the PCA. Sometimes the PCA maintains its origin from the ICA and the proximal PCA is atretic.

This is referred to as a "fetal" type PCA (Fig. 1-47B). On vertebral studies, unopacified blood may enter the PCA from the PCoA. The "washout" effect causes an apparent "filling defect" and should not be misinterpreted as an intraluminal clot.

The PCA continues to supply the medial portion and posterior tip of the occipital lobe. The two major branches to the occipital lobe are the more superior parieto-occipital branch and the inferior calcarine branch (Fig. 1-57). The PCA also gives off many branches to the medial and inferior surfaces of the temporal lobe, supplying the region of the hippocampus.

Displacement of these vessels is difficult to recognize, but they may be seen to be stretched in response to a mass in the occipital or posterior temporal lobe. Occlusive disease is common in this group, and one should look carefully for the absence of major branches.

Other important branches arise from the proximal PCA. The medial posterior choroidal artery arises from the proximal PCA, courses around the brain stem, and passes along the colliculi to reach the pineal. It then

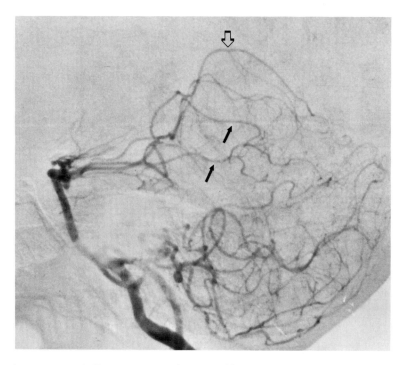

Fig. 1-58 Normal superior cerebellar arteries. Neither PCA fills from the vertebral injection. This view shows the SCA distribution to good advantage. The superior vermian branch (*open arrow*) and the hemispheric branches (*arrows*) are indicated.

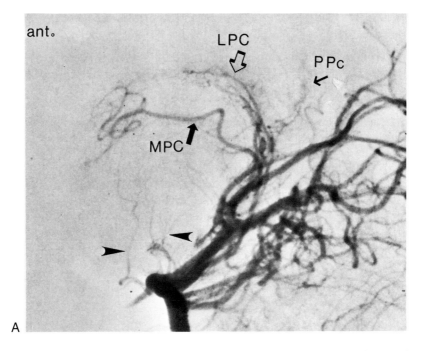

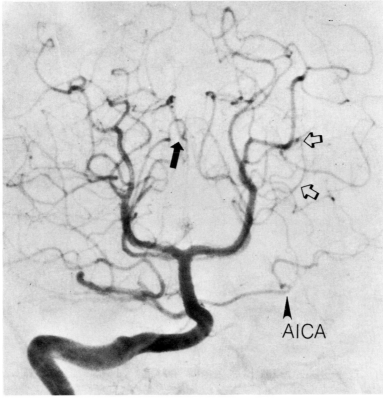

Fig. 1-59 Posterior choroidal vessels. (A) Lateral vertebral arteriogram shows the characteristic course of the medial posterior choroidal artery (MPC, *wide black arrow*). The sharp curves posteriorly are near the colliculi and pineal gland. The lateral posterior choroidal arteries (LPC, *open arrow*) curve broadly behind the pulvinar of the thalamic nuclei. The posterior pericallosal arteries (PPC, *thin black arrow*) course posterior to the splenium of the corpus callosum. The thalamoperforate arteries are seen *(arrowheads)*. **(B)** Towne view shows the medial posterior choroidal artery on the right *(arrow)* and the lateral posterior choroidal artery on the left *(open arrows)*. The characteristic loop of the AICA is seen as it enters the internal auditory canal *(arrowhead)*.

continues anteriorly over the roof of the third ventricle along with the internal cerebral veins (Fig. 1-59). This vessel is elevated by thalamic or pineal masses. The lateral posterior choroidal arteries arise from the proximal PCA and curve superiorly behind the pulvinar of the thalamus, supplying the choroid plexus of the atrium of the lateral ventricle and the posterior superior thalamus. These vessels are displaced and stretched posteriorly by thalamic masses and contribute to intraventricular tumor or vascular malformations in the region. The posterior pericallosal arteries arise from

the PCA and then course in the midline posterior to the splenium of the corpus callosum. They mark the posterior margin of the splenium and are displaced posteriorly by tumors extending into the splenium and by hydrocephalus (Fig. 1-60). The thalamoperforate arteries arise from the medial part of the PCA and fan out superiorly to supply the inferior thalamus, subthalamus, hypothalamus, and red nucleus. They are displaced by tumors in this region. They may also contribute vascular supply to tumors and vascular malformations.

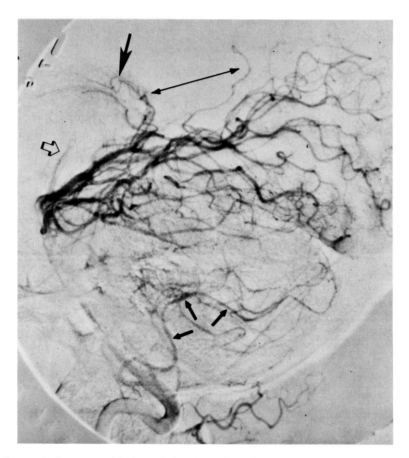

Fig. 1-60 Fourth ventricular mass and hydrocephalus. Lateral vertebral arteriogram shows widening of the choroidal loop by the intraventricular mass (*small arrows*). Hydrocephalus with lateral ventricular enlargement produces downward displacement of the choroidal vessels (*arrow*) and increased distance between the choroidal and posterior pericallosal arteries (*opposed arrows*). The thalamoperforate arteries are displaced and stretched by enlargement of the third ventricle (*open arrow*).

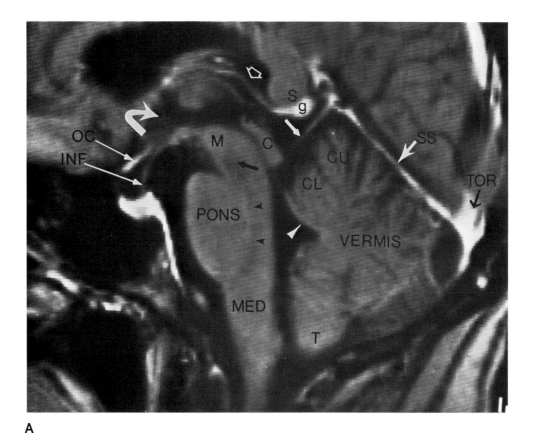

A

Fig. 1-61 Veins of the posterior fossa. (A) Sagittal T₁-weighted Gd-DTPA contrast-enhanced MRI shows the important anatomy of the posterior fossa. The precentral vein *(white arrow)* courses anterior to the superior vermis. It begins at the level of the superior medullary velum of the fourth ventricle *(white arrowhead)*. The internal cerebral vein is demonstrated *(open arrow)* in the roof of the third ventricle *(curved arrow)* as is the medial lemniscus *(small black arrowheads)* and decussation of the superior cerebellar peduncles *(black arrow)*. S, splenium; M, mesencephalon; C, collicular plate; MED, medulla; T, cerebellar tonsil; CL, central lobule; CU, culmen; SS, straight sinus; TOR, torcula; OC, optic chiasm; INF, infundibulum. (B) Lateral venous phase shows the normal course of the anterior pontomesencephalic vein (APM). The precentral vein (PC) is seen coursing anterior to the vermis. The inferior vermian vein (IV) runs behind the vermis inferiorly (arrow). The vein of Galen *(curved arrow)* is large and drains into the straight sinus (SS). Th, retrothalamic veins; SV, superior vermian veins; PV, petrosal vein. *(Figure continues.)*

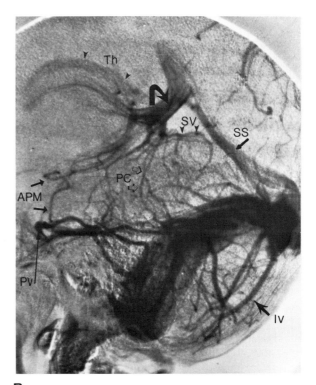

B

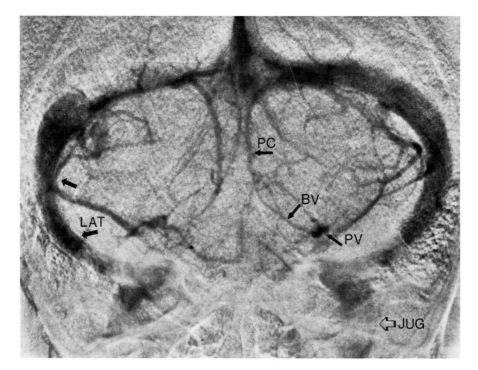

C

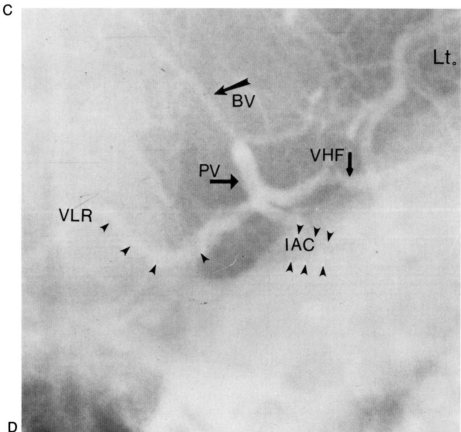

D

Fig. 1-61 *(Continued)*. **(C)** Towne projection. PC, precentral vein; BV, brachial vein; PV, petrosal vein; LAT, lateral sinus; JUG, jugular vein. Inferior vermian veins are not present in this patient. **(D)** Higher-power view of the Towne projection of the left petrosal vein. The petrosal vein (PV, *black arrow*) is short. The major tributaries are the vein of the lateral recess of the fourth ventricle (VLR, *arrowheads*), the brachial vein (BV, *sharp black arrow*), and the vein of the horizontal fissure (VHF, *blunt black arrow*). IAC, internal auditory canal, Lt, left side.

Veins of the Posterior Fossa

The posterior fossa contains only a few important veins (Fig. 1-61). The anterior pontomesencephalic vein is important as a marker of the position of the anterior margin of the pons. Because it is more closely applied to the pons than to the basilar artery, it more accurately defines displacement of the pons. Superiorly, the vein begins in the subthalamic region in the upper suprasellar cistern. It then courses posteroinferiorly in the midline within the interpeduncular sulcus on the anterior surface of the mesencephalon. At the superior belly of the pons, it abruptly curves anteriorly to round the pons and continue inferiorly on its anterior sur-

face. The vein is displaced forward by masses within the pons or cerebellum and backward by prepontine masses.

The precentral cerebellar vein courses superiorly, convex anteriorly, in the space immediately anterior to the superior vermis (Fig. 1-61A & B). It marks this structure and the posterior margin of the superior medullary velum of the fourth ventricle (the superior roof). The vein is hooked upward in a characteristic fashion by masses within the fourth ventricle (Fig. 1-62). It is displaced superiorly by upward transtentorial herniation, stretched anteriorly by masses within the vermis or cerebellar hemispheres, and stretched backward by masses within or in front of the pons.

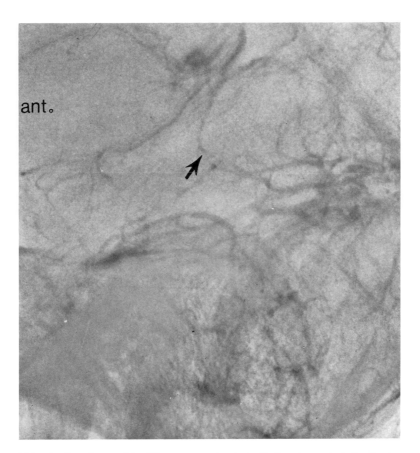

Fig. 1-62 Mass within the fourth ventricle. The precentral vein is displaced superiorly by the intraventricular mass (*arrow*), with a characteristic "hook."

The large paired inferior vermian veins run in the midline underneath the inferior margin of the vermis. Anteriorly, the superior and inferior tonsillar veins enter the vermian veins. The veins are displaced downward by inferior herniation of the cerebellum or vermis, pushed backward close to the inner table of the occipital bone by masses within the vermis, and shifted to one side by contralateral masses within the cerebellar hemisphere.

The petrosal vein is a short trunk that receives multiple veins from the anteroinferior cerebellum. It lies within the cerebellopontine angle cistern immediately above the porus acusticus and drains into the superior petrosal sinus. Its major importance is in the diagnosis of cerebellar pontine angle masses, particularly acoustic neurinoma. The vein is either displaced and stretched upward in an arc or occluded by the tumor.

Fortunately, vertebral angiography is no longer needed to diagnose mass lesions within the posterior fossa. This is done much better with MRI or CT scanning. The vertebral angiogram is now used for the reasons given in the box.

USES OF VERTEBRAL ANGIOGRAPHY

Atherosclerosis
  Basilar stenosis, thrombosis
  Branch occlusion
  Vertebral stenosis, occlusion
Aneurysm
  Berry
  Atherosclerotic
  Dissecting hematoma (dissection)
Vascular malformation
  Within the brain tissue
  Dural A-V fistula
Dural sinus thrombosis
  Lateral sinus
  Jugular vein
Arteritis
Hemangioblastoma, particularly following surgical
  excision of the tumor

MRI

MRI makes use of many parameters. It measures the concentration, density, and nature of chemical and molecular properties of tissue (spin-density and spectroscopy). Molecular behavior is manipulated through the T_1, T_2, and T_2^* relaxation phenomenon. It records motion both on a large scale (blood flow, CSF flow) and on a small scale (diffusion and perfusion). Many of these aspects are in their early stages of exploration and development. The ability to evaluate many different physical properties of tissue and fluids gives the MRI technique enormous flexibility in application and enormous potential for improved scanning in the future. This places a great responsibility upon the imager to understand the principles of MRI and the unique applications of each of the sequences.

Overall, MRI is the most sensitive imaging technique for the detection of CNS pathology (Fig. 1-63). Whenever possible, it is the preferred initial image examination, except for the detection of subarachnoid hemorrhage or bone lesions, and for the evaluation of acute trauma. The routine MRI examination uses spin echo technique (SE), including a short time of repetition of MRI sequence (TR), short time of echo (TE) T_1-weighted study (SE300–600/20 or inversion recovery (IR) 500), a long TR, long TE T_2-weighted study (SE2000–3000/60–100), and a long TR, intermediate TE proton density study SE (TR 2000–3000/25–50) (Table 1-5). Fast spin echo (FSE) T_2-weighted sequences may be substituted, saving time without losing sensitivity for the detection of pathology. Relatively thin (6- to 8-mm) slice thickness is used with as little interslice gap as possible. Transaxial sections are usually routine. Coronal sections have an advantage when the convexities, the temporal lobes, and the sella region are scanned. Sagittal sections are best for the subfrontal region, the colliculi, the corpus callosum, and the foramen magnum.

The T_1-weighted sequence produces a strong MRI signal that results in a clear anatomic image. It is relatively insensitive to artifact and motion degradation. This sequence is therefore best for evaluating the overall anatomic structure of the brain. Ventricular size and anatomic distortions are optimally displayed. The relatively short TR that is used also means that the total scanning time for the image is also relatively short. Fat is intensely bright on T_1 images; this is an advantage when the epi-

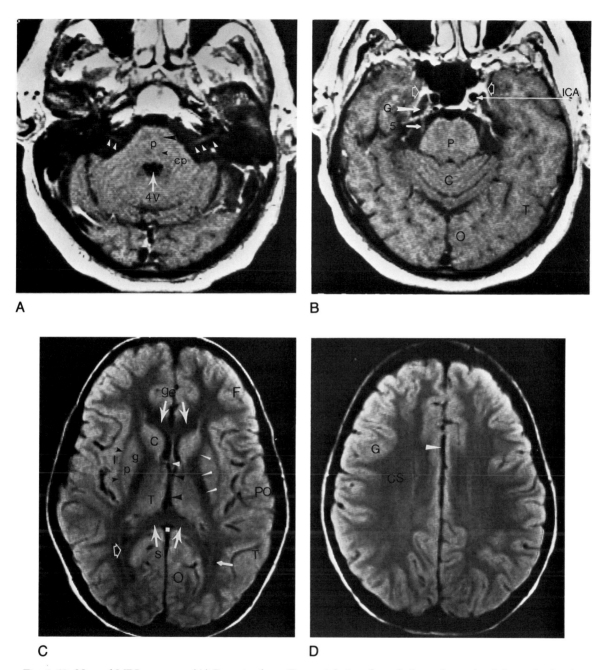

Fig. 1-63 Normal MRI anatomy. (A) Posterior fossa. Transaxial view through the mid-pons level shows the fourth ventricle *(white arrow)* behind the pons (p). Cortical spinal tracts *(large black arrowhead)* and the medial lemniscus *(small black arrowhead)* can be seen as regions of slight hypointensity. The cerebellar peduncles (cp) are large bilateral structures. The branches of CN VIII can be seen in the cerebellar pontine angle cistern entering the internal auditory canals *(white arrowheads).* **(B)** Transaxial view of a slightly higher level, Gd-DTPA enhanced. This view shows the level of the upper pons (P). CN V are seen extending anteriorly *(small arrow)* to enter Meckel's cave and the gasserian ganglion *(large arrow, G).* There is normal enhancement (high signal) within the cavernous sinus *(open arrow).* The internal carotid artery is seen as the round region of flow-void (ICA, *long arrow).* C, cerebellum; O, occipital lobe; T, temporal lobe. **(C)** MRI.SE2500/25 at the level of the foramina of Monro. The corpus callosum *(large white arrows)* is well seen. *(Figure continues.)*

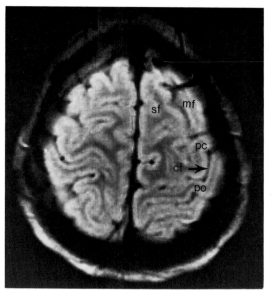
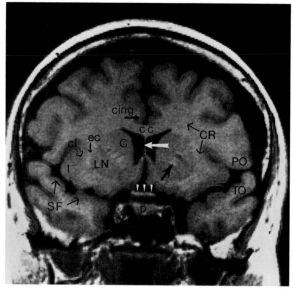

E   F

Fig. 1-63 *(Continued)*. The anterior portion is the genu (ge) and the posterior portion is the splenium(s). The frontal horns of the lateral ventricles lie just behind the genu and are separated by the septum pellucidum. At the posterior point of the septum pellucidum there is an enlargement, the fornix *(small white arrowhead)*. The paired internal cerebral veins run in the roof of the third ventricle *(black arrowheads)* and eventually empty into the vein of Galen *(white square)*. The basal ganglia consist of the caudate nucleus (C), putamen (p), and globus pallidus (g). The internal capsule *(thin white arrowheads)* is divided into the anterior limb, the genu, and the posterior limb. The external capsule *(small black arrowheads)*, which is lateral to the lenticular nuclei (globus pallidus and putamen), lies just deep to the insula (I). The atrium and occipital horn of the right lateral ventricle are seen *(open arrow)*. The optic radiation is just lateral to this area *(small white arrow)*. F, frontal lobe; PO, parietal operculum; T, temporal lobe; O, occipital lobe. **(D)** MR SE/2500/25 through the centrum semiovale (CS), the large central portion of the hemispheric white matter. The cortical gray matter is seen as slightly higher intensity (G). The pericallosal vessels run in the interhemispheric fissure *(white arrowhead)*. **(E)** MRI through the vertex shows the sulcal anatomy. The superior frontal gyrus (sf) and the middle frontal gyrus (mf) are separated by a fissure. The precentral gyrus (pc) and the postcentral gyrus (po) are separated by the central fissure (cf, *arrow*). **(F)** Coronal T_1-weighted MRI through the basal ganglia, showing the pituitary gland (p) and the optic chiasm *(white arrowheads)*. The frontal horns of the lateral ventricle are seen separated by the septum pellucidum *(white arrow)*. The internal capsule *(black arrow)* receives the fibers from the corona radiata (CR). The insula (I) lies just deep to the sylvian fissure (SF). The claustrum (cl) is a small region of deep gray matter just lateral to the external capsule (ec). LN, lenticular nuclei; C, caudate nucleus; PO, parietal operculum; TO, temporal operculum; cc, corpus callosum; cing, cingulate gyrus.

Table 1-5 T_1 and T_2 Weighting, SE Sequences

T_1 weighted	
TR short	< 600 msec
TE short	< 20 msec
T_2 weighted	
TR long	> 2,000 msec
TE long[a]	> 60 msec
Proton density	
TR long	> 1,000 msec
TE short	< 50 msec

[a] The longer the TE, the greater the T_2-based contrast.

dural space or marrow in the spinal column is being outlined. Subacute and chronic hematomas are seen as high density (see Ch. 2).

Differences of the *longitudinal relaxation* time within tissue determines the tissue contrast in the T_1-weighted image. This creates an image with relatively low contrast between most brain tissues and less sensitivity than T_2-weighted sequences to small changes in the water content (edema) of tissue. Therefore, small or early lesions of the brain that do not cause anatomic distortion are often invisible on routine nonenhanced T_1-weighted images. The inversion recovery (IR) technique with T_1-weighting enhances the T_1 contrast within tissues, and with specific adjustments suppresses high T_1 fat signal for better visualization of structures surrounded by fat tissue (e.g., optic nerve, and marrow lesions in the spine). T_1-weighted imaging detects the paramagnetic effect produced by gadolinium contrast enhancement (see below).

T_2-weighted images are much more sensitive than T_1-weighted images for the detection of pathology in the brain. This is because most pathology creates focal edema, and the T_2-weighted sequence is very sensitive to increases in water content of tissue. Because water has a long T_2 relaxation time, the T_2 signal of water lasts into the late echo times and shows after the signal of the background brain tissue has died away. Regions of increased water content (edema) show as regions of high signal superimposed on a dark background. The high sensitivity of T_2-weighted MRI for the detection of changes in tissue water content is the primary reason for the increased sensitivity of MRI over CT scanning in detecting brain and spinal cord pathology. However, because the signal of the background tissue is not strong, the T_2-weighted sequence is not as useful for displaying anatomy. The *transverse relaxation* of proton spins determines the T_2 contrast between tissues. Standard T_2-weighted SE sequences are time consuming and therefore relatively motion sensitive and inefficient. The standard SE MRI signal characteristics of various tissues is given in Table 1-6.

Fast Spin Echo

The FSE T_2-weighted sequence substitutes for standard SE T_2-weighted for examination of the brain and spine without significant penalty in most cases. FSE is now standard procedure. Instead of a dual echo, it uses a train of up to eight pairs of echoes, reducing scan time by the number of echoes/2. Higher-resolution scans can be done (512×512) without excessive time, and some motion artifacts are reduced by the quicker studies. The scans are much more accurate and more sensitive to pathology than gradient echo (GRE) images of the brain and spine. The FSE technique has limitation in the diagnosis of hemorrhage, cavernous hemangioma, metastatic melanoma, and neurodegenerative diseases that cause iron deposition.

The approximated TE of FSE, and the more rapidly driven readout gradient shifts, create different image characteristics compared with standard SE. Fat is much brighter on the later echo images of FSE. There also is some blurring, ghosting, loss of contrast between tissues, and less conspicuity of magnetic susceptibility effects compared with standard SE. However, edge enhancement may occur with some sequence patterns. FSE is

---

HYPERINTENSITY OF T_1-WEIGHTED MRI

Common
  Subacute and chronic hemorrhage
  Fat
  Mucin
  Very high protein fluid
Less common
  Some calcium deposits (with phosphates)
  Rare infarction (fogging effect, petichial hemorrhage)
  Hypercellular tumors
  Bone marrow following radiation therapy
  AIDS (calcium in PAIDS, gliotic nodules)

Table 1-6 MRI Signal Intensity of Cerebral Contents

Tissue	T_1	T_2
CSF	Low	High
White matter	Mod. high	Mod. low
Gray matter	Mod. low	Mod. high
Edema fluid	Mod. low	Very high
Fat	High	Mod. high
Tumors	Low	Mod. high
Fast-flowing blood	Very low	Very low
Slow-flowing blood	High	Variable
Bone	Very low	Very low
Bone marrow	High	Mod. low
Air	Very low	Very low
Cyst fluid, clear	Low	High
Cyst fluid, protein	Sl low	Very high
Cyst fluid, very high protein	High	Very high
Clotted blood	Complex (see text and Ch. 2)	Complex (see text and Ch. 2)
Calcium	Low, sometimes high	Very low

more susceptible to the adverse effects of motion compared with standard SE. But, long T_2 lesions within the white matter may have greater conspicuity. A few more "lesions" are seen within the white matter on FSE compared with SE (unidentified bright objects [UBOs]).

Gradient Echo

The GRE technique uses sequential pulsing but, instead of a 180-degree radiofrequency (RF) refocusing pulse (SE technique), opposing sequential gradients refocus the spins to produce an echo. The main advantage is the rapid speed of the technique, as the images are acquired in seconds rather than minutes. Flip angles of less than 90 degrees are employed with the initiating RF pulse. Because the gradient reversal does not recoup the dephasing produced by magnetic field inhomogeneity, the GRE technique is strongly T_2^* dependent. The echo signal strength follows free induction decay and falls off rapidly. Very short TEs of less than 12 msec are used to capture the signal before decay and limit the contribution of T_2^* dephasing. High-quality GRE imaging requires a very homogeneous static magnet field to minimize unwanted T_2^* dephasing effects.

Image contrast is controlled by altering the flip angle and the TR. T_1 weighting is accomplished with a short TR of less than 20 msec and a large flip angle of less than 60 degrees. Proton-density weighting is accomplished with a long TR of greater than 20 msec and a small flip angle of less than 30 degrees. Long TR of greater than 50 msec with a small flip angle of less than 30 degrees produce images with more T_2-weighting. The T_2-weighted sequences produce bright CSF, which is useful for spine imaging. True T_2 weighting is not present with GRE.

Sequences may be optimized for visualization of CSF or blood flow. Also, T_2^* sensitivity is exploited to produce images that are extremely sensitive to local field inhomogeneities from magnetic susceptibility effects within tissue. Hemosiderin and calcium show as regions of signal dropout. Here, long TEs are used to accentuate this T_2^*-dephasing effect. The strategy is robust for the detection of paramagnetic material deposited within lesions (hemosiderin, calcium) that occurs with chronic intraparenchymal hemorrhage, small occult vascular malformations, tuberous sclerosis, and other conditions. The technique is less sensitive for the detection of most other parenchymal pathology. GRE imaging exhibits greater signal with flow than the standard SE techniques. The time-of-flight (TOF) signal loss that occurs with SE does not occur as readily with GRE. The short TE also decreases dephasing effects from turbulence found with standard SE. This makes the GRE especially well suited for magnetic resonance angiography (MRA). Numerous types of gradient sequences are used (Table 1-7).

Echoplanar Imaging

The echoplanar imaging (EPI) technique uses a similar strategy to FSE but produces echoes by large numbers of gradient reversals centered around a single SE. The echo train length is vastly increased, so that many phase-encoded steps and echoes are contained within each TR.

Table 1-7 GRE Acronyms

Sequence	Names	TR/TE	Contrast	Use
Fourier steady state	P–Fast GE–Grass S–Fisp Ph–FFE	10–50/2–20	PD	MRA Axial spine CSF effect
GRE-spoiled	RF-Fast SPGR Flash FFE-T_1	10–35/2–20	T_1	Spine Gd studies MRA/MAST
Very fast GRE	Ram-Fast FSPGR Turbo- flash Turbo-FE	4–13/2–6	PD T_1	Rapid imaging
Fast spin echo	FSE Turbo-SE	Variable	T_2	Spine Head

Manufacturer abbreviations: P, Picker International; GE, General Electric; S, Siemens; Ph, Philips.

Each echo produces one line of raw data in K-space (the mathematical representation of the scanned volume). The entire K-space is sampled within one or two shots. Images take less than 1 second. This stops motion and allows a rapid number of sequential scans for dynamic studies. Because of the reliance on GRE and rapid shifting of strong gradients, the static magnetic field of the MRI unit must be extremely homogeneous and the gradient power supply robust and accurate. The images are predominantly T_2* weighted. They are strongly affected by chemical shift artifact and magnetic susceptibility, resulting in blurring of the image. EPI is not used for routine imaging at this time. EPI has promise for functional imaging of the brain.

Diffusion Imaging

Diffusion imaging is based on the molecular diffusion (motion) of protons within tissue. The SE T_2 decay time becomes shortened by diffusion of molecules within a studied volume. Inherent random molecular motion (Brownian movement) creates diffusion of water molecules within tissue. Although the molecular motion is random, zigzagging in all directions, a molecule within a given substance will travel an average linear distance over a given time period. The distance traveled is determined by the coefficient of diffusion for the medium and mechanical barriers presented to molecular motion. For the average measured time of diffusion in an MRI experiment, a molecule has a linear displacement of approximately 9 μm. This amount of linear travel is great enough to produce a loss of MRI signal from intravoxel dephasing.

A pulsed magnetic field gradient technique creates the condition to measure diffusion. With greater molecular diffusion, greater dephasing occurs, so that signal is lost for the readout. In the images, fast diffusion motion is represented by decreased signal intensity and slow motion by increased signal intensity. Patient motion significantly degrades the image. For this reason, ultrafast imaging techniques are normally used for diffusion imaging (EPI, Ram-fast). The pulsed magnetic field gradient sequence may be added to any MRI technique, either SE or GRE.

Diffusion imaging has an advantage for the study of acute cerebral infarction (see Ch. 2). The strategy is based on the changes in the apparent diffusion coefficient that occur with the edema and metabolic effects of infarction. Tissue planes and white matter tracks become

visible because of the distortion they produce in diffusion patterns.

Flowing Blood

The appearance of flowing blood depends on a number of factors, including the pulse sequences used, the order of slice acquisition, the velocity of flow, and the direction of flow relative to the plane of the imaging. In general, rapidly flowing blood appears dark (flow void) on SE sequences, a phenomenon termed *high-velocity signal loss*. Turbulence, also associated with high-velocity flow, contributes to the loss in signal by dephasing of spins. For most instances, signal void defines a vessel as patent. Rarely, deoxyhemoglobin in acute clot and hemosiderin with chronic clot within a vessel lumen produces a low signal, mimicking the signal void of flowing blood. This is most likely in the condition of dural venous thrombosis. Flow-related enhancement techniques or MRA must be used to diagnosis thrombosis in this instance.

A number of factors cause increased signal within a patent vessel. *Flow-related enhancement* occurs when unsaturated, fully magnetized blood enters the first few slices in a multislice series. Another potential cause of high signal within patent vessels is *even-echo rephasing*. Here the vessel lumen becomes bright on the second of two evenly spaced echoes. *Diastolic pseudogating* may also occur if, by chance, the repetition time of the sequence chosen approximates the cardiac rate of the patient. In this situation, slices that are consistently produced during the diastolic (slow-flow) phase of the cardiac cycle show high signal within arteries that normally would show high-flow signal loss. Care must always be taken to exclude these physical factors as the cause of a high intraluminal vascular signal before vessel thrombosis or slow flow is diagnosed (Fig. 1-64).

Intentionally produced flow-related enhancement has diagnostic value in proving the patency of a vessel. This effect results from a GRE sequence with single slice acquisition. Vessels with flowing blood become bright. This effect proves the patency of a vessel. Failure to produce a bright signal within a vessel with GRE confirms thrombosis or slow flow within that vessel.

Blood flow and CSF pulsation create image degrading artifacts. This is especially a problem on T_2-weighted sequences with longer echo times. The motion of blood within arteries and dural sinuses and the complex to-and-fro pulsating motion of the CSF are especially troublesome. Cardiac gating (to make the effect of flow con-

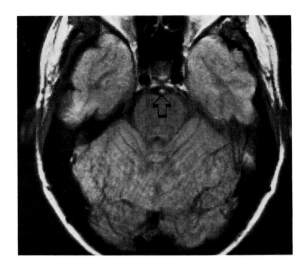

Fig. 1-64 Basilar artery thrombosis, MRI. The normal "flow void" is not present. High signal (*arrow*) is present within the lumen, indicating thrombosis.

sistent for each slice), gradient motion refocusing (motion artifact supression technique [MAST], to rephase spins perturbed by motion) and presaturation techniques (PRESAT, to reduce signal arising from regions of motion) reduce the artifacts. Suppression techniques are useful especially for spinal imaging and MRA.

MAGNETIC RESONANCE ANGIOGRAPHY

MRA depends on the ability to highlight flowing blood within vessels against a less prominent background. Two methods predominate: TOF and phase contrast. TOF MRA is the most commonly used.

TOF MRA depends on the inflow of blood into an imaging volume or scan slice. Using flow-related enhancement, optimized GRE sequences produce bright signal within vessels (bright blood) and suppress the signal from the background tissue so that the flowing blood stands out. In general, the shortest echo time possible (less than 8 ms) is used to reduce dephasing artifacts from T_2* effects on the GRE sequences. The slice thickness is also kept small (less than 2 mm) to decrease the intravoxel dephasing effects that produce signal dropout from turbulence. The format for collecting the information may be (1) two-dimensional (2D) TOF, or sequen-

tial contiguous slice acquisition; (2) three-dimensional (3D) TOF, or volume acquisition; and (3) multiple overlapping thin slab acquisition (MOTSA), a multiple overlapping 3D TOF technique. Each method has certain advantages.

2D TOF MRA uses GRE (RF-Fast, spoiled grass [SPGR], fast low angle shot [FLASH]) to obtain data by sampling each slice of the volume separately in a sequential fashion. It permits complete refreshment of unsaturated blood into each slice, a major difference from 3D TOF. There is no dropoff of blood signal intensity with penetration into the slab. Only vessels completely within the slice (parallel with the slice plane, in-plane flow) are not refreshed and lose signal. The in-plane flow problem is minimized with thin slices and a scan plane orthogonal to the major direction of flow of the vessels of major interest. This maximizes the signal within the vessels by exploiting through-plane flow. The usual sequence factors are TR = 50 msec, TE 8 msec, flip 30 to 45 degrees, and slice thickness 1.5 to 2.0 mm.

The signal from the inflowing blood is so strong that both arteries and veins are imaged. This produces confusion in interpretation.

To clarify the images, a selective presaturation pulse is applied close to the slice being imaged on the side of the inflowing blood targeted for cancellation. Thus, to eliminate the venous flow in the neck, the pulse is applied to the cranial side of the excited slice to be imaged. The presaturation pulses are walked along with the slice selection to produce maximum cancellation effect (walking PRESAT).

The 3D TOF method samples a defined volume of tissue. This allows for very thin slices, so the 3D TOF method is excellent for visualization of fine vessel detail within a thin volume of tissue. However, the bright signal within vessels falls off rapidly with increasing depth of penetration of the vessel into the tissue slab, and blood becomes invisible against the background tissue. The most useful strategy to counteract this effect uses a long TR (approximately 50 msec) and a small flip angle (less than 15 degrees). This produces a less bright signal at the entrance side of the volume slab but greater penetration of visible blood into the slab. There is little visualization of the vessel within deepest portion of a large volume slab regardless of the sequence strategy used.

MOTSA combines some of the advantages of both 2D and 3D TOF imaging. It employs 3D slabs that are thinner than those of standard 3D imaging and combines overlapping slabs to complete the entire volume desired.

The high spatial resolution of 3D technique is preserved, along with less signal dropoff in the total volume scanned.

Artifacts that affect all TOF MRA include loss of signal from (1) saturation of inflowing spins, (2) the dephasing effects of turbulent flow, (3) the outer margins of the vessel lumen because of slow laminar flow, and (4) intravoxel dephasing when there is nonorthogonal flow or curves in a vessel within the voxel. This last factor is especially a problem for the thicker voxel used with most 2D TOF MRA (slice thickness greater than 1 mm).

Paramagnetic Contrast with MRA

Paramagnetic aqueous contrast, such as gadolinium, when injected intravenously greatly reduces the T_1 relaxation time of blood. This increases the signal intensity of blood for T_1-weighted GRE sequences. The T_1 effect is confined to the intravascular blood by the blood-brain barrier. Contrast enhancement is especially advantageous with 3D TOF imaging, as the blood will remain brighter than the background brain tissue through the entire slab. The T_1 shortening effect is so great that slowly flowing blood in veins and sinuses becomes visible on the MRA. With 2D TOF imaging, in-plane flow becomes visible. Larger flip angles and shorter TRs may be used when intravascular contrast is employed.

Magnetic Transfer Contrast

Magnetization transfer contrast (MTC), or magnetization transfer saturation, is an RF strategy used with 3D TOF MRA. It suppresses the background signal of tissue as compared with blood. This renders the blood within vessels more conspicuous, and is especially useful for defining smaller vessels. There is some overall loss in the signal-to-noise (S/N) ratio, which limits very high-resolution thin section scanning. MTC technique is also applied for background suppression to increase the visualization of gadolinium enhancement with SE imaging.

Black Blood Technique

Black blood is produced by using techniques that reduce or eliminate the signal from blood. This can be done with T_2-weighted imaging. Here, the excited blood leaves the plane of excitation during the time between the 90- and 180-degree pulses and so there is no excited blood remaining to produce a signal. Black blood may also be produced by presaturation of blood entering a slice as with presaturation in 2D TOF MRA to eliminate venous flow. The advantage of the black blood technique is its decreased sensitivity to dephasing artifacts from flow or turbulence. However, other artifacts are introduced, as from bone and air. Vessels near sinuses are not visible.

Postprocessing Maximum Intensity Pixel

Most MRA images are obtained in a transaxial format (source images) and then reprocessed after acquisition into a more easily viewed format. The most common technique employs the maximum intensity pixel (MIP) projection algorithm. The volume of tissue is projected onto a 2D plane from any angle, using the single highest pixel along a line orthogonal to the chosen plane. Only the data from the pixels with the greatest value are used to render the final planar image that highlights the bright vessels. Multiple projection angles of view decrease overlap and improve visibility. Because this algorithm is inefficient and throws out much information, newer algorithms will undoubtedly be developed in the future.

Artifacts are inherent in the MIP technique. Overlapping margins of vessels are combined into one. The vessel imaged must be at least two times that of the background for the projection to be successful. The MIP projects vessels to be smaller than their actual size because of low signal along the vessel wall from slow laminar flow. Remote bright objects from within the background tissue may be projected along with the bright vessels causing spurious data. This may occur with subacute or chronic hemorrhage or fat within the scanned volume. Review of the source images may clarify these confusing pictures created by the MIP projection.

Phase-Contrast Magnetic Resonance Angiography

Phase-contrast magnetic resonance angiography (PC MRA) is based on the phase shift that occurs to spins with motion under defined gradient conditions. This phase shift occurs when protons with transverse magnetization move into regions of different field strength because of an applied gradient along the direction of flow. The magnitude of the phase shift over time is proportional to the velocity of the blood flow. The greater the velocity, the greater the distance spins travel over a given time period. The faster spins experience a greater change in magnetic field strength, and thus also a greater phase shift. This phase shift is measured to create images and quantitative data of vessel flow.

Motion is detected only in the direction parallel with the applied bipolar gradients. In order to visualize vessels that curve away from this direction, additional gradients are applied to produce three orthogonal planes. The information of the three data sets are combined into one planar (2D) image or one volume image (3D). 2D images are projected using MIP technique, as with TOF MRA. Variations in sequences and timing of data collection are used to reduce ghosting artifacts from pulsatile flow, which causes varying velocity within a vessel during a cardiac cycle. Cardiac gating is often employed. Data are collected from many points within the cycle and combined to average flow velocity. This averaging technique is called cine PC MRA.

PC MRA must be designed to read phase shifts from specific flow rates. This is called velocity encoding (VENC). This is necessary because of aliasing effects or cancellation of signals from too great a phase shift from applied gradients. Phase shift must be kept below 90 degrees, with the optimal maximum shift produced at about 57 degrees. VENC is done empirically. High gradient strength encodes for venous or slow arterial flow (approximately 20 cm/s). Low gradient strength encodes for fast arterial flow (approximately 100 cm/s). Arterial structures are better outlined because of better visualization of the slow blood flow that occurs along vessel walls.

PC MRA has some inherent advantages. The background tissue suppression is excellent, producing good visibility of blood motion within vessels. This background suppression will also suppress unwanted densities from hematomas and tumors, allowing visualization of the underlying vasculature. The technique may be optimized for slow flow and good venous visualization. Quantitative measurement of flow is possible. Image quality and vessel conspicuity is not affected by the size of the field of view (FOV) or slab volume. However, artifacts from turbulence occur and the images are more sensitive to these gradient produced errors than TOF MRA. The 2D technique suffers from artifacts of projection algorithms. In general, PC MRA is more time consuming than TOF MRA. It also requires an extremely homogeneous static magnetic field within the magnet bore and good control of eddy currents to limit dephasing artifacts.

MAJOR USES OF MRA

2D TOF
  Cervical carotid arteries
  Brachiocephalic arteries
  Venous dural sinuses
  Vertebral arteries
  More peripheral occlusions
  Arteriovenous malformation
3D TOF
  Circle of Willis
  Berry aneurysms
  Occlusion of the middle cerebral artery
MOTSA
  Cervical and cranial vessels
  All of above
PC MRA
  Venous flow
  Giant aneurysms
  True view of vessel diameter
  Cavernous arteriovenous malformation

MRI Contrast Enhancement

Contrast enhancement with MRI aims to deliver a paramagnetic agent to regions of abnormality within the brain and spinal cord. Here it is shown as a high signal on T_1-weighted images. Gadolinium is the major paramagnetic agent in use today.

In general, the contrast accumulates within regions of breakdown of the blood-brain barrier. Breakdown of the blood-brain barrier is a nonspecific alteration that occurs with pathology in the brain, including tumors, infection, and infarction. The use of contrast increases the sensitivity of MRI for detecting specific disease processes, particularly small tumors. The indications for contrast and the interpretation of the findings with MRI are essentially the same as for contrast enhancement with CT. Specific information about the use of enhancement is discussed under specific disease headings. The pattern of contrast enhancement with MRI may permit better tissue characterization, similar to the use of contrast with CT. Slow-flowing blood (veins, cavernous sinus) shows greatly enhanced T_1 signal after gadolinium infusion

INDICATIONS FOR CONTRAST ENHANCEMENT WITH MRI OR CT SCANNING

Signs/symptoms
    Focal neurologic signs and symptoms
        Hemiparesis
        Aphasia
    Focal sensory deficit
        Visual deficit
        Anosmia
        Hemisensory deficit
        Tic douloureux
        Hemifacial spasm
    Cranial nerve dysfunction
        Unilateral hearing loss
        Ophthalmoplegia
    New onset seizures, age > 20 years
    Ataxia
    Vertigo
    Behavior change, age < 50 years
Suspected pathology
    Primary brain tumor
    Metastatic brain tumor
    Meningeal carcinomatosis
    Pituitary adenoma
    Acoustic neurinoma
    Brain abscess or granuloma
    Epidural abscess
    Arteriovenous malformation
    Aneurysm
    Multiple sclerosis
    Infarction, not seen on nonenhanced examination
    Recurrent brain tumor
    To define mass lesions seen on the nonenhanced examination

(Fig. 1-61A). Dural membranes do not normally show enhancement. T_1-weighted imaging is used for the detection of contrast enhancement within tissue (Table 1-8).

MRI contrast agents consist of a paramagnetic agent chelated with a carrier substance. Gadolinium is the most useful paramagnetic agent. It is one of the rare earth metals of the lanthanide series. Paramagnetic agents produce the phenomenon called proton relaxation enhancement (PRE). That is, the paramagnetic agent accelerates T_1 (longitudinal) relaxation and T_2 and T_2* (transverse) relaxation of protons nearby. The degree of PRE is mostly proportional to the concentration of the paramagnetic within the tissues. Unlike radiopaque dyes used with radiography and CT, the paramagnetic contrast agent within tissue is not seen directly, but indirectly by its effects on immediately surrounding protons.

MRI chelated contrast agents are used similarly to the iodinated contrast agents. The agent is injected intravenously, ordinarily over a 1- to 2-minute period, for a total dose of 0.1 mmol/kg. The contrast remains within the vascular system of the CNS, except in regions in which there is breakdown in the blood-brain barrier. T_1-weighted scans are used to demonstrate the effect of the enhancement. After gadolinium injection, there is a general grayness to the normal brain with loss of tissue contrast, because of a general enhancement effect from contrast within capillaries. Normally the pituitary gland, the pineal, the pituitary infundibulum, the choroid plexuses, and the veins, including the dural sinuses, enhance. Rarely, other regions of the brain without a blood-brain barrier, such as the area prostrema, will enhance. Enhancement of arteries will occur only when there is slow flow within them. Fast flowing blood within arteries remains dark. The falx shows minimal or no enhancement, a difference from contrast CT scanning, where the falx is more prominently enhanced. The pia and ependyma do not normally enhance. The dura, pia, and ependyma will enhance following tumor invasion, infection, and trauma. The dural only will show enhance-

Table 1-8 Standard MRI Contrast Agents

Agent (Generic/Proprietary)	Formula	Osmolality (mOsm/kg)	Ion State
Gadopentetate Dimeglumine (Magnevist)	Gd-DTPA	1,940	Ionic
Gadoteridol (Prohance)	Gd-HP-DO3A	630	Nonionic
Gadodiamide (Omniscan)	Gd-DTPA-BMA	789	Nonionic

ment after shunting of hydrocephalus or intracranial surgery. Radiation therapy does not cause meningeal enhancement in the acute and subacute phase, but after 1 year, thin smooth pial enhancement may be seen in rare cases.

To be most effective, T_1 scanning is done immediately after the injection. Abnormal regions are bright. Postcontrast MRI is more sensitive than CT for the detection of regions of abnormal contrast accumulation. Overall, about 5 percent of cases show abnormal enhancement of lesions not seen with the nonenhanced T_1 and T_2 SE series.

A nonenhanced T_1 sequence is recommended before the enhanced sequence, to avoid misreading hemorrhage, fat, or artifact as abnormal contrast enhancement (Fig. 1-65). Fat suppression technique with the T_1 sequence significantly improves the conspicuity of an abnormally enhanced region, especially within the orbit or spinal canal. RF spoiled GRE (RF-Fast, FISP, spoiled GRASS) with T_1 weighting (large flip angle, short TR and TE) may be used for postcontrast studies when speed is needed. Contrast enhancement may be used with MRA to improve visibility of small vessels, vessels with slower flow, and aneurysms. However, care must be taken to avoid interpretation of accumulation within normal or abnormal structures (pituitary, choroid plexus, tumor, infarction) that may not be recognized as such on MRA projection images (MIP).

High dose (i.e., 0.3 mmol/kg) with nonionic low osmolar agents creates a greater enhancement effect. In some cases, lesions not seen with the lower dose may be seen after the higher dose. This is especially true for detecting metastases in the meninges (neoplastic meningitis), and additional brain metastases when one is found by routine technique. Lesions are rarely seen on high-dose scans if the routine dose scan is normal.

Magnetization transfer [(MT, magnetization transfer contrast MTC)] saturation is a selective RF saturation

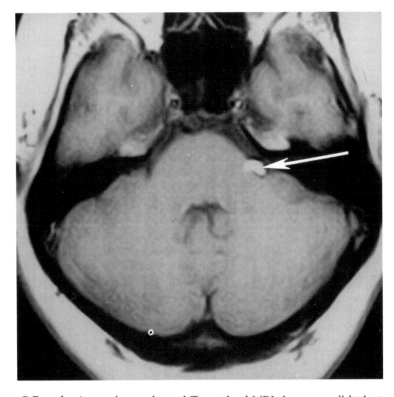

Fig. 1-65 Lipoma, C-P angle. An axial nonenhanced T_1-weighted MRI shows a small high-signal intensity lesion within the left C-P angle adjacent to the pons *(arrow)*. This is a rare lesion. A nonenhanced T_1-weighted MRI should be done before an enhanced sequence, so that a lipoma or other high-signal intensity lesion is not mistaken for a contrast-enhancing lesion.

ADVANTAGES OF CONTRAST-ENHANCED MRI

To improve detection sensitivity
    Metastatic disease of the brain, meninges, and spinal cord
    Extra-axial tumors (acoustic neurinoma, meningioma)
    Subacute infarction
    Infection (intracerebral, extra-axial)
    Multiple sclerosis
    Recurrent tumor after surgical excision
    Recurrent disc herniation in postsurgical spine
    Cranial nerve neuropathies
To improve diagnostic information
    Focal mass lesions
    Look for additional lesions to diagnose metastasis
    Gain information as to type of lesion
To enhance MRA
To evaluate cerebral blood flow

INDICATIONS FOR NONENHANCED MRI OR CT SCANNING

Symptoms/clinical setting
    Headache, nonspecific
    Chronic seizures
    Seizures, age < 20 years
    Immediate postoperative period
    Acute head trauma
    Dementia
Suspected pathology
    Hydrocephalus
    Shunt malfunction
    Intracranial hemorrhage
    Congenital malformation

---

pulse technique that suppresses the background brain signal intensity. It does not affect the signal from paramagnetic contrast accumulation within tissue, so contrast enhancement is made more conspicuous. MT application may be used to reduce the amount of contrast needed for adequate enhancement effect.

Paramagnetic agents may be used for perfusion studies of the brain. One method uses rapid-scanning GRE (Ram-fast, Turbo flash) for quantitative measurement of the diffusion of the agent into the brain tissues at sites of blood-brain barrier breakdown. Other methods are used for regions of intact blood-brain barrier. Here, the magnetic susceptibility effect of the intravascular contrast is exploited by measuring the regional increase in T_2^* relaxation rate using T_2^*-weighted fast GRE, or EPI. After the rapid injection of a bolus of agent, signal-versus-time data are recorded over a region of interest (ROI). Regions of flow will show low signal. These data can be converted into volume and flow maps of the cerebrum. Vascular regions may be separated from avascular regions.

Severe adverse reactions to the contrasts are extremely rare, but anaphylaxis has been reported (risk approximately 1 : 200,000). Most adverse reactions are minimal, the most common being dysgeusia (abnormal taste), nausea, urticaria or other rash, headache, pain at the injection site, and slight transient elevation in serum iron and bilirubin from hemolysis. Paramagnetic agents may be safely used in patients with renal failure. The agent is damaging to subcutaneous tissue, and sloughs have been produced in laboratory animals. The best treatment for extravasation is unknown. Careful injection is required. The contrast agent crosses the placenta and enters breast milk; it is to be used in pregnant or breast-feeding women only if absolutely necessary. Use

---

POTENTIAL INJURIOUS FACTORS IN MRI

Static magnetic field (B_o)
Gradient magnetic fields
Radiofrequency (RF) transmission
Cryogens
Monitoring difficulty

CONTRAINDICATIONS FOR MRI

Ferromagnetic aneurysm clips (check published lists)

Electronically, magnetically, or mechanically activated or electronically conductive devices

    Cochlear implants (magnetically activated)

    Cardiac pacemakers

    Implanted defibrillators

    Implanted electronic infusion devices

    Neural stimulators

    Magnetically activated dental devices

Internal or external conductive wire

    Implanted cardiac pacing wires

    Swan-Ganz thermodilution catheter

    Surface monitor wires

Starr-Edwards Heart valve, models < #6000

Fatio ocular implants

---

in children is generally restricted to the diagnosis of tumors, infections, and certain specific entities such as Sturge-Weber malformations. Delayed excretion of the agent is observed in neonates.

Hemorrhage

The MRI technique is generally sensitive for the detection of intracerebral hemorrhage, except subarachnoid hemorrhage and possibly hyperacute hematoma. The MRI scan varies with the age and cause of the hemorrhage. Its appearance changes with different pulse sequences. Higher field strength produces better images, but this statement is more theoretical than empirical. Subacute hematomas can be adequately imaged at low field strength. GRE techniques may also be used, as they have increased sensitivity to the magnetic susceptibility effect of hemosiderin. The theory of evolution of hematomas is complex (see Ch. 2).

CT SCANNING

CT has been the main neurodiagnostic imaging technique since its introduction in 1972. Because it is more readily available and less expensive than MRI, it will likely remain the most widely used modality for some time. It is accurate, fast, and versatile. It retains its advantage over MRI for detecting subarachnoid hemorrhage and bone destruction. It is much easier to use on patients who have difficulty remaining motionless or who require life support systems. Overall, CT is slightly less sensitive than MRI for detection of intracranial and spinal pathology, although it is sometimes capable of giving more specific tissue information.

The CT image is based on the same principles as all radiography. Tissue that has greater electron density and therefore greater radioabsorption is normally displayed as a high-density area on the CT image. Those tissues with little radioabsorption are displayed as a low-density area (Table 1-9). Relatively narrow windows are used for viewing brain tissue, and wide windows are used for viewing bone.

Transaxial 10-mm slices through the entire brain are performed for the routine scan (Figs. 1-66 and 1-67). Thin sections and coronal planes are used for detailed imaging, particularly of the posterior fossa and sella turcica. Intravenous contrast enhancement, using iodinated contrast agents, is frequently used for specific indications (see above). It highlights regions of breakdown in the blood-brain barrier. It not only increases the sensitivity for the detection of pathology, but often gives specific information for histologic diagnosis. The use of contrast enhancement is discussed in more detail under specific diseases.

CT scan has the advantage for the imaging of bone, much middle and inner ear pathology, acute brain trauma, skull fracture (particularly basal fractures), acute intracranial hemorrhage, subarachnoid hemorrhage, identification of calcification within lesions, the orbit, the paranasal sinuses, and spine fracture.

Table 1-9 Density of Cranial Tissue on CT Scans

Tissue	Appearance Relative to Water	HU
Bone	Very dense	> 500
Gray matter	Intermediate density	35–40
White matter	Intermediate density	28–35
CSF	Isodense	0–10
Fat	Hypodense	< −90
Air	Very hypodense	< −500
Tumors	Intermediate density	20–50
Cysts	Isodense	5–15
Calcification	Hyperdense	> 80
Clotted blood	Moderately hyperdense	50–100
Blood	Intermediate density	38–48

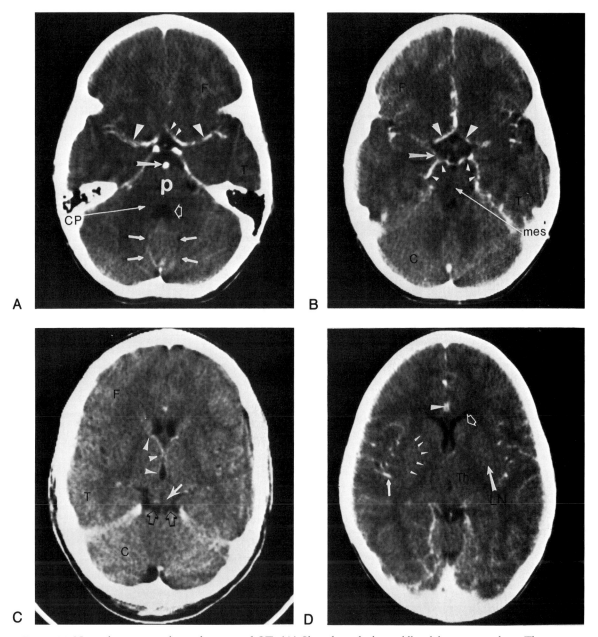

Fig. 1-66 Normal contrast-enhanced transaxial CT. (A) Slice through the middle of the posterior fossa. The vermis shows increased density both before and after contrast enhancement *(small white arrows)*. The fourth ventricle is a small trapezoidal cavity in the midline *(open arrow)*. It lies behind the pons (p). Contrast is seen in the basilar artery *(large white arrow)*. The horizontal portion of the MCA is seen bilaterally *(large arrowheads)*, as is the ACA *(small arrowheads)*. The cerebellar peduncle (CP) connects the pons with the cerebellar white matter. **(B)** Scan through the suprasellar cistern, which is seen as a pentagon at the center of the scan containing fluid of CSF density. It is surrounded by the ACA *(large arrowheads)*, posterior communicating artery *(large arrow)*, and posterior cerebral arteries *(small arrowheads)*. The more distal portions of the PCA surround the mesencephalon (mes). **(C)** Scan slightly higher shows the thalamo-striate vein draining into the internal cerebral vein *(arrowheads)*. The vein skirts the top of the third ventricle. The colliculi *(large arrow)* are seen protruding into the quadrigeminal plate cistern *(open arrows)*. C, cerebellum. **(D)** Scan slightly higher shows the basal ganglia. The caudate nucleus *(open arrow)* shows slight hyperdensity and forms the lateral border of the frontal horn of the lateral ventricle. The lenticular nuclei *(white arrow, LN)* are seen slightly posteriorly and laterally. The thalamus is a paired structure on either side of the third ventricle posteriorly (Th). The internal capsule *(small arrowheads)* separates the lenticular nuclei from the caudate nucleus anteriorly and the thalamus posteriorly. The pericallosal arteries *(large arrowhead)* and the middle cerebral arteries in the sylvian fissure *(small arrow)* are seen. *(Figure continues.)*

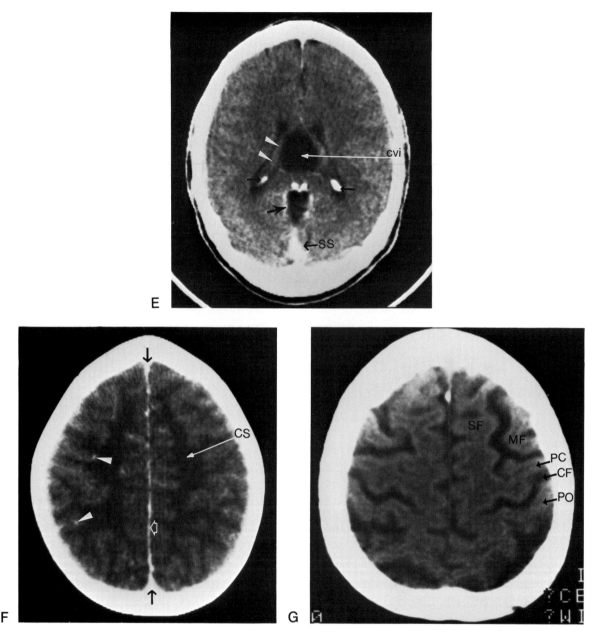

Fig. 1-66 *(Continued).* **(E)** Scan through the level of the bodies of the lateral ventricles show the anterior extension of the choroid plexus *(arrowheads)* and calcification within the glomus *(small black arrows).* The ventricular cavities are separated by an enlarged cavum velum interpositum (cvi), which is an extension of the subarachnoid space over the roof of the third ventricle. The supracerebellar cistern extends superiorly through a fenestration in the anterior tentorium *(black arrow).* The vein of Galen is divided into paired structures. The straight sinus (SS) runs posteriorly toward the torcula. **(F)** Scan through the centrum semiovale. The cortical gyral pattern is prominent. Cortical arteries are seen at the base of the sulci *(arrowheads).* The falx shows enhancement *(open arrow).* The sagittal sinus is seen as a triangular, contrast-filled structure *(black arrows).* CS, centrum semiovale. **(G)** Scan through the vertex shows the major posterior frontal and anterior parietal sulci. SF, superior frontal gyrus; MF, middle frontal gyrus; PC, precentral gyrus; CF, central fissure; PO, postcentral gyrus.

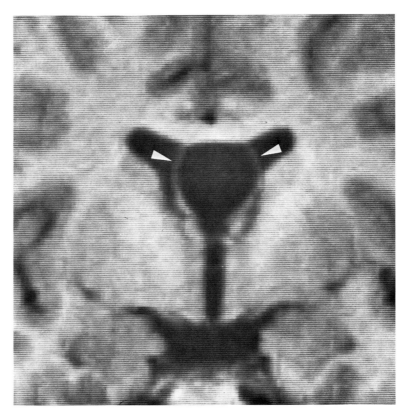

Fig. 1-67 Cyst of septum pellucidum, MRI. Fluid may separate the leaves of the septum pellucidum. If it communicates with the ventricular system, it is a cavum septum pellucidum. If no communication is present a cyst may form *(arrowheads)* that may obstruct the foramina of Monro.

Volumetric CT Scanning

Volumetric (spiral, helical) CT scans are acquired by continuously rotating the x-ray tube while continuously moving the patient through the scanner gantry. Rapid tube rotation is made possible by slip-ring power transmission to the x-ray tube, so that the tube gantry does not have to rewind after each scan cycle. The scan is said to have a pitch of 1, when the patient moves through the gantry a distance equal to the scan slice thickness during the time of one-tube revolution. The scan has a pitch of 2 when the patient is moved a distance of two times the scan slice thickness during one-tube revolution. The images are reconstructed using filtered backprojection, but with an interpolation algorithm that assigns raw data in such a way that nearly true axial scan slices are reproduced. The interpolations may be offset from the acquired scan slices by any amount to produce overlapping images. This increases the number of images ultimately viewed, but eliminates partial volume effects. The scan slice thickness is never altered by offsetting the level of the reconstruction.

The scan slice thickness is set by tube collimation. At a pitch of 1, the slice thickness of the reconstructed images equals the slice thickness set by collimation. The slice thickness may never be smaller than that set by collimation. With a pitch of greater than 1, the slice thickness increases proportionally, so that at a pitch of 2, the reconstructed slice thickness is two times that set by collimation. The slice sensitivity profile of a volume acquired scan is slightly larger than the nominal collimated slice thickness, but with modern algorithms using data offset by 180 degrees, the difference is insignificant. Volume scans can be noisy compared with conventional multislice CT if tube limitations restrict the scanning technique to lower milliampere levels. This is not a factor when scans are performed with high-capacity x-ray tubes.

ADVANTAGES OF VOLUME ACQUIRED VERSUS CONVENTIONAL CT SCANS

Thin overlapping slices eliminating partial volume effect
 Excellent CT angiography (cerebral, cervical carotid)
 Allows accurate densitometry of small lesions
 Pituitary studies
Elimination of motion artifact
 Infant heads
Allows use of decreased contrast volume
Excellent for MPR or 3D reconstruction (spine fracture, facial fracture)
Fast scanning of large area

Radiation and CT

CT scans expose the patient to about 1 to 2 rems of skin dose at each scan slice. The higher the resolution, the greater the radiation dose. The dose at the center of the scan circle may be higher, depending on the more peripheral resorption of the beam. Thin or overlapping slices increase the amount of radiation to any given tissue. It is important to keep the number of slices to the minimum necessary for diagnosis, particularly if the scan plane includes radiosensitive tissue, such as the lens of the eye.

SUGGESTED READINGS

General Texts

Atlas SW (ed): Magnetic Resonance Imaging of the Brain and Spine. Raven Press, New York, 1991

Barkovich AJ: Pediatric Neuroimaging. Raven Press, New York, 1990

Brant-Zawadzki M, Norman D: Magnetic Resonance Imaging of the Central Nervous System. Raven Press, New York, 1987

Burger PC, Scheithauer BW, Vogel FS: Surgical Pathology of the Nervous System and Its Coverings. 3rd Ed. Churchill Livingstone, New York, 1991

Burrows EH, Leeds NE: Neuroradiology; The Skull. Churchill Livingstone, New York, 1981

Davis RL, Robertson DM: Textbook of Neuropathology. Williams & Wilkins, Baltimore, 1991

Enzmann DR, DeLaPaz RL, Rubin JB: Magnetic Resonance of the Spine. CV Mosby, St. Louis, 1990

Heinz ER: Computerized tomography. In: The Clinical Neurosciences: Neuroradiology. Vol. 4. Churchill Livingstone, New York, 1984

Lee SH, Rao KCVG: Cranial Computed Tomography and MRI. 3rd Ed. McGraw-Hill, New York, 1992

Osborn AG: Introduction to Cerebral Angiography. Harper & Row, Hagerstown, MD, 1980

Osborn AG: Diagnostic Neuroradiology. CV Mosby, St. Louis, 1994

Patten J: Neurological Differential Diagnosis. Springer-Verlag, New York, 1987

Som PM, Bergeron RT (eds): Head and Neck Imaging. Mosby–Year Book, St. Louis, 1991

Stark DD, Bradley WG Jr: Magnetic Resonance Imaging. CV Mosby, St. Louis, 1988

Williams AL, Haughton VM: Cranial Computed Tomography. CV Mosby, St. Louis, 1985

Wolpert SM, Barnes PD: MRI in Pediatric Neuroradiology. Mosby–Year Book, St. Louis, 1992

Wong WS, Tsruda JS, Kortman KE, Bradley WG Jr: Practical MRI. Aspen, Rockville, MD, 1987

Yuh WTC, Tali ET, Afifi AK et al: MRI of Head and Neck Anatomy. Churchill Livingstone, New York, 1994

Additional Readings

Axel L: Blood flow effects in magnetic resonance imaging. In Kressel HY (ed): Magnetic Resonance Annual 1986. Raven Press, New York, 1986

Azar-Kia B, Naheedy MH, Elias DA et al: Optic nerve tumors: role of magnetic resonance imaging and computed tomography. Radiol Clin North Am 25:561, 1987

Berger B, Ortiz O, Gold A, Hilal SK: Total cerebrospinal fluid enhancement following intravenous Gd-DTPA administration in a case of meningiomatosis. AJNR 13:15, 1992

Bradley WG Jr: Magnetic resonance imaging in the central nervous system: comparison with computed tomography. In Kressel HY (ed): Magnetic Resonance Annual. Raven Press, New York, 1986

Brasch RC: Introduction to the gadolinium class. J Comput Assist Tomogr, suppl. 17:S14, 1993

Brooks RA, Di Chiro G, Patronas N: MR imaging of cerebral hematomas at different field strengths: theory and applications. J Comput Assist Tomogr 13:194, 1989

Burman S, Rosenbaum AE: Rationale and techniques for intravenous enhancement in computed tomography. Radiol Clin North Am 20:15, 1982

Bydder GM: Clinical application of gadolinium-DTPA. Radiology 189:587, 1993

Clark JA II, Kelly WM: Common artifacts encountered in magnetic resonance imaging. Radiol Clin North Am 26:393, 1988

Council on Scientific Affairs: Magnetic resonance imaging of the central nervous system. JAMA 259:1211, 1988

Daniels DL: Normal cerebral anatomy. In Williams AL, Haughton VM (eds): Cranial Computed Tomography. CV Mosby, St. Louis, 1985

Daniels DL, Haughton VM, Naidich TP: Cranial and Spinal Magnetic Resonance Imaging: An Atlas and Guide. Raven Press, New York, 1987

Debrun GM (ed): Interventional neuroradiology. Semin Intervent Radiol 4:219, 1987

Elster AD: Cranial MR imaging with Gd-DTPA in neonates and young infants: preliminary experience. Radiology 176:225, 1990

Erba M, Jungreis CA, Horton JA: Nitropaste for prevention and relief of vasospasm [during angiography]. AJNR 10:155, 1989

Finelli DA, Hurst GC, Gullapali RP, Bellon EM: Improved contrast of enhancing brain lesions on postgadolinium, T_1-weighted spin echo images with use of magnetization transfer. Radiology 190:553, 1994

Gado MH: Supratentorial anatomy. p. 269. In Stark DD, Bradley WG Jr (eds): Magnetic Resonance Imaging. CV Mosby, St. Louis, 1988

George AE: A systematic approach to the interpretation of posterior fossa angiography. Radiol Clin North Am 12:371, 1974

Grossman RI, Gomori JM, Goldberg HI et al: MR imaging of hemorrhagic conditions of the head and neck. Radiographics 8:441, 1988

Hudgins PA, Davis PC, Hoffman JC Jr: Gadopentetate Dimeglumine-enhanced MR imaging in children following surgery for brain tumor: spectrum of meningeal findings. AJNR 12:301, 1991

Hyman RA, Gorey MY: Imaging strategies for MRI of the brain. Radiol Clin North Am 26:471, 1988

Jones JP, Partain CL, Mitchell MR: Principles of magnetic resonance. In Kressel HY (ed): Magnetic Resonance Annual. Raven Press, New York, 1985

Jones KM, Mulkern RV, Schwartz RB et al: Fast spin-echo MR imaging of the brain and spine: current concepts. AJR 158:1313, 1992

Lane B: Erosions of the skull. Radiol Clin North Am 12:257, 1974

Lauffer RB: Magnetic resonance contrast media: principles and progress. Magnet Reson Q 6:65, 1990

Listerud J, Einstein S, Outwater E, Kressel HY: First principles of fast spin echo. Magnet Reson Q 8:199, 1992

Mafee MF, Putterman A, Valvassori GE et al: Orbital space-occupying lesions: role of computed tomography and magnetic resonance imaging. An analysis of 145 cases. Radiol Clin North Am 25:529, 1987

Makow LS: Magnetic resonance imaging: a brief review of image contrast. Radiol Clin North Am 27:195, 1989

Mathews VP, King JC, Elster AD, Hamilton CA: Cerebral infarction: effects of dose and magnetization transfer saturation at gadolinium-enhanced MR imaging. Radiology 190:547, 1994

Napel SA, Marks MP, Rubin GD et al: CT angiography with spiral CT and maximum intensity projection. Radiology 185:607, 1992

Oot RF, New PFJ, Pile-Spellman J et al: The detection of intracranial calcifications by MR. AJNR 7:801, 1986

Rubenfeld M, Wirtshafter JD: The role of medical imaging in the practice of neuroophthalmology. Radiol Clin North Am 25:863, 1987

Runge VM, Gelblum DY: The role of Gadolinium-Diethylenetriaminepetacetic acid in the evaluation of the central nervous system. Magnet Reson Q 6:85, 1990

Runge VM, Wood ML, Kaufman DM et al: The straight and narrow path to good head and spine MRI. Radiographics 8:507, 1988

Russell EJ, Schaible TF, Dillon W et al: Multicenter double-blind placebo-controlled study of gadopentate dimeglumine as an MR contrast agent: evaluation in patients with cerebral lesions. AJNR 10:53, 1989

Sage MR: Review: blood-brain barrier: phenomenon of increasing importance to the imaging clinician. AJNR 3:127, 1982

Sage MR, Wilson AJ: The blood-brain barrier: an important concept in neuroimaging. AJNR 15:601, 1994

Schnitzlein HN, Murtaugh FR: Imaging Anatomy of the Head and Spine. Urban & Schwarzenberg, Baltimore, 1985

Shellock FG, Kanal E: Policies guidelines and recommendations for MR imaging safety and patient management. J Magnet Reson Imaging 1:97, 1991

Shellock FG, Litwer CA, Kanal E: Magnetic resonance imaging: bioeffects, safety and patient management. Rev Magnet Reson Med 4:21, 1992

Shellock FG, Morisoli S, Kanal E: MR procedures and biomedical implants, materials, and devices: 1993 update.

Sze G, Soletsky S, Bronen R et al: MR imaging of the cranial meninges with emphasis on contrast enhancement and meningeal carcinomatosis. AJNR 10:965, 1989

Taveras JM: Interventional neuroradiology symposium on preoperative embolization. AJNR 7:926, 1986

Unsold R, Ostertag CB, de Groot J, Newton TH: Computer Reconstructions of the Brain and Skull Base. Springer-Verlag, Berlin, 1982

Villafana T: Fundamental principles of magnetic resonance imaging. Radiol Clin North Am 26:701, 1988

Wolf GL, Burnett KR, Goldstein EJ, Joseph PM: Contrast agents for magnetic resonance imaging. In Kressel HY (ed): Magnetic Resonance Annual. Raven Press, New York, 1985

Wong WS, Tsuruda JS, Kortman KE, Bradley WG Jr: Fundamentals of MR image interpretation. In: Practical MRI: A Case Study Approach. Aspen, Rockville, MD, 1987

Yock DH: Techniques in Imaging of the Brain. Part 1: The skull. In Heinz ER (ed): The Clinical Neurosciences: Neuroradiology. Vol. 4. Churchill Livingstone, New York, 1984

2
Cerebrovascular Disease

The term cerebrovascular disease refers to any process that results in abnormality of the cerebral blood vessels or cerebral blood flow. It includes both intracranial and extracranial vessels, as well as embolization, effects of hypertension, hyperviscosity, changes in global blood flow, and hypoxemia. Cerebrovascular disease is by far the most common of all neurologic problems, accounting for up to 70 percent of admissions to the neurologic service. Stroke is the most devastating consequence of cerebrovascular disease and is the third most common cause of death in the United States.

Stroke denotes the sudden development of a neurologic deficit, usually focal in nature. The neurologic manifestation of stroke depends on the volume and location of the tissue involved. About 85 percent of strokes result from ischemia and infarction caused by vascular occlusion, chiefly embolism from vessels or the heart. Hemorrhage and global ischemia account for most of the remaining strokes. Ischemia refers to a metabolically significant decrease in blood flow. Infarction refers to the death of tissue as a result of ischemia.

Because the brain cannot repair itself, medical and surgical intervention is directed primarily toward prevention. The aims are to remove an underlying pathologic process, to limit brain damage by improving perfusion to marginal regions, and to reduce risk factors. The radiologist is responsible for defining the cause and location of the infarction, demonstrating focal extracranial and intracranial vascular pathology, excluding other pathology that might mimic infarction, and evaluating the response to treatment.

CEREBRAL INFARCTION

Classification

A useful clinical categorization of cerebral ischemic events is outlined in the following sections.

TRANSIENT ISCHEMIC ATTACK

Transient ischemic attack (TIA) refers to a transient focal neurologic deficit of sudden onset, lasting a few seconds or minutes, and always clearing within 24 hours. Symptoms related to the internal carotid artery (ICA) distribution are hemiparesis, aphasia, and amaurosis fugax. Those relating to the vertebral basilar (VB) distribution are bilateral or alternating motor or sensory deficits, transient global amnesia, vertigo, diplopia, dysphagia, and ataxia. Hemianopsia and dysarthria may result from disease in either the ICA or VB circuits. TIAs are generally thought to be the result of transient ischemia and reversible damage in the brain, but small infarctions can also cause TIAs and may be detected with MRI. Rarely, what appears to be a TIA is caused by a tumor or chronic subdural hematoma.

A TIA may be thought of as the first manifestation of underlying cerebrovascular disease. About 15 percent of strokes are preceded by a TIA. It has significant prognostic implications, as persons with a history of TIA have a 5 percent per year incidence of a completed stroke. Additionally, there is a 5 percent per year incidence of myocardial infarction. Coronary heart disease is the most common cause of death in this group.

A major objective of screening imaging is the definition of a flow-limiting stenosis of a major cerebral vessel. About 50 percent of those with an ICA type of TIA have a demonstrable lesion in the ipsilateral ICA. Most lesions are atherosclerotic plaques at the origin of the ICA in the neck. Endarterectomy for stenosis of the ICA of greater than 70 percent will significantly decrease the incidence of TIAs and subsequent stroke. About one-half of the lesions of the ICA will have greater than 70 percent stenosis.

Persons with TIA undergo duplex color-flow Doppler ultrasonography as the initial screening study for cerebrovascular disease. Magnetic resonance angiography (MRA) substitutes for ultrasound carotid screening with about equal accuracy for the definition of carotid stenosis at the bifurcation. It adds evaluation of the intracranial circulation near the circle of Willis, but at considerable increase in cost. MRA cannot define precisely the lumen diameter and often overestimates the degree of stenosis. Ultrasound and MRA have an accepted role for the suspicion of carotid disease that is asymptomatic (carotid bruit); if normal, no further evaluation is necessary. If enough information is found with ultrasound or MRA for clinical decision making, angiography is unnecessary.

X-ray angiography (XRA) of the extracranial and intracranial circulation is still the most accurate test. It is recommended for the complete evaluation of symptomatic cerebrovascular disease (hemispheric and ocular TIA and stroke). The entire cerebrovascular system may be covered, including the aortic arch and vascular origins. It alone evaluates the patient for the broad spectrum of causes of TIA (see box). XRA most accurately defines the lumen diameter, collateral flow, tandem lesions, and intraluminal thrombus; it also differentiates complete occlusion from extremely tight stenosis.

THE PRIMARY OBJECTIVE OF SCREENING

To define:

    Normal, or near-normal vessel
    High-grade stenosis, > 70%
    Complete vascular occlusion.

CEREBROVASCULAR CAUSES OF TIA AND STROKE

Stenosis of the ICA at the carotid bifurcation
Stenosis of the distal ICA
Stenosis of the MCA or ACA
Stenosis of the CCA, VA and SA
Ulceration of the ICA at the bifurcation
Carotid and VA dissecting hematoma
Aneurysm, especially of cavernous ICA
Fibromuscular disease of the ICA
Arteritis
Migraine
Venous thrombosis

ASYMPTOMATIC CAROTID BRUIT

Controversy surrounds the prognostic implication of asymptomatic carotid bruit. Bruit occurs in about 5 percent of the population over age 50, but less than 1 percent of this population will have a carotid stenosis of greater than 75 percent. Carotid stenosis of greater than 75 percent is rare in asymptomatic individuals. Carotid bruit implies an increased risk of the complications of atherosclerosis, and an annual stroke incidence of 1 to 2 percent, with approximately 75 percent of the strokes occurring ipsilateral to the bruit. The annual death rate in those with carotid bruit is 4 to 6 percent, with approximately 80 percent of the deaths due to cardiac disease. Although neck bruits imply carotid atherosclerosis, they do not predict significant carotid stenosis, or the site of atherosclerotic disease.

REVERSIBLE ISCHEMIC NEUROLOGIC DEFICIT

Reversible ischemic neurologic deficit (RIND) refers to a focal neurologic deficit that is similar to a TIA but that lasts longer than 24 hours and subsequently clears. It has the same implications as a TIA.

PARTIALLY REVERSIBLE ISCHEMIC NEUROLOGIC DEFICIT

Partially reversible ischemic neurologic deficit (PRIND) is similar to RIND, except that there is not complete recovery of the function within the ischemic region. PRIND implies recovery within the penumbra region of reversible ischemic dysfunction that surrounds the profound central region of ischemia.

STROKE IN EVOLUTION

Stroke in evolution (SIE) refers to a stepwise worsening of a sudden neurologic deficit that occurs after 24 hours. It is thought to be a result of embolization, and secondary thrombosis about the initial clot or within more distal collateral vessels. Many neurologists recommend anticoagulation for this group of patients.

COMPLETED STROKE

A completed stroke is the sudden onset of a focal neurologic deficit that persists and is relatively stable. There may be clinical progression over the first 24 hours. It is the most common end result of cerebrovascular disease, caused by embolus in up to 80 percent of cases. It may be bland (ischemic) or hemorrhagic. Hemorrhage into the infarct may be delayed for up to 1 week or longer.

LACUNAR INFARCTION

Lacunar infarctions are relatively small (0.5 to 2.0 cm) and result from occlusion of the small penetrating end vessels (perforators) that supply the deep gray and white matter of the brain and brain stem. As there is little potential for collateral circulation in regions supplied by perforators, the infarcts develop pan-necrosis and progress to cysts, hence the name lacune or cyst. The major groups of perforating vessels are (1) lenticulostriate vessels, supplying the basal ganglia and internal capsule; (2) thalamoperforate vessels, supplying the thalamus; and (3) pontine perforate vessels, supplying the pons and mesencephalon.

Lacunar infarction is mainly the result of the effects of hypertension. Prolonged hypertension causes lipohyalinosis of the small perforating arteries. These changes result in narrowing of the vessels and their subsequent occlusions. Small microaneurysms form but are rarely seen with angiography.

Lacunar infarctions cause specific clinical syndromes. A pure motor deficit is the most common, resulting from infarction in the internal capsule or pons. Pure sensory deficit indicates infarction within the thalamus. Other syndromes include dysarthria with hand clumsiness (genu and anterior limb of internal capsule), ataxic hemiparesis (pons), hemiballismus (subthalamic), and mutism (bilateral nonsimultaneous lacunes). Most lacunar infarctions can be identified with MRI but are more difficult to detect in the early phase, compared with cortical infarction. Contrast enhancement is rare with lacunar infarction because of the lack of collateral circulation required to deliver contrast-filled blood to a region of ischemia.

NONLACUNAR INFARCTION

Nonlacunar infarction occurs in the cerebral hemispheres or cerebellum.

CHRONIC ISCHEMIA

Chronic ischemia implies multiple ischemic events from low blood flow, or vascular occlusion over a long period, most often affecting broad regions of the brain. Diffuse anatomic changes may be seen with imaging and is correlated with variable diffuse brain dysfunction. Binswanger's subcortical atherosclerotic encephalopathy is an example of chronic ischemia (see section below).

Etiology and Pathology

Thromboembolism and atherothrombosis account for about 85 percent of ischemic strokes in adults. Most occlusions are caused by arterioarterial thromboembolism. The embolus arises most often from the extracranial carotid arteries, but the vertebral arteries, aortic arch, and intracranial segments of cranial arteries may also be a source. The heart is a source of embolism, particularly with cardiac abnormality such as myocardial infarction, atrial fibrillation, prolapsed mitral valve, and endocarditis. Less commonly, embolism originates within aneurysms arising from cranial vessels.

Stroke is less common in those under 45 years of age. Nevertheless, the incidence is significant and approximates 10 per 100,000. The incidence is lower in those under 15, and higher in those aged 25 to 45. International variation in rates occur, and in some countries the rate among those aged 15 to 45 reaches 40 in 100,000. Most strokes in the young are also ischemic, although there is a higher incidence of hemorrhagic stroke. TIAs occur in the young, but the subsequent incidence of completed stroke is much lower than in adults. Women predominate in the under age 30 group, men in the over age 30 group. As with the adult population, the carotid circulation is involved in 60 percent, and the vertebrobasilar in 35 percent. The causes of stroke in the young are presented below.

CAUSES OF STROKE IN UNDER AGE 45 GROUP

Embolism
  Cardiac
    Valve disease
    Rhabdomyoma (tubular sclerosis)
    Myxoma
    Cardiomyopathy
  Arterial dissection
  Paradoxical pulmonary source
  Angiography
  Cardiac surgery
  Chagas's disease
Atherosclerosis
Arterial dissection
Arteriopathy
  Takayasu's
  Autoimmune, idiopathic
  Disseminated intravascular coagulation
  Infectious
    Meningitis
    Sinusitis
    AIDS
  Radiotherapy
  Peripartum angiopathy
Trauma
  Child abuse
  Increased intracranial pressure
Migraine

(Continued)

Hypertension
  Cocaine
  Amphetamine
  Phenylpropanolamine
  Toxemia of pregnancy
  Pheochromocytoma
Fibromuscular disease
Venous thrombosis
  Meningitis
  Sinusitis
  Trauma
  Peripartum
Hematologic abnormality
  Antithrombin III deficiency
  Protein C and S deficiency
  Hyperviscosity (myeloproliferative disease)
  Alcohol intoxication
Mitochondrial disorders

Pathophysiology

The brain has a constant demand for oxygen, but it has little oxygen reserve, so oxygen must be supplied continually in sufficient quantity. Total anoxia for 10 minutes results in total tissue death. Shorter periods may cause incomplete or reversible change. Neurons are more sensitive to anoxia than the supporting astrocytes. Some neurons are more sensitive than others to the effects of ischemia, a concept known as selective neuronal sensitivity (see below).

Normally, the cerebral blood flow (CBF) is maintained at an average of 50 ml/100 g/min. The gray matter receives 80 ml/100 g/min, whereas the white matter receives only 20 ml/100 g/min. Ischemia begins when the blood flow becomes less than 18 ml/100 g/min. Flow in the white matter is normally dangerously close to this level, and a slight decrease in flow results in white matter damage. This situation explains the greater damage done to the white matter in "watershed" regions after transient global ischemia. Conversely, gray matter is more sensitive to the effects of ischemia, once the critical level has been reached. However, the outer cortex is more resistant to ischemia and may be preserved even though the underlying tissue is severely damaged. Different neurons are more or less sensitive, depending on varying concentrations of neurotransmitters. Infarction occurs after short severe ischemia (< 5 ml/100 g/min), or more prolonged (30 to 60 minutes) less severe, but critical ischemia (< 15 ml/100 g/min). Cerebral ischemia is caused by a number of factors, with decreased regional cerebral blood flow (rCBF) the common denominator.

Autoregulation occurs, so that CBF is nearly constant between the arterial mean pressures of 60 and 180 mmHg. Above and below these levels, the regulation breaks down and flow increases or decreases. With chronic hypertension the window of autoregulation is set at a higher level. CBF is also proportional to the CO_2 content of the blood. Local increase in CO_2 and lactate cause an increase in regional CBF, resulting in "luxury perfusion."

CAUSES OF CEREBRAL ISCHEMIA

Embolism to cerebral arteries (approximately 80%)

Thrombosis of cerebral arteries

Narrowing of cerebral arteries from
    Atherosclerotic plaque
    Arteritis
    Vasospasm
    Regional edema
    Intracerebral hematoma
    Dissection

Global hypoperfusion
    Shock
    Cardiac arrest
    Increased intracranial pressure

Global anoxia
    Asphyxia
    CO poisoning
    Drowning
    Cyanide
    Severe anemia

Venous thrombosis
    Dural sinuses
    Cortical veins

Glucose is the major metabolic substrate of the brain delivered by the blood. Up to a point, with decreasing CBF, metabolism is maintained by increased extraction of glucose from the blood. Beyond this point, anaerobic metabolism of glucose occurs, leading to regional lactic acid buildup. If the ischemia is moderately severe (about 17 ml/100 g/min rCBF), electrical activity ceases, but cell membrane activity persists. With reperfusion after a short period of time, these cells may recover. This is the penumbra zone of misery perfusion, with nonfunctional cells kept alive by the collateral circulation. Cell death occurs after severe ischemia, when the membrane metabolism ceases. This mechanism is called cell death by energy failure and, if severe, leads to pan-necrosis of neurons, neuroglia, and vessels. The penumbra region may recover completely with timely re-establishment of good rCBF.

According to the theory of cell death by energy failure, a cascade of events is initiated when critical ischemia occurs. This cascade runs to completion regardless of re-establishment of flow. The shift to anaerobic metabolism causes an increase in the regional lactic acid and a decrease in the production of adenosine triphosphate (ATP). The integrity of the cell membrane and the function of the sodium pump are impaired, resulting in cytotoxic edema. An influx of calcium ions becomes toxic to mitochondrial activity. With cessation of mitochondrial activity, there is cell death.

There is a hierarchy to the vulnerability of neurons within the CNS. Neurons that are especially sensitive to the effects of ischemia may not survive within the penumbra regions of ischemia where other less sensitive neurons may survive. These sensitive cells may experience delayed death many days after the ischemic episode. Most of these cells receive fibers with glutamate-mediated neurotransmission at the synapse. This theory of selective vulnerability postulates that the glutamate persists at the synapse in regions of ischemia, causing prolonged stimulation of the neuron. The delayed death of the neuron is a consequence the metabolic demands and alterations of this prolonged stimulation. The ischemia alters DNA and RNA within these cells, interfering with subsequent protein synthesis and the ultimate integrity of the cell. Sublethal ischemia severe enough to cause transient alteration of the metabolism of a cell also increases its vulnerability to subsequent ischemia; repeated sublethal ischemia may itself cause cell death. The selectively most vulnerable neurons are found within Somer's sector of the hippocampal gyrus, the basal ganglia, and the third and fifth layers of the cerebral cortex, the inferior colliculus, and the substantia nigra, (mostly within the limbic lobe and basal ganglia). With cell death, postischemic edema begins during the first 3 hours as *cytotoxic edema* within the gray matter neurons. Later, lysis of cells causes extracellular edema within the gray matter.

After about 6 hours, edema continues within the white matter; this is known as *vasogenic edema*. Abnormal regional metabolites and chemical mediators from the neutrophilic inflammatory response cause vascular damage, resulting in loss of the blood-brain barrier. The transendothelial flow of water and plasma proteins into the extracellular space of the brain tissue is responsible for the edema. Loss of the blood brain barrier also allows leakage of intravenous contrast into the infarcted tissue. Vasogenic edema requires some blood flow into the infarcted region. Complete cessation of CBF causes cytotoxic edema only.

Postinfarction edema worsens progressively until reaching a maximum at about 4 days. If the infarction is large, the edema may be sufficient to cause craniocaudal herniation and death. Steroid treatment has little effect on the edema once the process has begun. A steroid given prior to an ischemic event decreases the amount of edema that may result from infarction.

After about 5 days, the edema begins to subside. Proliferation of microglia begins around the periphery and extends centrally into the infarcted tissue. These cells become lipid-laden macrophages. Within 6 weeks, the cellular debris is cleared. The region of the infarction becomes soft from a loss of cells (encephalomalacia). Complete loss of cells results in a fluid-filled cystic cavity, which is seen most commonly after lacunar infarction. Rarely, calcium is deposited within infarcted tissue.

In ischemic conditions, edema implies cell death. The edema, primarily vasogenic edema, is what is imaged with MRI and CT. The penumbra region of electronically nonfunctional but viable cells can only be detected by metabolically based tests, such as single photon emission computed tomography (SPECT), positron emission tomography (PET), and spectroscopic MRI.

Small hemorrhages are common within infarctions, particularly those caused by embolization. They usually occur sometime after the stroke, when blood flow to in-

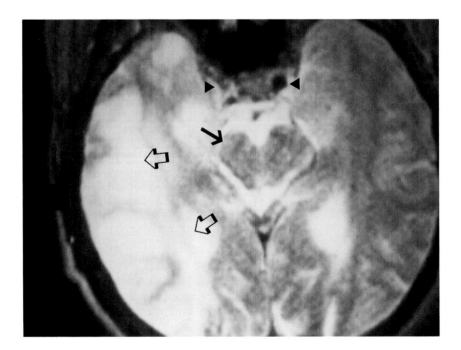

Fig. 2-1 Infarction; wallerian degeneration; ICA occlusion. Increased T_2 signal represents encephalomalacia from infarction in the right temporal lobe *(open arrows)*. The right crus cerebri is atrophic *(black arrow)*. There is no flow void in the right ICA. There is flow-void within the right ICA *(arrowheads)*. T_2-weighted axial MRI.

farcted tissue is re-established, but re-establishment of blood flow is not a necessary condition for hemorrhage to occur. The hemorrhage is usually petechial within the cortical gyri. Large hematomas are unusual.

Sodium accumulates in the extracellular fluid and in the intracellular fluid, so that the concentration of sodium within a volume of infarcted brain tissue is increased 300 percent over normal. This increase in sodium concentration can be imaged with sodium (^{23}Na) MRI, but this method currently has little advantage over proton MRI.

New capillaries are formed about the periphery and extend inward to a slight degree. Initially, these vessels do not have a blood/brain barrier. This allows leakage of injected contrast so there is often peripheral enhancement about infarctions beginning at the end of the first week and continuing for 1 month or more.

In the final stage, the infraction is sharply demarcated from the surrounding brain. Loss of brain tissue results in focal atrophy with dilatation of both the cortical sulci and the subjacent ventricle. Pan-necrosis results in vacuolization of tissue and the formation of cysts (encephaloclastic porencephaly). If the more resistant astrocytes survive, glia scar (gliosis) forms at the infarction site.

Wallerian degeneration of axons is a late consequence of cerebral infarction, becoming evident with imaging during the 6- to 12-week postinfarction period. This is a process of demyelination and degeneration of axons distal to a neuronal or proximal axonal injury. The changes are permanent. The demyelination and end-stage gliosis are best visualized with MRI as a high T_2 signal within the appropriate atrophic white matter tracts. For example, following a stroke in the motor region, the pyramidal tracts degenerate and show throughout the stack of slices as high T_2 signal intensity (Fig. 2-1). Magnetization transfer contrast MRI may show this degenerative process in its earlier stages.

It is important to recognize the common vascular territories of the brain, as a cerebral infarction most often corresponds with one of these territories. With imaging, territorial patterns help define the abnormality as infarction, rather than some other pathology. The average territories are shown in Figure 2-2. Because of the overlap at the territorial boundaries, there are territorial differences among normal persons.

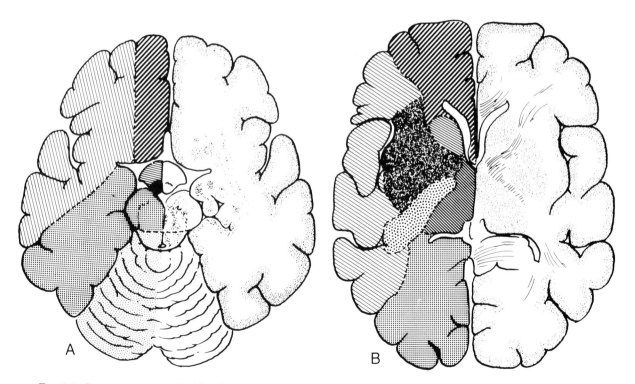

Fig. 2-2 Brain regions supplied by the major intracranial arteries. (From Drayer, 1984, with permission.) (*Figure continues.*)

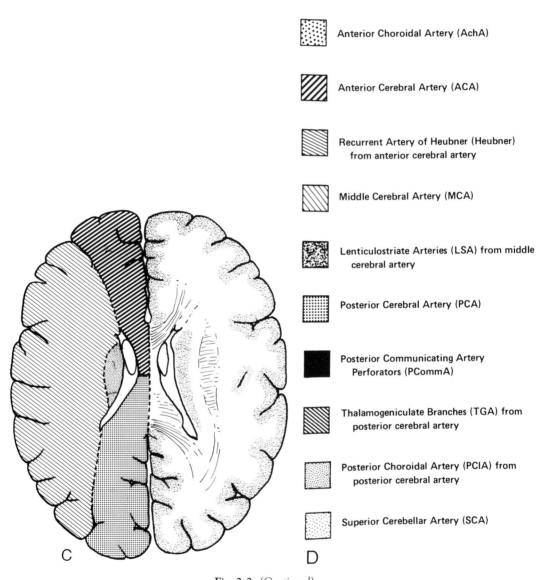

Anterior Choroidal Artery (AchA)

Anterior Cerebral Artery (ACA)

Recurrent Artery of Heubner (Heubner) from anterior cerebral artery

Middle Cerebral Artery (MCA)

Lenticulostriate Arteries (LSA) from middle cerebral artery

Posterior Cerebral Artery (PCA)

Posterior Communicating Artery Perforators (PCommA)

Thalamogeniculate Branches (TGA) from posterior cerebral artery

Posterior Choroidal Artery (PCIA) from posterior cerebral artery

Superior Cerebellar Artery (SCA)

C   D

Fig. 2-2 *(Continued).*

MRI

Overall, MRI is the most sensitive routine clinical imaging modality for the evaluation of cerebral infarction. The earliest change occurs within the blood vessels. Vascular occlusion or severe stenosis with slow flow nullifies the signal void effect of fast-flowing blood. The lumen with very slow or absent flow shows signal equivalent to tissue.

With gadolinium (Gd) enhancement, intraluminal high signal occurs within vessels with slow flow (Fig. 2-3). This condition occurs from more proximal stenosis or occlusion. Collateral circulation provides the necessary slow flow condition in segments distal to complete branch occlusions. This phenomenon may be seen immediately following the vessel occlusion, and so precedes by hours the edema and the swelling associated with the infarct. It persists for a few days, but occasionally lasts longer. Intraluminal vessel hyperintensity has been seen in up to 75 percent of Gd enhanced MRI studies during the acute stage of infarction. The territory of intravascular contrast enhancement sometimes extends beyond the region subsequently defined as infarction by high T_2 signal, probably representing the penumbra region of ischemia (misery perfusion). Intravascular enhancement is rare with basal ganglionic or brain stem infarctions.

Intravascular Gd enhancement occurs within both arteries and veins and is a nonspecific indicator of slow intravascular flow. Arterial and venous occlusive disease, meningitis, cervical artery stenosis or dissection, vasculitis, and slow blood flow from regional edema of any cause may produce this MRI finding.

The edema associated with infarction causes prolongation of T_1 and T_2 relaxation times seen best as bright signal on T_2 spin echo (SE) and fast spin-echo (FSE) sequences. It can be seen as early as 1 hour after experimental infarction, but clinically, it is not seen consistently until after 6 hours when vasogenic edema develops. The infarct becomes more conspicuous over the next 18 hours. About 80 percent of clinical strokes can be defined with MRI within the first 24 hours. Small infarctions or regions of reversible ischemia that do not produce vasogenic edema cannot be imaged with SE MRI. Infarction (cell death) is believed to be necessary to produce visible edema on MRI.

The infarction is best seen on T_2-weighted images as a region of high signal intensity (Fig. 2-4). FSE technique has the same sensitivity for the detection of the T_2 signal intensity increase from the edema of infarction. The region of edema defines cell death and corresponds with the region of infarcted tissue. Proton-density images will better define gyral hyperintensity as separate from the adjacent sulci or cisterna. T_1-weighted images show infarction as a region of low intensity which may be subtle. Swelling is recognized by compression and by a shift of the ventricles and sulci.

During the second and third weeks of the infarction, the T_2 signal within the infarct decreases often and may become isointense with the surrounding normal brain. This makes the infarct difficult or impossible to recognize, and is known as the "fogging effect" (see Fig. 2-17). The precise mechanism for this signal change is unknown, but is thought to be related to the influx into the infarcted tissue of macrophages, along with the subsidence of vasogenic edema. During this period, contrast accumulation within the infarct provides visualization on enhanced images (Fig. 2-5).

Hemorrhage within the infarction is best recognized during the acute phase with high field strength T_2-

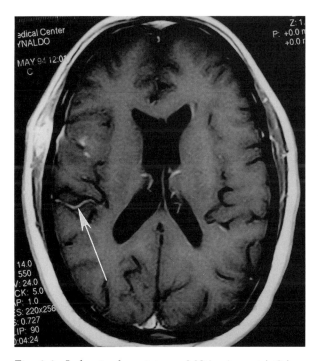

Fig. 2-3 Ischemia, hyperintense MCA. An axial Gd-enhanced T_1 MRI shows hyperintensity within an opercular branch of the right MCA *(arrow)*. This represents enhancement within the vessel that has slow flow. The vessel must be patent to exhibit contrast enhancement.

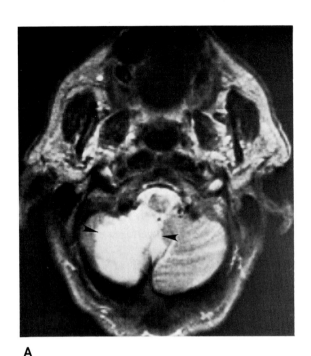

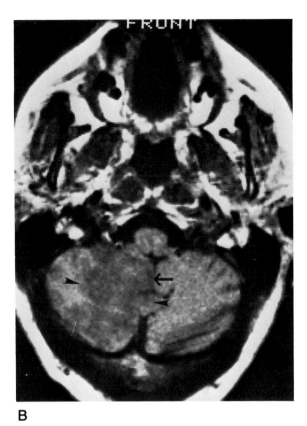

A   B

Fig. 2-4 Acute ischemic cerebellar infarction, MRI. (A) Transaxial T_2-weighted MRI shows the infarction as a sharply demarcated region of hyperintensity in the distribution of the right posterior inferior cerebellar artery (PICA). The hyperintensity represents cytotoxic and vasogenic edema that corresponds with the region of infarcted tissue *(arrowheads)*. **(B)** T_1-weighted image shows the infarction as a region of hypointensity *(arrowheads)*. Slight swelling is present with compression of the vallecula *(arrow)*. There are no signal changes of hemorrhage.

weighted SE or gradient echo recalled (GRE) images (see Ch. 1). Focal low signal intensity occurs within the high signal intensity of the edema because of the magnetic susceptibility of deoxyhemoglobin still within red blood cells (RBCs). Low field strength SE images and T_1-weighted images do not show these acute hemorrhages consistently. FSE sequences are also less sensitive to magnetic susceptibility effects, and may show gyral hemorrhage as high signal intensity. Subacute hemorrhage is well demonstrated on the T_1-weighted images as high signal within the low intensity signal of the edema. Low signal in the hemorrhage persists on the T_2-weighted images into the subacute phase (Fig. 2-6), then shifts to high signal. (For details of brain hemorrhage, see later section.)

Gd contrast enhancement demonstrates breakdown in the blood brain barrier from infarction. The paramagnetic contrast agent leaks through the vessels and creates high signal intensity within regions of vasogenic edema. Enhancement may be blotchy or may show a characteristic ribbon-like pattern of gyral enhancement similar to that seen on contrast-enhanced CT scans (Fig. 2-5). Gd enhancement within the infarcted tissue may be seen within the first 24 hours but more commonly is not found until day 2 or 3 after the event. Enhancement reaches peak incidence at about 7 days. When present, it helps differentiate infarction from tumor or abscess when the nonenhanced pattern is unclear. Enhancement may persist for 2 months.

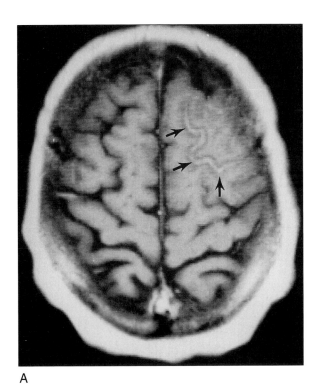

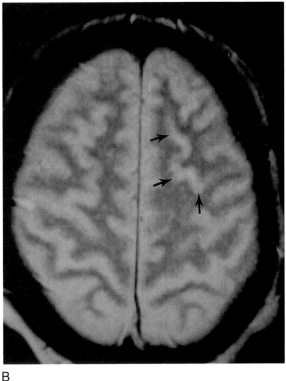

A   B

Fig. 2-5 Cortical ischemic infarction. (A) An axial Gd-enhanced T₁ MRI shows ribbon-like contrast enhancement within a left frontal cortical gyrus *(arrows)*. This pattern of enhancement is highly predictive of infarction. MTC background suppression improves sensitivity for the detection of enhancement and may show abnormal enhancement when the routine T₁ sequence is normal or equivocal. **(B)** Axial PD MRI through the same region as in Fig. A shows only subtle increase in the T₂ signal within the gyrus *(arrows)*. Contrast-enhanced MRI may indicate infarction when the nonenhanced does not.

Enhancement of the meninges accompanies infarction in up to 35 percent of cases, generally with larger supratentorial infarctions. It may precede the enhancement within the infarction itself and is best demonstrated with coronal imaging. The meningeal enhancement subsides prior to that within the infarcted brain parenchyma. The mechanism and the prognostic implication for the enhancement is unknown.

For detection of lacunar infarctions, MRI is considerably more sensitive than CT. The acute lacune is seen best as high signal intensity on T₂-weighted images (Fig. 2-7). After 1 week, lacunes are seen on T₁-weighted images as focal regions of hypointensity. During the chronic stage, the lacune has signal intensity equivalent to that of CSF.

Lacunes must be differentiated from enlarged Virchow-Robin spaces (Fig. 2-7). These spaces, present normally, may enlarge in older persons with chronic hyper-

tension. The mechanism and significance are unknown. The spaces appear as multiple small (less than 5 mm), round structures of CSF-equivalent intensity on all MRI pulse sequences, most prominently in the lateral portion of the putamen. Lacunar infarcts tend to be larger and

---

ONSET SEQUENCE OF MRI ENHANCEMENT WITH INFARCT

Intravascular

Meningeal

Parenchymal

(Text continues on page 104)

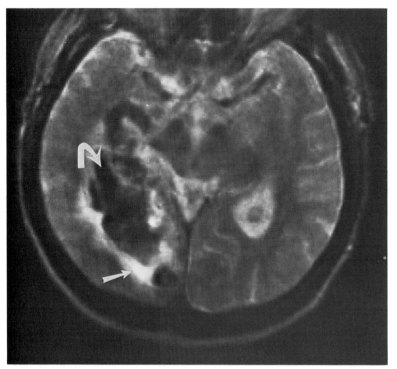

A

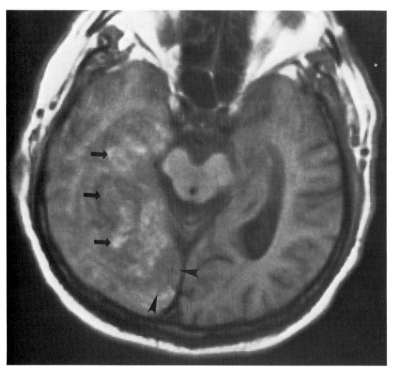

B

Fig. 2-6 Hemorrhagic infarction, MRI and CT scanning. (A) T$_2$-weighted MRI shows edema in the distribution of the right PCA within the occipital and medial temporal lobes *(arrow)*. The hemorrhage within the infarcted inferior gyri is demonstrated by central low signal intensity *(curved arrow)*. **(B)** T$_1$-weighted image shows the early subacute hemorrhage as regions of isointense and hyperintense signals *(arrows)*. The edema of the infarction is seen as slight signal hypointensity, which is extending to the cortical surface *(arrowhead)*. The MRI scan was obtained 4 days after the stroke. *(Figure continues.)*

Fig. 2-6 *(Continued)*. **(C)** CT scan performed on the day of the stroke shows acute serpiginous hemorrhagic high density in the right posterior inferior temporal and occipital lobe gyri *(open arrow)*. The edema of the infarction is imaged as low density *(arrowheads)*. **(D)** T$_2$-weighted MRI scan performed at a higher level demonstrates the edema of infarction of the right thalamus *(arrowheads)*. The central lower intensity may represent slight hemorrhage. Faint high signal intensity is seen in the upper right occipital lobe representing the edge of the infarction. Combination of an infarction involving the ipsilateral medial temporal lobe, occipital lobe, and thalamus indicates occlusion of the right PCA.

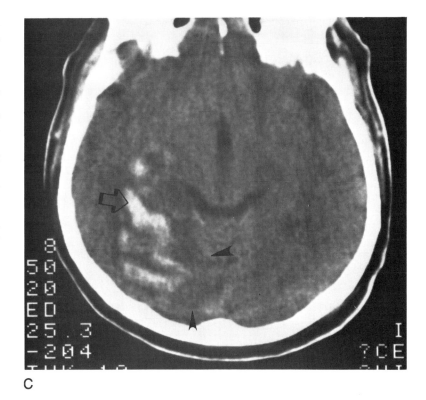

C

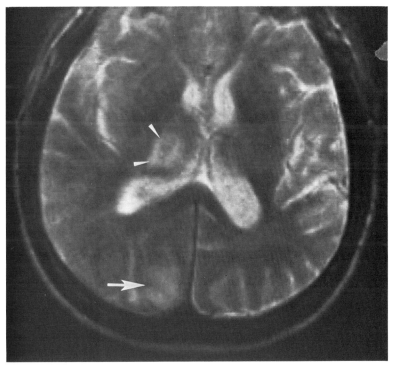

D

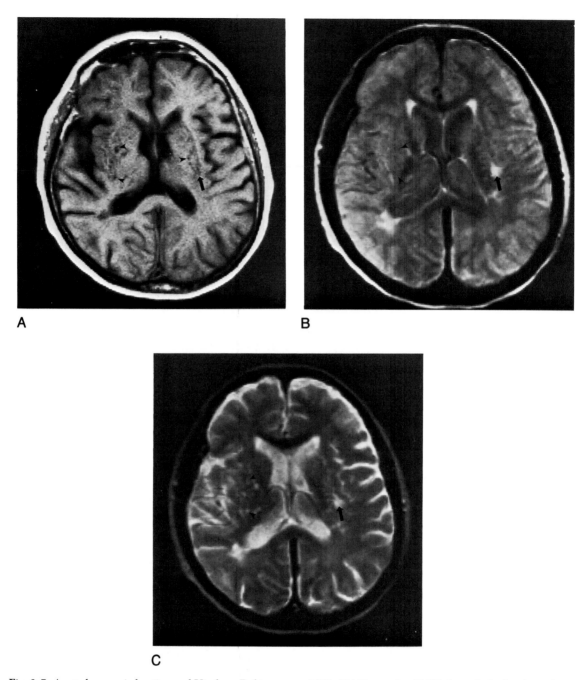

Fig. 2-7 Acute lacunar infarction and Virchow-Robin spaces, MRI. (A) T_1-weighted MRI through the basal ganglia shows multiple round hypointensities representing the dilated perivascular Virchow-Robin spaces around the perforating arteries *(arrowheads)*. The lacunar infarction in the posterior left lentiform nucleus is virtually impossible to diagnose on this sequence *(arrow)*. **(B)** Proton-density image (SE 2500/30) demonstrates the Virchow-Robin spaces as being slightly hypointense and equivalent with CSF *(arrowheads)*. The lacunar infarction is well demonstrated as a region of hyperintensity *(arrow)*. **(C)** T_2-weighted study (SE 2500/90) shows the Virchow-Robin spaces as rounded regions of high signal intensity equivalent with CSF. The lacunar infarction is also of high signal intensity but is larger and less rounded than the perivascular spaces *(arrow)*.

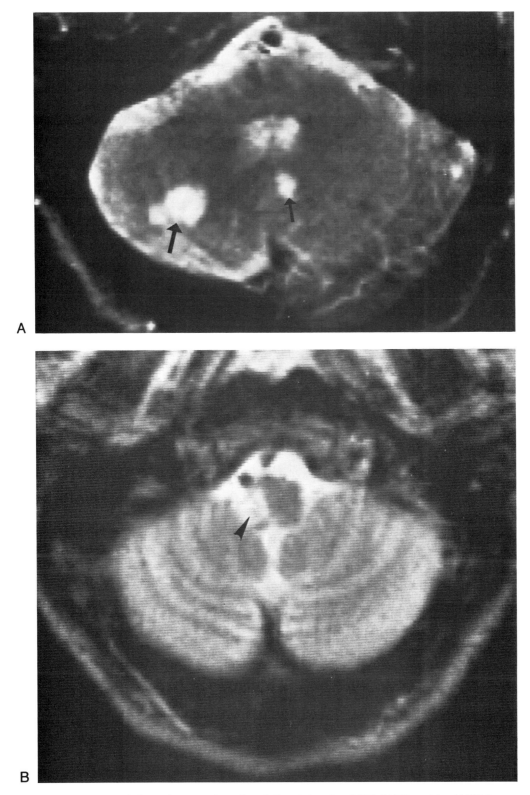

Fig. 2-8 Multiple acute cerebellar infarctions, lateral medullary infarction, MRI. (A) T_2-weighted MRI demonstrates two rounded regions of high signal intensity in the cerebellum *(arrows)*. Small infarctions in the cerebellum are frequently rounded in configuration and must be differentiated from metastases. **(B)** Lateral medullary infarction. T_2-weighted MRI shows high signal intensity in the right side of the medulla. This lesion is a relatively common infarction from a branch occlusion of the PICA *(arrowhead)*.

more oval in shape; during the acute phase, they show high signal on the proton density images (long time of repitition [TR] short time of echo [TE]).

The MRI technique is especially sensitive, compared with other modalities, for detecting infarction within the cerebellum (Fig. 2-5), pons, and medulla. The infarctions correspond with specific vascular territories. Pontine infarctions are almost always unilateral and paramedian or posterolateral, corresponding with the distribution of the paramedian perforating vessels from the basilar artery (see Fig. 2-15B). Bilateral abnormality is more likely to represent myelinolysis, the high signal intensity associated with aging (Fig. 2-9), or tumor, such as glioma, metastasis, or lymphoma.

It is difficult to determine the precise age of an infarction using MRI. With time, infarctions become soft (encephalomalacia) and relatively acellular. Cystic cavities may form and, when seen, indicate chronicity (Figs. 2-10 and 2-11). Gliosis within an old infarct has signal intensity between normal brain and CSF on T_1, and hyperintense signal intensity on T_2-weighted images. Cystic cavities from pan-necrosis have signal intensity similar to that of CSF on both T_1- and T_2-weighted images.

Although MRI is more sensitive than other methods in detecting infarction, it has some difficulty distinguishing infarction from other processes, particularly tumor and infection. The diagnosis of infarction is therefore made from the pattern of the edema on MRI, its correspondence with known vascular territories, and the appropriate clinical history. Gd-enhanced MRI may be more specific if gyriform enhancement is seen in the appropriate clinical setting.

Diffusion-Weighted MRI

Diffusion-weighted MRI is based on the difference in motion of protons within tissue in varying physiologic and anatomic states (see MRI in Ch. 1 for details for technique). Within a given tissue structure, there is an inherent diffusion coefficient for protons. That is, protons will diffuse farther in some tissue than in others within a given time interval. The influence of tissue on diffusion is complex, and not completely understood, so much of the observation is empirical. The diffusion coefficient is based on the cellular structure, the cellular integrity, the arrangement of cells within tissue, and the extracellular space (edema and substance). With imaging, motion of the patient and blood vessels contributes to the observed molecular motion, so that an apparent diffusion coefficient (ADC) is determined combining the effects of all the motion variables.

Diffusion MRI has particular advantage for the early evaluation of acute cerebral infarction. It is found that the ADC decreases in the region of ischemia within 20 minutes after the vascular occlusion. This is much earlier than the increase in T_2 signal on conventional SE MRI that occurs only after 3 to 6 hours from the onset of vasogenic edema. The region of decreased ADC correlates with regions of cytotoxic edema of infarction. By the theory of diffusion, extracellular water enters cells, causing cytotoxic edema; once within the cell, the protons are more constricted and diffuse less. It is not known whether the early decrease in the ADC in ischemia represents tissue that is doomed to necrosis or possesses potential for recovery. If the penumbra region of potential recovery from ischemia is also represented in the diffusion data, diffusion MRI will become a strong clinical tool for the evaluation of therapeutic intervention in ischemic stroke.

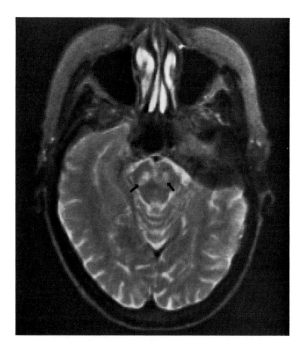

Fig. 2-9 High signal intensity within the pons associated with aging. Frequently there is irregular high signal intensity in the central portion of the pons and mesencephalon *(arrows)*. It probably represents a conglomeration of small microinfarctions, which must not be misinterpreted as acute brain stem infarction.

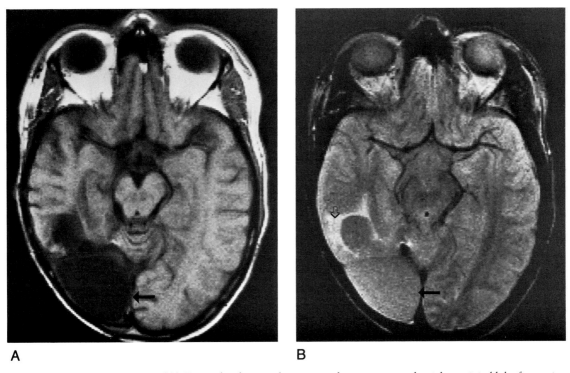

A   **B**

Fig. 2-10 Old infarction, MRI. (A) T_1-weighted image shows cystic degeneration in the right occipital lobe from prior ischemic infarction *(arrow)*. The region is of CSF intensity. **(B)** Proton-density MRI shows isointensity of the cystic cavitation *(arrow)*. Characteristically, there are regions of high signal intensity surrounding the cystic change representing gliosis *(open arrow)*.

After about 24 hours from the onset of infarction, increased ADC occurs in the cortex, in the subjacent white matter, and in the center of the lesion. The cortical and white matter change correlates with the development of extracellular edema (vasogenic edema). The central region of high ADC correlates with pan-necrosis and the complete disorganization of tissue structure. Within the regions of tissue damage, the ADC remains high. The precise mechanism for the observed change has not been worked out and is probably a complex addition of the effects of intracellular edema, extracellular edema, water volume, and metabolic function of cell membranes.

Perfusion Imaging

Perfusion contrast-enhanced MRI exploits the magnetic moment of a paramagnetic contrast agent and its magnetic susceptibility effect on T_2^* relaxation. Gd chelates have been used most often but, because of its much

greater magnetic moment, dysprosium chelate (e.g., Dy-DTPA-BMA) is preferred. The brain is scanned using an ultrafast GRE technique (turbo-GRE, echo-planar) before, during, and after intravenous contrast injection. The magnetic susceptibility effect of the magnetic moment creates loss of T_2^* signal and hypointensity on GRE images wherever contrast is delivered within the vessels. Regions of ischemia maintain normal signal intensity. To date, experimental studies in animals have been more effective than in humans in early detection of infarction and in estimation of CBF and volume.

MR Spectroscopy

MR spectroscopy (MRS) remains primarily a research tool. Advances in technology continue to make MRS more feasible, and it may eventually have important clinical use with infarction. Small region-of-interest capability to voxel level permits accurate spectral analysis

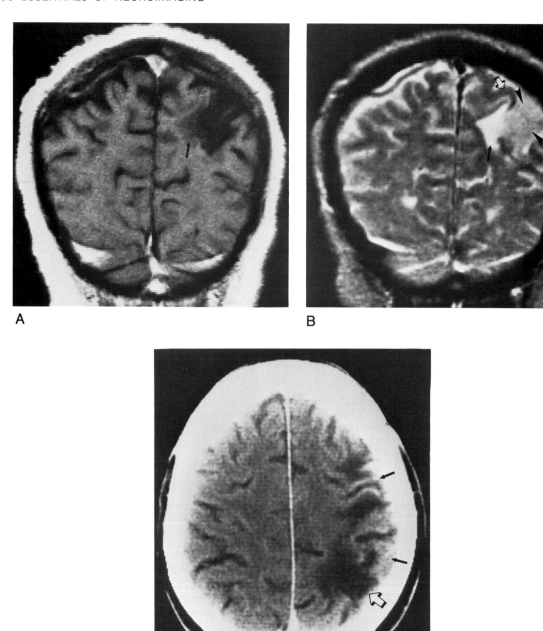

Fig. 2-11 Old hemispheric infarction, MRI and CT scans. (A) Coronal T_1-weighted MRI demonstrates a large hypointense region in the left midcortex. The cortex in this region has essentially dissolved, and there appears to be a cleft of CSF. Adjacent gliosis is represented by slight hypointensity *(arrow)*. **(B)** T_2-weighted image shows the region of brain dissolution as of nearly CSF intensity *(arrowheads)*. The adjacent gliosis is of greater intensity *(arrow)*. The cortex over the gliotic region is preserved *(open arrow)*. **(C)** Transaxial CT scan through the same region shows the cavitary dissolution as hypodensity *(open arrow)*. There is preservation of cortical stroma through much of the infarction *(arrows)*.

of tissue. Coalition of standard MRI images with spectral data provides anatomic localization. Proton MRS (H-1 MRS) defines the increase in choline, the decrease in pH, and changes in N-acetylaspartate (NAA) and γ-aminobutyrate glutamine (GABA) that reflect the metabolism of ischemia. Phosphorus (^{31}P MRS) spectroscopy is sensitive to the depletion of ATP, the changes in phosphocreatine (PCr). This information may allow timing of the ictus and the response to therapy.

CT Scanning

CT is still the most often used initial imaging technique for a person with an acute stroke. Although it is less sensitive than MRI for the detection of infarction, it is highly specific for excluding significant disease processes that mimic or complicate ischemic infarction, including hematoma, subarachnoid hemorrhage, chronic subdural hematoma, and tumor. It is excellent for the definition of hemorrhage prior to the initiation of anticoagulant therapy. CT defines very large infarcts in the early stage (within 6 to 12 hours) at a high risk of subsequent hemorrhage, poor outcome, and death. Overall, CT is much less sensitive than MRI for the definition of cerebellar, brain stem, and basal ganglionic infarctions.

The hyperdense vessel sign is the earliest abnormality that can be seen by CT with acute infarction (Fig. 2-12). It represents thrombosis within a major artery. It is present immediately after the ischemic event. It is seen most often within the horizontal portion of the middle cerebral artery (MCA). When seen, the sign is almost 100 percent accurate for the diagnosis of MCA occlusion. The finding has a 90 percent correlation with a large neurological deficit, and a less strong but positive correlation with poor outcome. The sensitivity of the finding for the diagnosis of MCA occlusion

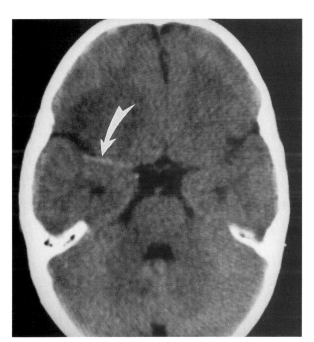

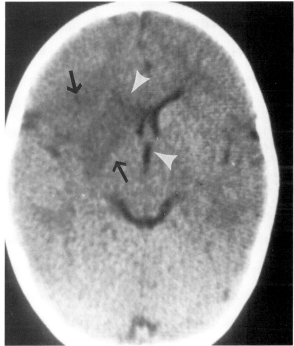

A   B

Fig. 2-12 MCA thrombosis, brain infarction. (A) Nonenhanced CT scan shows a hyperdense right MCA representing acute thrombosis *(arrow)*. **(B)** Hypodensity in the right caudate nucleus, lenticular nuclei, and the posterior frontal lobe represents the infarction *(black arrows)*. The swollen tissue compresses the right frontal horn of the lateral ventricle, and tilts the anterior third ventricle to the left *(white arrowheads)*.

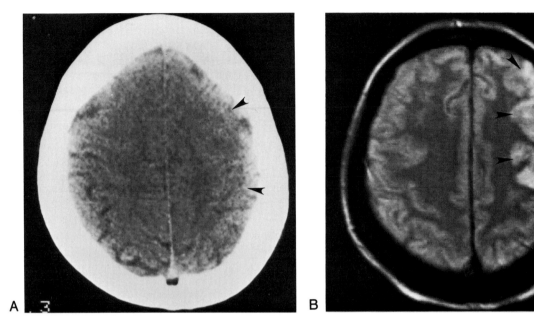

Fig. 2-13 Acute ischemic infarction, CT scan and MRI. (A) Nonenhanced CT scan shows subtle changes of a moderately large infarction of the left hemisphere. There is loss of the normally clear gray-white matter differentiation because of decreased density of the cortical gray matter from edema *(arrowheads)*. There is also minimal swelling with compression of the regional sulci. **(B)** MRI (SE 2500/30) shows hyperintensity of the cortical and subcortical tissues in the region of the infarction *(arrowheads)*.

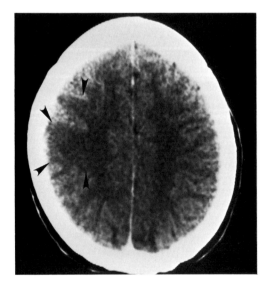

Fig. 2-14 Acute ischemic infarction, CT scan with contrast enhancement. Contrast-enhanced CT scan accentuates the early changes of acute infarction. The normal cortex shows increased density from perfusion with contrast-filled blood. The region of infarction does not show enhancement *(arrowheads)*. The low-density region goes to the cortical surface, a finding characteristic of infarction.

is unknown but is probably about 30 percent in MCA occlusion.

The diagnosis of the dense MCA is made by observing the vessel as more dense than other arteries of the same size. Overdiagnosis (false positive) occurs when the normally higher arterial CT density is misinterpreted for hyperdensity, especially if the surrounding brain is low density. Rapid-sequence contrast-enhanced thin-section helical CT of the basal arteries may prove more sensitive for the diagnosis of acute major vessel occlusion.

During the acute stage, CT scanning is dependent on the development of edema for detection of the infarcted tissue. The infarction is seen as a region of hypodensity. The quality and timing of the examination have a great effect on the sensitivity of the CT scan in detecting infarction. During the early phase after an infarct (less than 2 days), there may be insufficient edema for routine detection; after this time, most infarctions are seen. Large infarctions are more likely to be detected at an earlier time, some as soon as 3 hours after the ictus. Small infarctions of less than 2 cm are not seen consistently at any time. Infarctions within the cerebellum or pons are more difficult to image because of the artifacts inherent with CT scans of this region.

The first brain finding of an infarction is usually a subtle decrease in the density of the gray matter, so it becomes isodense with the underlying white matter (loss of gray and white matter differentiation) (Fig. 2-13). This change represents the cytotoxic edema of the gray matter. At this time there is little mass effect. If contrast enhancement is applied at this stage, there is normal enhancement of the adjacent normal cortex and accentuated hypodensity in the region of infarction because of ischemia (Fig. 2-14). If blood flow into an infarcted region has returned, this pattern may not be seen, and the infarcted region may enhance normally.

Over the next few hours, the edema increases, so there is more obvious well-marginated hypodensity of both the gray matter and the white matter. Typically, the infarction appears trapezoidal or triangular, more or less corresponding with a recognizable vascular territory (Fig. 2-15). Swelling increases and is first seen as a subtle compression of the cerebral sulci when compared with the corresponding opposite side (Fig. 2-13). The mass effect increases to a maximum at 4 to 5 days. Edema tends to become greater with infarctions caused by embolization. Lysis of the embolus results in reperfusion of the infarcted tissue, causing greater vasogenic edema and sometimes hemorrhage. The edema may be severe enough to cause herniation of brain or hydrocephalus (Fig. 2-16). During the 2- to 3-week period of the infarct, the infarct may become isodense with surrounding normal brain. This is known as the "fogging effect" (Fig. 2-17) (see above). Contrast accumulation in or around the infarct will provide diagnosis on enhanced CT.

Gradually, encephalomalacia or cystic cavitation develops. With CT scanning it is seen as a gradual decrease in the density of the infarcted tissue, until it becomes stable about 2 months after the stroke. At this stage, the infarct becomes sharply demarcated. Encephalomalacia with gliosis has a density slightly greater than that of CSF. Cystic cavitation from pannecrosis is isodense with CSF (Fig. 2-18). Because the outer layers of the cortex are more resistant to ischemia, sometimes a cortical ribbon of intact tissue is preserved (Fig. 2-11). Local atrophy causes dilation of the adjacent sulci and ventricle (Figs. 2-18 and 2-19).

Intravenous contrast enhancement may be used to define a region of infarction not seen with the nonenhanced examination or to differentiate an infarction from tumor. The enhancement with infarction may have many patterns. The principles are similar for both CT and MRI.

1. *Ribbon-like cortical gyral enhancement* (Fig. 2-20). This pattern is virtually pathognomonic of infarction. It is a result of a combination of loss of the blood-brain barrier and increased capillary filling of the gyri in the region of infarction ("luxury perfusion"). Only rarely does subarachnoid hemorrhage, meningitis, acute seizure, or hemiplegic migraine produce this type of enhancement.

2. *Irregular blotchy enhancement* (Fig. 2-21). Such enhancement may occur in any region of the infarction. It must be differentiated from the blotchy enhancement seen with grade III astrocytoma.

3. *Homogeneous enhancement throughout the region of the infarction.* This picture must be differentiated from the enhancement seen with lymphoma. Slight homogeneous enhancement may render a region of infarction isodense with the surrounding brain and invisible on enhanced CT scans.

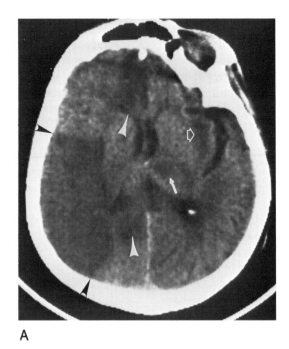

A

B

Fig. 2-15 Multiple ischemic infarctions, subacute phase. (A) A large infarction involves the right temporal lobe (*black arrowheads*). The margins of the infarct are sharp. There is mass effect with compression of the atrium of the right lateral ventricle. Additional infarcts are seen in the right pericallosal distribution (*white arrowheads*), left insula (*open arrow*), and left thalamus (*arrow*). **(B)** Pontine infarction. Axial T_2 MRI through the pons shows a large region of high signal in the right paramedian region. The infarction is sharply demarcated at the midline (*arrow*).

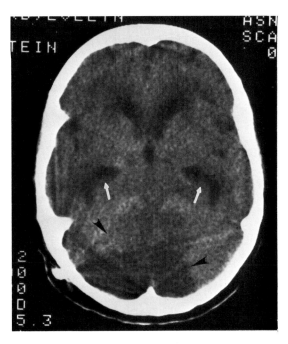

Fig. 2-16 Cerebellar infarction, acute. A large cerebellar infarction (*arrowheads*) has caused severe edema with obliteration of the basal cisterns and hydrocephalus (*white arrows*).

4. *Small, nonspecific, round, cortical, or subcortical enhancement* (Fig. 2-22). This picture may be indistinguishable from that produced by a small primary or metastatic tumor. Follow-up scans may be necessary to make the diagnosis.

5. *Peripheral enhancement outlining the boundaries of the infarction.* This pattern may be impossible to differentiate from a glioma, abscess, or prior hemorrhage. The enhancement may persist for a long time, contracting as the infarction matures. The clinical history, repeat scans, MRI, or a biopsy may be necessary to make the correct diagnosis. This type of enhancement is probably a result of the fibrovascular proliferation that occurs around infarctions as part of the healing process.

Hemorrhage may be seen as regions of hyperdensity within the hypodense infarcted tissue (Fig. 2-6C). Most commonly, the hemorrhage is petechial, but it may be linear within the gyri. The small petechial hemorrhages seen on pathologic examination are not demonstrated often with CT scanning. High field T_2-weighted or GRE MRI is more sensitive for demonstrating hemorrhage

within the infarction, especially during the subacute stage.

Lacunar infarctions are seen as regions of low density in the internal capsule (Fig. 2-18), basal ganglia, or thalamus (Fig. 2-15). CT scanning is not nearly as sensitive as MRI for detecting this type of infarction. Almost all lacunes progress to cystic change. After 2 months, they are seen as sharply defined, round or oval, CSF-equivalent-density lesions. Contrast enhancement is rare within a basal ganglionic or brain stem infarction because of the absence of collateral blood flow.

Angiography

The most reliable angiographic sign of cerebral infarction is the demonstration of an occluded artery. It usually appears as a gradual tapering of the lumen of the vessel until it is totally occluded. An avascular region appears distal to the occluded vessel and is best seen on the capillary ("brain stain") phase of the angiographic series. The circulation time through the infarcted region is usually slowed, with persistence of contrast in the proximal segments of the occluded vessels (Fig. 2-23). Angiography performed close to the time of the infarction (less than 8 hours) is most likely to define an occluded vessel. In recent trials of fibrinolytic therapy of stroke, early angiography demonstrated occluded vessels in 80 percent of persons with completed stroke. These trial also confirm the safety of cerebral angiography during the first 8 hours following the stroke. The overall incidence of complication is 4 percent, with a 0.6 percent of permanent CNS deficit.

ANGIOGRAPHIC SIGNS OF INFARCTION

Specific signs
 Occluded artery
 Collateral circulation
 Delayed washout of contrast from occluded
  artery
Nonspecific signs
 Avascular region in the brain (capillary
  phase)
 Early draining vein from "luxury perfusion"
 Local mass effect

(*Text continues on page 116*)

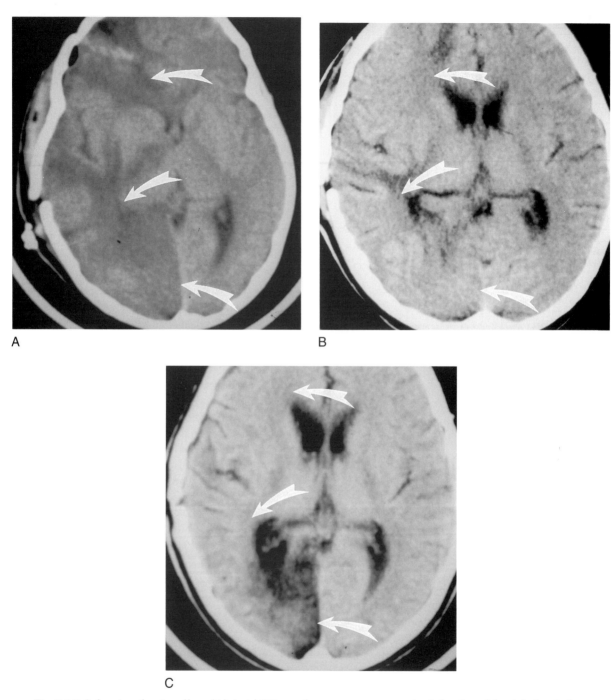

A

B

C

Fig. 2-17 Infarction; fogging effect. (A) Axial CT scan demonstrates an extensive infarction of the right hemisphere evidenced by the hypodensity of acute edema *(arrows)*. The patient had recent aneurysm surgery. **(B)** CT scan done 6 days later shows subsidence of the mass effect and the previously hypodense regions within the infarcted tissue are now isodense with the normal brain *(arrows)*. This isodensity is the "fogging effect" from cellular infiltration offsetting the edema during the subacute phase of the evolution of infarction. **(C)** Scan done 1 month later shows the hypodensity of encephalomalacia within the infarcted right occipital lobe. The right lateral ventricle dilates from atrophy in the infarcted right hemisphere. The tissue injury within the right frontal and temporal lobes is less than in the occipital lobe *(arrows)*.

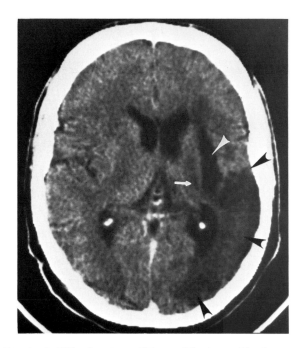

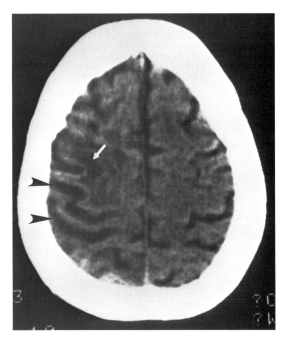

Fig. 2-18 Old infarctions, CT scan. The large old infarction in the region of the putamen (*white arrowhead*) has a density equivalent with that of CSF, representing cystic encephalomalacia. The large left posterior temporal infarction is hypodense but still has greater density than the CSF (*arrowheads*). It represents encephalomalacia and gliosis. Considerable atrophy is present, and there is dilatation of the left lateral ventricle and slight shift of the frontal horns to the left. Small lacunar infarctions are seen in the posterior limb of a left internal capsule (*white arrow*).

Fig. 2-19 Old parietal infarction. Nonenhanced CT scan shows regional enlargement of cerebral sulci (*arrowheads*). Hypodensity underneath the cortex represents gliosis (*white arrow*).

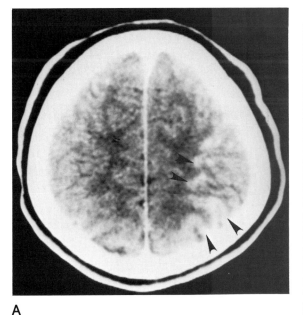

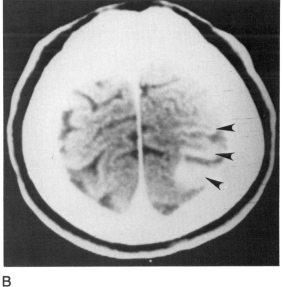

A   B

Fig. 2-20 Ischemic infarction, gyriform enhancement, CT scans. CT scans show the gyral pattern of enhancement associated with infarction (*arrowheads*). This pattern is almost pathognomonic of infarction or ischemia (see text).

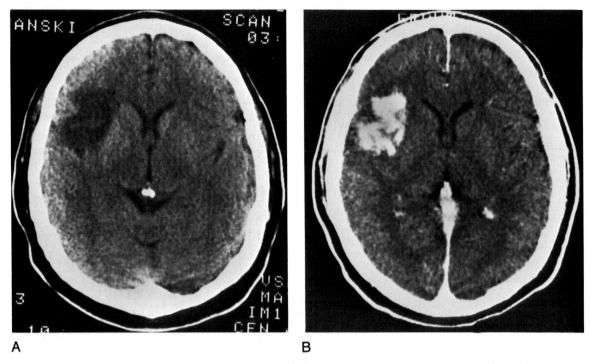

A   B

Fig. 2-21 Ischemic infarction, contrast enhancement. (A) Region of hypodensity is present in the right posterior lateral frontal lobe due to infarction. **(B)** Contrast CT scan shows dense, irregular enhancement throughout the region of infarction. It is difficult to differentiate this pattern of enhancement from that of an anaplastic astrocytoma (grade III).

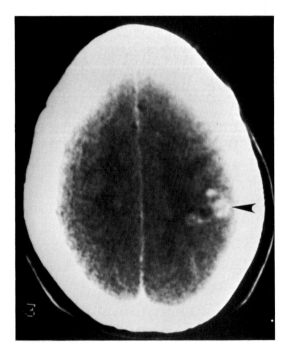

Fig. 2-22 Small infarction, cortical enhancement. Small region of nonspecific cortical enhancement is seen with this infarct. It is impossible to differentiate it from the early change of a peripheral glioblastoma *(arrowhead)* (cf. Fig. 5-21).

Fig. 2-23 Vascular occlusions from embolizations. (A) Supraclinoid ICA occlusion has occurred in a child secondary to embolization, with the clot outlined in the artery *(large arrowhead)*. A small trickle of contrast has passed by the clot to fill a few middle cerebral branches *(small arrowheads)*. **(B)** Multiple branch occlusions in the middle cerebral and anterior cerebral circulation has occurred secondary to embolization *(arrowheads)*. The occluded vessels are seen as an abrupt termination of the contrast column. A large region of the frontal and parietal lobes is void of contrast-filled vessels because of the occlusions *(curved arrow)*. Multiple occlusions in different vascular territories are characteristic of embolization.

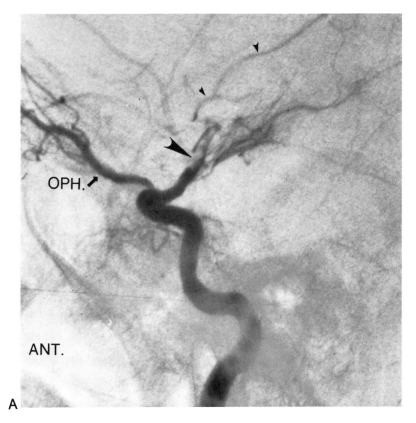

A

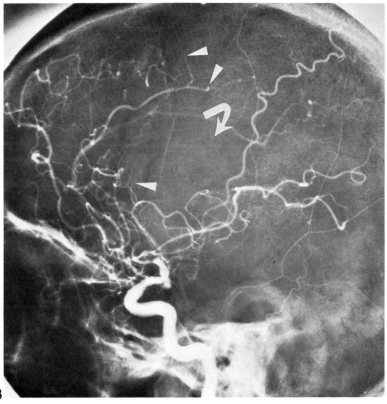

B

Collateral circulation is often seen in the region of the infarction, filling the vessels distal to the occlusion in a retrograde fashion. The flow into the infarcted region is delayed, so the retrograde contrast filling of arteries is seen in the capillary or venous phases of the angiography when other arteries have cleared of contrast. The demonstration of collateral circulation is specific for vascular occlusion (Fig. 2-24).

When the vascular occlusion is peripheral to the circle of Willis, the collateral flow is between major vascular territories through the normal leptomeningeal anastomotic channels (Fig. 2-25). For example, with MCA branch occlusion, the collateral flow into the peripheral middle cerebral branches beyond the occlusion comes from the ipsilateral pericallosal or posterior cerebral vessels passing through the small anastamotic channels over the convexity. The peripheral occluded branches fill from distal to proximal. Because of anatomic vascular variations and multiple sites of vascular occlusive disease, the collateral patterns are infinite and may become complex. The collateral pattern may become established within a matter of hours.

With occlusions proximal to the circle of Willis, the major vessels of the circle at the base of the brain normally provide the collateral flow (Fig. 2-26). The pattern depends on the point of obstruction of the major vessel and the configuration of the vascular ring. Variation in the circle of Willis is common (more than 50 percent) (see Ch. 1). Segments of the circle of Willis may not be completely formed. Other rare communications (e.g., a primitive trigeminal artery) and dural anastamoses may also provide collateral flow. Dural vessels may contribute to the collateral circulation of peripheral vessels by crossing the meninges to fill the cortical arteries.

If blood flow is decreased on one side, collateral unopacified blood may enter at anastamotic points, causing a confusing picture of apparent "filling defects" or "streaming" (Fig. 2-24A). These voids in the contrast column occur at the characteristic entry sites for collateral flow (posterior communicating artery [PCoA], anterior cerebral artery [ACA]) and must not be misinterpreted as atherosclerotic narrowing or intraluminal clot.

There may be increased circulation in regions of infarction ("luxury perfusion"), particularly at the periphery (Fig. 2-24B), as a result of local vasodilation in response to the ischemia. Arteriovenous shunting may be present. These perfusion changes are seen on the angiogram as a regional "blush" during the capillary phase commonly associated with one or more early filling veins, as noted below.

Radionuclide Scanning

The conventional radionuclide brain scan is no longer used to detect cerebral ischemia. The test depends on the accumulation of a labeled radiopharmaceutical in regions of blood-brain barrier breakdown. This technique is far less sensitive than CT scanning or MRI and has inferior spatial resolution. PET may be employed to give physiologic information and can define regions of ischemia in the brain. However, it requires the close proximity of a cyclotron for production of the positron-emitting radionuclides, as well as expensive scanning equipment; so far, the technique has been generally confined to research institutions.

SPECT is a recent practical advancement in imaging capability. It uses conventional nuclear cameras modified to rotate about the patient. In a fashion similar to CT scanning, the information is reconstructed into tomographic slices. Spatial resolution is better than that seen with conventional scanning, but the resolution is not nearly as fine as with CT scanning or MRI. SPECT is successfully applied to imaging the brain. It is used primarily to estimate brain metabolism by demonstrating rCBF using blood perfusion tracer agents. In general, cerebral metabolism is proportional to rCBF. Specific receptor site radiopharmaceuticals being developed promise to provide information about dopaminergic, adrenergic, and other neural receptor sites of interest to neurophysiologists and clinicians.

---

^{99m}Tc HMPAO SPECT SCANNING

Radionuclide distributed proportionally to
    Regional CBF
    Metabolic activity of neurons
Advantages of technique
    Earlier definition of infarction
    Definition of penumbra ischemia
    Definition of infarction when CT scanning is
        negative
    Assessment of metabolic state of brain distal
        to stenotic lesions

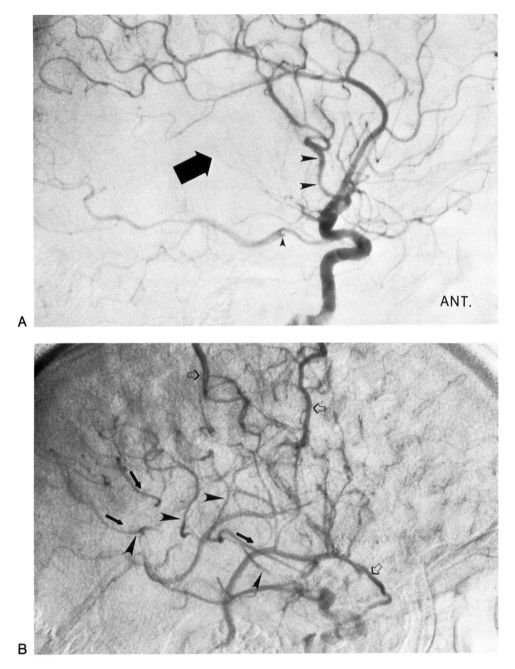

Fig. 2-24 Multiple branch occlusions of the middle cerebral artery. (A) Large vascular void is present in the arterial phase secondary to occlusion of most of the middle cerebral distribution *(large arrow)*. Some anterior branches remain patent but show variation in caliber secondary to vasomotor instability *(large arrowheads)*. It occurs frequently with infarction, particularly from embolization. Note the flow defect in the PCA at its junction with the PCoA *(small arrowhead)*. It represents unopacified blood entering from the basilar artery. Such defects, particularly at junctions between circulations, must not be misinterpreted as intraluminal clot. **(B)** Collateral circulation. The late capillary phase shows retrograde filling of distal portions of occluded MCAs *(arrowheads)*. These vessels have filled through pial anastomoses from the pericallosal vessels. *Black arrows* indicate the direction of flow. Adjacent regional hyperemia ("luxury perfusion") is indicated by the prominent early draining veins *(open arrows)*. A normal capillary phase is seen in the frontal lobe.

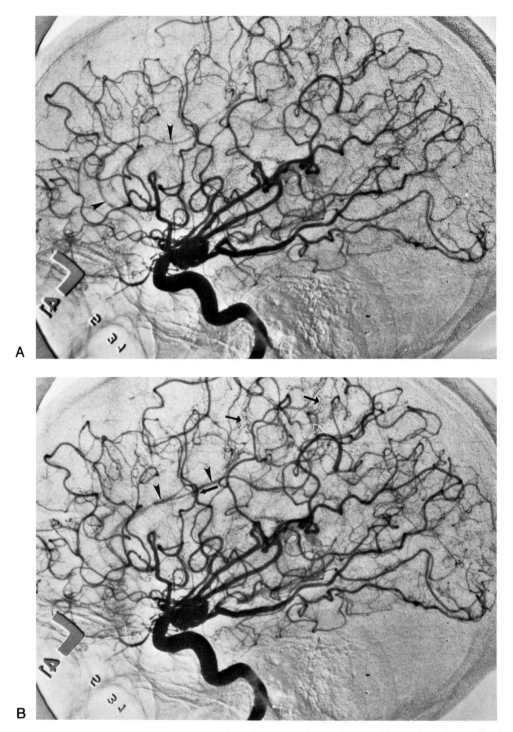

Fig. 2-25 Pericallosal occlusion. (A) Early arterial phase shows complete occlusion of the ipsilateral pericallosal artery. There is faint "flash filling" of the opposite pericallosal artery from cross filling through the anterior communicating artery *(arrowheads)*. **(B)** Delayed arterial phase shows opacification of distal branches of the ipsilateral pericallosal artery from retrograde filling *(arrowheads)*. Note the tangles of vessels representing the pial anastomoses between the middle cerebral and anterior cerebral circulations *(arrows)*. *(Figure continues.)*

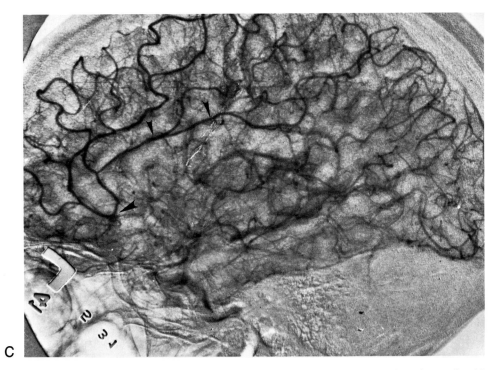

Fig. 2-25 *(Continued)*. **(C)** Capillary phase demonstrates extensive opacification of the ipsilateral pericallosal branches from retrograde filling *(arrowheads)*. Note the uniform "brain stain," indicating there is no ischemic tissue.

Technetium-99m hexamethylpropyleneamine oxime (HMPAO) is the most commonly used radionuclide blood tracer. It is lipophilic and diffuses easily through the cerebral endothelium into the brain parenchyma. Here, it is retained from backdiffusion into the bloodstream, probably by converting into a hydrophilic state. Technetium-99m-ethylcysteinate dimer (^{99m}Tc ECD) is a new more stable compound with similar distribution properties to ^{99m}Tc HMPAO, although like the older radiopharmaceutical ^{123}I IMP, its accumulation within the brain also reflects metabolic activity to some degree.

The scan shows the distribution of the radionuclide within the brain. The tracer amine distributes within the brain in proportion to the rCBF. It accumulates predominantly within the cerebral cortex, basal ganglia, thalamus, and cerebellum. If the patient's eyes are open during the injection, the visual cortex is highlighted. Essentially, no uptake occurs within the white matter. Ischemic and infarcted tissue show up as regions of photon deficiency (Fig. 2-27).

The major advantage of ^{99m}Tc HMPAO SPECT is its ability to define regions of nonlacunar cerebral infarction and ischemia earlier than with CT scanning. Infarcts may be defined consistently during the first day, whereas this is usually not possible with CT scanning.

It may also show ischemic regions when the CT scan is normal. In addition, the SPECT scan is more sensitive than the CT scan to regions of ischemia (penumbra of ischemia) that may surround the infarction. Thus, the region of abnormality may appear larger on the SPECT scan than on the CT scan, giving a better indication of the volume of tissue involved in the ischemic event. The volume often explains symptoms not explained by the distribution of abnormality seen with CT scans. Abnormality extending beyond the boundaries outlined by CT scans suggests that this tissue is ischemic but not infarcted. This finding has prognostic implications for recovery. SPECT scanning may be used to determine the metabolic activity of regions of the brain supplied by stenotic vessels. It may potentially be used to assess the

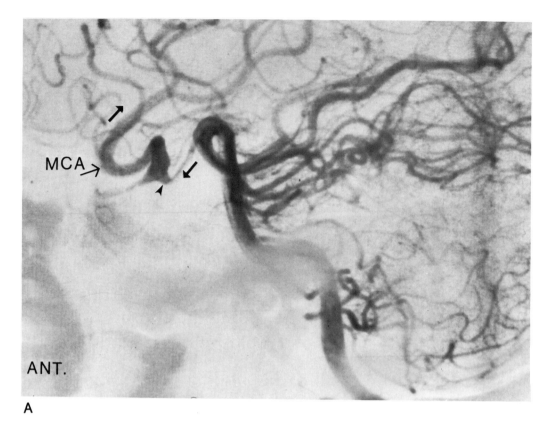

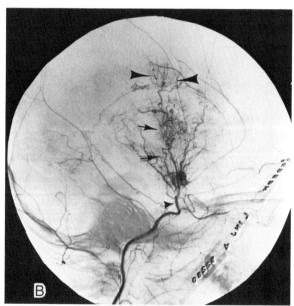

Fig. 2-26 Complete occlusion of the ICA. (A) The supraclinoid ICA fills via retrograde flow through the PCoA from the basilar circulation. The arrows indicate the direction of flow. Note the small infundibulum at the carotid origin of the PCoA *(arrowhead).* **(B)** Supraclinoid ICA occlusion, moya-moya. Gradual occlusion of the supraclinoid carotid artery has allowed development of extensive lenticulostriate collateral vessels that fill into deep medullary arteries. This deep form of collateralization, termed moya-moya, is usually found in children with multiple progressive intracranial arterial occlusions or neurofibromatosis 1. (Fig. B from Drayer, 1984, with permission.)

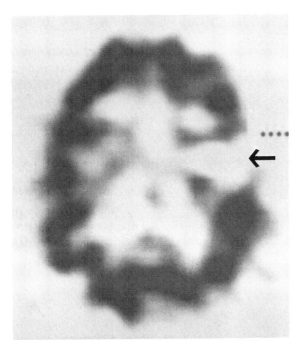

Fig. 2-27 SPECT radionuclide scan, cortical infarction. SPECT scan in the transaxial plane shows a photon-deficient region of the left parietal operculum representing infarcted tissue *(arrow)*.

adequacy of collateral circulation and cerebral flow as well as the need for surgical correction of stenotic lesions. After 5 days from the ictus, the accuracy of SPECT for the diagnosis of infarct decreases significantly. During this period, reperfusion masks the infarcted tissue. ^{99m}Tc ECD is less affected by the luxury perfusion because of its dependence on metabolic activity as well as rCBF. SPECT is of no value for the diagnosis of lacunar infarction.

SPECT brain imaging may be used to define rCBF following a TIA. Some evidence suggests that a persistent diminution of rCBF after recovery of clinical symptoms indicates a high risk of subsequent stroke. Also, SPECT is good for the evaluation of collateral circulation and the patency of major vessels after surgery.

Interventional Neuroradiology and Stroke

Thrombolytic (fibrinolytic) therapy attempts the rapid dissolution of intravascular embolus or thrombus before the occurrence of irreversible ischemic damage in the brain from stroke. Spontaneous intravascular thrombolysis takes place after stroke, but at a rate that is too slow to avoid ischemic damage. Streptokinase, urokinase, and tissue plasminogen activator (tPA) have been used. Delivery of the thrombolytic agents may be by peripheral intravenous infusion or by infusion directly into the clot (local intra-arterial fibrinolysis) after insertion of a microcatheter to the level of the clot. ICA and the MCA are the major targets, but clot lysis has been achieved in acute vertebrobasilar thrombosis with partially preserved brain stem function, occluded branches of the posterior circulation, and occlusion of the ophthalmic or central retinal arteries. The technique achieves thrombolysis in about 40 percent of cases. This must be accomplished within 6 to 8 hours after the ictus for the therapy to have any chance for clinical success. Fibrinolysis is not used for unilateral vertebral or chronic basilar occlusion, with cerebral aneurysm, arteriovenous malformation (AVM), and intracranial hemorrhage or in patients who have undergone surgery within the preceding 10 days.

Multicenter trials demonstrate that in response to thrombolytic therapy, vessels that recanalize do so within 1 hour about 50 to 80 percent of the time, but that the rapid recanalization cannot be equated to clinical benefit. Overall, clinical improvement rate is 30 percent. Hemorrhagic complications are significant with peripheral infusion, and infusion later than 6 hours after the ictus, where they occur in up to 35 percent of patients. These take the form of hemorrhagic infarction (small hemorrhage within the infarct) or parenchymal hematoma (large hematomas within the infarct or at other sites in the brain). Parenchymal hematomas are associated with significant clinical morbidity and mortality. Local delivery of low-dose tPA has a much lower hemorrhagic rate but delays the delivery of the thrombolytic to the clot. The ultimate value of intravascular thrombolysis is unknown, and the results of adequately controlled trials are not yet available.

Percutaneous Transluminal Angioplasty

Percutaneous transluminal angioplasty (PTA) is an alternative method to surgery for the treatment of significant extracranial carotid, vertebral, and subclavian artery stenosis. The technique has a low complication rate of about 5 percent, with few permanent sequelae. The indications for PTA are a surgical lesion in patients who cannot undergo surgery, failed anticoagulation therapy, and postendarterectomy restenosis from fibrous hyperplasia at the operative site. Patients are excluded from PTA if an acute infarction is seen with MRI, soft thrombus at the stenosis demonstrated with ultrasound, ulcerated plaque, and lack of hypoperfusion on ^{99m}Tc-HMPAO SPECT. All patients, except those with distal subclavian stenosis, are pretreated with anticoagulation for 1 week prior to PTA.

The PTA employs low-profile balloon catheters sized to the vessel and lesion (balloon is equal to vessel diameter) with a short tip. During the inflation of vessels supplying the brain, previously obtained heparinized arterial blood is infused through the lumen of the catheter into the distal artery during balloon inflation. This is to prevent ischemia during the time of inflation. Dilation of the stenosis is successful in 95 percent of cases. Distal vasospasm from PTA is unusual but is successfully treated with 100 μg of intra-arterial nitroglycerine. The long term success rate compared with surgery is unknown at this time.

ATHEROSCLEROSIS

More than 90 percent of infarctions are caused by atherosclerosis and its complications. The important mechanisms are vessel narrowing and occlusion from the atherosclerotic plaque, platelet emboli from ulcerated plaques and subendothelial hemorrhages, and embolism from endocardial infarction or cardiac arrhythmia.

Atherosclerosis is a diffuse disease with focal accentuations. Most often, it is the focal disease that is the cause of the infarction or cerebral symptoms. The significant focal lesion is the atherosclerotic plaque, which consists of subintimal proliferation of smooth muscle cells, excessive connective tissue matrix, and fat deposits in the intima. The plaque may grow large enough to cause complete occlusion of the vessel. Ulcers may form because of subintimal necrosis, resulting in the formation of fibrin-platelet thrombi on the ulcerated surface. Bits of these deposits may break loose and embolize to the cerebral vessels, causing TIA or infarction. Large thrombi

may form, causing local occlusion or embolism to a major cerebral vessel.

Almost always, atherosclerotic plaques occur at or near biburcations of vessels. Less commonly they occur at curves. In the carotid arteries, the lesions are most commonly found at the bifurcation of the common carotid artery (CCA), at the origin of the CCA from the aortic arch, in the cavernous portion of the ICA, and at the origin of the vertebral arteries.

Most of the attention is given to the carotid artery bifurcation. Persons with TIA and more than 80 percent stenosis of the origin of the ICA have been found to have a 5 to 10 percent risk per year of a stroke with permanent residual. The data also suggest that persons with asymptomatic internal carotid stenosis of more than 80 percent also have a much higher incidence of stroke in the distribution of that vessel compared with asymptomatic persons with less than 80 percent stenosis. Successful surgical removal of the plaque by endarterectomy decreases the risk of stroke to about 2 to 3 percent per year in both groups. Some researchers believe that there is also a positive correlation between surface ulceration and future stroke, but it has not been proved. Data are poor or lacking for other lesions, such as stenosis at the origins of vessels from the aortic arch. About 25 percent of the population of the United States over 60 years of age have an asymptomatic stenosis of 50 percent or greater.

Duplex Ultrasonic Scanning

Noninvasive tests have been developed to detect disease in the extracranial cerebral vessels, with the hope that it will lead to effective prophylactic therapy. Duplex scanning—the combination of B-mode imaging and pulsed Doppler flow detection—is now accepted as the standard noninvasive screening test for detection of atherosclerotic disease of the carotid bifurcation. It can reliably differentiate normal from abnormal and define the progression of disease. Oculoplethysmography is of limited value, as it depends on a significant decrease in blood flow or blood pressure for the detection of abnormality.

Duplex scanning of the carotid arteries is generally performed to select those patients who may be candidates for carotid angiography. Most commonly this test is performed in those who have an asymptomatic carotid bruit. Patients found to have more than 80 percent stenosis undergo angiography or MRA for confirmation of the lesion prior to possible carotid endarterectomy. Those found to have less than 80 percent stenosis do not undergo angiography but have follow-up duplex scans at 6-month intervals to detect any progression of disease. Those persons with definite TIAs usually have angiography without prior duplex scanning, as ultrasound is limited to evaluation of the carotid bifurcation. The sensitivity of duplex scanning for defining significant stenosis of the carotid arteries approaches 95 percent. The specificity for defining normal arteries is approximately 85 percent.

Most commonly, the B-mode scan is performed in both the longitudinal and the transverse planes. The lumen of the vessel is identified as an anechoic channel. The CCA is followed with the transducer along its course until the bifurcation is recognized, usually at the upper margin of the thyroid cartilage. The external carotid artery (ECA) normally branches anteromedially and the ICA posterolaterally. In 5 percent of persons, this relationship is reversed. The ICA is normally larger than the ECA. The ECA is reliably identified if the superior thyroid artery is seen to arise from its proximal portion. The waveform produced with the Doppler analysis also helps make this distinction. Because the ICA supplies a low-resistance intracranial circulation, it displays a diphasic pattern; in contrast, the ECA supplies a high-resistance circulation and displays a triphasic pattern.

The anatomic B-mode scan has not proved reliable for the identification of significant stenosis of the ICA. Ironically, the accuracy of the technique decreases as the severity of carotid stenosis increases. Thus, it can be particularly difficult to identify severe stenosis or occlusion. Heavy calcification within a plaque may obscure the vessel lumen. Attempts have been made to correlate the texture of the plaque with its composition and clinical prognosis, but the results are inconsistent. The more benign fibrofatty plaque is seen as a relatively homogeneous echogenic deposit with smooth margins bulging into the vessel lumen (Fig. 2-28A&B). Calcific plaques are strongly echogenic with irregular margins and acoustic shadowing (Fig. 2-28C). Ulcerations are seen as niches in the lumen, but they cannot be reliably identified.

Considerable attention has been given to the hemorrhagic plaque. Theoretically, ischemia of the vessel wall, produced as part of the atherosclerotic process, damages the vasa vasorum, which then ruptures, producing an acute hemorrhage. The hemorrhage enlarges the plaque, often causing stenosis or occlusion of the vessel lumen. The hemorrhage may rupture into the lumen, producing distal emboli and a residual ulcer crater. A plaque with a heterogeneous internal echo pattern and anechoic regions is thought to have a high probability of containing hemorrhage and surface ulceration.

The Doppler spectral analysis and blood flow rate calculations provide essential information about luminal stenosis. The examination is performed with a constant probe angulation. Ideally, the probe angulation is as close to parallel as possible with the vessel to be measured. Practically, about 60 degrees of angulation is possible and adequate. The ultrasound equipment compensates for the angulation when calculating the blood velocity within the vessel lumen. The pulsed technique allows sampling from a small volume of fluid at a specific point within the lumen of the vessel. The B-mode anatomic study is used to set the point of Doppler sampling. Bradycardia, hypotension, cardiac arrythmia, tandem lesions within the carotid artery, and hemodynamically significant stenosis of the opposite carotid adversely influence the spectral data degrading its accuracy.

Vessel narrowing increases the flow velocity through the region of stenosis. The increase becomes more pronounced with higher grades of stenosis but decreases as the stenosis exceeds 95 percent. The Doppler frequency shift is directly proportional to the change in velocity of the blood cells; that is, the observed peak frequency (in kilohertz) increases with the increase in blood velocity across the narrowed segment.

In addition, stenosis causes turbulence because of disruption of the normal laminar flow. The turbulence produces considerable variation in the speed of flow at different points in the cross section. Sometimes there is reversal of flow in focal segments producing a wide range of frequency shifts referred to as spectral broadening. This broadening generally becomes more pronounced with greater stenosis.

The spectrum of frequency within the vessel lumen is plotted graphically. By convention, the frequency is along the vertical axis, the time is along the horizontal axis, and the amplitude of echo at each frequency is represented by a gray scale point. The peak frequency increases and the spectral range broadens with carotid

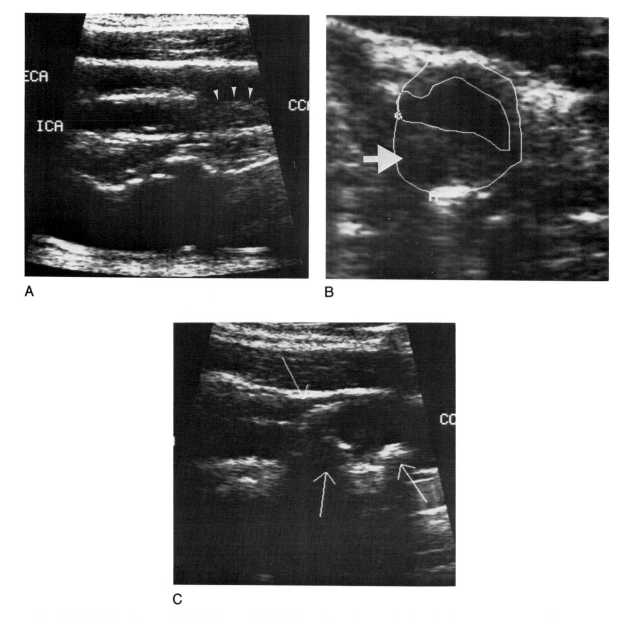

Fig. 2-28 B-Mode ultrasound, ICA plaque. (A) Fibrofatty plaque. Longitudinal study demonstrates a smoothly marginated, homogeneous, echogenic mound constricting the lumen of the distal CCA and the proximal ICA *(arrowheads)*. **(B)** Fibrofatty plaque. Transverse view of the distal CCA demonstrates the compromise of the cross-sectional area of the vessel lumen by the posterior plaque *(arrow)*. **(C)** Calcified plaque. Longitudinal study shows highly echogenic calcification within a posterior plaque *(arrows)*. Note the acoustic shadowing from the calcium. (Courtesy of Dr. Howard Raymond, Holyoke Hospital, Holyoke, MA.)

stenosis (Fig. 2-29). Color Doppler provides a color display of the spectrum of flow frequencies (Table 2-1). Colors are arbitrarily assigned to the frequency and direction of blood flow. Color display increases the conspicuity of turbulence and flow reversal within the vessel.

Ultrasonography may detect diseases processes other than atherosclerotic stenosis of the ICA. An intimal flap, thick hemorrhagic wall, and slow flow may be seen with dissection. Fibromuscular disease and Takayasu's arteritis show a thick wall. The vertebral artery is difficult to evaluate with ultrasound and with the exception of subclavian steal with flow reversal within the vertebral artery, ultrasound results are poor.

Transcranial Doppler

Transcranial Doppler (TCD) ultrasonography uses high-energy pulsed wave Doppler at low frequency (1 to 2 MHz). The probe is placed at the temporal bone, the orbit, or the neck, depending on the vessel to be imaged. The depth of the range-gated ultrasound transmission and the angle of the probe are set to optimize the signal from the vessel of interest. The spectral information obtained indicates the systolic, diastolic, and mean blood flow velocity. From Doppler spectrum, inference is made of the status of the vessel. The MCA, ACA, and the carotid artery at the base of the head are the most easily examined. The basilar and vertebral arteries may also be

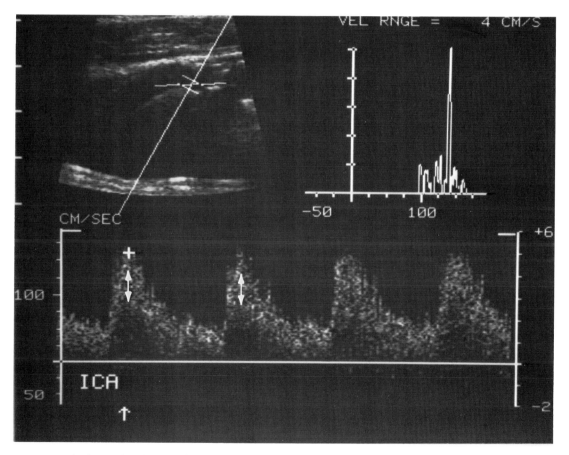

Fig. 2-29 Pulsed Doppler ultrasound. Spectral analysis was taken at the level indicated in the upper left of the figure. The plot shows spectral broadening (opposed arrows), higher peak frequency, and a prolonged peak, characteristic of significant carotid stenosis. (Courtesy of Dr. Howard Raymond, Holyoke Hospital, Holyoke, MA.)

Table 2-1 Spectral Analysis Criteria for ICA Stenosis

Stenosis (%)	Spectral Shift	
	kHz	cm/sec
0–30	< 3.5	< 110
31–50	3.5–4.0	110–125
51–90	4–8	125–250
91–95	> 8	> 250
95–99	Variable, may be < 4	
Occlusion		0

(Adapted from Grant et al., 1988, with permission.)

examined in some patients. Selective compression of a vessel is used to decrease flow in a defined circulation to confirm the vessel being imaged.

TCD is an indirect measurement of the anatomy of the vascular system and is a complementary examination to other imaging techniques. It is not accurate enough to be used as a sole screening test. With stenosis, peak velocity in the vessel increases over normal, until a critical stenosis of greater than 95 percent, when peak velocity decreases. Collateral flow is recognized by reversal of flow direction. Blood flow decreases in all vessels with increase in the intracranial pressure and brain death. TCD is used for the evaluation of a variety of intracranial and cervical vascular conditions (see box).

USES OF TRANSCRANIAL DOPPLER

Determination of severity of carotid or vertebral stenosis

MCA stenosis

Collateral intracranial circulation

Tolerance of carotid occlusion prior to surgery

Intraoperative CBF

Cervical carotid or vertebral dissection

Dolichoectatic arteries

Subclavian steal

Cerebral vasospasm from subarachnoid hemorrhage

Increased intracranial pressure

Brain death

Magnetic Resonance Angiography

MRA is not a single technique. It employs many variations of sequences, plane of view, voxel size, and echo times. Phase contrast or time-of-flight (TOF), two-dimensional (2D), or 3D must be chosen. Most techniques employ maximum-intensity pixel projection (MIP) display, which provides images in traditional form but has its own set of artifactual problems. Each technique selection has unique advantages, but no single one is optimal or solves all the imaging problems. In the medical literature, the panoply of MRA techniques and diagnostic criteria used make comparison of results and evaluation of utility difficult. Some studies evaluating MRA have used so many different sequences that the time of scanning approaches two hours, which is impractical. A more detailed discussion of the techniques of MRA is given in Chapter 1.

Cervical ICA

2D TOF MRA with MIP display is the most commonly used sequence for visualization of the cervical carotid artery and the bifurcation. The coverage includes the distal cervical ICA. Table 2-2 presents the average grading system for defining ICA stenosis and occlusion. The correlation with XRA of the ICA varies with the degree of disease measured, and the techniques used for MRA and XRA. It is best for near-normal and severe degrees of stenosis. Moderately severe stenosis is often erroneously graded as severe with MRA, and severe stenosis is often graded as occluded when the distal ICA is not visualized. Misreading results primarily from the technical limitations of the technique. The use of multiple sequences covering the same region in multiple planes increases the accuracy of MRA by nullifying limitations but requires a long and complicated examination.

Table 2-2 MRA ICA Grading[a]

Grade With XRA	Stenosis (%)	Criteria	Correlation (%)
Normal-mild	0–29	Min. narrowing	> 95
Moderate	30–50	More narrowing	60–85
Severe	50–99	Signal gap	90–95
		Distal flow signal	
Occluded	100	Signal gap	90–95
		No distal flow signal	

[a] Overall correlation 55–85%, depending on series reviewed.

Fig. 2-30 MRA, normal cervical carotid artery. Sagittal MIP of a 2D TOF MRA of the neck shows a normal common carotid bifurcation. The loss of signal within the posterior carotid bulb is from normal turbulence *(arrow)*.

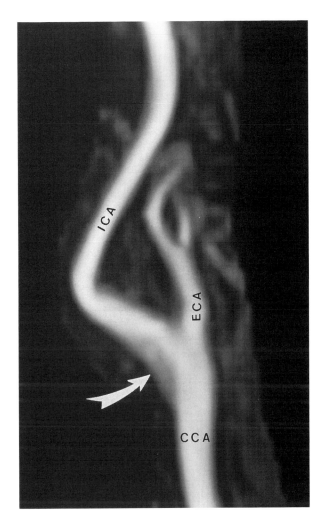

ADVANTAGE OF 2D TOF FOR CERVICAL ICA

Sensitive to slow flow

Less spin saturation with depth of volume

Less sensitive to motion affects (e.g., swallowing)

MRA provides a less detailed rendition of the carotid bifurcation compared with XRA (Fig. 2-30). It is insensitive to ulcerations and to the subtle changes in the vessel contour that occur with dissections and fibromuscular disease. The actual luminal diameter and stenosis cannot be measured. Grades of lesions are assigned by simple observation. A flow gap (lack of signal along the expected lumen) correlates with greater than 50 percent stenosis, although the anatomy of the actual lesion is invisible (Figs. 2-31 and 2-32). Lack of flow signal within the distal segment of the artery implies occlusion or a cessation of flow (Figs. 2-33, 2-34, and 2-35).

Artifacts plague MRA. Motion from swallowing causes misregistration. Turbulent flow causes intravoxel dephasing and loss of signal at, or just distal to, stenosis. Tortuous vessels and loops cause dephasing and flow gaps that spuriously mimic stenosis. Slow flow becomes saturated and appears the same as the background tissue.

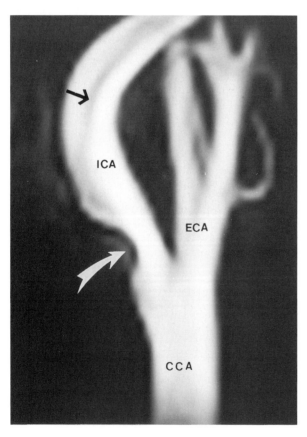

Fig. 2-31 ICA stenosis, MRA. MIP in the sagittal plane of a 2D TOF MRA of the neck shows the mild narrowing of the proximal ICA from atherosclerosis *(white arrow)*. There is low signal within the more distal ICA from flow variation within the bloodstream *(black arrow)*.

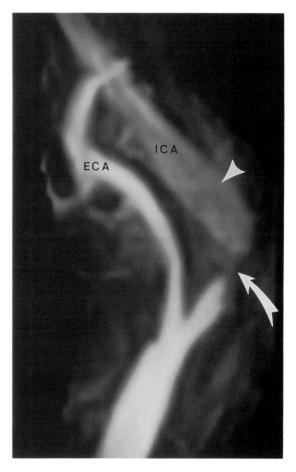

Fig. 2-32 Severe ICA stenosis, MRA. 2D TOF MRA shows loss of signal at the site of severe stenosis of the proximal ICA *(arrow)*. The intravascular signal is diminished distal to the stenosis from turbulance and slow flow *(arrowhead)*.

Fig. 2-33 Occlusion of the ICA, MRA. (A) Sagittal MIP of a 2D TOF MRA through the neck fails to produce signal in the region expected for the ICA *(large white arrows)*. This represents occlusion of the ICA. The ECA is well seen *(white arrowheads)*. The vertebral artery lies posterior *(open white arrows)*. "Notches" are common within the vertebral arteries, most of which represent a change in the course of the vessel within the foramen transversarium a signal void from turbulance *(thin white arrows)*. This vessel was normal at angiography. *(Figure continues.)*

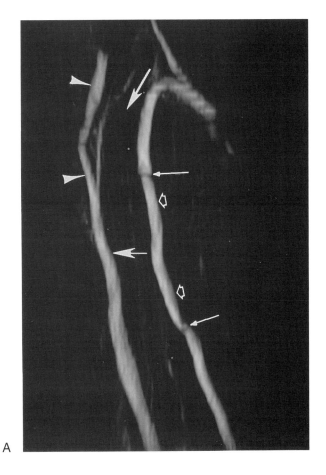

A

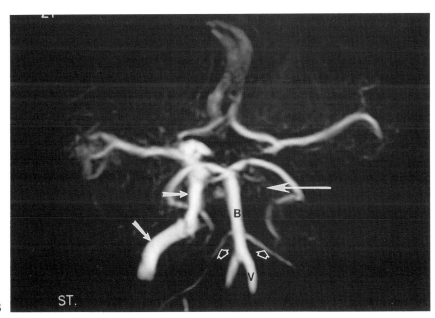

B

Fig. 2-33 *(Continued).* **(B)** Axial MIP of a 3D TOF MRA through the base of the skull demonstrates the absence of flow signal within the left ICA *(large white arrow)*. The normal ICA is seen on the right *(smaller white arrows)*. The vertebral (V), basilar (B), and AICA *(open arrows)* arteries show well.

129

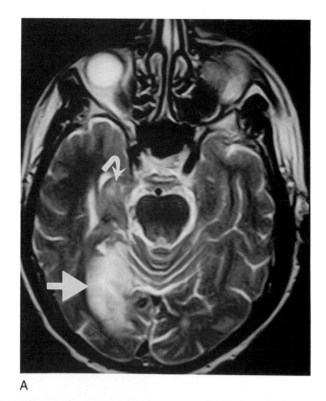

A

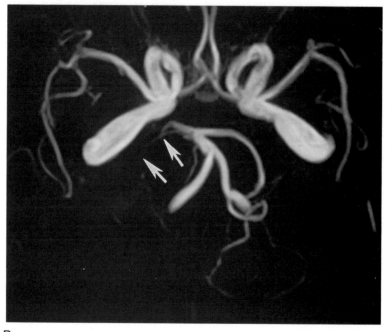

B

Fig. 2-34 Infarction right occipital and medial temporal lobes. PCA occlusion. (A) FSE (4/4) 96/3500 T_2 axial MRI shows homogeneous bright T_2 signal within the anteromedial right occipital lobe *(large arrow)* and to a lesser degree in the medial right temporal lobe *(curved arrow)*. The abnormal signal involves the cortex in most places, but only the subcortical region posteriorly in the occipital lobe. (B) 3D TOF RF-Fast MRA with motion artifact suppression technique (MAST) and MTC shows loss of normal signal in the right PCA just distal to its origin *(arrows)*.

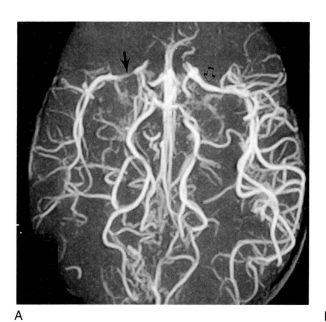

A

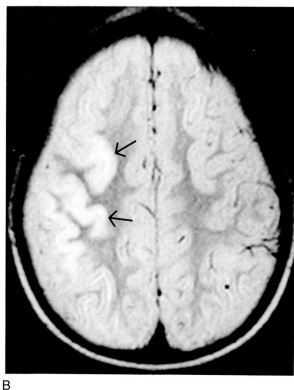

B

Fig. 2-35 **MCA stenosis; infarction, NF-1. (A)** Axial MIP of a 3D TOF MRA demonstrates the stenosis of the horizontal portion of the right MCA *(black arrow)*. The left MCA is normal *(open arrow)*. **(B)** Axial PD MRI through the hemispheres shows high signal within cortical gyri representing ischemic infarction *(arrows)*.

The MIP projection technique itself produces artifacts. The MIP depends on a high difference of signal intensity of the vessels at least two standard deviations (2 SDs) above the background tissue. With small vessels, or vessels with slow flow, this condition is not met, and the vessel drops from view. High signal intensities outside of vessels (e.g., blood and fat) can be projected along with the flowing blood when the background is not adequately suppressed. Review of the source images from which the MRA is made provides visualization of the spurious densities as artifacts. MIP always projects the vessel as narrower than normal because of the low signal produced along the vessel wall from laminar flow. The low signal is dropped out in the projection algorithm. View-to-view discrepancies form from inhomogeneities in the background tissue and asymmetric fields of view (matrix). Vessel overlap is a problem when edited volume boundaries are not used prior to projection.

Manipulation of the sequences and scanning parameters limits the artifacts. Signal from flowing blood is optimized with spoiled GRE at about 45/9/1, (TR/TE number of excitations [NEX]) flip 50 degrees. With this compromise sequence, there is good background tissue suppression with maintenance of strong blood flow signal. Voxels are kept small and the TE kept short to decrease intravoxel dephasing from turbulence and change in the direction of flow. Variable venous flow compensation (walking saturation [SAT] band) placed cranial to each excited slice eliminates the signal of flow within veins. The source images reveal the artifacts of MIP by showing the original information prior to projection.

The carotid MRA includes the cavernous portion of the ICA, the vertebral arteries, and the arteries of the circle of Willis at the base of the brain. Here, MRA has some difficulty. The curving nature of the carotid siphon and the air cavity of the sphenoid sinus produce artifacts that degrade the imaging of this segment of the vessel,

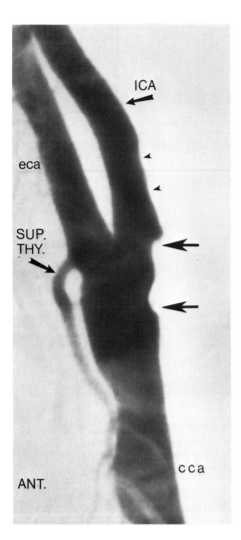

Fig. 2-36 Atherosclerotic plaque. A small atherosclerotic plaque is present that involves the posterior wall of the proximal ICA at the carotid bifurcation (*arrows*). Minimal plaque is seen more distally (*arrowheads*). SUP. THY, superior thyroid artery; cca, common carotid artery; eca, external carotid artery.

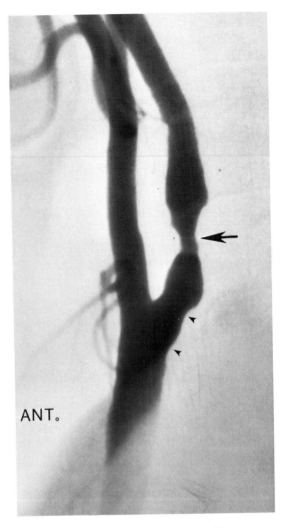

Fig. 2-37 Concentric atherosclerotic plaque. Lateral bifurcation angiography shows significant concentric stenosis of the proximal ICA just distal to the bifurcation (*arrow*). A small plaque is seen in the proximal bifurcation (*arrowheads*).

creating spurious stenosis. The base of the brain at the carotid canal presents similar difficulties. The origins of the vertebral arteries cannot be seen on most cervical studies without additional scans taken of the aortic arch. The basilar artery and proximal posterior cerebral arteries are well seen. Thin section 3D TOF MRA or 3D phase-contrast (3D PC) MRA has advantage over 2D TOF at the base of the skull. 3D TOF is relatively poor in evaluating more distal small intracranial arteries because of the saturation effects within the imaging slab that produce artifactual signs of occlusion. Tandem lesions of the ICA and stenosis of the MCA are visible with MRA,

although the sensitivity and clinical value of the technique are unknown. PC MRA and bolus tracking techniques provide flow velocity information which may have future clinical use.

Angiography (XRA, DSA)

Angiography remains the standard for detecting and evaluating focal atherosclerotic disease. Today digital subtraction angiography (DSA) provides equal spatial resolution compared with cut film XRA. However, DSA is susceptible to artifacts produced by motion and calcium within plaques.

The optimal examination is done with the catheter placed within the CCA about 2 cm proximal to the bifurcation. Care is taken not to pass the guidewire through the region of the bifurcation. Most lesions are located on the posterior or posterolateral wall of the ICA and are best seen on the lateral projection. Oblique or anteroposterior (AP) projections are needed to define the true size of the plaque and full extent of ulcerations, as the disease process is often asymmetric.

The plaque bulges into the vessel lumen (Fig. 2-36). The degree of stenosis of the vessel is determined by measuring the narrowest point seen on the two projections and comparing it with the normal portion of the ICA distal to the carotid bulb. The relationship is expressed as the percentage stenosis:

$$\frac{\text{Diameter of normal segment } (-\text{ Diameter of stenotic segment})}{\text{Diameter of normal segment}} \times 100$$

This technique is widely used, although there are some inherent inaccuracies. There is always the problem of which segment of the ICA to choose as a baseline and, with severe stenosis, decreased flow results in a smaller vessel lumen of the normal artery distal to the stenosis.

A stenosis of 50 percent causes a pressure gradient across the stenosis, but the flow rate remains normal. Stenosis of more than 70 percent causes a proportionately decreased flow rate (Fig. 2-37). The greater the length of the stenosis, the greater the effect on flow for a given degree of stenosis.

Special care must be taken during evaluation of the completely occluded ICA. Subtraction films are done so that a small trickle of contrast into a still patent ICA can be detected (Fig. 2-38). A prolonged injection into the CCA is best in this situation. Chronic occlusion usually shows a blunt stump. A high-grade stenosis that is still patent or a recent occlusion most often shows tapering to the level of occlusion. Even with the best technique, there is an occasional false-positive diagnosis of complete occlusion.

With complete occlusion, usually there is reconstitution of the distal intracranial ICA through collateral vessels. In most cases, it occurs via the ethmoidal branches of the ECA, filling the ophthalmic artery in a retrograde fashion, which in turn fills the supraclinoid ICA (Fig. 2-39). Also, the PoCA permits reversal of flow from the vertebral artery (Fig. 2-37).

This situation causes "streaming" of unopacified contrast into the supraclinoid ICA, which must not be interpreted contrast into the supraclinoid ICA, which must not be interpreted as intraluminal clot or stenosis. Sometimes there is enlargement of cavernous branches of the ECA filling the "siphon" or small branches from the neck filling the ICA artery within the carotid canal (Fig. 2-38). The site of re-entry of contrast into the ICA should be noted. With re-entry within the carotid canal or more proximally, endarterectomy is still possible, with complete occlusion of the ICA. With re-entry at the level of the cavernous portion of the ICA or beyond, endarterectomy is generally considered inadvisable.

Stenosis at the origin of the ECA is usually of little hemodynamic importance but may produce a bruit. It may be important, however, if it interferes with ECA collateral flow when the ICA is occluded.

PATTERNS OF COLLATERAL CIRCULATION WITH EXTRACRANIAL VASCULAR OCCLUSIONS

ICA occlusion
    Contralateral ICA–anterior communicating artery–MCA
    Posterior cerebral artery–PCoA
    Pterygopalatine branches of the internal maxillary artery–ethmoidal branches–ophthalmic artery
    Superficial temporal artery–ophthalmic artery
    Facial artery–inferior cavernous branches–ICA
    Vidian artery–ICA in the petrous portion
    Middle meningeal artery–ophthalmic artery along sphenoid ridge
CCA occlusion
    Vertebral artery–ECA–ICA
    Opposite ECA through midline anastomoses
Vertebral occlusion
    ECA–muscular neck branches–vertebral artery
Proximal subclavian occlusion
    "Subclavian steal" syndrome

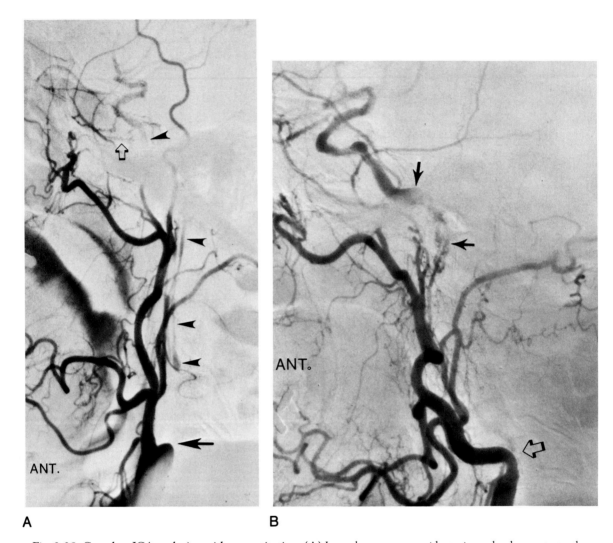

Fig. 2-38 Complete ICA occlusion with reconstitution. (A) Lateral common carotid arteriography demonstrates the stump of a complete ICA occlusion at its origin *(arrow)*. Distally, the cervical ICA is reconstituted *(arrowheads)*, in this case by flow through small collateral vessels from the ECA circulation. Additional collateral flow can be seen through cavernous anastomoses into the cavernous portion of the ICA *(open arrow)*. Patency of the distal portion of the cervical ICA makes endarterectomy technically possible. **(B)** Complete occlusion of the ICA is present *(open arrow)*. There is reconstitution in the petrous portion by way of anastomoses through the ascending pharyngeal artery *(arrows)*.

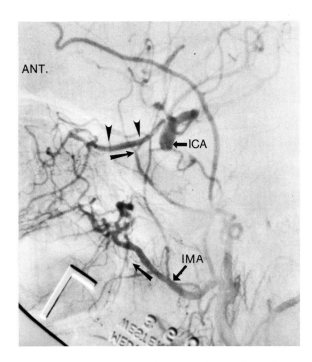

Fig. 2-39 Complete proximal ICA occlusion. Lateral common carotid arteriography demonstrates collateral circulation from the internal maxillary artery (IMA) to the ophthalmic artery (*arrowheads*) to fill the distal cavernous and supraclinoid portions of the ICA. *Arrow* below the ophthalmic artery indicates the direction of flow.

It is important to recognize "tandem" lesions, as they may be of hemodynamic significance and have bearing on the selection of patients for endarterectomy of a proximal ICA lesion. The most common lesions seen at sites other than the bifurcation appear at the origin of the CCA at the aortic arch (Fig. 2-40) and in the cavernous portion of the ICA (Fig. 2-41).

Stenotic atherosclerotic lesions or more than 70 percent often occur at the origins of the brachiocephalic and vertebral arteries (Fig. 2-42). These lesions are generally defined using aortic arch angiography. Severe stenosis of the proximal subclavian artery causes collateral circulation to develop from reversal of flow in the ipsilateral vertebral artery to fill the distal subclavian territory. This disorder is the subclavian steal syndrome, with most of the collateral blood supplied by the opposite vertebral artery through the junction at the basilar artery (Fig. 2-43). Subclavian steal is usually well tolerated by the patient, with claudication of the affected arm the predominant symptom. Only with stenosis of the origin of the opposite vertebral artery or with an inadequate circle of Willis are there signs of vertebrobasilar insufficiency. Patterns of collateral flow with extracranial vascular occlusion are given (see previous box).

Ulceration

Ulceration occurs commonly on the surface of the atherosclerotic plaque. Most ulcers are small and shallow, whereas others are large and extend deeply into the plaque to its base (Fig. 2-44). Platelet and fibrin thrombi form on the surface of the ulceration. It is estimated that more than 50 percent of ulcerations are not detectable with high-quality magnification angiography. Moreover, in about 30 percent of cases, surface irregularities thought to be ulcerations are found to be simply surface undulations and not ulcers. Therefore, except for larger craters, the accuracy of the diagnosis of surface ulceration is poor.

Surface fibrin and platelet thrombi are difficult to detect. An obvious intraluminal clot can sometimes be seen within the vessel lumen (Fig. 2-45). More commonly, however, thrombus cannot be differentiated from the underlying plaque. In general angiography significantly underestimates the amount of fibrin thrombus present. Atherosclerosis may recur or extend within vessels previously treated with endarterectomy (Fig. 2-46).

Dilation and Vessel Tortuosity

Atherosclerosis causes ectasia (dilation and elongation) of the vessel. Ectasia is most common in the aorta and the brachiocephalic vessels but may also be seen intracranially (Fig. 2-47). In the extreme, fusiform aneurysms form. There is a high association with hypertension.

Intracranial Atherosclerosis

Atherosclerosis also involves the intracranial vessels. The most common site is the cavernous portion of the ICA (Fig. 2-41). Severe cavernous segment disease indi-

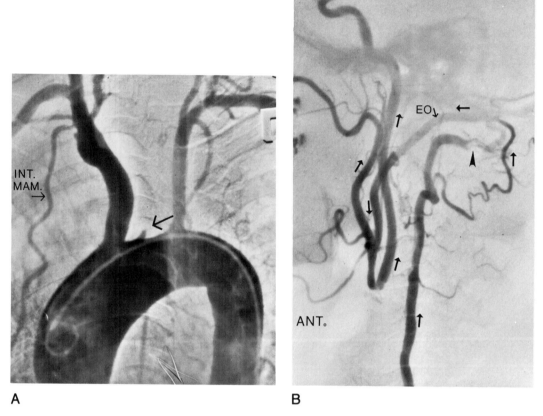

Fig. 2-40 Complete occlusion of the left common carotid artery. (A) Aortic arch examination shows the small "stump" of contrast at the base of a complete occlusion of the left ICA (*arrow*). INT. MAM, internal mammary artery. **(B)** Lateral neck view of vertebral injection shows collateral circulation from muscular branches of the distal vertebral artery to the external occipital artery (EO). Retrograde flow in the external occipital artery then fills the internal maxillary and ICAs. *Arrows* show the direction of flow. Note the "washout" defect in the distal vertebral artery from unopacified collateral circulation (*arrowhead*).

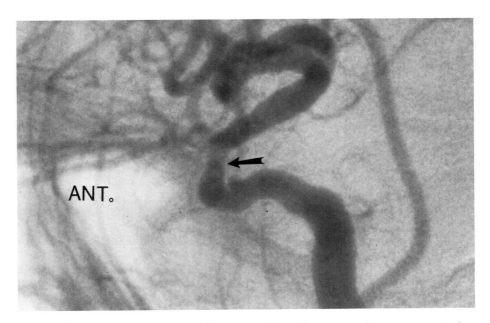

Fig. 2-41 Atherosclerosis, cavernous portion ICA. Atherosclerotic lesions may become severe in the cavernous portion of the ICA (*arrow*). This lesion is highly associated with death from stroke or myocardial infarction within 5 years.

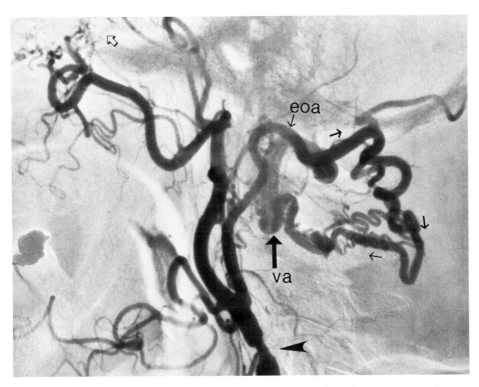

Fig. 2-42 Occlusion of the vertebral artery and ICA. Lateral neck angiography with common carotid injection shows complete occlusion of the ICA and stenosis of the ECA (*arrowhead*). Cavernous anastomosis branches can be seen from the distal internal maxillary artery (*open arrow*). Collateral flow occurs through the EOA into the occluded ipsilateral VA (*large arrow*). The small arrows indicate the direction of flow through the anastomotic muscular branches.

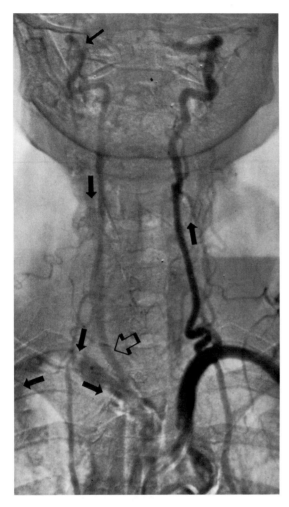

Fig. 2-43 Subclavian steal syndrome. There is high-grade stenosis of the innominate artery. Injection into the left subclavian artery shows antegrade flow into the left VA. Retrograde flow occurs down the opposite VA to fill the subclavian artery as well as the right CCA *(open arrow)*. Arrows indicate the direction of flow.

cates a poor prognosis, as 50 percent of patients die of severe stroke or myocardial infarction within 5 years. Other sites include the supraclinoid carotid artery, the proximal MCA, and less commonly any of the peripheral branches (Fig. 2-48) and the basilar artery. Branch narrowings tend to be focal at bifurcations and without intervening regions of dilatation as seen with vasculitis.

At times it is difficult to differentiate atherosclerotic change from arteritis or spasm.

EXTRACRANIAL NONATHEROSCLEROTIC ARTERIOPATHY

Dissection

Dissection (dissecting hematoma) of a major extracranial artery is an uncommon event. It may occur spontaneously or following neck trauma. Because the disease is generally not recognized unless angiography or MRI of the neck is performed, spontaneous dissection of the extracranial carotid and vertebral arteries is probably more common than is actually perceived. Patients with spontaneous dissection are usually younger than 50 years, and there seems to be a female predilection. It is a serious problem that frequently causes permanent neurologic deficits from embolization or vascular occlusion and may cause death from massive hemispheric infarction.

Two clinical syndromes are associated with dissection of the carotid arteries: (1) unilateral headache (hemicrania) and unilateral Horner syndrome (oculosympathetic palsy) with or without acute contralateral cerebral symptoms; and (2) unilateral headache with delayed contralateral cerebral symptoms. Rarely, the dissecting hematoma causes palsy of nearby cranial nerves (CN), particularly CN IX to XI. Frequently, the dissection occurs during strenuous exercise, such as marathon running or weight lifting, but it may occur at any time. There may be no predisposing factor, but there is a higher incidence of dissection in patients with connective tissue disorders or fibromuscular hyperplasia.

The dissection is an acute event. It usually regresses without therapy but occasionally progresses to cause permanent, complete occlusion of the vessel. It seldom recurs. Presently, angiography is the most accurate procedure for making the diagnosis and is performed in young patients who have the symptoms outlined above. Duplex ultrasonography and transcranial Doppler may suggest dissection, or exclude the diagnosis if flow and anatomy are normal. Presently, there are no studies regarding MRA and dissection of the cervicocranial vessels. MRA certainly demonstrates narrowing of the ves-

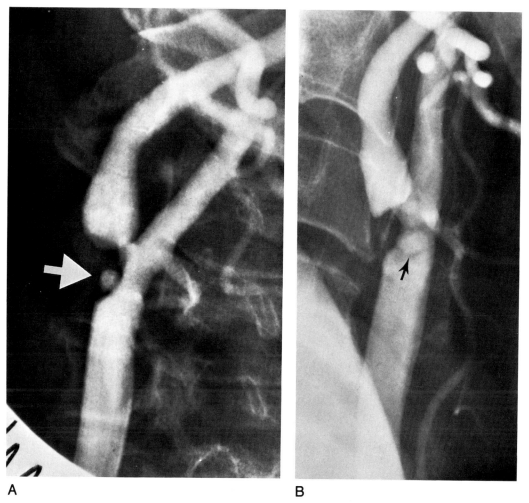

A   B

Fig. 2-44 Ulcerative plaque. (A) Oblique view of the carotid bifurcation shows a deep ulceration within a large posterolateral atherosclerotic plaque *(arrow)*. **(B)** On the direct lateral view it is difficult to identify the ulcer *(arrow)*, as it is seen more "en face."

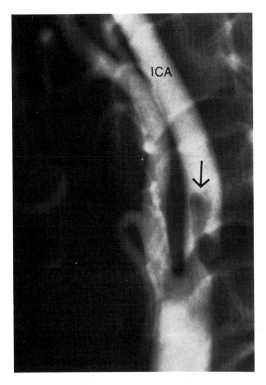

Fig. 2-45 Intraluminal clot. Lateral cervical CCA angiography demonstrates and intraluminal clot arising from a region of minimal disease in the proximal ICA *(arrow)*.

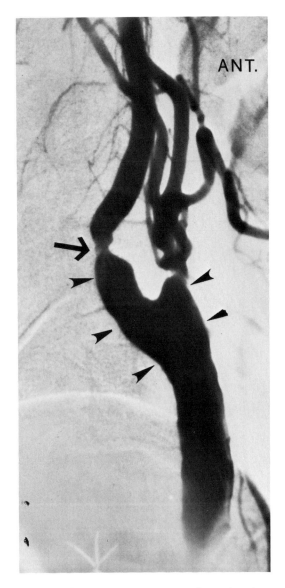

Fig. 2-46 Endarterectomy, recurrent atherosclerotic stenosis. Following endarterectomy, the vessel lumen becomes larger and has smooth margins *(arrowheads)*. If stenosis recurs, it is usually found at the distal end of the endarterectomy *(arrow)*. Less commonly it is found at the proximal margin.

sel (the string sign) and slow or absent flow, but its definition of dissections that cause little narrowing is unknown. The literature contains reports of nonocclusive dissections that have not been detected with MRA.

The most common angiographic finding is a tapered stenosis of the proximal ICA (Fig. 2-49). The stenosis may be asymmetric with a wavy configuration along one border of the artery (Fig. 2-50). Different projections may be necessary to demonstrate this important finding. Other possible patterns are a long, tapered stenosis (string sign) and total occlusion with a tapered proximal portion. A false lumen may be filled with contrast, and a saccular-appearing aneurysm may occur at either end of the dissected segment (Fig. 2-50). These saccular aneurysms may enlarge and must be followed with repeat angiography. The termination of the dissection is usually seen as an abrupt return to the normal caliber of the

vessel. Distal intraluminal thrombotic emboli may be seen at any level of the intracranial circulation. The dissection may extend into intracranial segments of the vessel and may be associated with subarachnoid hemorrhage. It is sometimes possible to identify the hemor-

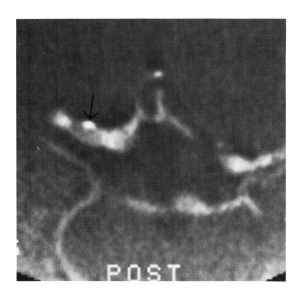

Fig. 2-47 Atherosclerotic ectasia. Contrast-enhanced CT scan through the base of the brain shows ectasia of the supraclinoid ICA and the proximal MCA *(arrow)*.

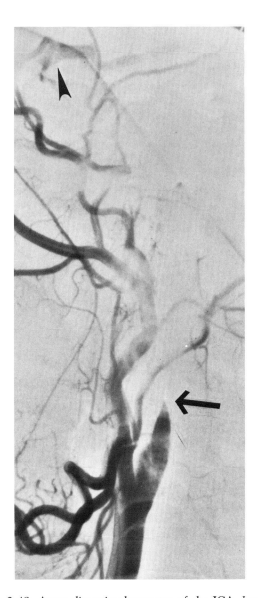

Fig. 2-49 Acute dissecting hematoma of the ICA. Lateral cervical projection of a common carotid injection shows tapered occlusion of the ICA. It is a nonspecific finding of occlusion, but in the appropriate clinical setting it is usually indicative of a dissection. Note the collateral filling into the supreclinoid ICA through the ophthalmic artery.

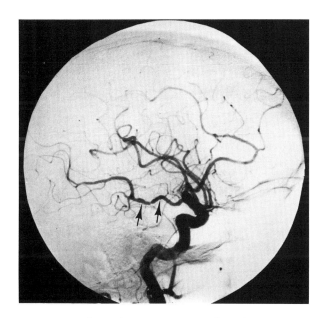

Fig. 2-48 Atherosclerotic narrowing. Irregular narrowing consistent with atherosclerosis *(arrows)* is seen in the posterior cerebral artery. Such an irregular, nontapered appearance of an artery distinguishes atherosclerosis from vasospasm. (From Drayer, 1984, with permission.)

rhage in the wall of the ICA with contrast CT scans. MRI may identify the dissecting hematoma in the vessel wall as high signal intensity on both T_1- and T_2-weighted images (Fig. 2-51) and as absence of flow void with vessel occlusion.

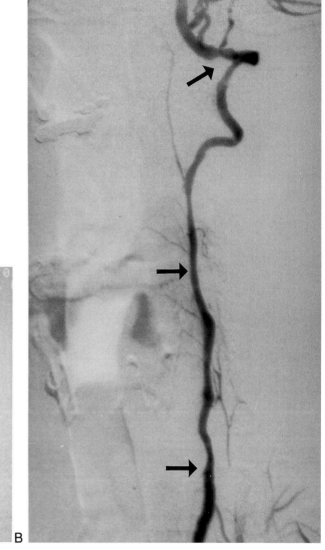

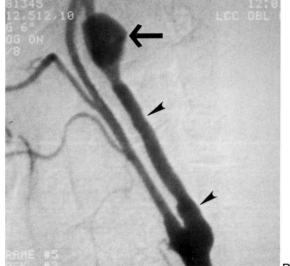

Fig. 2-50 Acute dissecting hematoma. (A) ICA. A dissected segment is patent but shows irregular, wavy "margins" *(arrowhead)*. An aneurysm has developed at the distal end of the dissection *(arrow)*. **(B)** VA. Left vertebral angiogram shows a long segment of the cervical vertebral artery compressed by dissecting hematoma. The margins of the vessel are wavy and the caliber is variable *(arrows)*. *(Figure continues.)*

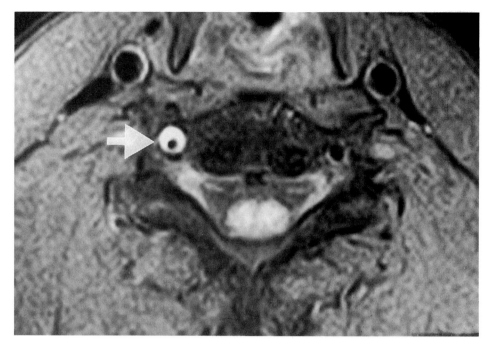

Fig. 2-50 *(Continued)*. VA. Axial T_1 fat suppressed (fat sat.) MRI through the cervical region demonstrates high signal within the right foramen transversarium representing dissecting hematoma within the right vertebral artery *(arrow)*. The central flow-void indicates that the vessel is patent but that the lumen is constricted.

The dissection most commonly involves the cervical and intracanalicular segments of the ICA (60 percent), the cervical segment only (30 percent), or a long segment extending from just above the carotid sinus through to the cavernous portion of the ICA (5 percent). The remaining dissections involve the vertebral artery or the MCA. The criteria for surgical intervention are not established. Most patients are treated with anticoagulation to decrease the risk of embolic complication.

Fibromuscular Dysplasia

Fibromuscular dysplasia (FMD) is an arteriopathy of unknown etiology. Segmental overgrowth of fibrous and muscular tissue occurs in the media leading to vessel stenosis. It occurs most frequently in the middle segments of the cervical ICA but may also involve the upper vertebral arteries and the branches of the ECAs. It is bilateral in more than 50 percent. There may be an associated intracranial berry aneurysm or involvement of the renal arteries with FMD, causing renovascular hypertension. The lesions may be found incidentally, but they are also associated with TIAs.

The diagnosis is made with angiography. The characteristic finding is an irregular string-of-beads configuration of the vessel lumen (Fig. 2-52), wherein the lumen is narrowed at multiple levels usually with intervening regions of dilated luminal diameter. There also may be only diffuse smooth narrowing of the vessel, which is difficult to differentiate from spontaneous carotid dissection. Vascular spasm must be differentiated from FMD. Spasm in the extracranial vessels is almost always caused by direct stimulation of the vessel wall by the catheter or the guidewire during the performance of angiography. With spasm, the vessel lumen has no regions of dilation but does have a more regular pattern of narrowings that change with time. MRA cannot reliably diagnose FMD. Slow flow in the periphery of the vessel

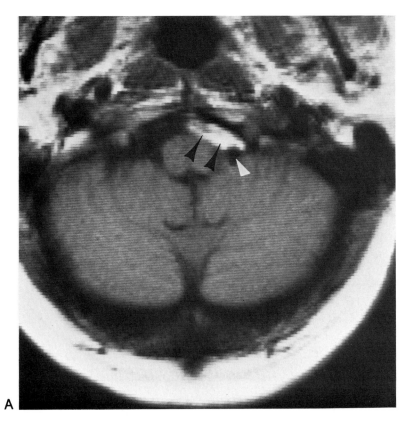

A

Fig. 2-51 **Subarachnoid hematoma appearing as a dissecting hematoma of the vertebral artery. (A)** T$_1$-weighted MRI shows the high signal intensity of subacute hemorrhage *(black arrowheads)* surrounding the distal VA *(white arrowhead)*. **(B)** Vertebral angiography shows a small aneurysm *(arrow)* and atherosclerotic ectasia of the VA *(arrowheads)*, but without findings of dissection. Angiography is necessary for definitive diagnosis in cases of suspected dissection.

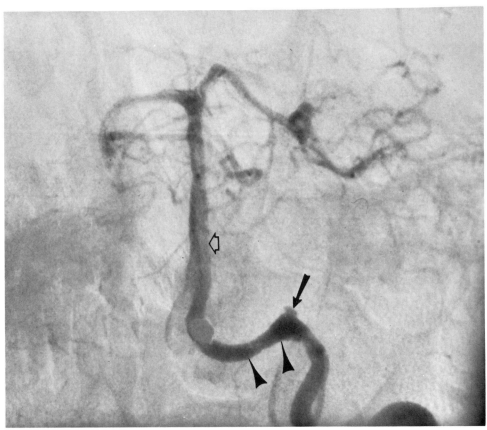

B

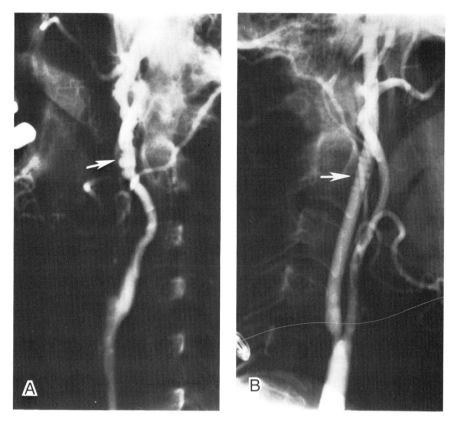

Fig. 2-52 Fibromuscular dysplasia and vascular spasm. (A) Fibromuscular dysplasia. Irregular dilatation and narrowing of the distal cervical segment of the ICA represents fibromuscular dysplasia *(arrow)*. **(B)** Vascular spasm. Regular concentric contractions are demonstrated in the cervical ICA *(arrow)* The regular spacing and lack of dilation differentiate spasm from fibromuscular dysplasia. (From Drayer, 1984, with permission.)

decreases signal so that the irregular luminal margins are not visible.

When associated with CNS syndrome, FMD can best be treated with percutaneous balloon angioplasty, which dilates the narrowed segment and usually stops the CNS symptoms. No treatment is given if the patient is asymptomatic and the lesion is found by chance.

Takayasu's Arteritis

Takayasu's disease is an acute inflammation of the aorta and brachiocephalic vessels that occurs in adolescent and young adult women. Systemic symptoms may be present that are similar to other collagen vascular diseases. Angiography demonstrates sooth, long segments of stenosis in the proximal subclavian and CCAs. The distal portions of these vessels are affected. Multiple vessel involvement occurs in 80 percent of cases. The stenosis may progress to complete occlusion. Calcification may be seen similar to that which occurs with atherosclerosis. The aorta usually appears normal, but during the later phases of the disease it may become aneurysmal predominantly in the ascending portion.

INTRACRANIAL NONATHEROSCLEROTIC ARTERIOPATHY

Numerous disease processes affect the intracranial vessels. These disorders are demonstrated by angiography as luminal narrowings, irregularities, dilations, and occlusions. The morphologic changes may be localized or diffuse. The vascular changes may be the result of an extrinsic process, such as meningitis or tumor, or of an intrinsic process, such as an autoimmune necrotizing angiitis. Because the arterial wall can react in only a limited number of ways to pathologic agents, most arteriopathies have a similar appearance on angiography. A few have characteristic narrowing patterns, and specific distributions can help limit the differential diagnosis, but the diagnosis is often made by exclusion of the various known etiologic factors or by biopsy of the meninges or involved vessels. The clinical history is important for defining the diagnosis. The disease may involve vessels that are too small to be visualized with XRA or MRA.

The arteriopathies may cause acute or chronic cerebral symptoms, such as headache, fever, personality change, TIAs, infarctions, and hemorrhage. If severe, the resulting ischemia may cause vast regions of brain destruction progressing to a vegetative state or death. Arteriopathies must be differentiated from vascular spasm, atherosclerosis, and dissection. The causes of angiopathy are listed below. Some of the categories are arbitrary, as the etiology and process are often poorly defined.

Collagen vascular disease
    Lupus
    Thrombotic thrombocytopenic purpura
    Scleroderma
    Rheumatoid arthritis
    Dermatomyositis
Necrotizing and granulomatous arteritis
    Polyarteritis nodosa
    Temporal arteritis
    Isolated intracranial arteritis
    Giant cell arteritis
    Wegener's granulomatosis
Amyloid angiopathy
Neoplasia
    Glioma
    Meningeal carcinomatosis
    Meningioma
    Atrial myxoma
Radiation vasculitis
Intravenous drug abuse

CAUSES OF ARTERIOPATHY IN ADULTS

Extracranial arteriopathy
    Atherosclerosis
    Spontaneous dissection
    Fibromuscular disease
    Takayasu's arteritis
    Temporal arteritis
Intracranial arteriopathy
    Hypertension
        Hematoma (microaneurysms of perforating arteries)
        Hypertensive encephalopathy
        Subcortical arteriosclerotic encephalopathy of Binswanger
    Infectious arteritis
    Bacteria
    Fungus
    Tuberculosis
    Syphilis
    Mycotic aneurysm

(Continued)

Hypertensive Arteriopathy; Intracerebral Hematoma

Hypertension is perhaps the major cause of cerebrovascular disease. It affects the middle-aged adult and elderly. Hypertension has a causative role in extracranial atherosclerotic plaques, lipohyalinosis of the perforating arteries at the base of the brain, and fusiform aneurysms of the aorta and intracranial vessels. The atherosclerotic lesions and their complications have been discussed in previous sections.

Hypertension is generally thought to be the cause of small (Charcot-Bouchard) microaneurysms in the perforating arteries of the brain, especially the lenticulostriate group. These aneurysms are occasionally seen with high-magnification selective intracranial angiography. They are susceptible to rupture, causing hematomas.

The putamen is the most common location of hemorrhage resulting from hypertension. The other locations are the caudate nucleus, thalamus, pons, dentate nuclei of the cerebellum, and subcortical regions of the cerebral hemispheres (see box). About 70 percent of all hypertensive hemorrhages are in the distribution of the lenticulostriate arteries (putamen and caudate). Hematomas

LOCATION OF HYPERTENSIVE INTRACEREBRAL HEMATOMA

Lenticulostriate arteries (70%)
    Putamen
    Caudate nucleus
Other sites (30%)
    Thalamus
    Subcortical hemisphere
    Brain stem
    Dentate nucleus

in these regions in patients over age 50 are almost invariably caused by hypertension, and no further angiographic investigation is necessary. Only rarely does a berry aneurysm rupture into the basal ganglia. When it does, it generally occurs in a young person and is associated with subarachnoid hemorrhage.

Hypertensive subcortical hemispheric hematomas occur in the white matter and have a lateral margin that conforms with the cortical gyri. They usually do not break into the subarachnoid space. In general, among the 50- to 70-year age group, these hematomas are considered secondary to hypertension. Contrast-enhanced CT scans and angiography, as well as careful dissection after surgical removal, have almost always failed to identify any underlying structural cause, such as a cryptic tumor or vascular malformation. In older patients (i.e., those over 70), these hematomas may be associated with amyloid angiopathy (see box). It is unusual for hemorrhage to be the presenting feature of a hemispheric tumor.

CT accurately detects intracerebral hematomas larger than 0.5 cm in diameter. Only the smaller petechial type cannot be seen reliably. The hematoma is represented as a region of nearly uniform increased density in the regions listed above (Figs. 2-53 and 2-54). Initially, during the acute phase, the hematoma is sharply demarcated from the surrounding brain parenchyma. At this time the density ranges from 50 to 90 Hounsfield units (HU). The

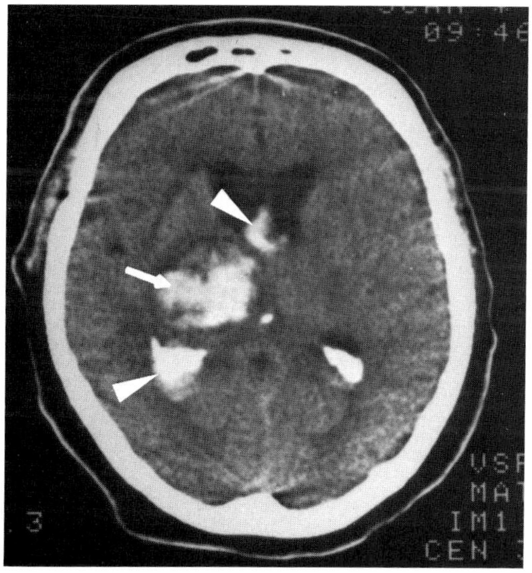

Fig. 2-53 Thalamic hematoma, hypertension. Large hyperdense region is present in the right thalamus representing the hematoma *(arrow)*. There is a minimal amount of surrounding edema. The hemorrhage has ruptured into the ventricular system *(arrowheads)*. Mild hydrocephalus is present.

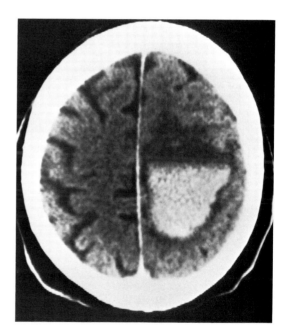

Fig. 2-54 Subcortical hematoma, hypertension. An acute subcortical hematoma is present in the left parietal white matter. The hematoma has not clotted and RBCs have layered dependently, producing a radiodense fluid level. In older persons, such hematomas are almost always secondary to hypertension or amyloid angiopathy.

high density is caused by the contraction of the clot increasing the concentration of the blood elements. The mass effect is from the blood volume itself. Within a few hours edema begins to surround the hematoma, becoming maximal after 3 to 4 days and increasing the overall mass effect. The ventricular system may be compressed or shifted, and the sulci are obliterated. Large hematomas may cause herniation and death. The mass effect from large hematomas may last as long as 4 weeks.

The hematomas have a propensity for dissection centrally into the ventricular system, which may be massive, forming a clot "cast" of the ventricle (Fig. 2-55). Only rarely do they break directly into the subarachnoid space. The prognosis is related to the size of the parenchymal hematoma and, to a lesser extent, to the amount of blood in the ventricular system. A small amount of blood within the ventricles does not have the grave prognosis once attributed to its presence.

The blood that diffuses from the ventricular system into the subarachnoid space may produce a communi-

cating hydrocephalus. The actual blood in the subarachnoid space is usually not seen on the CT scan. A hematoma in the posterior fossa can compress the fourth ventricle, and rarely a clot plugs the aqueduct resulting in noncommunicating hydrocephalus, which most often resolves spontaneously. The size of the ventricles must be followed by periodic CT scans, until the ventricular size is stable or decreasing. Because of the fibrinolysins in the CSF, blood within the ventricular system or subarachnoid space resorbs faster than the blood in the brain parenchyma.

Hematomas show a decreasing density of about 0.7 to 1.0 HU per day. The number of days until the hematoma becomes isodense with the surrounding brain can be predicted from the density of the fresh hematoma. For about 2 to 3 months, the density of the hematoma becomes progressively lower, until it is nearly that of the CSF (Fig. 2-56). During this time, the residual hematoma also becomes smaller. At about 6 months, the hematoma has completely resorbed, leaving a low density cavity that is

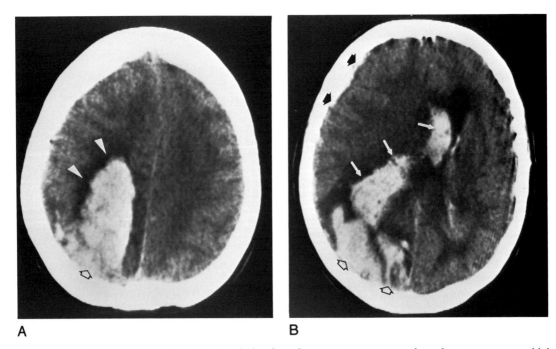

A   B

Fig. 2-55 Subcortical hematoma, hypertension. (A) A large hematoma is present in the right posterior parietal lobe. It is sharply demarcated but surrounded by moderate white matter edema (*arrowheads*). Hemorrhage has involved the cortical surface (*open arrow*). **(B)** The hematoma has extended down to the posterior parietal and upper occipital lobes (*open arrow*) and has dissected into the ventricular system, forming a cast of the right lateral ventricle (*white arrows*). Blood is also present in the subarachnoid or subdural space (*black arrows*). There is severe mass effect from hemorrhage and edema as well as subfalcial and craniocaudal herniation with ventricular dilation.

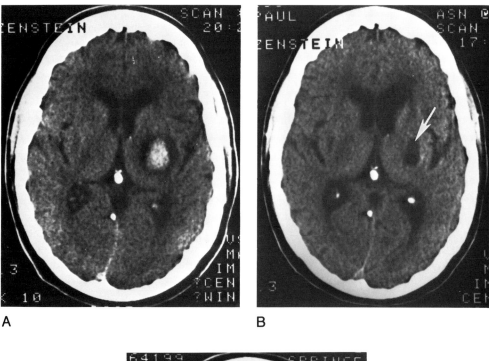

A   B

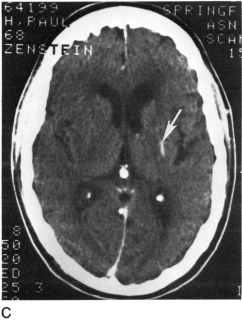

C

Fig. 2-56 Putaminal hematoma, hypertension (serial study). **(A)** Classic hypertensive hemorrhage in the left putamen is seen in the subacute phase. The hematoma has decreased slightly in density and is surrounded by a small amount of edema. **(B)** Scan 6 weeks later shows that the hematoma has become hypodense and smaller *(arrow)*, and the surrounding edema has resolved. The left frontal horn has dilated because of regional atrophy. Had contrast been used, ring enhancement about the hematoma would almost certainly be present, which might be misinterpreted as abscess or tumor. **(C)** Contrast-enhanced study performed 1 year later shows that the hematoma has completely resorbed and the cavity has collapsed to a small enhancing slit *(arrow)*.

smaller than the original hematoma. Small hematomas may leave no focal residual lesion at all. In addition to the focal cavity, the large lesions cause a more generalized regional atrophy, identified by focal dilation of the adjacent ventricular cavities and the cerebral sulci.

For their size and impressive appearance on the CT scan, hematomas cause relatively little damage to the brain. Hematomas dissect between the white matter fibers, causing compression but little destruction. The brain is able to withstand considerable compression without subsequent damage.

Approximately 7 to 10 days after the bleed, contrast enhancement may be demonstrated about the periphery of the hematoma (Fig. 2-57). The ring-like enhancement is usually separated from the hematoma and the surrounding parenchyma by a thin rim of low-density edema. This pattern is not seen with tumors, and a resolving hematoma may be suspected when such a pattern is recognized. With time, the intensity of the enhance-

ment increases until about 4 to 6 weeks and then gradually subsides over the next 3 to 4 months. The diameter of the ring decreases along with the hematoma. At the final stages, it appears as a small, round or linear density before its complete disappearance (Fig. 2-56). During the period when the enhancement appears ring-like or solid, the lesion may not be differentiated from a tumor or abscess (Fig. 2-57). MRI may show high signal on the T_1- and T_2-weighted images, indicating prior hemorrhage. In this situation, the clinical history and follow-up scanning may be necessary to make the correct diagnosis.

On MRI, the appearance of hemorrhage and hematoma is complex and time-dependent (Tables 2-3 and 2-4). Multiple factors are involved including the oxidative state of hemoglobin (oxyhemoglobin, deoxyhemoglobin, methemoglobin, and hemosiderin), the location of the hemoglobin (intracellular or extracellular), scan sequences, and the magnetic field strength. The various factors have different effects on proton relaxation. Al-

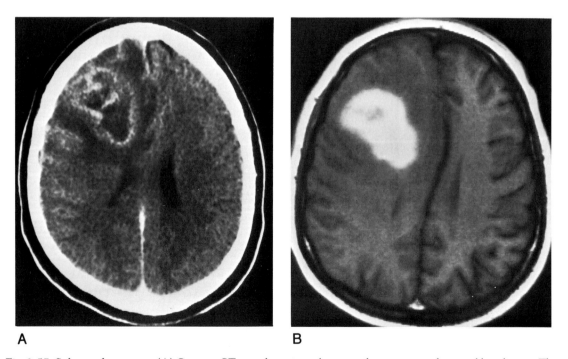

A   B

Fig. 2-57 Subacute hematoma. (A) Contrast CT scan shows irregular ring enhancement with central low density. This pattern is indistinguishable from that of a glioblastoma. There is surrounding edema. **(B)** T_1-weighted MRI shows the high signal intensity of methemoglobin within a resolving hematoma. This picture does not entirely exclude an underlying tumor, and follow-up scans are indicated if there is progression of the patient's symptoms.

Table 2-3 Phases of Hematoma for MRI Purposes

Phase	Time (days)	Condition
Hyperacute	< 12 hr	Intracellular oxyhemoglobin
Acute	1–5	Intracellular deoxyhemoglobin
Subacute	5–15	Methemoglobin—initially intracellular, then extracellular following lysis of RBCs
Chronic	> 15	Hemosiderin within macrophages around the periphery of the hematoma
		Gradual resorption of methemoglobin
		Conversion to cystic cavity
		Resorption of the cyst

though based on experimental data, the explanation for the appearance of a hematoma on MRI remains a theory that is not completely satisfactory. What follows is a simplified explanation of the present theory. Changes in the details of the theory may occur in the future.

The time phases of a hematoma can be divided into four periods: hyperacute (first few hours), acute (1 to 5 days), subacute (5 to 15 days), and chronic (more than 15 days). The borderline between each phase is not sharp and depends on the environment around the hematoma particularly the regional tissue oxygen tension. The

Table 2-4 MRI of Hematoma, High Field Strength (> 1.0 T)

Phase	Appearance
Hyperacute	
T_1	Isointense or slightly hypointense
T_2	Hyperintense
Acute	
T_1	Isointense or slightly hypointense
T_2	Hypointense hemorrhage surrounded by hyperintense edema
Subacute	
T_1	Hyperintensity of periphery of hematoma with gradual filling in of the center
T_2	Hyperintensity of periphery of hematoma with gradual filling in of the center—but all somewhat delayed compared with T_1 changes
	Thin peripheral rim of hypointensity from early hemosiderin deposition
Chronic	
T_1	Gradual decrease in signal intensity and size of the hematoma; will become near CSF intensity
T_2	Persistent high signal intensity until hematoma is completely resorbed
	Residual hypointensity of hemosiderin deposition

phases correspond with the various physical states and location of the hemoglobin molecule (Table 2-3). Field strength of the magnet affects the ability to observe magnetic susceptibility effects of intracellular deoxyhemoglobin and hemosiderin. The higher the field strength, the more sensitive the scanner is to magnetic susceptibility phenomena.

The hyperacute hematoma consists of oxyhemoglobin within intact RBCs. Oxyhemoglobin is not paramagnetic. Consequently, the hematoma appears as a protein solution with prolonged T_1 and T_2 relaxation times. The hematoma at this stage appears slightly hypointense on T_1-weighted images and hyperintense on T_2-weighted images. At this stage, it is nonspecific in appearance and cannot be differentiated from a neoplasm, except by history or the presence of a fluid level (Fig. 2-58). Within 12 hours, oxygen is depleted from the hemoglobin, producing deoxyhemoglobin; the hematoma is now said to have entered the acute phase, and its appearance changes.

During the acute phase, the hematoma consists of deoxyhemoglobin within intact RBCs. Deoxyhemoglobin is paramagnetic, but it is not available to interact with regional protons and so it has no effect on the T_1-weighted images. The hemorrhage remains of slightly low intensity, as during the hyperacute phase. However, the intracellular paramagnetic deoxyhemoglobin produces local variations (inhomogeneity) in the magnetic field within the hematoma. This causes both T_2 and T_2^* dephasing, resulting in a markedly shortened T_2 relaxation time. The actual hematoma or hemorrhage appears hypointense on the T_2-weighted images during this phase (Fig. 2-6). Edema develops around the hematoma, and clot retraction extrudes serum peripherally so the tissue surrounding the hematoma appears of high intensity on the T_2-weighted images.

The subacute phase is complex. It begins with the conversion of deoxyhemoglobin to methemoglobin, which usually occurs about 5 days after the bleed. The periphery of the hematoma converts first and the process progresses inward toward the center (centripetal) at the variable rate, until the entire hematoma is converted to methemoglobin. Methemoglobin is strongly paramagnetic and is available for proton interaction. This effect is especially strong when methemoglobin is free after cell lysis, but it has a moderately strong effect when methemoglobin remains within intact RBCs. Therefore, it produces a strong T_1 shortening effect, which is seen as high

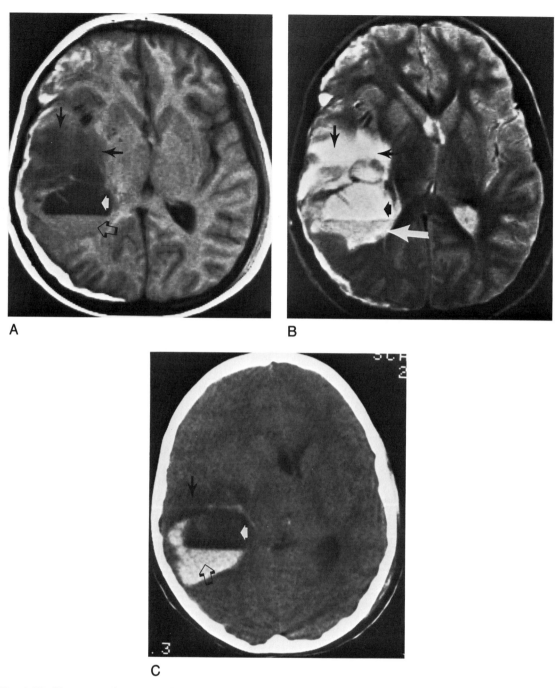

Fig. 2-58 Hyperacute hematoma. (A) T₁-weighted MRI shows the hematoma in the posterior right temporal lobe. RBCs with oxyhemoglobin are layered dependently and are nearly isointense with gray matter *(open arrow)*. The serum anteriorly is hypointense, but with just slightly greater intensity than that of CSF *(white arrow)*. Edema is present anteriorly *(black arrows)*. **(B)** T₂-weighted MRI shows the layered RBCs as moderately hyperintense *(white arrow)*. The serum is very hyperintense *(fat black arrow)*, as is the edema anteriorly *(black arrows)*. **(C)** CT scan shows the dependent RBCs as high density *(open arrow)*. The serum is low density, just slightly higher than that of CSF *(white arrow)*. The edema anteriorly is just slightly hypodense and is difficult to see *(black arrow)*. Note the subdural hemorrhage and right frontal contusion seen with MRI, but not with CT.

signal intensity on the T_1-weighted images. On T_1-weighted images, the hematoma becomes bright on the periphery and gradually fills in with brightness as the hematoma converts completely to methemoglobin (Fig. 2-59).

On the T_2-weighted images, the hematoma becomes of high intensity during the subacute phase. This change occurs primarily because of the strong T_1 shortening effect of methemoglobin and its carryover to T_2-weighted imaging. The net result is apparent T_2 prolongation. In fact, methemoglobin has little direct effect on T_2 relaxation. This T_1 shortening effect is strongest when methemoglobin is extracellular and is less strong when the methemoglobin remains within intact RBCs. Therefore, on the T_2-weighted images, the hematoma becomes of high signal intensity (bright) where cell lysis has occurred and remains of relative low intensity (dark) where RBCs remain intact (Fig. 2-59). The hematoma gradually fills in with high intensity throughout, as RBC lysis becomes complete, but this pattern usually lags behind the centripetal progression seen on the T_1-weighted images. The peripheral edema continues to be imaged as high signal intensity surrounding the hematoma nidus, although edema decreases during the late subacute phase.

After this point, the hematoma is considered to enter the chronic phase. A rim of low signal intensity appears around the hematoma and is thought to represent hemosiderin within macrophages (Fig. 2-60). The intracellular hemosiderin distorts the local magnetic field in a manner similar to the effect of intracellular deoxyhemoglobin. Therefore, it causes regional dephasing and a strong reduction of the T_2 and T_2^* relaxation time wherever hemosiderin is taken up within the macrophages. As the hematoma nidus becomes smaller, the thickness of the ring generally becomes thicker, filling in the region previously occupied by the hematoma. The finding of residual hemosiderin after the hemorrhage has resorbed can be useful for determining prior episodes of bleeding, as with an arteriovenous malformation (AVM).

On the T_1-weighted images, the hematoma gradually loses methemoglobin, resulting in gradual diminution of the T_1 signal intensity within the hematoma nidus. The nidus becomes isointense with the brain and eventually nearly isointense with the CSF as the hematoma resorbs and converts to a cystic cavity containing proteinaceous fluid. It may eventually disappear completely.

The central high T_2 signal intensity of the hematoma produced by methemoglobin persists for some time and becomes true T_2 prolongation as the hematoma liquefies into a high-protein cystic cavity (Fig. 2-61). It may slightly decrease in intensity, becoming nearly isointense with the CSF, depending on the amount of protein within the cystic fluid. The high T_2 signal disappears only with complete obliteration of the hematoma cavity. The low-intensity hemosiderin ring collapses with the residual cyst and may ultimately be seen as a small slit. This may be the only indication that a hemorrhage has been present. GRE sequences are the most sensitive for detection of hemosiderin deposits because of their greater sensitivity to the effects of magnetic susceptibility.

Angiography or MRA is necessary only if there is a reasonable probability that the hematoma may have been caused by an aneurysm, vascular malformation, or tumor. Etiologies other than hypertension are considered if the hematoma is inhomogeneous, has central low density on CT scans, occurs in a noncharacteristic location, or has associated calcium. Aneurysms of the horizontal portion of the MCA cause intratemporal hematoma from direct rupture into the brain parenchyma. Here, angiography is indicated. A blood-fluid level may occur with acute hematomas of any etiology, although this sign was once thought to represent hemorrhage into a tumor. This phenomenon occurs most often with anticoagulation, coagulopathy, or large post-traumatic hematomas.

The angiographic signs of intracerebral hematoma depend on the location of the hemorrhage. Local mass effect is seen, which is nonspecific and cannot be differentiated from other avascular mass lesions. Hemorrhages within the basal ganglia usually arise laterally in the putamen and external capsule, and therefore they almost always displace medially the lenticulostriate group of vessels. Aneurysm, vascular malformations, and tumors are recognized by their characteristic angiographic findings. Early draining veins may also be present in the region of the hematoma and do not specifically indicate a vascular lesion.

In cases of suspected hematoma or in patients with hypertension and stroke syndrome, a nonenhanced CT examination is performed as the first imaging procedure. MRI may be used and provides equal sensitivity but is generally unnecessary.

Contrast-enhanced CT or MRI may be used to define an underlying tumor, but it is often difficult to distinguish abnormal enhancement from the hematoma density until the hematoma has matured. On MRI, the usual

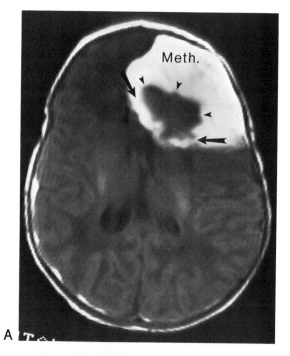

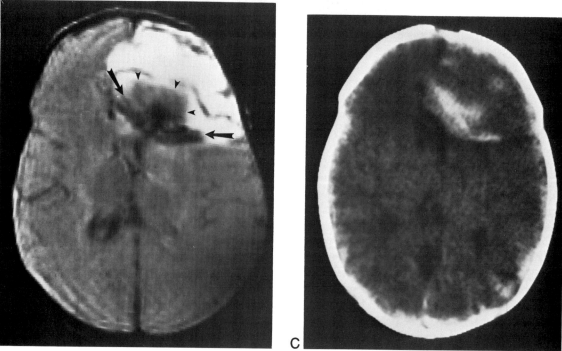

Fig. 2-59 Subacute hematoma, MRI and CT scan. The MRI scans are of a 10-day-old infant whose hemorrhage occurred at birth. **(A)** T$_1$-weighted (SE 600/15) scan shows a large hemorrhagic region in the left frontal lobe. The circumferential high signal intensity region represents extracellular methemoglobin (Meth.). The posteromedial hypointense region (*arrowheads*) represents persisting deoxyhemoglobin. Note that behind this region there is high signal from methemoglobin formation (*arrows*). **(B)** T$_2$-weighted (SE 2500/90) scan shows high signal intensity anteriorly, representing methemoglobin and edema. The posteromedial region of hypointensity represents deoxyhemoglobin (*arrowheads*). The far posterior region of hypointensity represents intracellular methemoglobin (*arrows*). Compare this region with the high signal intensity on the T$_1$-weighted study. **(C)** CT scan performed 1 day after the MRI shows patchy hemorrhagic contusion of the left frontal lobe with a more consolidated hematoma in the posteromedial aspect. It is this hyperdense region of hematoma that shows persisting deoxyhemoglobin and methemoglobin in the intracellular state.

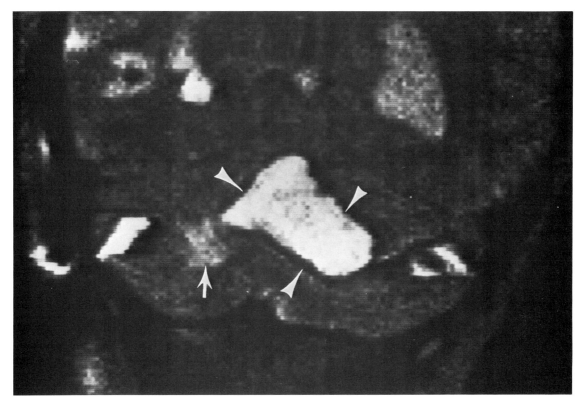

Fig. 2-60 Late subacute hematoma. T_2-weighted MRI shows the hyperintense hematoma in the central portion of the cerebellum. The hematoma has filled in almost completely with high signal as RBC lysis has become complete. A surrounding rim of low signal intensity is developing and represents hemosiderin with marcophages *(arrowheads)*. A small amount of edema persists in the right cerebellar hemisphere *(arrow)*. Other small hemorrhages are present more laterally.

temporal evolution of hemorrhage within tumor is often delayed. The intracellular deoxyhemoglobin phase may persist for weeks, and there may be little peripheral hemosiderin deposition. Hemorrhage within tumor may appear less homogeneous than nontumoral hemorrhage because of the underlying structure of the tumor and changes from prior hemorrhages. A Gd-enhanced MRI with T_1-weighted sequence may define a tumor during the hyperacute and acute phases as a region of high signal intensity within the low-signal hemorrhage.

Hypertensive Encephalopathy

Hypertensive encephalopathy is an acute neurologic syndrome consisting of severe hypertension initially associated with headache, nausea and vomiting, and, later, seizures, visual disturbances, stupor, and coma. Papilledema, retinal hemorrhages, and cardiac and renal failure are usually present. The syndrome requires several days of increasing blood pressure to develop. Complete or partial recovery occurs with early control of blood pressure.

The pathogenesis of the syndrome has not been established. The disorder is probably a result of a breakdown of autoregulation of CBF in the face of the extremely high blood pressure. The latter appears to cause intermittent regions of vasodilatation (regions of lost autoregulation) alternating with regions of preserved autoregulation (vasospasm). The loss of the autoregulation of the flow results in reversible hyperemia and cerebral vasogenic edema, in the posterior cerebral hemispheres, particularly the occipital lobes. Pregnancy-induced hyperten-

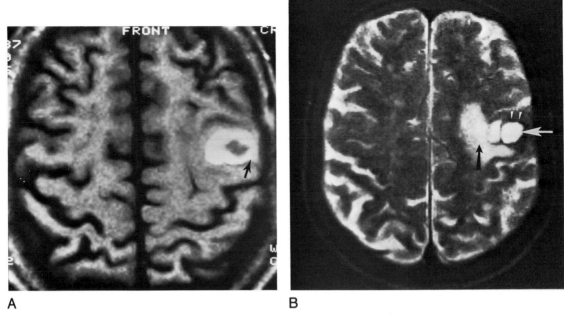

A   **B**

Fig. 2-61 Hemorrhagic infaction. (A) T_1-weighted image shows a hemorrhage in the subacute phase with peripheral high signal intensity of methemoglobin and central low density of persistent deoxyhemoglobin. The hemorrhage goes right to the cortical surface in what was thought to be hemorrhagic infarction. **(B)** In this case, the T_2-weighted MRI shows diffuse hemogeneous central high T_2-signal intensity *(white arrow)*, which persists until the hematoma is completely resorbed. A minimal amount of peripheral hemosiderin is seen *(arrowheads)*. The white matter high signal intensity represents subcortical edema and later gliosis from the infarction.

sion (pre-eclampsia/eclampsia) is essentially the same process, although petechial hemorrhages are more common.

The CT findings correlate well with pathologic studies. Diffuse, symmetric, well-demarcated, low-density regions in the white matter occur, particularly within the occipital regions, resolving after the blood pressure is lowered. The diffuse brain swelling causes ventricular compression with sulcal and cisternal obliteration. The central cerebellum and basal ganglia are involved in the most severe cases, and hydrocephalus may occur. MRI shows high T_2 signal in the white matter, often with patchy low signal from petechial hemorrhage. Gd enhancement occurs within the abnormal regions. SPECT radionuclide scanning with ^{99}Tc-HMPAO demonstrates hyperemia in the occipital lobes. Angiography sometimes shows the varying caliber of the vessels but is not done for diagnosis. The imaging changes revert to normal after successful control of the blood pressure. Focal infarction is rare.

Acute Severe Hypertension

Acute severe hypertension results in intracerebral hemorrhage and infarction, distinct from the pattern of the more slowly developing syndrome of hypertensive encephalopathy. This occurs most commonly after cocaine use or over exposure to sympathomimetic drugs. The hemorrhages develop most commonly from pre-existing vascular lesions such as aneurysms and AVMs but may occur from rupture of normal vessels. Infarctions probably result from regional vasospasm.

Migraine

Migraine is a common clinical constellation, with headache as the central feature. It affects up to 25 percent of the adult population and may occur in adolescents. It is more frequent in women. The cause is unknown.

The syndrome is subclassified as classic migraine, complicated migraine, and common migraine. Classic migraine consists of focal CNS abnormalities (prodromes) preceding and blending into the onset of severe headache, which is most often unilateral and throbbing. The prodromes usually are various scotomata but may include aphasia, hemiparesis and hemisensory deficits as well. The focal symptoms last only a short time, up to 30 minutes. If the focal symptoms last longer, or result in permanent residua, the syndrome is called complicated migraine. Common migraine refers to a migraine-type headache without the prodromes. Vertebrobasilar migraine refers to prodromes involving structures within the posterior fossa. Ophthalmoplegic migraine refers to oculomotor palsy that occurs during the headache.

Migraine is thought to be caused by a cerebrovascular disturbance. According to the current theory, the process begins with vasoconstriction of both the intracranial and extracranial vessels. The vasoconstriction phase may produce significant and measureable cerebral ischemia resulting in the prodromes or focal CNS deficits. The ensuing vasodilation phase is associated with the headache. A decrease in glucose metabolism and CBF has been documented with PET scanning and Xenon-133 blood flow studies.

In general, patients with migraine do not require CT or MRI scanning for diagnosis or management. However, certain abnormalities have occasionally been seen with imaging. On MRI, focal high T_2 signal white matter abnormalities have been demonstrated in a small number of patients. The abnormalities vary from a few small punctate high signal lesions to more extensive confluent involvement in both the periventricular and subcortical regions. The lesions resemble those seen with multiple sclerosis or Binswanger's subcortical arteriosclerotic encephalopathy. Mild diffuse atrophy is common, especially with migraine that has been present for more than 5 years. These findings have unknown clinical significance. Less commonly, cortical infarction may be seen. Very rarely, migraine has been associated with cerebral AVM.

Cerebral angiography has on rare occasions demonstrated cerebral vasospasm with migraine. However, it should not be performed for diagnosis, as cerebral angiography poses a significant risk to the patient with migraine, especially during the prodromal phase. Cerebrovasospasm may be triggered or exacerbated and result in permanent CNS deficits. The test should be avoided when possible and should never be performed during the migraine attack. If the angiography is essential for evaluation of an unrelated problem, pretreatment with 40 mg of propanalol, or other β-blockers, may decrease the risk of iatrogenic vasospasm.

Subcortical Arteriosclerotic Encephalopathy of Binswanger

Binswanger's encephalopathy is discussed in Chapter 10.

Bacterial Arteritis

Intracranial arteries are susceptible to surrounding inflammatory reaction or pus, particularly meningitis. An arteritis occurs from outside in, with edema, inflammatory cells, and bacteria traversing inward across the vessel wall to the intima. The vessel lumen is narrowed by the effects of the swelling in the vessel wall, and thrombosis may occur on an inflamed intima. The arteries at the base of the brain are usually the most severely affected, probably because pus tends to accumulate in the basal cisterns. *Haemophilus influenzae* and *Staphylococcus* cause the most severe changes of arteritis and often involve the cortical vessels as well.

There is a general correlation between the severity the meningitis and the arteritis. However, even in relatively mild infections or infections that are treated early and appropriately, the arteritis may be severe and cause extensive brain damage from infarction. Venous occlusive disease may also occur. Inflammatory arteritis cannot be differentiated from other types of arteritis by angiography. The infarctions and subsequent encephalomalacia may be imaged with MRI and CT.

Fungal Arteritis

Arteritis may be seen after a large number of fungal infections of the meninges. Rarely, arteritis occurs from hematogenous intraluminal spread. Today, fungal infections are seen more commonly because of the relatively large number of people who are immunodeficient and thus susceptible to these opportunistic infectious agents. The mechanism of the arteritis is the same as for the bacterial infections described above. The most common infectious agents are *Cryptococcus, Nocardia, Coccidioides, Aspergillus, Candida, Mucor,* and, rarely, *Actinomyces* and *Histoplasma*. The angiographic findings are nonspecific and similar to those of bacterial arteritis.

Tuberculous Arteritis

Tuberculous meningitis is generally seen in children and is associated with a primary infection outside the intracranial structures. The CSF contains high levels of protein, mononuclear leukocytes, and low glucose and chloride levels. The disease primarily involves the basal cisterns (pontine and suprasellar cisterns), in which a thick, gelatinous, inflammatory exudate forms. In time, the exudate becomes hard and often calcifies. The engulfed vessel walls are involved with the inflammatory process and are mechanically constricted by the fibrosis. The vessel lumens are severely narrowed, and the obliterative endarteritis leads to total occlusions. The supraclinoid portions of the ICA, the proximal MCAs, the lenticulostriate arteries, and the basilar artery are most severely involved (Fig. 2-62). Vascular insufficiency syndromes and stroke may occur. Peripheral involvement of the cortical vessels has not been seen following tuberculosis.

Syphilitic Arteritis

A severe arteritis may result from tertiary meningovascular syphilis. CNS symptoms may occur within months or may not become apparent for up to 5 years. There is a predilection for involvement of the proximal portions of the MCA, although any vessel, large or small, may be affected. The angiographic picture is that of arteritis, although there is a greater tendency for aneurysm formation, particularly if the large vessels (i.e., basilar artery) are involved. The CSF serology is always positive.

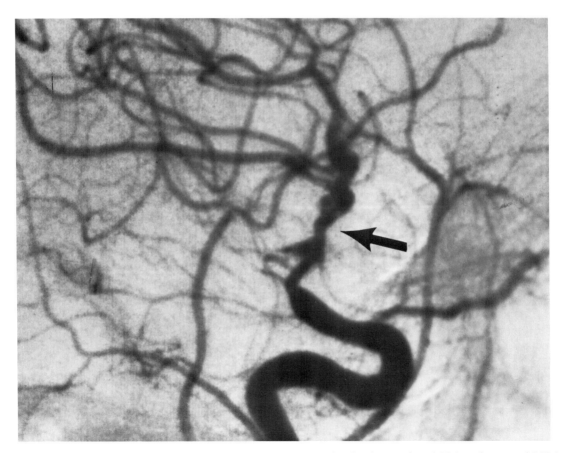

Fig. 2-62 Tuberculous arteritis. Constriction is demonstrated in the distal supraclinoid ICA and proximal MCA (*arrow*). This pattern may also be seen with syphilitic or fungal arteritis.

Mycotic Aneurysm

Mycotic aneurysms result from septic embolizations, usually associated with subacute bacterial endocarditis. They are rarely associated with acute bacterial endocarditis. Almost all of the aneurysms occur in the more peripheral branches of the intracranial vessels, most often the MCA. They may be small or large, and they may rupture, often with lethal consequences. Because there is usually an inflammatory reaction adjacent to the aneurysm, adhesions form, directing the extravasated blood into either the brain parenchyma or the subdural space. The hemorrhages may be small or large. Free subarachnoid hemorrhage is unusual.

Rarely, a mycotic aneurysm occurs in the thoracic aorta, particularly after thoracic cardiovascular surgery. These aneurysms are frequently associated with infections, especially those caused by *Aspergillus* or *Salmonella*. Large vegetations may form on the intima of the proximal aorta and result in distal septic embolization to the brain, causing arteritis and occasional aneurysm formation.

It is usually not possible to diagnose an intracranial mycotic aneurysm directly with CT scanning or MRI. The findings are nonspecific: infarction, hemorrhagic infarction, intracerebral hematoma, subdural hematoma, and sometimes subarachnoid hemorrhage. When found in the presence of subacute bacterial endocarditis, a mycotic aneurysm is strongly considered and angiography performed.

Collagen Vascular Disease

Systemic lupus erythematosus (SLE) is the most common collagen vascular disorder. This generalized disorder predominantly affects women (8:1), most commonly between the ages of 15 and 40. As many as 50 percent of persons with the disease have lesions in the CNS, although not all are symptomatic.

A diffuse vasculitis occurs that involves both intracranial and extracranial vessels. Angiography shows the typical, but nonspecific, narrowings and dilations of the involved vessels (Fig. 2-63). MRA is not as sensitive as XRA for the demonstration of arteritis, especially if the vessel changes are not so severe. Symptoms include seizures, mental changes, headache (migraine type), and focal neurologic signs from small infarctions. Large infarctions are rare.

Most commonly, CT scans and MRI show generalized cerebral volume loss, which results from atrophy caused by microinfarctions and from the effects of steroid therapy. MRI is more sensitive than CT scanning for detecting focal abnormalities of the brain in patients with SLE. The lesions have prolonged T_1 and T_2 relaxation times. Three patterns are usually seen: (1) cerebral infarction with a relatively small area of increased intensity; (2) multiple small regions of increased intensity, probably representing microinfarctions; and (3) focal ribbon-like areas of increased intensity involving the cortical gray matter that regress in about 2 weeks. These lesions may represent areas of ischemia without actual infarction. Only the larger lesions may be seen on high-resolution CT scanning. Contrast enhancement may show lesions that would otherwise be occult.

Necrotizing Angiitis

The vascular changes of necrotizing angiitis are similar to those of collagen diseases, although they tend to be somewhat more severe, and there is more necrosis within the vessel walls. Polyarteritis nodosa is the characteristic disease, but similar necrotizing changes occur with thrombotic thrombocytopenic purpura, isolated intracranial arteritis, temporal arteritis, and Wegener's granulomatosis.

Angiography may show regions of focal narrowing, but biopsy is necessary for a precise diagnosis. When possible, the biopsy is performed on these narrowed regions, as the diseases have a multifocal distribution with skip areas. The ophthalmic artery is frequently involved in temporal arteritis (Fig. 2-64). The erythrocyte sedimentation rate is almost always elevated in the presence of active arteritis.

Amyloid Angiography

Cerebral amyloid angiography is becoming recognized as a common cause of subcortical hematomas in the elderly. The disease is generally not seen in persons under 60 years of age. Over this age, the incidence proportionately increases, with as many as 50 percent of people over the age of 80 years showing the amyloid changes. These changes are unrelated to the effects of hypertension, and the cause is unknown.

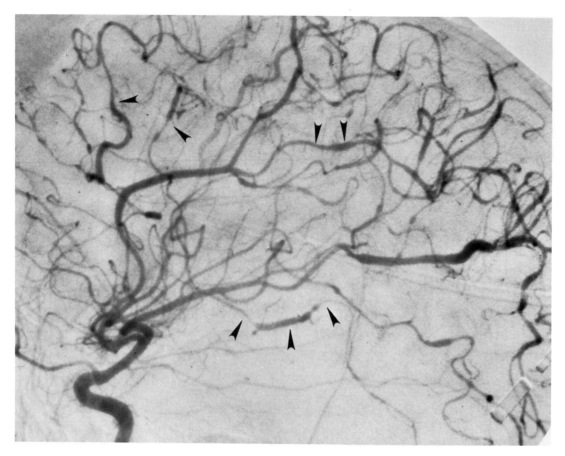

Fig. 2-63 Arteritis. ICA angiography shows diffuse segmental narrowing and dilation of multiple intracranial vessels (*arrowheads*). This pattern is typical of arteritis, whatever the cause.

The amyloid infiltration affects only medium-size and small arteries in the cerebral cortex and leptomeninges. The involved vessels rupture easily after minor trauma or mild elevations of blood pressure. They may also rupture spontaneously. CT scanning easily demonstrates the usually large peripheral subcortical hematomas, which occur most often in the more posterior parts of the brain, particularly the occipital lobes. Multiple hematomas may occur, sometimes simultaneously. Changes of arteritis and occlusion are unusual on angiography. The definitive diagnosis is made by biopsy of the meninges, usually at removal of the hematoma.

Neoplasia

In certain situations, tumors severely affect the vessels in the brain. Most commonly, vessel damage is secondary to constriction by a meningioma at the base of the brain (see Ch. 5). Parasellar meningiomas may constrict or totally occlude the cavernous portion of the ICA, its supraclinoid portion, or the proximal portions of the MCA. In these instances, it is usually impossible for the neurosurgeon to remove the tumor without causing either rupture or occlusion of the major vessel involved. The diagnosis of meningioma is usually clear from the characteristic vascular blush of the tumor. Rarely, an ex-

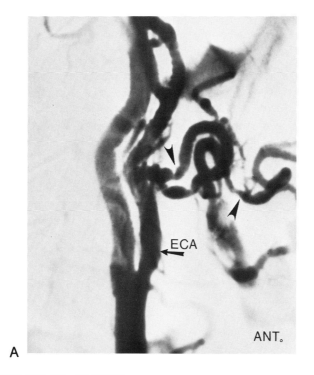

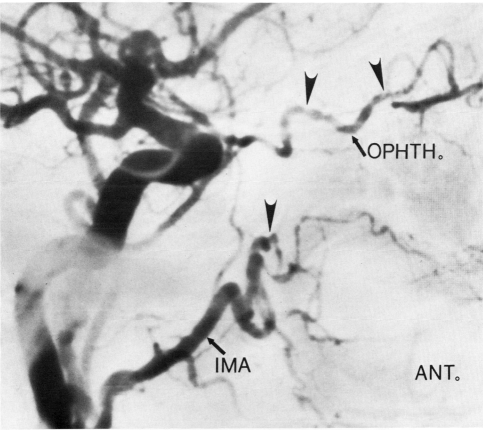

Fig. 2-64 Necrotizing angiitis; temporal arteritis. (A) Common carotid arteriography in the neck shows diffuse changes of arteritis involving the ECA and its branches (*arrowheads*). Change in the ECA distribution may be the only finding with arteritis. (B) Lateral intracranial view shows the changes of arteritis in the ophthalmic and internal maxillary arteries (*arrowheads*).

ceptionally virulent glioblastoma multiforme invades a cortical vessel, resulting in its occlusion and leading to the stroke syndrome. Atrial myxomas embolize to the peripheral cortical branches, particularly at bifurcations, invade the vessel wall, and cause a focal aneurysm. Sporadically, other tumors follow this sequence as well.

Radiation Arteriopathy

High doses of radiation (more than 5,000 R) may cause an obliterative arteritis. The risk of radiation-induced angiopathy increases in an exponential fashion with doses over 6,000 R. On angiography, the vessels may show diffuse narrowing of the lumen within the fields of irradiation. High-dose brachytherapy has a higher incidence of radiation necrosis than conventional therapy. The vasculopathy usually occurs 1 to 3 years after cessation of irradiation, but the latent period may be as long as 10 years. Mild involvement causes demyelination. More severe involvement produces focal tissue necrosis. Tissue necrosis causes striking and confusing changes on the CT scan with large regions of intense, irregular contrast enhancement, edema with mass effect, and increased intracranial pressure (Fig. 2-65). MRI shows a high T_2 signal within white matter from demyelination (Fig. 2-66).

Radionuclide imaging helps differentiate radiation necrosis from recurrent tumor. With FDG PET, the radionuclide lable accumulates within the metabolically ac-

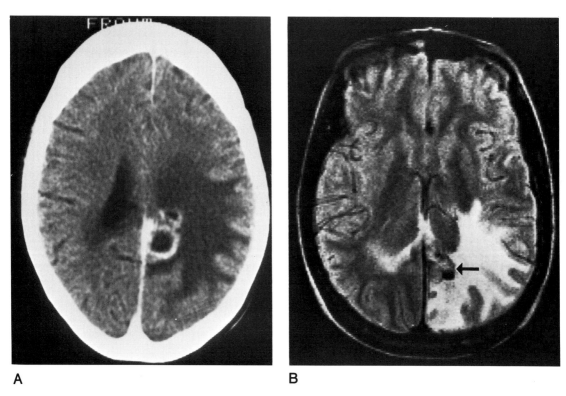

A   B

Fig. 2-65 Radiation necrosis. (A) Contrast-enhanced CT scan shows ring-like enhancement deep in the posterior left hemisphere surrounded by considerable white matter edema. This material was removed and was found to be entirely necrotic tissue. **(B)** T_2-weighted MRI shows the necrotic region as one of hypointensity *(arrow)*. It is surrounded by high signal intensity, representing edema and gliosis. Slight white matter changes are seen on the opposite side. This patient had radiation implantation. Postirradiation changes are confined to the field of exposure.

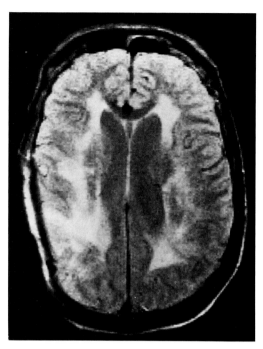

Fig. 2-66 Postirradiation change, MRI. Bilateral high T_2 signal intensity is seen in the white matter, representing demyelination. This patient had undergone whole-brain irradiation.

tive regions of viable tumor, but usually not within regions of radiation necrosis. Dual-isotope SPECT using Thallium-201 (^{201}Tl), a potassium analog, and ^{99m}Tc-HMPAO, achieves good differentiation of recurrent tumor from necrosis. Tumor commonly exhibits high uptake of thallium. Necrosis shows little or moderate uptake. When the uptake is moderate, the perfusion study demonstrates the high vascularity of the tumor differentiating it from the avascular region of necrosis. Biopsy makes the more definitive diagnosis in unresolved cases. The symptoms of the increased intracranial pressure are greatly relieved by removal of the enhancing nidus of necrotic tissue, which causes the edema. See also Chapter 9 for more discussion on radiation brain changes.

Drug Abuse

Intravenous amphetamines may cause the typical changes of arteritis described earlier in the chapter. Mycotic aneurysms may occur, not always associated with subacute bacterial endocarditis. Both of these processes,

along with the induced hypertension, may cause intracranial hemorrhages of almost any type. Today's mercurial fashion in the use of drugs means that any combination of agents may be encountered, and so the changes seen must be considered in the appropriate clinical setting. Chronic use of foreign substances of almost any type can produce brain cell loss and atrophy that is visible on the CT scan. Cocaine use is associated with cerebral hematomas and infarctions. Changes consistent with vasospasm, but not with vasculitis, have been seen on angiography. Subarachnoid hemorrhage after cocaine use is usually due to rupture of a pre-existing berry aneurysm.

THROMBOSIS OF THE CEREBRAL VEINS AND DURAL SINUSES

Cerebral venous thrombosis is usually not an independent disease but occurs as the result of various pathologic factors:

1. Local effects of infection, particularly sinusitis
2. Trauma, including surgical trauma
3. Hypercoagulable states, as with pregnancy
4. Congestive heart failure
5. Local tumor, particularly meningioma
6. Following intracranial surgery

The symptoms depend on the location and extent of the brain infarctions. Headache is almost always present, and there may be papilledema from increased intracranial pressure. Hemiparesis, the most common deficit, frequently fluctuates in severity. Seizures may occur. These symptoms are somewhat different from those usually associated with arterial occlusion and so help with the differential diagnosis. Cranial nerve palsy of CN III, IV, and VI may be present with cavernous sinus thrombosis.

The infarctions caused by venous thrombosis are often hemorrhagic. The hemorrhages are almost always confined to the cortical areas in the region of the thrombosis and vary in size from small, punctate collections to widespread cortical involvement. With thrombosis of the superior sagittal sinus, hemorrhages frequently occur bilaterally in the high parasagittal convexities. Deep vein thrombosis can cause symmetric deep hemorrhagic necrosis within the thalamic nuclei or basal ganglia.

On noncontrast CT examination, a linear increased density within a cortical vein (cord sign) or within a dural sinus is sometimes seen, probably representing a clot. There can be considerable swelling of the brain from hyperemia. Subarachnoid hemorrhage is frequent, but the amount of blood is small and is not usually seen on the CT scan.

Abnormal contrast enhancement occurs in the cortical "ribbon" pattern. A defect in the contrast-filled sagittal sinus may represent a clot and can best be seen in the region of the torcula (delta sign) (Fig. 2-56). Tentorial or falx enhancement may be unusually prominent from collateral flow. Ischemic infarct, without hemorrhage, occurs in up to 40 percent, and the diagnosis of venous thrombosis is suggested only by the clinical symptoms listed above. Dilated contrast filled collateral deep medullary veins is an indirect sign of sinus thrombosis.

MRI is the most useful technique for the diagnosis. Routine SE scans may show the clotted sinus with absent flow-void. The thrombus within the sinus shows the usual features of hemorrhage. Initially, there is isointensity within the sinus on the T_1-weighted images which converts to hyperintensity with change to methemoglobin. The T_2-weighted images will show the clotted sinus as low intensity in the early stages, which should not be mistaken for flow-void. Later, this is also converted to high signal. The infarctions and parenchymal hemorrhages are better demonstrated compared with CT and are seen as low intensity within the infarct on the T_2-weighted images. However, care must be taken to be certain that the increased intensity within a sinus is not flow-related enhancement.

Both 2D TOF MRA and PC MRA reliably define dural sinus thrombosis and is the accepted procedure today. The 2D TOF sequence is done in the coronal plane to optimize the flow-related enhancement effect of blood within the sagittally oriented sinuses. A spoiled gradient echo sequence with a 60-degree tip angle is used with a TR/TE of 50 msec/10 to further enhance inflow effect. An arterial presaturation is used to suppress the arterial signals. Bright signal within the sinus indicates flow and no thrombosis. No signal indicates thrombosis. A frayed appearance indicates late thrombosis with recanalization. Indirect signs include collateral venous flow over the cortex and into the deep medullary veins. PC MRA may also be used with similar findings. With PC MRA, there is excellent background tissue suppression, so the high signal of methemoglobin is not a prob-

lem. Cut film or DSA is done only if the MRA is negative when there is high suspicion for sinus thrombosis.

Angiography remains a definitive diagnostic test (Fig. 2-67C) but is done only if MRA is nondiagnostic or unavailable. The diagnosis is made when a sinus or vein is not opacified. An occluded vein may be seen, although a single cortical vein thrombosis is almost impossible to recognize because of the great variation in venous anatomy. The circulation is slowed in the region of the sinus occlusion, and there is prolongation of the "brain blush" of the capillary phase of the angiogram. Irregular and tortuous, almost corkscrew-like, collateral veins may be seen, and the direction of the flow in these veins is reversed from normal. Enlarged deep medullary veins can be seen with extensive sagittal sinus thrombosis. To best visualize the sagittal sinus, slightly oblique lateral and AP projections are used to distinguish the sinus from the density of the inner table of the calvarium and eliminate superimposition. Angiography is usually done with a unilateral injection of contrast, and there is unopacified blood from the opposite side flowing into the sinuses, which may be misinterpreted as thrombosis. If there is any question about the diagnosis, bilateral simultaneous carotid injection is done.

Dural sinus venography followed by direct infusion of a thrombolytic agent into the thrombosed sinus may improve the outcome in venous sinus thrombosis. The jugular bulb is catheterized in a retrograde femoral vein approach. Using coaxial technique, a 2.7 Fr. Tracker catheter is manipulated through the sigmoid sinus, transverse sinus, and torcula into the sagittal sinus. Digital subtraction sinus venography is accomplished with 4 ml of 60 percent low osmolar contrast agent (LOCA) to prove the diagnosis. Bolus infusion of 80,000 U of urokinase is given over 5 minutes every 15 minutes until thrombolysis is achieved.

GLOBAL ISCHEMIA OR HYPOXEMIA

Global ischemia or hypoxemia occurs when either the cerebral perfusion falls below the critical level or the blood is not adequately oxygenated. Both factors may be present simultaneously. The same cellular effect results from failure of cell metabolism from any reason. In adults, global ischemia or hypoxia results most commonly from the circulatory arrest that occurs with hemorrhagic or cardiogenic shock, carbon monoxide poison-

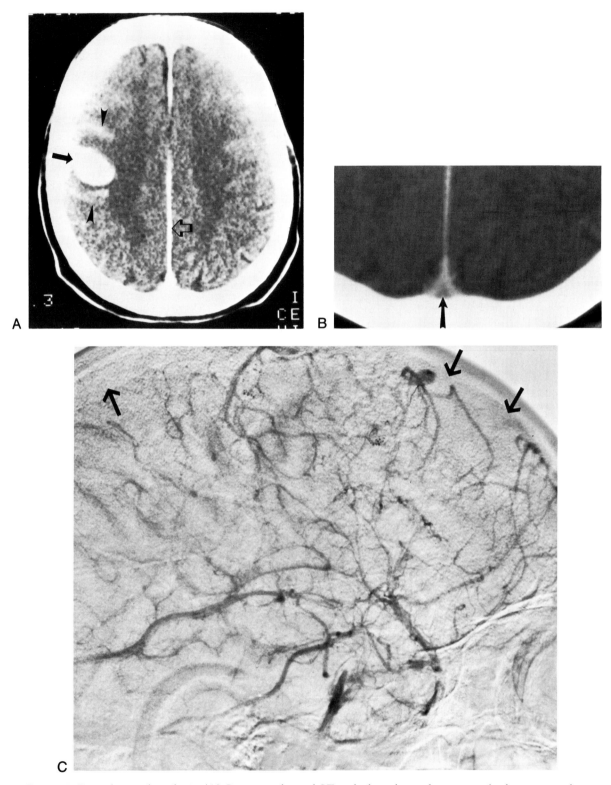

Fig. 2-67 Sagittal sinus thrombosis. (A) Contrast-enhanced CT study shows hemorrhagic cortical infarction over the right parietal lobe *(arrow)* with adjacent cortical gyral enhancement *(arrowheads)*. The falx shows striking enhancement *(open arrow)*. **(B)** Wide window view of the posterior sagittal sinus shows the delta sign with nonopacification of the sinus cavity *(arrow)*. Wide windows are necessary to observe this sign. **(C)** Venous phase angiography demonstrates nonopacification of the entire sagittal sinus *(arrows)*. There is prominent opacification of the inferior cortical veins draining to patent basal sinuses. *(Figure continues.)*

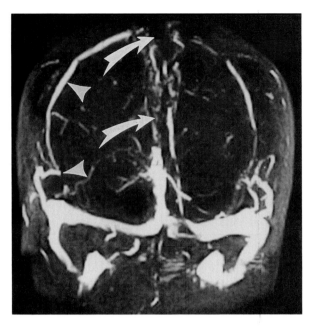

Fig. 2-67 *(Continued)*. **(D)** Coronal 3D TOF venous MRA shows absence of high signal within the sagittal sinus *(arrows)*. There is a prominent cortical vein acting as collateral *(arrowheads)* and tiny collateral veins adjacent to the falx.

most commonly the result of prematurity, perinatal asphyxia, status epilepticus, near-drowning, child abuse, and mitochondrial disorders. The effects on the brain vary with the age of the patient, selective neuronal susceptibility, the rapidity of onset of the hypoxia–ischemia, and whether deoxygenation or low perfusion is predominant. Combination patterns are common.

Global Hypoxemia

When deoxygenation is the primary factor (global hypoxemia), the major abnormality occurs bilaterally in the globus pallidus, anterior thalamus, hemispheric white matter, and cerebellum. Acute asphyxia, poisoning from carbon monoxide, cyanide, and manganese exposure, near-drowning, severe hypoglycemia, and lightning strike cause this pattern of pathology. It can also be seen in full term newborns with acute extremely severe asphyxia. Parkinsonism often follows.

Cell necrosis, edema, and hemorrhage occurs within the affected regions. On CT scans, the edema causes bi-

lateral hypodensity in the globus pallidus nuclei (Figs. 2-68 and 2-69) and cerebral white matter. With MRI, the edema is hypointense on T_1- and hyperintense on T_2-weighted images. The lesions may be invisible in the subacute phase from the fogging effect. Hemorrhage shows the high signal of methemoglobin or low signal of hemosiderin. An hypoxic-ischemia event precipitated by one of the deoxygenating factors may produce the pattern of global ischemia described below as the hypoxia often leads to hypoperfusion as well. Mitochondrial disorders and hypoglycemia produce unpredictable patterns of infarction.

Global Ischemia

Global ischemia results most commonly from cardiac arrest and hemorrhagic shock. When global low perfusion is the major factor, the ischemia generally creates one, or more of infarction(s) within the "watershed"

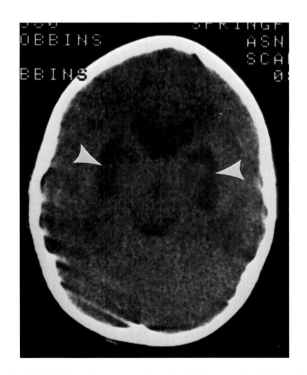

Fig. 2-68 Hypoxic encephalopathy. CT scan of a child in a near-drowning accident shows bilateral symmetric regions of hypodensity involving the lentiform nuclei *(arrowheads)*. This pattern is characteristic after pure hypoxemia.

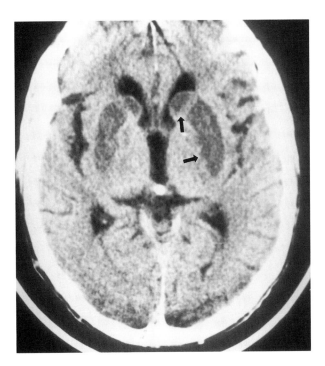

Fig. 2-69 Hypoxic encephalopathy. Hypodensity occupies the head of the caudate and lenticular nuclei bilaterally (*arrows*). Axial CT scan.

regions between main cerebrovascular territories, cortical laminar necrosis, diffuse cortical and white matter hemispheric infarction or global infarction. The "watershed" abnormalities are seen on CT scans as hypodensity in the deep parasagittal frontal and parietal white matter. With more severe ischemia, the cortex becomes hypodense in this distribution as well. Later, atrophy develops with dilatation of the lateral ventricles, particularly the atria. MRI shows high T_2 signal within the deep white matter (see Fig. 2-72). Laminar necrosis shows best with MRI as high T_2 signal in the gyri, which enhance with Gd.

Global infarction shows involvement of the entire cerebrum on both CT and MRI. The basal ganglia, thalami, and cerebellum are typically spared unless extreme ischemia has occurred. The region along the course of the main branches of the MCA tend to be less affected than other portions of the hemispheres. Mild damage from global ischemia may not be seen during the acute phase but only after 1 to 2 months, when slight dilatation of the atria of the lateral ventricles indicates the mild deep posterior parietal lobe injury. Focal infarctions sometimes follow ischemia as the only finding.

Hypoxic Ischemic Encephalopathy in the Newborn

Intrapartum factors account for most of the hypoxic ischemic encephalopathy (HIE) seen in the newborn. Some of these factors may be antepartum, such as maternal diabetes, eclampsia, placental abruption or placental insufficiency. More often, the insult is intrapartum from cord compression, prolonged labor, and other difficult labors. About 10 percent of HIE results from postpartum events. Adverse antepartum factors make the fetus more vulnerable to intrapartum adverse events. In some cases, the cause of HIE is unknown.

The pattern of neonatal HIE is variable. The most common finding is diffuse or multifocal infarction of the cortex and white matter of the cerebral hemispheres. This commonly takes watershed parasagittal pattern (Fig. 2-70). The basal ganglia, thalamus, and posterior fossa structures appear normal in all but the most severe cases. The basal ganglia may be selectively involved in extremely severe, acute, complete asphyxia. An unusual

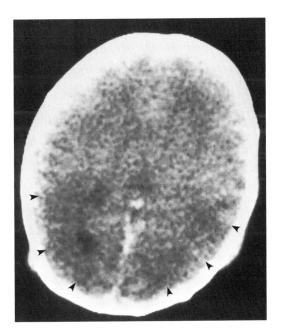

Fig. 2-70 Moderate ischemic encephalopathy. CT scan shows hypodensity in the posterior parietal lobes representing ischemic change (*arrowheads*).

form of hypermyelination develops in some of these cases, seen after about 6 months as irregular hyperdensity on CT, a condition known as status marmoratus. However, in milder cases, selective neuronal necrosis may be present in the basal ganglia and Somer's sector of the hippocampus and may not become visible until later as subtle atrophy. Multicystic encephalomalcia forms in the hemispheric white matter in severe cases.

CT shows the involved regions as hypodensity often with varying parenchymal and intraventricular hemorrhage. The CT may be normal in the first 24 hours after the insult. With the most severe cases, it may become abnormal as early as 12 hours (Fig. 2-71). The edema becomes maximum at about 4 days. Hemorrhage within the parenchyma frequently lags after the changes of edema, becoming apparent most often after the second day. The hemorrhage may be found within the parenchyma, the ventricles or the subarachnoid space. Dissolution of the brain tissue seems more rapid in infants than in adults, occurring within 1 to 2 weeks in regions of pan-necrosis (Fig. 2-72). Hydrocephalus occurs commonly from the associated hemorrhages and may require shunting. Ultrasound demonstrates the edema and hemorrhages as increased echogenicity within the brain parenchyma. Loss of pulsations of the arteries of the circle of Willis indicates severe brain swelling. MRI is generally less useful than CT during the acute phase because of the immature white matter and its normally high T_2 signal, but it is very useful in the later stages in assessing the full extent of brain injury. In milder cases, there is less global damage. Multifocal

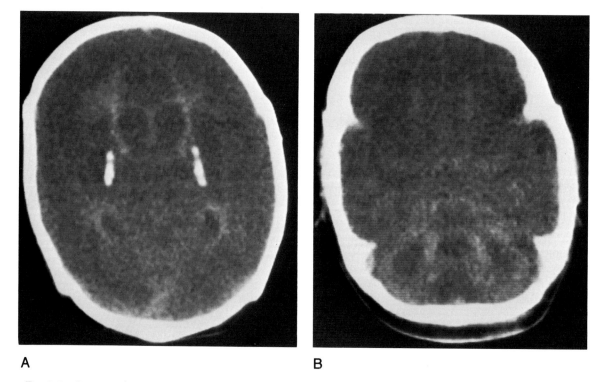

A   B

Fig. 2-71 Severe ischemic encephalopathy. (A) Scan through the level of the ventricles shows hypodensity involving all portions of both cerebral hemispheres including the basal ganglia and the thalamus. The ependyma shows preserved density. Calcifications developed in the globus pallidus nuclei bilaterally. **(B)** Scan through the posterior fossa shows preservation of cortical and periventricular tissue density in the cerebellum. Only the most severe ischemic events produce changes in the cerebellum and thalamic nuclei.

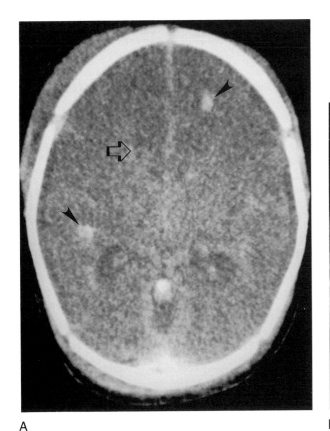

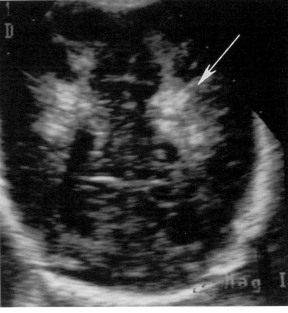

A   B

Fig. 2-72 Severe hypoxic-ischemic encephalopathy (HIE, neonatal asphyxia). **(A)** Axial nonenhanced CT scan of the middle portion of the brain shows severe hypodensity of brain tissue from edema so that the CSF within the ventricles is barely identifiable *(open arrow)*. There is loss of the gray–white matter differentiation and hypodensity within the basal ganglia and thalamus. Small hemorrhages occurred within the damaged tissue *(arrowheads)*. **(B)** Neonatal head ultrasonogram shows high echogenicity within the brain tissue representing brain edema *(arrow)*.

PATTERNS OF NEONATAL HIE IN TERM INFANTS

Global ischemia

Focal or multifocal ischemia

Multicystic encephalomalacia

Watershed ischemia

Selective neuronal necrosis

Status marmoratis

infarctions are common, ultimately causing the ulegyria, a condition of loss of the gray matter at the depths of gyri (Fig. 2-73). MRI is also better equipped to identify congenital structural abnormalities, as compared with CT.

Periventricular leukomalacia (PVL) is a pattern of HIE peculiar to premature infants. Here, infarction injures the germinal matrix in the immediate paraventricular white matter, causing cystic change or atrophy. If large enough, the hemorrhage breaks into the ventricular system. When severe, the intraventricular hemorrhage (IVH) forms a cast of the ventricle. It may spread laterally to the more peripheral hemispheric white

Table 2-5 Grading of PVL-IVH

Grade	Location of Hemorrhage
I	Germinal matrix only
II	IVH, < 50% of volume
III	IVH, > 50% of volume
IV	Parenchymal hemorrhage

matter (Table 2-5). Hydrocephalus is common after intraventricular bleed. Neonatal ultrasonography is very useful during the postnatal period (Fig. 2-74). After 1 year, MRI shows the periventricular leukomalacia to be a high T_2 signal within the paraventricular white matter, which may remain immature and atrophic (Fig. 2-75).

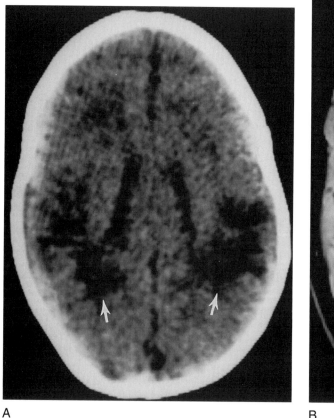

A

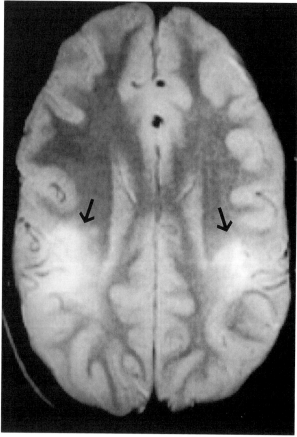

B

Fig. 2-73 Hypoxic–ischemic encephalopathy, ulegyria. (A) Axial CT scan demonstrates the bilateral regions of subcortical and deep cortical hypodensity representing infarctions in an infant with perinatal asphyxia (*white arrows*). Multifocal infarction is one of the patterns of damage occurring in full-term infants after asphyxia. **(B)** Axial T_2 MRI shows high signal in the posterior parietal subcortical white matter and at the base of the gyri (*arrows*).

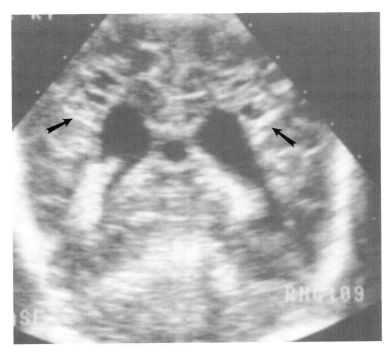

Fig. 2-74 Paraventricular leukomalacia (PVL), prematurity. Cranial ultrasound demonstrates increased echogenicity in the paraventricular regions associated with rounded echo-deficient locations representing multiple cysts *(arrows)*. The ventricles are dilated from white matter atrophy. This is typical of damage within the germinal matrix, found with prematurity or insults to the fetus at 7 to 8 months gestation. Scan done at age 2 months.

Fig. 2-75 Paraventricular leukomalacia, mild hypoxic ischemic encephalopathy. T_2-weighted MRI examination shows high signal in the frontal and parietal white matter. The white matter is also atrophic in these regions *(arrows)*, a residual of mild hypoxic–ischemic encephalopathy.

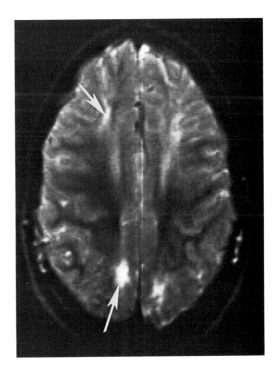

BRAIN DEATH

Brain death is the result of irreversible global cerebral circulatory arrest. In most cases, this type of arrest is the result of a marked increase in the intracranial pressure from brain edema or hemorrhage, so that the pressure is higher than the circulatory perfusion pressure. Clinical criteria have been established to define brain death, including the absence of cerebral electrical activity as defined by electroencephalography. When there is any doubt about meeting the clinical criteria, definite proof of absent cerebral circulation is considered equivalent to brain death. There must be absent flow in both internal carotid and vertebrobasilar circulations.

Selective cerebral angiography is the most accurate test. The demonstration of absent CBF with preserva-tion of normal extracranial circulation is accepted as proof of brain death. All the cerebral vessels must be studied.

Because of the cost and difficulty of performing cerebral angiography, other methods have been used to define brain death. The radionuclide brain scan is most commonly employed and can be performed with portable equipment. Brain death is diagnosed when there is no evidence of intracranial tracer but normal tracer is seen in the superficial scalp tissue (Fig. 2-76). Dynamic contrast-enhanced CT scans, transcranial Doppler, and MRA may also be used. With dynamic CT contrast infusion, there is no measured increase in density of the brain, whereas there is normal increase in density of the soft tissues over the calvarium.

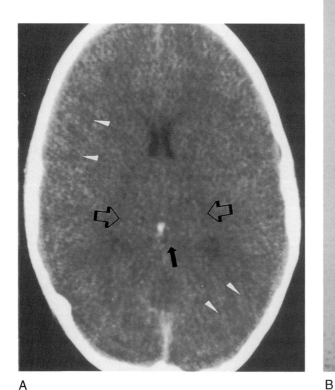

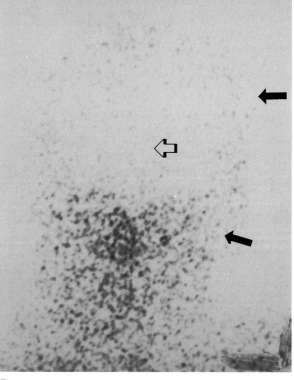

A

B

Fig. 2-76 Brain death. (A) An axial CT scan early after the onset of coma shows a subtle decrease in density of the basal ganglia and thalami *(open black arrows)*. The cistern of the quadrigeminal plate is obliterated from brain swelling *(closed black arrow)*. There is loss of the gray-white matter differentiation *(white arrowheads)*. **(B)** Radionuclide brain scan illustrates normal activity of the face and scalp *(closed arrows)*, but no intracranial activity *(open arrow)*.

SUGGESTED READING

Ackerman RH, Candia MR: Assessment of carotid artery stenosis by MR angiography, commentary. AJNR 13:1005, 1992

Akers DL, Markowitz IA, Kerstein MD: The value of the aortic arch study in the evaluation of cerebrovascular insufficiency. Am J Surg 154:230, 1987

Anderson CM, Saloner D, Lee RE et al: Assessment of carotid artery stenosis by MR angiography: comparison with x-ray angiography and color-coded Doppler ultrasound. AJNR 13:989, 1992

Asato R, Okumura R, Konishi J: "Fogging effect" in MRI of cerebral infarct. J Comput Assist Tomogr 15:160, 1991

Barkovich AJ: MR and CT evaluation of profound neonatal and infantile asphyxia. AJNR 13:959, 1992

Barkovich AJ, Atlas SW: Magnetic resonance imaging of intracranial hemorrhage. Radiol Clin North Am 26:801, 1988

Barnett HJM, Mohr JP, STein BM, Yatsu FM (eds): Stroke: Pathophysiology, Diagnosis and Management. Churchill Livingstone, New York, 1986

Bluth EI, Wetzner SM, Stavros AT et al: Carotid duplex sonography: a multicenter recommendation for standardized imaging and Doppler criteria. Radiographics 8:487, 1988

Bousser MG, Chiras J, Bories J, Castaigne P: Cerebral venous thrombosis—a review of 38 cases. Stroke 16:199, 1985

Braffman BH, Zimmerman RA, Trojanowski JQ et al: Brain MR: Pathologic correlation with gross and histopathology. 1. Lacunar infarction and Virchow-Robin spaces. AJNR 9:621, 1988

Brant-Zawadzki M: Ischemia. In Stark DD, Bradley WG Jr (eds): Magnetic Resonance Imaging. CV Mosby, St Louis, 1988

Brant-Zawadzki M, Weinstein P, Bartowski H et al: MR imaging and spectroscopy in clinical and experimental cerebral ischemia: a review. AJNR 8:39, 1987

Brooks RA, Di Chiro G, Patronas N: MR imaging of cerebral hematomas at different field strengths: theory and applications. J Comput Assist Tomogr 13:194, 1989

Brown JJ, Heslink JR, Rothrock JF: MR and CT of lacunar infarcts. AJNR 9:477, 1988

Caplan LR, Wolpert SM: Angiography in patients with occlusive cerebrovascular disease: views of a stroke neurologist and neuroradiologist. AJNR 12:593, 1991

Carroll BA: Carotid sonography. Neuroimag Clin North Am 2:533, 1992

Crain MR, Yuh WTC, Greene GM et al: Cerebral ischemia: evaluation with contrast enhanced MR imaging. AJNR 12:631, 1991

Drayer BP: Diseases of the cerebral vascular system. p. 247. In Heinz ER (ed): Neuroradiology. Vol. 4. The Clinical Neurosciences. Churchill Livingstone, New York, 1984

Earnest F, Forbes G, Sandok BA et al: Complications of cerebral angiography: prospective assessment of risk. AJNR 4:1191, 1983

Elster AD: Magnetic resonance contrast enhancement in cerebral infarction. Neuroimag Clin North Am 4:89, 1994

Erickson SJ, Middleton WD, Mewissen MW et aL: Color Doppler evaluation of arterial stenoses and occlusions involving the neck and thoracic inlet. Radiographics 9:389, 1989

Estol C, Claassen D, Hirsch W et al: Correlative angiographic and pathologic findings in the diagnosis of ulcerative plaques in the carotid artery. Arch Neurol 48:692, 1991

Fishman MC, Naidich JB, Stein HL: Vascular magnetic resonance imaging. Radiol Clin North Am 24:485, 1986

Fujita N, Hirabuki N, Fujii K et al: MR imaging of middle cerebral artery stenosis and occlusion: Value of MR angiography. AJNR 15:335, 1994

Furie DM, Tien RD: Fibromuscular dysplasia of arteries of the head and neck: imaging findings. AJR 162:1205, 1994

Goes E, Janssens W, Maillet B et al: Tissue characterization of atheromatous plaques: correlation between ultrasound image and histologic findings. J Clin Ultrasound 18:611, 1990

Golden GS: Stroke syndromes in children. Neurol Clin 3:59, 1985

Gomori JM, Grossman RI, Hackney DB et al: Variable appearances of subacute intracranial hematomas on high-field spin-echo MR. AJNR 8:1019, 1987

Gomori JM, Grossman RI: Head and neck hemorrhage. p. 71. In Kressel HY (ed): Magnetic Resonance Annual. Raven Press, New York, 1987

Gonzalez CF, Doan HT, Han SS et al: Extracranial vascular angiography. Radiol Clin North Am 24:419, 1986

Grant EG, Wong W, Tessler F et al: Cerebrovascular ultrasound imaging. Radiol Clin North Am 26:5, 1122, 1988

Hankey GJ, Warlow CP, Sellar RJ: Cerebral angiographic risk in mild cerebrovascular disease. Stroke 21:209, 1990

Hayman LA, Taber KA, Jhingran SG et al: Cerebral infarction: diagnosis and assessment of prognosis by using [123]IMP-SPECT and CT. AJNR 10:557, 1989

Hecht-Leavett C, Gomori JM, Grossman RI et al: High-field MRI of hemorrhagic cortical infarction. AJNR 7:581, 1986

Hennerici M, Rautenberg W, Sitzer G, Schwartz A: Transcranial Doppler ultrasound for the assessment of intracranial arterial flow velocity. I. Examination technique and normal values. Surg Neurol 27:439, 1987

Hennerici M, Rautenberg W, Schwartz A: Transcranial Doppler ultrasound for the assessment of intracranial arterial flow velocity. II. Evaluation of intracranial arterial disease. Surg Neurol 27:523, 1987

Holman BL (ed): Radionuclide Imaging of the Brain. Churchill Livingstone, New York, 1985

Houser OW, Campbell JK, Baker HL Jr et al: Radiologic evaluation of ischemic cerebrovascular syndromes with emphasis on computed tomography. Radiol Clin North Am 20:123, 1982

Jacobs IG, Roszler MH, Kellt JK et al: Cocaine abuse: neurovascular complications. Radiology 170:223, 1989

Johnson BA, Fram EK: Cerebral venous occlusive disease. Neuroimag Clin North Am 2:769, 1992

Kilgore BB, Fields WS: Arterial occlusive disease in adults. p. 2310. In Newton TH, Potts DG (eds): Radiology of the Skull and Brain: Angiography. CV Mosby, St Louis, 1974

Kissel JT: Neurologic manifestations of vasculitis. Neurol Clin 7:655, 1989

Lande A, Berkman YM: Aortitis. Radiol Clin North Am 14:219, 1976

Laster RE, Acker JD, Halford HH III, Nauert TC: Assessment of MR angiography versus arteriography for evaluation of cervical carotid bifurcation disease. AJNR 14:681, 1993

Mayberg MR, Wilson SE, Yatsu F et al: Carotid endarterectoma and prevention of cerebral ischemia in symptomatic carotid stenosis. JAMA 266:3289, 1991

Mitchell DG: Color Doppler imaging: principles, limitations, artifacts. Radiology 177:1, 1990

Savoiardo M, Bracchi M, Passerini A et al: The vascular territories of the cerebellum and brainstem: CT and MRI study. AJNR 8:199, 1987

Schwartz RB, Carvalho PA, Alexander E III et al: Radiation necrosis vs high-grade recurrent glioma: differentiation by using dual-isotope SPECT with ^{201}Tl and ^{99m}Tc-HMPAO. AJR 12:1187, 1992

Schwartz RB, Jones KM, Chernoff DM et al: Common carotid bifurcation: evaluation with spiral CT. Radiology 185:513, 1992

Schwartz RB, Jones KM, Kalina P et al: Hypertensive encephalopathy: findings on CT, MR imaging and SPECT imaging in 14 cases. AJR 159:379, 1992

Smullens SN: Surgical treatable lesions of the extracranial circulation, including the vertebral artery. Radiol Clin North Am 24:453, 1986

Soges LJ, Cacayorin ED, Petro GR: Migraine: evaluation by MR. AJNR 9:425, 1988

Takahashi S, Higano S, Ishii K et al: Hypoxic brain damage: cortical laminar necrosis and delayed changes in white matter at sequential MR imaging. Radiology 189:449, 1993

Tomsick TA: Sensitivity and prognostic value of early CT in occlusion of the middle cerebral artery trunk, commentary. AJNR 15:16, 1994

Tratnig S, Schwaighofer B, Hübsch P et al: Color coded Doppler sonography of vertebral arteries. J Ultrasound Med 10:221, 1991

Tsai FY, Higashida RT, Matovich V, Alfieri K: Acute thrombosis of the intracranial dural sinus: direct thrombolytic treatment. AJNR 13:1137, 1992

Tsai FY, Higashida R, Meoli C: Percutaneous transluminal angioplasty of extracranial and intracranial arterila stenosis in the head and neck. Neuroimag Clin North Am 2:371, 1992

Tsuruda JS, Kortman KE, Bradley WG Jr: Radiation effects on cerebral white matter: MR evaluation. AJNR 8:431, 1987

Tsuruda J, Saloner D, Norman D: Artifacts associated with MR neuroangiography. AJNR 13:1411, 1992

Vieregge P, Klostermann W, Blumm RG, Borgis KJ: Carbon monoxide poisoning: clinical, neurophysiologic and brain imaging observations in acute disease and follow-up. J Neurol 236:478, 1989

Vogl T, Bergman C, Villringer A et al: Dural sinus thrombosis: value of venous MR angiography for diagnosis and follow-up. AJR 162:1191, 1994

Volpe JJ: Neurology of the Newborn. WB Saunders, Philadelphia, 1987

Volpe JJ: Value of MR in definition of the neuropathology of cerebral palsy in vivo. AJNR 13:79, 1992

von Kummer R, Meyding-Lamadé U, Forsting M: Sensitivity and prognostic value of early CT in occlusion of the middle cerebral trunk. AJNR 15:9, 1994

Warach S, Chien D, Li W et al: Fast magnetic resonance diffusion-weighted imaging of acute human stroke. Neurology 42:1717, 1992

Weingarten K, Barbut D, Filippi C, Zimmerman RD: Acute hypertensive encephalopathy: findings on spin-echo and gradient echo MR imaging. AJR 162:665, 1994

3
Intracranial Aneurysm and Subarachnoid Hemorrhage

An aneurysm is an abnormal focal bulge of the vessel wall. More than 99 percent arise from arteries. Intracranial aneurysms occur in approximately 2 percent of the population of the United States. More than 90 percent are congenital, or "berry," aneurysms that occur in characteristic sites around the circle of Willis. Atherosclerotic fusiform aneurysms of the larger intracranial arteries account for 5 percent. The remaining rare types are infectious (mycotic), post-traumatic, dissecting, or neoplastic. Venous aneurysms occur as part of arteriovenous malformations (AVMs). Rupture of an intracranial aneurysm will occur in less than 1 percent of the population.

LOCATION OF "BERRY" ANEURYSMS

Common (90%)
    ACoA (35%)
    PCoA (30%)
    MCA (25%)
Less common (10%)
    Bifurcation of the ICA
    Tip of the basilar artery
    Origin of the PICA
    Supraclinoid ICA
    Miscellaneous sites

CONGENITAL ("BERRY") ANEURYSMS

Congenital aneurysms are thought to result from a structural weakness in the media of the arterial wall, usually small fenestrations at the junctions of arteries. Hypertension and atherosclerosis may contribute to the process. A familial tendency is present. The incidence of aneurysms occurring at similar sites is twice that of the remaining population.

Ninety-five percent of aneurysms occur in the internal carotid distribution; only 5 percent occur in the vertebrobasilar system. The junction of the anterior cerebral artery (ACA) with the anterior communicating artery (ACoA) is the most common site. The next most common sites are the internal carotid artery (ICA) at the level of the posterior communicating artery (PCoA) and the middle cerebral artery (MCA) at the bifurcation (Fig. 3-1). Aneurysms are multiple in

175

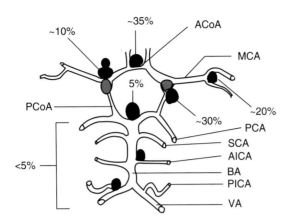

~10%
~35%
ACoA
MCA
5%
~20%
PCoA
~30%
PCA
SCA
AICA
BA
PICA
<5%
VA

Fig. 3-1 Location of aneurysms.

about 20 percent of cases. Most aneurysms are small (less than 1 cm). Giant aneurysms (more than 2.5 cm) are most common in the extradural segments of the ICA, the bifurcation of the MCA, or at the tip of the basilar artery. Intracranial "berry" aneurysms may also be associated with AVMs and other vascular anomalies, including the primitive trigeminal artery, fibromuscular hyperplasia, coarctation of the aorta, polycystic kidney disease (approximately 10 percent incidence), and sickle cell anemia. The smaller aneurysms primarily present with subarachnoid hemorrhage (SAH). The giant aneurysms act as mass lesions and have a low incidence of rupture. The abrupt onset of a unilateral third nerve (CN III), palsy may be an indication of the presence of an expanding internal carotid aneurysm at the level of the PCoA.

Subarachnoid Hemorrhage

SAH is the hallmark presentation of the small congenital "berry" aneurysms. It usually occurs in patients 40 to 50 years old. Seventy-five percent of adults with nontraumatic SAH have a demonstrable aneurysm by angiography, 10 percent have a primary intracerebral hemorrhage as the cause without a demonstrable aneurysm, and only 3 to 5 percent have an AVM. Other rare causes make up 1 to 2 percent. About 10 percent have no demonstrable cause—presumably an occult (nonimaged) aneurysm or AVM. SAH in persons under 25 years of age is more likely to be secondary to an AVM. Aneurysms are infrequent in children.

SAH presents with the abrupt onset of a severe headache, usually the worst experienced by the patient. Nausea, vomiting, and a decrease in mental status and consciousness frequently occur. Prior to a major hemorrhage, the patient may experience small "sentinel" hemorrhages that cause relatively mild headache that may be dismissed.

In addition to the subarachnoid bleeding, about 25 percent of patients have an intraparenchymal hematoma. It occurs most commonly in the deep medial temporal lobe due to rupture of an MCA aneurysm at its bifurcation and in the medial inferior frontal lobes, the corpus callosum, and the septum pellucidum due to rupture of an ACoA aneurysm. When large, the hematomas are surgically removed to control intracranial pressure and herniation. Small focal perianeurysmal clots may form.

Intraventricular hemorrhage also occurs after rupture of aneurysms, most commonly from rupture of an ACoA aneurysm through the adjacent thin lamina terminalis into the third ventricle. Less commonly, aneurysms in other locations rupture into the ventricles. Intraventricular bleeding may leave hemosiderin deposits along the ependymal lining.

The prognosis after SAH can be related to the "grade" of the patient at initial evaluation by a physician. The most common grading system consists of the five categories listed above.

SUBARACHNOID HEMORRHAGE PROGNOSTIC GRADING SYSTEM

Grade 1: Either no symptoms or only mild signs of meningeal irritation. The patient is fully alert and the headache is mild.

Grade 2: The patient is still alert but has a moderate or severe headache. A minimal neurologic deficit may be present.

Grade 3: Mental function changes are present such as lethargy or confusion. A mild to moderate degree of neurologic deficit may be present.

Grade 4: Patient is stuporous.

Grade 5: Coma is present with or without signs of decerebration.

The grading system is useful for defining the prognosis for the patient or evaluating surgical or medical therapeutic regimens. Patients in the grade 1 or 2 category have a good prognosis for successful corrective surgery, whereas those with grade 4 or 5 disease have a poor prognosis, with mortality approaching 100 percent. Grade 3 is intermediate, with about 30 to 40 percent overall mortality. There is general correlation with the amount of SAH and the clinical grade of the patient.

Complications After Subarachnoid Hemorrhage

Delayed complications have a strong effect on the eventual outcome after SAH. The most important are given below. There is a 35 percent risk of major rebleeding during the first three weeks after a SAH due to aneurysm. Afterward, this risk gradually decreases. Rebleed significantly worsens the prognosis. To decrease this risk of rebleed, some neurosurgeons recommend early aneurysm surgery, especially in patients of low clinical grade. The use of antifibrinolysis therapy has been discontinued.

Delayed vasospasm is an additional significant risk factor. This complication occurs in more than 50 percent of persons with SAH. It is known that the degree of spasm correlates positively with the amount of the SAH, but the mechanism of its production is unknown. Most likely, it is caused by the chemical breakdown products of the subarachnoid blood. Vasospasm is not present immediately but becomes evident 5 to 10 days after the hemorrhage. The vascular narrowing may last for weeks. In the case of severe spasm, inflammatory changes develop in the walls of the affected arteries, so the narrowing becomes fixed for some time. These narrowed segments have been dilated with some success using small balloon angioplasty catheters.

The vasospasm may become severe enough to cause brain ischemia and infarction, although the relationship between the spasm and ischemia is not proved conclusively. Additionally, with vasospasm the intima may be affected, so there is secondary formation of local thrombus and distal embolization. If surgery is performed in the presence of spasm, there is a high rate of subsequent infarction and a poor outcome. Multiple angiography or transcranial Doppler ultrasonography is used to define the status of the vasospasm prior to surgery. Calcium channel blockers have been used with some success in an attempt to decrease the amount of vasospasm after SAH. Controlled hypertension is also used to maintain blood flow through regions of spasm and within collateral channels.

Some degree of communicating hydrocephalus almost always accompanies a SAH no matter how small the

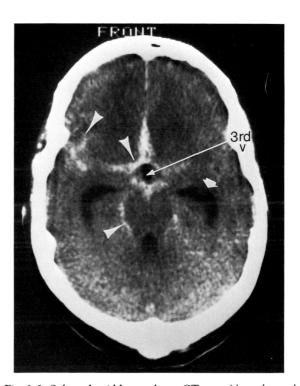

Fig. 3-2 Subarachnoid hemorrhage. CT scan. Nonenhanced transaxial CT scan shows diffuse increased density in the suprasellar cistern, the right sylvian fissure, and the perimesencephalic cisterns (*arrowheads*). It represents the subarachnoid blood. Moderate hydrocephalus is present with dilatation of the temporal horns (*arrow*).

DELAYED COMPLICATIONS OF SAH

Vasospasm and infarction

Rebleeding

Hydrocephalus

bleed. It is thought to be caused by decreased resorption of CSF by mechanical plugging of the arachnoid villi by the subarachnoid red blood cells. The hydrocephalus occurs almost immediately and can usually be seen on the initial CT or MRI scan. Sometimes there is only slight dilation of the temporal horns. In most cases it resolves spontaneously, but occasionally the ventricular system becomes large, necessitating surgical ventricular decompression. Rarely, a large amount of blood within the ventricular system blocks the ventricular foramina, resulting in an entrapped ventricle or noncommunicating hydrocephalus.

CT Scanning

The CT scan is the most useful imaging technique for defining acute SAH. It has greater sensitivity than MRI. When the nonenhanced CT examination is performed within 48 hours of the hemorrhage, the subarachnoid blood can be defined in about 90 percent of cases. The CT scan is negative only in those patients with low clinical grade and a small amount of subarachnoid blood.

The hemorrhage is seen as high density within the basal cisterns and the cerebral sulci (Fig. 3-2). Specific patterns of bleeding frequently make it possible to locate the aneurysm that has bled (Figs. 3-3 and 3-4). This ability may be of great importance when multiple aneurysms are found by subsequent angiography. The patterns are given in Table 3-1. When a large SAH has occurred, blood diffusely floods the subarachnoid space, and localization is not possible (Fig. 3-2). Small hemorrhages may raise the density of the basal cisterns just enough that they become isodense with the brain, and the diagnosis may be difficult to make (Fig. 3-5). In this situation, contrast infusion may show diffuse enhancement of the meninges (Fig. 3-6). Contrast enhancement may also demonstrate the aneurysm itself, although only large aneurysms (more than 1 cm) are routinely seen (Figs. 3-7 and 3-8). Because patients with SAH undergo angiography, it is not necessary to try to demonstrate aneurysms. Routine CT scanning cannot be used to screen for aneurysm.

CT angiography (CTA) employing helical (spiral, volume) scanning technique and rapid intravenous contrast infusion (more than 5 ml/sec) of 120 ml of 60 percent contrast is able to demonstrate most intracranial aneurysms. Thin 1 mm sections, a 1 : 1 pitch and 0.4-mm interpolation reconstruction are used to improve the spatial resolution and decrease problem partial volume effects. Multiplanar reconstruction with cine loops, and three-dimensional (3D) technique are used for display. The aneurysm is seen as a contrast-filled structure arising from the main vascular channel. Aneurysms greater than 5 mm are routinely diagnosed. It is not known whether CTA will replace conventional angiography for the diagnosis and surgical management of patients with suspected intracranial aneurysm.

Communicating hydrocephalus may be subtle but is almost always present with SAH (Fig. 3-2). It is sometimes recognized on CT scans only by the slight dilation of the temporal horns of the lateral ventricles (Fig. 3-5A). If hydrocephalus is moderate, serial scans are performed every few days until the ventricles stabilize or become smaller. Hydrocephalus can become a significant problem and causes an increase in the intracranial pressure, which can further compromise the cerebral blood flow in the patient with vasospasm. There are no CT criteria for deciding at what point the ventricular enlargement becomes significant. The decision for shunting is a matter of clinical judgment.

Focal regions of low density, mostly conforming to known vascular patterns, are indicative of cerebral ischemia or infraction. These low-density regions usually appear after 1 week and are considered a consequence of vasospasm. When a low-density region is first seen, it is impossible to determine whether it represents reversible ischemia or infarction. Regions of ischemic edema without infarction regress in time, whereas a completed infarction shows persistent hypodensity. The severity of the ischemic change on CT and the vasospasm seen on angiography generally correlate with the severity of clinical cerebral dysfunction.

MRI

MRI is not as useful as CT scanning for the diagnosis of SAH. The SAH is much more difficult to demonstrate with MRI. A possible explanation is the high PO_2 of the CSF, which delays the formation of deoxyhemoglobin. Although MRI can define infarction and hydrocephalus, it does not have clinical advantage in this setting. If a clot forms within the cisterns and

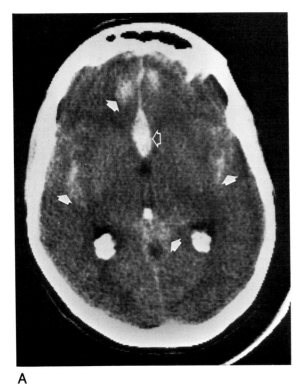

A

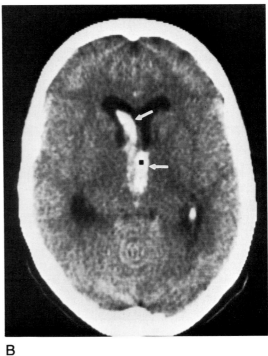

B

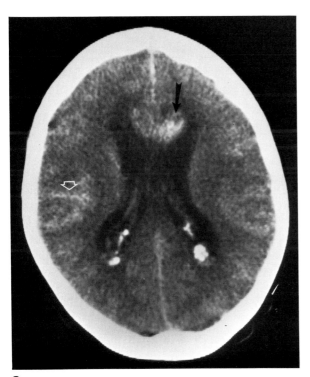

C

Fig. 3-3 ACoA aneurysm rupture. (A) CT scan shows a focal rounded clot in the posterior frontal interhemispheric fissure, possibly extending into the base of the septum pellucidum *(open arrow)*. SAH is also seen diffusely within the cerebral sulci and basal cisterns *(solid arrows)*. **(B)** Intraventricular rupture has occurred, with blood clot seen in the right frontal horn and third ventricle *(arrows)*. Mild hydrocephalus is present. Subarachnoid blood might not be present when the aneurysm ruptures directly into the ventricular system. **(C)** ACoA aneurysm has ruptured into the corpus callosum with blood dissecting through the genu *(arrow)*. Minimal subarachnoid blood is seen in the sylvian fissure *(open arrow)*.

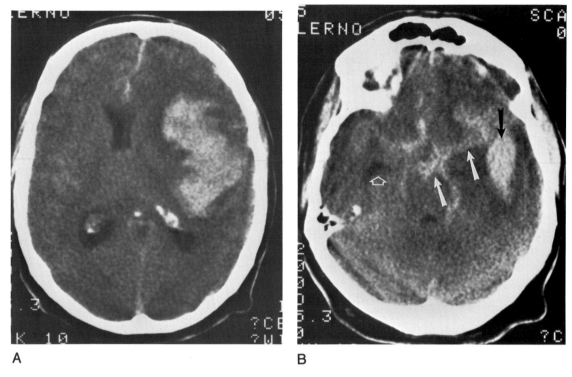

A   B

Fig. 3-4 MCA aneurysm rupture. (A) Large blood clot is seen in the region of the left sylvian fissure and temporal lobe. It could also represent a subcortical intraparenchymal spontaneous hemorrhage. Mass effect is present with subfalcial herniation and mild hydrocephalus of the right lateral ventricle. **(B)** CT scan at a lower level shows the base of the hematoma within the left sylvian fissue *(black arrow)*. In addition, there is subarachnoid blood in the horizontal sylvian fissure, suprasellar cistern, and perimesencephalic cisterns *(white arrows)*. The left perimesencephalic cistern is larger than the right one because of the contralateral shift of the pons from the temporal hematoma. The right temporal horn is slightly dilated from hydrocephalus *(open arrow)*.

Table 3-1 Characteristic Patterns of Hemorrhage With Location of Aneurysm

Site	Hemorrhage
ACoA	Frontal interhemispheric fissure
	"Flame shape" into frontal lobe
	Anterior suprasellar cistern
	Corpus callosum
	Intraventricular blood
PCoA	Ipsilateral suprasellar cistern
MCA	Ipsilateral sylvian fissure
	Medial temporal lobe/sylvian fissure
Basilar artery	Posterior suprasellar cistern
	Prepontine cistern
PICA	Foramen magnum

persists for a few days, it can be imaged with MRI (Fig. 3-9).

The aneurysm itself may be identified with standard spin echo MRI as a round region of abnormal flow-void adjacent to a major artery at the base of the brain (Fig. 3-10). The signal intensity within the aneurysm may be variable from the dephasing effect of turbulent flow or the variable signal intensity of intraluminal blood clot. An aneurysm with subacute thrombus may show high signal intensity. These effects are more pronounced within larger aneurysms. A flow-enhancing radiofrequency (RF)-spoiled gradient echo recalled MRI (GRE) single-slice sequence with a large tip angle, short time

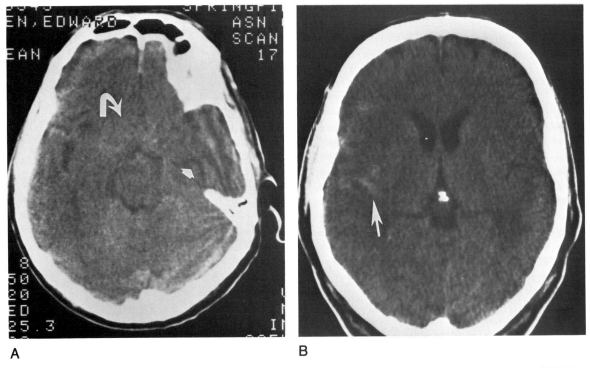

A   B

Fig. 3-5 Subtle SAH. (A) Transaxial view through the basal cisterns shows isodensity of the cisterns from a small SAH (*curved arrow*). Minimal hydrocephalus is present, indicated by slight dilatation of the left temporal horn (*arrow*). **(B)** Subtle increased density is seen in the right sylvian fissure from SAH (*arrow*).

Fig. 3-6 SAH, meningeal enhancement. Contrast-enhanced CT scan of a patient with no demonstrable SAH shows diffuse "smudgy" enhancement of the meninges (*arrows*). It is a nonspecific finding and may also be seen with meningitis or meningeal carcinomatosis.

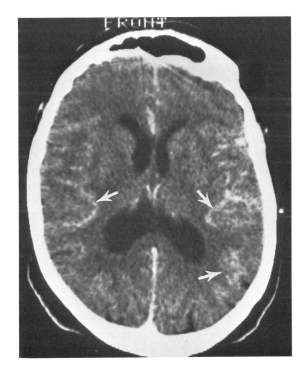

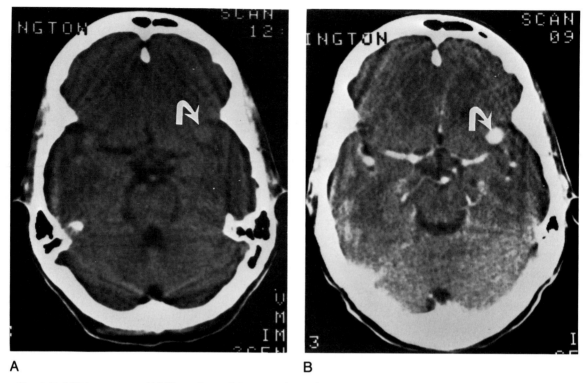

Fig. 3-7 MCA aneurysm (A) Nonenhanced CT scan shows the moderate-size left MCA aneurysm as a focal found region of slight hyperdensity within the sylvian fissure *(curved arrow)*. **(B)** Contrast-enhanced examination shows the aneurysm lumen as a homogenous dense contrast collection *(curved arrow)*. A clotted aneurysm would not fill with contrast, which is common with giant aneurysms. Small aneurysms cannot be routinely detected with a contrast CT scan.

of repetition (TR), and short time of echo (TE) will show bright signal within aneurysms with flow and may aid in the diagnosis. However, the routine MRI techniques are not reliable for the detection or definition of small aneurysms (less than 1 cm) and cannot be used as a screening examination or in place of angiography prior to surgery.

Intravenous contrast usually does not produce high signal enhancement within the aneurysm because of the relatively fast flow and so it does not improve the detection rate. The aneurysm wall may enhance with contrast suggesting a neoplastic lesion. The changes of subacute or chronic hemorrhage are sometimes seen adjacent to an aneurysm that has bled further, confusing the pattern on the image.

The ependyma of the ventricles or the brain surface may be outlined on T_2 or T_2* sequences as a rim of very low signal intensity. This results when hemosiderin is deposited following SAH. Low-signal-intensity structures adjacent to the circle of Willis, such as a pneumatized anterior clinoid process or a calcified meningioma, may mimic a flow-void and lead to a false-positive diagnosis of aneurysm.

MRA

Magnetic resonance angiography (MRA) is a promising technique for the demonstration of smaller aneurysms. Aneurysms 3 mm or greater have been demon-

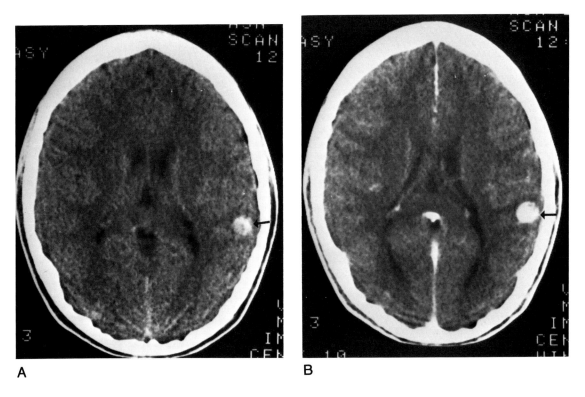

A   B

Fig. 3-8 Peripheral aneurysm of MCA.
(A) Nonenhanced CT scan shows a round
hyperdense lesion in the region of the left
posterior sylvian fissure *(arrow)*. The hyperdensity represents either a small amount
of calcium or acute clot. **(B)** Contrast-enhanced study shows the densely enhancing
lumen of the aneurysm. **(C)** ICA angiography demonstrates the aneurysm sac on the
angular branch of the MCA *(open arrow)*.
Peripheral aneurysms may due to infection
(mycotic) or trauma, or they may be idiopathic.

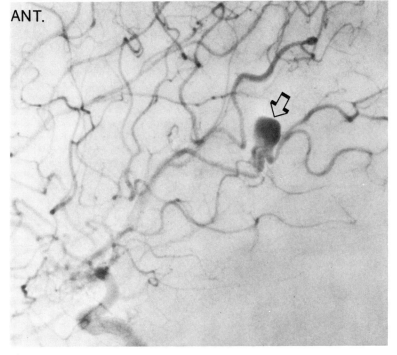

C

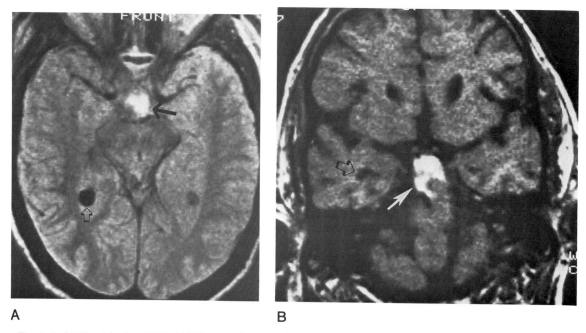

A   B

Fig. 3-9 SAH, with clot; MRI. (A) Transaxial proton density MRI through the suprasellar cistern shows a hyperintense blood clot *(arrow)*. A calcified glomus is seen in the atrium of the right lateral ventricle *(open arrow)*. **(B)** Coronal T$_1$-weighted MRI demonstrates the hyperintense clot in the suprasellar cistern extending inferiorly on the right in the prepontine cistern *(arrow)*. It was from a rupture of a PCoA aneurysm. The right temporal horn is slightly dilated from hydrocephalus *(open arrow)*.

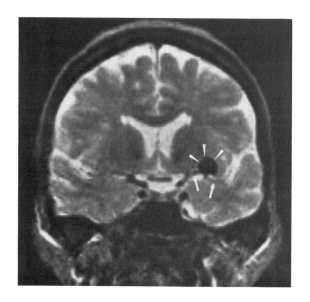

Fig. 3-10 Left MCA aneurysm, MRI. T$_2$-weighted MRI shows a rounded region of flow-void, representing a moderately large aneurysm of the bifurcation of the left MCA *(arrowheads)*. The MCA can be seen within the sylvian fissure because of the flow-void within the hyperdense cisterns *(arrows)*.

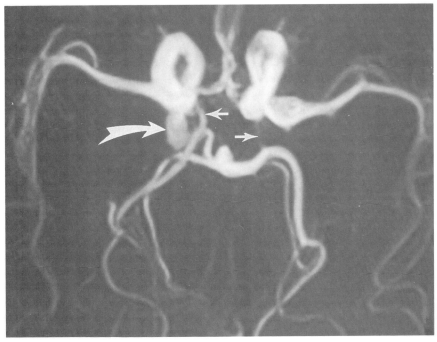

A

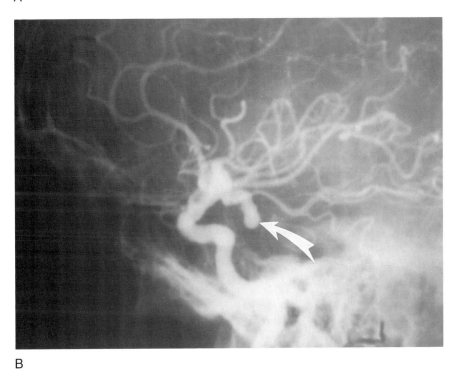

B

Fig. 3-11 Berry aneurysm at the right PCoA. (A) 3D TOF MRA through the base of the brain shows the PCoA aneurysm on the right *(large white arrow)*. This view is similar to a submentovertex x-ray projection. Note the PCoAs *(small white arrows)*. Aneurysms as small as 4 to 5 mm may be seen with 3D thin-section (< 1 mm) MRA. **(B)** Lateral projection XRA of the same person showing the large aneurysm with multiple bulges pointing backward and downward from the ICA at the origin of the PCoA *(arrow)*.

strated with sensitivity reported at about 60 to 95 percent. The 3D time-of-flight (TOF) sequence is the most useful because of the ability to obtain thin sections of less than 1 mm in multiple projections. The scan slab may be appropriately positioned on the region of the circle of Willis, so the blood saturation effect in the depth of the slab is not a problem. This technique may replace angiography as a screening test for asymptomatic persons at high risk of intracranial aneurysm (Fig. 3-11). Phase-contrast (PC) MRA is advantageous in demonstrating giant aneurysms or aneurysms with high signal within an intraluminal clot. This is because of its much greater background signal suppression, which eliminates confusing signals in nonvascular tissue.

Angiography

Cerebral angiography remains the most reliable means of identifying cerebrovascular aneurysms. It is performed with selective ICA and vertebral injections, using magnification rapid serial filming technique (2 to 3 films/sec). A very high-resolution digital subtraction technique angiography (DSA) may be used. This has the advantage of direct rotational angiography of display in a cine loop. These techniques provide alterations in projection that clearly resolve overlapping vessels and provide some depth perception.

The intra-arterial contrast injection rate is reduced slightly (4 to 5 ml/sec) to decrease the potential for aneurysm rupture due to pressure changes downstream. The angiogram may be performed at any time after the aneurysm rupture. Aspirin should not be given prior to the angiography. Hypertension and patient agitation should be controlled with medication. Oblique and submental vertex views are frequently necessary to define the details of the aneurysm. When possible, a complete angiogram is performed to identify all potential aneurysms. Both vertebral arteries may need to be injected to opacify the origin of the posterior inferior cerebellar arteries (PICAs) sufficiently. Occasionally, cross-carotid compression is used to identify an aneurysm at the origin of the ACoA. If no aneurysm or other source of the SAH can be found, selective external carotid angiography is performed to search for dural AVMs. Occasionally, an aneurysm is seen on follow-up angiography that was not seen initially. The goals of angiography are as follows:

1. Identification of an aneurysm or multiple aneurysms
2. Determination of which aneurysm has bled

3. Definition of the neck of the aneurysm for treatment planning
4. Relationship of the aneurysm to the surrounding vessels and brain structures
5. Detection of vasospasm and ischemia
6. Definition of the flow patterns of the circle of Willis

The aneurysm is identified as a rounded, contrast-filled bulge originating from the adjacent artery at characteristic sites (Figs. 3-12 and 3-13). The density of the contrast within the aneurysm is the same as that within the normal arteries. A vessel seen "on end" or a tortuosity of a normal vessel may simulate an aneurysm but its observed contrast density is greater than that of the adjacent normal vessels. Oblique views are almost always necessary for better definition of the aneurysm.

When multiple aneurysms are present, the size of the aneurysm is the most reliable sign for determining which aneurysm has bled. In 95 percent of cases of multiple aneurysms, it is the largest one that has bled. Aneurysms smaller than 5 mm almost never bleed. Because ruptured aneurysms have almost always expanded near the apex of the dome and bleed from this expansion, the dome usually shows a focal, rounded, secondary bulge, or at least some irregularity of the margin (Fig. 3-12). There may be multiple rounded focal expansions, appearing as either a segmented worm or a small cluster of berries. Local vasospasm also points to the nearest aneurysm as the source of the SAH but is less reliable (Fig. 3-14). Demonstration of extravasation of contrast from the aneurysm is clearly definitive of rupture, but this problem almost never presents. A local hematoma may indicate the site of rupture (Figs. 3-3, 3-4, and 3-15) (see below).

A small aneurysm in the region of the PCoA must be differentiated from an infundibulum of the PCoA itself.

INDICATIONS OF WHICH ANEURYSM HAS BLED

Large size
Multiple small aneurysms on dome
Local hematoma
Local vasospasm

Fig. 3-12 Aneurysms: PCoA, supra-clinoid ICA. Lateral ICA angiogram shows a typical aneurysm of the ICA at the level of the PCoA *(black arrow)*. Note the small "daughter" bulges at the apex of the aneurysm *(small open arrow)*, which is a sign highly associated with recent bleeding or expansion of the aneurysm is present just distal to the ophthalmic artery *(large open arrow)*. Note that the contrast density within the aneurysm is similar to that of the adjacent vessel.

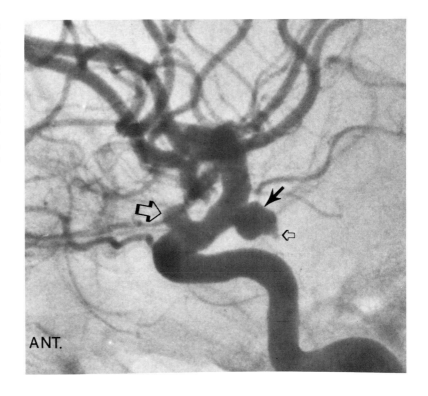

Fig. 3-13 Aneurysm of the tip of the basilar artery. A small aneurysm is seen at the tip of the basilar artery *(curved arrow)* on a Water's projection of vertebral angiography. This view shows the entire basilar artery to good advantage and is excellent for aneurysm detection. Both vertebral arteries need to be opacified either by bilateral selective injection or by reflux from unilateral injection.

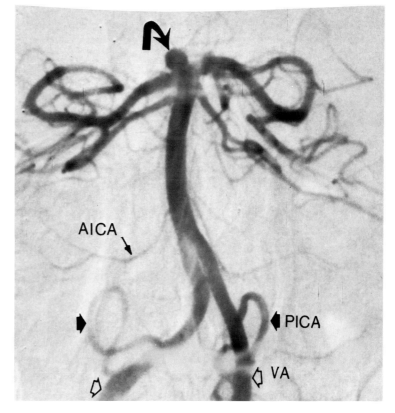

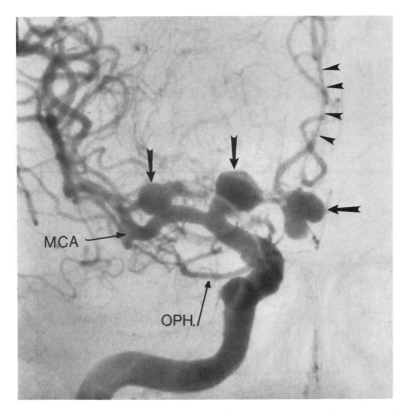

Fig. 3-14 Multiple aneurysms. Aneurysms are present at the right MCA, the proximal anterior cerebral artery, and the level of the ACoA (*arrows*). The aneurysms are similar in size. However, the lobular shape of the ACA aneurysm and the regional spasm of the pericallosal artery (*arrowheads*) make it likely that this aneurysm is the one that has ruptured.

An aneurysm is more than 3 mm and has a rounded, dome-like configuration. An infundibulum is smaller than 3 mm, is triangular in shape, and frequently has a visible vessel emerging from the dome (see Fig. 2-26A).

Vasospasm is represented by narrowing of the vessel lumen. It has a multifocal distribution (Fig. 3-16). Vasospasm usually occurs near the region of the aneurysm but may be seen at some distance. When severe, it causes a demonstrable decrease in the cerebral circulation. It begins 5 to 10 days after the SAH but may persist for weeks.

Angiography following SAH does not define the cause of the hemorrhage in about 10 percent of cases. MRA has identified an aneurysm previously undiagnosed by high-quality angiography. At most centers, if the source of bleeding has not been found, repeat angiogram is performed after 2 weeks, although the likelihood of discovering the origin of the bleed is reported to be low (0 to 10 percent). Often those with SAH seen by CT to be localized solely within the prepontine cistern and around into the ambient cisterns will have normal cerebral angiography. This group reportedly has a better prognosis.

GIANT ANEURYSM

Only 2 percent of intracranial aneurysms are giant aneurysms. These lesions are arbitrarily defined as those more than 2.5 cm in diameter. Giant aneurysms seldom rupture; rather, they become apparent because of local mass effect. They are recognized predominantly in the older age groups. About 75 percent have peripheral calcification that can be seen on plain films and by CT

Fig. 3-15 Intraparenchymal hemorrhage from peripheral aneurysm rupture. (A). CT scan shows a large hematoma involving the corpus callosum and extending into the deep frontal white matter. This hemorrhage is deeper than the average subcortical hemorrhage associated with hypertension. Angiography should be performed for hemorrhages such as this one to detect a possible aneurysm or vascular malformation. **(B)** Lateral ICA angiography demonstrates an aneurysm of the pericallosal artery at the origin of the callosal marginal artery *(large arrow)*. The pericallosal artery is stretched. The hematoma has depressed the sylvian vessels *(small arrows)* and is stretching a posterior frontal opercular branch *(open arrow)*.

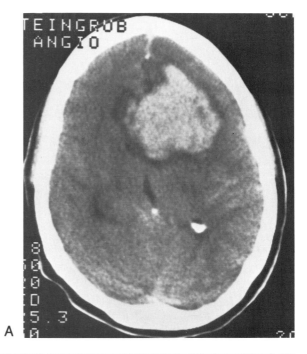

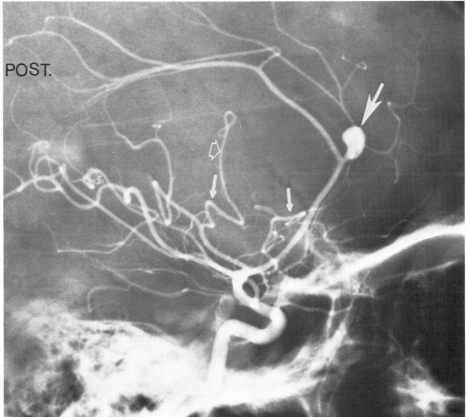

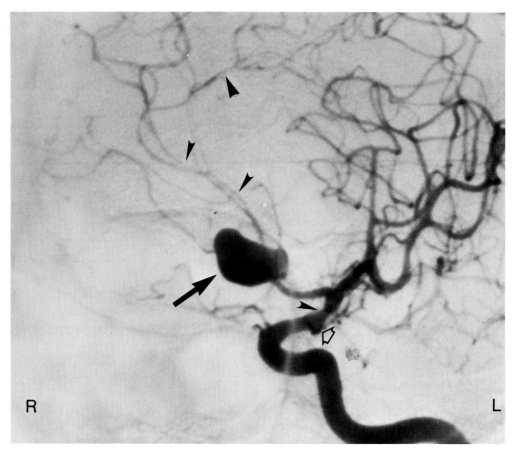

Fig. 3-16 ACoA aneurysm, with spasm; oblique projection anteriography. A large aneurysm is present at the ACoA (*arrow*). There is diffuse severe spasm of the supraclinoid ICA and the pericallosal vessels (*arrowheads*). Spasm is multifocal with varying degrees of vasoconstriction along the vessel. A small aneurysm is present at the PCA (*open arrow*).

scanning. Thick walls and mural thrombus are common. They may cause bone erosion, particularly in the region of the sella turcica.

Most giant aneurysms arise in the intracavernous portion of the ICA (Fig. 3-17) and are at least partially extradural. These lesions may cause compression of CN III to VI, which course in the lateral dural wall of the cavernous sinus. When they are very large, the aneurysm erodes and enlarges the sella from the side. The predominant unilateral expansion indicates an aneurysm, rather than an intrasellar tumor. Proptosis can occur with expansion forward into the posterior orbit.

The next most common location is the basilar artery, from either the tip (Fig. 3-18) or along the main portion. These lesions can compress the posterior third ventricle or the brain stem. Rarely, giant aneurysms arise from the MCA at the bifurcation or from the ACA. Intradural giant aneurysms present with SAH about 40 percent of cases. Peripheral embolization can occur from the mural thrombus. The parent vessel may be atheromatous and friable in the region from which the aneurysm arises and may be incorporated into the wall of the aneurysm sac. The neck of the aneurysm is usually small.

Fig. 3-17 Giant aneurysm of the ICA. (A) A giant aneurysm is present arising from the cavernous portion of the right ICA *(arrowheads)*. A small amount of calcification is present in the wall *(arrow)*. Erosion of the dorsum and the right side of the sella has occurred *(open arrowhead)*. This aneurysm shows nearly uniform contrast enhancement. **(B)** Lateral common carotid angiography demonstrates a contrast-filled giant aneurysm of the cavernous segment.

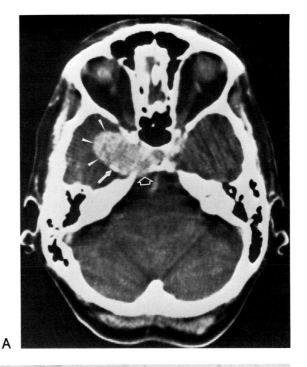

A

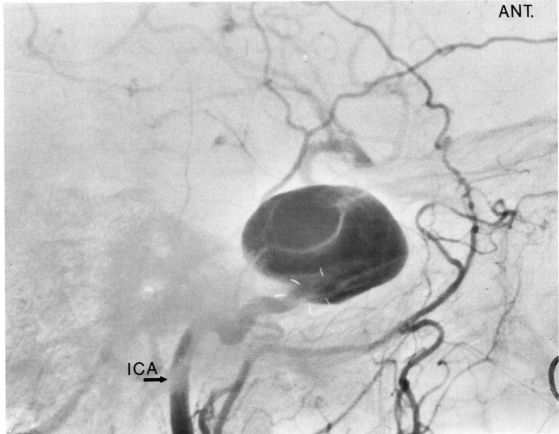

B

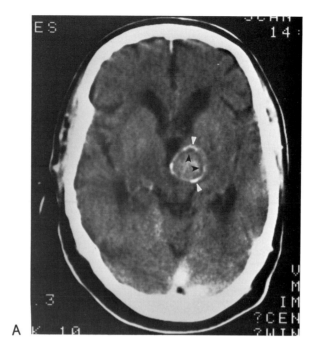

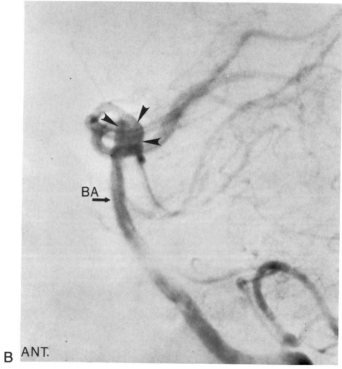

Fig. 3-18 Giant aneurysm from the tip of the basilar artery. (A) Transaxial contrast-enhanced CT scan shows a spherical mass extending up into the base of the thalamus, distorting the posterior third ventricle. The rim of the mass has irregular calcification (*white arrowheads*). The central portion of the sphere has inhomogeneous hyperdensity with only minimal if any contrast enhancement. This aneurysm has caused mild hydrocephalus. Such aneurysms are often misdiagnosed as tumor. The concentric ring of hypodensity just inside the calcified wall represents chronic clot (*black arrowheads*). (B) Lateral vertebral angiography demonstrates contrast filling only the base of the aneurysm (*arrowheads*). The remainder of the aneurysm is filled with clot. BA, basilar artery.

Angiography demonstrates most giant aneurysms as large abnormal collections of contrast (Fig. 3-17). A swirling effect of the contrast is often seen at the time the contrast bolus reaches the aneurysm. However, the actual size of the aneurysm is often underestimated because of the large amount of clot lining the periphery of the cavity. Occasionally, only a small part of the aneurysm fills with contrast (Fig. 3-18).

The CT scan better demonstrates the size of the aneurysm, as the entire structure can be visualized (Fig. 3-18). The appearance of the aneurysm depends on the amount of clot within the lumen. If the entire lumen is patent, on the nonenhanced CT scan the lesion appears as a slightly hyperdense mass, often with rim density representing the calcification in the wall. The contrast-enhanced CT scan shows uniform hyperdensity of the opacified blood within the aneurysm (Fig. 3-17). The homogeneity and hyperdensity may be identical to some tumors, especially pituitary adenomas and parasellar meningiomas. In this instance, angiography, MRI, or MRA may be necessary to make the correct diagnosis.

Partially thrombosed aneurysms have a different appearance. On the nonenhanced CT scan, the lesion appears as a mass with a central or eccentric round region of slight hyperdensity with peripheral hypodensity representing the chronic clot (Fig. 3-18). With contrast enhancement, the central region becomes very hyperdense owing to opacified flowing blood. Little change occurs in the other portions of the lesion except the wall, which may enhance.

A completely thrombosed aneurysm is nearly homogeneously isodense because of the lumen-filled clot. Slight peripheral contrast enhancement may occur. These lesions may be difficult to diagnose and are often mistaken for a tumor, particularly meningioma or craniopharyngioma.

The characteristic signs of a giant aneurysm can be seen with MRI. If the cavity is patent, there is almost always a central flow-void, although inhomogeneity of signal or even hyperintensity may occur because of turbulence, flow-related enhancement, or the echo-rephasing phenomenon. The peripheral clot shows inhomogeneous hyperintensity. Calcium in the rim is not reliably seen with MRI. There may be a small amount of subacute hemorrhage adjacent to the aneurysm, and sometimes edema of adjacent brain parenchyma. The finding of flow-void reliably differentiates giant aneurysm from tumor. Thrombosed aneurysms show vary-

ing intensity depending on the age of the clot. The wall of the aneurysm may demonstrate contrast enhancement.

ENDOVASCULAR TREATMENT OF ANEURYSMS

It is now possible to treat some intracranial aneurysms safely using endovascular technique. Aneurysms that cannot be surgically obliterated may be successfully obliterated by placement of fine wire coils within the lumen of the aneurysm. Catheterization of the intracranial vessel and aneurysm is accomplished with very small flexible wire and catheter coaxial systems (e.g., the Tracker). Extreme care is taken to avoid pressure on the dome of the aneurysm, to prevent inadvertent rupture. Once the aneurysm is catheterized, a relatively long length of platinum wire is coiled within the lumen. The coil is attached to a proximal length of stainless steel wire guide. A small current of 9 V and 10 mA is applied. This produces a positive charge in the coil, attracting the negatively charged blood elements, including platelets and fibrinogen. The blood elements thromboses and form a stable clot. A short (less than 3-mm) length of the most distal stainless steel wire guide is left uninsulated. Electrolysis occurs and erodes the short segment of wire until it detaches from the intra-aneurysmal coil. The catheter system is removed, leaving the coil and clot within the aneurysm (Fig. 3-19). Special skill and training are needed for the performance of these delicate procedures.

About 200 cases have been done in this fashion with a morbidity of less than 10 percent, a mortality of less than 5 percent, and a thrombosis rate of 70 to 98 percent of the aneurysm volume. Few rebleeds have been reported, although the follow-up period is short. Aneurysms with a narrow neck hold the wire better and have a greater success rate. The size of giant aneurysms is not reduced by this technique.

ATHEROSCLEROTIC FUSIFORM ANEURYSMS

Fusiform aneurysms represent the progression of atheroectasia of the vessel. The aneurysms may become large, with a diameter of more than 2.5 cm, and are

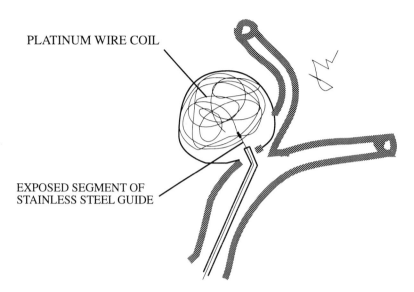

PLATINUM WIRE COIL

EXPOSED SEGMENT OF
STAINLESS STEEL GUIDE

Fig. 3-19 Endovascular treatment of aneurysms.

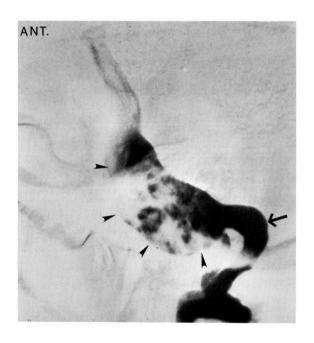

ANT.

Fig. 3-20 Atherosclerotic giant basilar artery aneurysm. Lateral vertebral angiogram shows a giant fusiform atheroslerotic aneurysm of the basilar artery *(arrowheads)*. Atherosclerotic ectasia of the distal vertebral artery is present *(arrow)*.

therefore classified as a form of giant aneurysm. Most commonly they involve the basilar artery (Fig. 3-20), but they may also involve the upper terminal portion of the ICA, with extension into the proximal segments of the MCA and ACA. The vessel is not only dilated but elongated and tortuous. The aneurysms may compress CNs; occasionally, a basilar artery aneurysm compresses the posterior third ventricle and aqeduct, causing hydrocephalus. These aneurysms seldom rupture. Because small perforating branches arise from the aneurysmal sac, they are almost impossible to resect safely. Some have been treated successfully by endovascular technique. Except for the shape of the aneurysm, both the CT and angiographic findings follow the same principles of diagnosis as described above for giant saccular aneurysms.

DISSECTING ANEURYSM

A dissecting aneurysm, or dissecting intramural hematoma of the vertebral, ICAs, or MCAs may be associated with SAH. The dissection may occur spontaneously, secondary to vascular pathology such as Marfan syndrome, or secondary to trauma. SAH results only from those dissections that extend to the subadventitial layer external to the muscular layer of the vessel. The more common dissection between the intima and muscularis causes ischemia from vessel narrowing or thrombosis and does not rupture into the subarachnoid space.

DURAL VASCULAR MALFORMATION

Dural vascular malformations, or fistulas, are usually fistulas that develop between dural arteries and a dural venous sinus. They are diagnosed almost solely by external carotid angiography. Drainage from the fistula may occur into cortical veins which may rupture into the subararachnoid space. Dural fistula is discussed in more detail in Chapter 4.

MYCOTIC ANEURYSM

Mycotic aneurysms are discussed in Chapters 2 and 8.

SUGGESTED READING

Alcock JM: Aneurysms. p. 2435. In Newton TH, Potts DG (eds): Radiology of the Skull and Brain: Angiography. CV Mosby, St. Louis, 1974

Atlas SW: Intracranial vascular malformations and aneurysms. Radiol Clin North Am 26:821, 1988

Davis KR, Kistler JP, Heros RC et al: A neuroradiologic approach to the patient with a diagnosis of subarachnoid hemorrhage. Radiol Clin North Am 20:87, 1982

Drayer BP: Diseases of the cerebrovascular system: aneurysms. In Heinz ER (ed): The Clinical Neurosciences: Neuroradiology. Churchill Livingstone, New York, 1984

Guglielmi G: Endovascular treatment of intracranial aneurysms. Neuroimag Clin North Am 2:269, 1992

Huston J III, Rufenacht DA, Ehman RL, Weibers DO. Intracranial aneurysms and vascular malformations: comparison of time-of-flight and phase-contrast MR angiography. Radiology 181:721, 1991

Rinkel GJE, Wijdicks EFM, Vermeulen M et al: Nonaneurysmal perimesencephalic subarachnoid hemorrhage: CT and MR patterns that differ from aneurysmal rupture. AJR 157:1325, 1991.

Sasaki O, Koike T, Tanaka R et al: Subarachnoid hemorrhage from dissecting aneurysm of the middle cerebral artery: case report. J Neurosurg 74:504, 1991

Silver AJ, Pederson ME Jr, Ganti SR et al: CT of subarachnoid hemorrhage due to ruptured aneurysm. AJNR 2:13, 1981

4

Arteriovenous Malformations

Congenital cerebrovascular malformations have been classified by Rubenstein into four groups: (1) arteriovenous malformation (AVM), (2) cavernous hemangioma (angioma), (3) venous angioma, and (4) capillary telangiectasis. The AVM is by far the most common type, occurring in about 3 percent of persons with subarachnoid hemorrhage (SAH) or intracerebral hematoma. AVMs are uncommon in the general population, occurring in about 0.1 percent. Ninety percent of AVMs occur in the cerebral hemispheres. The other types are rare. Even more rare are the malformation syndromes discussed in this chapter. Overall, AVMs account for less than 1 percent of all strokes.

ARTERIOVENOUS MALFORMATION

An AVM is a developmental anomaly of blood vessels, consisting of a tortuous tangle of arteries and veins that are abnormal in structure, caliber, and length. The malformation may be small, with a single feeding artery-to-vein fistula, or large, with multiple feeding arteries from multiple vascular territories feeding large varicose deep and superficial veins. The arteries and veins communicate directly without an intervening capillary network, allowing the shunting of blood from artery to vein. They occur predominantly in the parietal lobe, but may be found in any region of the brain. The malformations are further subclassified according to their vascular supply. Pial malformations (75 percent) are supplied solely by cerebral or cerebellar cortical arteries. Dural malformations (10 percent) are supplied solely by dural menin-

---

CEREBROVASCULAR MALFORMATIONS

Common
    Arteriovenous malformation—pial, dural, mixed
Uncommon
    Venous angioma
    Cavernous hemangioma
    Capillary telangiectasis
Rare
    Carotid–cavernous fistula
    Vein of Galen aneurysm
    Traumatic A-V fistulas
    Encephalofacial angiomatosis (Sturge-Weber)
    Hereditary telangiectasis (Osler-Weber-Rendu disease)
    Wyburn-Mason syndrome

geal arteries. Mixed malformations (15 percent) are supplied by both pial and dural arteries. Posterior fossa malformations are of the mixed type in 50 percent of cases.

The brain parenchyma in and around the malformation degenerates, resulting in atrophy and gliosis, so that the vessels of the malformation become compact. Calcification occurs in almost all malformations and may be minimal or heavy. Chronic changes from small subclinical hemorrhages are seen both within the lesion and in the adjacent brain. Hemosiderin deposits are found particularly around the periphery. Areas of cerebral infarction may be found outside the region of the lesion and are thought to result from ischemia caused by a steal of blood flow to larger AVMs. The draining veins may become large with the development of deep venous varices. It is generally accepted that AVMs hemorrhage from rupture of these dilated thin-walled veins.

The initial clinical presentation is varied in persons with an AVM. Fifty percent present with either an intraparenchymal hematoma or SAH. The peak incidence is during the 20- to 30-year age period. Ironically, it seems to be the smaller AVMs that are at greater risk of hemorrhage. In contrast to SAH due to berry aneurysms, the mortality rate from the first bleed is relatively low, about 10 percent, probably because the hemorrhage is of venous origin. Some persons present with a subdural hematoma near the region of the malformation, most likely attributable to dural malformation. When considered as a group, all malformations have a yearly 1 to 4 percent risk of hemorrhage and a yearly rebleed risk of 3 percent. The risk is slightly higher during the first year after a hemorrhage from the AVM.

About 33 percent of persons with an AVM present with a seizure disorder that may be either focal or generalized. Headache resembling migraine is the presenting symptom in 9 percent of patients. In later life, presenile dementia may occur, the cause of which is unknown.

The consensus is that AVMs should be resected if they have bled at least once and are in a location in which they can be totally resected without causing neurologic damage. Incomplete resections are of limited value. In rare cases, an AVM is removed in an attempt to relieve intractable seizures, but results have been inconsistent. Headache is generally not an indication for surgical removal. AVMs that have not bled are usually not removed.

Various methods are used to treat AVMs. Treatment aims to reduce the threat of future hemorrhage, control vascular steal, or reduce headache. Surgical removal is the most common and most successful treatment for lesions near the cortical surface limited to one arterial vascular distribution. Complete treatment of the AVM entails obliteration of the actual tangle ("sponge"). If an AVM is incompletely removed or obliterated and a fistula remains, new vessels contribute to the lesion, and the AVM may grow again. Only complete obliteration of the AVM eliminates the threat of repeat hemorrhage.

Interventional neuroradiologists use superselective microcatheter techniques and transcatheter embolization to obliterate feeding arteries. Up to 30 percent of AVMs initially thought to be surgically unresectable are subsequently totally resected. Partial obliteration can reduce surgical morbidity and the duration of the operation. In about 20 percent, embolization alone obliterates the entire AVM. Transcatheter embolization successfully completes obliteration after partial treatment by other methods.

AVMs are treated with various radiotherapy regimens. X-rays, γ-rays (γ-knife, radiosurgery), and Bragg-peak proton therapeutic techniques have all been tried with variable success. The aim is to produce fibrosis of the vessels, ultimately leading to thrombosis of the malformation. The smaller AVMs (less than 3.5-cm-diameter) have the greatest potential for obliteration. Larger malformations are only incompletely obliterated. The AVM may continue to decrease in size or to thrombose up to 2 years after the completion of radiotherapy. After that time, it is stable. Surgery or transcatheter embolization may safely obliterate AVMs that have been incompletely treated with radiotherapy.

IMAGING TECHNIQUES

Skull Radiography

About one-third of persons with an intracranial AVM show some characteristic skull abnormality, most commonly evidence of large feeding arteries or veins, causing erosion of the bone. This includes enlargement of the carotid groove in the paracavernous region, enlarged meningeal grooves on the inner table, enlarged or increased numbers of holes for perforating vessels (particularly in the region of a dural AVM), and inner table thinning from dilated cortical draining veins. Calcifica-

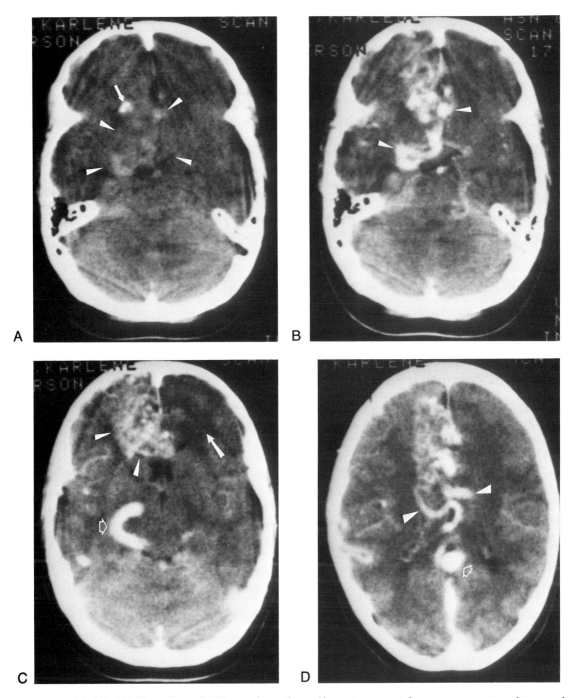

Fig. 4-1 AVM, CT. (A) Nonenhanced CT scan shows the malformation as an inhomogeneous region of increased density representing blood-filled vascular channels and adjacent gliosis *(arrowheads)*. Calcification is present *(arrow)*. **(B)** Contrast-enhanced CT scan at the same level as the scan in Fig. A shows the contrast-filled, grossly dilated vessels, mostly enlarged veins *(arrowheads)*. **(C)** Scan at a higher level shows the "sponge" of the malformation *(arrowheads)*. It extends across the corpus callosum but causes little mass effect. There is a huge draining vein on the right, probably the basal vein *(open arrow)*. Region of hypodensity in the left frontal lobe represents ischemia or infarction from "steal" *(arrow)*. **(D)** Scan through a higher level shows that the malformation is large and extends posteriorly in the corpus callosum and cingulate gyrus. Note large dilated draining veins *(arrowheads)* and a moderately enlarged vein of Galen *(open arrow)*.

199

tion is seen in only 20 percent with plain film skull radiography, whereas it is seen in most AVMs with CT scanning.

CT Scanning

The CT appearance of an AVM is characteristic. On the nonenhanced CT scan, the lesion is most commonly seen as a region of mixed high and low density (Fig. 4-1A). The margins are poorly defined, calcifications are frequently observed, and there is almost always some degree of local atrophy, which may be indicated by subtle focal dilation of the nearby ventricular cavities or sulci. Minor degrees of focal encephalomalacia may be seen adjacent to the nidus. It is rare to see large regions of ischemia or infarction. When close observation is made of these findings, most AVMs that are larger than about 1 cm can be suspected on the nonenhanced CT examination.

The contrast-enhanced CT examination shows nonhomogeneous increased density of the lesion and better demonstrates the malformation. The enhanced regions represent contrast-filled vessels with flowing blood, and the lower density indicates clotted channels and atrophic tissue. Large feeding arteries and veins are shown as serpiginous linear channels (Fig. 4-1). It is important to recognize the vascular territories involved and the location of large draining veins. If there is recent hemorrhage or excessive calcification, the abnormal enhancement may not easily be seen. However, the recognition of dilated arteries at the base of the brain and large draining veins allow one to make the diagnosis of AVM.

CT remains a primary means of detecting the complications associated with vascular malformations of the brain. SAH is recognized by the increased density in the subarachnoid cisterns (see Ch. 3). There is more likely to be blood within the sulci over the convexities than in the basal cisterns. A brain hematoma is more likely to be seen with a ruptured AVM than with a ruptured aneurysm (Fig. 4-2). The hematoma may be large and cause brain herniation. Small hematomas may cause little permanent damage, as the bleed may occur into the already atrophied brain tissue. The hematoma may be indistinguishable from hematomas of other causes, such as hypertension or tumor. The hemorrhage may be the only indication of a cryptic (nonvisible) vascular malformation.

Hydrocephalus may occur after rupture of an AVM, caused acutely by the mass of the hematoma, SAH, or intraventricular blood. The size of the ventricles must be followed, as shunting may be required. An entrapped ventricle is also possible.

MRI

Routine spin echo (SE) MRI distinguishes a vascular malformation from other abnormalities, primarily by the recognition of flow-void within the serpiginous abnormal feeding and draining vessels of the malformation. The dilated vascular channels of the AVM appear dark on the MRI image (Fig. 4-3). When there is difficulty determining whether low signal is representative of blood flow or paramagnetic effect, gradient echo (GRE) imaging may be used. Most commonly, vessels with flow-void on the SE images appear hyperintense with GRE sequences. Calcium or hemosiderin remain of low intensity. Gadolinium shows variable enhancement of the vessels, depending on the speed of flow. Slower flow shows greater enhancement. In general, contrast enhancement is not necessary for the diagnosis. As with CT scanning, a large hematoma may obliterate visualization of the AVM.

The MRI scan shows changes of subacute and chronic hemorrhage, as well as variable hemosiderin deposition within and near the AVM. A thin hemosiderin deposit may be seen over the ventricular and cortical surfaces. Calcium is poorly recognized and may be difficult to differentiate from hemosiderin within the malformation. Local gliosis surrounding the AVM appears as high signal intensity on the T_2-weighted images (Fig. 4-3C). Usually there is no mass effect or edema with an AVM unless there has been recent hemorrhage. Focal atrophy may be seen.

Magnetic resonance angiography (MRA) is capable of identifying AVMs. However, only the larger vessels with more rapid flow are reliably identified. Smaller, but important, arterial feeders are not identified. Presently, MRA demonstrates much of the malformation and its feeding and draining vessels, but it cannot substitute for cerebral angiography for the preoperative management of AVM.

Angiography

Cerebral angiography remains the most precise technique for the anatomic definition of the AVM. The study must be complete and of high technical quality. It is essential to selectively opacify all the vessels that may

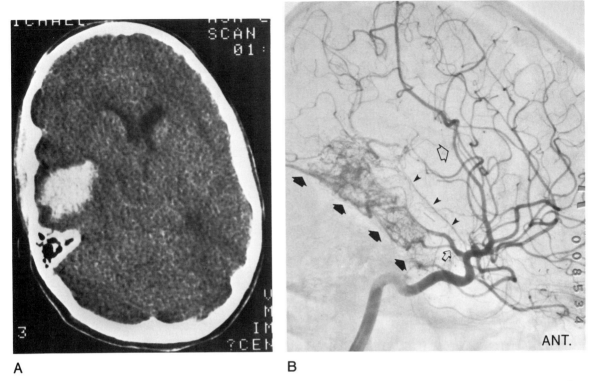

Fig. 4-2 AVM, intraparenchymal hematoma. (A) Moderate-size subcortical hemorrhage is present in the right middle temporal lobe. There is mild mass effect. The basal cisterns and sulci are not well seen, possibly indicative of a small SAH making the CSF isodense with the brain. **(B)** Lateral ICA angiography demonstrates a large vascular malformation involving the medial and inferior temporal lobe *(black arrows)*. It is supplied by the temporal branches of the posterior cerebral artery, which has a fetal type communication with the ICA *(bottom open arrow)*. The temporal hematoma has caused elevation of the sylvian vessels *(top open arrow)* and delayed filling of temporal opercular arteries over the lateral cortex. The anterior choroidal artery is stretched because of the temporal lobe mass *(arrowheads)*.

be contributing to the malformation (Fig. 4-4). Rapid serial filming must be done through the venous phase to determine the flow patterns of both the arteries and the veins. In some cases, three or four films per second are required. Selective bilateral external carotid angiography is routinely performed when the AVM is of the posterior fossa or in the peripheral portion of the cerebral hemispheres so as to detect dural vascular supply (Fig. 4-4C). Contralateral contribution from the external or internal carotid arteries (ICAs) to the malformation can be seen in up to 30 percent of cases. Large doses of contrast may be necessary to adequately visualize a large malformation. This method can be done safely, as most of the contrast passes through the AVM, not the normal cerebral vascu-

larity. The angiographic criteria of an AVM are as follows (Fig. 4-4 to 4-6).

1. A *"sponge" of abnormal vessels, representing the nidus of the malformation*—The vessels are tightly packed together, as the intervening brain tissue is atrophic. Tumor neovascularity is not packed nearly as tightly as with an AVM.

2. *Large feeding arteries, with uniform enlargement from their origin*—Vessels feeding tumors may enlarge, but more peripherally near the tumor and not from the origin of the vessel.

3. *Large rapidly filling veins with frequent varix formation*—Early draining veins are also seen with high grade

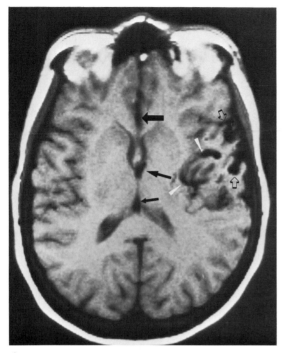

A

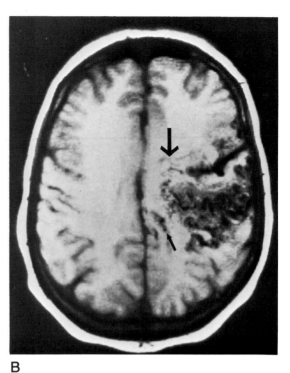

B

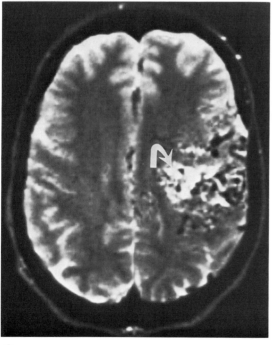

C

Fig. 4-3 AVM, MRI. (A) T$_1$-weighted MRI through the ventricular system shows the flow-void hypointensity of enlarged sylvian vessels *(white arrowheads)* and large cortical veins over the temporal lobe *(open arrows)*. The pericallosal arteries are enlarged *(large arrow)*, indicating that they also supply the malformation. Enlargement of the internal cerebral vein *(small arrows)* indicates that the malformation has deep venous drainage as well. There are no changes of hemorrhage. **(B)** Scan through the hemisphere shows the main nidus of the large parietal malformation with irregular dilated vascular channels. Enlarged deep medullary veins drain to the deep venous system *(large arrow)*. Enlarged cortical arteries in the pericallosal distribution also feed the malformation *(small arrow)*. **(C)** T$_2$-weighted MRI shows the high intensity regions of gliosis within the malformation *(arrow)*. There is no hemosiderin deposition associated with this AVM.

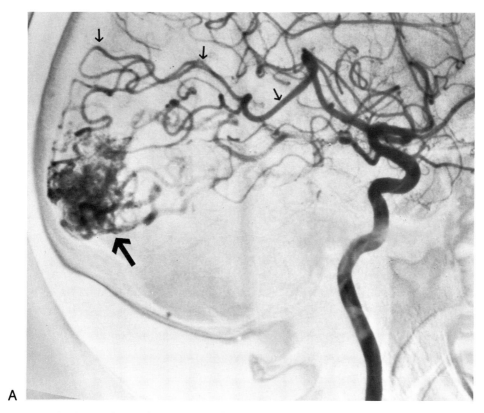

A

Fig. 4-4 AVM, multiple vessel contribution. (A) Nidus of the AVM (*large arrow*) is fed by the angular branch (*small arrows*) and posterior temporal branches of the middle cerebral artery. (*Figure continues.*)

tumors, but the veins tend to be of normal size and do not fill as early as with the AVM.

4. *Mass effect if the AVM has recently ruptured*—This effect also occurs with tumors. When the AVM is small, it may be difficult to distinguish from a tumor by angiography.

Pseudoaneurysm represents a nonclotted region of an acute hematoma that communicates with a vascular lumen. Angiography demonstrates it as a smooth collection of contrast, most often adjacent to a feeding artery, but sometimes related to a vein. It represents the site of hemorrhage from the AVM. This must be recognized prior to interventional embolization or surgery, as the pseudoaneurysm must be obliterated first to avoid rupture from change in the dynamics of the AVM. MRI

identifies the pseudoaneurysm as a swirling inhomgeneous flow void within, or at the margins of, the hematoma. Pseudoaneurysm must be differentiated from a venous varix, which may have a similar angiographic appearance, but on the venous side of the AVM. MRA is poor for evaluation of most AVMs.

DURAL ARTERIOVENOUS FISTULA

A dural arteriovenous fistula (DAVF) (dural vascular malformation) accounts for about 15 percent of all intracranial vascular malformations. It may be confined entirely to the dura and fed by dural vessels, or it may involve both the dura and the brain and be fed by both

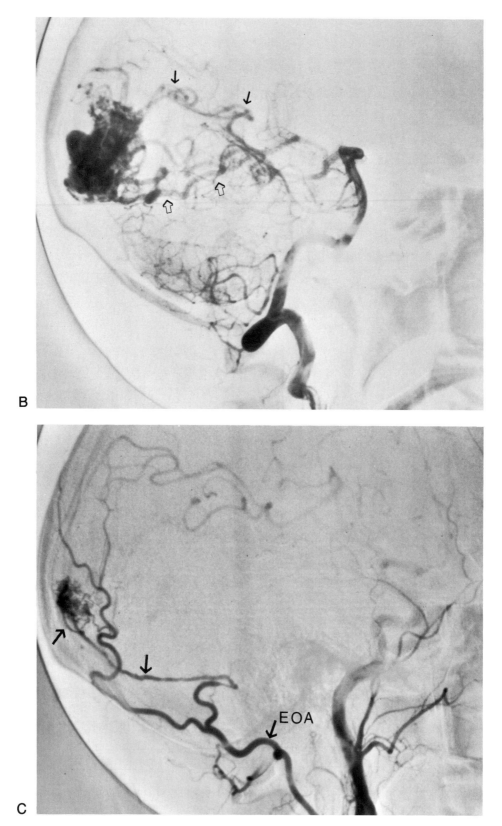

Fig. 4-4 *(Continued)*. **(B)** Vertebral angiography shows that the malformation is also supplied by the parieto-occipital branches *(arrows)* and the medial temporal branches *(open arrows)* of the posterior cerebral artery. **(C)** Selective external carotid angiography. The AVM also receives supply from a dural vessel rising from the external occipital branch of the external carotid artery *(arrows)*. EOA, external occipital artery.

Fig. 4-5 AVM, arteriography. (A) Lateral
ICA angiography in the early arterial phase
shows a large feeding middle cerebral vessel
(*small arrows*) supplying the nidus of the malformation (*open arrows*). Vessel shows uniform enlargement throughout. Extremely
early draining veins are already seen in this
early arterial phase (*large black arrows*). Note
that the nidus appears compact. **(B)** Late arterial phase shows large superficial draining
veins (*large black arrows*). A vein drains inferiorly (*small arrow*) into the cavernous sinus
(CA. SIN.) and the inferior petrosal sinus
(IPS).

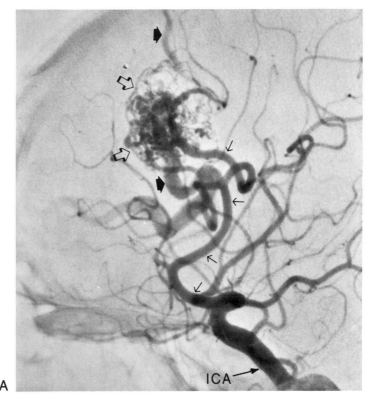

A

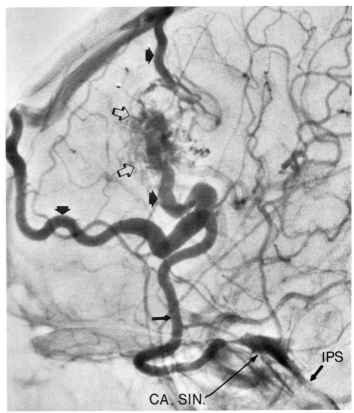

B

205

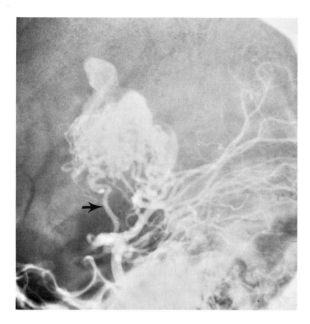

Fig. 4-6 AVM, within the thalamus. Lateral vertebral angiography shows a large vascular malformation within the thalamus. It is fed by enlarged thalamoperforate branches *(arrow)*.

dural and pial vessels. A typical DAVF is made up of a nidus of numerous small fistulas within the dural wall of a venous sinus. Dural arteries feed the fistulas, which drain directly into a dural venous sinus. Some have a direct arterial feeder into a dural sinus, a carotid-cavernous fistula being a striking example. The dural sinus is the sole drainage in most cases, but cortical veins may participate. The most common sites of dural AVMs are the lateral or sigmoid sinus of the posterior fossa (Fig. 4-7), the cavernous sinus, and the sagittal sinus, but they may involve any segment of the dura, including the tentorium. Tentorial DAVFs have a high incidence of cortical venous drainage, as well as associated infarction and hemorrhage. Selective internal ICA and external carotid artery (ECA) arteriography are usually necessary to define the vascular supply of these unusual malformations. The occipital artery of the ECA is the most common supplier. Supply from contralateral vessels is unusual. Most DAVFs are now thought to be acquired lesions, probably secondary to thrombosis or to injury to a dural venous sinus and

the subsequent development on neovascular connections.

These malformations are most frequently found incidentally. The most common DAVF of the sigmoid sinus may cause tinnitus or bruit near the mastoid. Those that are demonstrated to have prominent cortical venous drainage have a higher incidence of SAH, cerebral hemorrhage, or infarction. Subdural hematoma may also occur from their dural location. Spinal cord infarction has been seen from a spinal dural AVM. Those involving the cavernous sinus produce ophthalmoplegia, proptosis, and chemosis of the globe. Angiography is the only means for their diagnosis, as the lesions cannot be seen with CT scans or MRI.

Transarterial superselective embolization of the feeding vessel may obliterate or significantly decrease the size of the smaller DAVFs. Obliteration of the dural sinus controls the DAVF and is safely employed when the sinus is stenotic or otherwise contributes little to local cortical venous drainage. Occlusion of a dural sinus that diverts drainage retrogradely into cortical veins creates a dangerous condition with the risk of infarction or hemorrhage. Coils delivered through microcatheters placed retrogradely through the inferior petrosal sinus successfully occlude the cavernous sinus DAVF. Complications here include ophthalmoplegia from injury to the vasa nervosum and stroke from interference of cortical venopus drainage through the cavernous sinus. Coils are the preferred embolic material for larger fistulas and the sinus, but polyvinyl alcohol particles (Ivalon), silk suture wads, and cyanoacrylates are successful alternatives. Most tentorial DAVFs must be treated by surgical approach.

VEIN OF GALEN ANEURYSM

The vein of Galen aneurysm is a rare type of vascular malformation that is almost always diagnosed during the newborn period. It is comprised of direct fistulous communications of arteries with the deep veins. A large single feeding artery may be the cause. There is high flow into the vein, resulting in the aneurysmal dilatation. The high flow may cause congestive heart failure and a loud intracranial bruit. The vein of Galen aneurysm may compress the aqueduct, causing hydrocephalus (Fig. 4-8).

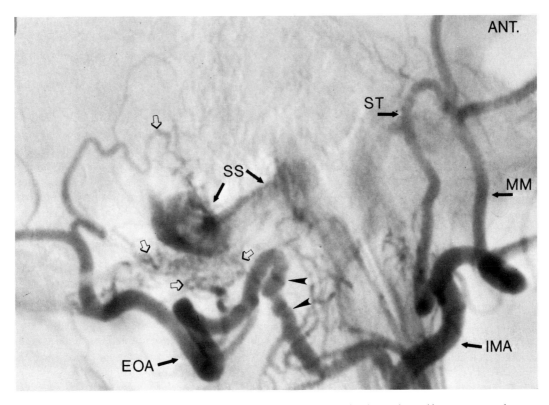

Fig. 4-7 Dural AVM. Lateral selective ECA angiography demonstrates a dural vascular malformation into the sigmoid sinus (SS). Multiple feeders supply the fistula *(open arrows)*. These feeders arise from the external occipital artery (EOA). The irregularity of the EOA most likely represents fibromuscular dysplasia *(arrowheads)*. IMA, internal maxillary artery; ST, superficial temporal artery; MM, middle meningeal artery.

Fig. 4-8 Vein of Galen aneurysm. Huge aneurysm of the vein of Galen is filled with contrast. It empties into a large, straight sinus posteriorly. This aneurysm is a result of the thalamic vascular malformation demonstrated in Figure 4-6.

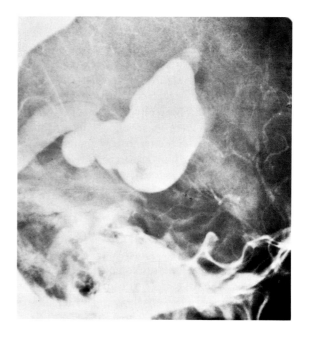

VENOUS ANGIOMA

Cerebral venous angioma is an uncommon form of vascular malformation that is composed solely of veins. On angiography, it is characterized by a group of enlarged deep medullary veins that converge to drain into a large intraparenchymal vein that usually courses through the central white matter to the cortical surface to drain into a dural sinus (Figs. 4-9 and 4-10). The veins fill at the normal time, and there is no shunting. It sometimes has a "head of Medusa" pattern (Fig. 4-9B). It occurs most often in the frontal portions of the cerebral hemispheres but may be seen in any location, including the cerebellum, brain stem, and spinal cord. Venous angioma is considered an anomalous venous drainage pattern, rather than a true angioma.

On the contrast-enhanced CT scan, the large intracerebral vein is often seen as a curvilinear density coursing through the brain to the cortical surface. There is no edema, and the lesion does not calcify. The nonenhanced examination is almost always normal. On MRI, the large anomalous vein is seen as tubular structure with flow-void (Fig. 4-9). Gadolinium contrast enhancement is prominent because of the slow venous blood flow. Rarely, some associated adjacent parenchymal intensity changes, both high and low signal, possibly representing local ischemia, gliosis, or hemosiderin deposition. The clinical significance of these changes is unknown.

The clinical presentation is varied. Most of the lesions are asymptomatic and are found incidentally at CT scanning or angiography. They have been associated, however, with hemorrhage, seizures, headaches, or ataxia, although their causal role in these processes is unknown. A large retrospective evaluation of 100 persons with venous angioma found symptoms with hemorrhage in only one. Surgical excision of a venous angioma has resulted in infarction in the territory of the venous drainage of the angioma, supporting the theory that these structures represent aberrant but otherwise normal deep veins.

CAPILLARY TELANGIECTASIS

Capillary telangiectases are composed of dilated capillaries that vary in caliber. Most commonly seen as incidental findings at autopsy, they are occasionally associated with a small intracerebral hemorrhage. These lesions are "cryptic," as they cannot be demonstrated by CT scan, MRI, or angiography.

CAVERNOUS HEMANGIOMA

The cavernous hemangioma is the rarest type of vascular malformation, but it is clinically important because it can hemorrhage or cause seizures. It is potentially totally resectable. These lesions most commonly occur in the subcortical regions of the cerebral hemispheres, but they may also lie deep within the brain, particularly near the region of the pineal. Calcification is present in 30 percent of lesions.

The CT scan demonstrates a well-circumscribed, slightly hyperdense, often calcified, lesion that shows minimal homogeneous contrast enhancement (see Fig. 6-21). There is no mass effect or surrounding edema. A hematoma may be present and, in this situation, the lesion usually cannot be recognized. The lesion has characteristics of a meningioma or lymphoma and, depending on the location of the malformation, it may be difficult to differentiate between these possibilities. Because the blood flow is so slow through the hemangioma, it is almost never detected with angiography. With MRI, the lesion is slightly hypointense on both T_1- and T_2-weighted images, but there may be circumferential changes of low intensity from hemosiderin and high T_2 signal representing gliosis and chronic hemorrhage (Fig. 4-11).

"CRYPTIC" VASCULAR MALFORMATION

A "cryptic" malformation is defined as one that cannot be observed with angiography. It is thought to be small, possibly obliterated by hemorrhage or thrombosis. The lesion may be suspected with MRI or CT scanning. The subacute and chronic changes of multiple hemorrhages are detected, and it is assumed that they are due to an underlying occult vascular malformation.

With MRI, the T_2-weighted image is the most useful, particularly with high field strength units. The lesion is seen as inhomogeneous, with a "target" configuration

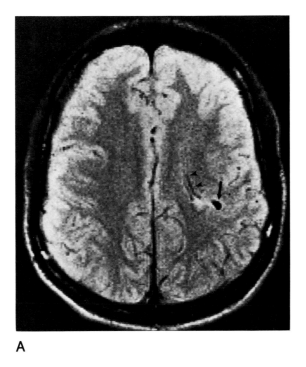

A

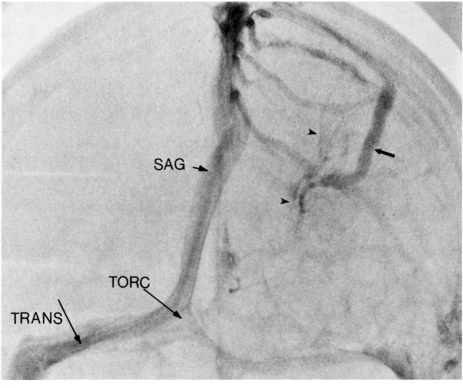

B

Fig. 4-9 Venous angioma. (A) Proton-density MRI in the transaxial plane shows the anomalous medullary veins (*arrowheads*) draining into a large vein (*arrow*), which traverses the white matter to the cortical surface. (B) AP angiography of the same patient shows the medullary veins (*arrowheads*) converging to drain into the large anomalous vein (*arrow*), which courses through the white matter to the cortical surface. SAG, posterior sagittal sinus; TORC, torcula; TRANS, transverse sinus.

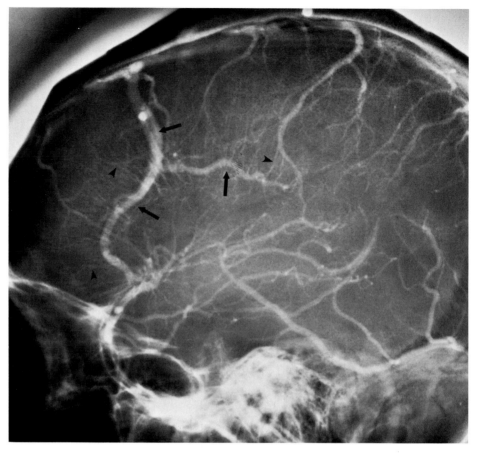

Fig. 4-10 Large venous angioma. Multiple enlarged deep medullary veins *(arrowheads)* drain into large anomalous veins coursing through the white matter to the cortical surface *(arrows)*. This "angioma" is the predominant drainage of the entire frontal lobe in this patient. Subependymal veins of the lateral ventricles are underdeveloped.

(Fig. 4-12). Characteristically, there is a peripheral rim of low signal intensity, thought to represent hemosiderin deposits within macrophages. Centrally, there is mixed high signal from chronic hemorrhage and low signal from calcium and iron. MRI is more sensitive than the CT scan in detecting cryptic AVMs. Few of the lesions described in the literature as cryptic AVMs have been histologically proved.

The CT scan generally displays cryptic malformations as foci of calcium or iron without mass effect. Contrast enhancement may be minimal, but it is inconsistent. On CT scans, the lesion cannot be differentiated from the early changes of a calcified glioma or the rare calcified infarction.

MALFORMATION SYNDROMES

Encephalofacial (Trigeminal) Angiomatosis

Encephalofacial (trigeminal) angiomatosis, or Sturge-Weber syndrome, is a congenital anomaly usually associated with a hemangioma or nevus of the face, most commonly in the distribution of the trigeminal nerve. The anomaly is a phakomatosis. Affected persons may have seizures, hemiparesis, mental retardation, glaucoma, and buphthalmos. Intracranially, a complex venous and capillary malformation of the leptomeninges (pial angioma) occurs over the cortex of a hemisphere.

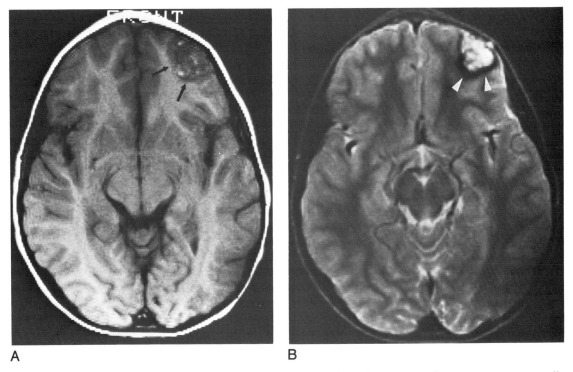

Fig. 4-11 Cavernous hemangioma, MRI. (A) T_1-weighted MRI shows the cavernous hemangioma as a generally low-intensity lesion on the cortical surface of the left frontal lobe *(arrows)*. A few focal high signal intensity regions represent chronic hemorrhage. **(B)** T_2-weighted MRI shows high signal intensity through the lesion representing gliosis. The lesion is surrounded by a rim of extreme hypointensity representing hemosiderin deposition.

The cortical veins are absent in the region of the angioma, and there are large collateral deep medullary veins in the subjacent white matter. Calcification occurs in the cortex of the brain underneath the malformation, creating a typical gyriform pattern that can be seen on skull radiography (Fig. 4-13) and CT scans. Local atrophy is present that may be severe enough to cause calvarial changes of the Dyke-Davidoff-Mason syndrome, characterized by thickening of the calvarium and unilateral volume loss of the cranial cavity. The size of the malformation over the cortex correlates with the severity of clinical symptoms and degree of mental retardation. The ipsilateral choroid plexus is enlarged. Rarely, all changes may be bilateral. MRA and x-ray angiography (XRA) demonstrate occluded arteries and veins in the region of the malformation and the enlarged deep medullary veins draining deep to the subependymal veins.

Gadolinium enhanced MRI consistently produces hyperintensity over the abnormal cortex on T_1-weighted images, most likely representing enhancement within the angioma. The CT scan shows the calcium and atrophy, but little or no contrast enhancement. GRE MRI sequences show the cortical calcifications as serpiginous signal dropout (Fig. 4-14).

Klippel-Trenaunay-Weber syndrome is a similar phakomatosis, consisting of port-wine hemangioma, venous varicosities, and a grotesque hypertrophy of bones and soft tissue, particularly the feet. The hemangiomas may be bilateral and involve the skin of any part of the body. Intracranially, there is bilateral cerebral atrophy, bilateral cortical calcification, cortical angiomas, and large choroid plexi. Glaucoma, hemimegalencephaly, and vascular anomalies at the base of the brain occur as well. Mental retardation is common.

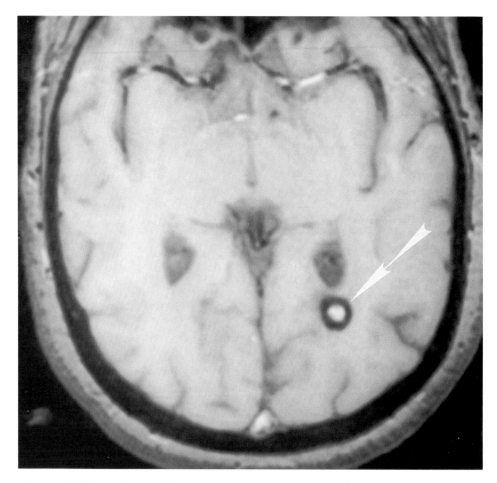

Fig. 4-12 Cryptic AVM. Axial unspoiled GRE MRI 13/200, 60° shows the characteristic circle of low signal from hemosiderin surrounding higher signal of hemorrhage found with cryptic arteriovenous malformations (*arrow*). This finding may also be seen with certain metastases, especially melanoma. Slight dilation of the atrium of the left lateral ventricle indicating minimal regional atrophy favors the diagnosis of AVM. GRE MRI technique increases the sensitivity of MRI for the detection of the magnetic susceptibility effect of hemosiderin.

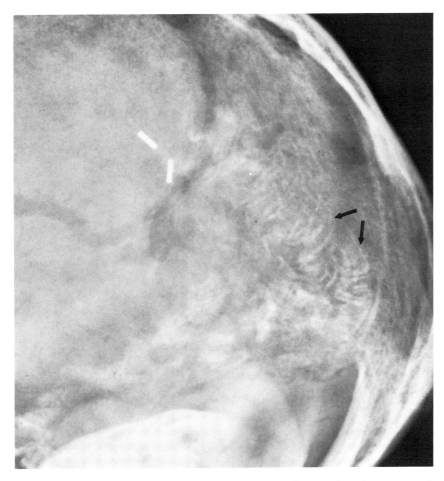

Fig. 4-13 Sturge-Weber syndrome. Plain skull radiograph demonstrates the typical gyriform pattern of cortical calcification (*arrows*) indicative of the angiomatous malformation.

Meningioangiomatosis is another, more rare, phakomatosis with findings similar to those associated with the Sturge-Weber syndrome. This hamartomatous plaque-like mass lesion over the cortex shows serpiginous calcification on CT scan. The absence of facial hemangioma, local cortical atrophy, and cortical contrast enhancement on MRI differentiates the lesion from Sturge-Weber syndrome.

Wyburn-Mason Syndrome

In its complete form, Wyburn-Mason syndrome consists of cutaneous, retinal, cerebral, and mandibular or maxillary AVMs. Any portion of the brain may be involved, but most often it is the optic pathways. The malformations may be large and appear as typical AVMs, as described above. The facial malformations may cause serious bleeding into the nasal cavities or mouth.

Sinus Pericranii

Sinus pericranii (sagittal venous varix) is an unusual benign malformation of the sagittal dural venous sinus. The skull is deficient in a portion of the midline, allowing the dural venous sinus to bulge outward. It presents as a soft midline calvarial defect with intact skin (Fig. 4-15).

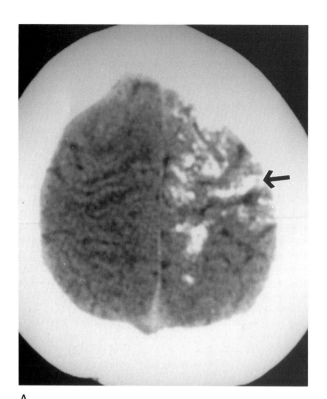

A

Fig. 4-14 **Sturge-Weber encephalotrigeminal angiomatosis.** (A) Axial nonenhanced CT scan demonstrates the typical gyral calcification just below the cortical surface beneath the angioma *(arrow)*. (B) Axial Gadolinium-enhanced T_1 MRI shows the absence of visible cortical veins in the region of the angioma *(open arrows)*. Normal cortical veins show on the right *(solid arrows)*. *(Figure continues.)*

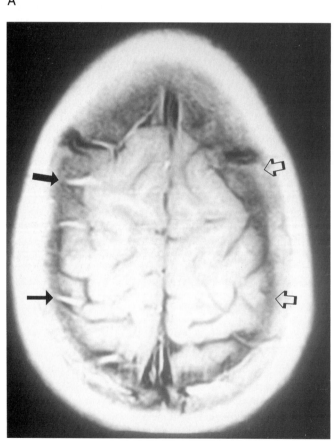

B

Fig. 4-14 *(Continued).* **(C)** Axial MRA through the level of the ventricles demonstrates dilated deep medullary veins on the left *(white arrows)* acting as collaterals. These drain to the subependymal veins of the lateral ventricle *(curved white arrow). (Figure continues.)*

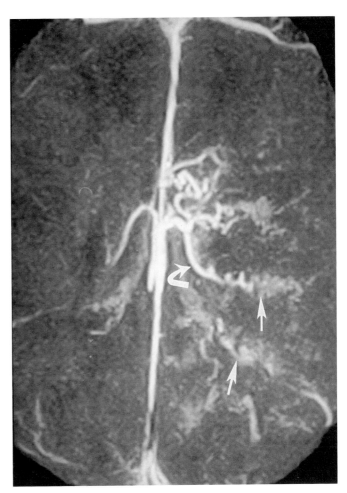

C

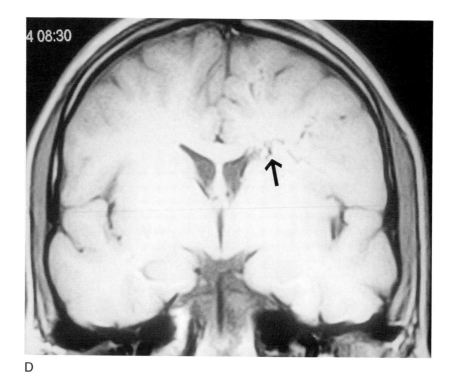

D

E

Fig. 4-14 *(Continued)*. **(D)** Coronal T_1 MRI shows dilated deep medullary veins acting as collateral conduits and draining to the subependymal veins of the lateral ventricles *(arrow)*. **(E)** Coronal T_1 Gadolinium-enhanced MRI of another patient, with Sturge-Weber syndrome showing diffuse enhancement of the subcortical white matter and gyri beneath the malformation on the cortical surface *(arrows)*. Contrast enhancement better defines the full extent of the malformation.

Fig. 4-15 Sinus pericranii. Axial contrast enhanced CT scan shows focal enhancement of the sagittal sinus *(arrow)*. This enlargement protrudes through a defect in the skull, creating a palpable soft "bump" beneath the scalp.

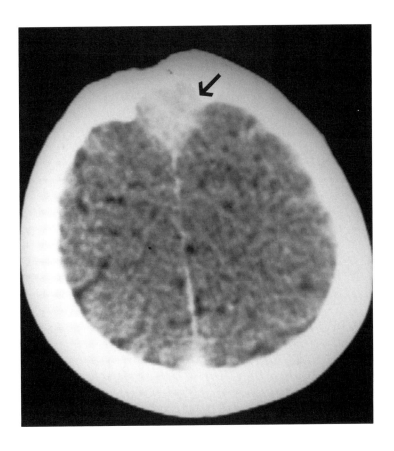

SUGGESTED READING

Atlas SW: Intracranial vascular malformations and aneurysms: current imaging applications. Radiol Clin North Am 26:821, 1988

Duckwiler G: Dural arteriovenous fistula. Neuroimag Clin North Am 2:291, 1992

Eskridge JM: Interventional neuroradiology: state of the art. Radiology 172:991, 1989

Fults D, Kelly DL Jr: Natural history of arteriovenous malformations of the brain: a clinical study. Neurosurgery 15:658, 1984

Garcia-Monaco R, Rodesch G, Alvarez H et al: Pseudoaneurysms within ruptured intracranial arteriovenous malformations: diagnosis and early endovascular management. AJNR 14:315, 1993

Garner TB, Curling OD Jr, Kelly DL Jr, Laster RW: The natural history of intracranial venous angiomas. J Neurosurg 75:715, 1991

Lemme-Plaghos L, Kucharczyk W, Brant-Zawadzki M et al: MR imaging of angiographically occult vascular malformations. AJNR 7:217, 1986

Pelz DM, Fox AJ, Vineula F et al: Preoperative embolization of brain AVMs with isobutyl-2 cyanoacrylate. AJNR 9:757, 1988

Rubinstein LJ: Tumors of the Central Nervous System. p. 241. Armed Forces Institute of Pathology, Washington, DC, 1972

Smith HJ, Strother CM, Kikuchi Y: MR imaging in the management of supratentorial intracranial AVMs. AJNR 9:225, 1988

Tien RD, Osumi A, Oakes JW et al: Meningioangiomatosis: CT and MR findings. J Comput Assist Tomogr 16:361, 1992

Viñuela F: Update on intravascular functional evaluation and therapy of intracranial arteriovenous malformations. Neuroimag Clin North Am 2:279, 1992

Vogl TJ, Stemmler J, Bergman C et al: MR and MR angiography of Sturge-Weber syndrome. AJNR 14:417, 1993

5
Cranial Tumors

Tumors of the brain and its linings account for about 1 percent of all deaths, and for about 9 percent of all neoplastic disease. The tumors are generally classified according to cytogenetic lines. The classification system presented here is a compilation of those proposed by the World Health Organization (WHO) and by Rubinstein in the Armed Forces Institute of Pathology fascicle, Tumors of the Central Nervous System. The tumor types and approximate frequency of occurrence are given in the following box. The frequency of tumor types actually observed varies according to referral pattern.

FREQUENCY OF ADULT CEREBRAL NEOPLASMS ACCORDING TO TUMOR TYPE

Neuroglial tumors (40%)
    Glioblastoma (55%)
    Astrocytoma (22%)
    Ependymoma (6%)
    Medulloblastoma (6%)
    Oligodendroglioma (5%)
    Mixed and gliosarcoma (4%)
    Others (2%)
Metastic tumor (35%)
    Lung (adenocarcinoma, small cell carcinoma) (40%)
    Breast carcinoma (40%)

(Continued)

Others (kidney, colon, prostate, melanoma) (20%)
Mesodermal tumors (10%)
    Meningioma (common)
    Sarcoma (rare)
        Hemangiopericytoma
Pituitary tumors (5%)
    Microadenoma
    Macroadenoma
Cranial nerve tumors (3%)
    Schwannoma (neurinoma): CN VIII (common), CN V (uncommon)
Lymphoreticular system tumors (2%)
    Lymphoma
    Leukemia (chloroma)
Others (5%)
    Pineal region tumors
    Craniopharyngioma
    Hemangioblastoma
    Embryonic tumors
    Neuronal tumors
        Gangliocytoma
        Ganglioglioma
        Ganglioneuroblastoma
        Neuroblastoma
    Chordoma
    Central neurocytoma
    Dermoid, epidermoid, teratoma

Table 5-1 Neoplasm Type According to Age and Location[a]

Site	Adults	Children
Cerebral hemispheres, including deep structures	**Glioma**	**Astrocytoma**
	Metastasis	**Ependymoma**
	Neuronal tumors	Mixed tumor
	Mixed tumors	Metastasis
	Lymphoma	Primitive neuronal tumor
Meninges	**Meningioma**	**Leukemia**
	Metastasis	**Metastasis**
Sella, juxtasellar area	**Pituitary adenoma**	**Craniopharyngioma**
	Meningioma	Optic glioma
	Craniopharyngioma	Hypothalamic glioma
	Hypothalamic glioma	(Hamartoma)
	Epidermoid, teratoma	
	CN V neurinoma	
	Chordoma	
	Sphenoid sinus carcinoma	
	Nasopharyngeal carcinoma	
Pineal region	**Metastasis**	**Dysgerminoma**
	Glioma, usually tectal	Pinealoma
	Subependymoma	Pineoblastoma
		Embryonal yolk sac tumor
		Teratoma
		Glioma
Intraventricular area	**Meningioma**	**Ependymoma**
	Exophytic glioma	Choroid plexus papilloma
	Colloid cyst of third ventricle	
	Ependymoma	
	Choroid plexus papilloma	
	Epidermoid	
	Metastasis	
	Xanthogranuloma	
Cerebellar hemisphere	**Metastasis**	**Astrocytoma**
	Hemangioblastoma	Medulloblastoma (> 12 years of age)
	Astrocytoma (rare)	
	Neuronal tumors	
Midline cerebellum	**Metastasis**	**Medulloblastoma**
	Meningioma of fourth ventricle	**Ependymoma**
		Astrocytoma
Brain stem	**Metastasis**	**Astrocytoma**
	Glioma	
Cerebellar pontine angle	**Acoustic neurinoma**	Epidermoid
	Meningioma	Acoustic neurinoma (NF-2)
	Epidermoid	Meningioma
	Metastasis	
Skull	**Metastasis**	Metastasis
	Chordoma	Epidermoid
	Sinus carcinoma	Histiocytosis X
	Direct invasion	Sarcoma
	Glomus tumor	
	Sarcoma	

[a] Boldface letters indicate the most common tumors.
(Adapted from Dubois, 1984, with permission.)

Nearly 95 percent of tumors are accounted for by glioma, metastasis, meningioma, pituitary adenoma, and acoustic neurinoma. All the others are uncommon. Brain tumors characteristically occur within given age ranges and have strong predilections for certain intracranial regions (Table 5-1). Although some tumors have specific imaging characteristics, it may be impossible to identify a specific tumor type correctly solely with imaging. The imaging factors used to identify specific tumors are given below.

IMAGING OF TUMORS

The main diagnostic imaging modalities for evaluation of neoplastic disease of the brain are MRI and CT scanning. MRI has the greatest sensitivity for detection of

---

FACTORS USEFUL FOR IDENTIFICATION OF BRAIN TUMORS

Location
    Site
    Surface of brain
    Gray-white junction
    White matter
Age
Imaging characteristics of the tumor
    Relative CT density
    Intensity of MRI signal
    MRI signal change with T_1 and T_2
Contrast enhancement
    Presence
    Pattern
Edema
    Pattern
    Amount
Calcification
Cystic change
Shape
Pattern of growth
Adjacent bone changes
Number of lesions

---

tumors and is the preferred initial examination. CT scanning is best for defining bone destruction or sclerosis associated with metastatic tumors, pituitary adenomas, meningiomas, acoustic neurinomas, adjacent carcinomas from the sinuses or pharynx, and glomus tumors. Radionuclide scanning has little diagnostic role here. CT also better defines calcification associated with tumors. Metabolic nuclide imaging with positron emission tomography (PET) and single photon emission computed tomography (SPECT) may have future roles in determining the true breadth of the tumor, its response to therapy, and its differentiation from similar nonmetabolically active lesions such as postradiation necrosis.

Contrast Enhancement

Intravenous contrast enhancement is basic to the evaluation of neoplastic disease with both MRI and CT scanning. Gadolinium (GD) contrast chelates are used with MRI and aqueous iodinated agents with CT scans.

Most tumors show contrast enhancement. This abnormal accumulation of contrast in the tumor results primarily from leakage of contrast into the tumor interstitium because of the absence of a blood-brain barrier within the tumor neovascularity. Contrast enhancement is seen as hyperdensity on CT scans (Fig. 5-1) and hyperintensity on T_1-weighted MRI (Fig. 5-2).

The region of contrast enhancement corresponds well with the main tumor mass. However, malignant tumor cells are commonly found beyond the enhanced portion of the tumor, particularly with glioma.

Enhancement correlates with tumor grade (degree of malignancy) only for adult glial tumors. PET scans with agents that define increased metabolic activity may be capable of defining the penumbra of infiltrating active tumor adjacent to the main mass.

Contrast enhancement indicates tumor viability. Therefore, whenever a biopsy is to be performed, it is best done in a region of the tumor that demonstrates contrast enhancement. Ring enhancement implies central necrosis, cyst formation, or abscess. Some tumor types show characteristic enhancement, discussed below. The diagnostic implications of contrast enhancement in tumors is the same for both CT scanning and MRI.

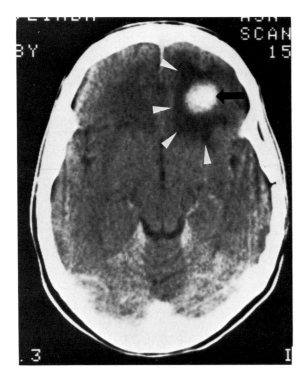

Fig. 5-1 Metastatic adenocarcinoma of the lung, CT scan with contrast enhancement. Note the typical features of a cerebral neoplasm. There is dense contrast enhancement of the tumor nidus *(black arrow)* surrounded by white matter vasogenic edema *(white arrowheads)*.

MRI

On MRI, neoplastic lesions are detected because of their long T_1 and T_2 relaxation times compared with those for brain parenchyma. This prolongation results from the increased water content of the neoplastic tissue. Therefore, the tumor is seen as low signal intensity of the T_1-weighted images and as high signal intensity on the T_2-weighted images. Similarly, MRI is particularly sensitive to the detection of surrounding vasogenic edema, which also exhibits prolonged T_1 and T_2 relaxation times and is also apparent as low intensity on T_1-weighted images and high intensity on T_2-weighted images.

The tumor nidus can be difficult to differentiate from the surrounding edema, as both regions exhibit prolonged T_1 and T_2 relaxation times. Commonly, because

there is less prolongation of T_1 and T_2 within the tumor than in the surrounding edema, on the T_1-weighted images the tumor is differentiated by its slightly greater intensity than that of the surrounding edema and on the T_2-weighted images by its slightly less intensity than the surrounding edema (Fig. 5-3).

Contrast-enhanced MRI is the most sensitive imaging technique for the detection of almost all intracranial tumors, particularly glioblastoma, metastasis, meningioma, acoustic neurinoma, and those in the posterior fossa in children. Enhanced MRI provides accurate localization of the tumor nidus.

MRI contrast enhancement is accomplished by intravenously administering Gd-chelate. The current recommended dose is 0.1 mmol/kg body weight, but newer hypo-osmolar and nonionic compounds permit doses up to 0.3 mmol/kg, the triple dose scan. High-dose enhanced MRI is most useful for defining additional lesions within the brain when one lesion is already found with single dose, and for the detection of neoplastic infiltra-

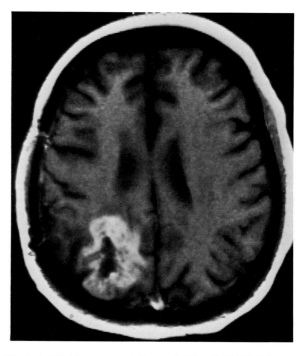

Fig. 5-2 Glioblastoma multiforme. MRI with Gd-DTPA contrast enhancement shows the striking hyperintensity of the tumor nidus on T_1-weighted images.

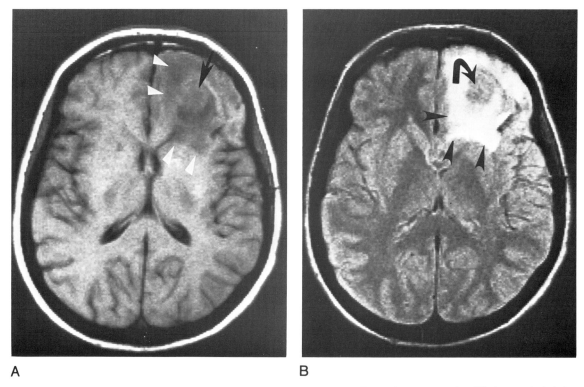

A **B**

Fig. 5-3 Metastasis, carcinoma of the lung. (A) T_1-weighted image shows the tumor nidus *(black arrow)* slightly hyperintense to the surrounding hypointense edema *(white arrowheads)*. **(B)** Lightly T_2-weighted image shows the tumor nidus *(curved black arrow)* to be relatively hypointense to the surrounding hyperintense edema *(black arrowheads)*. The CT scan for this patient is shown in Figure 5-1.

tion of the meninges and CSF (neoplastic meningitis). The post-contrast MRI scan is best performed immediately after the injection, but it may be performed up to 1 hour later. T_1-weighted technique is used to visualize contrast enhancement. It clearly distinguishes the high-intensity enhancement from the surrounding lower-intensity edema and brain tissue. Magnetization transfer contrast (MTC) background suppression may improve the conspicuity of smaller lesions or of lesions that enhance only faintly. More detail of MR contrast is given in Chapter 1.

Specific identification of tumor type is often difficult with MRI. The specific diagnosis is based primarily on the statistics of location, image pattern, clinical history, and age of the patient. Specific T_1 and T_2 relaxation rates have been of little use in defining specific tumor types.

Some helpful additional signal characteristics are discussed under specific subsections.

CT

The contrast-enhanced scan in the transaxial plane is the most common CT examination for the evaluation of tumors. A bolus or bolus-drip infusion of contrast is given using 27 to 42 g of iodine. When subtle metastatic disease is sought as part of staging, the high-dose (80 g of iodine) drip technique, followed by a 1.0- to 1.5-hour delayed scan, is the most sensitive CT technique for the detection of the lesions. Other projections, particularly the coronal view, are useful in evaluating tumors at the base of the skull (sella, planum sphenoidale) and those near the tentorium. Thin sections (less than 5 mm)

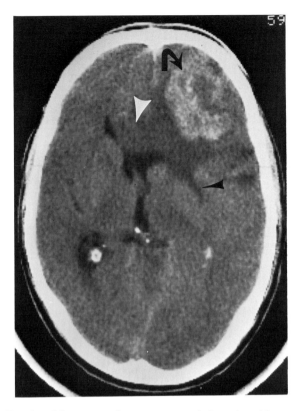

Fig. 5-4 Metastatic adenocarcinoma of the colon. Nonenhanced CT scan shows a hyperdense tumor nidus in the left frontal lobe *(curved black arrow)*. Note the extensive white matter edema extending into the external capsule *(black arrowhead)* and across the corpus callosum *(white arrowhead)*. There is subfalcial herniation and compression of the frontal horns of the ventricular system.

HYPERDENSE TUMORS, NONENHANCED CT SCANS

Adults
    Meningioma
    Metastatic tumors
        Melanoma
        Colon
        Kidney
        Choriocarcinoma
        Osteogenic sarcoma
    Lymphoma
    Colloid cyst
    Cavernous hemangioma
Children
    Medulloblastoma
    Hamartoma

---

through the posterior fossa and temporal regions reduce any artifact due to bone asymmetry.

Tumors that do not enhance are recognized (1) directly by their different absorption density compared with the surrounding brain, or (2) indirectly by peritumoral edema and mass effect. Additional differential diagnostic information is sometimes gained with a nonenhanced scan (calcium, tumor density). Nonenhanced lesions are described according to the following terminology.

Hyperdensity: Tumors that have a generally homogeneous greater absorption density compared with the surrounding brain (Fig. 5-4). This may result from fine calcification, hemorrhage, or a high nuclear-to-cytoplasmic ratio, as shown in the box.

Hypodensity: Tumors that have a lower absorption density than that of the surrounding brain. It is accounted for by the water content of the tumor tissue, cysts, necrosis, or fat (extreme hypodensity). Low-grade astrocytoma is a common example.

Isodensity: Tumors that have the same absorption density as that of the surrounding brain and therefore are at the same gray scale density as the brain. This pattern is the most common one for a tumor nidus.

Mixed density: Tumors that are inhomogeneous with regions of hyperdensity and hypodensity, resulting from a mixture of necrosis, hemorrhage, cysts, calcification, and viable tumor nidus (Fig. 5-5).

Angiography

The role of angiography in the evaluation of brain tumors has decreased considerably. The technique is now used infrequently to (1) differentiate tumor from infarction in difficult cases, (2) differentiate glioma from metastasis when a biopsy cannot be performed, (3) define vascular anatomy for large meningiomas prior to surgical removal and preoperative embolization, and (4) differentiate an aneurysm from a pituitary tumor (magnetic resonance angiography [MRA] is now the preferred method). Specific angiographic signs are discussed in the following sections.

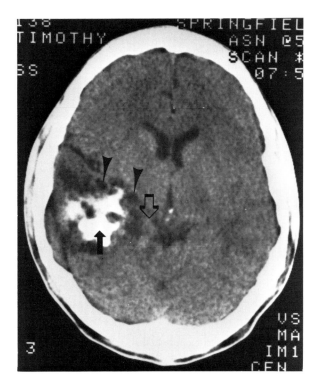

Fig. 5-5 Astrocytoma. Nonenhanced CT scan shows a mixed density tumor in the right posterior temporal lobe. Features include dense calcification *(black arrow)*, isodense solid tumor *(open black arrow)*, and low-density cysts *(black arrowheads)*. An oligodendroglioma or ganglioneuroma could have this same appearance.

Brain Edema

One of the most serious side effects of brain tumors is cerebral edema. Brain tumors produce varying amounts of edema, usually related to their speed of growth and neovascularity. The edema is of the vasogenic type; that is, it results from the absence of the blood-brain barrier in tumor neovascularity, which allows leakage of proteins and other solutes into the surrounding extracellular space of the white matter. The edema may spread along white matter tracts far away from the tumor mass, sometimes crossing the midline (Fig. 5-4). As a rule, tumors produce relatively little edema in the gray matter.

Brain edema may occur with both intracerebral and extracerebral tumors, although the extracerebral tumors (meningiomas) tend to produce less. Edema is particularly frequent and severe with metastatic carcinomas and may be excessive from a small, sometimes invisible,

focus. As edema is a primary cause of increased intracranial pressure and mass effect with tumors, it should be quantified in the report. On CT scans, edema is seen as a region of hypodensity (Fig. 5-4). On MRI, it is hypointense on T_1-weighted images and hyperintense on T_2-weighted images (Fig. 5-3). Brain edema generally decreases following glucocorticoid therapy.

Cyst Formation

Cysts occur within tumors and may be small (microcysts) or large. Large cysts are most often seen with low-grade astrocytomas and craniopharyngiomas, but they may also be seen with a pituitary adenoma, meningioma, acoustic neurinoma, and hemangioblastoma. Microcysts are most commonly seen in the more malignant primary gliomas but imply a better prognosis for the malignant tumor. Cystic cavities may result from necrosis or hemorrhage, and they may be difficult to differentiate from true cysts.

Cysts may contain nearly pure water or considerable protein or other debris, often from prior hemorrhage. Most large cysts have homogeneous contents, whereas solid tumor tends to be more inhomogeneous. Some cysts and solid tumors will have similar imaging characteristics and cannot be differentiated. A fluid level caused by the heavier, dependent proteinaceous material is diagnostic of a cyst when seen on CT scans or MRI.

On MRI, the signal characteristics of cysts are determined predominantly by the characteristics of the contained water. The water may be "pure" and free of protein, and therefore "unbound." In this state, the cyst fluid has the same signal characteristics as CSF. With increased protein content, protons become "bound" in a hydration layer adjacent to the protein, significantly decreasing the T_1 relaxation time of the water solution. The net effect is to increase the signal intensity of the cyst on both the T_1- and T_2-weighted images; thus, the contents of the cyst appear brighter than the CSF on both sequences (see Fig. 5-9). When the cyst is extremely proteinaceous (greater than 25 g/dl), there is a paradoxic decrease in the MRI signal intensity, presumably from the extreme viscosity of the cyst contents. In general, the most benign tumors have cyst fluid whose intensity approximates that of CSF. MRI is the most accurate imaging study used to differentiate a cyst from a solid tumor. Intraoperative ultrasonography may also be used.

On CT scans, cysts appear as a low-density area, resembling CSF. Higher protein density is reflected in greater CT density of the cyst (compared with CSF) and may appear the same as solid tumor. Because of decreased contrast resolution, the CT scan is not as accurate as MRI for defining the contents of a cyst or for differentiating a cyst from a solid or necrotic tumor (see Fig. 5-9).

Hemorrhage

A large hemorrhage from a tumor is unusual; when it does occur, it is generally impossible to identify a tumor immediately as the cause. Only later does the tumor become apparent on MRI or contrast-enhanced CT scans. Large hemorrhages occur most commonly within metastases and malignant hypervascular tumors (glioblastoma), but they may occur within any tumor, including meningioma. Small hemorrhages are more frequent within tumors, most commonly with metastasis from melanoma, hypernephroma, and choriocarcinoma. These are seen as a slight hyperdensity of the tumor on CT scans. Because the changes associated with hemorrhage are usually chronic, on MRI there is focal hyperintensity within the tumor on T_1- and T_2-weighted images. Late paramagnetic effects of hemosiderin sometimes produce focal low signal within or around hemorrhagic sites on T_2-weighted spin echo (SE) or gradient echo (GE) images.

Calcification

Calcification occurs within many tumors and has diagnostic significance (see box below). It may be punctate or diffuse and is best seen as high density with CT scanning. Calcium produces signal void on MRI, which is more difficult to identify by that technique. GRE images are somewhat more sensitive to calcium and may define the deposits when SE MRI cannot. Calcium cannot be differentiated from hemosiderin by MRI. Calcification frequently occurs within tumors after radiotherapy, and this must not be mistaken for contrast enhancement on CT scans.

Metastasis of Primary CNS Tumors

Metastasis of primary intracerebral tumors is relatively unusual. The most frequent mode of spread is through the CSF to the leptomeninges or into the ventricular system (see Fig. 5-17C). Tumor contact with the surface of the brain or ependyma is necessary for spread through the CSF. Medulloblastoma is the tumor most likely to

---

TUMORS THAT COMMONLY CALCIFY

Meningioma
Craniopharyngioma
Oligodendroglioma
Astrocytoma
Ependymoma
Choroid plexus papilloma
Ganglioglioma
Dysgerminoma (pineal)
Chordoma
Central neurocytoma
All tumors after irradiation
Pleomorphic xanthoastrocytoma
Meningioangiomatosis

---

spread in this fashion, although it can occur with other types, particularly ependymoma, germinoma, neuroblastoma, glioblastoma, and hemangioblastoma.

Distant somatic metastasis occurs in rare cases by hematogenous spread. Metastasis outside the CNS almost always occurs after a surgical procedure that allows the tumor access to the vascular system or other body cavities. Glioma is the most likely to metastasize in this fashion, being most often found in the regional lymph nodes or lungs. Medulloblastoma most frequently goes to the bones. Tumors may metastasize through shunt catheters into distant body cavities.

SPECIFIC TUMORS

The specific cranial tumors discussed here are organized into broad anatomic categories. For adults, the anatomy has been divided into the supratentorial region (cerebrum, meninges, calvarium), the sella and base of the skull, and the posterior fossa. Tumors more common to children and adolescents are discussed in Chapter 6.

Supratentorial Tumors

TUMORS OF NEUROGLIAL ORIGIN

Glial tumors are by far the most common primary brain neoplasms. Current concepts ascribe the origin of glial tumors to the three types of neuroglial cell normally

Table 5-2 Glial Tumors, Grading[a] and CT/MRI Characteristics

Histology	Contrast Enhance	Edema	Angiography
"Benign"			
Grade I (astrocytoma)	No	No	Avascular mass
Grade II (astroblastoma)	No	No	Avascular mass
"Malignant"			
Grade III (anaplastic astrocytoma)	Usually (80%)	Yes	Avascular mass or sometimes "blush"
Grade IV (glioblastoma)	Yes	Yes	Highly vascular mass

[a] Grading according to Kernohan and Sayre (1952). Corresponding classification according to Rubinstein (1972) is shown in parentheses.

found in the brain and spinal cord: astrocytes (the most common origin for glioma), oligodendrocytes, and ependymal cells. These cells constitute the supporting stroma of the brain. The medulloblastoma, generally considered a glial tumor, arises from a primitive cell with potential to differentiate along multiple cell lines.

Grading of glial tumors arising from the fibrillary astrocyte (accounting for 90 percent of gliomas) is somewhat confused because of the inherent inhomogeneity of the tumor and the use of many nosologic systems (Ringertz, 1950; Kernohan, 1952; Rubinstein, 1972). All are based on defining nuclear atypia, mitosis, vascular proliferation, and necrosis. The more within each of these categories, the higher the grade and the more malignant the tumor. Table 5-2 gives a simplified classification that compares the two commonly used systems. The Ringertz system, which with modification is used by the WHO, is given below (see box).

The histologic determination of malignancy is made by quantifying the mitotic figures, necrosis, and vascular proliferation. Tumor grade is assigned according to the most malignant portion of the tumor. Glial tumors are notoriously pleomorphic, and grading errors can occur with biopsy if only small portions of the tumor are sampled. A glial tumor may also change with time, so a once benign-appearing tumor may change into a more malig-

nant variety. When there are nearly equal numbers of two cell lines (i.e., glioma and fibrosarcoma), the tumor is called gliosarcoma. Lesser amounts of mixing of cell types are called "mixed" tumors. Glial tumors rarely demonstrate a pure cell line.

In general, the higher the malignant grade of glioma, the greater the amount of abnormal vascular proliferation within the tumor nidus. This situation results in greater contrast enhancement and greater surrounding vasogenic edema. The neovascularity can be seen with angiography. As a rule, grade III and IV gliomas demonstrate abnormal contrast enhancement, surrounding vasogenic edema, and neovascularity on angiography. Grade I and II tumors show no enhancement, little edema, and no neovascularity on angiography (Table 5-2).

The grade of astrocytoma and the age of the patient are closely correlated. As a general rule in adults, well-differentiated astrocytoma occurs at age below 40 years, anaplastic astrocytoma at age below 50 years, and glioblastoma at age above 50 years. Age is a powerful predictor of astrocytic tumor grade, with little crossover.

Glial tumors most commonly arise within the white matter of the cerebral hemispheres but may be deep within the internal capsule, thalamus, or brain stem. In rare cases, glial tumors arise within the gray matter or the subpial space and reach the cortical surface. Characteristically, glial tumors occur within the cerebral hemispheres in adults and in the cerebellum, pons, hypothalamus, and optic chiasm in children. These tumors tend to be varied in shape and often resemble the structures they replace. Although the contrast-enhanced margin may appear relatively sharp, there is always a gradual histologic transition from tumor to normal brain. Even the most benign gliomas do not have a true tumor capsule, although some have a pseudocapsule of compressed gliotic brain. Concurrent malignant degeneration may

---

RINGERTZ CLASSIFICATION OF ASTROCYTIC NEOPLASMS

Well-differentiated astrocytoma

Malignant, anaplastic astrocytoma

Glioblastoma multiform

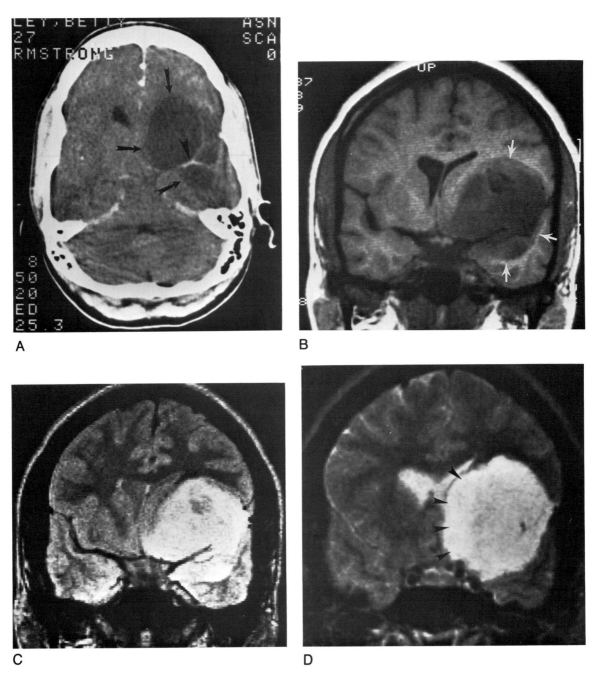

Fig. 5-6 Astrocytoma, grade I. (A) Contrast-enhanced CT study shows a hypodense mass in the left posterior frontal and anterior temporal regions *(arrows)*. The density of the tumor is greater than that of the CSF. There is no contrast enhancement. The mass surrounds the left middle cerebral artery *(arrowhead)*. **(B)** Coronal T_1-weighted image shows the tumor mass in the left posterior frontal lobe extending into the superior medial portion of the temporal lobe *(arrows)*. The mass had greater intensity than CSF and is inhomogeneous, indicating that it is solid rather than cystic. **(C)** Lightly weighted T_2-weighted image shows the mass to be inhomogeneous and of greater signal intensity than CSF, which is typical for solid astrocytomas. **(D)** Heavily T_2-weighted image shows the mass to be nearly isointense with the CSF. There is slightly greater intensity along the medial margin of the tumor *(arrowheads)*. *(Figure continues.)*

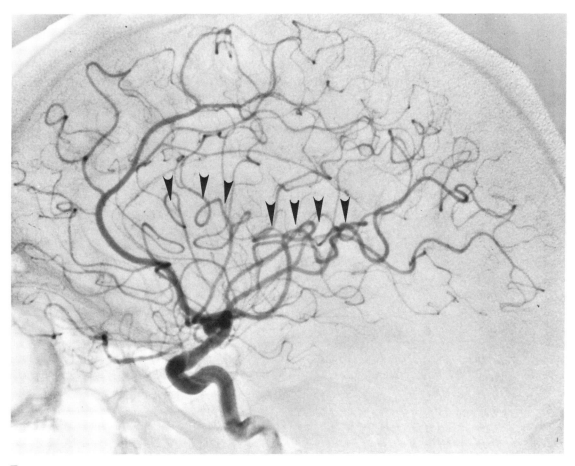

E

Fig. 5-6 *(Continued)*. **(E)** Left internal carotid artery angiography shows a subtle avascular mass distorting the anterior sylvian vessels *(arrowheads)*. The posterior frontal cortical branches are stretched.

take place at distant sites within the brain, creating multicentric gliomas. The most extreme example of this pattern of growth is gliomatosis cerebri, in which nearly the entire brain exhibits low-grade neoplastic transformation. Delayed development of other gliomas may occur after removal of the initial tumor. This pattern of growth has grave implications for the treatment of glioma and explains the seemingly inevitable recurrence of the tumor after resection.

Astrocytoma (Grades I and II
Well-Differentiated Astrocytoma)
In the adult, low-grade astrocytoma (well-differentiated astrocytoma) is nearly always a solid tumor of the cerebral hemispheres. Cystic astrocytomas are much less common. When discovered, the tumors may be of any

size and superficial or deep. Diffuse growth may occur within the white matter, especially extending from the frontal into the temporal lobe, a characteristic location. Little or no surrounding edema is present, although the tumor itself has a markedly increased water content when compared with normal brain tissue. The mass effect may be slight and is a direct result of the tumor itself. Neovascularity is minimal (Fig. 5-6).

Calcification may occur within the tumor in either small flecks or dense hunks (Fig. 5-5). Occasionally, these inclusions are the only indication that a tumor is present. Similar calcification occurs within oligodendrogliomas and gangliogliomas. Long-term serial examination at yearly intervals may be necessary to distinguish the calcification of tumor from focal benign idiopathic calcification, sometimes called a *brain stone* (Fig. 5-7).

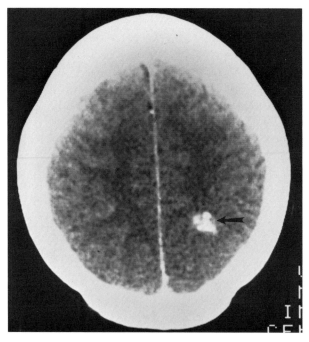

Fig. 5-7 **"Brain stone."** CT scan shows a calcific lesion in the left parietal subcortical region *(arrow)*. It is impossible to determine the nature of such a lesion. However, this finding could be the only sign of a low-grade glioma, and serial scanning is necessary to detect possible change.

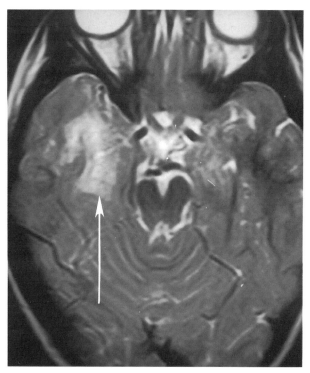

Fig. 5-8 **Astrocytoma, low grade.** Axial FSE (4'4) 96/3500 T_2 MRI shows homogeneous high signal intensity within the anteromedial right temporal lobe white matter *(arrow)* of a young adult—a typical finding of low-grade astrocytoma. Contrast enhancement does not occur. Herpes simplex encephalitis may also produce this picture but tends to exhibit Gd enhancement, and sometimes hemorrhage.

Because of increased water content, low-grade astrocytomas exhibit relatively low intensity on T_1- and high intensity on T_2-weighted MRI. Most often the tumor is well circumscribed, but it may be poorly marginated. Most solid tumors demonstrate some internal inhomogeneity (Figs. 5-6 and 5-8). Cystic tumors are homogeneous, with MRI signal characteristics of pure or proteinaceous water (see section above). In some cases, it is impossible to differentiate a cystic tumor from a solid tumor. Low-grade astrocytomas do not enhance with Gd. Calcifications may be seen as regions of signal dropout.

With CT scanning, low-grade astrocytomas appear as well-circumscribed regions of hypodensity (18 to 24 HU) (Fig. 5-6). Calcification occurs in 10 percent of tumors and may have any form. Contrast uptake does not occur with grade I and II astrocytoma, with rare exceptions. However, the lack of tumor enhancement is no guarantee of low-grade histology, as 20 percent of anaplastic (grade III) astrocytomas show no enhancement either. Sometimes there is increased density after con-

trast in the compressed brain tissue or gyri adjacent to the tumor, but this increased density must not be misinterpreted as abnormal enhancement of the tumor itself.

At best, angiography demonstrates a nonspecific mass effect, and it may show no abnormality (Fig. 5-6). Low-grade astrocytoma does not demonstrate neovascularity or early draining veins. Angiography has little role in the evaluation of low-grade glioma, except perhaps in difficult cases, in which it may differentiate tumor from infarction with vascular occlusion.

Grade III Anaplastic Astrocytoma
Anaplastic astrocytoma contains any number of cells displaying more malignant change but falls short of the high-grade malignancy of glioblastoma. It exhibits mitosis, necrosis, vascular proliferation, and cyst formation,

particularly microcysts. Significant edema is usually present around the tumor. These tumors are inhomogeneous, with regions of both low-grade and high-grade change. Clinically, the tumors are moderately aggressive, but not as much as the glioblastoma.

On MRI, the tumor shows hypointensity on T_1-weighted images and hyperintensity on T_2-weighted images, but usually with considerable inhomogeneity. Surrounding edema is usually present. Gd enhancement occurs in a large number of the tumors and is indicative of anaplasia (Fig. 5-9). The patterns are similar to those observed by CT scanning (see below). It is usually not possible to identify microcystic change and necrosis specifically within the tumor. Large cysts may be seen (Fig. 5-10).

As expected from the histology, these tumors also have a variable CT appearance. Solid tumors may be hypodense, isodense, or hyperdense. Inhomogeneity is common. Sharply outlined cysts occur that have a CT density of about 12 to 25 HU. Calcification is rare. It may be impossible to differentiate tumor from edema. About 80 percent of anaplastic astrocytomas demonstrate contrast enhancement. Some tumors have such a small region of anaplasia that contrast uptake is nondetectable, and the tumor appears identical to a low grade astrocytoma. The pattern of contrast uptake varies considerably and may show any of the following patterns.

1. Small blotch of enhancement at the periphery of a cyst or within the solid portion of the tumor (Fig. 5-9)
2. Fine rim enhancement of a cyst with or without an enhancing tumor nodule (Fig. 5-10)
3. Several somewhat thick ring-enhancing structures
4. Homogeneous enhancement throughout the entire solid portion of the tumor
5. Dense, thick, irregular enhancement, with a ring-like pattern indistinguishable from that of glioblastoma

In glial tumors, the slightest amount of contrast enhancement is enough to make the diagnosis of anaplasia (malignancy) likely.

On angiography, a mass lesion can be seen that may be avascular or show a small vague tumor "blush." Discrete neovascularity is generally not seen (see Fig. 1-13).

Grade IV Glioblastoma

Glioblastoma is the most common primary tumor of the brain. It accounts for up to 20 percent of all intracranial tumors and for 50 percent of glial tumors. The tumor is

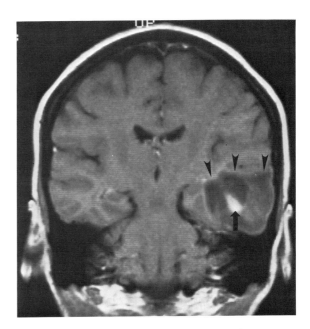

Fig. 5-9 Anaplastic astrocytoma, grade III. Gd-DTPA contrast-enhanced T_1-weighted MRI shows an inhomogeneous low-intensity tumor mass in the left temporal lobe (*arrowheads*). There is a focal region of enhancement within the tumor (*arrow*). Note that the tumor extends to the cortical surface.

highly anaplastic and grows in an infiltrative and destructive manner. It induces striking neovascularity. Central necrosis is an almost constant feature. Multiplicity of tumor sites is present in about 5 percent of cases. Left untreated, the usual survival is 3 to 5 months after diagnosis. With surgical decompression followed by radiation treatment, the survival may be extended 6 to 12 months. The tumor is almost always fatal.

The tumor is most commonly found in the cerebral hemispheres but may also be within the deep structures and the brain stem. Cerebellar glioblastoma is rare. It typically grows as an irregularly shaped mass in the white matter, but it may reach the cortical surface, where it invades surface arteries and produces infarction. The tumor frequently involves the corpus callosum, crossing the midline and producing the characteristic "butterfly" pattern. White matter edema occurs with more than 90 percent of tumors, and is extensive in 75 percent. The edema significantly adds to the mass effect produced by the tumor. It may be diminished by systemic glucocorticoid therapy.

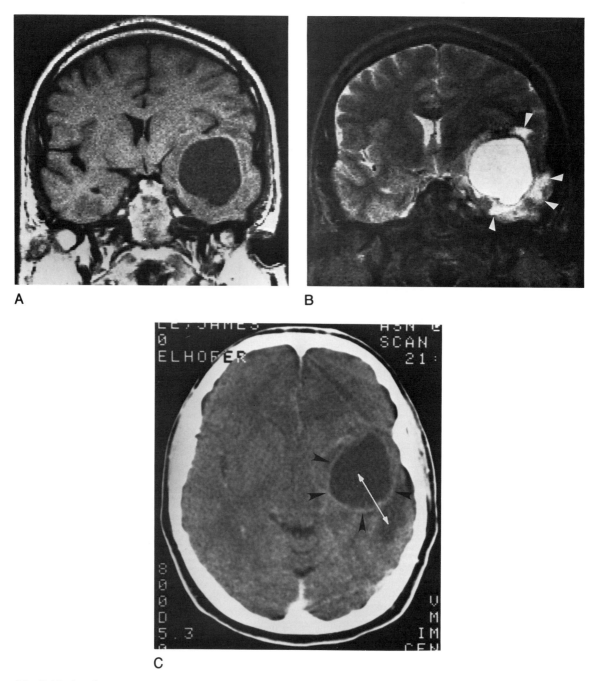

Fig. 5-10 Anaplastic astrocytoma, grade III, cystic. (A) Coronal T_1-weighted MRI shows the well-circumscribed left temporal lobe mass. It is uniformly hypointense but of slightly greater intensity than the CSF. This picture is typical of a cyst with proteinaceous fluid. **(B)** T_2-weighted MRI showing a hyperintense cystic mass in the left temporal lobe. Note that the cyst fluid has higher signal intensity than the surrounding edema *(arrowheads)*. **(C)** Contrast-enhanced CT scan shows enhancement of the cyst capsule *(arrowheads)*. Note that the cyst fluid is of lower density than the edema fluid *(opposing arrows)*. For an angiogram of a temporal lobe mass, see Figure 1-44.

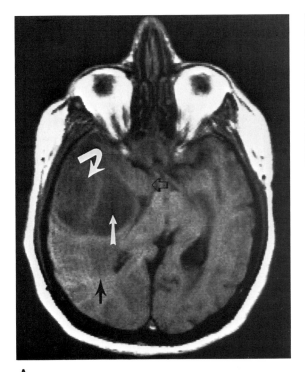

A

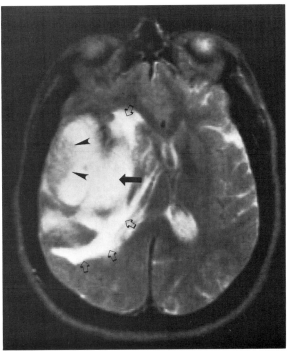

B

Fig. 5-11 Glioblastoma, grade IV. (A) T_1-weighted MRI shows an inhomogeneous hypointense mass in the right temporal lobe. The more medial portion of the mass (*white arrow*) is of lower intensity than the lateral portion (*curved arrow*), indicating greater necrosis. Surrounding edema is seen as slight hypointensity (*black arrow*). Note the uncal herniation (*open arrow*). **(B)** T_2-weighted MRI showing the medial portion of the tumor as high intensity (*arrow*) and a lower intensity, more solid portion laterally (*arrowheads*). The surrounding white matter edema is of greater intensity than the necrotic portion of the tumor (*open arrows*). **(C)** Contrast-enhanced CT scan shows the typical irregular ring-like enhancing mass with central low-density necrosis (*arrowheads*). Edema is seen as low density in the white matter surrounding the contrast-enhanced tumor nidus (*curved arrows*).

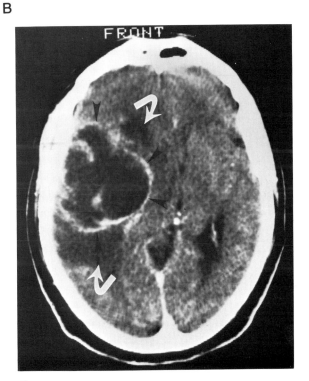

C

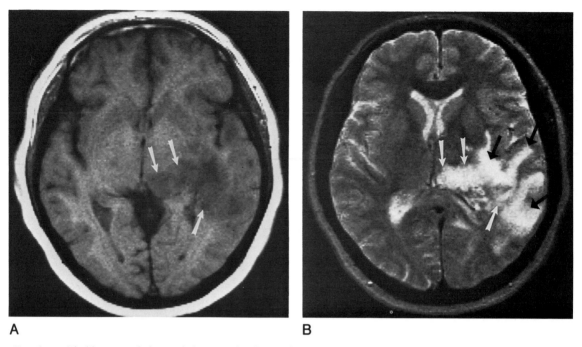

A   B

Fig. 5-12 Glioblastoma, thalamic. (A) T_1-weighted MRI shows a hypointense lesion within the posterior left thalamus and appearing to extend into the medial left temporal lobe (*arrows*). **(B)** T_2-weighted image shows the tumor in the thalamus to be of high intensity (*white arrows*) but slightly less intense than the surrounding edema (*black arrows*).

The diagnosis is based on the pattern of contrast enhancement, the shape of the lesion, its location within the white matter (especially the corpus callosum), and the presence of central necrosis.

On MRI, the tumor shows varying degrees of hypointensity on T_1- and hyperintensity on T_2-weighted images (Fig. 5-11). There is inhomogeneity of the tumor mass, representing necrosis, cystic change, hemorrhage, neovascularity. The high T_2 signal of vasogenic edema is striking and extends far beyond the margin of the main tumor mass. On the T_2 images, the tumor mass is usually slightly less intense than the surrounding edema (Fig. 5-12). Focal low intensity on T_2-weighted images usually represents acute hemorrhage but may rarely be due to calcium (Fig. 5-13).

Gd enhancement is almost always present. It is usually ring-like, with thick, irregular walls that have a garland-like inner pattern (Fig. 5-2). The central nonenhancing region represents necrosis or less commonly cyst formation. The enhancement clearly outlines the main tumor mass, distinguishing it from low-intensity surrounding edema on T_1-weighted images. Contrast enhancement is often necessary to help differentiate glioblastoma from other tumors, abscess, and infarction.

On CT scans, contrast uptake in demonstrated in more than 95 percent of glioblastomas, and the use of contrast is essential for the evaluation of this tumor (Fig. 5-11). With early tumors, the enhancement is minimal and nonspecific (Fig. 5-14), or even absent. The contrast outlines the region of neovascularity of the tumor and accurately indicates the size of the main tumor mass. However, the tumor is always beyond the perimeter of the enhancement. Similar to Gd-enhanced MRI, glioblastomas almost always appear as irregular ring-enhancing lesions with a variable shape and ring thickness. Homogeneous tumor enhancement is rare. The central low-density nonenhancing segment of the tumor represents necrosis in 95 percent of cases. It may fill in somewhat with contrast over time. Rarely, contrast accumulates in the dependent portion of a cyst. Calcification is rare.

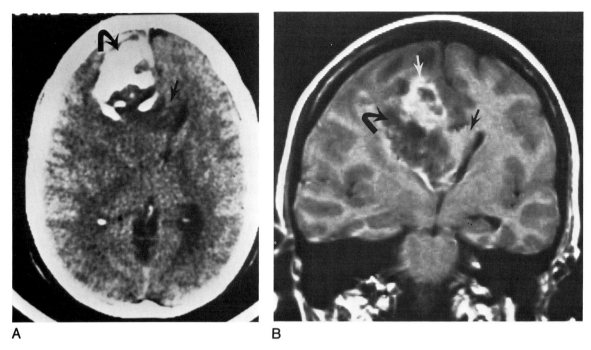

Fig. 5-13 Glioblastoma, with calcium. (A) Nonenhanced CT scan shows the tumor mass in the right frontal lobe with a large amount of dense calcification *(curved arrow)*. Noncalcified hypodense tumor is seen crossing the corpus callosum *(arrow)*. **(B)** T₁-weighted MRI with Gd-DTPA shows the typical enhancement pattern of a glioblastoma *(white arrow)*. More homogeneous enhancement is seen in the portion of the tumor that crosses the corpus callosum *(black arrow)*. The calcium is seen as regions of signal loss *(curved arrow)*. This growth pattern is typical for glioblastoma. The presence of calcium suggests that originally this tumor was a predominantly lower grade astrocytoma or oligodendroglioma. The tumor is invading the ventricular cavity.

On angiography, a typical vascular pattern is demonstrated in about 50 percent of cases (Fig. 5-15). When present, it is accurate for the diagnosis, differentiating glioma from metastasis or abscess. Therefore, angiography is still useful in confirming the diagnosis of glioblastoma when biopsy cannot be done. The typical diagnostic pattern is as follows:

1. Abundant irregular neovascularity, with sites of dilation and narrowing of the vessels, contribution from multiple cortical branches, and rarely meningeal contribution

2. Large early-appearing draining veins, representing arteriovenous shunting

3. Drainage by way of enlarged deep medullary veins to the subependymal veins

Drainage by deep medullary veins reflects the white matter location of the tumor. The deep veins of the white matter can be identified by their characteristic straight radial course toward the subependymal veins of the lateral ventricles. Under normal circumstances, these veins are seen late in the venous phase as fine vessels, visible no more than 1 cm into the paraventricular white matter. When enlarged, they provide strong evidence of glioblastoma. These veins may also be enlarged with other white matter pathology, such as lymphoma, multifocal leukoencephalopathy, fulminant multiple sclerosis, and deep arteriovenous malformations (AVMs). However, the clinical setting and the type of neovascularity usually permit distinction of these other entities. Deep medullary vein enlargement is not seen with metastasis, an important differential angiographic finding.

The combination of the age of the patient and the characteristic findings on MRI, CT scans, and angiography, if necessary, is essentially pathognomonic of the diagnosis of glioblastoma multiforme.

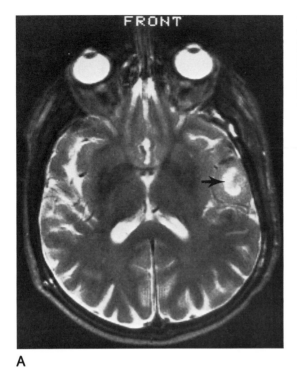

A

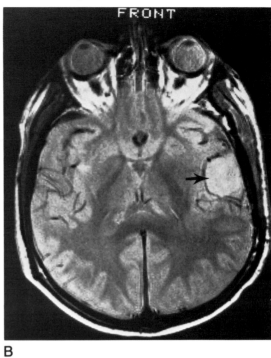

B

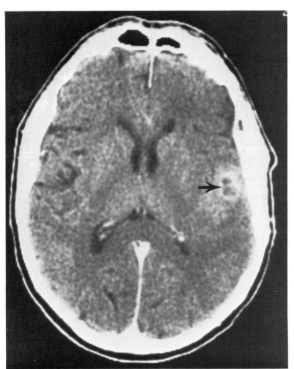

C

Fig. 5-14 Early glioblastoma. (A) T_2-weighted MRI shows a signal intensity tumor nidus in the left temporal lobe. **(B)** Lightly T_2-weighted image shows a nearly homogeneous region of high signal intensity in the left middle temporal lobe *(arrow)*. **(C)** Contrast-enhanced CT scan shows a small ring-enhancing tumor nidus in the left temporal subcortical region *(arrow)*.

Gliomatosis Cerebri

Gliomatosis cerebri is the term applied to the rare diffuse infiltration of the cerebral hemispheres with undifferentiated astrocytic cells. One or both hemispheres are involved, along with the corpus callosum. The tumor commonly infiltrates the optic chiasm, the brain stem, and the cerebellum. The affected tissue is expanded, and the ventricles, sulci, and cisterns are compressed. CT scan usually shows an isodense expansion of one entire hemisphere, with little or no contrast enhancement. MRI shows a uniform high T_2 signal within the involved white matter, and variable contrast enhancement within the tumor or gyri (Fig. 5-16). The diagnosis can be made only by biopsy. Clinical signs and symptoms reflect the diffuse process.

Differential Diagnosis

Metastasis. It is important to differentiate metastasis from glioblastoma. Occasionally, on MRI or CT scanning, a metastatic tumor has a configuration and enhancement pattern similar to that of the classic pattern of glioblastoma. The differential points are given below. It is important to emphasize that it may be impossible to differentiate a glioblastoma from a metastasis by imaging, and biopsy may be required (Fig. 5-17).

Brain Abscess. The pattern of a brain abscess may mimic that of a glioblastoma. The diagnosis of abscess must be strongly considered when there is a relatively thin, regular rim of contrast enhancement around a central cavity, although a glial tumor occasionally has this

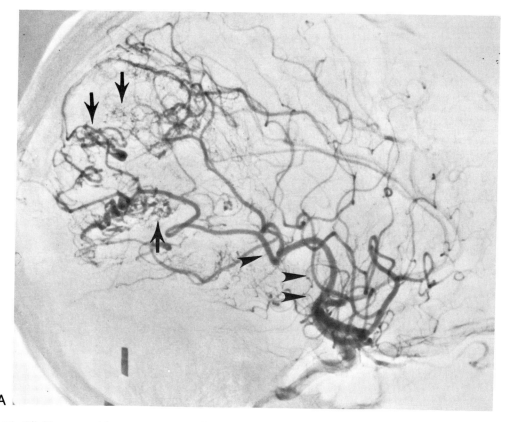

A

Fig. 5-15 Glioblastoma, right posterior parietal lobe. (A) Internal carotid artery angiography shows a large right posterior parietal mass with irregular neovascularity *(arrows)*. The sylvian vessels are displaced forward *(arrowheads)*. *(Figure continues.)*

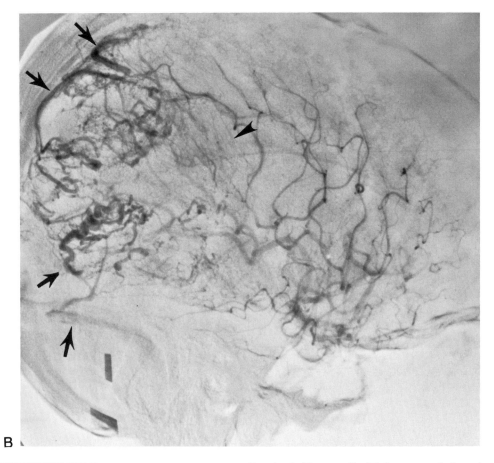

B

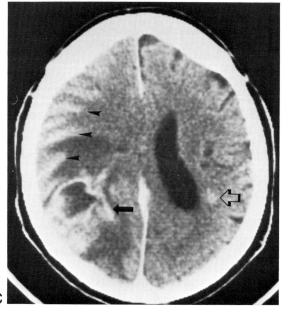

C

Fig. 5-15 *(Continued).* **(B)** Late arterial phase shows multiple early cortical veins draining into sinuses *(arrows)*. Small straight medullary arteries and veins are seen *(arrowhead)*. **(C)** Contrast-enhanced CT scan shows the typical irregular enhancing tumor nidus in the right posterior parietal lobe *(black arrow)*. There is surrounding edema and "pseudoenhancement" of compressed gyri *(arrowheads)*. A second glioma shows early development in the left hemisphere *(open arrow)*.

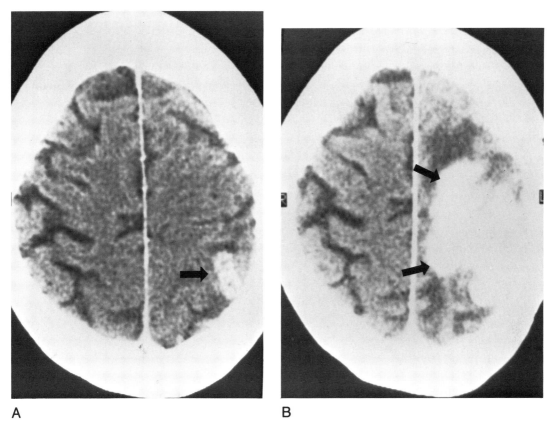

A   B

Fig. 5-16 Early peripheral glioblastoma. (A) Small cortical enhancing lesion is present in the left posterior parietal lobe. At this stage it is indistinguishable from the gyral enhancement of infarction *(arrow)*. **(B)** CT scan 6 weeks later shows gross enhancement of the tumor *(arrows)*.

pattern (see Ch. 8). An abscess may also have a characteristic capsular blush on angiography. A thick enhancing wall with the inward-curving garland pattern is generally not present with abscess.

Infarction. Rarely, a peripheral glioblastoma is examined during its early stage of development. There may be only edema of the gray matter or a small ring or gyral enhancement on MRI or CT scans (Fig. 5-18). It may resemble infarction. Only the history or repeat scanning in a few weeks can ensure the correct diagnosis.

ADDITIONAL TUMORS OF NEUROGLIAL ORIGIN

Oligodendroglioma

Oligodendroglioma is a rare tumor that accounts for about 5 percent of intracranial gliomas. Its peak incidence is in persons 30 to 55 years of age. These tumors

---

DIFFERENTIAL DIAGNOSIS OF SOLITARY RING ENHANCING BRAIN LESIONS

Metastasis
    Clinical findings
        Known primary malignancy that frequently metastasizes to the brain
    CT/MRI findings
        Multiple enhancing lesions of varying size; may be solitary in 50 percent
        At corticomedullary junction
        Excessive surrounding edema
    Angiography
        Drainage to single cortical vein

(Continued)

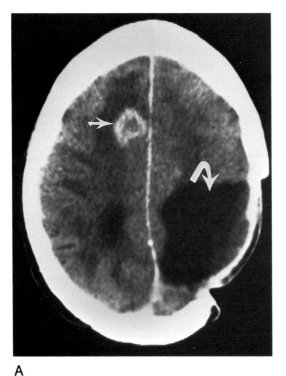

A

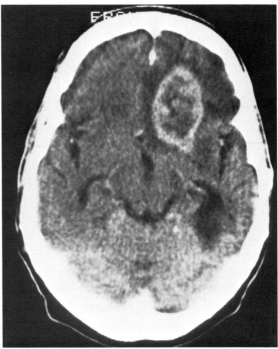

B

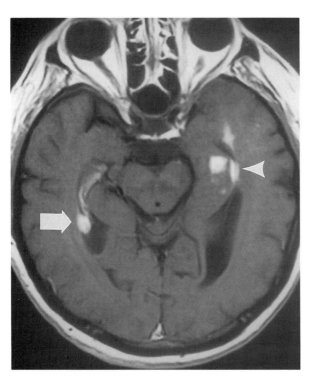

C

Fig. 5-17 Multicentric glioma, metastatic glioma. (A) Contrast-enhanced CT scan shows a large region of porencephaly from prior excision of a glioblastoma *(curved arrow)*. Now, a year later, the patient has a new lesion in the right posterior frontal lobe *(arrow)*, which proved to be a glioblastoma. **(B)** A second, larger, glioblastoma is present in the left frontal lobe. Without biopsy, it would be impossible to differentiate these lesions from metastasis. **(C)** Anaplastic astrocytoma in hippocampus, with metastasis to ependyma. Axial T_1 MRI after Gd shows irregular enhancement within the astrocytoma of the left anterior hippocampus and temporal lobe *(arrowhead)*. A metastatic tumor nodule arises from the ependyma of the atrium of the right lateral ventricle *(arrow)*. CSF spread of malignant glioma is a common pathway for metastasis.

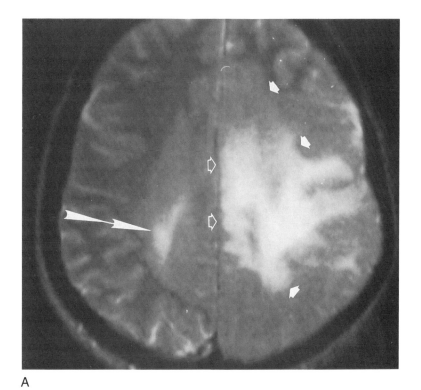

A

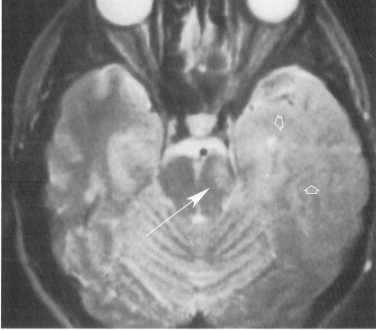

B

Fig. 5-18 Gliomatosis cerebri. (A) Axial T$_2$ MRI illustrates diffuse high signal intensity within the left hemispheric white matter, the centrum semiovale *(solid arrowheads)*. In some places it extends through the cortex to the brain surface *(open arrowheads)*. The tumor is contiguous across the corpus callosum into the right hemisphere *(arrow)*. **(B)** Axial T$_2$ MRI through the mesencephalon shows tumor extending through white matter tracts downward into the left peduncle *(arrow)*. The tumor is within the left temporal lobe *(open arrowheads)*.

241

Single feeding artery
Well-organized neovascularity or diffuse "stain"
Glioblastoma
    CT/MRI finding
        Single, irregularly shaped enhancing lesion, central necrosis
        Located within white matter
    Angiography findings
        Grossly irregular neovascularity
        Large multiple early draining veins
        Deep medullary venous drainage
Abscess
    Clinical findings
        Fever
    CT/MRI findings
        Thin-walled homogeneous ring-enhancing lesion
    Angiography findings
        Faint, round "blush" from uptake by the abscess capsule
    Radionuclide study
        Indium-111-labeled leukocyte uptake

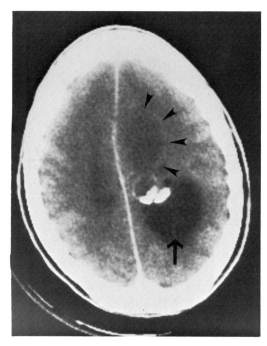

Fig. 5-19 Oligodendroglioma. Contrast-enhanced CT scan shows the typical findings of oligodendroglioma. There is no contrast enhancement. Dense calcification is present in the central portion of the tumor. There is a large cystic lesion posteriorly (*arrow*). A barely identifiable solid portion of the tumor is seen more anteriorly (*arrowheads*). The faintly identifiable rim represents a "pseudocapsule."

most often occur in the frontal or anterior temporal lobes and rarely in the occipital lobes. Calcification is characteristic of the tumor, occurring in about 75 percent of persons. Most of these tumors are low grade; they exhibit little or no contrast enhancement on CT or MRI, and only minimal if any peritumoral edema (Fig. 5-19). The imaging characteristics of an oligodendroglioma are the same as those of low-grade astrocytoma. Noncalcified oligodendroglioma cannot be differentiated from low-grade astrocytoma. As with all gliomas, prominent contrast enhancement indicates anaplasia.

Ependymoma
Ependymoma is primarily a tumor of childhood and is discussed in that section. Uncommonly, cerebral hemispheric ependymoma occurs in the young adult, usually at the level of the junction of the parietal, temporal, and occipital lobes. They are characteristically located adjacent to the lateral ventricle and may have an intraven-

tricular component (see Ch. 6). Cysts, calcification, and surrounding edema are common. The tumor shows variable contrast enhancement on CT or MRI scan. When there is no calcification, ependymoma appears similar to anaplastic astrocytoma or glioblastoma, although it occurs in a younger age group.

Subependymoma
Subependymoma is a benign usually slow-growing astroglial tumor that is small; most are discovered incidentally at autopsy. Occasionally, a subependymoma becomes large and obstructs the ventricular system, causing hydrocephalus. When symptomatic, it presents in persons aged 10 to 40 years. This tumor arises most commonly at the level of the foramina of Monro, where it is attached to the septum pellucidum, the periaqueductal

region, and the inferior floor of the fourth ventricle. The tumor is characteristically small, but it may become bulky. Calcification and cysts are common and are easily seen with CT. They demonstrate slight contrast enhancement. Surgical removal is successful in treating the obstruction.

Mixed Glioma

Mixed glioma accounts for about 5 percent of glial tumors and, except for the gliosarcoma, are usually low grade. They are indistinguishable from the other glial tumors, particularly the oligodendroglioma. Gliosarcoma appears as a peripheral tumor with an imaging pattern of glioblastoma. Angiography usually shows some meningeal vascular contribution to the tumor. Mixed low-grade tumors are common in the region of the hypothalamus in children and tend to display uniform contrast enhancement.

Pleomorphic Xanthoastrocytoma

Pleomorphic xanthoastrocytoma has only recently been recognized and classified. The pathology suggests an aggressive tumor, but the clinical history suggests a more benign course. Because it has so recently become recognized, its clinical behavior is incompletely known. The tumor presents at ages 25 to 40 and is located predominantly within the lateral temporal lobe or the peripheral parietal lobe. The lesion is cystic and calcified and has a mural nodule that is commonly contiguous with the leptomeninges. With Gd, the nodule enhances, but the cyst wall does not.

Subependymal Giant Cell Tumor

Subependymal giant cell tumor is a rare glial tumor that occurs exclusively in the bodies of the lateral ventricles in patients with the tuberous sclerosis complex. Some patients will show minimal or no stigmata of the phakomatosis. This tumor is discussed in more detail in the section on tuberous sclerosis complex.

OTHER TUMORS

Central Neurocytoma

Central neurocytoma is a rare benign neuroepithelial tumor that occurs within the body of the lateral ventricles near the foramina of Monro and adjacent to the septum pellucidum. It presents most commonly in young

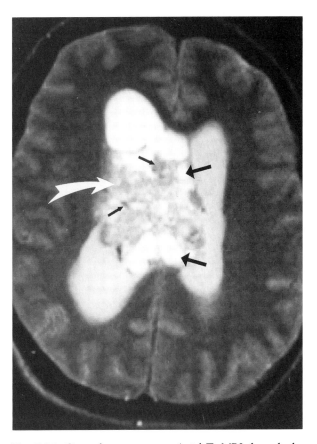

Fig. 5-20 Central neurocytoma. Axial T$_2$ MRI through the lateral ventricles shows the large inhomogeneous large mass within the anterior body of the right lateral ventricle (*large white arrow*). Round bright regions represent cysts (*larger black arrows*), and the focal darker areas represent calcium (*smaller black arrows*).

adults, aged 25 to 35, and may cause unilateral hydrocephalus. The tumor contains calcium and cysts. It is an isodense, or hyperdense lobular cystic mass on CT, and slightly hyperintense on T$_2$-weighted MRI (Fig. 5-20). Contrast enhancement is variable, but when present is generally less than with meningioma or choroid plexus papilloma. The tumor cannot be differentiated from a calcified intraventricular oligodendroglioma. An extremely rare malignant form of this tumor is intraventricular neuroblastoma.

Colloid Cyst

A colloid cyst is a rare benign tumor that probably originates from ectopic endoderm incorporated in the roof of the anterior third ventricle at the level of the foramina of

Monro. Previously, colloid cysts were classified as neuroepithelial in origin. It is composed of a dense fibrous capsule filled with a fluid that varies from a thick mucoid material with debris to clear fluid. It commonly appears on CT scans as a round, hyperdense lesion less than 2 cm in diameter (Fig. 5-21); uncommonly, it is isodense or hypodense. On MRI, it is usually isointense on the T_1- and T_2-weighted images but may be hyperintense or hypointense on T_2, depending on the amount of protein contained within the cyst cavity. Occasionally, the capsule enhances with contrast, but contrast is neither necessary nor useful in the diagnosis, if the characteristic findings are present.

These cysts are congenital but slow-growing, hence usually do not cause symptoms until adulthood. Colloid cysts obstruct the foramina of Monro and cause hydrocephalus, which is sometimes intermittent. Sudden death from acute hydrocephalus is reported but is rare. Meningioma and ependymoma rarely occur in this region and may be differentiated from a colloid cyst by their more uniform contrast enhancement or hypointense T_1 signal on MRI.

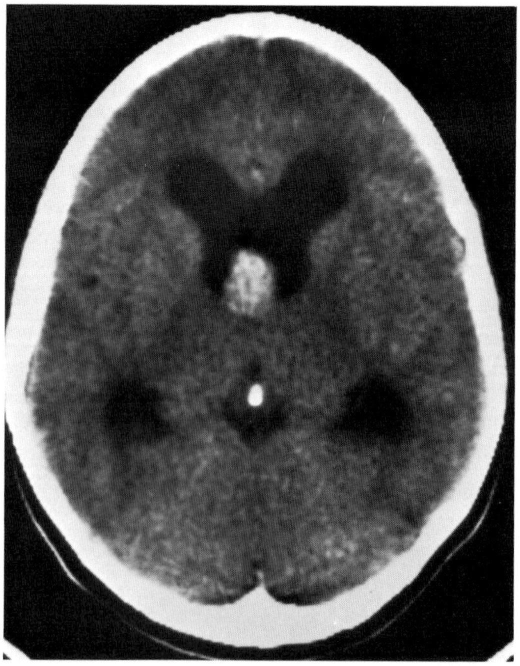

Meningioangiomatosis

Meningioangiomatosis is a rare benign hamartomatous meningovascular fibroblastic proliferation of the leptomeninges. It occurs almost exclusively in children and young adults under age 25, with a history of seizures. Fifty percent of those with this tumor have stigmata of neurofibromatosis-1 (NF-1). The lesion extends over the cortex in a thick plague-like fashion but may have a more focal configuration, extending from the cortex deeper into the brain. Characteristically, the hamartoma heavily calcifies or ossifies, easily seen with CT. T_2-weighted MRI shows the tumor as isointense or hypointense from calcium, with surrounding high-intensity edema. Contrast enhancement is variable, but a gyriform pattern is frequently seen. These are slow-growing tumors and amenable to surgical removal. They must be differentiated from oligodendroglioma, astrocytoma with calcification, meningioma, granulomatous meningitis (sarcoid, tuberculosis), and Sturge-Weber syndrome.

SECONDARY NEOPLASMS (METASTASIS)

Cerebral metastatic tumor is the second most common type of intracerebral neoplasm, accounting for about 35 percent of all intracranial tumors. The incidence of metastasis to the brain is rising, so that in some series metastasis is the most common type of cerebral tumor found. Most are hematogenous metastases. The most common types are lung tumor (small cell carcinoma, adenocarcinoma) and breast carcinoma. Less frequent metastatic tumors arise from renal cell carcinoma, colon carcinoma, and melanoma. However, any advanced or aggressive tumor may metastasize to the brain. The meninges of the brain or spine are only rarely involved, and then most commonly with breast carcinoma or small cell carcinoma of the lung. Metastasis to the spinal cord is uncommon.

Most commonly, hematogenous metastases are seen as multiple lesions located at the gray–white corticomedullary junction. The lesions are usually spherical but may assume any shape. Some appear to be entirely cystic. Neovascularity is present, accounting for vasogenic white matter edema and contrast enhancement of the

Fig. 5-21 Colloid cyst. Nonenhanced CT scan shows homogeneously dense, round lesion at the level of the foramina of Monro. The foramina are obstructed, causing hydrocephalus.

tumor. The amount of edema is variable, but characteristically it is extensive for the size of the tumor nidus. In general, the specific histology of the metastasis cannot be determined from its shape or pattern of contrast enhancement.

Cerebral metastasis is the initial presentation of neoplasia in about 10 percent of patients. The metastatic tumor deposit is solitary in 30 to 50 percent of cases, depending on the tumor type. Carcinoma of the lung is the most likely to be solitary. Headache, seizures, and progressive focal neurologic symptoms are the most common presenting complaints. Ninety percent of patients with cerebral metastasis have a history of a carcinoma known to metastasize to the brain. Differentiation of a solitary metastasis from a primary brain tumor may not be possible without biopsy or discovery of the likely primary tumor.

Gd contrast-enhanced MRI is the most sensitive examination for the detection of metastatic disease in the brain. Additional lesions may be seen with enhancement that were not apparent on the nonenhanced scan. High-dose Gd enhancement (0.3 mmol/kg) and the use of MTC with T_1-weighted imaging shows additional small parenchymal metastases that are not seen with the standard dose and SE technique. However, this shows only rarely a single lesion when none is seen with the standard techniques. The detection of a single tumor in a patient at risk of metastasis requires additional MRI scanning with augmented Gd enhancement to detect additional occult lesions. Implications for treatment are much different if multiple lesions are found.

With nonenhanced MRI, the lesions are ordinarily hypointense on T_1- and hyperintense on T_2-weighted images (Fig. 5-3). However, melanoma characteristically shows high intensity on T_1- and hypointensity on T_2-weighted images because of the paramagnetic effects of both melanin and the acute and chronic hemorrhage that is almost always present in these tumors (Fig. 5-22). Some mucinous cystadenocarcinomas and choriocarcinoma may show this pattern as well. Edema is a prominent accompaniment to metastasis and is strikingly seen with MRI. Sometimes with small tumors only the edema is seen on the scan, and the edema may be difficult to differentiate from infarction when it involves the cortical surface.

The post-contrast CT scan detects about 95 percent of metastatic tumors in the brain that are 5 mm or larger. It is sometimes possible to detect the presence of smaller lesions by the observation of surrounding edema or by the use of very high-dose contrast enhancement and delayed scanning. Characteristically, the metastatic tumor deposits are seen as multiple spherical contrast-enhancing lesions with sharp outlines (Fig. 5-23). Edema is present in about 90 percent of metastatic tumors. Small lesions tend to have uniform complete enhancement, whereas larger lesions are often ring-like with central necrosis.

In general, it is not possible to determine the type of tumor metastasis from the appearance on CT. Hyperdensity from either hemorrhage or calcium suggests that the tumor is a melanoma, hypernephroma, choriocarcinoma, mucinous cystadenocarcinoma (Fig. 5-4), or lymphoma. A large solitary metastatic lesion may be indistinguishable from a glioblastoma (Fig. 5-24). Some large cystic-appearing lesions with focal contrast enhancement are identical to anaplastic (grade III) astrocytoma. In a small percentage (less than 5 percent), only the edema or necrosis of a tumor situated on the cortex is seen (Fig. 5-25), and differentiation from infarction is sometimes difficult. CT scanning for metastatic disease includes bone window images. CT scanning is generally superior to MRI for the evaluation of osseous metastasis.

Neoplastic Meningitis

Meningeal carcinomatosis (neoplastic meningitis, carcinomatous meningitis) is metastatic spread of tumor in the leptomeninges. The tumor may be focal and nodular or diffuse and sheet-like. In adults, carcinoma of the breast, small cell carcinoma of the lung, and melanoma account for most cases of meningeal carcinomatosis. In children, CSF spread from primary CNS tumor and neuroblastoma accounts for most metastatic meningeal tumors. Meningeal metastasis may occur without brain metastasis.

Post-contrast CT and MRI make the diagnosis by demonstrating the presence of diffuse contrast enhancement of the meninges or ependyma (Fig. 5-26). The enhancement may be diffuse (linear and even) or nodular. When both pial and dural enhancement occurs, the diagnosis is more secure. Contrast-enhanced MRI is much more sensitive than CT scanning for the detection of meningeal carcinomatosis (Fig. 5-27 and 5-28). High-dose Gd may show enhancement of the meninges with meningeal metastasis when the standard dose (0.1 mmol/kg) fails to do so.

The sulci and cisterns may be obliterated with extensive disease, but meningeal metastatic tumor cannot be

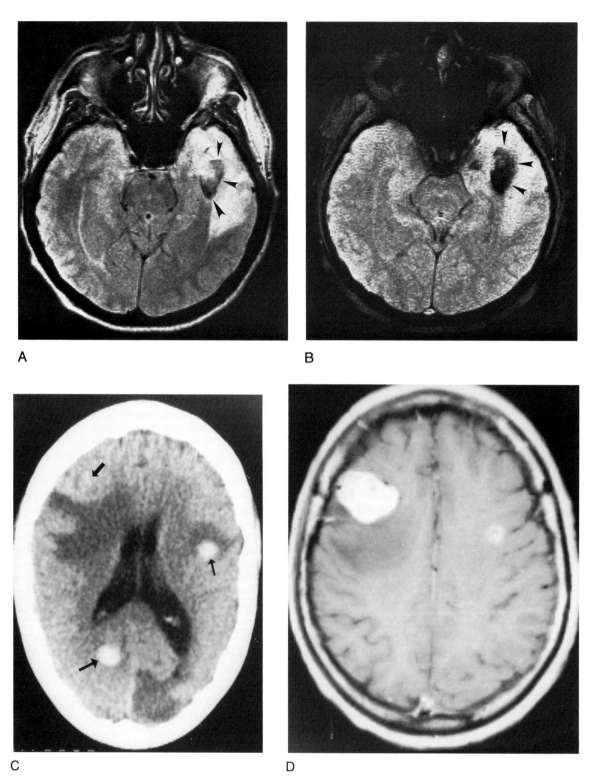

A

B

C

D

Fig. 5-22 Metastatic melanoma. (A) T_2-weighted SE MRI shows the hypointense tumor nodule in the medial left temporal lobe *(arrowheads)*. The surrounding high intensity signal represents edema. **(B)** Fisp 20;300/20 GRE MRI shows extreme hypointensity of the tumor nodule because of accentuation of paramagnetic effects with the GRE images *(arrowheads)*. **(C)** Axial nonenhanced CT scan shows the multiple melanoma deposits as high-density lesions surrounded by edema *(arrows)*. **(D)** Axial Gd-enhanced MRI demonstrates the intense enhancement of the tumor deposits. Some of the lesions were also hyperintense on the nonenhanced T_1 MRI.

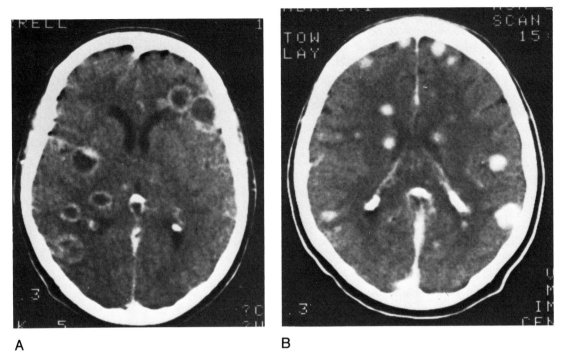

Fig. 5-23 Metastasis, adenocarcinoma of the lung. (A) Multiple metastases with ring-like contrast enhancement. **(B)** Multiple metastases with homogeneous contrast enhancement. Multiple metastases tend to have varying sizes of lesions, whereas multiple abscesses tend to be more uniform in size.

seen reliably without enhancement. Mild hydrocephalus is usually present from cisternal obstruction to CSF flow. CSF cytology confirms the diagnosis. Meningeal enhancement from carcinomatosis must be differentiated from leptomeningeal enhancement of other causes (see box).

Cerebral angiography is no longer used for the detection of cerebral metastasis. However, it can be helpful for differentiation of metastasis from primary tumor infarction or abscess and for localization prior to surgical excision. Almost all metastatic tumors produce enough mass effect to be visualized with angiography. About 50 percent exhibit some hypervascularity.

Four patterns of hypervascularity have been described:

CAUSES OF GADOLINIUM ENHANCEMENT OF THE LEPTOMENINGES

Neoplastic meningitis
    Metastasis (carcinoma)
    Lymphoma
    Leukemia
    CSF spread of CNS tumor
Granulomatous meningitis

(Continued)

Prior cranial surgery
Shunting of hydrocephalus
Infarction
    Cortical, ischemic
    Venous thrombosis
Meningioma (tail sign)
Subarachnoid hemorrhage
Acute meningitis
Trauma
Idiopathic

(Text continued on page 251)

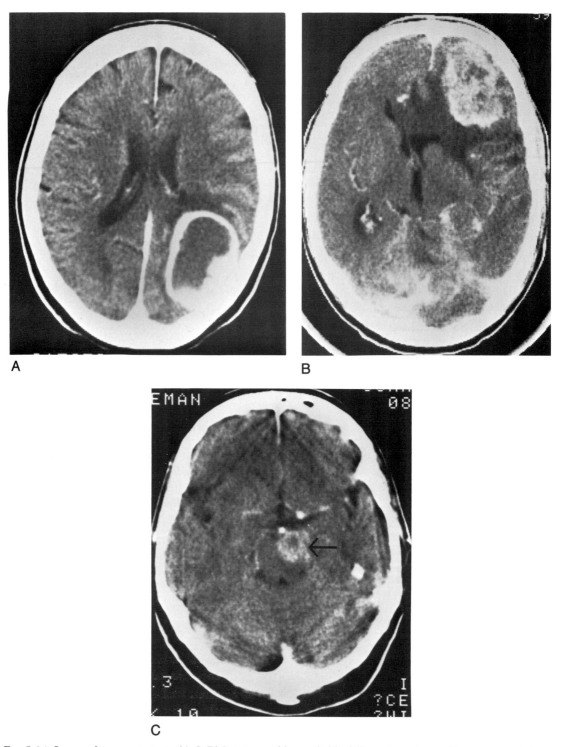

Fig. 5-24 Large solitary metastasis. (A & B) It is impossible to reliably differentiate a large solitary metastasis from a glioblastoma. **(C)** This contrast-enhanced lesion in the pons represents a metastasis *(arrow)*. Although less common, a glioma could also have this appearance.

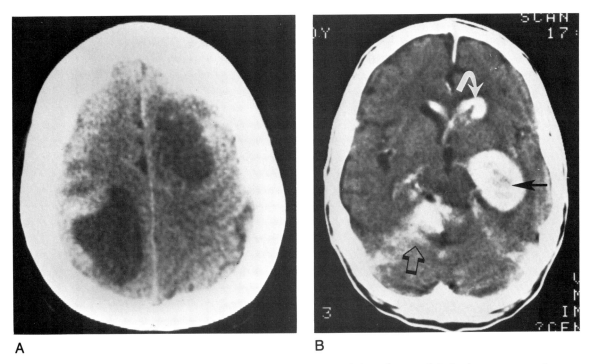

A   B

Fig. 5-25 Metastatic tumor. (A) Metastases sometimes appear entirely hypodense with little if any contrast enhancement. They can mimic cerebral infarction. **(B)** Small cell carcinoma of the lung. Contrast-enhanced CT scan shows metastatic tumor deposits in the parenchyma *(black arrow)*, along the ependyma of the ventricular cavities *(curved arrow)*, and in the meninges of the cisterns *(open arrow)*.

Fig. 5-26 Meningeal carcinomatosis. Contrast-enhanced CT study shows intense contrast enhancement of the meninges of the basal cisterns, sylvian fissures, and interhemispheric fissure *(arrows)*. Communicating hydrocephalus is also common because of cisternal blockage.

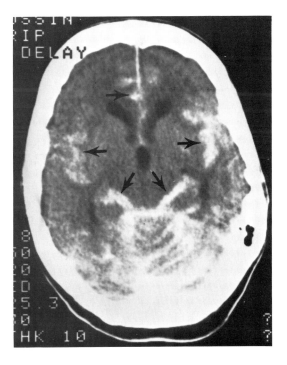

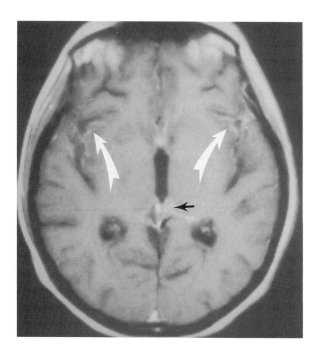

Fig. 5-27 Neoplastic meningitis, melanoma. Axial T₁ triple-dose Gd-enhanced MRI illustrates diffuse enhancement of the pia arachnoid from metastatic melanoma *(white arrows)*. The abnormality was just barely identified with single-dose contrast study. The tumor has a more solid mass appearance near the pineal gland *(black arrow)*.

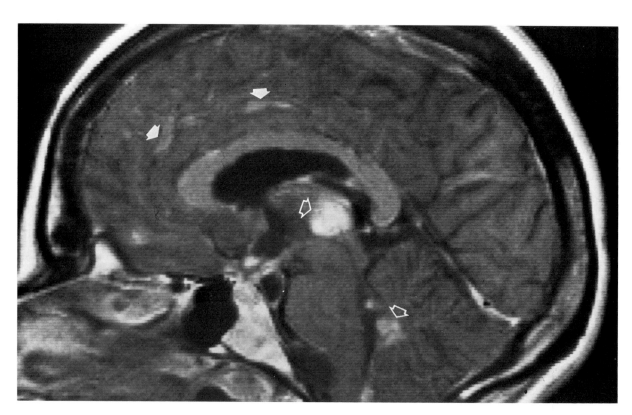

Fig. 5-28 Meningeal metastasis, astrocytoma of the spinal cord. Parasagittal Gd-DTPA-enhanced MRI demonstrates multiple enhancing lesions along the meningeal surface of the interhemispheric fissure *(solid arrows)*. Deposits are also present within the ventricles *(open arrows)*. Gd-DTPA-enhanced MRI is the most sensitive technique for the detection of meningeal metastasis.

1. Network of thin, regular, tortuous vessels, with early venous filling; frequently shows a single feeding artery and one, or perhaps two, cortical draining veins—a pattern essentially pathognomonic of a metastatic tumor (Fig. 5-29)
2. Diffuse blush, simulating a meningioma but without supply from meningeal vessels and a shorter duration of the "stain" than a meningioma
3. Ring-shaped lesion, formed by a vascular periphery and an avascular central region of necrosis
4. Irregular, "wild" neovascularity with prominent early draining veins, indistinguishable from glioblastoma

The histology of these tumors correlates poorly with the pattern of neovascularity, except for the hypernephroma, which is highly vascular with a generally homogeneous persisting stain; it mimics hemangioblastoma when seen in the posterior fossa. Metastatic tumors almost always drain to the surface cortical veins, whereas glioblastomas nearly always have at least some of their drainage deep into the ventricular system, often with demonstrable enlargement of the deep medullary veins.

Radionuclide brain scanning has been supplanted by contrast CT scanning or MRI. The sensitivity of radionuclide scanning of the brain is about 85 percent—less than that of contrast-enhanced CT scanning and MRI. The radionuclide study is performed with the chelates of technetium, diethylenetriamine pentaacetic acid (DTPA), or glucoheptinate. Metastatic tumors are detected as regions of radionuclide accumulation and are generally round and discrete, sometimes exhibiting a doughnut pattern. Single lesions are nonspecific, and identical patterns may be seen with primary tumors, abscess, and sometimes infarction. However, an indium-111-labeled white blood cell examination may be used to differentiate an abscess from a tumor.

Calvarial metastasis occurs, but with less frequency than cerebral metastasis. Focal head pain is common, caused by the lesion. By far the most common calvarial metastasis is carcinoma of the breast (see box). Most commonly, these metastatic lesions are lytic (Fig. 5-30), but blastic lesions or a combination of the two forms (mixed) may occur. On skull radiographs, the margins of metastatic lesions are usually irregular and lack definition. A small central bone sequestrum may remain. Beveled margins may be present, particularly when only one table of the calvarium is involved. Blastic metastasis does not cause skull thickening, differentiating it from Paget's

disease. Calvarial metastatic tumor may grow inward into the epidural space or through the dura, to involve the subdural space and sometimes the cortical surface of the brain (Fig. 5-31). Skull lesions may also be seen with bone windows on CT scans or with MRI (Fig. 5-30).

When multiple and small, lytic metastasis may not be distinguishable from multiple myeloma. Myeloma lesions involve the diploë, are usually small and generally uniform in size, and have rather sharp margins compared with metastatic tumors. The clinical history, age, and evaluation of the serum protein are important for making the differential diagnosis. The small lytic lesions of hyperparathyroidism must be differentiated from metastatic disease, but they tend to be uniform in size and relatively sharply demarcated.

Isolated metastatic deposits must be differentiated from the lucency produced by venous lakes. These normal dilations of the diploic veins do not cause erosion of the inner or outer tables, which may be evaluated by tangential radiography. Also, a draining diploic vein may be seen entering the region of lucency. The radionuclide bone scan is negative.

Secondary neoplasm may involve the base of the skull as well. Most commonly, this is a metastasis from carcinoma of the prostate (usually blastic) or breast. Bone destruction from direct invasion occurs from carcinoma of the nasopharynx and sphenoid sinus and glomus

---

MULTIPLE LYTIC LESIONS OF THE CALVARIUM

Metastatic
    Adults
        Breast
        Lung
        Kidney
        Thyroid
    Children
        Neuroblastoma
        Leukemia
        Histiocytosis X
Myeloma (other macroglobulinemias)
Hyperparathyroidism
Sarcoidosis and other granulomas
Venous "lakes"

(*Text continued on page 256*)

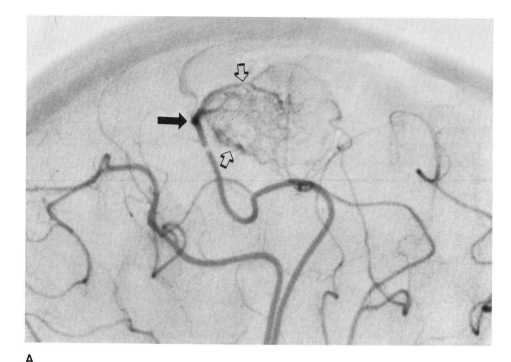

A

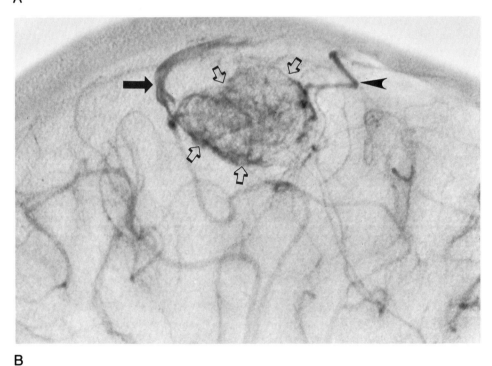

B

Fig. 5-29 Metastatic adenocarcinoma. (A) Early arterial phase shows the fine tumor neovascularity *(open arrows)* being filled primarily by a single feeding artery *(solid arrow)*. **(B)** The tumor nodule is well circumscribed with uniform neovascularity *(open arrows)*. There is a primary early filling cortical draining vein *(solid arrow)* and a small secondary cortical draining vein *(arrowhead)*. This pattern is typical for cerebral metastasis.

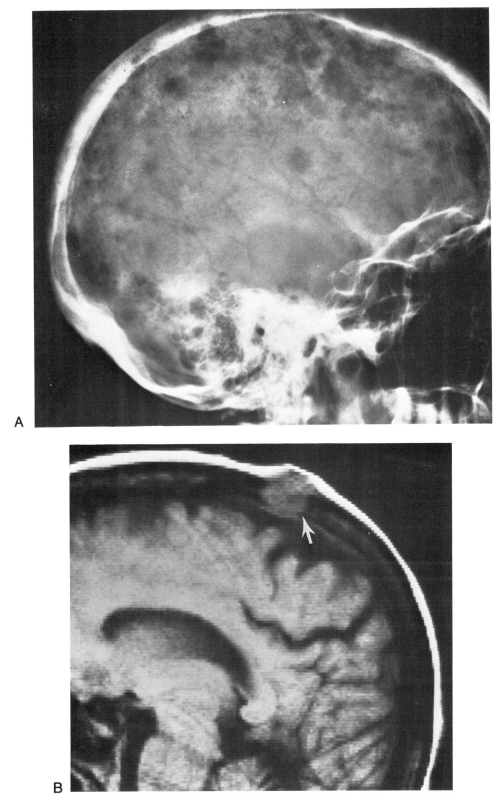

Fig. 5-30 Calvarial metastases, breast carcinoma. (A) Multiple lytic metastases are seen in the calvarium. The lesions are of many sizes with indistinct margins. **(B)** MRI demonstrates an isodense mass within the calvarium *(arrow)*. It displaces the marrow and has grown outward into the subcutaneous fat of the scalp.

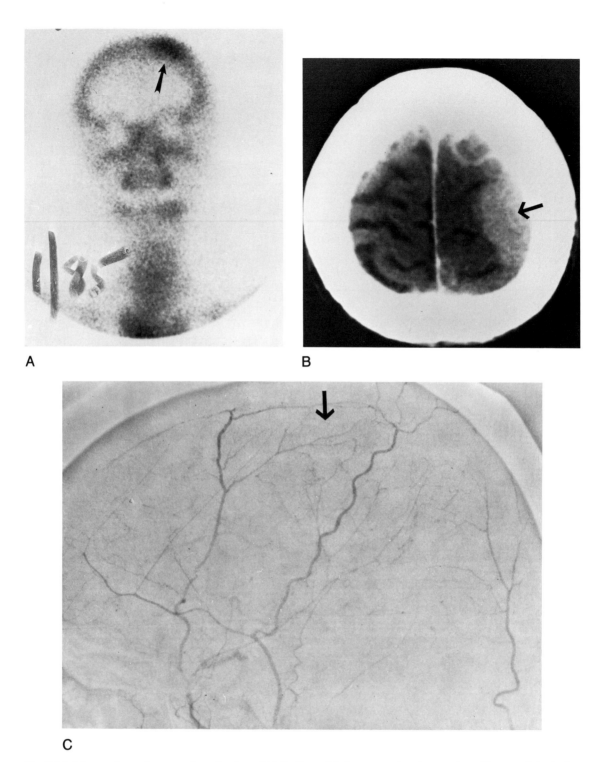

Fig. 5-31 Metastatic carcinoma to the calvarium. (A) Radionuclide bone scan showing a region of hyperactivity in the left parietal bone *(arrow)*. **(B)** Contrast-enhanced CT scan shows an enhancing extracerebral mass underneath the left calvarium, causing compression of the cerebral cortex (arrow). This presentation mimics that of a meningioma. **(C)** Selective external carotid arteriography shows the tumor to be essentially avascular, differentiating it from a meningioma *(arrow)*. Metastasis was to the calvarium with secondary inward growth. *(Figure continues.)*

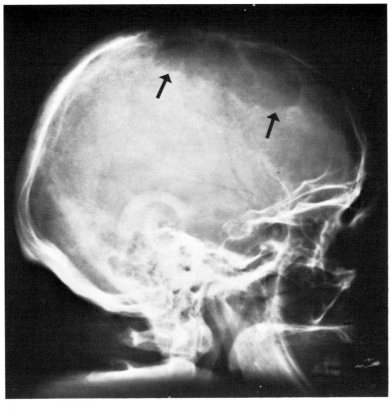

D

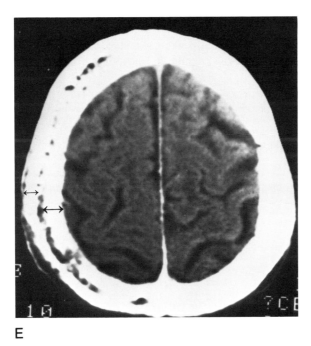

E

Fig. 5-31 *(Continued)*. **(D)** Osteoporosis circumscripta. An extensive lytic process involves the calvarium. It may be indistinguishable from metastatic or primary tumor without biopsy. **(E)** Paget's disease. CT shows the typical changes of thickening of the outer and inner tables with overall enlargement of the calvarium *(opposed arrows)*. There is irregular density within the diploic space. Blastic metastasis would cause increased bone density, but not thickening.

tumors. CT with bone windowing is the best modality with which to visualize these lesions, although MRI shows better the full extent of the soft tissue component of the tumor mass.

MENINGIOMA

Meningioma is the third most common intracranial tumor, accounting for about 10 percent of tumors involving the CNS. It is the most common extracerebral tumor. The tumor originates from cell elements within the meninges, usually those packing the arachnoid villi. As a result, the tumor is usually attached to the dura. The preferential sites for meningioma correspond with the distribution of the arachnoid villi along the major dural sinuses and at the exits of the spinal nerves from the meningeal sleeves. However, because meningioma may arise from any arachnoidal site (see box), it is seen over the convexities or at sites of infolding of the arachnoid into the brain (choroid plexus). Rarely, it occurs entirely within the brain along the pia extension around cortical vessels.

The histopathologic classification of meningioma is in transition. The classic classification divided the tumor into meningothelial, transitional, and fibroblastic types. Attempts at differentiating these types with MRI have produced conflicting and indefinate results. The newer classification divides meningioma into (1) classic benign

meningioma, with slow growth, low cellularity, and no mitosis; (2) atypical meningioma, with signs of more rapid growth, increased cellularity, and some mitosis; and (3) anaplastic meningioma, with rapid growth, high cellularity, and prominent mitosis, anaplasia, and necrosis. The angioblastic meningioma of the older classification is more commonly labeled hemangiopericytoma of the meninges and is no longer considered part of the meningioma series. MRI studies concerning the newer classification are not available at this time.

Most meningiomas are "classic" benign, slow-growing, well-encapsulated tumors that cause symptoms by indentation of the brain or compression of cranial nerves. About 5 percent of meningiomas show signs of aggressive growth or malignancy (sarcoma) and may actually invade brain tissue. Meningiomas may infiltrate and destroy adjacent bone. True hyperostosis with an increase in the thickness of bone is common adjacent to the tumor and has diagnostic importance. The bone expansion may cause compression of exiting cranial nerves. Meningiomas may hemorrhage spontaneously and appear as a large hematoma. The morbidity and mortality of this complication is high, at greater than 50 percent.

Meningiomas may grow as a diffuse sheet of tumor over the brain or at the base of the skull. This development is referred to as *en plaque growth*. These tumors are difficult to image and sometimes are suspected only because of the presence of hyperostosis of the adjacent bone.

Calcification is common within meningiomas. It is usually diffuse and fine but may occur in large focal clumps. Small densely calcified meningiomas are slow-growing and may remain the same size for years. Necrosis and cyst formation are rare. Multiple meningiomas occur in about 2 percent of cases, most often in association with neurofibromatosis (NF-2).

In general, the prognosis for meningioma is excellent. Many tumors can be completely removed; however, local tumor recurrence is a problem. Periodic follow-up imaging is essential for the timely detection of tumor recurrence, and the tumor may be much more aggressive after recurrence. Contrast-enhanced scans are essential. MRI may better show tumor recurrence when multiple surgical clips are present.

Stroke may result from arterial occlusion, when large tumors encase major vessels. Venous sinus thrombosis may result from local tumor invasion, particularly at the sagittal sinus. Rarely, a meningioma erodes into a paranasal sinus, causing pneumocranium or CSF fistula.

SITES OF CRANIAL MENINGIOMA (IN ORDER OF FREQUENCY)

1. Parasagittal, falx, or sagittal sinus
2. Cerebral convexity
3. Sphenoid ridge
4. Subfrontal, olfactory groove
5. Parasellar
6. Tentorial
7. Posterior fossa (CPA cistern and foramen magnum)
8. Temporal fossa
9. Intraventricular
10. Intraorbital, along the optic nerve sheath

Table 5-3 Characteristic Clinical Symptoms Produced by Meningioma

Symptom	Location
Spastic paraparesis	Large parasagittal
Spastic monoparesis of the leg	Unilateral, parasagittal
Seizure, focal	Convexity
Anosmia	Subfrontal
Cranial nerve paresis	Parasellar, sphenoid ridge, tentorium, CPA
Exophthalmos	Sphenoid ridge
Medullary compression with focal neck pain	Foramen magnum
Organic brain syndrome	Large frontal tumor

Meningioma is sought when certain characteristic clinical findings are present (Table 5-3).

The diagnosis of meningioma is generally made with contrast-enhanced MRI or CT scanning. The tumor always exhibits strong contrast enhancement because the neovascularity derived from the meninges is somatic in nature. Only heavily calcified tumors will not show enhancement. Angiography also has an important role in the evaluation of meningioma, namely to confirm the diagnosis by demonstrating the characteristic meningeal vascular supply. It is also important in mapping the vascular supply and demonstrating the relationship of the tumor to major cerebral vessels prior to surgical removal or preoperative embolization.

With MRI, meningioma characteristically appears as a slightly hypointense or isointense extracerebral mass on T_1-weighted images (Fig. 5-35). It is moderately hyperintense on lightly T_2-weighted (proton density) images but decreases in intensity on the heavily T_2-weighted images. A thin rim of water density may outline the mass and is thought to represent entrapped CSF. After Gd injection, meningioma shows strong hyperintensity on the T_1-weighted images (Fig. 5-32). An enhancing "dural tail" often extends along the inner surface of the skull a short distance away from the main tumor mass. This is most likely a reactive change within the dura and does not imply spread of tumor along the dura. Once thought specific for the MRI diagnosis of meningioma, the dural tail sign, while most commonly seen with meningioma, has been found with a large spectrum of disease, including peripheral astrocytoma, metastasis, lymphoma, and sarcoid. The meninges may enhance in the region of acute cerebral infarction.

Small meningiomas are easily seen after enhancement (Fig. 5-33). Without Gd enhancement small tumors may be difficult to recognize and then often only by a subtle distortion of the adjacent brain anatomy. Bone invasion is seen as loss of the normally high marrow signal at the tumor site. Cysts are sometimes seen within the tumor. Calcium is more difficult to recognize than with CT scans and is represented by signal void within the mass. MRI is generally not as reliable as MRA for determining the patency of adjacent venous sinuses, although GRE imaging sometimes helps by showing increased intravascular signal in patent vessels.

Most meningiomas exhibit a characteristic pattern on CT scans (Fig. 5-34). On the precontrast CT scan, 80 percent of the tumors are identified as a homogeneous, slightly hyperdense mass with its base on a dural surface. Isodense tumors occur in about 15 percent of cases. Rarely, the tumor is hypodense or cystic and mimics an anaplastic astrocytoma. Cysts may be within or adjacent to the tumor. Occasionally, there is a thin rim of CSF trapped between the tumor and the cortical surface.

The tumor may show fine or conglomerate calcification (20 percent). Edema is common (60 percent) and occasionally severe. In general, for the size of the tumor, the edema is less severe than that seen with glioma or metastasis. Infarction may be seen in the distribution of vessels occluded by the tumor.

Bone window images are essential for the CT evaluation of meningioma and identify hyperostosis or bone destruction of the adjacent calvarium. The hyperostosis may be just a small region of enostosis (Fig. 5-35), or it may cause gross enlargement over a large region of the bone (Fig. 5-36). Tumor is usually present within hyperostotic bone.

Except for those that are entirely calcified, almost all meningiomas (98 percent) demonstrate intense and generally homogeneous contrast uptake. Uptake may not be identifiable in tumors that are thin (en plaque) or obscured by adjacent bone artifact. The contrast-enhanced region represents the actual tumor.

Malignant sarcomatous change is suggested by blurred borders, inhomogeneous enhancement due to regressive changes within the tumor, and adjacent satellite lesions. In these cases, it may be impossible to differentiate the malignant meningioma from a glioblastoma or gliosarcoma.

Angiography is useful for the diagnosis and mapping of the vascular supply prior to surgery. The angiographic hallmark of meningioma is the meningeal vascular supply of the tumor. It is normally demonstrated by selective

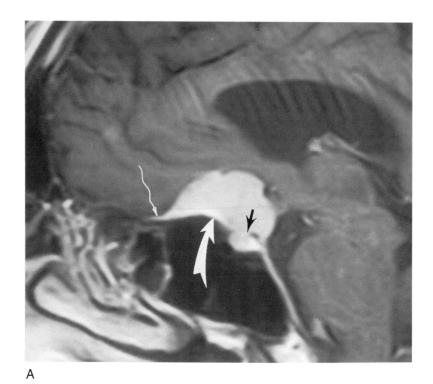

A

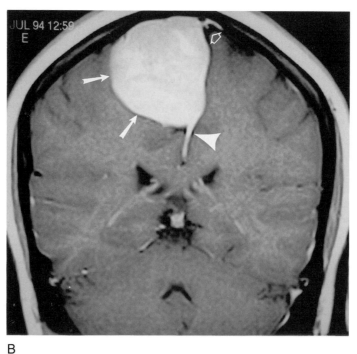

B

Fig. 5-32 Meningioma. (A) Sagittal T₁ Gd-enhanced MRI shows the bright homogeneous enhancement that is typical with meningioma *(large white arrow)*. This arises from the planum sphenoidale. The tumor extends into the suprasellar cistern indenting the hypothalamus and chiasm. It crosses over the sella turcica but does not enter it *(black arrow)*. A small dural "tail" extends from the anterior edge of the tumor *(white squiggle arrow)*. **(B)** Coronal T₁ Gd-enhanced MRI shows a parasagittal meningioma arising from the falx *(small white arrows)*, showing the typical homogeneous contrast enhancement. The sagittal sinus is displaced to the left but is not occluded *(open white arrow)*. The falx enhances away from the bulk of the tumor, the "tail sign" *(arrowhead)*. The tail of contrast does not mean tumor extends into this region of the falx.

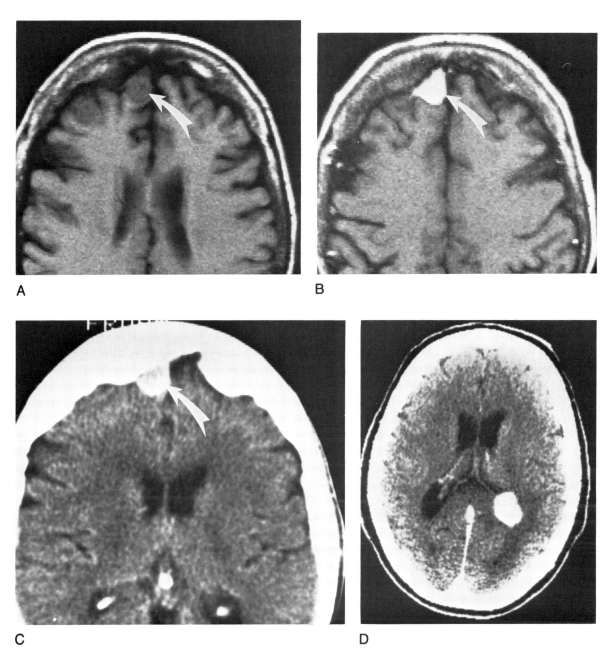

Fig. 5-33 Meningioma. (A) T_1-weighted MRI shows the subtle changes of a small meningioma *(arrow)*. **(B)** Gd-DTPA-enhanced contrast enhancement of the meningioma shows strong enhancement on T_1-weighted images *(arrow)*. **(C)** Contrast-enhanced CT study easily identifies the strongly enhancing meningioma *(arrow)*. **(D)** Intraventricular meningioma. A densely enhancing round mass is present within the atrium of the left lateral ventricle attached to the glomus. A benign xanthogranuloma of the glomus would have the same appearance.

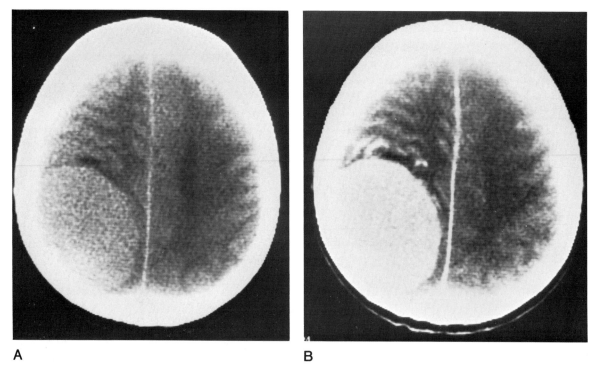

Fig. 5-34 Meningioma, CT scans. These scans show the classic appearance of a large meningioma. **(A)** Nonenhanced CT scan shows the large right posterior parietal convexity meningioma as being hyperdense. **(B)** With contrast enhancement there is intense homogeneous enhancement of the meningioma.

cerebral angiography, particularly selective external carotid injection. The tumor receives vascular supply from regional meningeal vessels at the site of its attachment to the dura (Fig. 5-37). Intraventricular tumors are supplied by the regional choroidal arteries. When the tumor is large, its inner portion may receive additional vascular supply from the pial vessels of the cerebral arteries, but most of the tumor is still supplied by the meningeal vessels. The superficial temporal artery or other scalp vessels may contribute to tumors, especially those that have grown through the calvarium (Fig. 5-35F). The location of meningiomas and their characteristic vascular supplies are given in Table 5-4.

The meningeal vascular supply is more or less radially oriented from the base into the tumor. The tumor shows an intense homogeneous blush that begins in the later arterial phase and persists throughout the venous phase (Fig. 5-38). The mass is sharply outlined. Early venous drainage is rare. Except for those that are small and cal-

cified, all meningiomas show significant meningeal vascular supply and blush. If the meningeal supply is little or absent, the diagnosis of meningioma is unlikely (see differential diagnosis below, and Fig. 5-31).

Selective angiography is preferred. For convexity tumors, external and internal carotid angiographic runs are appropriate for clear differentiation of the meningeal and cortical supply to the tumor. Bilateral studies are performed, as frequently there is important supply from contralateral vessels. Subtraction films are basic.

Special note is made of the patency of the dural venous sinuses (Fig. 5-39), the position of major vessels relative to the tumor, and any vascular constriction or occlusion. The last point is particularly important for tumors at the base of the brain near the internal carotid artery (ICA) and middle cerebral artery (MCA). MRA is now the first imaging study performed for the evaluation of the arteries and veins in relationship to meningioma (Fig. 5-40).

(*Text continued on page 264*)

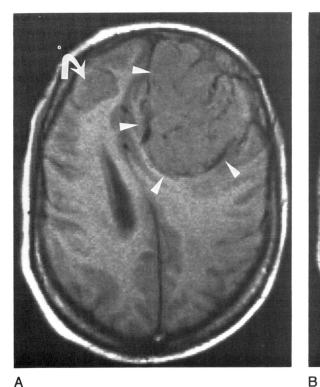

A

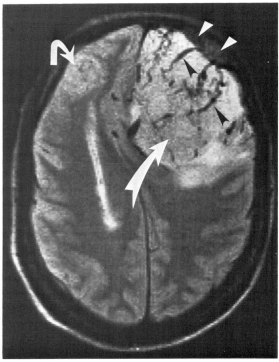

B

Fig. 5-35 Meningioma. (A) T_1-weighted MRI shows a large, slightly hypointense mass in the left frontal fossa (*arrowheads*). The brain is compressed and displaced away from the mass. A second, smaller meningioma is present in the right frontal region (*curved arrow*). **(B)** Lightly T_2-weighted MRI shows the meningioma as hyperintense (*large white arrow*). Flow-voids of large vascular channels are present within the tumor (*black arrowheads*). The smaller, right frontal meningioma is also hyperintense (*curved white arrow*). Note the thickened calvarium (*white arrowheads*). **(C)** Heavily T_2-weighted image shows the meningioma as being less intense, a characteristic finding (*large white arrow*). Right frontal meningioma shows the same signal change (*curved white arrow*). Also note the slight edema in the adjacent brain (*open arrow*). (*Figure continues.*)

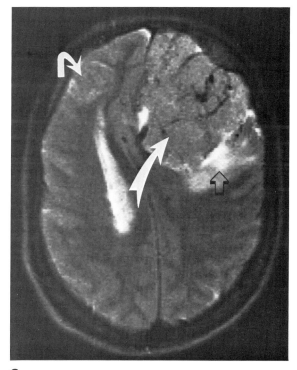

C

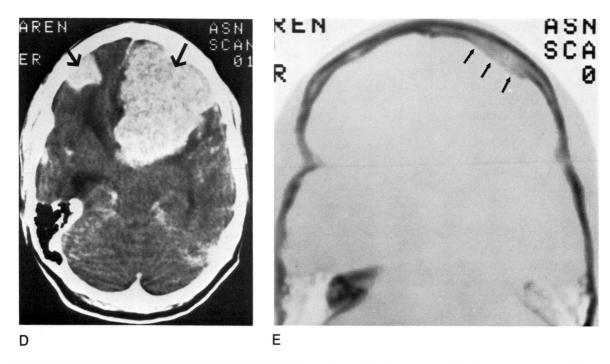

D   E

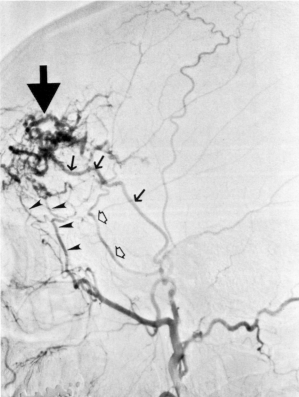

F

Fig. 5-35 *(Continued).* **(D)** Contrast-enhanced CT scan shows the homogeneously enhancing meningiomas *(arrows).* **(E)** Bone windows show local hyperostosis of the inner table of the left frontal bone *(arrows).* **(F)** Selective left external carotid angiography shows the marked hypervascularity of the tumor *(large arrow).* It is supplied by anterior branches of the superficial temporal artery *(small arrows),* ethmoidal branches of the internal maxillary artery *(arrowheads),* and (minimally) the middle meningeal artery *(open arrows).*

Fig. 5-36 Meningioma, hemangioma.
(A) Extensive hyperostosis of the frontal and parietal bones from a large meningioma. Note the enlarged vascular grooves in the calvarium caused by both meningeal and superficial vessels *(arrowheads)*. **(B)** The hyperostosis of meningioma must be distinguished from other lesions that expand or thicken the inner table. The typical hemangioma expands the calvarium without causing hyperostosis *(open arrows)*. The diploic space is replaced with enlarged vascular channels and intervening radially oriented bone trabeculae *(arrow)*. *(Figure continues.)*

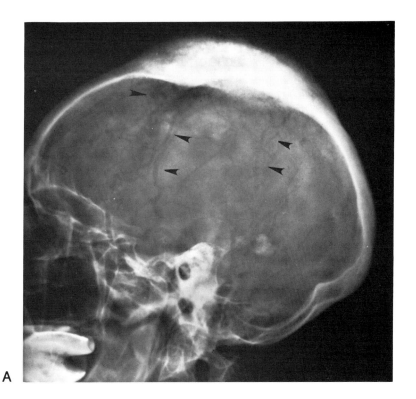

A

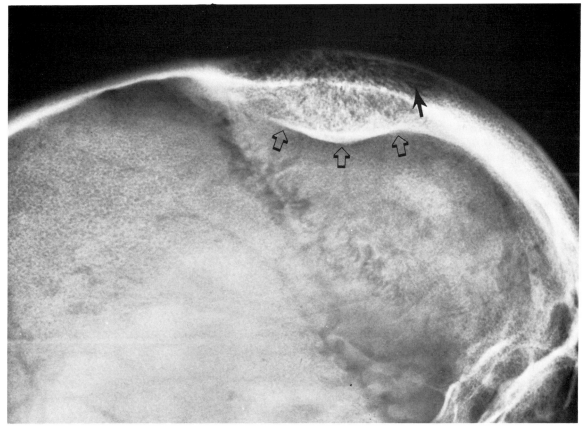

B

263

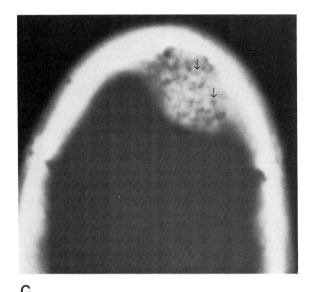

C

Fig. 5-36 *(Continued)*. **(C)** CT scan of the hemangioma demonstrates the radially oriented vascular channels *(arrows)* with intervening bone trabeculae.

Large vascular tumors may be embolized before surgical removal to decrease the vascularity of the tumor so that surgical removal is easier. Gelfoam, Ivalon, coils, or other material is injected into the external carotid artery or subselectively into the arterial branches supplying the tumor. Proximal balloon occlusion catheters must be used to avoid inadvertent backflow embolization into the internal carotid circulation.

Differential Diagnosis
Although most meningiomas are obvious, the tumor may appear similar to other neoplastic masses. For example, it may be especially difficult to differentiate a meningioma from a peripheral lymphoma, as the two tumors often have similar MRI and CT scanning characteristics. Some cerebral metastases and gliomas have an appearance similar to that of meningioma, especially small peripheral enhancing lesions or cystic masses. If the dura is invaded, these tumors may have slight meningeal vascular supply, seen with angiography, but it is always less than with meningioma. It may be particularly difficult

to differentiate blastic calvarial metastasis (prostate or breast carcinoma) from meningioma, especially if there has been inward tumor invasion through the dura (Fig. 5-31).

A suprasellar meningioma must be differentiated from a pituitary macroadenoma. Both tumors show homogeneous enhancement and have similar characteristics on MRI and CT scans. Meningioma usually arises from the posterior planum sphenoidale and lies entirely above the sella within the suprasellar cistern (Fig. 5-41). Meningeal vascular supply is seen with angiography. Pituitary adenoma is almost always an intrasellar tumor, enlarging the sella and then extending into the suprasellar cistern. Coronal MRI and CT scans usually differentiate between the two tumors. Pituitary adenoma does not ordinarily show significant neovascularity on angiography.

A cerebellopontine angle meningioma must be differentiated from an acoustic neurinoma, exophytic pontine astrocytoma, and metastatic tumor, particularly melanoma. The acoustic neurinoma enlarges the internal auditory canal; a meningioma does not. Meningioma also shows meningeal vascular supply on angiography, whereas an acoustic neurinoma shows little neovascularity. On CT scan, large aneurysms occasionally simulate a meningioma in the suprasella region or cerebellar pontine angle cistern. MRI or angiography makes the correct diagnosis.

LYMPHOMA

Non-Hodgkin's B-cell lymphoma is an uncommon tumor of the CNS, accounting for about 2 percent of all intracranial tumors, although the incidence is increasing. Primary CNS lymphoma is much more common than metastatic lymphoma. The peak incidence is at about 50 years of age, although the tumor may occur at any age after 10 years. Persons who are immunodeficient, particularly those with T-cell deficiency, are at far greater risk of CNS lymphoma (e.g., congenital deficiencies, those with acquired immunodeficiency syndrome (AIDS), allograft recipients, cancer victims, and those undergoing cancer chemotherapy). CNS lymphoma afflicts 6 percent of those with AIDS.

Most primary lymphomas involve the deep supratentorial structures (basal ganglia, thalamus, corpus callosum) and the deep white matter of the cerebrum, al-

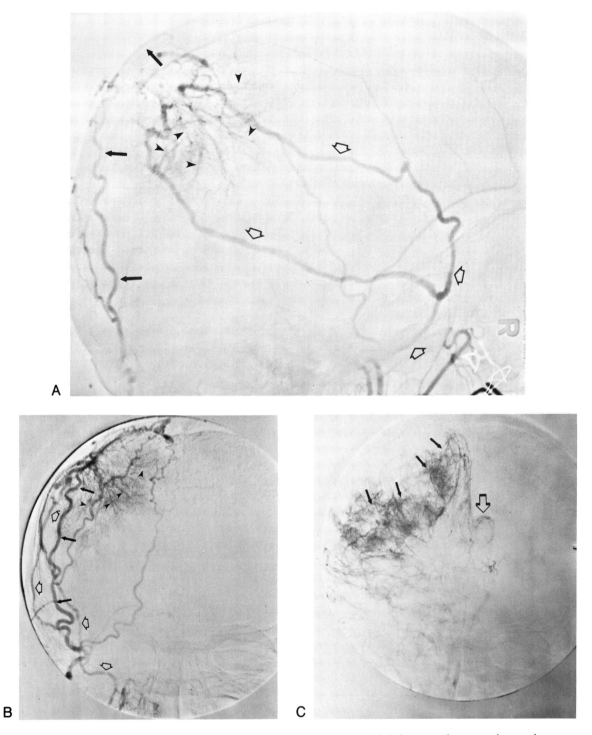

Fig. 5-37 Meningioma. CT scan of this case is shown in Figure 5-34. **(A)** Selective right external carotid artery arteriography shows gross enlargement of the posterior branches of the middle meningeal artery *(open arrows)*. This artery is the primary supply to the fine radial vessels of the tumor *(arrowheads)*. The posterior occipital scalp branch also supplies the tumor *(arrows)*. **(B)** AP view of Fig. A. The vessels are labeled with the same types of arrow. **(C)** Internal carotid artery injection in the AP projection. It shows the pial vascular supply to the inner portion of this large meningioma *(black arrows)*. Note the subfalcial herniation *(open arrow)*. *(Figure continues.)*

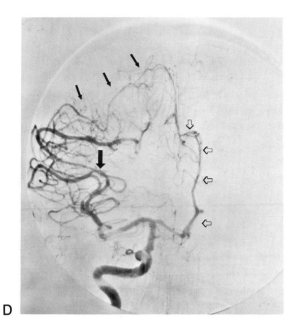

D

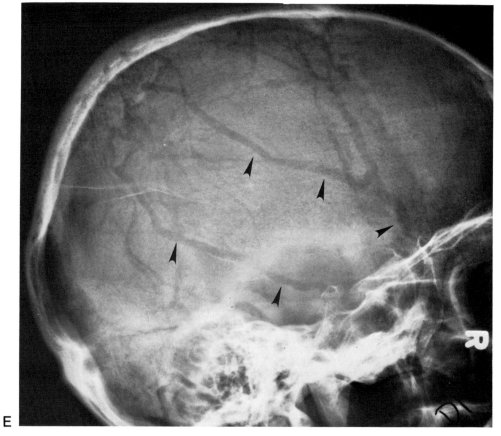

E

Fig. 5-37 *(Continued).* **(D)** Internal carotid artery injection in the AP projection. The arterial phase shows the displacement of the cortical vessels away from the inner table *(small black arrows)*, downward displacement of the sylvian point *(large black arrow)*, and subfalcial herniation with a "square" type shift of the pericallosal artery *(open arrows)*. The square shift is commonly seen with posterior masses. **(E)** Skull radiograph showing abnormal vascular channels corresponding to the enlarged meningeal vessels demonstrated in Fig. A *(arrowheads)*.

266

Table 5-4 Characteristic Meningeal Vascular Supply to Meningioma

Location	Arteries
Convexity, parasagittal	Middle meningeal artery
Falx	Superficial temporal artery Anterior falx artery
Subfrontal	Penetrating dural arteries from the ethmoidal branches
Parasellar	Meningohypophyseal vessels Cavernous dural branches
Tentorial	Tentorial branch of meningohypophyseal trunk
CPA	Penetrating branches from the ascending pharyngeal artery
Posterior fossa	Posterior meningeal branches from the vertebral artery
Intraventricular	Choroidal artery (anterior, posterolateral, PICA)

though the tumor may be found anywhere in the brain, including the retina and vitreous of the eye. The disease may be confined to the meninges and the cranial nerves and may mimic meningioma. Paraventricular location is characteristic. Single and multiple lesions are present with about equal frequency. With multiple lesions, the deep structures are almost always involved. Basal ganglionic lymphoma with central necrosis is characteristic of lymphoma with AIDS. There is usually less mass effect than would be expected for the size of the tumor, because of relatively little surrounding edema. Response to glucocorticoid therapy and radiation therapy may be dramatic, with the tumor disappearing within days. However, considerable clinical disability remains despite apparent disappearance of the tumor. Guided needle biopsy is excellent for diagnosis, immunologic typing for therapy, and for finding associated infection, such as toxoplasmosis, in immunodeficient patients.

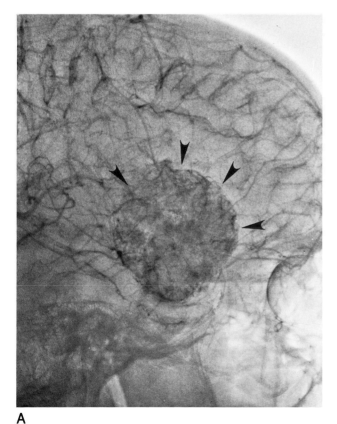

A

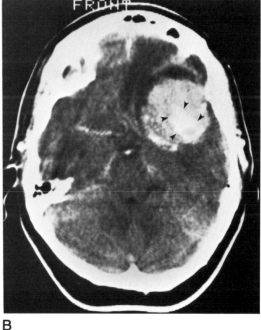

B

Fig. 5-38 Meningioma, sphenoid ridge. (A) Angiographic stain of the meningioma normally becomes homogeneously hyperintense during the capillary and venous phases *(arrowheads.)* **(B)** Contrast-enhanced CT scan. Note the dense calcium that is barely seen within the enhanced tumor *(arrowheads).*

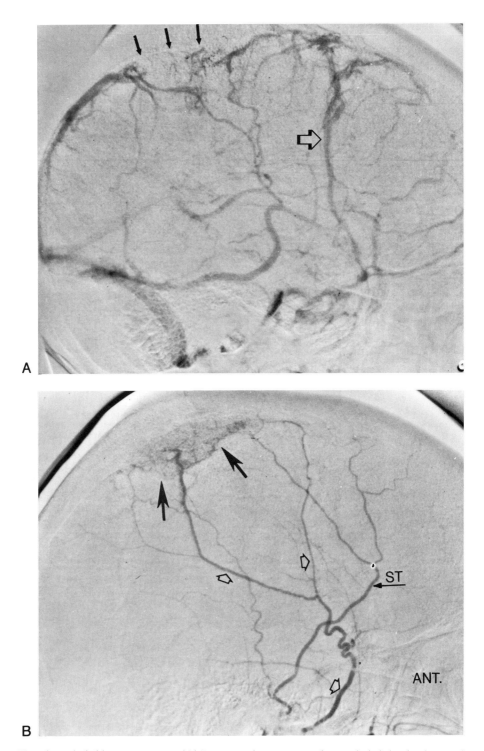

Fig. 5-39 Vessels occluded by meningioma. (A) Parasagittal meningioma has occluded the dural sagittal sinus (*black arrows*). Note the collateral veins draining the frontal region (*open arrow*). **(B)** Lateral selective external carotid artery arteriography. The parasagittal meningioma (*black arrows*) is supplied by branches of the middle meningeal artery (*open arrows*). The "blush" outlines the tumor. ST, superficial temporal artery. (*Figure continues.*)

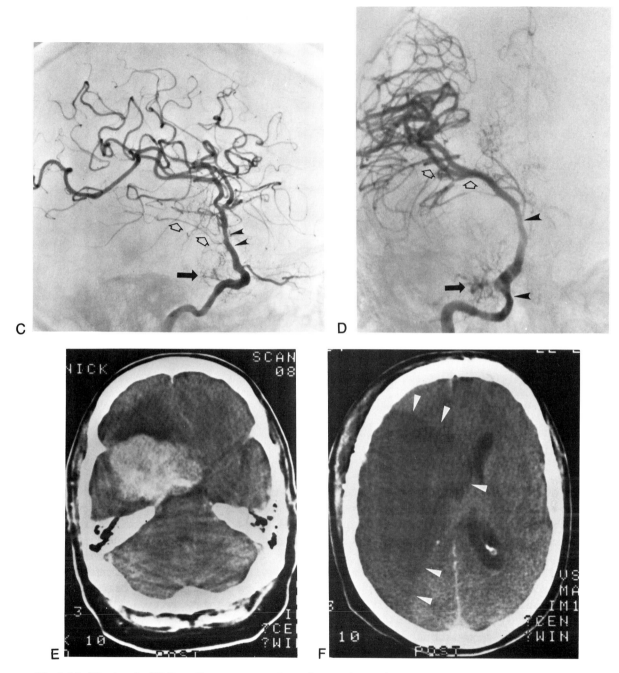

Fig. 5-39 *(Continued)*. **(C)** Parasellar meningioma. Lateral internal carotid artery injection shows elevation of both the supraclinoid internal carotid artery and horizontal middle carotid artery. Vessel irregularity indicates tumor encasement *(arrowheads)*. Hypertrophied branches from the meningohypophyseal trunk supply the tumor *(arrow)*. There is gross elevation of the sylvian vessels and of the anterior choroidal artery *(open arrows)*. **(D)** AP view of the same injection showing constriction of the internal carotid artery *(arrowheads)*, elevation of the middle carotid artery *(open arrows)*, and hypertrophy of the meningohypophyseal vessels *(black arrow)*. **(E)** Contrast-enhanced CT scan of the meningioma of the right sphenoid wing and parasellar region. **(F)** Postoperative infarction of the right hemisphere from right middle carotid artery occlusion *(arrowheads)*. Infarctions may result from thrombosis caused by the tumor constriction or by manipulation at surgical excision.

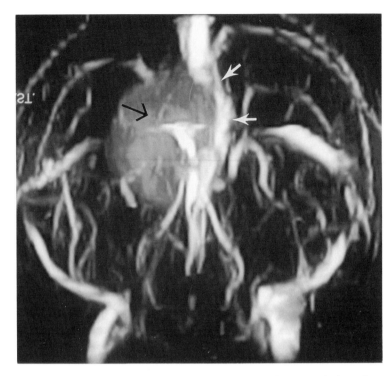

Fig. 5-40 Meningioma, compression of sagittal sinus. Venous MRA in the coronal plane shows a parasagittal falx meningioma *(black arrow)* (see Fig. 5-32B). The dural sagittal sinus is deviated toward the left and compressed but is not occluded *(white arrows)*.

On the nonenhanced CT scan, the tumors are usually seen as large (greater than 4 cm), homogeneous, isodense, or slightly hyperdense, often well-demarcated lesions. There is intense homogeneous enhancement with contrast in 70 percent of lymphomas (Fig. 5-42). In 20 percent, particularly those with AIDS, ring-like enhancement occurs and is identical to that toxoplasmosis or abscess. Surrounding edema is variable but is often minimal or absent. The tumor may infiltrate the subependymal regions diffusely, outlining the lateral ventricles.

The full extent of the edema is shown better on MRI than on CT scanning. The tumor is slightly hypointense ion T_1- and slightly hyperintense on T_2-weighted images (Fig. 5-42). It may be difficult to define the tumor within the edema without Gd enhancement. Rare gyriform enhancement of peripheral tumor mimics infarction or meningitis.

Metastatic non-Hodgkin's lymphoma from somatic sites occurs only rarely. It tends to have a peripheral location, having infiltrated the brain from the meninges. Hodgkin's lymphoma rarely involves the brain parenchyma, although it may involve the meninges. When peripheral, lymphoma in the brain resembles meningioma. Biopsy or angiography may be necessary to differentiate between the two tumors.

PRIMARY NERVE CELL TUMORS

Primary nerve cell tumors arise from neurons within the brain or ganglion cells outside the brain. These tumors account for fewer than 0.5 percent of brain tumors and are found most commonly in persons under 30 years old. They account for 4 percent of CNS tumors in children. Seizures are the most common clinical symptom, and neurologic deficit is rare. Most tumors grow slowly

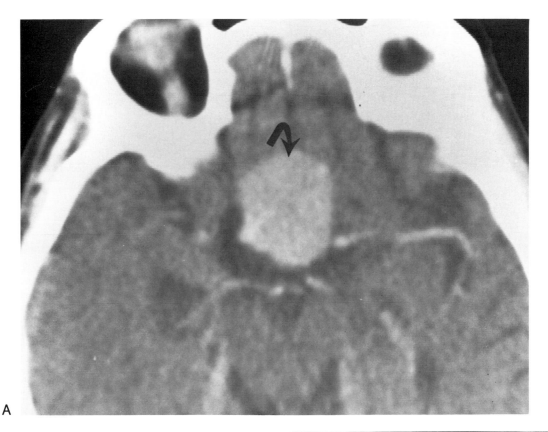

A

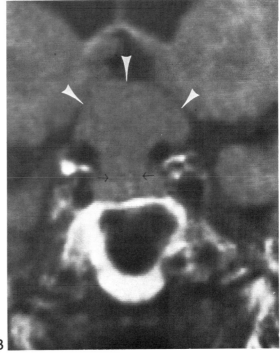

B

Fig. 5-41 Meningioma, suprasellar. (A) Transaxial CT scan shows a homogeneously enhancing mass within the suprasellar cistern *(arrow)*. This mass cannot be differentiated from a pituitary adenoma with suprasellar extension. **(B)** Coronal T$_1$-weighted MRI through the sella turcica demonstrates the homogeneous isointense mass within the suprasellar cistern *(arrowheads)*. The mass is separated from the slightly more hyperintense normal pituitary gland by the diaphragma sellae *(arrows)*. The sella is normal in size. In this case, MRI makes the correct diagnosis. With a pituitary macroadenoma with suprasellar extension, the sella is almost always enlarged (see also Fig. 5-32B).

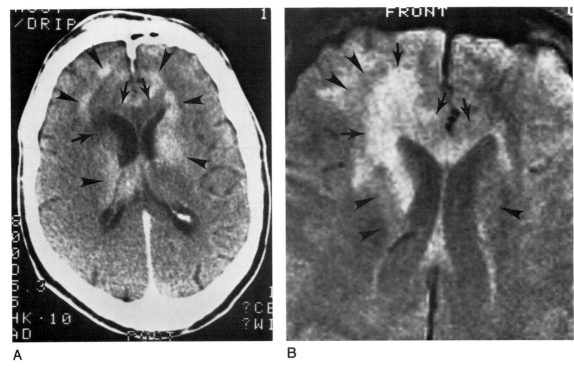

A   **B**

Fig. 5-42 Lymphoma. (A) Contrast-enhanced CT scan shows homogeneous enhancement of lymphoma involving the deep paraventricular structures bilaterally, including the basal ganglia and thalamus *(arrowheads)*. Edema is present in the corpus callosum and adjacent to the right frontal horn *(arrows)*. **(B)** Lightly T_2-weighted MRI shows the edema in the corpus callosum and adjacent to the right frontal horn as regions of high signal *(arrows)*. The tumor itself shows only minimally increased T_2 signal and is difficult to identify *(arrowheads)*.

and may have been present for years before presentation. They are named according to an ascending grade of malignancy: gangliocytoma, ganglioglioma, ganglioneuroblastoma, and neuroblastoma. Neuronal or glial ectopia and hypoplasia of the corpus callosum often accompany these tumors. The more malignant primitive type of neuronal tumors are discussed in Chapter 6.

Most of these tumors arise within the medial portion of the temporal lobe adjacent to the atrium of the lateral ventricle, although a few may also be found within the cerebellum or deep structures, particularly the floor of the third ventricle. Characteristically, the tumor is moderately large, with large cysts, calcification, and a peculiar tendency to produce local atrophy and adjacent ventricular dilatation (Fig. 5-43). Contrast enhancement may be seen within the solid portions of the tumor, but it does not correlate with the degree of malignant change. The tumor appears similar to oligodendroglioma and ep-

endymoma. A calcified tumor in a cerebellar hemisphere of an adult is usually one of the neuronal series.

EPIDERMOID TUMOR

An epidermoid tumor is a congenital lesion. It is a cyst lined by squamous epithelium and filled with the debris of desquamation. Most commonly it occurs in the cerebellopontine angle (CPA) cistern, the suprasellar cistern, and the temporal fossa. It may also occur, however, within the calvarium, where it appears on a skull radiograph as a well-outlined round lytic lesion with a slightly sclerotic margin (Fig. 5-44A).

On CT scan, an epidermoid tumor is a well-circumscribed mass that is usually hypodense and equivalent to CSF and that shows no contrast enhancement. It is occasionally isodense or even hyperdense (Fig. 5-44B,C,D). It may resemble an arachnoid cyst. Rarely, there is a rim of calcification.

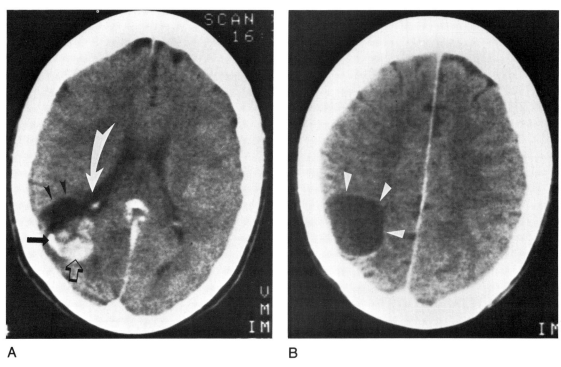

A   B

Fig. 5-43 Ganglioglioma. (A) Tumor is present in the medial posterior temporal lobe. Note cyst *(arrowheads)*, contrast enhancement *(open arrow)*, calcification *(black arrow)*, and local atrophic change with dilation of the atrium of the right lateral ventricle *(large white arrow)*. **(B)** With neuronal tumors, the cyst is characteristically large *(arrowheads)*.

With MRI, it is hypointense (close to CSF intensity) on T_1-weighted images and moderately hyperintense to CSF on T_2-weighted images. Inhomogeneity within the cyst is common. T_2-weighted MRI characteristics differentiate the lesion from an arachnoid cyst. The tumor may become large without causing symptoms (Fig. 5-44E,F).

Sella and Juxtasella Tumors; The Pituitary

PITUITARY ANATOMY

The pituitary gland is composed of two main parts: the adenohypophysis (anterior pituitary lobe), and the neurohypophysis (the posterior pituitary lobe). The adenohypophysis consists of a number of different groups of cells that secrete the trophic hormones of the gland. The neurohypophysis comprises the median eminence of the hypothalamus, the infundibulum (pituitary stalk), and the posterior lobe. The hypothalamus controls the output of the anterior lobe through a portal venous system

within the pituitary stalk. Interruption of the stalk causes hypopituitarism. The poor arterial blood supply to the anterior lobe makes it, and tumors within it, susceptible to hemorrhage. The posterior lobe is a direct extension of the brain.

PITUITARY HORMONES

Adenohypophysis
    Prolactin
    Thyroid-stimulating hormone (TSH)
    Follicle-stimulating hormone (FSH)
    Adrenocorticotropic hormone (ACTH)
    Growth hormone (GH)
Neurohypophysis
    Vasopressin (ADH)
    Oxytocin

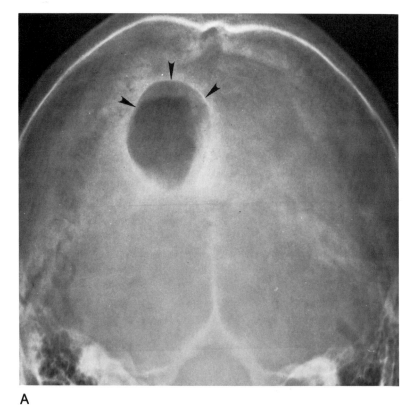

A

Fig. 5-44 Epidermoid. (A) Skull radiograph demonstrates a purely lytic lesion of the calvarium with sharply defined sclerotic margins. It is a characteristic picture. Compare with Figure 6-26B. **(B)** Epidermoid within the left sylvian fissure. CT scan typically shows an epidermoid as very hypodense. There is minimal mass effect, as with most slowly enlarging tumors or cysts. **(C)** Epidermoid. Rarely, an epidermoid has high density contents. This one extends through the tentorium into the cerebellopontine angle (CPA) cistern (arrow). *(Figure continues.)*

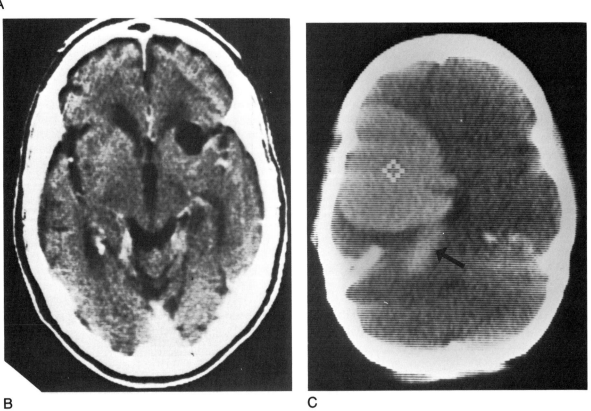

B   **C**

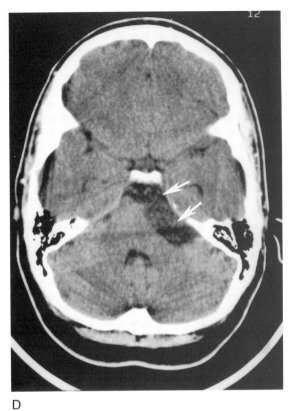

D

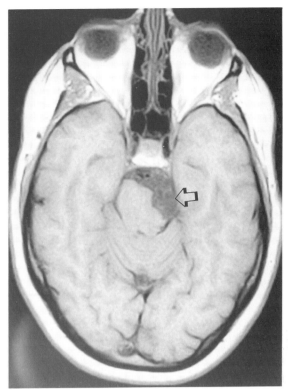

E

Fig. 5-44 *(Continued)*. **(D)** Axial CT scan shows a left CPA epidermoid tumor as a slightly inhomogeneous mass *(arrows)*. **(E)** T₁ MRI shows the epidermoid as a hypointense mass nearly equivalent to CSF *(open arrow)*. **(F)** FSE T₂ MRI shows the epidermoid as generally bright with a slightly inhomogeneous internal matrix of varying hypointensity *(arrow)*. The inhomogeneous matrix differentiates the epidermoid from an arachnoid cyst. (Fig. 5-44B from Dubois, 1984, with permission.)

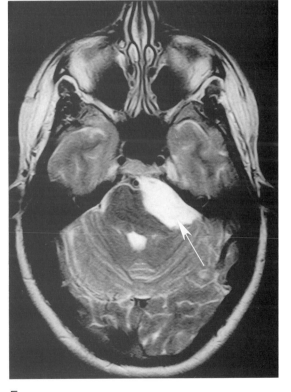

F

From birth to 2 months of age, the anterior pituitary gland has hyperintensity on T_1-weighted images thought to reflect hypersecretion. From then on, the normal anterior pituitary gland has signal intensity approximating that of the pons, although occasionally it will become bright again and larger during normal pregnancy (Fig. 5-45). The gland occupies the inferior sella turcica, averaging 4 to 8 mm in height. It is a dynamic organ, so its size increases up to 12 mm in height with puberty, pregnancy, precocious puberty, and, to a lesser degree, the middle of the menstrual cycle. It tends to grow larger in women and may protrude slightly into the suprasellar cistern. Its upper border may be straight or it may have slight upward or downward convexity. The pituitary stalk is seen as a thin midline structure, extending downward from the infundibulum of the hypothalamus into the posterior portion of the gland. It may enlarge with the gland up to 4 mm in width. The stalk may normally angulate slightly away from the midline, and its position cannot be used to indicate a pituitary tumor.

With bolus injection of iodinated or Gd contrast, there is immediate strong enhancement of the infundibulum and posterior lobe. The anterior lobe, which has a poorer arterial blood supply, enhances about 1 minute later. Small tumors within the gland (microadenomas) enhance little and later. Small nonenhancing foci that resemble microadenomas are found within the gland in about 20 percent of clinically normal people. The precise nature of these is often unknown unless defined at sur-

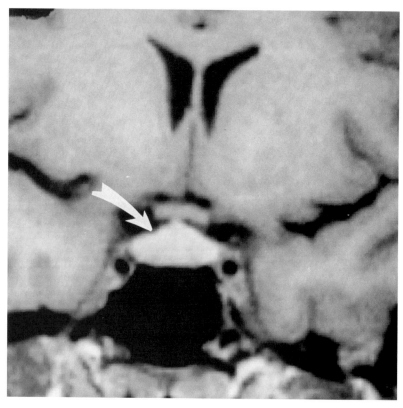

Fig. 5-45 Normal pituitary gland, pregnancy. Coronal nonenhanced CT scan through the sella illustrates an enlarged, homogeneous, hyperdense pituitary gland. This is normal in some persons during pregnancy and reflects normally increased pituitary activity. The other time of hyperdense pituitary gland is during the postnatal period. In adolescent girls, the pituitary gland becomes slightly enlarged, but it is not usually hyperdense.

gery or autopsy. Some of the hypodensities may be image artifacts. Autopsy examinations of the pituitary have shown small nonfunctioning adenomas, cysts, focal infarctions, or occult metastases in 25 to 50 percent of cases. This creates a dilemma for management. Most are ultimately ignored unless something develops clinically.

The normal pituitary is bright in the posterior portion on routine T_1-weighted images—referred to as the neurohypophyseal "bright spot." This hyperintensity reflects the normal metabolic activity of the neurohypophysis, although the precise molecular structures responsible are unknown. It is seen in about 90 percent of routine scans. The high signal may be displaced to the region of the median eminence whenever there is interruption of the axonal flow of neuropeptides in the hypothalamic–hypophyseal tract. This occurs with tumors in the suprasellar cistern and congenital absence or injury to the stalk. The high signal in the neurohypophysis is absent in patients with diabetes insipidus. However, technical failure is the most common cause of nonvisualization of the bright spot with MRI.

Diabetes insipidus is the excessive clearance of water through the kidneys. It is called nephrogenic diabetes insipidus when there is a defect within the kidney, so that it does not respond to the influence of antidiuretic hormone (ADH). It is called central diabetes insipidus when there is failure of adequate secretion of ADH into the bloodstream. Central diabetes insipidus is caused most commonly by head trauma (50 percent) and by local tumors or infarcts (25 percent). Twenty-five percent of cases are idiopathic. The bright spot is absent in most patients with central diabetes insipidus. If the bright spot persists, the patient most likely has primary polydipsia, a condition that may clinically mimic true diabetes insipidus or, less likely, a very rare form of diabetes insipidus (less than 1 percent of cases).

PITUITARY ADENOMA

Pituitary adenoma is the most common tumor of the sella turcica and suprasellar cistern. This benign, slow-growing tumor arises from the adenohypophysis (anterior lobe). Only rarely is it invasive or malignant. In adults, pituitary tumors account for about 10 percent of primary intracranial tumors. The tumor is rare in children. Seventy percent of these tumors produce a hormone. Small tumors (less than 10 mm) are called microadenomas. Large tumors (more than 10 mm) are called macroadenomas. The sella turcica is usually normal with a microadenoma but is expanded with a macroadenoma.

A small pituitary tumor (i.e., confined within the sella) is usually suspected by the clinical syndrome of hormone excess. A prolactin-secreting adenoma (prolactinoma) is the most common; it produces amenorrhea, galactorrhea, and decreased libido in women and hypogonadism and decreased libido in men. The tumor is 10 times more common in women. Hyperprolactinemia with the serum prolactin level more than 100 ng/ml indicates a high probability of a pituitary tumor. Excessive growth hormone (GH) production produces acromegaly in the adult and gigantism in adolescents. Excessive adrenocorticotropic hormone (ACTH) or thyroid-stimulating hormone (TSH) produces Cushing's disease or hyperthyroidism. Small nonsecreting tumors are clinically unrecognized.

Nonsecreting tumors are almost always large at diagnosis, expanding the sella (Fig. 5-46) and frequently extending into the suprasellar cistern. Within the suprasellar cistern, the tumor may compress the optic chiasm, producing variable visual-field deficits, with bitemporal hemianopsia the most common. Much less commonly, there may be cranial nerve (CN) III, IV, V, and VI abnormalities from lateral invasion or extension into the cavernous sinus. In rare cases, a sudden hemorrhage into the tumor causes severe headache, often with acute chiasmal compression or ocular nerve palsy from rapid expansion of the tumor. This condition is referred to as *pituitary apoplexy*. Large tumors may extend high in the brain to obstruct the foramina of Monro causing hydrocephalus. Headache is common and may be the only presenting symptom with pituitary tumor.

Highly aggressive tumors erode the floor of the sella and infiltrate the dural sinuses or upper clivus. Some break through the dura into the temporal or posterior fossa. These invasive adenomas are refractory to definitive treatment by either surgery or irradiation.

The previous histologic classification of the tumor as chromophobic, eosinophilic, or basophilic has generally been replaced by classification based on the presence and type of hormonal secretion. Classification is not important for prognosis. Metastasis or spread into the subarachnoid spaces has been reported, but it is rare.

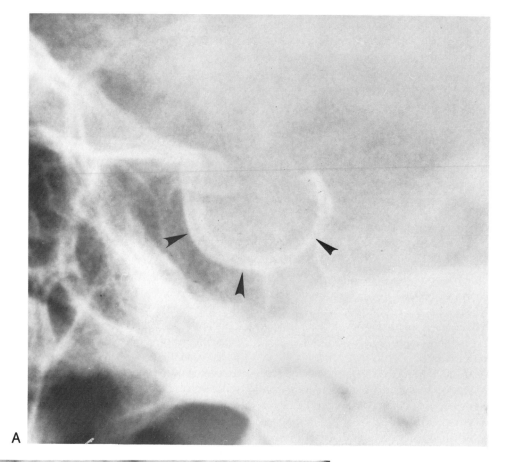

A

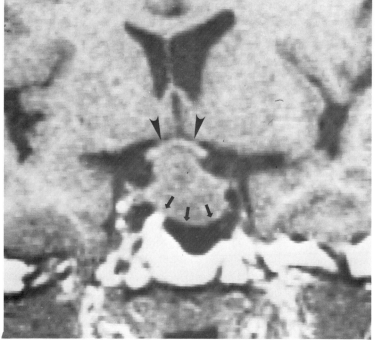

B

Fig. 5-46 Pituitary macroadenoma. (A) Lateral skull radiograph showing spherical enlargement of the sella turcica (*arrowheads*). The dorsum is thinned. The "double floor" appearance is caused by asymmetric enlargement. **(B)** Coronal T_1-weighted MRI shows a homogeneous isointense mass within the sella turcica causing its enlargement (*small arrows*). Suprasellar extension of the tumor causes upward displacement of the optic chiasm (*arrowheads*).

Fig. 5-48 Pituitary adenoma with chronic hemorrhage. (A) T$_1$-weighted MRI in the coronal plane shows a pituitary macroadenoma with suprasellar extension (*arrowheads*). The remains of a hemorrhage into the tumor or cyst is seen as high signal intensity (*curved arrow*). **(B)** Contrast-enhanced CT scan of the same patient shows the suprasellar extension of the contrast-enhancing tumor (*arrowheads*). The low-density region represents the old hemorrhage or hemorrhagic cyst (*curved arrow*). The floor of the sella has resorbed and is expanded downward. Laterality is reverse of MRI image.

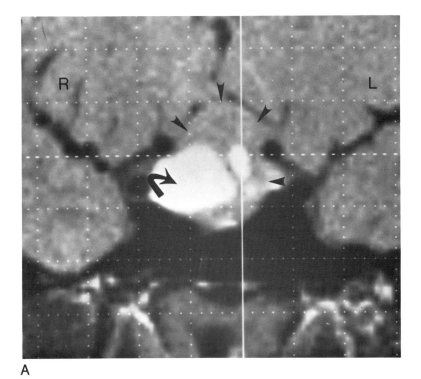

A

B

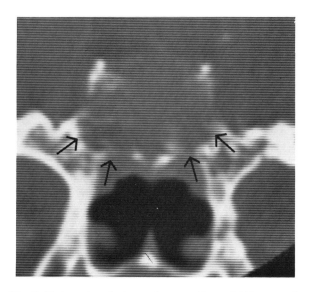

Fig. 5-49 Invasive pituitary adenoma. Coronal CT scan with a wide window shows that the tumor has extended inferiorly to fill the sphenoid sinus and cause destruction of the base of the skull (*arrows*). A sphenoid sinus carcinoma might produce the same findings.

CT adequately demonstrates the important features of macroadenoma. Coronal contrast-enhanced thin section imaging is the most useful. On the noncontrast CT scan, a pituitary macroadenoma usually appears as a sharply marginated, homogeneous, moderately hyperdense (35 to 50 HU) mass within the sella and suprasellar cistern. The sella is enlarged. Ten percent of these tumors are hypodense or of mixed density because of the presence of cysts or old hemorrhage within the tumor. Density greater than 50 HU implies recent hemorrhage (apoplexy). Calcification within the tumor is rare but may occur as either a thin rim along the tumor margin or diffusely within the tumor.

Most commonly, there is uniform contrast enhancement of the adenoma, with characteristics similar to those of a meningioma. Mixed-density enhancement may occur (Fig. 5-48). The nonenhancing regions within the tumor usually are cysts or regions of resolving hemorrhage. There is no correlation between enhancement and hormonal function or malignancy of the tumor.

Invasive tumors cause either erosion of the floor of the sella or more diffuse destruction of the clivus, simulating a sphenoid sinus carcinoma (Fig. 5-49). CT scanning with bone window imaging is superior to MRI for detecting invasion of the clivus. Invasion of the cavernous sinus is difficult to define with CT scans, but it can be diagnosed when there is enlargement of a cavernous sinus or obvious tumor beyond the lateral or posterior wall of the cavernous sinus (Fig. 5-50B).

After surgical resection of a pituitary macroadenoma, often considerable non-neoplastic tissue remains in the region of the original tumor that slowly resolves over 2 to 3 months (Fig. 5-51). Therefore, the amount of residual tumor cannot be accurately evaluated until after this period.

Tumor recurrence is seen in about 12 percent of patients and usually becomes apparent after 4 to 8 years. The sella is often packed with fat or muscle at surgery, making recognition of recurrent tumor more difficult. Initially, the fat displays high intensity on MRI; the intensity decreases with time, however, so that after 1 year it is isointense with other intrasellar tissues and cannot be differentiated from recurrent tumor. Baseline postoperative scans are obtained 3 months after surgery so that future changes in the amount of intrasellar tissue will be apparent.

Microadenoma
The preferred technique for detecting a microadenoma is MRI. The examination is performed in the coronal plane using thin-section (3-mm) T_1-weighted imaging or T_1-weighted Gd-enhanced imaging. T_2-weighted images are not as sensitive for this lesion.

A pituitary microadenoma is seen as a small (less than 10 mm) round region of hypointensity within the pituitary gland on T_1-weighted sequences. It is isointense or slightly hyperintense on T_2-weighted sequences. Prolactin-secreting microadenomas are most commonly posterolateral, often enlarging the ipsilateral portion of the gland. It may displace the pituitary stalk contralaterally, but this is not a reliable diagnostic sign.

Gd-contrast enhancement increases the sensitivity of the MRI when the nonenhanced T_1-weighted images are nondiagnostic. The scan is performed as soon as possible after infusion of contrast. The normal gland and the cavernous sinus show increased signal intensity. The tumor is seen as a small round region of nonenhanced

Fig. 5-50 Pituitary adenoma, invasion of the cavernous sinus and upper clivus. (A) Gd-DTPA contrast-enhanced T$_1$-weighted MRI shows an enhancing adenoma invading laterally into the cavernous sinus. The sinus is enlarged, and the tumor surrounds the internal carotid artery *(black arrow)*. The right cavernous sinus *(black arrowheads)* is normal. **(B)** Contrast-enhanced CT scan in the coronal plane also shows the tumor invading and enlarging the right cavernous sinus *(black arrows)* surrounding the contrast-filled internal carotid artery *(white arrow)*. The normal size and enhancement of the right cavernous sinus is seen *(arrowheads)*. This lesion is an invasive adenoma with inferior extension and destruction of the upper portion of the left side of the clivus *(open arrow)*. Laterality of CT is reverse of MRI.

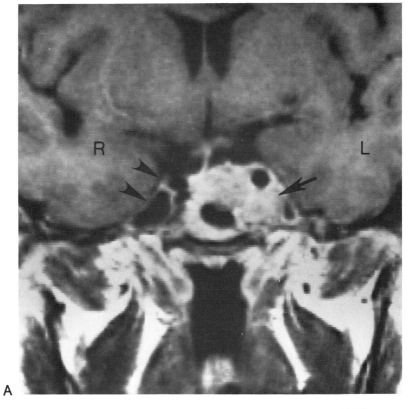

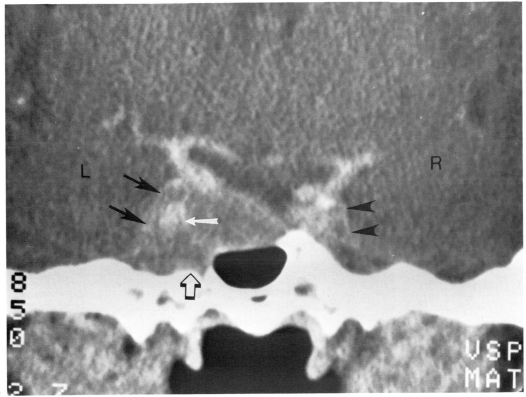

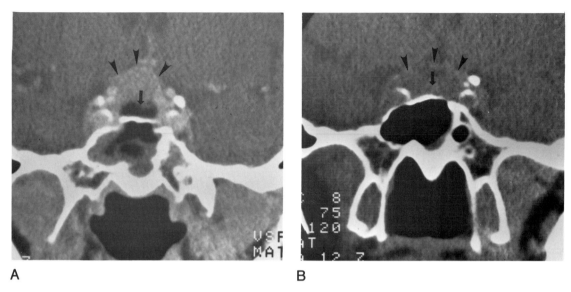

A   B

Fig. 5-51 Pituitary adenoma, postoperative evaluation. (A) Postoperative contrast-enhanced CT scan in the coronal plane 2 weeks after transsphenoidal hypophysectomy shows considerable abnormal enhancing tissue remaining within the sella and suprasellar cistern *(arrowheads)*. Fat packing is seen in the inferior sella turcica *(arrow)*. There are postsurgical changes within the sphenoid sinus. **(B)** CT scan without contrast done 3 months after surgery shows that the tissue mass has decreased considerably in size without treatment *(arrowheads)*. The density of the fat has increased *(arrow)*. The changes within the sphenoid sinus have resolved.

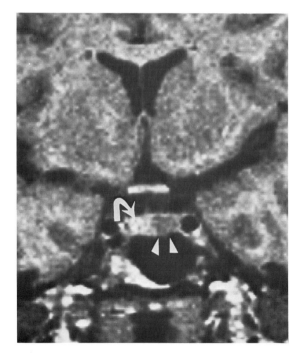

Fig. 5-52 Pituitary microadenoma. Coronal Gd-DTPA-enhanced T_1-weighted MRI shows the nonenhancing adenoma *(arrowheads)* displacing the intensely enhanced normal pituitary gland *(curved arrow)*.

low signal intensity within the intensely enhancing gland (Fig. 5-52). With time, the enhancement within the pituitary and the microadenoma equalize, and the tumor becomes invisible. Rarely, a microadenoma will enhance greater than the gland on the early postinjection images. Microadenomas frequently invade the dura of the medial margin of the cavernous sinus. However, this dural membrane is so thin that it is not reliably imaged with MRI, and the invasion cannot be defined.

CT is an acceptable technique for evaluating microadenomas, although compared with MRI it is slightly less sensitive for detection of the tumor. Thin-section (1.5- to 2-mm) high-resolution coronal imaging is used. The patient lies prone, with the neck extended as far as possible. The gantry is angled so the plane of the scan is as close as possible to true coronal. The scan sections must fall away from any dental hardware. Bone images are also obtained. High-dose (40 to 60 g of iodine) intravenous contrast is essential and is best given as a rapid bolus just before rapid-sequence scanning. This method densely opacifies the pituitary gland, cavernous sinus, and internal carotid arteries.

Microadenoma most commonly appears as a focal low-density nonenhancing region within the densely enhancing pituitary gland (Fig. 5-53). The gland may be enlarged to more than 9 mm in height, with displacement of the infundibulum away from the side of the tumor. There is so much variation in the position of the stalk, however, that its position cannot be used to indicate the presence of an adenoma. A small, focal, downward bulge of the sella floor is sometimes seen adjacent to the tumor but by itself cannot be used to indicate a microadenoma. Rarely, microadenomas calcify, resulting in a "pituitary stone." The sensitivity of contrast-enhanced thin-section CT scanning is reported to be 80 to 85 percent, depending on the study. The criteria and approximate correlation with surgical findings are given in Table 5-5.

Because most prolactin-secreting pituitary adenomas are successfully treated with bromocriptine, identification of the adenoma is often unnecessary. The imaging is done primarily to exclude a macroadenoma or lesion of the hypothalamus. The diagnosis and follow-up can be done by clinical measurements. There is no longer any indication for plain-film tomography in the diagnosis of macroadenoma, and angiography is of no use.

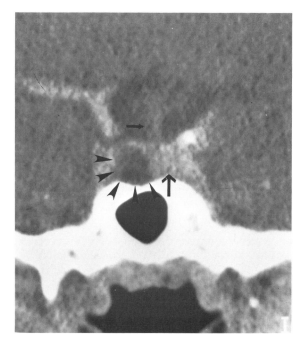

Fig. 5-53 Pituitary microadenoma, CT scan. Coronal contrast-enhanced CT scan shows the nonenhancing hypodense adenoma (*arrowheads*) displacing the normally enhanced pituitary gland (*arrow*). The height of the gland is increased. The pituitary stalk is displaced to the opposite side (*small arrow*).

Cushing's Disease

Pituitary Adenoma-Secreting Adrenocorticotropic Hormone

MRI detects a pituitary tumor in about 50 percent of those patients who are cured following hypophysectomy. MRI is superior to CT in the detection of microadenoma. Tumor detectability is related to size. MRI will detect 100 percent of adenomas 7 mm or greater in diameter, 80 percent at 4 mm, and a few at 3 mm or

Table 5-5 Criteria for CT Diagnosis of Microadenoma

Finding	Frequency of Finding (%)	Correlation with Histology (%)
Discrete hypodense lesion	50	90
Focal floor erosion or bulge	40	70
Focal upward convexity of gland	30	80
Pituitary height > 9 mm	10	100
Infundibular displacement	20	80

less. Contrast enhancement adds little increase in sensitivity, as only a few tumors are seen following contrast injection that are not seen on nonenhanced study. Some tumors seen on nonenhanced scans become invisible following contrast. When tumors are not visualized, bilateral simultaneous inferior petrosal sinus (IPS) sampling both before and after corticotropin-releasing hormone may differentiate Cushing's disease from Cushing syndrome (ectopic adrenocorticoid secretion). IPS corticoid levels two times that of peripheral vein concentration is diagnostic of a pituitary tumor when corticotropin-releasing hormone is not used and is 95 percent sensitive. Following corticotropin-releasing hormone challenge, a three times IPS-to-peripheral vein ratio is diagnostic and is 100 percent sensitive for the diagnosis of Cushing's disease in appropriately performed studies. The tumor may be lateralized by IPS sampling in 70 percent of cases. The technique of IPS sampling is safe when done skillfully. Knowledge of the variable anatomy of the IPS and of its anastamosis with the jugular vein is essential.

CRANIOPHARYNGIOMA

Craniopharyngioma has its peak incidence during childhood and adolescence, with a second peak in adults aged about 40 to 50 years. This tumor is discussed in Chapter 6.

DIFFERENTIAL DIAGNOSIS

Meningioma

Meningioma occurs in the sellar region, arising from the dura of the posterior portion of the planum sphenoidale, tuberculum sellae, diaphragm sellae, dorsum sellae, and lateral dura of the cavernous sinus. Meningioma in the suprasellar region may mimic a pituitary adenoma (Fig. 5-41). However, with meningioma, the sella is not enlarged and there is often characteristic sclerosis of the adjacent bone. Angiography shows the characteristic meningeal vascular supply.

Paraganglioma

Paraganglioma is an unusual tumor arising within the sella or suprasellar cistern from neural crest tissue incorporated within the pituitary gland. It shows intense contrast enhancement and frequently prominent intratumor vascularity seen as flow-void phenomenon.

LESIONS OF THE SELLA AND JUXTASELLA REGION

Common lesions
  Pituitary adenoma
  Meningioma
  Empty sella
  Giant aneurysm
  Invasive tumors of the sphenoid sinus and nasopharynx
Uncommon lesions
  Arachnoid cyst (suprasellar)
  Epidermoid
  Rathke's pouch cyst
  Metastasis
  Trigeminal schwannoma (Meckel's cave)
  Chordoma/chondrosarcoma
  Granuloma
  Cysticercosis (common in endemic regions)
  Paraganglioma

Giant Aneurysm

A giant aneurysm of the internal carotid artery may expand into the sella and suprasellar cistern. It may mimic a pituitary adenoma or meningioma. Contrast enhancement on CT scans is intense and occurs early. On MRI, flow-void is usually seen within the aneurysm. However, the signal intensity within an aneurysm may be inhomogeneous because of high signal clot or slow related enhancements. The sella is usually eroded and enlarged from one side, rather than the more central enlargement seen with a pituitary adenoma. Rim calcification is common with giant para sellar aneurysms. When there is any question of an aneurysm, MRI or bilateral carotid angiography is performed (see Ch. 3).

Empty Sella Syndrome

The empty sella refers to herniation of the suprasellar arachnoid into the sella turcica through an opening in an incompletely formed diaphragma sella. The sella then becomes filled with CSF. The CSF pulsations may enlarge the sella; on plain-film radiography, it mimics the changes of a pituitary adenoma. CT scanning and MRI can identify the fluid within the sella. Defining the pituitary stalk within the fluid is diagnostic of empty sella. The stalk is not seen with an intrasellar cyst or cystic tumor (Fig. 5-54). If necessary, CT positive-contrast cisternography can differentiate between cyst and empty sella. The empty sella fills with the contrast; the intrasellar cyst does not.

An empty sella is common, found in 10 to 20 percent of the general population, and is usually asymptomatic. Rarely, there is minor hypopituitarism due to posterior compression of the gland, visual field deficit due to "falling" of the optic chiasm into the sella, and a CSF fistula into the sphenoid sinus. It has been reported secondary to hydrocephalus or pseudotumor cerebri. Because it is so common, an empty sella may be found in association with other lesions, such as pituitary microadenoma.

Rathke's Pouch Cyst

Rathke's pouch cyst is a rare congenital endodermal cyst within the sella turcica, or suprasellar cistern. It may be of any size and has the typical appearance of simple cystic lesions on both MRI and CT scans. When large, these cysts are impossible to differentiate from a cystic macroadenoma.

Meckel's Cave Tumor (Trigeminal Neurinoma)

Meckel's cave is an invagination of the dura at the posterior cavernous sinus, adjacent to the posterior sella turcica. It contains the gasserian ganglion of the trigeminal nerve (CN V). Trigeminal neurinoma (schwannoma) is a rare tumor that occurs most often in the gasserian gan-

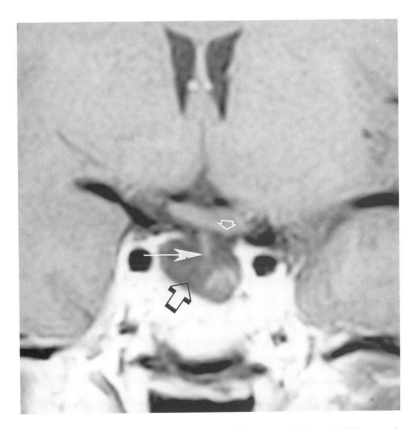

Fig. 5-54 Empty sella. Coronal T$_1$ MRI shows the expanded sella turcica filled with CSF equivalent intensity signal (*open arrow*). The pituitary stalk is outlined by the CSF as it enters the sella (*white arrow*). This finding differentiates an empty sella from one filled with a cyst. The chiasm is in a low position (*small open white arrow*), which is relatively common with empty sella. Other brain tissue may herniate into the sella as well.

glion within Meckel's cave. The tumor may become large, erode the lateral sella, and displace the proximal cavernous portion of the internal carotid artery anteriorly. On MRI, these tumors are hypointense on T_1-weighted images and hyperintense on T_2-weighted images. CT scans show intense homogeneous enhancement typical of a neurinoma. Angiography shows forward displacement of the posterior segment of the cavernous portion of the internal carotid artery, as well as neovascularity and blush in the tumor fed by vessels from the meningohypophyseal trunk (Fig. 5-55). It may be impossible to differentiate this tumor from a meningioma in the same region. Meckel's cave and tumors of the cavernous sinus are discussed in greater detail in Chapter 13.

Metastasis

Metastasis to the pituitary gland is uncommon. The deposit most often involves the posterior lobe, and commonly extends into the suprasellar cistern and anterior recesses of the third ventricle. It may grow entirely within the pituitary stalk in the suprasellar cistern. Breast and lung carcinomas are the most frequent tumors to metastasize to the pituitary gland. Diabetes insipidus and general pituitary hypofunction are the most common clinical abnormalities. The pituitary stalk may fail to enhance in those who have diabetes insipidus or in those whom its development is imminent. The metastatic tumor appears similar to a macroadenoma on CT scanning and MRI, but the sella does not expand as with

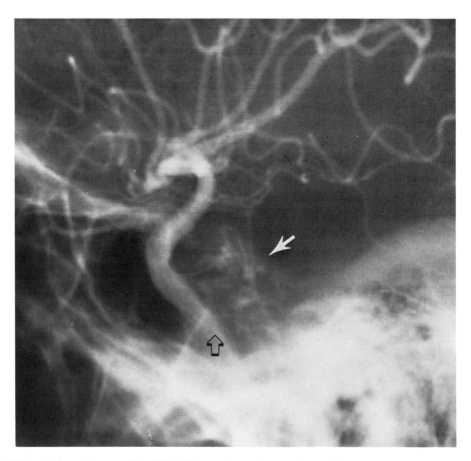

Fig. 5-55 Trigeminal neurinoma within Meckel's cave. Lateral internal carotid artery angiography demonstrates the characteristic anterior displacement of the proximal cavernous segment of the internal carotid artery *(open arrow)*. Meningohypophyseal vessels supply the tumor *(arrow)*.

macroadenoma. Metastasis to the sphenoid bone is more common, sometimes simulating an invasive adenoma; it is most commonly from breast or prostate carcinoma (Fig. 5-56A). Pain and CN abnormalities may result from metastatic tumor to the base of the skull. Blastic metastasis must be differentiated from fibrous dysplasia, which tends to thicken the bone (Fig. 5-56B). Langerhans cell histiocytosis (histiocytosis X) involves the anterior third ventricle and pituitary axis in children.

Sphenoid sinus carcinoma and nasopharyngeal carcinoma may spread into the sella and sphenoid bone. Lytic bone destruction occurs, which may be difficult to differentiate from a large invasive adenoma. A retropharyn-

geal or sinus mass suggests the correct diagnosis (Fig. 5-57).

Chordoma and Chondrosarcoma

Chordoma is a rare tumor that arises from notochord remnants in the middle clivus, just below the sella turcica. The tumor is slow-growing, but there is always extensive bone destruction and often invasion upward into the sella and cavernous sinuses. Calcification within the tumor is common. On MRI, the tumor shows prolonged T_1 and T_2 relaxation times. The presence of calcium is less reliably demonstrated. MRI is excellent for defining associated carotid artery constriction or invasion into the

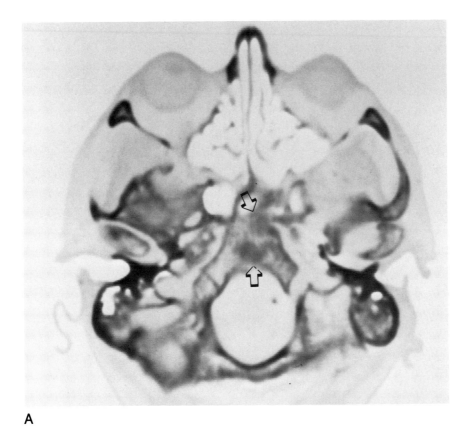

A

Fig. 5-56 Metastatic carcinoma, fibrous dysplasia. (A) Metastatic prostate carcinoma to the clivus. CT scan with bone windowing and reversed gray scale demonstrates the blastic reaction to the metastatic deposits (*arrows*). This technique is the most sensitive for demonstrating metastatic disease to the base of the skull. Prostate carcinoma frequently involves the base of the skull but only rarely involves the calvarium. (*Figure continues.*)

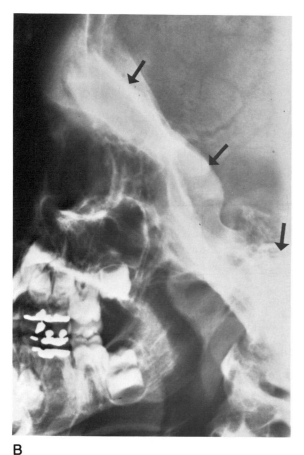

Fig. 5-56 *(Continued).* **(B–D)** Fibrous dysplasia. **(B)** Lateral skull radiograph shows dense sclerosis of the base of the skull *(arrows)*. Such sclerotic density may also be found with meningioma and prostatic metastasis. **(C)** Transaxial CT scan through the skull base shows the diffuse involvement. The bone is thickened, causing constriction of the foramen magnum *(arrow)*. Typical "ground glass" increased medullary density is present with fibrous dysplasia *(open arrow)*. **(D)** CT scan through the calvarium shows the typical features of fibrous dysplasia. The calvarium is thickened. The external table is expanded outward *(arrows)*, but there is little change of the internal table *(open arrows)*. The tables are not thickened. Between the tables the dysplasia produces fine bone spicules, which create the homogeneous increased density. Some regions may show greater density *(curved arrow)*.

B

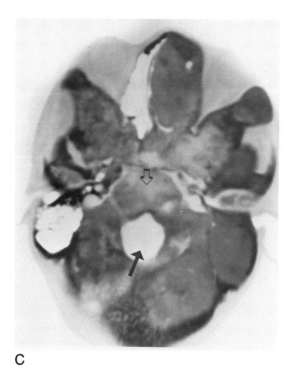

C

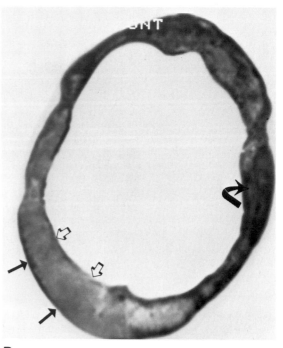

D

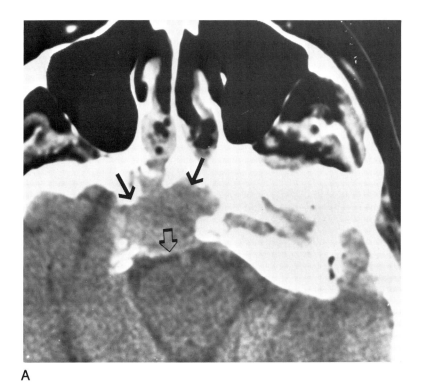

A

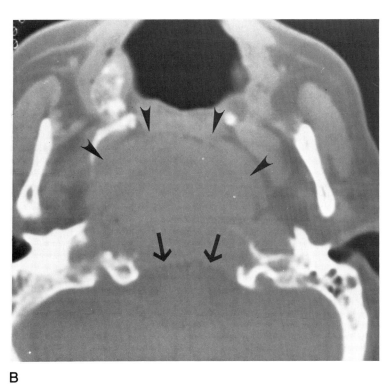

B

Fig. 5-57 **Nasopharyngeal carcinoma.** (A) Transaxial CT scan through the upper clivus and sphenoid sinus shows a large soft tissue tumor mass (*black arrows*). The mass has eroded posteriorly through the clivus into the prepontine cistern (*open arrow*). (B) Scan at a lower level shows the large nasopharyngeal mass (*arrowheads*). It has completely eroded the clivus posteriorly (*arrows*).

posterior fossa. The tumor cannot be differentiated from the even more rare chondrosarcoma, although chondrosarcoma tends to arise more laterally (see also Ch. 14).

Granuloma and Cysticercosis

Sarcoid, tuberculosis, and giant cell granuloma may occur within the sella and extend into the suprasellar cistern, but these lesions are rare. The lesions show uniform contrast enhancement on CT and may be indistinguishable from an adenoma. Cysticercosis appears as any cystic lesion of the sella and is suspected only in the appropriate clinical context.

POSTERIOR FOSSA TUMORS IN ADULTS

Cerebellar (Intra-axial) Tumors

METASTASIS

Metastasis is by far the most common neoplastic lesion seen in the posterior fossa in adults. It occurs primarily within the cerebellar hemispheres (Fig. 5-58) but is also seen within the pons and subarachnoid space. Hydrocephalus may result from compression of the fourth ventricle. The characteristics of a metastasis are the same as described in the section on metastasis above.

Hemangioblastoma

Hemangioblastoma, a low-grade tumor arising in the blood vessels, almost always occurs in the cerebellar hemispheres. It is histologically similar to hemangiopericytoma (angioblastic meningioma). Accounting for only about 7 percent of posterior fossa tumors in adults, hemangioblastoma is the most common primary cerebellar tumor. The peak incidence is at age 30 years. There is a 2 : 1 male predominance. The tumor may be found in the spinal cord but almost never in the cerebral hemispheres. About 10 percent of hemangioblastomas are seen as part of the Von Hippel-Lindau syndrome (see Ch. 12). Erythrocythemia is commonly produced by the tumor and may serve as a marker for recurrence.

On CT scan, the tumor most commonly appears as a hypodense cyst, with one or more solid intensely enhancing mural nodules or regions of cyst wall thickening. Prominent feeding arteries may be seen. Sometimes the

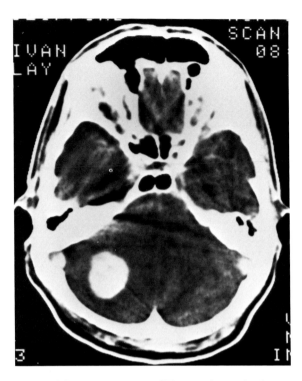

Fig. 5-58 Metastasis, contrast CT scan. A round enhancing lesion in the cerebellum in an adult is almost always a metastasis.

tumors are more solid with multiple cysts (Fig. 5-59) or ring-like in appearance with low-density central necrosis. Small tumors show homogeneous contrast enhancement. In these instances, the tumor is difficult to differentiate from metastasis or glioma by CT scanning. Because the tumor is highly vascular, a dynamic contrast examination through the lesion shows rapid uptake of contrast, followed by a rapid decrease, while remaining hyperdense. Multiple lesions may occur, particularly with Von Hippel-Lindau syndrome or as recurrent tumor after surgical excision.

The MRI procedure has a slightly greater sensitivity than CT scanning for the detection of these tumors. The cyst is isointense with CSF on both T_1- and T_2-weighted images. On the T_2-weighted images, the mural nodule is slightly hyperintense to the brain but hypointense to the cyst fluid. Prominent vascular channels with flow-void are characteristic in the solid portions (Fig. 5-60).

Fig. 5-59 Hemangioblastoma. (A) Contrast-enhanced CT scan shows an intensely enhancing large mass in the right cerebellar hemisphere. Multiple cysts are within the tumor. The mass has caused hydrocephalus with dilation of the lateral ventricles and temporal horns *(arrowheads)*. An occipital craniectomy defect is present from prior surgery *(arrows)*. **(B)** Lateral vertebral angiography demonstrates the very hypervascular mass with large irregular vessels *(arrowheads)*. The PICA is displaced downward and forward *(arrow)*. *(Figure continues.)*

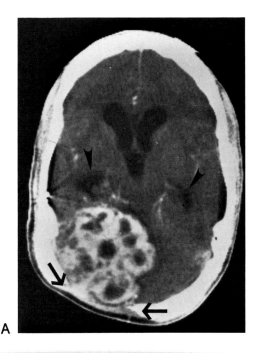

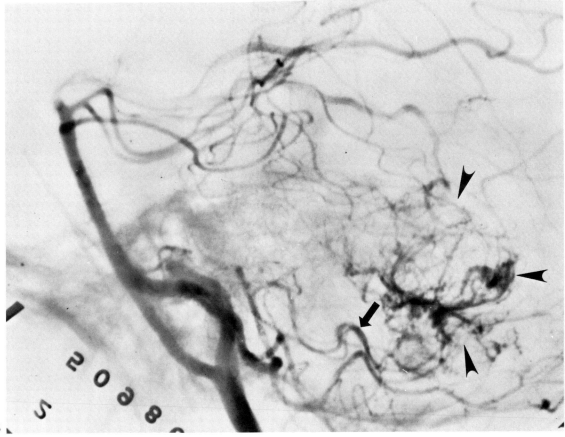

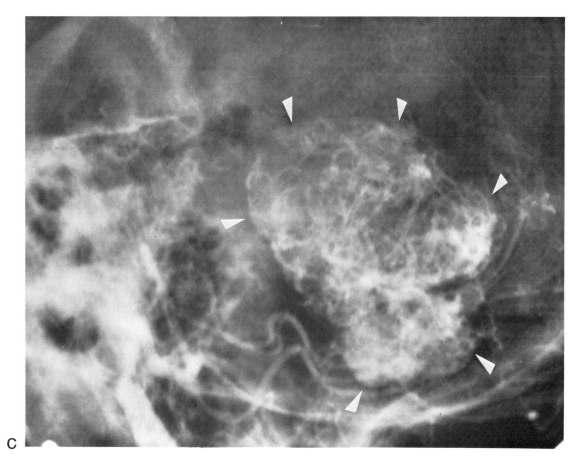

C

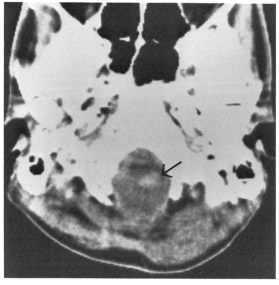

D

Fig. 5-59 *(Continued)*. **(C)** There is an intense delayed stain *(arrowheads)*. Often there are prominent early draining veins. **(D)** Small tumor is present at the foramen magnum, representing a meningeal metastasis *(arrow)*.

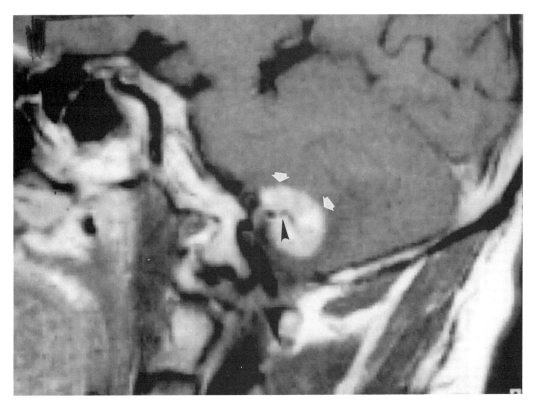

Fig. 5-60 Hemangioblastoma. Sagittal T_1 Gd-enhanced MRI shows the intensely enhancing tumor nodule of hemangioblastoma in the inferior cerebellum *(white arrows)*. Hypervascularity is a characteristic feature of this tumor. Here, a flow-void is seen within a large feeding vessel *(arrowhead)*.

Angiography demonstrates the vascular component of the tumor as a homogeneously dense nodule surrounded by mass effect representing the cyst. Occasionally, there is a tangle of vessels with rapid shunting to early-draining veins, simulating a vascular malformation or vascular metastasis, such as renal carcinoma (Fig. 5-59). Angiography may detect lesions that cannot be seen with CT scans (Fig. 5-61). It is unclear whether angiography or MRI is more sensitive for the detection of multiple tumors.

Other Intra-Axial Tumors
All other posterior fossa intra-axial tumors are rare. Glioma occurs most commonly within the pons and almost never in the hemispheres in adults. Tumors of the neuronal series (ganglioglioma) occur within the hemi-spheres and are often cystic and calcified, producing little or no mass effect.

Tumors Outside the Cerebellum (Extra-axial Tumors)

Most of the extra-axial tumors in the posterior fossa occur in the cerebellopontine angle cistern. These tumors produce tinnitus, unilateral neurosensory hearing loss, and sometimes vertigo. Large tumors produce hydrocephalus, cerebellar dysfunction, or contralateral hemiparesis.

SCHWANNOMA (NEURINOMA)

Neurinomas arise from the Schwann cells that envelop the cranial nerves. Sensory nerves are much more frequently involved than motor nerves. Most originate in

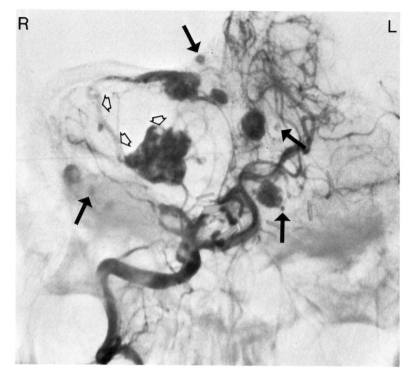

Fig. 5-61 Hemangioblastoma, recurrent, with metastasis. Right vertebral angiography in the Towne projection demonstrates a typical hemangioblastoma nodule with intense stain, early draining vein, and surrounding mass from cyst formation *(open arrows)*. Small metastatic nodules *(arrows)* are best demonstrated with angiography because of the high spatial resolution of the technique.

the acoustic nerve (CN VIII), particularly the vestibular branch. Acoustic neurinoma is the most common extra-axial tumor of the posterior fossa.

Less commonly, a neurinoma arises from the trigeminal nerve at the entrance to Meckel's cave. Tumors of the other cranial nerves are rare. The peak age range is 35 to 60 years, but rarely it is found in adolescents, especially those with NF-2. The female/male ratio is 2:1. Multiple tumors of the cranial nerves, including bilateral acoustic neurinomas, are found with NF-2.

At discovery, an acoustic neurinoma may be small and entirely within the internal auditory canal (intracanalicular), or it may be large, indenting the pons and cerebellar hemisphere. Rarely, it produces hydrocephalus owing to compression of the fourth ventricle. The tumor may have considerable cystic change but no calcification. Enlargement of the internal auditory canal is almost always present with larger tumors and is a hallmark of the diagnosis (Fig. 5-62).

EXTRA-AXIAL POSTERIOR FOSSA LESIONS IN ADULTS

Common lesions
    Acoustic neurinoma (80%)
    Meningioma (10%)
    Epidermoid (4%)
Uncommon lesions
    Metastasis
    Invasive nasopharyngeal carcinoma
    Trigeminal neurinoma
    Chordoma/chondrosarcoma
    Glomus tumor (Fig. 5-69)
    Arachnoidal cyst
    Sarcoma of the petrous pyramid
    Aneurysm
    Basilar artery ectasia

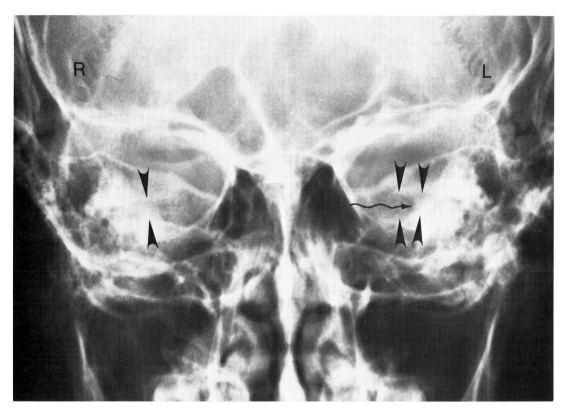

Fig. 5-62 Acoustic neurinoma. AP skull radiography shows enlargement of the porus acusticus on the left compared with that on the right *(arrowheads)*. There is erosion of the posterior wall of the internal auditory canal *(arrow)*.

The most sensitive imaging technique for the detection of an acoustic neurinoma is MRI. Thin-section transaxial slices are preferred. Most tumors can be seen on the nonenhanced T_1-weighted images but are better defined after Gd enhancement (Fig. 5-63). The tumor obliterates the normal outline of the acoustic nerve within the internal auditory canal. The mass is hyperintense to the CSF and slightly hypointense to the adjacent pons. Lightly weighted T_2 (proton-density) images may show the tumor as slightly hyperintense to CSF, especially if there is cystic change. On late echo T_2-weighted images, the tumor is slightly hypointense to CSF. Large tumors exhibit considerable signal inhomogeneity, sometimes with demonstrable cysts (Fig. 5-64). Almost never is any edema produced in the adjacent pons.

On the nonenhanced CT scan, these tumors are almost always isodense and, unless large, difficult to identify. Almost all neurinomas require contrast enhancement for visualization (Fig. 5-65). These tumors show characteristically intense, usually homogeneous, enhancement. Inhomogeneity occurs with central cystic change. Only tumors that are larger than 1 cm are routinely detectable on 5-mm transaxial scans of the posterior fossa. Bone window images show enlargement of the internal auditory canal with all but the smallest tumors.

Small intracanalicular tumors present a special problem for diagnosis with CT scanning. A contrast CT cisternogram must be performed if these tumors are to be seen; the safest and easiest is the air cisternogram. It is performed by introducing about 5 cc of air into the lumbar subarachnoid space with the patient horizontal and the side of interest up. The air is manipulated blindly into the cerebellopontine angle cistern and internal auditory canal by slight elevation of the upper torso. A

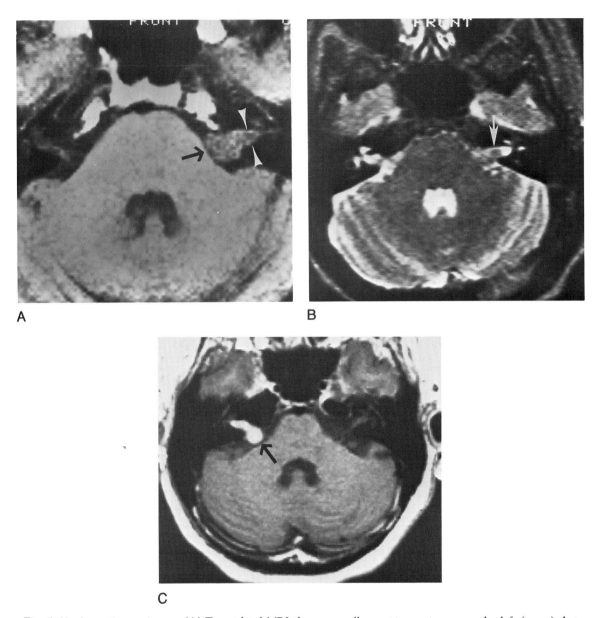

Fig. 5-63 Acoustic neurinoma. (A) T_1-weighted MRI shows a small acoustic neurinoma on the left *(arrow)* that extends into the IAC *(arrowheads).* **(B)** Heavily T_2-weighted MRI shows the tumor to be hypointense relative to the CSF *(arrow).* **(C)** Gd-DTPA-enhanced T_1-weighted MRI shows intense enhancement of a small right-sided acoustic neurinoma with intracanalicular extension *(arrow).* This method is the best imaging technique for demonstration of small acoustic neurinomas.

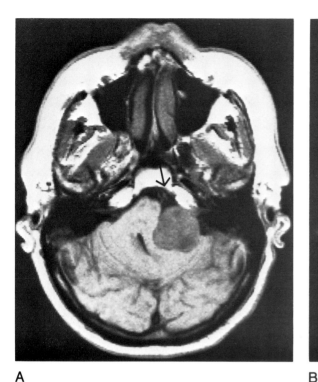

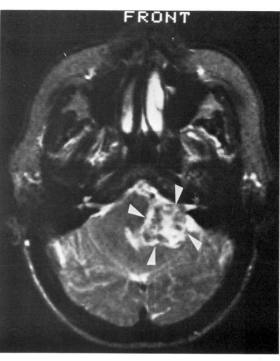

A   B

Fig. 5-64 Large acoustic neurinoma. (A) T_1-weighted MRI shows the slightly hypointense mass in the left CPA cistern. There is sharp demarcation from the distorted pons. The ipsilateral peripontine cistern is widened, confirming the extra-axial position of the tumor (*arrow*). **(B)** Heavily T_2-weighted MRI shows considerable inhomogeneity within the tumor (*arrowheads*).

small tumor is seen as a filling defect in the internal auditory canal, sometimes bulging slightly into the adjacent cistern (Fig. 5-66). Occasionally, adhesions prevent the air from entering the canal, resulting in a false-positive diagnosis of tumor. Gd-enhanced MRI is the more accurate technique for the diagnosis of these small tumors.

Angiography is no longer necessary for the diagnosis of acoustic neurinoma, although sometimes it is requested by the neurosurgeon to identify the position of the regional arteries. Usually, the anterior inferior cerebellar artery (AICA) is displaced upward and anteriorly over the tumor, and the petrosal vein is often compressed or occluded. Large tumors displace the basilar artery (Fig. 5-67).

Thin-section tomography is no longer used for the diagnosis of acoustic neurinoma. This technique relies on enlargement of the internal auditory canal to indicate the presence of tumor. A more than 2-mm difference in the diameter of the porus acousticus (medial entrance to the canal) is strongly suggestive of the diagnosis in the appropriate clinical setting. The posterior wall of the canal may be shortened (Fig. 5-62). The test is much less sensitive than MRI or CT scanning.

A neurinoma of the trigeminal nerve is rare. It most often occurs at the level of the gasserian ganglion within Meckel's cave, but it can be seen in the posterior fossa along the nerve adjacent to the pons. It has the same imaging characteristics as those of an acoustic neurinoma but is more superior and anterior in location.

(*Text continued on page 303*)

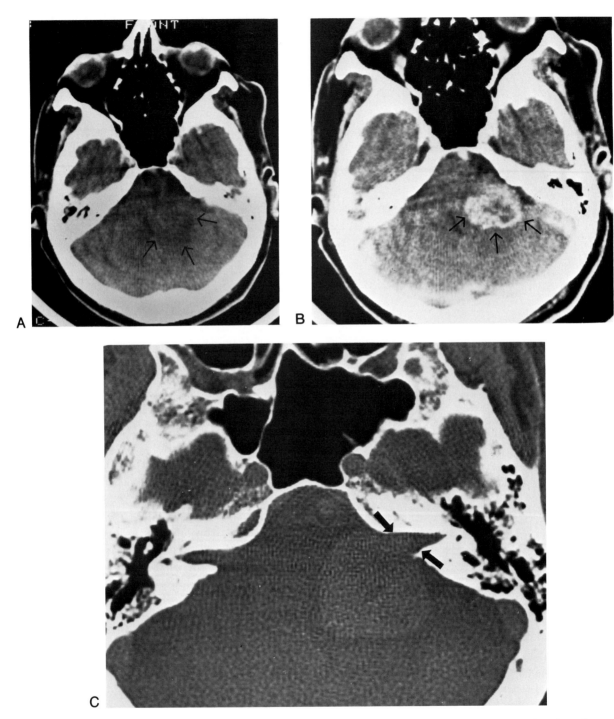

Fig. 5-65 Large acoustic neurinoma. (A) The tumor is nearly isodense with the brain and is difficult to identify on the nonenhanced CT scan *(arrows)*. **(B)** Contrast-enhanced CT scan shows intense enhancement of the tumor, with the central nonenhanced regions most likely representing cysts *(arrows)*. **(C)** High-resolution bone window study shows characteristic enlargement of the left internal auditory canal *(arrows)*.

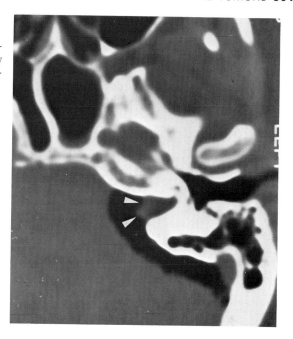

Fig. 5-66 Intracanalicular acoustic neurinoma. CT air cisterno-gram shows the small tumor mass within the internal auditory canal (*arrowheads*). A normal internal auditory canal would com-pletely fill with air.

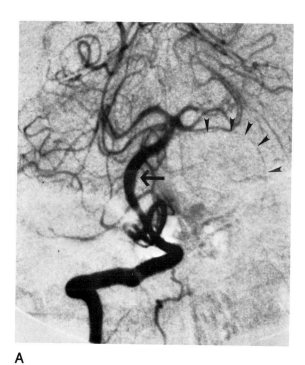

A

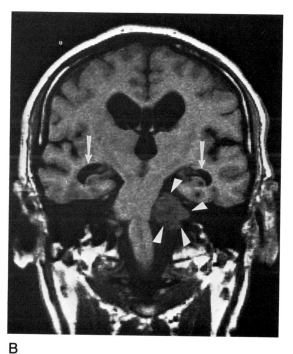

B

Fig. 5-67 Acoustic neurinoma. (A) AP vertebral angiography shows a large left-sided acoustic neurinoma. The AICA is elevated (*arrowheads*), and the basilar artery is displaced to the right (*arrow*). **(B)** Coronal T$_1$-weighted MRI showing the left-sided tumor (*arrowheads*) with displacement of the pons. Note the hydrocephalus and dilation of the temporal horns (*arrows*).

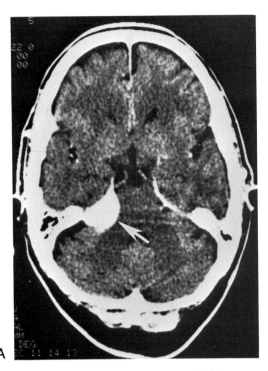

A

Fig. 5-68 Meningioma within CPA cistern. (A) Contrast-enhanced CT scan shows homogeneous enhancement of the tumor in the right CPA cistern (*arrow*). It lies slightly higher than the usual acoustic neurinoma. **(B)** High-resolution bone window CT scan shows a normal internal auditory canal (*arrowheads*). The outline of the tumor can barely be seen (*arrows*). (*Figure continues.*)

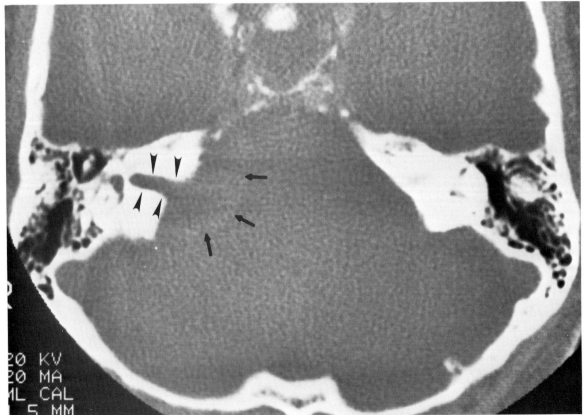

B

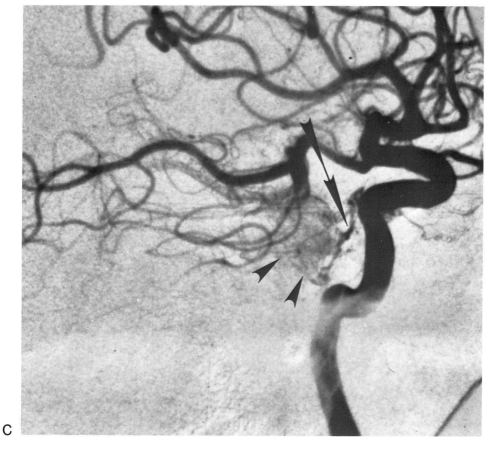

C

Fig. 5-68 *(Continued).* **(C)** Internal carotid artery angiography shows the characteristic blush of the meningioma *(arrowheads)* supplied by meningohypophyseal vessels *(arrow).* An acoustic neurinoma shows minimal if any neovascularity on angiography.

It may be difficult to differentiate from a meningioma (Fig. 5-55).

MENINGIOMA

Meningioma is the second most common extra-axial tumor of the posterior fossa. It occurs within the CPA cistern, along the clivus, or at the level of the foramen magnum. The imaging characteristics were described earlier.

Meningioma may appear similar to an acoustic neurinoma (Fig. 5-68). However, the internal auditory canal does not enlarge. There are almost never any bone changes with meningioma in the posterior fossa. Calcification may be present. Angiography usually shows enlargement of regional meningeal vessels and a "blush" that is much more intense than with the usual acoustic neurinoma. The common carotid injection most commonly demonstrates the meningeal vascular supply to CPA tumors.

EPIDERMOID TUMORS

Epidermoid tumors may be found in the posterior fossa, particularly within the cerebellopontine angle cistern. They are discussed in a previous section.

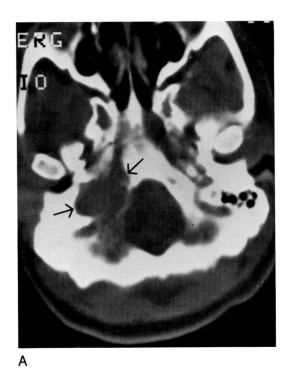

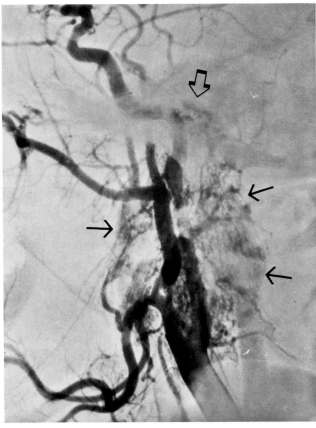

A   B

Fig. 5-69 Glomus vagale tumor. (A) CT scan through the base of the skull shows bone destruction on the right *(arrows)*. **(B)** A large hypervascular glomus vagale tumor is present within the neck *(arrows)*. It has extended superiorly to erode the base of the skull *(open arrow)*. Glomus tumors may be found within the tympanic cavity, at the jugular bulb, in the upper neck, or in the carotid body.

METASTASIS

Metastasis occurs in the region of the CPA. It is usually the result of meningeal involvement with carcinoma of the breast, small cell carcinoma of the lung, and melanoma. Metastasis may rarely involve the petrous pyramid. Tumors from the nasopharynx can invade the posterior fossa and enter the cerebellopontine angle after having eroded through the base of the skull.

Other uncommon extra-axial tumors (Fig. 5-69) that can extend into the posterior fossa are listed above. Their characteristics are the same as discussed in their respective sections elsewhere in the text.

SUGGESTED READING

Aizpuru RN, Quencer RM, Norenberg M et al: Meningioangiomatosis: clinical, radiologic and histopathologic correlation. Radiology 179:819, 1991

Buchfelder M, Nistor R, Falbusch R et al: The accuracy of CT and MR evaluation of the sella turcica for detection of adrenocorticotropic hormone-secreting adenomas in Cushing disease. AJNR 14:1183, 1993

Chakeres DW, Curtin A, Ford G: Magnetic resonance imaging of pituitary and parasellar abnormalities. Radiol Clin North Am 27:265, 1989

Curnes JT: MR imaging of peripheral intracranial neoplasms: extraaxial vs intraaxial masses. J Comput Assist Tomogr 11:932, 1987

Davis JM, Zimmerman RA, Blaniuk LT: Metastasis to the central nervous system. Radiol Clin North Am 20:417, 1982

Davis PC, Hudgins PA, Peterman SB et al: Diagnosis of cerebral metastasis: double dose delayed CT vs. contrast-enhanced MR imaging. AJNR 12:293–300, 1991

Destian S, Sze G, Krol G et al: MR of hemorrhagic intracranial neoplasms. AJNR 9:1115, 1988

Dorne HL, O'Gorman AM, Melanson D: Computed tomography of intracranial gangliogliomas. AJNR 7:281, 1986

Dropcho EJ: The remote effects of cancer on the nervous system. Neurol Clin 7:579, 604, 1989

Dubois PJ: Brain tumors. p. 365. In Heinz ED (ed) The Clinical Neurosciences: Neuroradiology. Vol. 4. Churchill Livingstone, 1984

Elster AD: Modern imaging of the pituitary. Radiology 187:1, 1993

Ganti SR, Hilal SK, Stein BM et al: CT of pineal region tumors. AJNR 7:97, 1986

Hasso AN, Fahmy JL, Hinshaw DB: Tumors of the posterior fossa. In Stark DD, Bradley WG (eds): Magnetic Resonance Imaging. CV Mosby, St. Louis, 1988

Haughton VM, Daniels DL: The posterior fossa. In Williams AL, Haughton VM (eds): Cranial Computed Tomography. CV Mosby, St. Louis, 1985

Jack CR, Bhansali DT, Chason JL et al: Angiographic features of gliosarcoma. AJNR 8:117, 1987

Kazner E, Wende S, Grumme T et al: Computed Tomography of Intracranial Tumors. Springer-Verlag, Berlin, 1982

Kernohan JW, Sayre GP: Tumors of the Central Nervous System. Fascicle 35. Atlas of Tumor Pathology. Armed Forces Institute of Pathology, Washington, DC, 1952

Kortman KE, Bradley WG: Supratentorial neoplasms. In Stark DD, Bradley WG (eds): Magnetic Resonance Imaging. CV Mosby, St. Louis, 1988

Kulkanni MV, Lee KF, McArdle CB et al: 1.5-T MR imaging of pituitary microadenomas: technical considerations and CT correlation. AJNR 9:5, 1988

Lee SR, Sanches J, Mark AS et al: Posterior fossa hemangioblastomas: MR imaging. Radiology 171:463, 1989

Lee Y, Van Tassel P: Intracranial oligodendrogliomas: imaging findings in 35 untreated cases. AJNR 10:119, 1989

Lee Y-Y, Bruner JM, VanTassel P, Libshitz HI: Primary central nervous system lymphoma: CT and pathologic correlation. AJNR 7:599, 1986

Manelfe C, Lasjaunias P, Ruscalleda J: Preoperative embolization of intracranial meningiomas. AJNR 7:963, 1986

Miller DL, Doppman JL: Petrosal sinus sampling: technique and rationale. Radiology 178:37, 1991

Miller DL, Doppman JL, Peterman SB et al: Neurologic complications of petrosal sinus sampling. Radiology 185:143, 1992

Moses AM, Clayton B, Hochhauser L: Use of T_1-weighted MR imaging to differentiate between primary polydipsia and central diabetes insipidus. AJNR 13:1273, 1992

Naheedy MH, Haag JR, Azar-Kia B et al: MRI and CT of sellar and parasellar disorders. Radiol Clin North Am 25:819, 1987

Oct RF, Melville GE, New PFJ et al: The role of MR and CT in evaluating clival chondromas and chondrosarcomas. AJNR 9:715, 1988

Press GA, Hesselink JR: MR imaging of cerebellopontine angle and internal auditory canal lesions at 1.5T. AJNR 9:241, 1988

Rippe DJ, Boyko OB, Radi M et al: MRI of temporal lobe pleomorphic xanthoastrocytoma. J Comput Assist Tomogr 16:856, 1992

Rubinstein LJ: Tumors of the Central Nervous System. Second series. Fascicle 6. Atlas of Tumor Pathology. Armed Forces Institute of Pathology, Washington, DC, 1972

Russell EJ, Geremia GK, Johnson CE et al: Multiple cerebral metastasis: detectability with Gd-DTPA enhanced MR imaging. Radiology 165:609, 1987

Schubiger O, Haller D: Metastasis to the pituitary-hypothalamic axis: an MR study of 7 symptomatic patients. Neuroradiology 34:131, 1992

Schwaighofer BW, Hesselink JR, Press GA et al: Primary intracranial CNS lymphomas: MR manifestations. AJNR 10:725, 1989

Tampieri D, Melanson D, Ethier R: MR of epidermoid cysts. AJNR 10:351, 1989

Teng MMH, Huang C, Chang T: Pituitary mass after transsphenoidal hypophysectomy. AJNR 9:23, 1988

Tien RD: Intraventricular mass lesions of the brain: CT and MR findings. AJR 157:1283, 1991

VanTassel P, Lee Y-Y, Bruner JM: Symchronous and metachronous malignant gliomas. AJNR 9:725, 1988

Whelan MA, Reede DL, Meisler W et al: CT of the base of the skull. Radiol Clin North Am 22:177, 1984

Wichmann W, Schubiger O, von Deimling A et al: Neuroradiology of central neurocytoma. Neuroradiology 33:143, 1991

Williams AL: Tumors. In Williams AL, Haughton VM: Cranial Computed Tomography. CV Mosby, St. Louis, 1985

Wilms G, Lammens M, Marchal G et al: Prominent dural enhancement adjacent to nonmeningiomatous malignant lesions on contrast enhanced MR images. AJNR 12:761, 1991

Yanaka K, Kamezaki T, Kobayashi E et al: MR imaging of diffuse glioma. AJNR 13:349, 1992

Yuh WT, Engelken JD, Muhonen MG et al: Experience with high dose gadolinium MR imaging in the evaluation of brain metastasis. AJNR 13:335, 1992

Yuh WTC, Wright DC, Barloon TJ, et al: MR imaging of primary tumors of trigeminal nerve and Meckel's cave. AJNR 9:655, 1988

6
Cranial Tumors in Children and Adolescents

Of all neoplasia in children, brain tumors are second in frequency only to leukemia. Twenty percent of all brain tumors occur in persons less than 20 years old, with the peak age range of 5 to 10 years in this group. About 50 percent of the tumors occur in the posterior fossa. There is a definite tendency for childhood tumors to occur toward the midline, which accounts for the associated high incidence of hydrocephalus. Headache, vomiting, and ataxia are the usual presenting symptoms. In most cases, skull radiography shows changes of increased intracranial pressure (Fig. 6-1).

In general, MRI is preferred for the detection of childhood tumors. The tumors exhibit prolonged T_1 and T_2 relaxation times. The anatomic location and spread of the tumor are well demonstrated. Transaxial and parasagittal planes are most useful in the posterior fossa; the coronal plane often helps in the supratentorial region. Hydrocephalus is easily identified. However, most of these tumors have similar MRI characteristics; thus, histologic diagnosis is often not possible. The pattern on CT scans, with and without contrast enhancement, often gives more information about the specific tissue type. Plain films are of little specific use, except for calvarial lesions. Angiography is seldom required for diagnosis or presurgical evaluation. Virtually all tumors, except for the rare choroid plexus papilloma or meningioma, are avascular and are seen only indirectly as a mass effect.

The classification of pediatric brain tumors is not precise, nor are the tumors static entities. Variations occur within the same tumor, and over time with genetic mutation. Degrees of malignancy are difficult to determine on the basis of histology alone. Classification of the "small blue cell tumors," the primitive neuroectodermal tumor (PNET) is especially in flux. The classification used in this chapter is current, but it may change in the future with newer methods of tumor analysis, particularly DNA analysis.

INFRATENTORIAL TUMORS

INFRATENTORIAL TUMORS
IN CHILDREN

Cerebellar astrocytoma (33%)
Medulloblastoma (33%)
Pontine astrocytoma (20%)
Ependymoma (10%)
Others (4%)
    Choroid plexus papilloma
    Dermoid, epidermoid
    Neuroectodermal tumor
    Meningioma
    Neurinoma

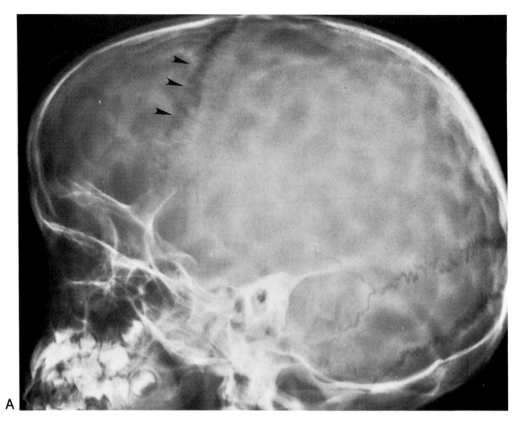

A

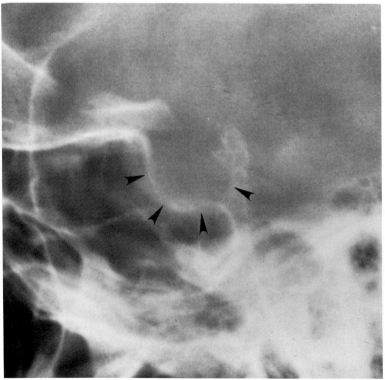

B

Fig. 6-1 Increased intracranial pressure. (A) Plain skull radiograph shows separation of the intracranial sutures (*arrowheads*) and increased convolutional markings. **(B)** Lateral view of the sella shows demineralization of the floor of the sella (*arrowheads*), which has occurred as a result of increased intracranial pressure.

Cerebellar Astrocytoma

Cerebellar astrocytoma is one of the two most common posterior fossa tumors (Table 6-1). Sixty percent of all intracranial astrocytomas arise within the cerebellum. It is usually the pilocytic type, named for the elongation of the cells found on histologic examination. The tumor characteristically has both small and large cysts. Pilocytic astrocytoma almost always has a benign clinical course. The tumor may be cured by surgical excision alone, but recent reports indicate that it occasionally metastasizes within the CSF pathways, showing up many years later.

Pilocytic astrocytoma grows slowly and is often large at clinical recognition (Fig. 6-2). Almost invariably, the presenting signs and symptoms are those of hydrocephalus, caused by compression of the fourth ventricle. The peak age of occurrence is 5 to 10 years. Pilocytic astrocytoma shows no sex preference. Most tumors straddle the boundary between the vermis and the adjacent cerebellum, although some are entirely within the vermis, cerebellar hemisphere, or fourth ventricle. Fifty percent have a large central cyst with a focal mural nodule. The rim of the cyst is most often composed of compressed cerebellar tissue, rather than tumor. The remaining 50 percent are partially cystic (Fig. 6-3). Anaplasia rarely occurs. When anaplastic tumors occur in the midline, they are difficult to differentiate from medulloblastoma and ependymoma, using either MRI or CT scanning (Table 6-2).

On MRI, the tumor is hypointense on T_1 and hyperintense on T_2-weighted images. Solid tumors are relatively homogeneous. Cysts are recognized as round regions of CSF intensity within the tumor mass. Peripheral edema is uncommon. Characteristically, intense enhancement occurs within the solid portions of the tumor and is not indicative of malignant potential in these tumors. The rim may enhance slightly, indicating tumor in the cyst wall.

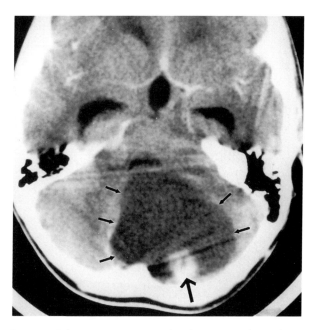

Fig. 6-2 Pilocytic astrocytoma. Axial contrast-enhanced CT scan through the posterior fossa illustrates a large cystic mass within the cerebellum (*small black arrows*) with enhancing mural nodule arising from the posterior wall (*larger black arrow*). This is the classic appearance of a pilocytic type of astrocytoma.

Diffuse fibrillary astrocytoma is a less common form of astrocytoma arising from the cerebellum. These tumors tend to be solid without cysts (Fig. 6-4). It is more infiltrative than the pilocytic astrocytoma and may grow outward through the lateral foramina of the fourth ventricle into a cerebellopontine angle cistern, in a pattern that mimics ependymoma. Spread to the CSF pathways is much more common than with pilocytic astrocytoma, and the cure rate for this tumor is much less than with pilocytic astrocytoma. Fibrillary astrocytoma may have anaplastic histologic features.

On CT scan, any solid astrocytoma is slightly hypodense to brain tissue (Figs. 6-3 and 6-4). Only rarely is an astrocytoma hyperdense or of mixed density. Calcification is seen in 10 percent, usually as a conglomeration. When calcification is seen in a childhood posterior fossa tumor eccentrically positioned in the cerebellum, the diagnosis of fibrillary astrocytoma is virtually certain. Contrast enhancement may be absent but becomes more intense with the more malignant tumors. Enhancement is usually inhomogeneous. On MRI, the tumor is hypointense to isointense on T_1 and hyperintense on T_2-

Table 6-1 Intracranial Location of Astrocytoma in Children

Location	Overall (%)	Common Type
Cerebellum	60	Pilocytic
Cerebral hemispheres	20	Fibrillary
Brain stem	17	Fibrillary
Optic chiasm—hypothalamus	3	Pilocytic

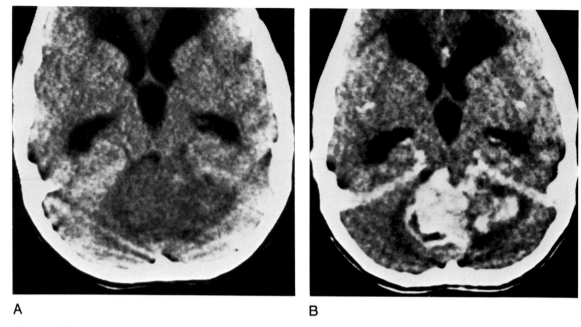

A   B

Fig. 6-3 Pilocytic astrocytoma. (A) Nonenhanced CT scan shows a large hypodense mass involving the medial left cerebellar hemisphere and vermis. There is secondary hydrocephalus. **(B)** Contrast-enhanced study shows enhancement throughout the solid portions of the tumor. Note minimal surrounding edema and cysts.

weighted images. Hemosiderin indicates prior hemorrhage and does not correlate with malignant potential of the tumor.

Medulloblastoma

The medulloblastoma is a highly malignant devastating tumor of the midline of the posterior fossa. It is now classified as a PNET. Its peak incidence is at 5 to 10 years of age, but it may occur in newborns and adolescents. When it occurs in adolescents, the tumor tends to be more peripheral in the cerebellar hemispheres. The male-to-female ratio is 2 : 1.

The tumor arises from the velum (roof) of the fourth ventricle and spreads in all directions. It frequently occupies the fourth ventricle, spreading through the foramina into the pericerebellar cisterns and into the aqueduct. Hydrocephalus is common. The tumor can metastasize, most commonly by seeding to the entire subarachnoid space, including the ventricular system and spinal region. Shunting of hydrocephalus into the abdomen may spread tumor to the peritoneum. Occasionally, hematogenous metastases occur in the lungs or bones.

Table 6-2 CT Differential Diagnosis of Common Posterior Fossa Tumors in Children

Parameter	Medulloblastoma	Astrocytoma	Ependymoma
CT density	Hyperdense	Hypodense	Isodense
Enhancement	Homogeneous (except for necrosis)	Homogeneous (in solid portion)	Heterogeneous
Cysts	Microcysts	Common, large	No
Necrosis	Common	Uncommon	Sometimes
Calcification	Rare	Sometimes	Common

See also Table 5-1.

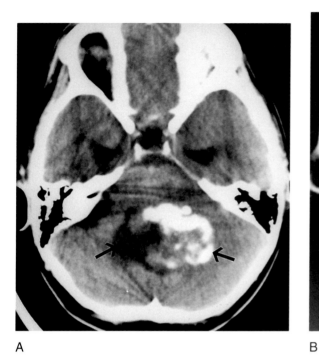

A

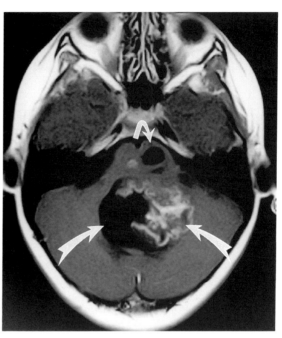

B

Fig. 6-4 Fibrillary astrocytoma, cerebellum. (A) Nonenhanced axial CT scan through the posterior fossa identifies an inhomogeneous calcified solid and cystic mass in the central cerebellum occupying the fourth ventricle *(black arrows)*. Calcified mass in the cerebellum is most commonly an astrocytoma. (B) Gd-enhanced T₁ MRI shows the enhancing left side solid portion of the tumor. The cyst fills the fourth ventricle *(large white arrows)*. The tumor also involves the pons *(curved white arrow)*. (C) T₂ MRI shows the tumor as inhomogeneous high signal intensity. The tumor extends laterally through the foramen of Luschka into the left C-P angle cistern *(arrow)*. *(Figure continues.)*

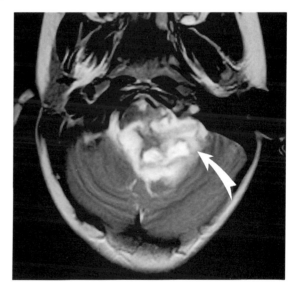

C

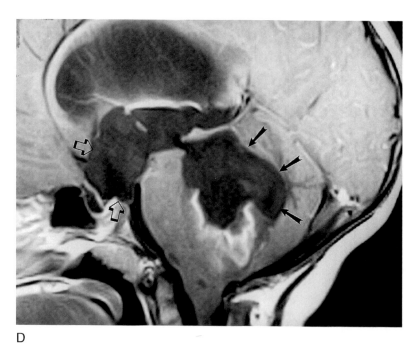

D

Fig. 6-4 *(Continued).* **(D)** Cystic tumor fills the fourth ventricle *(small black arrow)* and causes severe hydrocephalus and expansion of the third ventricle *(open arrows).* Enhancing irregular mass and invasive nature of the tumor are characteristic of the fibrillary type of astrocytoma in children.

On MRI, medulloblastoma appears as a well-defined midline vermian and fourth ventricular mass of low intensity on T_1- and high intensity on T_2-weighted images (Fig. 6-5). The presence of hemorrhage, necrosis, and microcysts is common. Surrounding edema is often seen. Enhancement within the tumors with gadolinium (Gd) tends to be relatively homogeneous. Enhanced MRI is the best method for diagnosis of meningeal metastasis to the subarachnoid spaces and ventricles.

On CT scans, most medulloblastomas appear as well-delineated, slightly hyperdense, tumors in the midline of the posterior fossa, occupying the region of the vermis and fourth ventricle (Fig. 6-6). Mixed density is a result of cysts, necrosis, and hemorrhage within the tumor. Iso-dense and hypodense medulloblastomas are unusual. Calcification is rare and, if present, is of the diffuse, fine variety. Frequently, there is surrounding edema.

Contrast enhancement, a constant feature, may be slight to intense. Focal regions of necrosis and microcysts create defects in the otherwise uniform enhancement pattern. Metastatic tumor in the meninges also enhances.

Ependymoma

Ependymoma is much less common, accounting for about 10 percent of the posterior fossa tumors in children. These lesions are intimately attached to the floor of the fourth ventricle, making complete removal impossible. They are malignant and are classified as grade III or IV. In the posterior fossa, they have a peak incidence at 3 to 7 years of age but may be encountered at any age.

The tumor grows to fill the fourth ventricle, causing hydrocephalus. Growth through the foramina into the cisterns is a characteristic feature. The tumor may entrap CSF around the periphery, creating a halo effect. Metastasis to the meninges of the brain and spinal column occurs, particularly after surgical resection.

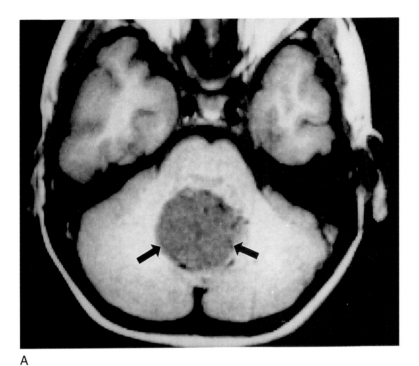

A

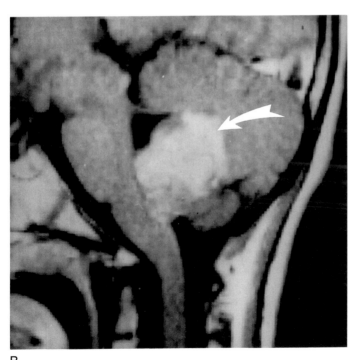

B

Fig. 6-5 Medulloblastoma. (A) Axial T_1 nonenhanced MRI shows a nearly homogeneous mass hypointense to brain within the inferior vermis and protruding into the fourth ventricle *(arrows)*. **(B)** Sagittal Gd-enhanced midline parasagittal MRI demonstrates the nearly homogeneous intense contrast enhancement of the tumor arising within the anterior inferior vermis *(arrow)*.

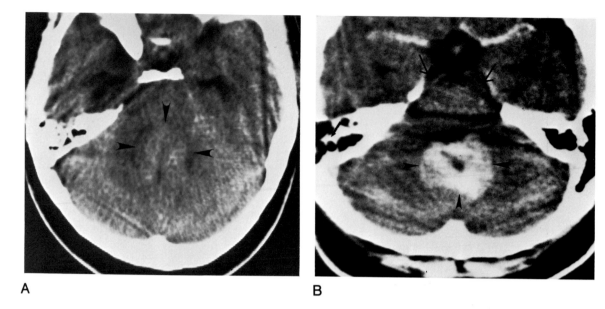

A

B

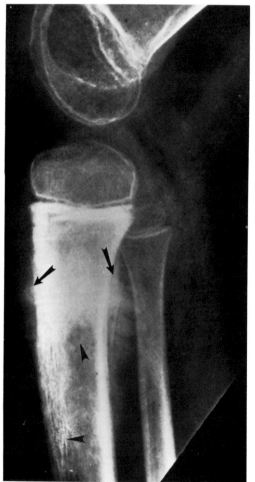

C

Fig. 6-6 Medulloblastoma. (A) Nonenhanced CT scan shows a slightly hyperdense midline mass within the vermis and posterior fourth ventricle (*arrowheads*). (B) There is blotchy enhancement of the solid tumor (*arrowheads*). Note the anterior displacement of the pons and obliteration of the prepontine cistern (*arrows*). Hydrocephalus is present. (C) Metastatic medulloblastoma to bone. The metastatic deposit has caused permeative change (*arrowheads*) and periosteal new bone formation (*arrows*). Pure lytic lesions are more commonly seen.

On MRI, the tumor is seen as an irregular, inhomogeneous fourth ventricular mass, hypointense on T_1- and hyperintense on T_2-weighted images (Figs. 6-7 and 6-8). Cystic change is relatively uncommon. The widespread growth is well defined with MRI. Calcification may be seen as signal void, especially on gradient echo imaging. Gd enhancement occurs within the tumor to a variable degree.

On the precontrast CT scan, ependymomas are usually isodense or slightly hyperdense. Calcification is especially common and may be fine or coarse. A densely calcified midline posterior fossa tumor is almost always an ependymoma. Accompanying hydrocephalus is the rule. The outline of the tumor is commonly irregular. However, a large calcified fibrillary astrocytoma may be indistinguishable from an ependymoma. Contrast enhancement, almost always present, is usually inhomogeneous because of necrotic change (Fig. 6-9). Angiography shows an avascular intraventricular posterior fossa mass.

Brain Stem Glioma

Brain stem gliomas mostly occur in children and adolescents, with a peak incidence at 3 to 7 years. The histology of the astrocytoma is variable, although most of these tumors are low grade of the diffuse fibrillary type. Almost all occur in the pons but may spread in any direction into the mesencephalon or medulla or show exophytic growth that may encase and elevate the basilar artery. Exophytic growth may occur posteriorly into the fourth ventricle. Cysts are common within the tumor. The usual presentation is gradually progressive cranial nerve dysfunction, often asymmetric. Hydrocephalus is uncommon, even with large tumors.

Although other lesions, such as lymphoma, granuloma, or angioma, can lead to brain stem enlargement in children, they are rare. Therefore, almost all solid or cystic masses in the pons are considered gliomas; many are treated with irradiation without a definitive tissue diagnosis. It is important to recognize cysts, as decompression

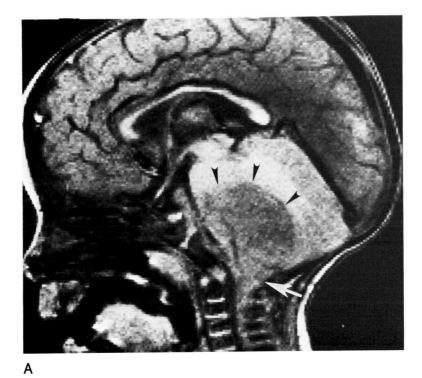

A

Fig. 6-7 Ependymoma. (A) T_1-weighted gradient echo image shows the hypointense large mass in the central portion of the posterior fossa occupying the fourth ventricle *(arrowheads)*. Note the inferior extension into the foramen magnum and upper cervical canal *(arrow)*. *(Figure continues.)*

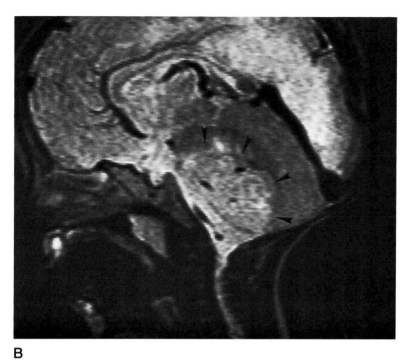

B

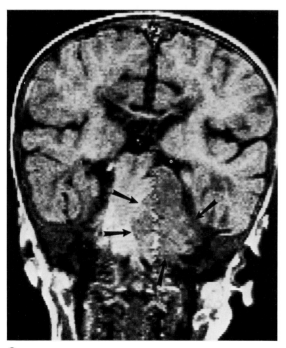

C

Fig. 6-7 *(Continued).* **(B)** T$_2$-weighted image showing the tumor as slightly hyperintense *(arrowheads).* **(C)** Coronal T$_1$-weighted image shows the hypointense tumor with exophytic growth to the left into the peripontine and basal cisterns *(arrows).*

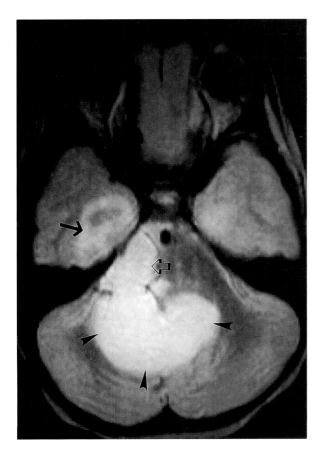

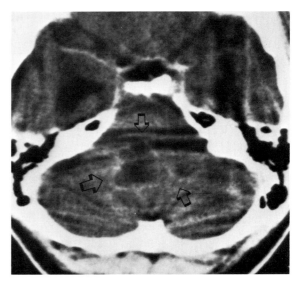

Fig. 6-9 Ependymoma, intraventricular tumor. Contrast-enhanced CT study shows irregular ring-like enhancement surrounding regions of necrosis (*arrows*). The tumor lies within the fourth ventricle.

Fig. 6-8 Ependymoma of the fourth ventricle. Axial proton-density (lightly weighted T₂) MRI shows this tumor with homogeneous high signal intensity occupying the fourth ventricle and central cerebellum (*arrowheads*). The tumor extends laterally through the right foramen of Luschka, to fill the cerebellopontine angle cistern (*open arrow*). It grows superiorly through the tentorial hiatus into the inferior right temporal fossa (*black arrow*).

cisterns, but the tumor is much more difficult to recognize with CT scanning than with MRI. The fourth ventricle is characteristically displaced posteriorly, so it becomes slit-like. About 50 percent of pontine gliomas are hypodense; the remainder are isodense or of mixed density. About 50 percent show some degree of contrast enhancement in the solid portions of the tumor (Fig. 6-11). The enhancement may be uniform, patchy, or ring-like. Calcification does not occur, except after radiation therapy.

can be performed. Postirradiation calcification of the tumor is common.

MRI is the most effective imaging method for the diagnosis of pontine glioma. The tumor appears hypointense on T₁- and hyperintense on T₂-weighted images (Fig. 6-10). The pons is enlarged and distorted. Cysts are identified by round regions of CSF density. Hemorrhage may be seen within the tumor. Gd enhancement is variable and generally not of prognostic value.

The precontrast CT scan may show some degree of pontine enlargement and obliteration of the peripontine

Other Tumors of the Posterior Fossa

Acoustic neurinomas are benign tumors rarely seen in children. When they do occur, they are often bilateral, and almost always in children with neurofibromatosis (NF-2). When discovered, these tumors are usually large, with associated enlargement of the internal auditory canals. Hearing loss is the first symptom (see Ch. 5). Meningioma is very rare.

Epidermoid tumors and arachnoid cysts are found within the posterior fossa, often in the cerebellopontine

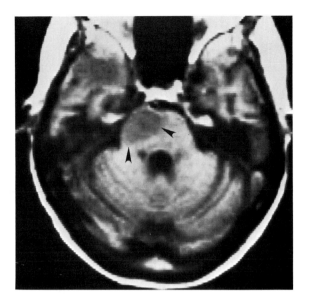

Fig. 6-10 Pontine glioma. T$_1$-weighted MRI shows a focal low-intensity mass within the right anterolateral aspect of the pons (*arrowheads*). The contour of the pons is locally expanded. On T$_2$-weighted imaging the tumor would be hyperintense.

SUPRATENTORIAL TUMORS

Astrocytoma

Astrocytoma is the most common cerebral hemispheric tumor in children. The tumors are usually of the low-grade diffuse fibrillary type, but almost all are locally invasive to some extent. Almost all are found within the parietal or temporal lobes. Headache and seizures are the most common presenting symptoms.

These tumors may be solid or cystic and exhibit MRI and CT patterns similar to those seen with low-grade astrocytomas in adults (see Ch. 5). Astrocytomas vary from small to very large at recognition. When a small cystic tumor is first seen, it is impossible to differentiate it from the rare benign cyst of the brain (Fig. 6-12). All cystic lesions within the brain parenchyma must be considered a tumor until proved otherwise. Follow-up scans must be obtained. Some may arise deep within the white matter, extending into a lateral ventricle (Fig. 6-13).

angle cistern. These entities may be difficult to differentiate, as both have CSF characteristics. Epidermoid may be slightly more intense than CSF on T$_1$ and T$_2$-weighted MR images.

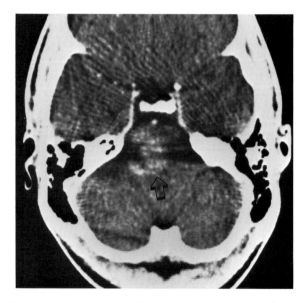

Fig. 6-11 Pontine glioma, CT scan. Contrast CT scan shows diffuse enhancement of a pontine astrocytoma (*arrow*).

**SUPRATENTORIAL TUMORS
IN CHILDREN**

Common tumors
    Astrocytoma (50%)
        Hemispheric
        Hypothalamic and optic chiasm
    Ependymoma
    Craniopharyngioma
    Pineal region tumors
    Choroid plexus papilloma
    Leukemia
Uncommon tumors
    Meningioma
    Hamartoma
    Histiocytosis X
    Cavernous hemangioma
    Neuroblastoma
    Primitive neuroectodermal tumor
Other mass lesions
    Dermoid
    Epidermoid
    Arachnoid cyst

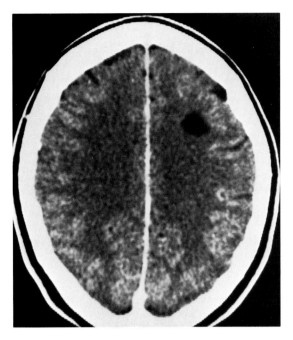

Fig. 6-12 Hemispheric cyst. Contrast CT scan shows a non-enhancing hypodense cyst-like lesion in the left posterior frontal subcortical white matter. It did not change over many years and presumably is a benign idiopathic cyst. A low-grade astrocytoma might also have this appearance.

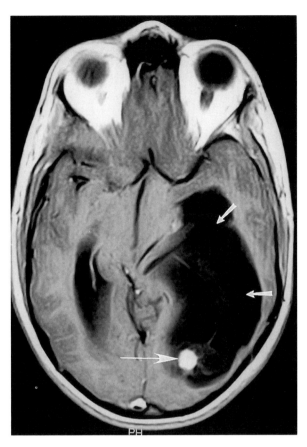

Fig. 6-13 Pleomorphic xanthoastrocytoma, intraventricular. Axial T_1 Gd-enhanced MRI illustrates a large intraventricular cystic tumor with an enhancing nodule *(large arrow)*. The cystic portion of the tumor completely fills the atrium and temporal horn of the left lateral ventricle *(small arrows)*. The signal intensity within the cyst is slightly greater than that of the CSF within the ventricle. This tumor was initially thought to be a pilocytic astrocytoma. (See p. 330.)

Some tumors contain varying amounts of calcification. Sometimes the only evidence of the tumor is a small dot of calcium on the CT scan. In this situation, follow-up scanning or MRI is necessary for a final diagnosis of the lesion as a tumor that may take many years to show any growth. Contrast enhancement indicates a higher grade of histologic malignancy.

Hypothalamic and Optic Chiasm Astrocytoma

Astrocytomas of the hypothalamus or optic chiasm are almost always benign grade I, pilocytic type tumors. In the optic chiasm, they most commonly occur with neurofibromatosis (NF-1). In the hypothalamus, they produce the diencephalic syndrome, with signs of cachexia, particularly in infants. Diabetes insipidus, precocious puberty, and narcolepsy may be the presenting symptoms or signs.

On MRI, astrocytomas show slight hypointensity on T_1- and slight hyperintensity on T_2-weighted images (Fig. 6-14). On CT scans, the tumors are either isodense or hyperdense and are occasionally calcified. Most of these tumors are solid, although sometimes those within the hypothalamus have a few cysts. MRI and CT scanning demonstrate strong homogeneous contrast enhancement of the solid portions of the tumor. Astrocytomas cause partial obliteration of the suprasellar cistern. It is impossible to differentiate a hypothalamic astrocytoma from mixed glioma, Langerhans cell histiocytosis, mid-

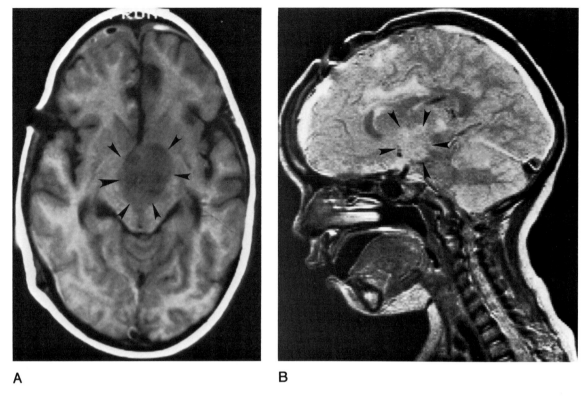

A   **B**

Fig. 6-14 Hypothalamic astrocytoma. (A) T_1-weighted MRI shows a hypointense mass in the hypothalamus extending upward to fill the third ventricle *(arrowheads)*. **(B)** Parasagittal lightly T_2-weighted MRI shows a round, slightly hyperintense, midline mass in the region of the hypothalamus and third ventricle *(arrowheads)*.

line germinoma (ectopic pinealoma), and hamartoma (Fig. 6-15), which also occur in this region.

Ependymoma

Ependymoma arises from ependymal cells lining the walls of the ventricles or from ependymal rests that occur in the paraventricular white matter. The tumors may therefore be found entirely within the ventricle or entirely within the white matter adjacent to the ventricle. Frequently, both ventricle and white matter are involved. The most common location is adjacent to the atrium of the lateral ventricle. Peak incidence is during adolescence and young adulthood. All carry a poor prognosis. An especially malignant form of the tumor, called ependymoblastoma, may grow rapidly and attain large

size. This aggressive tumor has a strong tendency to spread within the ventricular system and may form a tumor cast of the ventricular cavity.

The supratentorial tumors appear similar to posterior fossa ependymomas on both MRI (Fig. 6-16) and CT scanning. Calcification and cysts are common. About 50 percent are homogeneous and solid. Contrast enhancement is usually present.

Craniopharyngioma

Craniopharyngioma is a tumor with a peak incidence in children and adolescents and a second peak incidence in adults aged about 40 to 50 years. This tumor exhibits male predominance. The tumor usually arises in the suprasellar cistern within the pituitary stalk at the level of

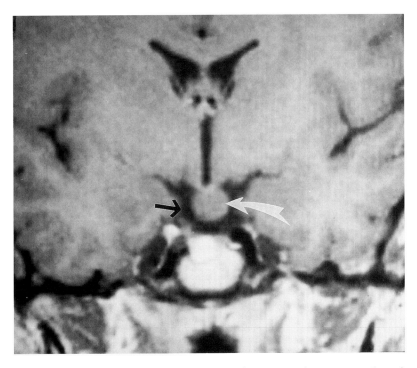

Fig. 6-15 Hamartoma of the hypothalamus. Coronal T$_1$ MRI shows a round mass arising from the tuber cinereum beneath the third ventricle *(white arrow)*. It protudes into the upper suprasellar cistern *(black arrow)*. These tumors usually exhibit uniform contrast enhancement.

the tuber cinereum, although in 20 percent of cases it is within the sella turcica (Fig. 6-17). Fifty percent of tumors in the suprasellar–hypothalamic region in children are craniopharyngiomas. According to the most accepted theory, craniopharyngiomas develop from a remnant of Rathke's pouch, an invagination of epidermal tissue of the buccal mucosa, but the origin is still controversial. The suprasellar location of the tumor produces visual-field deficits, diabetes insipidus, or hydrocephalus. These tumors have both solid and cystic components and may become large enough to occupy more than one-half of the brain. The epidermal tissue secretes predominantly keratin (protein).

Enlargement of the sella occurs only with an intrasellar location of the tumor. With downward growth of the suprasellar mass, truncation of the dorsum sellae may be found without change in its normal forward-directed axis. Calcification is common in children (70 percent) but is less common in adults (30 percent). The calcifica-

tion is variable in appearance and may be densely globular, diffusely speckled, or thin and rim-like. Craniopharyngiomas are generally histologically benign. When large, however, they may become impossible to remove, and recurrence is common. Spillage of contents into the subarachnoid space may cause arachnoiditis.

Variable signal intensity is seen on MRI with this tumor (Fig. 6-18). The cystic components may have either high or low signal on T$_1$-weighted images, depending on the amount of protein within the cystic contents. With low protein concentration, the cyst appears nearly CSF equivalent. Moderately high concentrations show high T$_1$ signal intensity. With very high concentrations of protein (inspissation), the cyst becomes dark on T$_1$- and T$_2$-weighted sequences and is difficult to distinguish from calcification. T$_2$-weighted images show variable intensity as well. Gd enhancement may occur within solid portions of the tumor but is not necessary for diagnosis and has no prognostic value.

(Text continued on page 325)

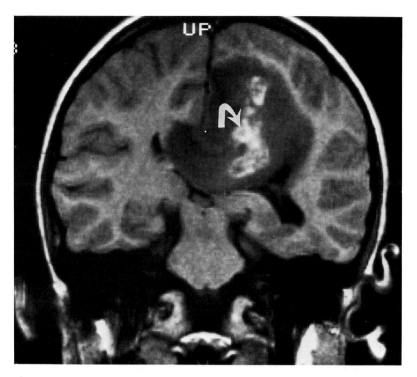

A

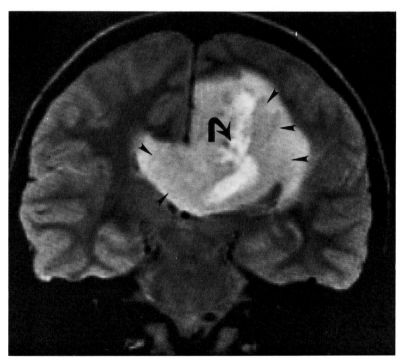

B

Fig. 6-16 Ependymoma. (A) T_1-weighted MRI shows a hypointense mass in the medial left posterior parietal lobe adjacent to, and extending into, the atrium of the left lateral ventricle. The tumor crosses the midline through the splenium of the corpus callosum. The high-intensity region *(arrow)* represents hemorrhage. **(B)** Lightly T_2-weighted MRI shows the mildly hyperintense tumor mass *(arrowheads)* surrounded by a thin rim of edema with greater hyperintensity. The high intensity in the central portion of the tumor *(curved arrow)* represents hemorrhage.

Fig. 6-17 Craniopharyngioma. (A) Coronal contrast-enhanced CT scan shows homogeneous nonenhancing mildly hypodense mass within the sella turcica, with just slight suprasellar extension *(arrowheads)*. There is minimal erosion of the floor of the sella *(arrows)*. In an adult, pituitary adenoma is the most common tumor with this appearance. **(B)** Intrasellar and suprasellar craniopharyngioma. The sella is grossly enlarged with undercutting of the anterior clinoid processes *(arrowheads)*. The dorsum is truncated *(arrow)*. Faint suprasellar calcification is seen within the tumor *(open arrows)*.

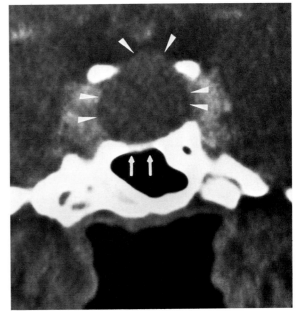

A

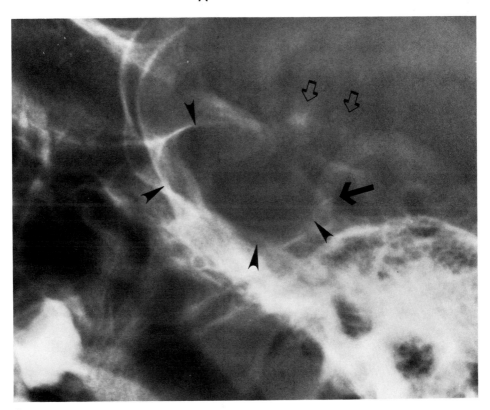

B

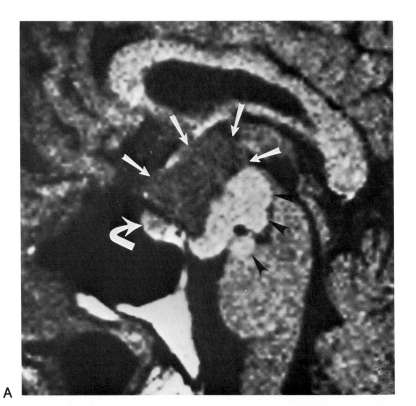

A

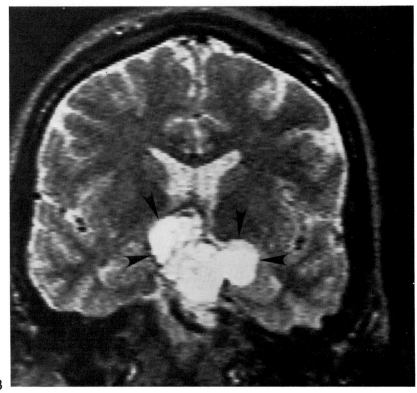

B

Fig. 6-18 Craniopharyngioma. (A) Parasagittal T$_1$-weighted MRI shows the large tumor within the suprasellar cistern and extending posteriorly into the interpeduncular and prepontine cisterns. The anterior portion is solid and is slightly hypointense *(arrows)*, whereas the posterior portion is cystic and hyperintense *(arrowheads)*. Note the normal pituitary gland and sella turcica *(curved arrow)*. **(B)** Coronal T$_2$-weighted MRI shows the very high signal intensity of the cystic component of the tumor *(arrowheads)*. The solid portion is moderately hyperintense. The tumor has extended superiorly to occupy the third ventricle and displace the basal ganglia.

The CT appearance is also variable, with solid, cystic, and calcified portions (Fig. 6-19). The cystic regions are usually of a density approximately equal to that of the CSF, but they can be hypodense or hyperdense, as a function of the protein content. About 66 percent show some contrast enhancement within the solid portions of the tumor.

Pineal Region Tumors

Most tumors in the region of the pineal gland occur in persons 10 to 20 years old. The presentation is generally that of hydrocephalus (i.e., headache, nausea, vomiting, confusion), often accompanied by Parinaud syndrome (supranuclear impairment of upward gaze, defective convergence, and slow pupillary reaction to light). Most tumors in this region are classified as either of germ cell origin or of neuroectodermal origin from the pineal parenchyma. Some of these tumors arise from glial cells.

Germ cell tumors (GCTs) are the most common, accounting for more than 50 percent. They are similar to GCTs occurring in the gonads, retroperitoneum, and mediastinum. Overall, these tumors are rare, and accounting for no more than 3 percent of CNS tumors. GCTs display a wide range of histology and malignant potential. The mature teratoma contains three embryonal cell lines and is the most benign and uncommon of the group. The germinoma consists of two primitive cell lines and is comparable to testicular seminoma and ovarian dysgerminoma. It is highly radiosensitive. The malignant non-germinoma germ cell tumors (NGGCTs) are a group consisting of endodermal sinus tumor (yolk sac tumor), embryonal carcinoma, choriocarcinoma, and immature teratoma. This aggressive group is insensitive to radiotherapy but may be responsive to chemotherapy. Mixed tumors containing cells of all the above groups are common, accounting for about 40 percent of cases. Embryonal carcinoma and choriocarcinoma often secrete human chorionic gonadotropin (hCG); embryonal carcinoma and endodermal sinus tumor secrete α-fetoprotein (AFP). Survival for the GCT group is good (>80 percent); survival for the NGGCT is poor (<10 percent). Teratoma and choriocarcinoma predominates in childhood. The other types predominate in adolescents and young adults.

GCT of the pineal occurs almost exclusively in males, with a peak frequency at 15 years. Most are mixed (germinoma plus NGGCT) or pure NGGCT. A suprasellar tumor in the region of the anterior third ventricle in a female is almost always a pure germinom (GCT). These tumors are generally slow growing and well circumscribed. Hydrocephalus and Parinaud syndrome (failure of upward gaze, nystagmus with convergence, and disordered light response of the pupils) is the most common presentation for pineal region tumors; diabetes insipidus and growth failure or precocious puberty is the most common for suprasellar GCT.

On CT scans, GCTs are displayed as a homogeneous, slightly hyperdense mass, usually without calcification and showing intense uniform contrast enhancement (Fig. 6-20). On MRI, they tend to be isointense with the brain on both T_1- and T_2-weighted images. Uniform Gd enhancement occurs. Teratoma is usually inhomogeneous with cystic, lipomatous, and

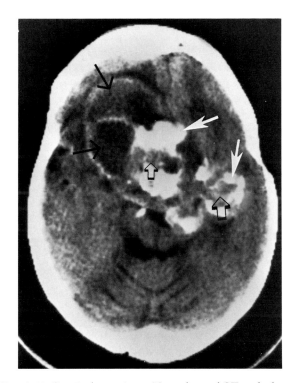

Fig. 6-19 Craniopharyngioma. Nonenhanced CT study shows a large tumor mass with cysts containing variable-density material *(black arrows)*, solid tissue *(open arrows)*, and dense calcification *(white arrows)*. (Courtesy of Dr. Sam Mayerfield, UMDNJ.)

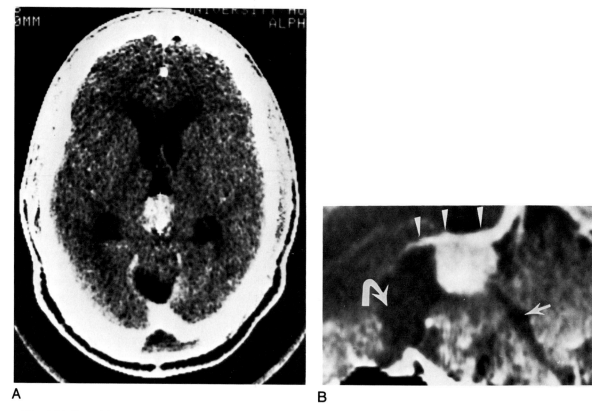

Fig. 6-20 Pineal dysgerminoma. (A) Transaxial contrast-enhanced CT scan shows an intensely enhancing mass in the posterior third ventricle in the region of the pineal gland, causing hydrocephalus. **(B)** Sagittal reconstruction shows the enhancing mass in the posterior third ventricle occluding the aqueduct. It lies underneath the internal cerebral veins and the vein of Galen *(arrowheads)*. Note the dilated third ventricle extending into the sella turcica *(curved arrow)*. The fourth ventricle is small *(arrow)*. (Courtesy of Dr. Sam Mayerfield, UMDNJ.)

calcific change, which can be seen with CT scanning and MRI. It shows irregular contrast enhancement. Endodermal sinus tumor has a nonspecific appearance but tends to be somewhat more aggressive, with poorly defined margins.

Ectodermal pineal tumors are either pinealoma (pineocytoma) or pineoblastoma. Both tumors are either isodense or slightly hyperdense on CT scans, with intense contrast enhancement indistinguishable from that of germinoma. They occur with equal frequency in males and females. The pinealoma can be found at any age and is slow growing. By contrast, the pineoblastoma behaves similar to a medulloblastoma, with rapid growth and the propensity for seeding in the subarachnoid space. The pineoblastoma is considered one of the PNET series. (Fig. 6-21). Initially, the two tumors appear similar on all imaging studies and may be diagnosed only on the basis of clinical findings, CSF cytology, or biopsy.

Non-neoplastic pineal cysts are common. Most are small and are found incidentally, either with imaging or at autopsy. Occasionally, these cysts are large enough to cause mass effect, with hydrocephalus and Parinaud syndrome. Most small cysts do not change over time. Follow-up imaging is currently not recommended for a cyst found incidentally. Other tumors or masses that occur in this region are listed in the box (Fig. 6-22).

Fig. 6-21 Pineoblastoma. Axial Gd-enhanced MRI illustrates an intensely and uniformly contrast enhancing mass in the region of the pineal *(black arrow)*. The mass is nonspecific and could represent pineocytoma, or germ cell tumor, although the irregular margin suggests a more aggressive type of tumor.

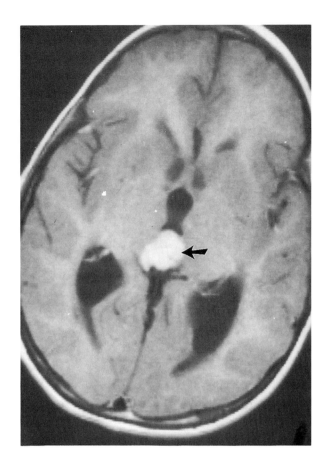

Fig. 6-22 Pineal cyst and glioma. (A) Nonenhanced CT scan shows a cyst in the region of the pineal with contents of just slightly higher density than CSF *(arrowheads)*. Small uncomplicated cysts generally do not cause any problem. This cyst appears to have hemorrhage within it *(arrow)* and is causing hydrocephalus. *(Figure continues.)*

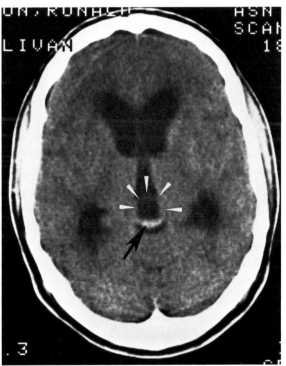

A

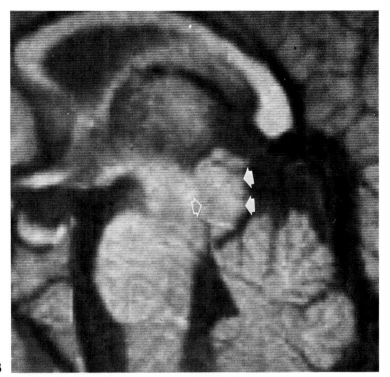

B

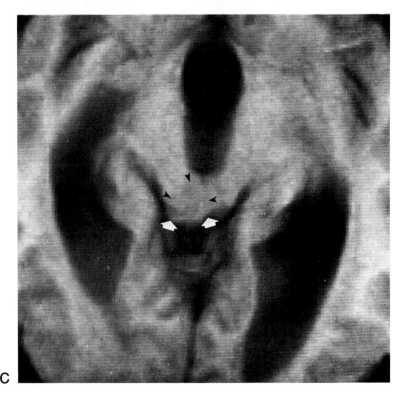

C

Fig. 6-22 *(Continued)*. **(B–C)** Glioma of the quadrigeminal plate. **(B)** Midline parasagittal T₁-weighted MRI shows enlargement of the quadrigeminal plate expanding posteriorly into the cistern *(arrows)*. The aqueduct is obliterated *(open arrow)*. **(C)** Transaxial T₁-weighted MRI shows the enlargement of the right side of the quadrigeminal plate *(arrows)*. Subtle hypointensity represents the glioma (arrowheads). Hydrocephalus is present from obstruction of the aqueduct.

Choroid Plexus Papilloma

Choroid plexus papillomas are rare intraventricular tumors that arise from the ependymal layer of the choroid plexus. They occur most frequently in infants at the site of the glomus within the atrium of the lateral ventricle. Rarely, they are seen in the roof of the third or fourth ventricle. These tumors cause hydrocephalus from overproduction of CSF, which may be asymmetric. Hemorrhage sometimes occurs. Complete surgical removal is curative.

On precontrast CT scan, the tumor is homogeneously isodense or hyperdense, with sharp undulating margins lying within a dilated ventricle. Calcification is common. Contrast enhancement is intensely homogeneous. Malignant change is suggested when there is edema in the adjacent brain parenchyma.

Angiography is important for defining the vascular pedicle before attempting surgical removal. The tumor itself shows an angiographic "blush," similar to the common pattern of meningioma. The anterior choroidal artery enlarges, contributing significantly to a tumor of the glomus.

Leukemia

CNS leukemia is occurring with increasing frequency, chiefly because of the increased survival of children with leukemia. Therefore, CNS involvement develops at some time in 25 to 50 percent with leukemia, which can occur during peripheral remission. CNS leukemia always carries a poor prognosis.

The meninges are the usual sites of involvement, with sheets of tumor cells spreading diffusely throughout the subarachnoid space. This situation blocks the flow of CSF, resulting in mild hydrocephalus. Rare cases exhibit focal peripheral infiltration of the brain with the formation of a tumor mass (chloroma).

CT scanning almost always demonstrates some degree of hydrocephalus when meningeal leukemic infiltration is present. The hydrocephalus may be minimal, indicated only by a slight increase in the size of the temporal horns. When there is extensive infiltrate, the meninges, and sometimes the ependyma of the ventricles, show diffuse contrast enhancement. Focal chloroma or infiltration in the brain is identified by a region of homogeneous enhancement (Fig. 6-23). The lesions tend to be isodense or slightly hyperdense on the nonenhanced scan. Hemorrhage may occur in the brain in the absence

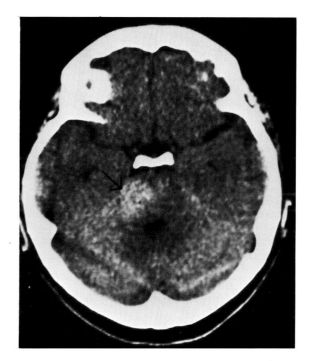

Fig. 6-23 Leukemia (chloroma). Contrast-enhanced CT study shows diffuse homogeneous enhancement in the right side of the pons and cerebellar peduncle. It was thought to represent leukemic infiltration of the brain and disappeared after treatment.

of leukemic infiltration, usually as a result of thrombocytopenia. Cerebral infections may occur secondary to pansinusitis and retrograde contamination through the veins into the cavernous sinus, the brain, and the subdural space.

RARE TUMORS

Meningioma

Meningioma and meningeal sarcoma occur in children; 80 percent of these rare tumors are supratentorial and 20 percent intraventricular. En plaque growth is common and often causes skull sclerosis. The most common sites are along the sphenoid ridge or the fronto-parietal convexity, although the lesions may be found at any of the characteristic locations. Bone destruction may occur. The CT and angiographic characteristics are the same as those described in adults (see Ch. 5).

Hamartoma

Hamartomas are slow-growing gliomatous lesions. They are most common in the temporal lobe, where they may cause psychomotor seizures. Another common location is the hypothalamus near the tuber cinereum. On CT scan they are seen as calcified, cystic, slightly contrast-enhancing lesions. They cannot be accurately differentiated from other low grade gliomas. They are hyperintense on T_2-weighted MRI (Fig. 6-15). The dysembryoblastic neuroepithelial tumor is probably a hamartomatous type of slow-growing tumor that arises in the cerebral hemispheric cortex. It is nodular, and its appearance resembles an oligodendroglioma.

Pleomorphic Xanthoastrocytoma

Pleomorphic xanthoastrocytoma is a rare calcified cystic tumor, usually of the temporal and parietal lobe cerebral cortex. Most patients with this very slow-growing tumor have had a long history of seizures. The characteristic appearance with imaging is of a superficial contrast-enhancing nodule within a calcified cystic mass (Fig. 6-13). It is rare within the ventricle. The tumor

appears malignant on histologic examination, but its clinical behavior is of a benign tumor.

Cavernous Hemangioma

Cavernous hemangiomas are rare malformations of the brain. These rare lesions can occur anywhere, but most are hemispheric. They appear hyperdense on the precontrast CT scan and show varying degrees of contrast enhancement (Fig. 6-24). Most commonly, seizures result, but intracerebral or subarachnoid hemorrhage may ensue. The peak incidence of presentation is during adolescence or young adulthood. Angiography fails to exhibit these lesions because of the slow blood flow through the malformation (see Ch. 4).

Primitive Neuroectodermal Tumors

PNET is a broad category of tumors (see box) that includes all those tumors composed mostly of undifferentiated small primitive cells (small blue cell tumor). It encompasses many tumors that have traditional different names related to their site of origin (e.g., medulloblastoma in the vermis, pineoblastoma in the pineal). These tumors tend to be highly malignant. Small cell carcinoma of the lung and Ewing sarcoma are often considered PNETs.

Primary hemispheric PNETs are highly aggressive rapidly growing lesions that occur during the first decade of life. The tumors are poorly marginated and often infiltrate, but cause little mass effect. Cystic change is frequently present. The tumors are hypodense or isodense on the precontrast CT scan and show moderate enhancement on the postcontrast scan.

**PRIMITIVE
NEUROECTODERMAL TUMORS**

Primitive neuroectodermal tumor (PNET)
Medulloblastoma
Pineoblastoma
Neuroblastoma
Ependymoblastoma
Primitive polar spongioblastoma

Fig. 6-24 Cavernous hemangioma. (A)
Nonenhanced CT scan shows an irregular,
slightly hyperdense lesion in the region of
the pineal gland *(arrowheads)*. The lesion
had hemorrhaged and caused hydrocephalus.
There is a shunt catheter in the right frontal
horn *(arrow)*. **(B)** Lesion shows diffuse homogeneous contrast enhancement *(arrowheads)*.

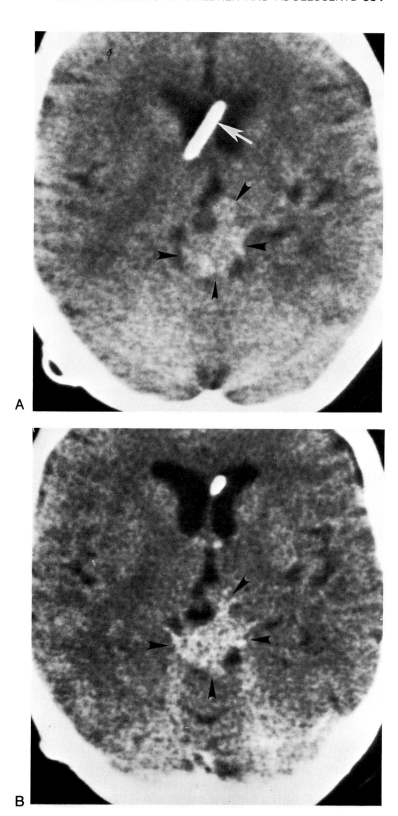

Neuroblastoma

Intracranial neuroblastoma may be metastatic or primary. Metastatic neuroblastoma involves the calvarium as destructive lesions invading the orbit and intracranial epidural space; only rarely does it involve the brain parenchyma.

Primary neuroblastoma usually arises within the cerebral hemispheres, most often the medial posterior temporal lobe in children under 2 years of age. It is related to the PNET group of tumors. On CT scan, it appears hypodense, isodense, or of mixed density, sometimes with dense calcification and cysts. Contrast enhancement is present (Fig. 6-25). Hemorrhage occurs relatively frequently, a finding that helps differentiate the tumor from an astrocytoma, which it otherwise closely resembles. Peritumoral edema is common.

Esthesioneuroblastoma is the term given to neuroblastomas originating from the olfactory nerve. This rare tumor penetrates the anterior frontal fossa and the superior nasal cavity. It is slow-growing but infiltrates and is locally destructive. The great vascularity of this tumor is well demonstrated on angiography. The CT findings are nonspecific, resembling those of other invasive tumors (see Ch. 13).

Primitive Polar Spongioblastoma

Primitive polar spongioblastoma is a rare tumor with a predilection for the cerebellum and spinal cord in children. It has the appearance of an astrocytoma and cannot be specifically distinguished with imaging.

Langerhans Cell Histiocytosis

Langerhans cell histiocytosis (histiocytosis X) comprises a group of disorders characterized by abnormal proliferation of histiocytes in multiple organs, including

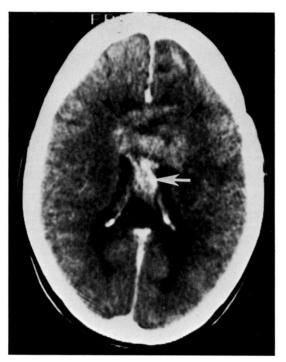

A

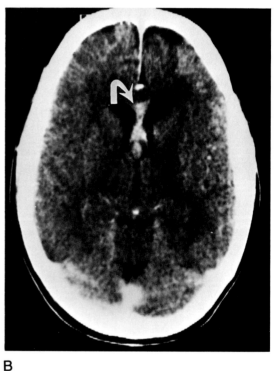

B

Fig. 6-25 Primary neuroblastoma. (A) Contrast-enhanced CT scan shows diffuse enhancement of a tumor mass within the corpus callosum and hemispheric white matter bilaterally (*arrowheads*). Calcification is present in the posterior portion (*arrow*). **(B)** Tumor has extended into the septum pellucidum (*arrow*).

the skull and brain. The disease may occur at any age but is seen primarily in children and young adults. The brain is only rarely involved, but when it is, there is usually only a single lesion in the hypothalamus or sella turcica. Rarely, a lesion occurs in the cerebrum, cerebellum, optic chiasm, or spinal cord. It is associated with diabetes insipidus in about 50 percent of cases. Typical purely lytic bone lesions are present in the skull (Fig. 6-26) and mandible.

On CT scan, these lesions are displayed as a homogeneously enhancing mass, surrounded by edema. On MRI, the lesion shows nonspecific prolongation of the T_1 and T_2 values.

Dermoid

Dermoids and epidermoids occur as a result of embryonic rests resulting from the infolding of the neural tube. Dermoids are most commonly seen in children, and epidermoids are discovered more often during adulthood. Both lesions tend to be in or near the midline.

Dermoid tumors are most often found in the posterior fossa. They assume a variety of configurations, from a small cyst in the occipital subcutaneous tissue to a large cyst that occupies the fourth ventricle, associated with a sinus tract connected to a skin pit. About 50 percent are associated with a sclerotic tract within the calvarium. In infants or children with an occipital skin pit or cyst, a CT scan of the posterior fossa is obtained to look for an intracranial component of the lesion. The dermoid causes problems by its mass effect, infection, or rupture into the subarachnoid space. When rupture occurs, fat density material is seen scattered in the cisterns and leads to a chemical meningitis.

Dermoids contain partially saponified fatty tissues and are therefore seen as very low density, 0 to −150 HU, on CT scan. Rarely, a tooth or calcification is present. There is generally no contrast enhancement. Dermoids have variable intensity on MR images (see section on epidermoid in Ch. 5).

Arachnoid Cyst

Most arachnoid cysts are thought to result from a development disorder of the formation of the meninges that creates pockets within the arachnoid membrane with accumulation of CSF between the layers. Some cysts may result from adhesions that occur following meningitis, subarachnoid hemorrhage, or trauma. Suprasellar cysts are thought to be unique, occurring as a result of upward expansion of an imperforate membrane of Lilliquist into the hypothalamus. Frequently, there is atrophy or distortion of the underlying adjacent brain.

Arachnoid cysts occur most commonly in the temporal fossa or sylvian fissure, posterior fossa (CP angel, and cisterna magna), and suprasellar cistern. On CT scan, they are homogeneous and of CSF density. Calcification or contrast enhancement does not occur. Large cysts in the suprasellar region expand into the third ventricle, obstructing the foramina of Monro, causing hydrocephalus. Detection of these cysts on the initial CT scan may be difficult, as they may mimic a dilated third ventricle. However, widening of the prepontine cistern is almost always associated with the cyst. Lateral cysts may expand the calvarium locally. The adjacent atrophy produces less distortion of the brain than would be expected for the size of the cystic mass. MRI shows the cysts as CSF-intensity masses within the subarachnoid space (Fig. 6-27). Benign intraventricular cysts form, some of which can be large filling or expanding a ventricular cavity. Most arise from the ependyma of the choroid plexus (Fig. 6-28).

CONSEQUENCES OF TREATMENT FOR BRAIN TUMORS

An undesirable result of the combination of cranial radiotherapy and intrathecal methotrexate to treat CNS leukemia is a significant disorder known as subacute necrotizing leukoencephalopathy. This disorder is characterized by demyelination, axonal destruction, and coagulation necrosis, often with fibrinoid vascular change. The greater the dose of the radiotherapy and chemotherapy, the more likely the development of destruction. Total cranial irradiation of more than 2,000 cGy and total intrathecal methotrexate of more than 50 mg leads to the leukoencephalopathy in about 45 percent of patients. The changes are irreversible. Damage can be minimized if therapy is modified during early development of the lesions.

The initial CT finding is a diffuse homogeneous decrease in the density of the paraventricular white matter, which is difficult to recognize in the early phase of the

(*Text continued on page 337*)

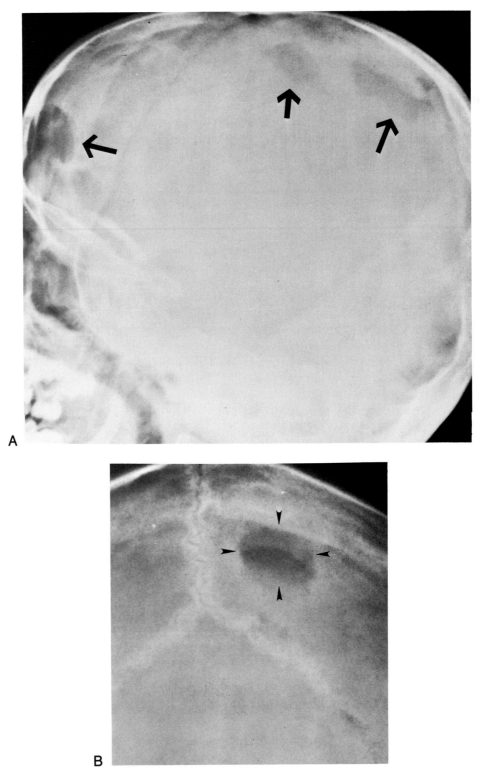

Fig. 6-26 **Langerhans cell histiocytosis and eosinophilic granuloma. (A)** Histiocytosis X. Plain skull radiograph shows multiple lytic lesions of the calvarium *(arrows)*. **(B)** Eosinophilic granuloma. Skull radiograph shows a well-circumscribed solitary lytic lesion of the calvarium *(arrowheads)*. On tangential view, the margins may appear beveled.

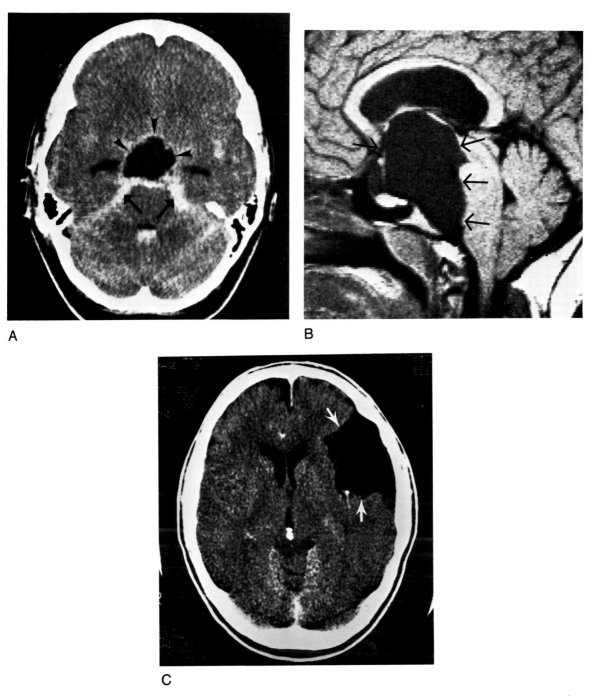

Fig. 6-27 Arachnoid cyst. (A) Transaxial CT positive-contrast cisternogram shows a noncommunicating cyst with CSF-density fluid *(arrowheads)*. Aqueous contrast is seen in the cisterns *(arrows)*. **(B)** Parasagittal T_1-weighted MRI shows a large cyst in the suprasellar prepontine cistern causing displacement of the brain *(arrows)*. **(C)** Arachnoid cyst of the temporal region. CT scan shows the typical trapezoidal configuration with straight margins *(arrows)*. Slight mass effect is present with contralateral shift of the frontal horns. The density of the contained fluid is equivalent to that of CSF. Arachnoidal cysts may be discovered in any age group.

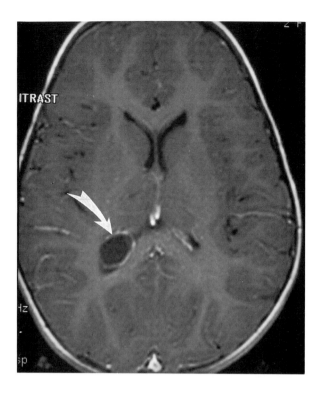

Fig. 6-28 Benign ependymal cyst of the choroid plexus. T_1 MRI after Gd identification of an oval cyst within the atrium of the right lateral ventricle *(arrow)*. The cyst contains fluid with signal equivalent to CSF. The cyst capsule enhances, but this is not a sign of malignancy or tumor. Pathologic proof of diagnosis following surgical removal.

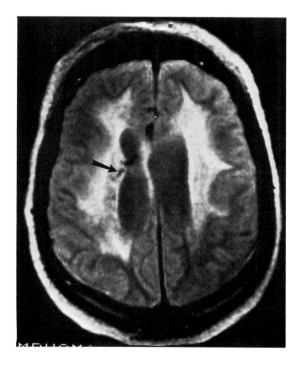

Fig. 6-29 Subacute necrotizing leukoencephalopathy. T_2-weighted MRI shows high signal intensity in the periventricular white matter throughout both hemispheres. Note dilation of the lateral ventricular system due to atrophy; the low signal region *(arrow)* probably represents calcium.

disease. MRI is much more sensitive in detecting early disease and shows paraventricular high signal intensity on T_2-weighted images (Fig. 6-29). Calcification may occur in the blood vessels and is seen as clumps within the white matter. Later, atrophy leads to ventricular dilation. In some cases, small transient high T_2 signal white matter lesions develop with radiation or chemotherapy. The true significance of these is unknown.

EFFECTS OF CNS TUMOR THERAPY

Decrease in intellect
Endocrinopathy
  GH deficiency
  Hypothyroidism
Visual loss
  Injury to optic nerves
Oncogenesis (5 to 25 years latency)
  Glioma, usually malignant glioblastoma multiform (GBM)
  Meningioma
  Sarcoma
  Leukemia
  Thyroid carcinoma
Leukoencephalopathy

SUGGESTED READING

Ahmadi J, Destian S, Apuzzo MLJ et al: Cystic fluid in craniopharyngiomas: MR imaging and quantitative analysis. Radiology 182:783, 1992

Barnes PD, Lester PD, Yamanshi WS et al: Magnetic resonance imaging in childhood intracranial masses. Magn Res Imag 4:41, 1986

Batnitzky S, Segall HD, Cohen ME: Radiologic guidelines in assessing children with intracranial tumors. Cancer 56:1756, 1985

Brandt-Zawadzki M, Kelly W: Brain tumors. In Brandt-Zawadzki M, Norman D (eds): Magnetic Resonance Imaging of the Central Nervous System. Raven Press, New York, 1987

Burger PC, Fuller GN: Pathology-trends and pitfalls in histologic diagnosis. Neurol Clin 9:249, 1991

Bydder GM: Magnetic Resonance Imaging of the Posterior Fossa. Magnetic Resonance Annual. Raven Press, New York, 1985

Curtin HD: CT of acoustic neuroma and other tumors of the ear. Radiol Clin North Am 22:77, 1984

Gentry LR, Smoker WRK, Turski PA et al: Suprasellar arachnoid cysts. I. CT recognition. AJNR 7:79, 1986

Hasso AN, Fahmy JL, Hinshaw DB Jr: Tumors of the posterior fossa. p. 425. In Stark DD, Bradley WG Jr (eds): Magnetic Resonance Imaging. CV Mosby, St. Louis, 1988

Haughton VM, Daniels DL: The posterior fossa. In Williams AL, Haughton VM (eds): Cranial Computed Tomography. CV Mosby, St. Louis, 1985

Lee Y-Y, VanTassel P, Bruner J: Juvenile pilocytic astrocytomas: CT and MR characteristics. AJNR 10:363, 1989

Myers SP, Kemp SS, Tarr SW: MR imaging of medulloblastomas. AJR 158:859, 1992

Pusey E, Kortman KE, Flannigan BD: MR of craniopharyngiomas. AJNR 8:439, 1987

Segall HD, Batnitzky S, Zee C-S et al: Computed tomography in the diagnosis of intracranial neoplasms in children. Cancer 56:1748, 1985

Yock DH Jr: Techniques in imaging of the brain. Part 1. The skull. p. 1. In Heinz ER (ed): Neuroradiology. Churchill Livingstone, New York, 1984

7
Head Trauma

Accidental injury is the most common cause of death and disability among adolescents and young adults. Fifty percent of the injuries are the result of vehicular accidents, and at least 40 percent of the deaths result from head injury, a percentage that is increasing. Many of the deaths inevitably result from severe impact injury; with lesser injuries, however, rapid resuscitation, diagnosis, and treatment of intracranial pathology can produce a better outcome.

Post-traumatic events in the brain can occur with surprising rapidity. The gravity of the intracranial pathology does not necessarily correlate either with the severity of the initial trauma or with the condition of the patient immediately after the accident.

The changes demonstrated with CT scans or MRI may presage clinical events while there is still opportunity for effective therapy, especially generalized brain swelling and impending herniation. It is not enough for the radiologist simply to diagnose the presence of a hematoma or swelling. Rather, it is necessary also to assign some value regarding its severity and importance. Recommendations for the timing of follow-up scans must be given.

Skull radiographs have limited use in the evaluation of head trauma. The main indication is the clinical suspicion of depressed or basal skull fracture or of a gunshot wound of the head. A depressed fracture results from a blow with a relatively small-faced object, such as a hammer. The clinical signs of a basal skull fracture are (1) hemotympanum, (2) hearing loss, (3) vestibular dysfunction, (4) peripheral type of facial nerve palsy, (5)

Battle's sign (ecchymosis over the mastoid), (6) anosmia, or (7) otorrhea or rhinorrhea.

Life-threatening injury to organs outside the CNS take precedence; in these cases, CT of the head is deferred. Otherwise, CT scanning or MRI is done for persons with head injury that results in any neurologic symptoms or signs, including minor changes in mental status. Persons with minor head injury without neurologic dysfunction, loss of consciousness, or decrease in sensorium do not require a scan of the brain. The clinical severity of the injury is determined with the Glasgow Coma Scale. This method evaluates the eye response, speech, and movement of the patient. The scale consists of 3 to 15 points. The lower the score, the more severe the effect of the injury.

The CT scan remains the preferred initial examination, as all immediately important post-traumatic lesions can be detected. CT is easier, faster, and less expensive than MRI, particularly if the patient requires life-support systems. A nonenhanced CT scan of the whole head is obtained using 10-mm-scan slice intervals. Intravenous

GLASGOW COMA SCALE	
13–15	Mild injury
9–12	Moderate injury
<9	Severe injury

contrast is of little use for acute head trauma, indeed, it may be dangerous. Bone window images are routine. Chest, abdominal, and spine imaging can follow as needed. Helical scanning technique may reduce motion artifacts and permit faster scanning in certain circumstances. It has an advantage for obtaining artifact free contiguous images for CT angiography of cervical or intracranial vessels, and multiplanar reconstruction or 3-D rendering.

MRI demonstrates most important acute lesions just as well as CT, including acute hematomas. Spinecho (SE) or fast spinecho (FSE) T_1- and T_2-weighted and gradient echo (GRE) transaxial images are used. The edema associated with axonal shear injury is better seen with MRI SE T_2-weighted MRI. Coronal plane images may better demonstrate hematomas in the subfrontal and subtemporal regions. Small extracerebral hematomas may be seen with MRI, and not with CT scans, but the hematomas found are not immediately important, as they do not require surgical removal. T_2*-weighted GRE is advantageous for the detection of acute hemorrhages during the deoxyhemoglobin phase (1 to 3 days). The T_1-weighted radiofrequency (RF)-spoiled GRE sequence is a fast method designed to show anatomy of the brain; a selected single slice may be done in a few seconds. However, MRI does not reliably demonstrate subarachnoid hemorrhage (SAH). The primary role of MRI is the diagnosis of parenchymal lesions in moderately or severely injured patients when CT scanning does not positively correlate with the clinical condition, or when more prognostic information is needed for clinical decision making.

Traumatic lesions can be characterized as primary or secondary, intracerebral or extracerebral. Primary lesions are those that occur at the instant of injury. They include axonal (shear) injuries, contusions, hematomas, lacerations, and extracerebral hematomas. Except for the removal of hematomas, little can be done to alter the severity of the primary damage. The severity and type of primary injury are the major determinants of the final outcome.

Secondary responses may occur within minutes of the injury or may not become apparent for days or weeks. Intracranially, these responses include brain swelling, delayed hemorrhage, herniation, infarction, infection, and atrophy. Sometimes, particularly in children, the severity of subsequent swelling is greater than would be expected for relatively minor injuries. Significant hydrocephalus after trauma is rare.

TRAUMATIC LESIONS OF THE HEAD

Primary lesions
 Intracerebral lesions
  Contusion
  Axonal injury (shear)
  Hematoma
 Extracerebral lesions
  Subdural hematoma
  Epidural hematoma
  Subdural hygroma
  Subarachnoid hemorrhage
 Skull fracture
  Calvarium
  Basal
  Sinuses
 Vascular lesions
  Laceration
  Pseudoaneurysm
Secondary lesions
 Brain swelling
  Edema
  Hyperemia
  Increased intracranial pressure
 Delayed hemorrhage
 Herniation
 Infarction
 CSF fistula
 Infection
  Intracranial
  Sinus
  Mastoid
 Atrophy

PRIMARY INJURY

Cerebral Contusion

Cerebral contusion is a common post-traumatic brain lesion. It consists of varying amounts of necrosis, hemorrhage, and edema. Primarily a cortical lesion, it may be nothing more than a small, superficial bruise. In its more severe forms, the necrosis and hemorrhage are extensive, not only involving a large portion of the cortex but extending deep into the subjacent white matter. Almost all of the damage is done at impact.

Most contusions occur at sites distant from the point of impact and represent the classic contrecoup lesion. Here the contusion is produced by acceleration and sudden deceleration of the head. With the sudden halt of the calvarial motion, the nonrigid brain continues its forward motion, riding over internal skull protuberances, particularly the orbital roof, sphenoid ridge, and petrous pyramid. Therefore, the inferior surface of the frontal lobe and the anterior and inferior portions of the temporal lobes are the most commonly contused portions of the brain. Oblique motions can cause contusions of the medial surfaces of the cerebral hemispheres, as they are momentarily displaced beneath the falx.

Occipital, parietal, and cerebellar contusions are unusual, presumably because of the smoothness of these re-gions of the calvarium. A depressed skull fracture may penetrate the brain, causing cortical laceration.

Multiple lesions occur in more than 50 percent of persons with contusions, with subdural hematoma the most common associated lesion. Skull fractures are frequently associated with contusions. The prognosis is generally worse if a fracture is present.

The CT appearance of a cerebral contusion depends on its size, the amount of hemorrhage present, and the interval between the injury and the scan. Some small contusions do not hemorrhage and show only edematous changes. A small peripheral bruise may not be detected on the initial CT scan, and only a small amount of focal atrophy may be detected after about 1 month. However, MRI can display the edema and hemorrhage of these small contusions (Fig. 7-1).

With more severe contusions, CT scanning shows a heterogeneous abnormality composed of (1) low-density

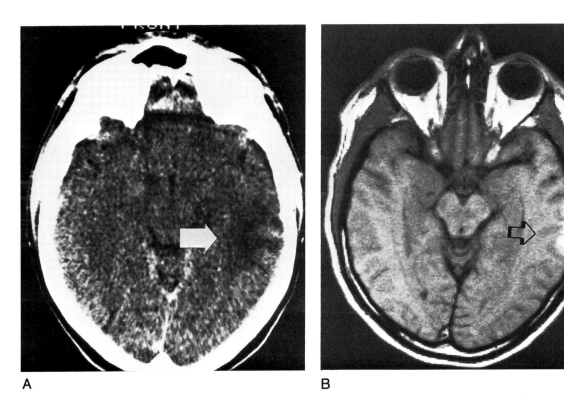

A   B

Fig. 7-1 Cerebral contusion. (A) Cortical contusion is present in the lateral left temporal lobe *(arrow)*. The findings are subtle on the CT scan, with just minimal cortical hemorrhage and edema seen. **(B)** On T$_1$-weighted MRI, the high-intensity change of subacute cortical hemorrhage is easily seen *(arrow)*.

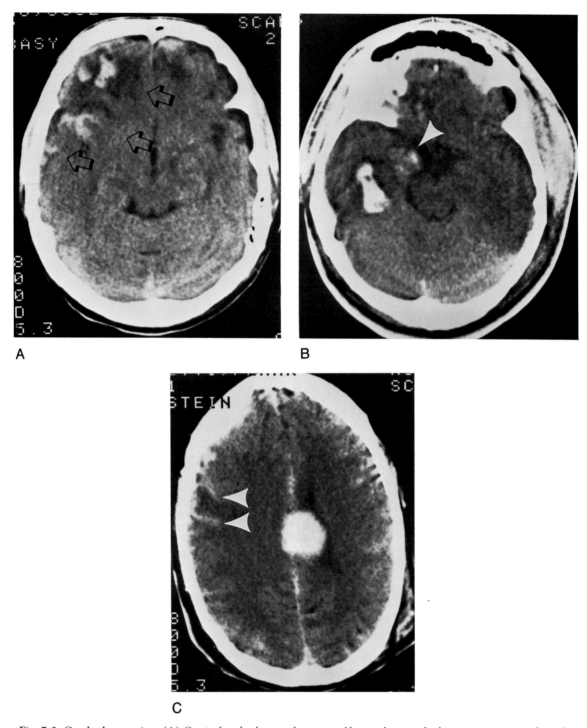

Fig. 7-2 Cerebral contusion. (A) Cortical and subcortical regions of hemorrhage and edema are present in the right frontal and right temporal lobes *(arrows)*. **(B)** Right inferior temporal contusion adjacent to petrous ridge and a small contusion of the hippocampal gyrus *(arrowhead)* are seen. **(C)** Hematoma within the corpus callosum and subarachnoid hemorrhage *(arrowheads)*.

tissue due to necrosis and edema, and (2) regions of multifocal increased density, representing hemorrhage (Fig. 7-2). The mass effect is usually small during the initial phases of the lesion and is proportional to the size of the contusion. It becomes maximum at 4 to 7 days. If the hemorrhage is severe, the cortex may be completely replaced by homogeneous blood density. Hematomas may occur within deep structures (Fig. 7-2C).

MRI scan shows contusions well. The regions of hemorrhage are seen as very low-intensity signal on T_2-weighted SE and GRE images. The hemorrhage is surrounded by high signal intensity edema (Fig. 7-3). During the acute phase, the hemorrhage is isointense on T_1-weighted images. Over a 3- to 10-day period, the hemorrhages begin the resorption process, losing density and sharpness on the CT scan. On MRI, the hemorrhage first becomes bright on T_1-weighted images and then later on T_2-weighted images.

Delayed hemorrhage may occur into regions of contusion, usually within 48 hours of the injury. It may occur spontaneously or after surgical removal of an extracerebral hematoma. Presumably, surgery results in removal of a tamponade effect. Delayed hemorrhage can result in a significant increase in the intracranial pressure and deterioration of the patient. The prognosis thus becomes worse.

CT examination 2 weeks to 6 months after the injury demonstrates the gradual development of gliosis and encephalomalacia, which may become cystic. The density of any hemorrhage disappears, and there is a gradual decrease in the CT tissue density of necrotic tissue, until it approximates that of CSF. On MRI, the late appearance of the contused tissue depends on the amount of hemorrhage that was present initially. For nonhemorrhagic contusions, as gliosis occurs there is a gradual decrease in the signal intensity of the tissue on T_1-weighted images

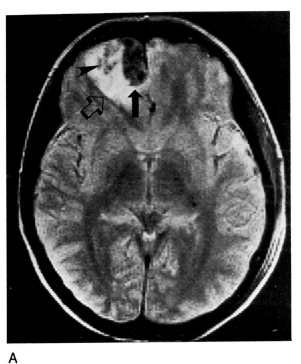

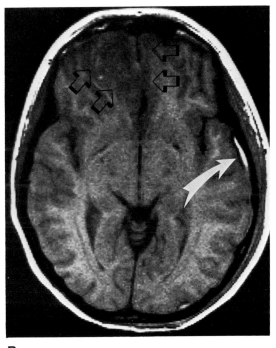

A   **B**

Fig. 7-3 Frontal contusion, MRI. (A) Proton-density MRI shows the hemorrhage as low signal intensity in the medial frontal pole *(solid black arrow)*. Petechial-type hemorrhages are seen laterally *(black arrowhead)*. The edema is shown as high signal intensity *(open arrow)*. **(B)** T_1-weighted MRI shows the hemorrhage to be isotense as it is imaged acutely *(open black arrows)*. A small left temporal extracerebral blood collection has higher intensity than the intracerebral hemorrhage *(white arrow)*.

and an increase in the signal intensity on T_2-weighted images. Hemorrhage within the contusions is high intensity on T_1-weighted images, and this picture persists for months or even years after the injury. On T_2-weighted images, the hemorrhage also becomes bright and persists indefinitely. Very low-intensity hemosiderin rings surround the high-intensity hemorrhages. Focal atrophy is present, and at this stage the contusion may mimic old infarction. Depending on the depth of the contusion injury, there may be associated focal dilation of the adjacent lateral ventricle (Fig. 7-4).

As with infarction, after about 5 days there is proliferation of peripherally located granulation tissue. These new vessels do not possess a blood-brain barrier, so there is ring-like or gyral enhancement on contrast CT scans that persists for many months, although gradually decreasing in area and intensity. If a contrast CT examination is performed and the history of recent trauma is unknown, it is possible to misdiagnose the lesion as a tumor, infarction, or infection.

The signs and symptoms of cerebral contusion are related to the size and location of the contusion. Small contusions may resolve without causing any deficit. Lesions in the motor or speech regions generally result in focal neurologic deficit.

Axonal Injury

Diffuse axonal injury, or shear, occurs as an immediate consequence of the rotational forces associated with impact. Because the gray and white matter have different mass densities, with rapid rotation these two tissues deform at different rates, causing shearing of axons at gray–white matter interfaces. The lesions are characteristically found within the white matter at the corticomedullary junction and in the corpus callosum. They may also occur deep within the brain or in the brain stem when especially severe forces have occurred. Small intraventricular hemorrhages may be seen as an indirect sign of deep axonal injury (see Fig. 7-6C).

Most axonal injuries are small, widely distributed, and nonhemorrhagic, producing only a small amount of local edema. About 20 percent have a small associated hemorrhage. The amount of axonal injury generally correlates with neurologic dysfunction. This correlation is greatest with more severe injury. With lesser degrees of axonal injury, the correlation with neurologic function is unpredictable. When severe, the summation of multiple lesions may produce significant diffuse brain swelling with compression of ventricles and sulci. Axonal injury produces significant loss of white matter, causing ventricular enlargement after 1 month.

T_2-weighted MRI is superior to CT scanning for detecting axonal injury. The injury is best seen during the acute phase. The lesions are small (less than 1-cm) foci of high signal intensity within the white matter, representing the focal edema (Fig. 7-5). Small acute hemorrhages are seen on T_2-weighted images as low-intensity regions within the high-intensity edema. In the subacute and chronic phase, hemorrhages are best detected as high signal on the T_1-weighted images. CT scans demonstrate only the hemorrhagic lesions (Fig. 7-6), as the small nonhemorrhagic regions of subcortical edema are usually too small for routine CT scan detection. MRI findings of axonal injury must be differentiated from other multifocal white matter diseases, particularly multiple sclerosis, diffuse infection (fungal infection, acquired immunodeficiency syndrome [AIDS]), metastasis, and small subcortical infarctions that might pre-exist in the trauma patient.

Intracerebral Hematoma

Intracerebral hematomas are well-defined homogeneous hemorrhages in the brain parenchyma, usually the white matter. They are formed when a major perforating artery ruptures from an acceleration–deceleration injury or when there is profuse hemorrhage in a region of contusion, causing the hemorrhages to coalesce. The differentiation of hemorrhagic contusion from hematoma can be difficult and somewhat arbitrary. Like contusions, 80 to 90 percent of hematomas occur most commonly in the frontal and temporal lobes. Hematomas also occur underneath depressed skull fractures or may be associated with penetrating wounds. They are rare in the cerebellum. CT and MRI examinations show the findings of intracerebral hematoma (see Ch. 2).

Extracerebral Subdural Hematoma

A subdural hematoma (SDH) is a collection of blood within layers of the dura. Post-traumatic SDH is thought to result from rupture of surface cortical veins or arteries

(Text continued on page 348)

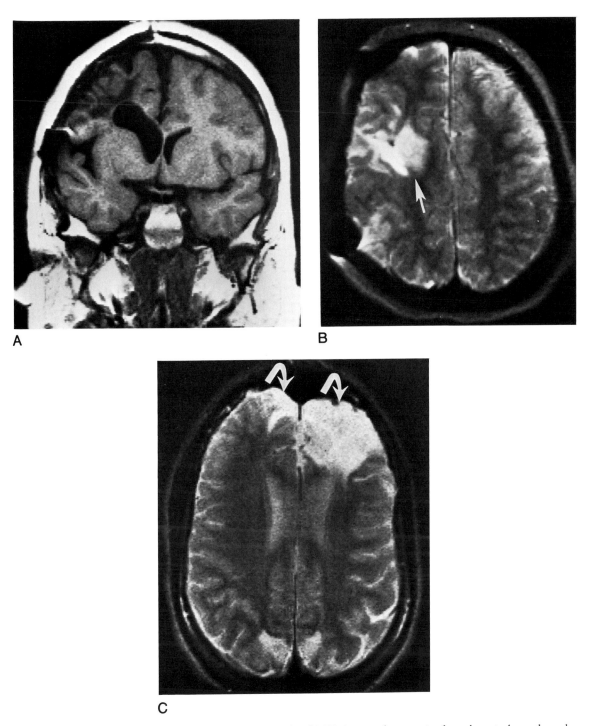

Fig. 7-4 Post-traumatic encephalomalacia. (A) T_1-weighted MRI shows right posterior frontal cortical atrophy, subcortical gliosis, and porencephalic dilation of the adjacent lateral ventricle. **(B)** Gloisis is seen as high signal intensity on the T_2-weighted images *(arrow)*. **(C)** Bifrontal contusion in another patient. T_2-weighted MRI shows gliosis and demyelination as regions of high signal intensity extending to the cortical surface *(arrows)*.

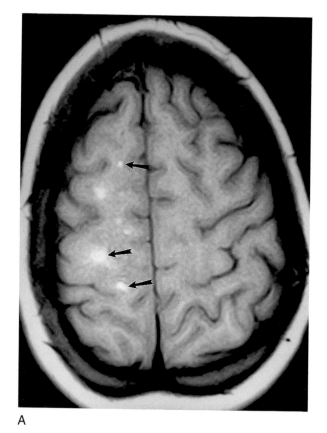

A

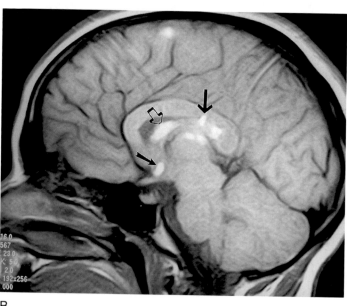

B

Fig. 7-5 Axonal injury. (A) Axial T_1 MRI shows multiple small hemorrhages at the corticomedullary junction, the typical location for axonal shear injuries (*arrows*). T_2 MRI may show additional lesions, evident from local edema without hemorrhage. **(B)** Sagittal T_1 MRI shows hemorrhage shear injuries within the hypothalamus (*small black arrow*), the fornix (*open arrow*), and the corpus callosum (*larger black arrow*). Additional lesions occur within the thalamus and the subcortical posterior frontal lobe.

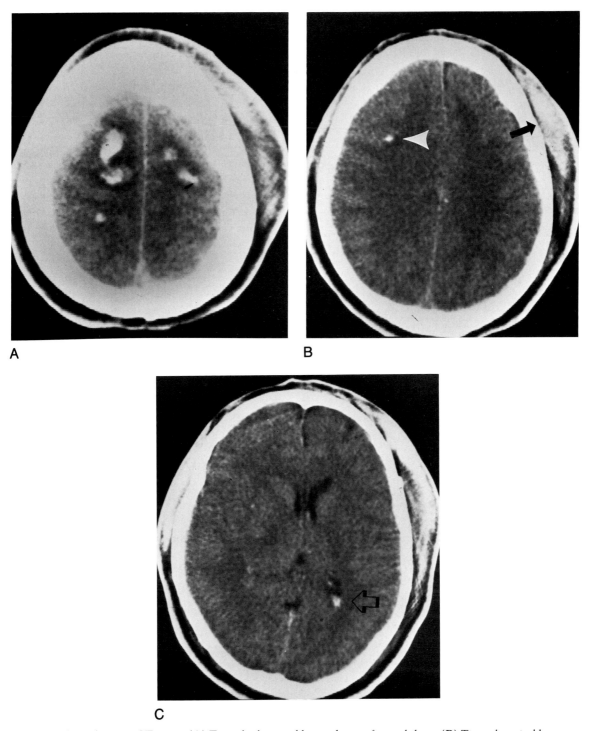

A

B

C

Fig. 7-6 Axonal injury, CT scans. (A) Typical subcortical hemorrhages of axonal shear. **(B)** Tiny subcortical hemorrhage representing axonal shear *(white arrowhead)*. Subcutaneous scalp hematoma *(black arrow)* is also seen. **(C)** Small intraventricular hemorrhage secondary to deep axonal shear. This picture may be the only CT indication of deep white matter injury.

at sites of brain contusion or from tearing of the bridging veins that drain into the dural venous sinuses. The blood collects within the dural membrane at the junction of the meningeal and periosteal layers. These two layers separate as the blood accumulates (Fig. 7-7). It almost always occurs as a result of trauma. These lesions have been arbitrarily classified according to their age relative to the trauma. An acute SDH is one that clinically manifests within the first 72 hours after the injury. Subacute SDH refers to one that is 3 to 20 days old. Chronic SDH is one that is more than 20 days old. The CT and MRI appearances, their pathogenesis, and their epidemiology differ among the three groups, so they are discussed separately.

ACUTE SUBDURAL HEMATOMA

Acute SDH almost always occurs as a result of significant head trauma, but it may occur spontaneously from dural arteriovenous malformation (AVM) or mycotic aneurysm or after the most minimal injury when a coagulopathy is present. Rarely, it occurs after rupture of a berry-type aneurysm. In most cases, it occurs over the cerebral convexity, but it may extend beneath the brain, over the tentorium, or into the interhemispheric fissure. A posterior fossa SDH is unusual. Most patients with acute post-traumatic SDH have an underlying brain contusion that is often severe. The mass effect and elevation of the intracranial pressure may be predominantly due to the

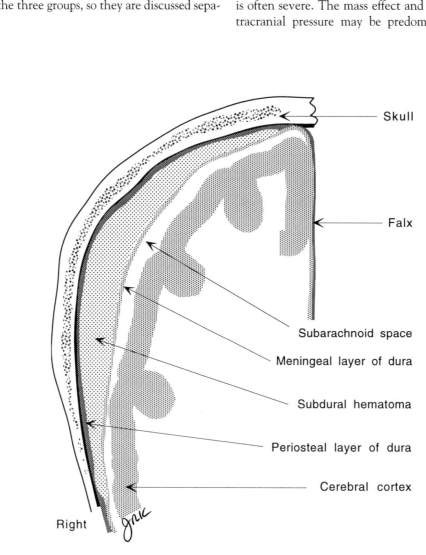

Fig. 7-7 Subdural hematoma. Coronal drawing shows collection of subdural hematoma between the meningeal and periosteal layers of the dura.

changes associated with the contused brain. The severity of the associated brain injury is the primary determining cause of the outcome of patients with SDH, even after its timely surgical evacuation.

In the hyperacute phase, MRI shows the SDH as a high-protein solution, with isointensity to brain on T_1- and high intensity on T_2-weighted images. The first day, the SDH demonstrates low intensity on T_2-weighted images.

On CT scan, an acute SDH appears as a high-density (50- to 90-HU) crescent-shaped extracerebral mass, most commonly over the frontal or parietal convexity (Fig. 7-8). Its medial border may be slightly angulated into the region of the sylvian fissure. SDH may also be found in the subtemporal, suboccipital, and interhemi-spheric regions (Figs. 7-9 and 7-10). The clot density may be uniform or may have low-density regions within it, representing either unclotted blood or continued bleeding during CT scan. With unclotted blood, a density level may be seen from red blood cells settling in the dependent region. It is possible for the SDH to have a relatively low density (30 to 40 HU) in patients with severe anemia (hemoglobin usually less than 9 mg/dl). The hematoma decreases in density with time, so that at some point between 1 and 3 weeks, the SDH is isodense with the adjacent brain. After 3 weeks, the SDH becomes hypodense.

Cerebral angiography demonstrates the SDH in a high percentage of cases. It is seen as an avascular crescent-shaped mass over the frontal and parietal convexity, compressing the brain and displacing the cortical arteries and veins away from the inner table of the skull (Fig. 7-11D).

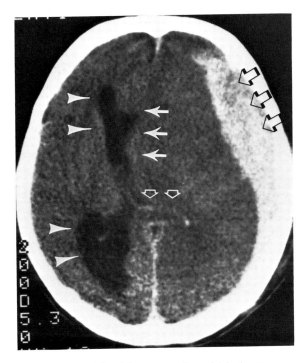

Fig. 7-8 Acute subdural hematoma. Large high-density extra-cerebral fluid collection over the left frontal and parietal con-vexities represents an acute subdural hematoma (*large open arrows*). There is subfalcial herniation (*white arrows*) and tran-stentorial herniation, represented by enlargement of the right lateral ventricle (*white arrowheads*) and obliteration of the qua-drigeminal plate cistern (*small open arrows*).

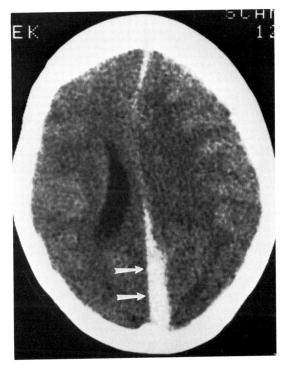

Fig. 7-9 Acute interhemispheric SDH. Interhemispheric SDH is seen with a flate medial border (*arrows*) and a convex lateral border. The shift of the brain is due to contusion.

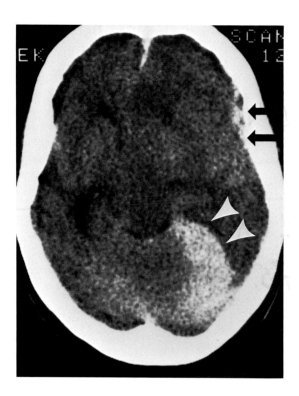

Fig. 7-10 Acute suboccipital SDH. The high density over the left tentorium represents a suboccipital SDH *(white arrowheads)*. A small left convexity SDH also is seen *(black arrows)*, and there is subfalcial herniation.

SUBACUTE SUBDURAL HEMATOMA

An SDH seen 3 to 20 days after injury is called a subacute SDH. During this period, MRI easily detects a SDH. With the development of methemoglobin, T_1-weighted images show the extracerebral blood collection as hyperintense, first at the periphery and then filling in centrally. The same change occurs with the T_2-weighted imaging, although at a slightly later time, depending on the field strength used (Fig. 7-11).

During this time, the hematoma is more difficult to detect with CT scanning. An SDH decreases in CT radioabsorption properties with time, so that at some point it appears isodense with the adjacent brain. Nevertheless, it is possible to detect nearly all collections without resorting to the use of intravenous contrast. The extracerebral collection causes a mass effect that compresses the brain, which can be detected by recognizing (1) obliteration of the cerebral sulci; (2) inward displacement of the cortex, seen as "buckling" of the ipsilateral gray–white matter junction (Fig. 7-12A); (3)

ipsilateral compression of the ventricles; and (4) subtle mottled extracerebral density adjacent to the inner table of the skull, where normal brain sulci are usually seen. Bilateral subacute SDHs are more difficult to recognize because of symmetry.

In difficult situations, CT scanning with intravenous contrast helps detect the subacute SDH. The contrast enhances the hypervascular inner membrane that forms around the SDH after 1 week. It defines the border between the SDH and the cortex of the brain (Fig. 7-12B). In addition, the contrast may outline cortical veins and the brain cortex itself, further defining the margin of the SDH.

CHRONIC SUBDURAL HEMATOMA

By definition, a chronic SDH is more than 20 days old. In the elderly, an SDH frequently goes unrecognized until it has reached the chronic stage. Up to 50 percent of these patients cannot even recall a specific head

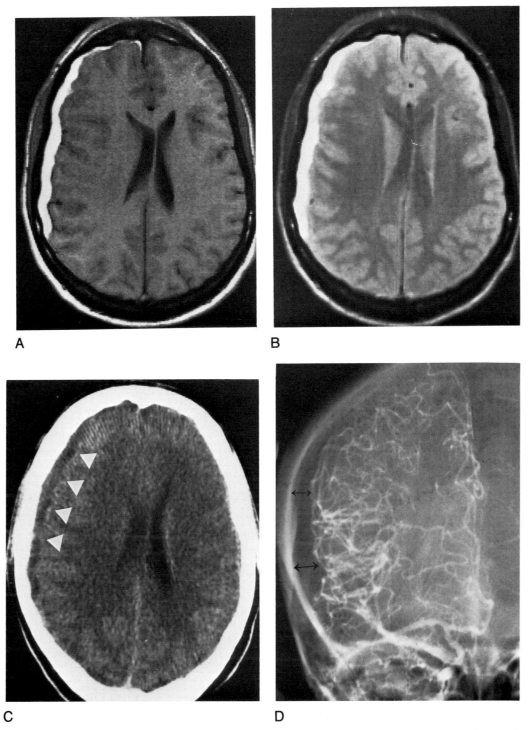

Fig. 7-11 Late subacute phase SDH. (A) T₁-weighted MRI shows hyperintense subacute subdural blood accumulation over the right convexity. **(B)** Proton-density MRI shows high-intensity fluid accumulation over the convexity. **(C)** CT scan shows slightly hypodense fluid accumulation over the right convexity. **(D)** Angiography shows cortical arteries displaced away from the inner table *(opposed arrows)*.

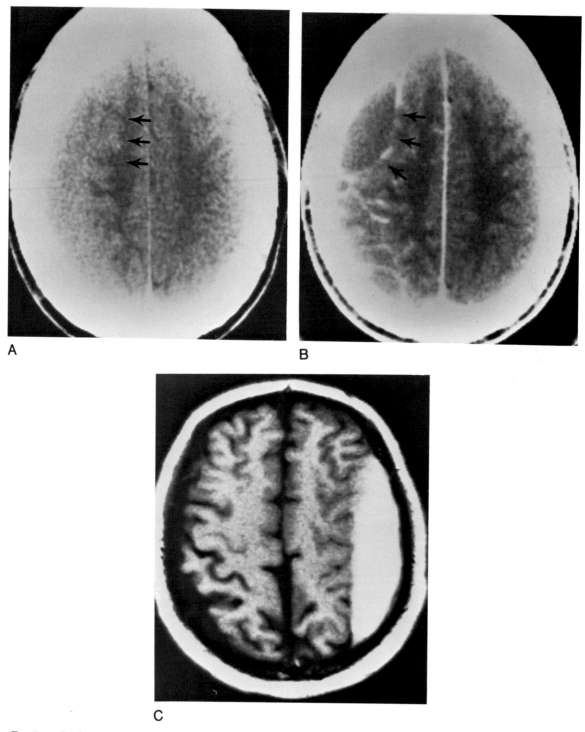

Fig. 7-12 Isodense subacute SDH, CT scans. (A) Inward displacement (buckling) of the gray–white matter junction on the right *(arrows)* secondary to the invisible SDH. **(B)** Contrast enhancement outlines the dural membrane *(black arrows)* and compressed cerebral cortex by the now visible SDH. **(C)** Subacute SDH is easily seen with T$_1$-weighted MRI.

injury, and those who can recall a mild injury. Fifty percent of persons who present with a chronic SDH are chronic alcoholics.

Most SDHs that present in the chronic stage produce few symptoms. They enlarge slowly and are not associated with underlying brain injury. They may be asymptomatic for weeks, probably at least partly a consequence of the greater compliance of an atrophic brain to accommodate an intracranial mass lesion. Bilateral chronic SDH is common, with an incidence of up to 50 percent in some series.

There are multiple clinical patterns of presentation: (1) dementia or inappropriate behavior; (2) acute motor or sensory loss mimicking a stroke; (3) transient neurologic deficit, mimicking a transient ischemic attack (TIA); (4) seizure; (5) increased intracranial pressure with headache and vomiting; and (6) gradual progression of neurologic signs, mimicking a tumor. Headache is a cardinal symptom and may be present without associated neurologic abnormality. Therefore, in the elderly, headache is a definite indication for the performance of CT scanning or MRI.

In the elderly, SDH is thought to result from rupture of the fragile bridging veins that connect the cortical veins with a dural sinus. These veins are under greater stretch in persons with cerebral atrophy. The hemorrhage begins as a small, silent, acute SDH. Ingrowth of fibroblasts into the hematoma from the inner surface of the dura leads to the formation of a thick lateral membrane over the subdural blood collection. A thin membrane also forms along the medial margin of the hematoma. This formation occurs within the first 3 weeks after the injury, and the membrane can be imaged by either contrast-enhanced CT scans or radionuclide brain scanning because of its lack of a blood-brain barrier. The hematoma increases in size, usually because of repeated hemorrhages. It may take any shape; the most common is concave medially, similar to an acute SDH (Fig. 7-13). A biconvex shape is less common and usually indicates multiple bleeding episodes (Fig. 7-14A).

With MRI, chronic SDH appears as an extracerebral high-intensity collection on both T_1- and T_2-weighted images. Inhomogeneity may be present from loculated hemorrhages of differing ages. Because of contained protein, chronic SDH is very hyperintense on T_2-weighted images. After a few months, the hematoma may become hypointense on T_1-weighted images, although it remains of high intensity indefinitely on T_2-weighted images. Peripheral hemosiderin deposits are much less common

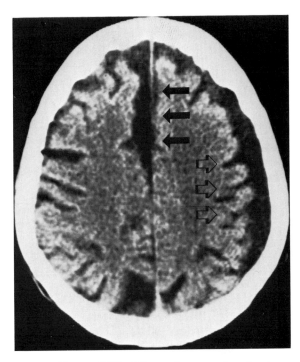

Fig. 7-13 Chronic SDH. CT scan shows very low-density extracerebral fluid collection on the left causing compression of the cerebral sulci (*open black arrows*) and narrowing of the ipsilateral interhemispheric fissure (*solid black arrows*).

with extracerebral hemorrhages than with parenchymal hemorrhages.

With CT scanning, the collection is primarily hypodense (Fig. 7-13). Multiple loculations of differing densities may be present, sometimes with a compartmentalized acute hemorrhagic density (Fig. 7-14A). Fresh blood elements may gravitate to the dependent portion of the collection. If rebleeding has not occurred for some time, CT density of the SDH is very low, just slightly above that of the CSF, and may appear as enlargement of the subarachnoid space due to severe atrophy. However, mass effect is usually present with an SDH (Fig. 7-13). T_2-weighted MRI always shows a chronic SDH as high signal intensity, easily differentiating SDH from the enlarged subarachnoid space associated with severe atrophy.

It is generally not possible to evacuate a chronic SDH completely at surgery in elderly patients because of underlying brain atrophy and the inability of the brain to

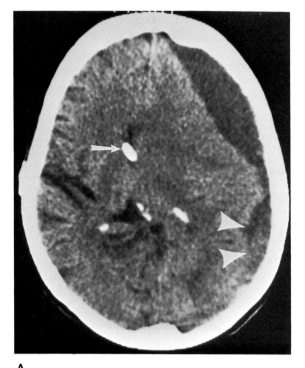

A

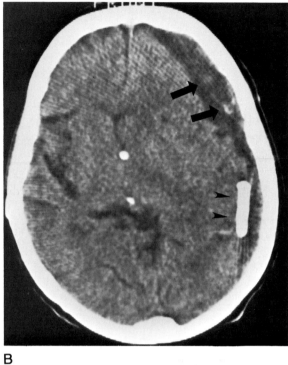

B

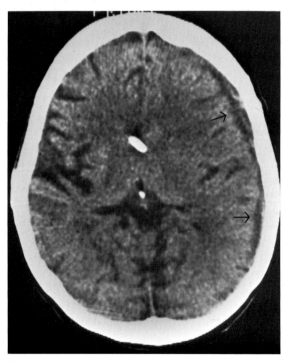

C

Fig. 7-14 Chronic SDH, surgical evacuation. (A) CT scan shows a moderately hypodense multiloculated chronic SDH on the left, with a convex inner margin. Slight increased density is seen in the posterior loculation from recent hemorrhage *(arrowheads)*. The brain is compressed and shifted. A shunt tube is present in the right frontal horn *(arrow)*. **(B)** Subtotal removal of the SDH on the left. A small amount of rebleeding has occurred *(black arrows)*. A drain is in place posteriorly *(arrowheads)*. **(C)** One month later, near-complete resorption of the extracerebral fluid collection has occurred *(arrows)*.

re-expand to fill the extracerebral void. Serous fluid usually fills the remaining space. However, over weeks this fluid gradually resorbs (Fig. 7-12).

Subdural Hygroma

Subdural hygromas, collections of CSF in the subdural space, may occur following trauma or rapid decompression of the ventricular system after shunting. There may also be a small amount of blood within the fluid with acute head injury.

These peculiar collections probably result from a tear in the arachnoid and meningeal layer of the dura, with escape of CSF into the subdural space. Because the fluid may not be able to escape from the subdural pocket, it can act as a mass effect. Small hygromas resorb on their own, but those associated with deteriorating neurologic signs may benefit from surgical drainage. However, hygromas that develop after ventricular overshunting are treated by placement of a higher-pressure shunt valve,

rather than removal of the extracerebral fluid (see Ch. 11). On CT, a hygroma may appear as a chronic SDH but is seen in a different clinical context (Fig. 7-15).

Subarachnoid Hemorrhage

SAH is common after head injury. It is caused by rupture of small perforating vessels. It can be seen on CT scans as increased density within the cisterns and sulci (Fig. 7-2C). SAH is usually not detected with MRI. Although it does not cause a mass effect, SAH may cause a mild communicating hydrocephalus and contribute to increased intracranial pressure.

Epidural (Extradural) Hematoma

Acute epidural hematoma refers to a blood collection that occurs between the dura and the inner table of the skull. It is relatively infrequent—about one-tenth as common as SDH. Epidural hematoma is rare in those

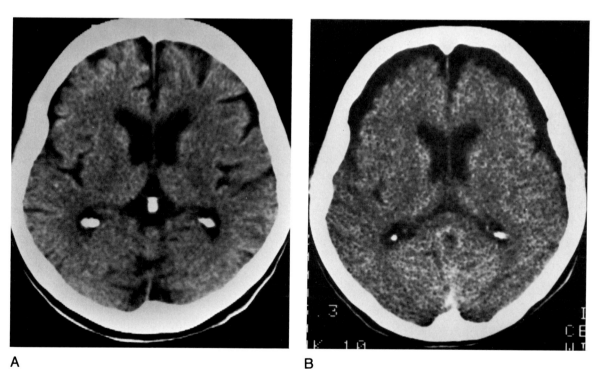

A   B

Fig. 7-15 Acute subdural hygroma. (A) CT scan of the brain obtained for other reasons just prior to a head injury. **(B)** Acute bilateral subdural low density fluid collections found at surgery to be CSF. Note the compression of the ventricles and cerebral sulci.

under age 2 or over age 60, because in these two groups the dura is firmly adherent to the undersurface of the skull. It occurs most commonly in the supratentorial region and only rarely in the posterior fossa.

Epidural hematoma is due to laceration of the meningeal arteries and veins that are on the calvarial surface of the dural membrane. Less commonly it is caused by laceration of a dural sinus. Eighty-five percent are associated with a local skull fracture, generally one that crosses a vascular groove or dural sinus. The force of the trauma produces inward bending of the calvarium, which in turn strips the dura from the inner table. This action creates an epidural pocket into which bleeding may occur. Venous bleeding remains localized to the original pocket. Arterial bleeding may exert enough force to enlarge the pocket. If the associated fracture is large, the hemorrhage may decompress outward into the subgaleal space, instead of into the epidural space.

An epidural hematoma may enlarge rapidly, requiring immediate surgical evacuation. Because there is usually relatively little underlying brain injury, the results of surgical removal are much better than with acute SDH. Small epidural hematomas (less than 1 cm thick) tend not to enlarge and may be followed with serial CT scans. They resorb without surgical evacuation.

With CT scanning, the acute epidural hematoma appears as a homogenous high-density (40 to 90 HU) biconvex mass underneath the inner table (Fig. 7-16). It causes local displacement of the brain, with compression of the sulci and the ventricles, and midline shift, but underlying brain injury is uncommon. A skull fracture can often be demonstrated with the use of bone window images. Hematomas caused by a tear of the sagittal sinus are seen high on the convexity and often cross the midline. Small epidural hematomas that are not removed gradually decrease in density and size, eventually becoming hypodense lesions and disappearing. There may be slight thickening of the inner table of the skull with healing. Epidural hematomas appear on MRI with the intensity characteristics of any hematoma. The shape is the same as that described with CT scans (Fig. 7-17).

SECONDARY INJURY

Intracranial Pressure

The intracranial pressure is normally below a mean of 15 mmHg. This steady state is maintained by a balance of the formation of CSF, resorption of CSF, resistance to

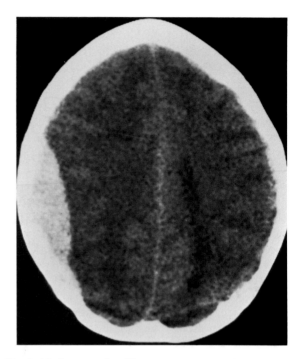

Fig. 7-16 Acute epidural hematoma. CT scan shows the typical high-density extracerebral collection on the right with a convex medial border. The right lateral ventricle is compressed, and there is a slight shift of the falx.

the flow of CSF through the ventricles and cisterns, and compliance of the intracranial fluid volume to accommodate increased cerebral mass. After cranial trauma, the normal mechanisms controlling intracranial pressure may be disturbed or overwhelmed, so there is a devastating increase in intracranial pressure.

It is important to understand the concept of intracranial compliance and the nature of the CSF pressure–volume curve. Because of the buffer of the vascular system, an increase in CSF volume or the formation of brain edema is compensated for by a decrease in the intracerebral vascular volume. Similarly, the CSF spaces surrounding the brain and the ventricular cavities can accommodate changes in brain size by displacement of fluid. Clearly, these mechanisms can work only to a point. If the brain volume (e.g., from edema) becomes so great that all the blood is displaced from the vascular system and the CSF is displaced from the ventricles and cisterns, the intracranial pressure rises rapidly to the level of the blood pressure. The intracranial blood flow ceases, and brain death ensures.

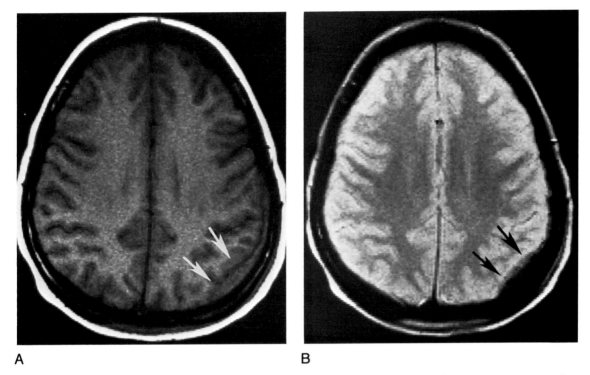

Fig. 7-17 Acute epidural hematoma. (A) Small epidural left posterior parietal blood collection is isointense on the T$_1$-weighted image *(arrows)*. **(B)** T$_2$-weighted image shows the acute small collection as extreme low signal intensity *(arrows)*.

At conditions close to normal, small changes in the cerebral volume (brain swelling) can be accommodated with little increase in the intracranial pressure. However, when the capacity of the cisterns and the vascular system is approached, an equally small incremental increase in the intracranial volume results in a large, possibly detrimental, increase in intracranial pressure. At this point, intracranial compliance has become markedly reduced. The converse is also true, and a slight decrease in cerebral volume due to therapeutic maneuvers results in a significant decrease in intracranial pressure. The brain volume–intracranial pressure curve is shown in Figure 7-18.

The nature of the pressure–brain volume curve has important imaging implications. From the curve alone, one can predict that a considerable amount of brain swelling can be present with little increase in intracranial pressure. It is also true that it is impossible to determine where the patient is on the curve simply by measuring intracranial pressure. Therefore, serial CT or MRI scans become important for management of the trauma victim, as the findings of cerebral swelling may be seen on CT examination before there is a significant rise in intracranial pressure or a change in the patient's clinical state. The scans can also predict the amount of compliance present by determining the size of the cisterns. If the cisterns are small or obliterated, intracranial compliance for an increase in brain volume is low.

Other factors reduce the compliance of the brain to absorb changes in intracranial volume. Compliance is reduced by the presence of intracranial hematoma, hydrocephalus, SAH, transtentorial or tonsillar herniation (increased resistance to flow of CSF in the cisterns), systemic hypertension, or fluid overload. These factors must be considered when evaluating the CT or MRI scan.

Brain Swelling

Brain swelling is a poorly understood phenomenon that may accompany any type of head injury. It may be severe after either major and surprisingly minor injuries. The effects of the swelling are additive to the effects of

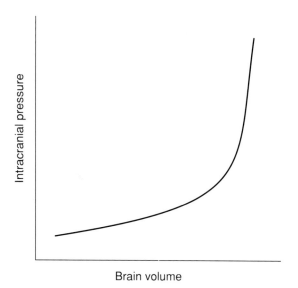

Fig. 7-18 Relationship of cerebral volume to intracranial pressure. Intracranial pressure rises only slightly until a critical brain volume is reached. The pressure then increases greatly with slight increases in brain volume.

primary injury itself and may become the most significant factor determining outcome.

Acute brain swelling is usually the result of increased intracerebral blood volume brought about by changes in the autoregulation of the cerebral vessels. Delayed swelling is usually due to brain edema. The mechanism of the swelling has important therapeutic ramifications. Hyperventilation therapy through vasoconstriction is aimed at decreasing intracranial vascular volume. Mannitol therapy is aimed at decreasing actual cerebral edema. Severe post-traumatic brain swelling is difficult to control by any means.

Brain swelling is identified on the CT scan by a decrease in the size of the ventricles, cortical sulci, and basal cisterns. Special attention is given to observing the size of the cisterns. If the supratentorial and incisural cisterns are totally obliterated, cranial–caudad transtentorial herniation is either imminent or has already occurred (Fig. 7-8). This sign indicates extreme danger and may be recognized on the CT scan before clinical signs of herniation have occurred. When this degree of swelling is present, even small changes in intracranial volume due to edema or bleeding result in a rapid increase in the intracranial pressure. In severe situations, the posterior fossa cisterns are also obliterated (Fig. 7-19).

The CT density of the swollen brain following trauma is variable and depends on a complex relationship between blood volume, edema, and microhemorrhages. The brain may be any density, from slightly hypodense to slightly hyperdense; it is unpredictable.

Herniation

Internal herniations are common with severe head injuries. There are many types. Subfalcian herniation occurs when the brain shifts across the midline underneath the relatively rigid falx (Fig. 7-6). Such a shift results from unilateral hemispheric or extracerebral lesions. The frontal horns of the ventricles are distorted. Infarctions may result in the distribution of the compressed pericallosal arteries. Cranial–caudad transtentorial herniation is due to a large unilateral or bilateral supratentorial mass lesion. Here the base of the brain is forced downward into the tentorial hiatus, which obliterates the perimesencephalic and quadrigeminal plate cisterns. Acute hydrocephalus results from compression of the aqueduct. The ventricular dilation may be asymmetric, depending on the side of the maximal lesions (Fig. 7-8). There may be unilateral or bilateral third cranial nerve (CN III) compression.

Uncal herniation is the result of lesions within or adjacent to the temporal lobe. Here, the medial portion of the temporal lobe (uncus) is forced medially and downward into the ipsilateral tentorial hiatus, and the brain stem is displaced contralaterally. There is ipsilateral compression of CN III.

Infarction

Focal cerebral infarctions may result from injured or compressed vessels. Most commonly they occur in the following regions: (1) pericallosal region, due to compression by subfalcial herniation; (2) occipital lobes, due to compression of the posterior cerebral arteries against the tentorium with transtentorial herniation (Fig. 7-20); (3) posterior parietal "watershed" region; (4) middle cerebral territory, due to compression of the supraclinoid internal carotid artery against the anterior clinoid processes; and (5) global infarction, due to a substantial increase in intracranial pressure.

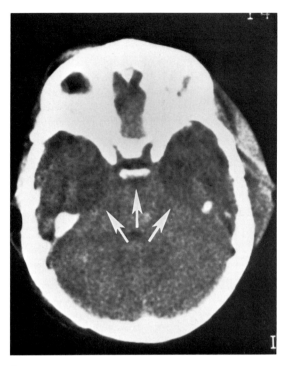

A

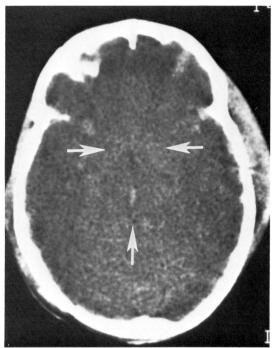

B

Fig. 7-19 Severe acute brain swelling. (A) Obliteration of the prepontine and cerebellopontine angle cisterns *(arrows)*. **(B)** Obliteration of the suprasellar and quadrigeminal plate cisterns *(arrows)*. **(C)** Compression of the lateral ventricular system *(arrows)*. All scans show loss of gray–white matter differentiation, obliteration of cerebral sulci, and small contusions at typical sites. The findings on this scan indicate increased intracranial pressure and low compliance.

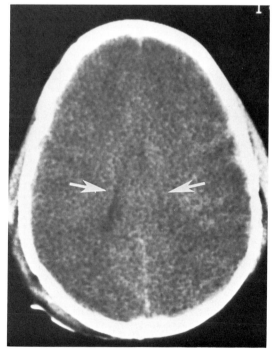

C

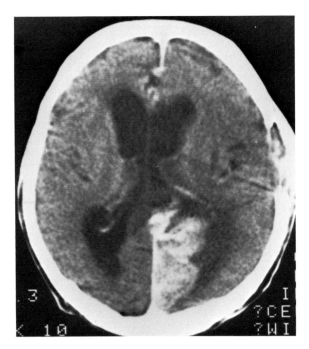

Fig. 7-20 Occipital infarction after transtentorial herniation.
This contrast-enhanced CT study shows diffuse enhancement
in a left occipital infarct. It is the postoperative CT scan of the
acute SDH shown in Figure 7-8.

Atrophy

Atrophy is common after head injury. It may be mild
or severe and focal or diffuse. After the subsidence of
swelling, it progresses in a linear fashion, becoming maximal about 2 months after the injury. The white matter is
usually most affected, resulting in dilation of the ventricles that can be considerable (Fig. 7-4A). The cerebral
sulci show less enlargement. The diagnosis of ventricular enlargement from atrophy should not be confused
with hydrocephalus, an uncommon finding after head
injury.

Pneumocephalus (Pneumocranium)

A careful search is made for air within the calvarium.
Air most frequently collects in the anterior portion of the
frontal fossa, best seen on the transaxial CT scan, but
also on a cross-table lateral skull radiograph (Fig. 7-21).

The presence of air implies rupture of the dura, most
commonly due to fracture through the posterior wall of
the frontal sinus or the ethmoidal plate. The finding of
parasellar air suggests a fracture through the sphenoid
sinus. Fracture of the petrous bone only rarely produces
pneumocranium. Air may collect anywhere after a compound depressed fracture. The presence of intracranial
air indicates a risk of subsequent meningitis.

Infections

Intracranial infections are surprisingly rare after head
injury. Most occur as a postoperative complication or
with dural tears. Paranasal sinus infection is a problem in
the immobilized head-injured patient and may cause
sepsis. It can be identified as fluid in the paranasal sinuses
on head scans (Fig. 7-22). Not all fluid collections within
a sinus represent infection, however. Fluid is common
within the sphenoid sinus with nasotracheal intubation.

SKULL RADIOGRAPHY AND SKULL FRACTURES

Key observations to make from the skull radiograph of
the person with head trauma are pineal shift, fracture,
pneumocranium, and facial fracture.

Pineal Shift

A shift of the pineal gland of more than 3 mm is considered significant. The presence of a shift denotes an
emergency, and the shift is believed to be caused either
by brain swelling or by hematoma. A CT scan or MRI
must be done immediately to determine the nature of the
problem.

Because the pineal is positioned directly above the
odontoid process and is therefore at the center of rotation of the skull, rotation of up to 20 degrees has little
effect on the measured position of the gland. Calcific
densities in the tentorium and choroid must not be misinterpreted as representing the pineal gland. To be visible on the frontal projections, the pineal must also be
visible on the lateral projection. Atrophy of the brain
and posterior fossa masses generally does not produce
pineal shift.

Fig. 7-21 Pneumocranium. (A) Air behind dorsum sellae *(arrow)*. **(B)** Plain film showing air within frontal cortical sulci *(arrow)*. A compound depressed fracture is seen posteriorly *(arrowheads)*.

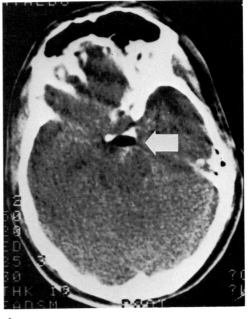

A

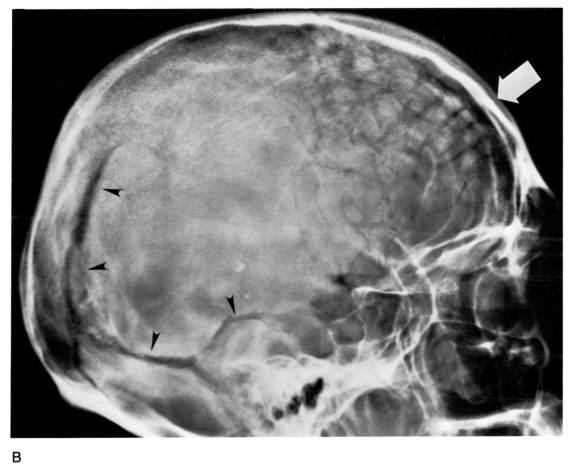

B

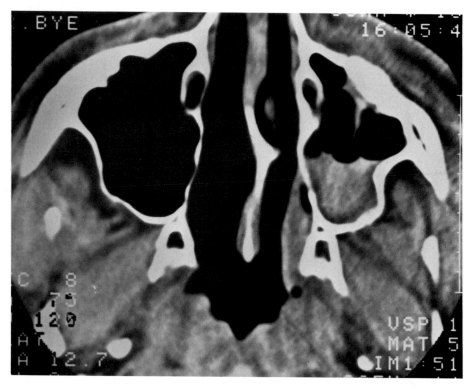

Fig. 7-22 Acute sinusitis. Fluid level is seen in the left maxillary sinus. This infection was the source of a troublesome fever after severe head injury.

Depressed Fracture

A depressed fracture is considered significant if the fragment is displaced inward a distance equal to, or greater than, the width of the calvarium. There is a high incidence of an associated dural tear with communication of the subarachnoid space with the outside air (compound fracture) and cortical laceration. Pneumocranium—air within the cranial cavity—indicates a compound fracture (Fig. 7-21). An estimated 85 percent of depressed fractures are compound. The development of post-traumatic seizures is thought to occur with greater frequency if completely displaced fragment is not elevated.

A depressed fracture may be recognized by observing the inward displacement of bone fragments at the tangential portions of the calvarium or the increased density of overlapping bone margins when the fracture is viewed en face (Fig. 7-23). Special note is made of a depressed fracture that overlies a dural sinus. Magnetic resonance angiography (MRA) or x-ray angiography (XRA) may be necessary to determine precisely the relationship of a fracture to a sinus. CT scanning with bone windows is also an excellent means of evaluating depressed fractures. Unconventional scan planes may be needed for an optimal view of the fracture.

Linear Fracture

Linear fractures are the most common type of fracture and require the least amount of kinetic energy. They are of little consequence unless they are diastatic or cross-meningeal vessels or dural sinuses. When this situation is present, they may be associated with a dural tear or epidural hematoma. A normal suture line or vascular groove must be differentiated from a fracture. A fracture line appears more radiolucent than a vascular groove and does not have sclerotic margins. Fractures also tend to be straight or to make angular changes in direction. Occa-

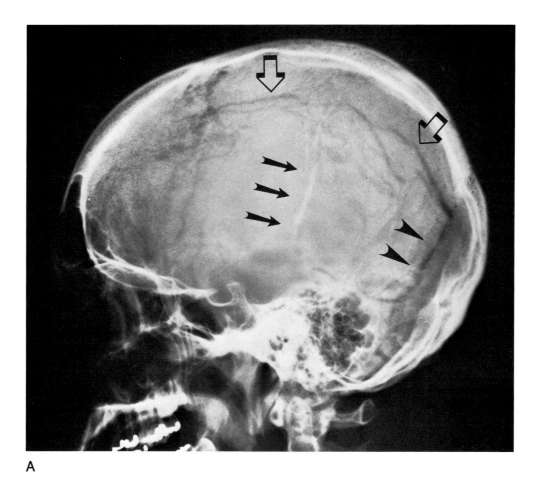

A

Fig. 7-23 Depressed skull fracture. (A) Lateral skull radiograph shows a linear region of increased density over the right parietal bone representing the overlapping bone margins of the depressed skull fracture (*arrows*). Note diastatic linear parietal fracture (*open arrows*) and diastasis of the lambdoid suture (*arrowheads*). (B) CT scan shows overlapped edges of the right parietal depressed skull fracture. There is also an acute left SDH and subfalcial herniation.

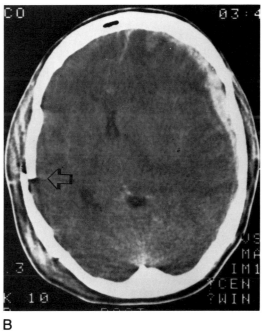

B

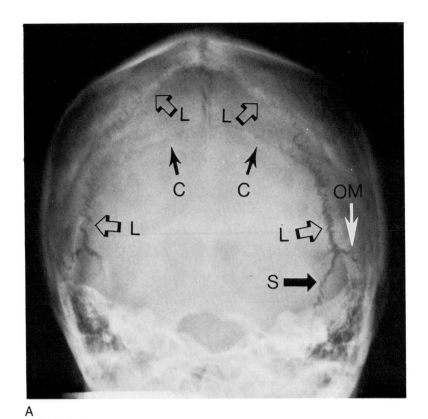

A

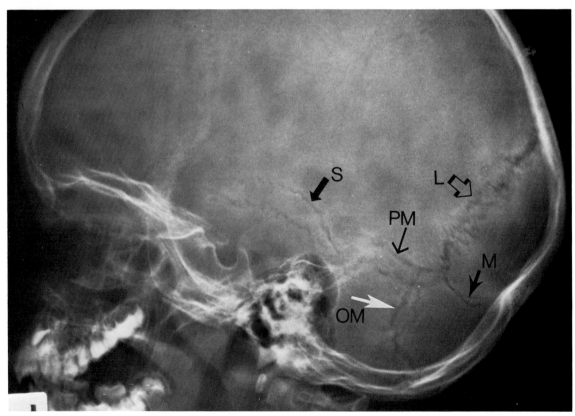

B

Fig. 7-24 Cranial sutures. (A) Towne view. (B) Lateral view. L, lambdoid; C, coronal; S, squamosal; OM, occipitomastoid; PM, parietomastoid; M, mendosal.

sionally, sutures are mistaken for fractures. The common suture lines are shown in Figure 7-24. Linear fractures heal within 1 to 2 years in adults.

Diastatic Fractures

A diastatic fracture is a widened linear fracture or a separated sutural fracture. These fractures may be associated with dural tears. In young children, under 2 years of age, the separated fracture may fail to heal because of interposition of meninges into the fracture site. This problem may cause progressive expansion of the fracture —the "growing fracture" (post-traumatic leptomeningeal cyst) (Fig. 7-25). All diastatic fractures of the calvarium in young children require follow-up radiographs to ensure proper healing of the fracture.

Basal Fractures and CSF Fistula

Basal skull fractures are those that involve one of the five bones that make up the base of the skull: the cribriform plate of the ethmoid bone, the orbital plate of the frontal bone, sphenoid bone, petrous and squamous portions of the temporal bone, and occipital bone. Basal fractures are of considerable significance because the regional dura may be disrupted, allowing communication of the subarachnoid space with an underlying paranasal or mastid sinus. A dural disruption here places the person at increased risk of the development of meningitis or of a CSF fistula.

Fractures through the petrous bone are either horizontal or vertical. Horizontal (longitudinal) fractures are caused by blows to the temporal or parietal region. The fracture may be in any plane. It is best viewed with coronal or transaxial high-resolution CT scanning. More than one plane of scanning may be required for the detection of these fractures. Helical technique is useful to avoid partial volume artifacts. There may be tearing of the membranes of the external auditory canal and of the tympanic cavity or ossicular dislocation. Patients may experience bleeding from the external canal, otorrhea, or conductive hearing loss (Fig. 7-26).

Vertical (parasagittal) fractures of the petrous bone result from an occipital blow. The fracture is orthogonal to the petrous pyramid, usually in the sagittal plane. It can be identified with tomography or CT scans in either the frontal or transaxial plane. The tympanic membrane is usually intact; therefore, otorrhea does not occur. If the fracture passes through the cochlea and vestibular structures, the result is sensorineural hearing loss. CN VII and VIII may be damaged.

A severe complication of a basal fracture is a CSF fistula into the sinuses. It is recognized clinically by the presence of rhinorrhea or otorrhea. The site of the fistula must be determined for successful surgical repair. A basal fracture may cross several sinus cells, and the actual site of the fistula may be indeterminate. Fluid within a sinus is an indirect sign (Fig. 7-27) but may be present owing to other causes, such as hemorrhage, a prolonged recumbent position, or a nasotracheal tube. Special studies may be necessary.

Positive-contrast CT cisternography, performed with the patient in the position that produces a leak, most frequently demonstrates the fistula. Accumulation of contrast within a sinus cell is diagnostic. Repeated studies are sometimes necessary, occasionally with the use of positive pressure in the subarachnoid space. The demonstration of herniation of brain into the sinus is rare but diagnostic. Radionuclide cisternography with nasal pledgets can be used to lateralize the fistula if other studies are not informative.

Frontal Sinus Fractures

Fractures through the anterior wall of the frontal sinus, if depressed, may need surgical cosmetic repair; however, these fractures do not affect the intracranial structures. By contrast, it is important to recognize a fracture of the posterior wall of the frontal sinus, as these fractures present a risk of development of meningitis or CSF fistula (Fig. 7-28). Trauma to the orbit and a discussion of the temporal bone appears in Chapter 13.

VASCULAR INJURIES OF THE HEAD AND NECK

Intracranial and extracranial vascular injuries are relatively rare but critical. Because angiography is no longer routinely performed in the acute trauma situation, vessel injuries may go undiagnosed. These injuries are impor-

(Text continued on page 369)

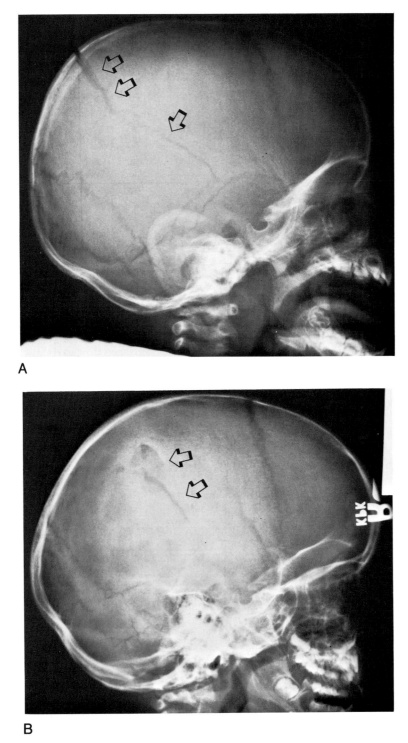

A

B

Fig. 7-25 **Diastatic fracture, leptomeningeal cyst. (A)** Parietal diastatic linear skull fracture *(arrows)*. **(B)** Radiograph 6 month later shows persistence of a portion of the fracture line as well as erosion of the inner table along the posterior portion *(arrows)* *(Figure continues.)*

Fig. 7-25 *(Continued)*. **(C)** CT scan shows persistence of the fracture line with erosion of the inner table and some local thickening of the calvarium *(arrow)*.

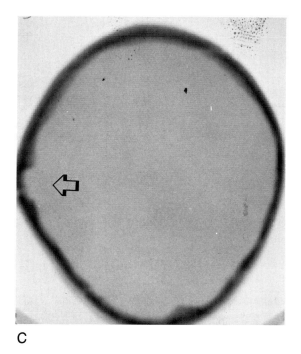

C

Fig. 7-26 Fracture of the temporal bone. Oblique fracture line is seen in the temporal bone passing through the middle ear cavity and into the lateral margin of the jugular foramen. High-resolution CT scanning with bone windows is the best method for demonstrating basal skull fractures.

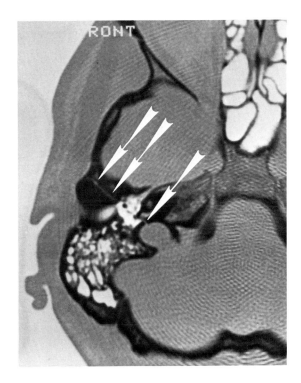

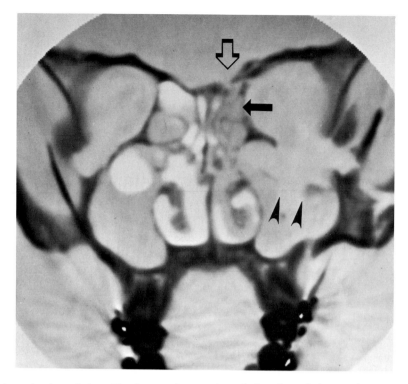

Fig. 7-27 Fracture through ethmoid plate. The diastatic fracture through the ethmoid plate can be seen on the coronal CT scan *(open arrow)*. Fluid is present in an ethmoid air cell *(arrow)*. A blow-out fracture of the orbit is also seen *(arrowheads)*.

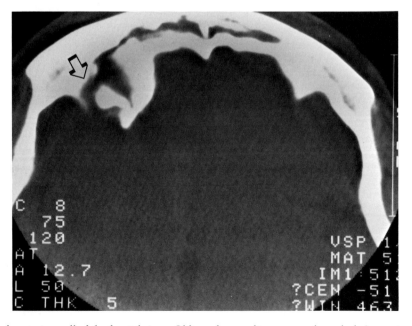

Fig. 7-28 Fracture of posterior wall of the frontal sinus. Oblique fracture line courses through the posterior wall of the right frontal sinus and orbital roof *(arrow)*.

tant because of their associated high mortality rate and the potential for repair. Vascular injury should be suspected and angiography considered in the presence of (1) a history of a blunt or penetrating injury to the neck, (2) a hematoma in the neck, (3) a focal neurologic deficit suggesting a stroke or TIA, or (4) a local bruit.

Extracranial vascular injury may be the result of direct laceration of the vessel against a bone, especially the transverse process of C2. Dissection of the internal carotid artery may result from stretching. Intimal injury may lead to thrombosis, which can embolize or progress to vessel occlusion. Adventitial injury results in the formation of a pseudoaneurysm. The apparent trauma that causes the vascular injury may be minimal.

Selective angiography of the extracranial vessels is generally necessary for the definitive diagnosis. MRA or CT angiography (CTA) may have use here to select patients for XRA. The usual findings are pseudoaneurysm at the injury site (Fig. 7-29), mural hematoma with vessel constriction (dissection), or total occlusion from intimal injury or thrombosis. Two orthogonal views are needed to exclude injury of a vessel. If routine MRI is performed on a patient with head trauma, the absence of flow-void in the major vessels at the base of the brain suggests acute vascular injury.

Intracranial vascular injury may result from a basal or depressed fracture, internal herniation of the brain, or a direct penetrating injury (bullet or knife wound). Dissection may occur, particularly of the middle cerebral artery (MCA). Traumatic aneurysms may be found in the cortical vessels over the convexities or at the base of the skull. Depressed fractures in the region of the posterior portion of the sagittal sinus or the transverse sinuses often cause excessive hemorrhage from dural sinus laceration or occlusion of the sinus with resultant hemorrhagic brain infarction, frequently accompanied by intractable seizures. Dural arteriovenous fistula may be seen. Intracranial angiography is performed for infarctions not directly explained by the trauma.

Injury to the cavernous portion of the internal carotid artery is particularly apt to result from severe head trauma. Often there is a fracture of the sphenoid bone, but not necessarily. A carotid–cavernous fistula occurs when laceration of the internal carotid artery or one of its small cavernous dural branches allows communication between the carotid artery and the venous channels of the cavernous sinus. If the rupture is contained by a

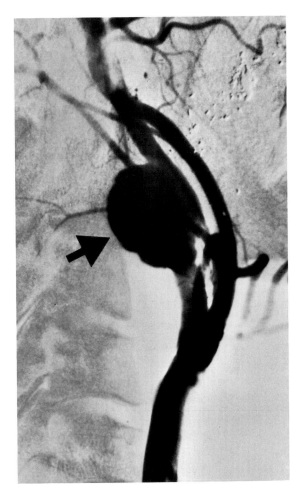

Fig. 7-29 Pseudoaneurysm of the internal carotid artery. Vascular injury *(arrow)* may follow blunt trauma, gunshot wounds, or knife wounds.

noncommunicating compartment of the cavernous sinus, a pseudoaneurysm results (Fig. 7-30).

The clinical diagnosis of a carotid–cavernous fistula is usually straightforward. The rush of blood to the cavernous sinus generally drains anteriorly to the ophthalmic veins, causing proptosis, chemosis, bruit, ophthalmoplegia, loss of vision, and headache. Despite the large amount of blood flow through the fistula, focal cerebral neurologic deficit is rare. Angiography in patients with a carotid–cavernous fistula is done with rapid serial filming so the size and location of the fistula can be determined. The fistula may be successfully treated by catheter placement of detachable balloons.

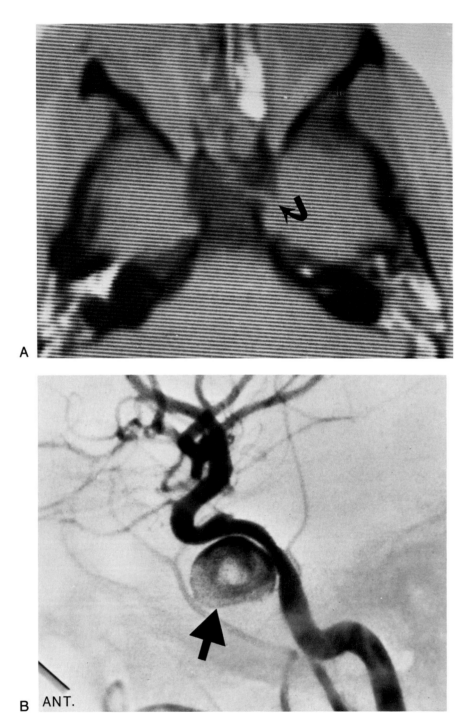

Fig. 7-30 Fracture of the sphenoid bone, pseudoaneurysm of the internal carotid artery. (A) CT scan with bone windowing shows a fracture through the sphenoid bone displaced into the left cavernous sinus *(arrow)*. **(B)** Internal carotid artery angiography demonstrates a pseudoaneurysm of the cavernous segment *(arrow)*. If the internal carotid artery rupture had been into a different compartment of the cavernous sinus, a carotid–cavernous fistula might have resulted.

Gunshot Wounds

Gunshot wounds of the head are becoming more common among civilians, especially in the United States. Military and civilian wounds generally differ, mainly because of the slower bullet velocity of the common civilian weapons. The higher-velocity missiles used by the military are much more lethal, and a direct hit in the head almost always produces death. Unfortunately, military weapons are becoming much more available for civilian use.

The extent of injury produced in the brain following penetration by a bullet is directly proportional to the amount of kinetic energy deposited from the bullet into the tissues. The amount of kinetic energy of a moving bullet is proportional to the mass of the bullet and the square of its velocity. The impact and penetration of the bullet cause a shock wave that produces an instantaneous rise in tissue pressure. Although the shock wave lasts only several microseconds, the pressure is high and is transmitted to the central axis of the brain, often causing deep irreversible pontine damage, or even herniation of brain tissue through the tentorium. Thus, a relatively peripheral bullet tracking can produce immediate death if enough energy is deposited.

Cavitation of the tissue occurs along the missile path. A permanent cavity is produced that nearly corresponds with the width of the bullet fragment. Also, a transient cavitation is produced more peripheral to the permanent tract, the width of which is proportional to the velocity of the bullet; it may be up to 30 times the width of the central cavity. Because the missile slows as it traverses tissue, the path of destruction is more or less conical in shape, being widest at the site of entry. Deformation of the bullet or bullet yaw may alter the shape of the path of damage in an unpredictable way.

When the bullet enters the skull, bone fragments are driven inward, but always to a lesser distance than the final resting place of the bullet. The bone fragments are therefore embedded in the part of the brain that is closest to the site of entry. If the bullet exits the opposite side of the skull, the path of the bullet is nearly a straight line from the entry to the exit site. If the bullet does not possess enough kinetic energy to penetrate the opposite skull table, the bullet may ricochet backward and come to rest away from the expected path. In this situation, the path of the bullet cannot be determined with certainty. Normally, in the civilian situation, only large-caliber bullets fired at close range exit the skull. The site of the bullet fragments must be reported with accuracy, as they must be removed if possible (Figs. 7-31 and 7-32).

Hemorrhage occurs along the track of the bullet because of laceration of the large and small vessels. Large vessels at the base of the skull may be lacerated completely, resulting in massive hemorrhage and infarction. Incomplete lacerations of major arteries may result in aneurysms or pseudoaneurysms. Special note is made of bullet wounds in the region of the sagittal or transverse sinuses, as there may be life-threatening hemorrhage or thrombosis. Subdural or epidural hematoma may result from rupture of bridging veins or laceration of cortical or dural arteries.

It is important to identify the site of entry, so the surgeon may be directed to the site of dural tear. This point is especially true for a bullet that has passed through a paranasal sinus, as the risk of subsequent infection is increased.

Plain-skull radiographs and CT scanning of the brain are the basic neuroradiologic examinations. The prognosis of intracranial gunshot wounds is poor and for the most part depends on the neurologic conditions and the state of consciousness at evaluation.

CHILD ABUSE

In the United States, more than one million cases of child abuse are reported annually, about 50 percent of which involve inflicted trauma. It is the group with bodily injury that comes to the attention of the radiologist. Most children with head injuries are under 2 years of age. Cerebral damage can result from many factors, including direct impact, shaking, whiplash, asphyxia, and vascular occlusion due to strangulation. Clinically visible signs of injury and a history of injury are often lacking, so the responsibility may fall to the radiologist to suggest child abuse. A key component to diagnosis is the "discrepant history," i.e., the description of an injury that does not match the clinical or radiographic findings.

The triad of imaging modalities used to evaluate possible child abuse include plain-film radiography, CT scanning, and MRI. Ultrasonography and radionuclide scanning are sometimes of additional value.

The primary imaging technique here is the CT scan, as it is sensitive for detecting intracranial hemorrhage and skull fracture. The most common finding is SAH, seen in about 75 percent of abused children with intracranial head injury. It may be subtle, seen only as a slight in-

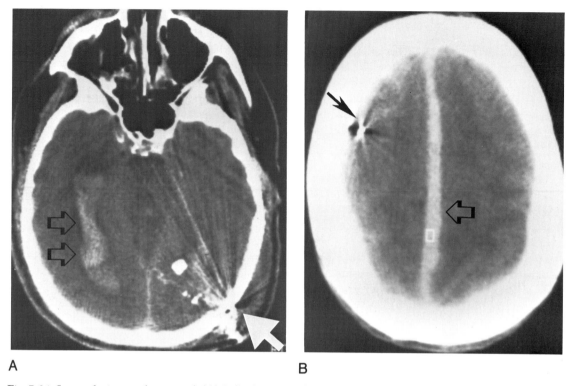

A   B

Fig. 7-31 Low-velocity gunshot wound. (A) Bullet has entered the left occipital region *(white arrow)*. Intraventricular blood is seen in the right temporal and occipital horn *(open black arrow)*. **(B)** Bullet fragment has come to rest against the inner table of the right posterior frontal bone *(black arrow)*. Interhemispheric blood is also present *(open arrow)*, and there is diffuse brain edema.

crease in the density of the subarachnoid space. SAH in the interhemispheric fissure is common, but blood may also be seen in the basal cisterns, sylvian fissures, and cerebral sulci. CT scanning is superior to MRI for detecting SAH. In infants, SAH is far more common in the child-abuse situation than with trauma due to other causes, such as falls or car accidents. Whenever SAH is seen in a child under 2 years old, child abuse must be considered. MRI is used to detect small extracerebral hematomas that are not seen with CT in the appropriate clinical setting. MRI also differentiates chronic SDH from normally widened subarachnoid spaces. Here, MRI shows high T_1 and T_2 signal from blood and increased protein content with SDH (Fig. 7-33).

Interhemispheric SDH is also a common finding associated with child abuse (Fig. 7-34), seldom found with other causes of head injury. It appears as a unilateral interhemispheric region of high attenuation, usually posteriorly. This hematoma has a flat medial border but a slightly convex lateral border. It may extend over the posterior convexity of the brain or the tentorium. In the chronic phase, a small residual low-density region confirms the diagnosis. It may be impossible to differentiate an interhemispheric SDH from SAH. However, this point is not a particularly important distinction, as both blood collections have the same diagnostic implication.

It is important to differentiate interhemispheric blood from a normally dense falx. The normal falx may appear especially dense if there is adjacent brain edema. The falx is generally much thinner than blood collections, but repeat scans may be necessary to make this crucial distinction. It is equally important to avoid overdiagnosis of child abuse.

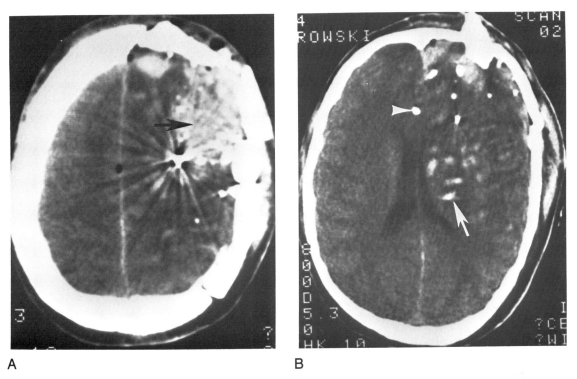

A   B

Fig. 7-32 High-velocity gunshot wound. (A) CT scan superiorly near the tract of the bullet demonstrates the explosive force produced by a high-velocity missile. Comminuted fracture fragments are displaced outward, and there is a large hemorrhage *(arrow)* in severely contused brain. (B) CT scan at a lower level, some distance from the bullet tract, demonstrates the punctate hemorrhages of multiple axonal injuries *(arrow)* and severe acute brain swelling. Small bullet fragments are scattered widely *(arrowhead)*.

Fig. 7-33 Child abuse, subdural collection. Nonenhanced CT scan shows a prominent dural membrane separating a subdural fluid accumulation from the subarachnoid space *(arrow)*. This membrane is not seen with most subdural collections. MRI may help in identifying the higher intensity of the subdural collections (containing blood) as separate from the CSF within the subarachnoid space. In children under 2 years of age, the subarachnoid space may be widened normally. Identification of subdural collections is important for the diagnosis of injury from child abuse.

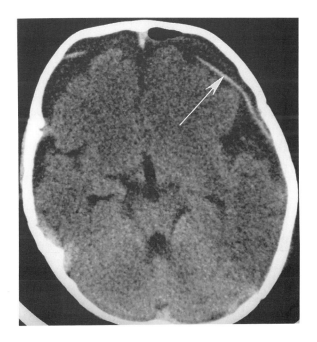

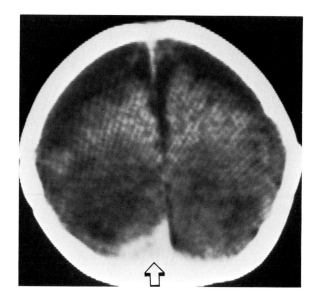

Fig. 7-34 Child abuse, interhemispheric subdural blood collection. CT scan shows high-density blood collection in the posterior interhemispheric fissure extending over the right convexity (*arrow*). This finding has a high association with inflicted injury.

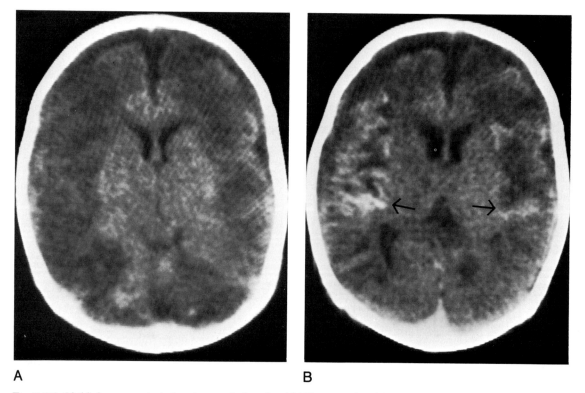

A

B

Fig. 7-35 Child abuse, anoxic-ischemic encephalopathy. (A) Note peripheral low density with general preservation of normal density of the deep basal ganglionic and posterior fossa structures. Extracerebral fluid collections are also present. **(B)** A few days later, there is hemorrhage into the ischemic tissue (*arrows*).

Cerebral anoxic or ischemic encephalopathy may occur as a result of child abuse. Resulting from coma, seizures, aspiration, or strangulation, it is reflected in the brain as diffuse cerebral edema, similar to global ischemic encephalopathy due to any cause. Most commonly, the cerebral hemispheres have uniform low density with loss of gray–white matter distinction. The thalami, cerebellum, and brain stem are usually spared and remain of relatively normal density (Fig. 7-35). Sometimes the pattern is that of near-drowning or carbon monoxide poisoning, with low density seen in the basal ganglia (see Ch. 2).

Focal infarction may result from strangulation caused by occlusion of the carotid artery in the neck. The infarction usually involves multiple vessel territories supplied by the internal carotid artery. Spontaneous hemiplegia of childhood due to cerebrovascular disease usually involves only the internal capsule or a single cerebral vessel distribution.

Focal cerebral parenchymal hemorrhages occur, most of which represent contusion. A well-circumscribed tri-angular hemorrhage pointing toward the ventricle indicates laceration of the brain. Low-density extracerebral fluid collections represent hygromas, which are common following child abuse (Fig. 7-36).

Skull fractures result from direct-impact trauma. The more severe or comminuted a fracture in a child, the more likely it is the result of inflicted injury (Fig. 7-37). Radiography of other regions is worthwhile and may reveal additional fractures of different ages. This examination includes the ribs, clavicles, spine, and long bones. Radionuclide bone scanning can identify otherwise occult regions of injury, but it is unable to give the age of the lesion. Organ injury also may occur. MRI is used in cases of suspected child abuse when the CT scan is nondiagnostic. Small subdural blood collections or regions of infarction, axonal injury, or contusion may be seen with MRI that cannot be seen with the CT scan.

Atrophy is often seen after head injury caused by abuse. It is particularly severe after anoxia or infarction. The ventricles may become enlarged. The brain damage

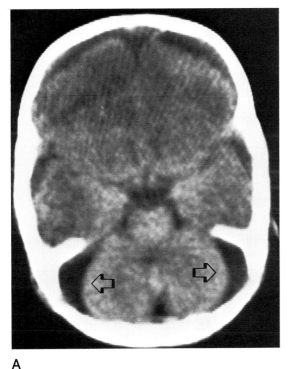
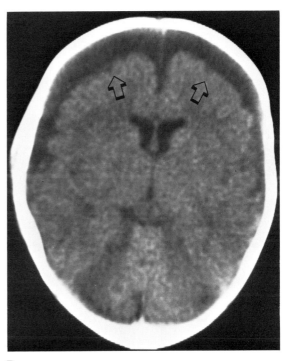

A

B

Fig. 7-36 Child abuse, low-density extracerebral collections. (A) Low density fluid collections surround the cerebellum *(arrows)*. **(B)** Frontal extracerebral fluid collections compress the frontal lobes *(arrows)*.

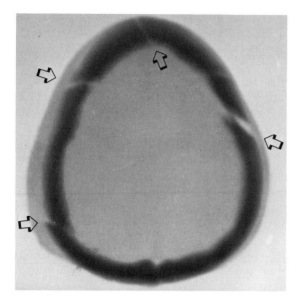

Fig. 7-37 Child abuse, multiple skull fractures. Multiple skull fractures *(arrows)* have a high association with abuse in infants.

of inflicted injury is often irreparable, and post-traumatic neurologic dysfunction is common and often devastating.

SUGGESTED READING

Allen WE III, Kier EL, Rothman SLG: Pitfalls in the evaluation of skull trauma: a review. Radiol Clin North Am 11:479, 1973

Cooper PR (ed): Head Injury. Williams & Wilkins, Baltimore, 1982

Eskridge JM: Interventional neuroradiology: state of the art. Radiology 172:991, 1989

Fobben ES, Grossman RI, Atlas SW et al: MR characteristics of subdural hematomas and hygromas at 1.5 T. AJNR 10:687, 1989

Genieser NB, Becker MH: Head trauma in children. Radiol Clin North Am 12:333, 1974

Gentry LR: Imaging of closed head injury. Radiology 191:1, 1994

Gentry LR, Godersky JC, Thompson B: MR imaging of head trauma: review of the distribution and radiopathologic features of traumatic lesions. AJNR 9:101, 1988

Gentry LR, Godersky JC, Thompson BH: Traumatic brain stem injury: MR imaging. Radiology 171:177, 1989

Gentry LR, Godersky JC, Thompson B et al: Prospective comparative study of intermediate-field MR and CT in the evaluation of closed head trauma. AJNR 9:91, 1988

Kelly AB, Zimmerman RD et al: Head trauma: comparison of MR and CT experience. AJNR 9:699, 1988

North CM, Ahmadi J, Segall HD, Zec C-S: Penetrating vascular injuries of the face and neck: clinical and angiographic correlation. AJNR 7:855, 1986

Williams A: Trauma. In Williams A, Haughton VM (eds): Cranial Computed Tomography. CV Mosby, St. Louis, 1985

Wright JW Jr: Trauma of the ear. Radiol Clin North Am 12:527, 1974

Zimmerman RA: Vascular injuries of the head and neck. Neuroimag Clin North Am 1:443, 1991

Zimmerman RD, Balaniuk LT: Head trauma. In Heinz RE (ed): The Clinical Neurosciences: Neuroradiology. Churchill Livingstone, New York, 1984

Zimmerman RD, Danziger A: Extracerebral trauma. Radiol Clin North Am 20:105, 1982

8

Infections

PYOGENIC INFECTIONS

Pyogenic (pus-forming) bacterial infections of the CNS occur by direct spread from contiguous structures, hematogenous seeding, or trauma. The infection may be seen in many forms, such as meningitis, ventriculitis, cerebritis, brain abscess, and subdural and epidural empyema. Although the incidence of pyogenic infections of the CNS has decreased, the disease is still important and causes significant morbidity and mortality.

Meningitis

Meningitis is an inflammation of the arachnoid, pia, and CSF. It is the most common form of pyogenic infection of the CNS. Most often it is a result of hematogenous seeding, but it may result from trauma, direct extension from a sinus infection or pre-existing brain abscess, or a congenital fistula. The etiologic agent of the infection depends on the age of the victim and the clinical setting in which the infection occurs. Neonates are most commonly infected with *Escherichia coli* or a group B streptococcus. Infants up to age 5 are most often infected with *Haemophilus influenzae*. If infection occurs with these agents later in life, it suggests a congenital neurologic defect that permits direct infection of the CSF. In those over age 5 and throughout adulthood, *Neisseria meningitidis* and *Streptococcus pneumoniae* are the predominant agents. After trauma and surgery, *Staphylococcus aureus* and gram-negative organisms are most commonly found.

The purulent material is widely distributed throughout the subarachnoid space and ventricles (ventriculitis). Ventricular empyema (pus-filled ventricle) is rare. Brain abscess does not result from meningitis, as the pia is an effective barrier to the spread of infection. However, underlying brain edema may be present and can become severe enough to cause increased intracranial pressure and herniation. A cortical thrombophlebitis, or secondary arteritis may result in brain infarction. This problem is particularly likely with *H. influenzae* and may progress to gross devastation of the cerebral hemispheres despite appropriate treatment of the infection. Hydrocephalus may result from obstruction by pus, but this condition tends to be mild; it occurs almost exclusively in children. Sterile subdural effusions may occur, most frequently in infants and most often after *H. influenzae* infection. Subdural empyema as a result of meningitis is rare.

The diagnosis of meningitis is best accomplished by lumbar puncture and CSF analysis. Imaging during the acute phase of the infection is of little value, as the results are usually normal or demonstrate only a subtle hydrocephalus. If contrast enhancement is used with MRI or CT scanning, there may be enhancement of the meningeal surface, particularly at the base of the brain. This picture is indistinguishable from the enhancement that may be seen with acute subarachnoid hemorrhage or meningeal carcinomatosis (Fig. 8-1). With severe infections, the cerebral sulci may appear widened, owing to gross pus.

Imaging primarily defines complications and the underlying etiologic process. Studies are performed in pa-

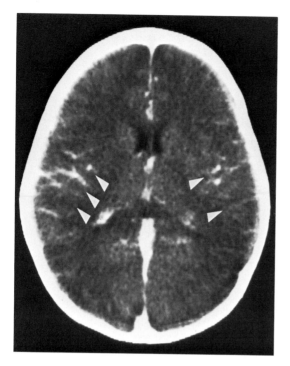

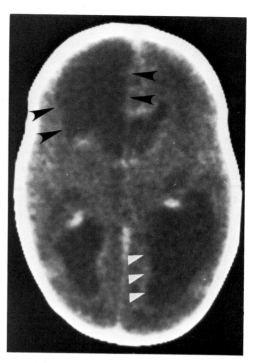

Fig. 8-1 Meningitis, meningeal contrast enhancement, CT scan. Note enhancement of the meningeal surface of the sulci (*arrowheads*). This pattern of enhancement can also be seen with subarachnoid hemorrhage and meningeal carcinomatosis.

Fig. 8-2 Meningitis, infarction, and hydrocephalus. Large right frontal region of hypodensity, representing infarction (*black arrowheads*). Large ventricles represent hydrocephalus and developing porencephaly (*white arrowheads*).

tients in whom seizures, focal neurologic signs, persistent fever, an enlarging head, or signs of increased intracranial pressure develop. A nonenhanced MRI or CT scan is preferred and should include the paranasal and mastoid sinuses. Hydrocephalus is detected by the enlargement of the ventricles. Special attention is directed toward observing any septations within the ventricles, as they may loculate fluid and cause persistent infection. A ventricle may become completely isolated by adhesions and remain enlarged after routine lateral ventricular shunting. Focal brain abnormalities almost always represent ischemia or infarction and may be global and severe (Fig. 8-2). Contrast is used to better define lesions if the diagnosis is uncertain. Enhancement may occur in the gyri, representing loss of the blood-brain barrier, but in the presence of meningitis it does not necessarily represent infarction or cortical infection (Fig. 8-3). By contrast, ependymal enhancement does represent significant ventriculitis or ventricular empyema (Fig. 8-4).

Subdural effusions appear as crescent-shaped fluid collections nearly equivalent to CSF density over the frontal and parietal lobes (Fig. 8-5). They occur almost exclusively in children. Those that are infected (rare empyema) show contrast enhancement along the inner thickened membrane. Severe edema with herniation is recognized by the compression of the lateral ventricles and by obliteration of the basal cisterns. MRI shows simple effusions as CSF-equivalent intensity and empyema as a high-protein solution with very high T_2 signal.

CT and MRI are both useful for defining a possible cause of the meningitis. Multiplanar scanning through the paranasal sinuses, inner ear, or mastoids may detect sinusitis, CSF fistula, encephalocele, otitis media, or cholesteatoma. Bone windows are essential. MRI may better show small effusions and ischemia. Skull radiography is useful in post-traumatic situations. Angiography may show the changes of vasculitis but has little practi-

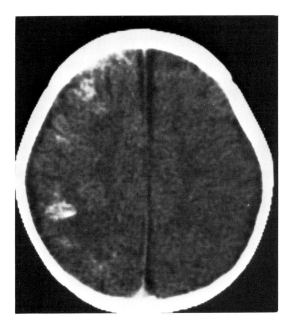

Fig. 8-3 Meningitis, gyral enhancement. Damage to the blood-brain barrier is indicated, but not necessarily infarction. It must not be interpreted as cerebritis.

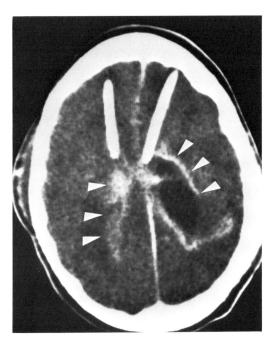

Fig. 8-4 Ventriculitis. Contrast enhancement of the ependyma in a patient with ventriculitis *(arrowheads)*. Bilateral shunt catheters are present, but the left shunt is malfunctioning.

Fig. 8-5 Subdural effusion. Bilateral low-density fluid accumulations of CSF-equivalent density, representing effusions *(arrowheads)*. Additionally, note the low-density abnormalities present bilaterally in the posterior temporal and occipital lobes, representing ischemia *(open arrows)*.

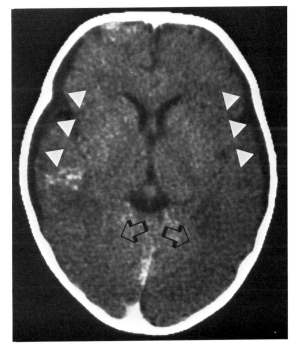

cal value. Mycotic aneurysm following meningitis is rare. CT scans and radionuclide cisternography may help in the diagnosis of CSF fistula (see CSF rhinorrhea, Ch. 7).

Cerebritis and Brain Abscess

Cerebritis is a region of infection and inflammatory response but without associated necrosis. Brain abscess constitutes encapsulated or free pus within the brain parenchyma. The abscess may vary in size from a microscopic focus to a large region of suppurative necrosis. The abscess may be single or multiple and always occurs within the white matter, completely sparing the gray matter. It is most common in the cerebral hemispheres, though it may also occur in the cerebellum (here almost always a direct extension from otologic infection), thalamus, pons, and pituitary gland. MRI and CT scans are the primary imaging modalities used.

The infection begins with local inflammation (cerebritis) limited to the white matter and surrounded by vasogenic edema. CT scanning and MRI with contrast show enhancement in the region of infection, (Fig. 8-6). MRI, without or with contrast, is more sensitive for detection of this phase. The edema of inflammation is seen as high signal intensity on T_2-weighted images. Small hemorrhages may be detected on the T_1-weighted images.

With progression, central necrosis develops in regions of cerebritis, to create an abscess. After about 10 days, the abscess becomes walled off by a reticular-collagen capsule, formed by fibroblasts migrating into the region. With virulent infection or immunocompromise, the encapsulation takes longer and is less complete or absent. The capsule is always thinner on the ventricular side of the abscess, and infection may escape medially, forming daughter abscesses or reaching the ventricle, and causing ependymitis. Ventriculitis results if pus spills into the ventricular cavity.

On CT scan, the capsule is represented by a ring of contrast enhancement (Fig. 8-7). It is impossible to determine the maturity of the encapsulation solely by the nature of the contrast-enhanced ring, but a mature capsule is presumed to develop sometime during the 14- to 21-day stage of the abscess. The central low density accurately represents the necrotic abscess cavity. Ependymitis is indicated by contrast enhancement of the ven-

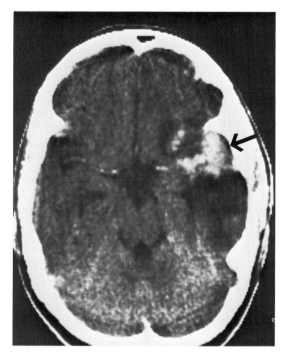

Fig. 8-6 Cerebritis. Region of contrast enhancement present in the left anterior temporal and posterior frontal lobe (*arrow*). Note considerable surrounding edema, particularly in the left temporal lobe.

tricular surface. Ventriculitis from actual spillage of inflammatory cells and bacteria into the ventricular CSF is indicated by ventricular enlargement and intense enhancement of the choroid plexus. The ventricular fluid may become more dense, owing to accumulation of pus.

T_1-weighted MRI shows the capsule as an isointense ring separating the central low-intensity abscess cavity and surrounding edema (Fig. 8-7). On the T_2-weighted images the pus of the central abscess cavity is seen as very high intensity separated from the surrounding edema by the hypointense capsule. Gadolinium (Gd) enhancement demonstrates very high signal intensity of the capsule on T_1-weighted images following the principles of contrast enhancement in CT scanning.

The peripheral edema is maximum at the beginning of the abscess stage but subsides with the formation of the capsule. A decrease in the diameter of the ring and in the

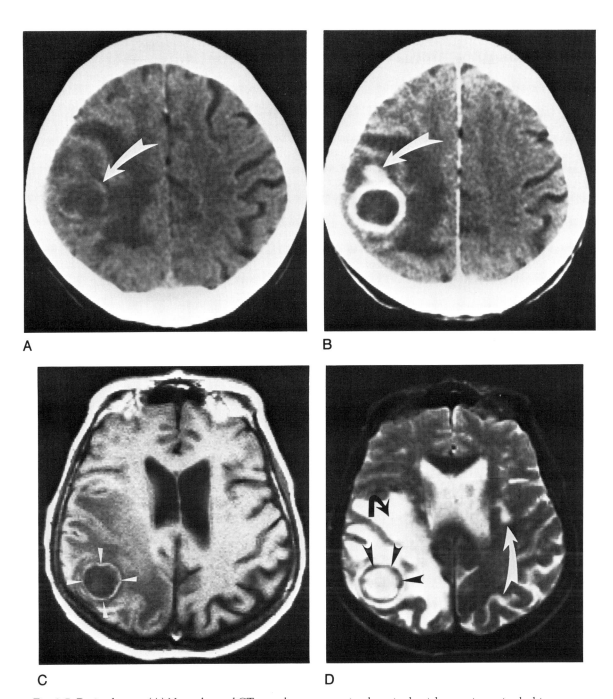

Fig. 8-7 Brain abscess. (A) Nonenhanced CT scan shows vasogenic edema in the right anterior parietal white matter. The isointense ring of the abscess capsule can be seen *(white arrow)*. **(B)** Contrast-enhanced study shows a thick but uniformly enhancing capsule with the beginnings of a daughter abscess anteriorly *(arrow)*. The enhancement of abscess is typically circular and nearly uniform, except along the medial surface. **(C)** T_1-weighted MRI shows the abscess cavity as hypointense, surrounded by an isointense capsule *(arrowheads)*. **(D)** T_2-weighted MRI shows high-intensity white matter edema *(black arrow)* and high-intensity central pus collection. The capsule is hypointense *(arrowheads)*. A small region of cerebritis is present in the left hemispheric white matter *(white arrow)*.

size of the central cavity indicates a favorable response to treatment. An abscess may recur after complete CT scan disappearance, necessitating follow-up scans for up to 2 years.

Cerebral angiography shows an abscess as a well-circumscribed ring-like "blush," without tumor neovascularity (Fig. 8-8). The primary use of angiography was previously used to differentiate abscess from tumor, but this task is better done with needle biopsy. Radionuclide brain scanning has no inherent advantage over the other modalities, although gallium or indium-labeled white blood cell studies may differentiate abscess from tumor.

An abscess may occur by direct extension from cranial sinuses (Fig. 8-9), trauma, or metastatic seeding (Fig. 8-10), particularly from infection in the lungs. In a few cases, no source of infection is found. Brain abscess occurs relatively infrequently with bacterial endocarditis. The most common organisms isolated are aerobic and anaerobic streptococci. Staphylococcal infection occurs most often after trauma but may be seen in intravenous drug users as well. Mixed infections are common.

Percutaneous needle aspiration of the abscess may be done with CT or MRI guidance. Material can be obtained for culture through a needle as small as a 22 gauge. Anaerobic organisms must be handled efficiently in airless containers so they will grow in vitro in the laboratory. Ideally, the microbiologist is present to transfer the aspirate, so all significant organisms will be later identified. Complete aspiration of the abscess cavity may be curative, obviating the need for surgical excision (Fig. 8-11).

Differentiation between abscess, glioblastoma, and metastasis is at best difficult, if not impossible, with ei-

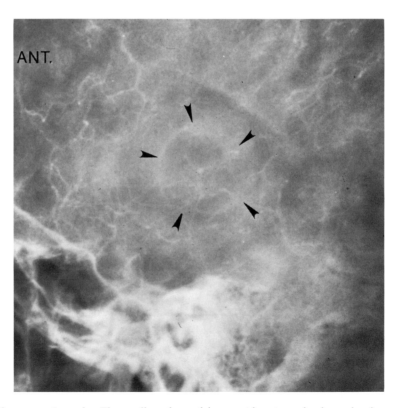

Fig. 8-8 Brain abscess, angiography. The capillary phase of the carotid angiography shows the abscess capsule as a faint ring-like "blush" (*arrowheads*) with surrounding edema.

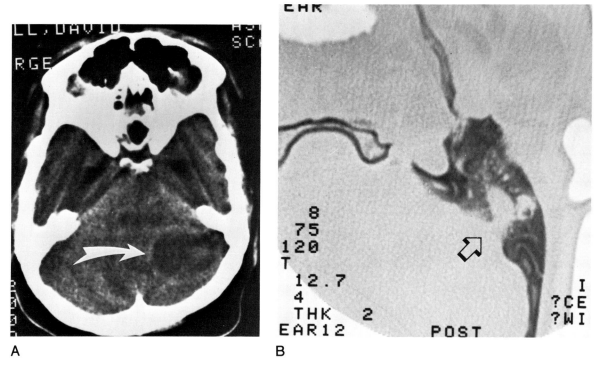

A   B

Fig. 8-9 "Otitic" cerebellar abscess. (A) Contrast-enhanced CT scan shows faintly enhancing ring of a large left cerebellar abscess (*arrow*). **(B)** Destruction of the temporal bone from purulent mastoiditis (*arrow*).

Fig. 8-10 Multiple metastatic brain abscesses. Various stages. Multiple metastatic tumors could also produce this same picture.

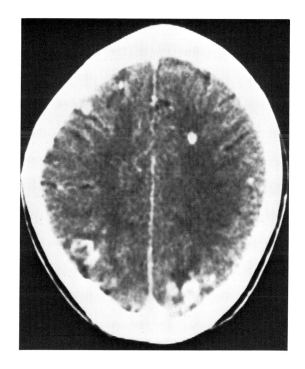

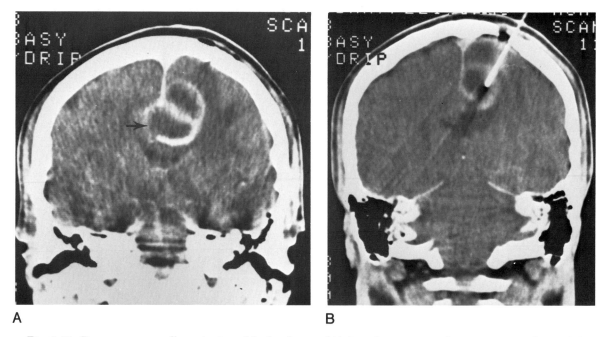

A   B

Fig. 8-11 **Percutaneous needle aspiration of brain abscess. (A)** Irregular contrast-enhancing mass in the medial parietal lobe extending into the corpus callosum. Note the deficient capsule *(arrow)*. It is impossible to differentiate between an abscess and glioma in this case. **(B)** Needle aspiration of the central cavity.

ther CT scans or MRI. The shape of the lesion, the thinness of the ring enhancement and its regularity, and the density of the central fluid may suggest an abscess, but the diagnosis cannot be made with certainty. The presence of gas within the central cavity is the only pathognomonic finding of abscess. The diagnosis of brain abscess must be considered whenever ring-like enhancing lesions are seen.

Septic Embolization

Septic thromboembolization to the brain commonly occurs in association with subacute bacterial endocarditis (SBE). About 15 percent of those with SBE are affected. The embolus causes infarction that is sometimes hemorrhagic, but abscess sometimes follows. About 5 percent of those with septic embolization develop a mycotic aneurysm, most commonly in a peripheral branch of the middle cerebral artery (MCA). Hemorrhage into the brain,

or subarachnoid or subdural space suggests its presence (Fig. 8-12). To identify the aneurysm prior to rupture, angiography is recommended in all those who have diagnosed septic emboli. Magnetic resonance angiography (MRA) demonstrates many of these aneurysms, but its sensitivity compared with x-ray angiography (XRA) is unknown. MRA sequences must be designed to create good slab penetration of unsaturated intravascular spins, so that peripheral vessels are well demonstrated.

Subdural Empyema

Subdural empyema is a rare but grave form of intracranial infection, with pus collecting within the subdural space. The infection is rapidly progressive, generally requiring emergency surgical drainage. Most commonly, it results from infection within the frontal sinuses or middle ear. Other causes include trauma, surgery, septicemia, and osteomyelitis of the calvarium.

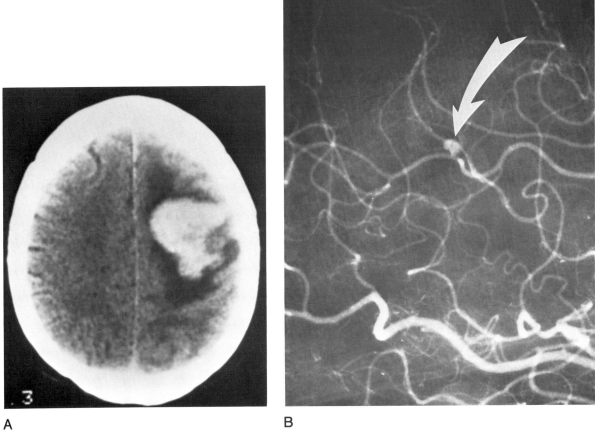

A

B

Fig. 8-12 Septic embolization, mycotic aneurysm, and brain hemorrhage. (A) Subcortical hematoma is present in the left anterior parietal lobe. **(B)** Small mycotic aneurysm in a peripheral left middle cerebral artery branch (*arrow*).

On CT scan, the empyema appears as an extracerebral hypodense collection of fluid over the convexity or within the interhemispheric fissure of slightly greater density than CSF. Rarely, it is along the tentorium. Contrast enhances the inner membrane (Fig. 8-13). Over the convexity, large collections may have the same appearance as a chronic subdural hematoma or epidural abscess. Bone windows may show osteomyelitis. On MRI, the collection is seen as low intensity on T_1-weighted images and as very high intensity on T_2-weighted images. The signal characteristics are those of a proteinaceous fluid.

Epidural Empyema

The findings for epidural empyema by CT scan are similar to those for subdural empyema, except that there is usually a convex margin medially (Fig. 8-14). The epidural collections do not extend into the interhemispheric fissure. The etiologic factors are the same as for subdural empyema, although the epidural infection most commonly occurs following craniotomy (Fig. 8-15). Angiography may define the epidural location of extracerebral fluid if the dural sinuses are elevated away from the inner table of the calvarium. Epidural infections do not

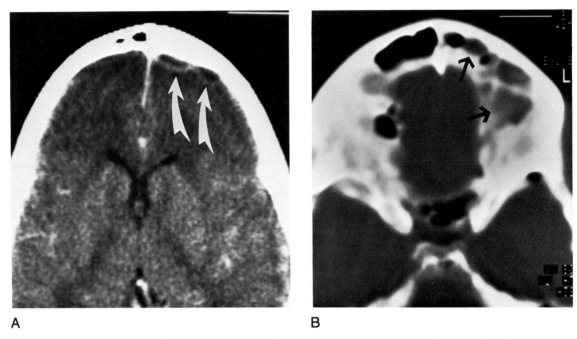

A   B

Fig. 8-13 Subdural empyema. (A) Small extracerebral fluid collection is seen over the left frontal lobe with membrane enhancement *(arrows)*. The findings are subtle but important. **(B)** Left frontal sinusitis *(arrows)* cause the subdural empyema.

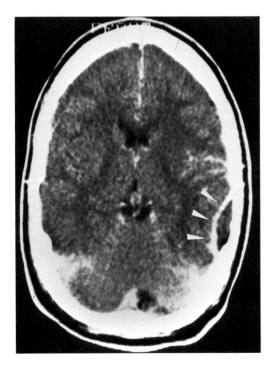

Fig. 8-14 Epidural empyema. Low-density extracerebral fluid collection is seen over the left posterior temporal lobe. There is a convex medial border with membrane enhancement *(arrowheads)*. This collection was secondary to mastoiditis.

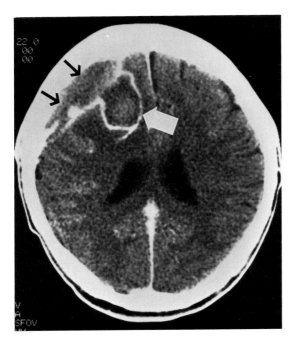

Fig. 8-15 Epidural abscess, postcraniotomy. Relatively high-density extracerebral fluid collection is present over the right frontal lobe with an enhancing medial margin *(black arrows).* Abscess was also found within the tumor bed *(white arrow).* This same appearance can be found with sterile hemorrhagic postoperative collections.

advance nearly as rapidly as those within the subdural space.

Major Dural Sinus Thrombosis

The dural sinuses may become thrombosed as a result of infection. This situation occurs when the sinuses are adjacent to epidural or subdural empyema. The venous sinus thrombosis produces local brain infarction that is usually hemorrhagic. On CT scan, the sinus does not fill normally with contrast; on MRI, there is signal intensity within the thrombosed vein instead of the usual flow-void. Care must be taken with MRI diagnosis of thrombosis, as flow-related enhancement effects can produce the same appearance. Angiography or MRA optimized for venous flow shows sinus occlusion and collateral venous flow (see Ch. 2).

Cavenous sinus thrombosis may be caused by facial, orbital, or paranasal sinus suppuration. In most cases, the afflicted person is severely ill with high fever, edema, and cyanosis of the upper face and eyelid. There may be ophthalmoplegia. MRI shows signal intensity within the cavernous sinus instead of the low signal of flow-void. The cavernous sinus does not become hyperintense after GD enhancement. Dynamic contrast CT scanning also shows an absence of contrast within the cavernous sinus. Venous MRA will show the lack of flow within the thrombosed sinus.

VIRAL INFECTIONS

Acquired Immunodeficiency Syndrome

Acquired immunodeficiency syndrome (AIDS) was first described in 1981. Since that time the disease has exploded to epidemic proportions and is now one of the major health problems of the world. At this writing, most persons with AIDS in the United States are homosexual or bisexual men or intravenous drug abusers, but heterosexual AIDS is increasing, particularly among female prostitutes and among the sexual partners of those with human immunodeficiency virus (HIV) infection. The remaining 10 percent are infants of infected mothers, persons who have received contaminated transfusions, or health workers accidentally infected by needle puncture or splashes of contaminated blood or body fluid. A few individuals with AIDS have no identifiable risk factor. HIV infection is pandemic in much of Africa, Southeast Asia, and India.

AIDS is caused by a retrovirus—human T-cell leukemia virus (HTLV III)—now called HIV-I. Other subtypes of retrovirus are also responsible for the disease, and HIV-II has now been found associated with AIDS. These specific RNA retroviruses primarily infect the T lymphocyte. Containing reverse transcriptase, the virus is capable of replicating its code into DNA and then inserting the DNA into the genetic code of the host cell. The end result is destruction of the T-cell-mediated defense functions within the body. Once immunocompromised, the host develops unusual infections and tumors, some of which are far more common than others. The severity of the immunodeficiency among patients with AIDS is variable.

Neurologic involvement is common, found clinically in about 50 percent of patients with AIDS, and in about

80 percent at autopsy. Neurologic complications occur earlier in those with more severe immunodeficiency. Neurologic involvement of any kind portends a severe clinical course and poor prognosis. About 10 percent of persons with AIDS present with a neurologic complaint. Neurologic complication is seen less commonly with the AIDS-related complex (ARC) (lymphadenopathy syndrome).

The neurologic complications of AIDS are protean. They are best studied with MRI. The more common afflictions are discussed separately below. Detailed accounts for some of the infections or tumors may be obtained in their respective sections elsewhere in this book. Multiple processes within the same patient are common, and in many instances it is impossible to diagnose a specific lesion on the basis of the imaging, serologic study, or clinical symptoms. Therefore, biopsy of the CNS lesions becomes important for accurate diagnosis and timely treatment of the complications.

The AIDS–dementia complex, or subacute encephalitis, is the most common neurologic complication of AIDS, occurring in more than 30 percent. Minimal dementia (depression, social withdrawal) is the usual presenting symptom. The dementia advances relentlessly to a state of global mental impairment with behavioral and motor dysfunction, often leaving the patient with rudimentary intellect as death draws near. Rarely, a myelitis accompanies the encephalitis.

The syndrome correlates positively with the finding of HIV within monocytes in the cerebral parenchyma. Microglial nodules of giant cells form and are associated with vacuolar lesions, primarily in the subcortical white matter. These lesions are variably associated with an inflammatory reaction. The subcortical gray matter (basal ganglia) may be involved.

The most common finding with both MRI and CT scans in AIDS–dementia complex is generalized atrophy. The ventricles show progressive enlargement from the predominant white matter loss. Sulcal enlargement is usually a later finding. In addition, MRI often demonstrates multifocal patchy or punctate regions of increased signal intensity on T_2-weighted images (Fig. 8-16), most likely representing the glial nodules. Later, the lesions become more confluent and bilateral in the subcortical white matter (Fig. 8-17). These signal abnormalities represent the demyelination caused by the HIV encephalitis. CT scanning has not been able to demonstrate these lesions consistently, but when severe enough, the diffuse demyelination is seen as hypodensity within the

white matter. High T_2 signal abnormality attributable to HIV infection may also be seen in the basal ganglia and spinal cord with MRI (Fig. 8-17). The lesions regress following azidothimidine (AZT) therapy. Rarely, CT scans show these lesions as regions of abnormal enhancement.

Cytomegalovirus (CMV) may produce similar findings within the brain. It is associated with HIV infection in less than 5 percent of cases. It sometimes produces a thickened rind of tissue seen as hyperintensity along the subependymal tissue of the lateral ventricles on T_2-weighted MRI that enhances with Gd in a nodular paraventricular pattern. Calcifications do not occur.

NEUROLOGIC COMPLICATIONS WITH AIDS

Viral infections
- Common
  - HIV subacute encephalitis (30%)
  - Cytomegalovirus (CMV) encephalitis (5%)
  - HIV atypical meningitis (8%)
  - HIV vacuolar myelopathy
- Uncommon
  - Herpes simplex encephalitis (3%)
  - Progressive multifocal leukoencephalopathy (4%)
  - Varicella-zoster encephalitis (2%)

Nonviral infections
- Common
  - *Toxoplasma gondii* encephalitis (25%)
  - *Cryptococcus neoformans* meningitis (12%)
  - Tuberculosis
  - Syphilis
- Uncommon
  - Other fungi, mycobacteria, bacteria (2%)

Neoplasia
- Lymphoma, primary CNS (5%)
- Kaposi sarcoma, metastatic (1%)

Other complications
- Infarction, hemorrhage, and unknown lesions (2%)

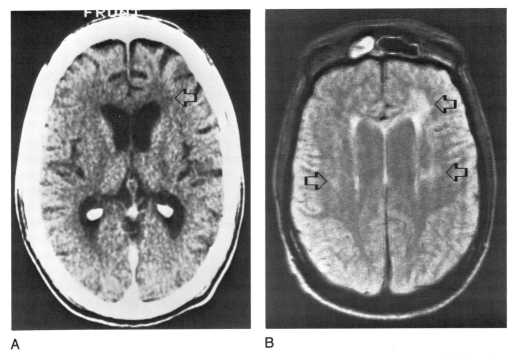

A   B

Fig. 8-16 AIDS, subacute encephalitis. (A) CT scan shows generalized cerebral atrophy. Note subtle low density in the white matter adjacent to the left frontal horn *(arrow)*. **(B)** Lightly T_2-weighted MRI shows patchy paraventricular high signal intensity in the white matter *(arrows)*.

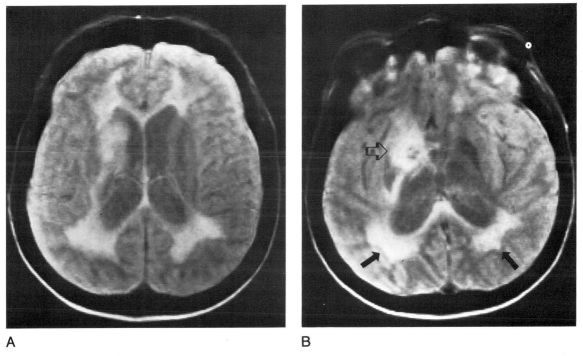

A   B

Fig. 8-17 AIDS, subacute encephalitis. (A) Large confluent regions of high T_2 signal are seen in the white matter with MRI. **(B)** High T_2 signal may also be seen in the basal ganglia *(open arrow)*. Confluent high T_2 signal is seen about the ventricular atria *(black arrows)*. *Toxoplasma* can produce identical basal ganglionic lesions.

Herpes simplex virus (HSV) infection is not common with HIV infection. It may assume its characteristic appearance on CT scans with slightly hemorrhagic, contrast-enhancing, low-density mass lesions bilaterally within the temporal or frontal lobes. However, because the immune response of the brain is diminished, often little necrosis is associated with the infection, and the CT scan may show only atrophy. MRI also shows edema and hemorrhage. Varicella-zoster virus (VZV) has been isolated in autopsy series, but there are no reports of its detection by imaging.

Progressive multifocal leukoencephalopathy (PML), with its characteristic findings (see later discussion) occurs in about 4 percent of those with AIDS. The high T_2 signal intensity lesions on MRI begin as small lesions within the subcortical white matter anywhere within the hemispheres, often bilateral and asymmetric. The lesions become larger and coalesce, spreading over time to the deeper paraventricular white matter. There is no mass effect or contrast enhancement.

Toxoplasma gondii causes, by far, the most common secondary nonviral infection of the brain. Depending on the series cited, this infection occurs in 10 to 40 percent of patients with AIDS. The trophozoites invade the brain, producing regions of necrosis that are often large. Most cases are thought to represent reactivation of latent disease. The lack of immune response allows the infection to persist, even with appropriate antimicrobial therapy, so recurrence is common even after apparent clearance of the lesions with imaging.

The contrast-enhanced CT examination detects all but a few cases of cerebral toxoplasmosis. A diffuse *Toxoplasma* encephalitis without necrosis may occur and escape detection with CT scan. A high-dose drip/delayed examination may detect some lesions when the routinely enhanced study is negative. The imaging test is important, as the diagnosis on the basis of serologic data is unreliable and is frequently negative when infection is raging.

The most common CT pattern is a 1- to 4-cm low-density region, with mass effect, showing ring-like contrast enhancement. Occasionally, the enhancement is uniform (homogeneous) throughout the lesion. The lesions are most commonly found at the corticomedullary junction or within the basal ganglia (75 percent of cases). It does not involve the corpus callosum or spinal cord, and only rarely the cerebellum. Mild hydrocephalus may result from third ventricular compression. The lesions are most often multiple (66 percent). Small hemor-

rhages may occur. About 33 percent of lesions either fail to show enhancement or demonstrate nonspecific enhancement of the cortical surface near more peripheral lesions. *Toxoplasma* encephalitis and lymphoma commonly coexist in the same region. The lesions may regress but frequently remain visible, even with adequate antimicrobial therapy (Fig. 8-18). Calcification may occur within the lesions following therapy.

MRI is the most sensitive imaging study for the detection of toxoplasmosis. The actual lesions are isointense with white matter on T_2-weighted imaging, surrounded by hyperintense edema. They may be single or multiple (Fig. 8-19). MRI may show abnormality when the CT scan is entirely negative, except for atrophy. Gd enhancement increases the sensitivity on MRI, with the principles similar to CT scanning (Fig. 8-20). If the MRI does not show focal disease, toxoplasmosis infection in the brain is unlikely. Toxoplasmosis is difficult to differentiate from lymphoma within the brain (Fig. 8-21). One distinguishing clue is that with lymphoma, the lesions tend to be larger, single, and paraventricular.

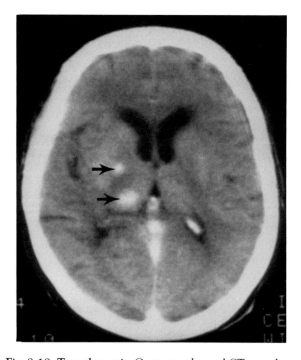

Fig. 8-18 Toxoplasmosis. Contrast-enhanced CT scan shows two densely enhancing solid-appearing lesions in the thalamus and lenticular nuclei (*arrows*). This appearance is characteristic of acquired toxoplasmosis.

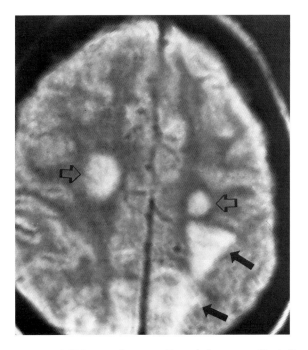

Fig. 8-19 AIDS, toxoplasmosis. Rounded regions of high T_2 signal in the white matter represent toxoplasmosis (*open arrows*). The more confluent regions of high T_2 signal posteriorly (*black arrows*) are undiagnosed. These findings could just as well represent lymphoma or Kaposi sarcoma.

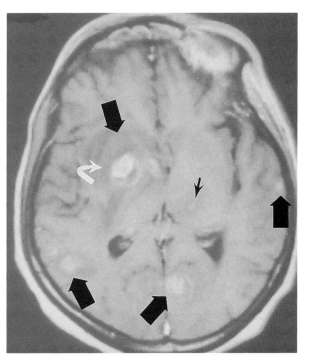

Fig. 8-20 Toxoplasmosis, AIDS. Axial Gd-enhanced MRI shows multiple lesion of variable size and location (*large arrows*). Small lesion is seen in the left thalamus (*small arrow*). Lesions of toxoplasmosis may appear solid and cystic with rim enhancement or complex. They are most common within the basal ganglia and thalamus but may be found anywhere both within the gray or white matter. Edema is variable around the lesions but can be significant (*curved white arrow*).

Cryptococcus meningitis is common, occurring in about 5 to 10 percent of those with AIDS. Meningitis is the most common form of the infection. Meningeal inflammatory reaction is minimal, and contrast enhancement of the meninges is rare. When enhancement occurs, it is most commonly nodular in form. The meningeal infection has a characteristic tendency to invade into the Virchow-Robin spaces of the basal ganglia, forming gelatinous nonenhancing pseudocysts at the base of the brain. Sometimes, the meningitis is associated with *Cryptococcus* encephalitis. Focal cryptococcoma may occur within the brain and are indistinguishable from other abscesses.

Other opportunistic infections found in association with AIDS are caused by *Mycobacterium*, *Aspergillus fumigatus*, *Syphilis*, and *Candida albicans*, as well as coccidioidomycosis and pyogenic abscess. Most of these lesions are indistinguishable from those of toxoplasmosis. Syphilis tends to present as infarction from the associated vas-

culitis (see the specific headings for these infections in this chapter).

Non-Hodgkin's B-cell lymphoma is the most common CNS neoplasm with AIDS, occurring in 5 percent of cases. It is almost always primary within the CNS and only uncommonly arises outside the brain. When it does arise outside of the CNS, the paranasal sinuses and nasal cavity are likely to be involved (Fig. 8-22). On CT scan, CNS lymphoma may be hypodense and may demonstrate ring-like enhancement, hence indistinguishable from the lesions of *T. gondii* infection, or the lesions may be hyperdense and show homogeneous enhancement (Fig. 8-23). In about 25 percent of cases, the enhanced CT scan fails to display the lesions. MRI shows greater sensitivity for detecting the process, but with nonspecific high-density signal abnormality on T_2-weighted images.

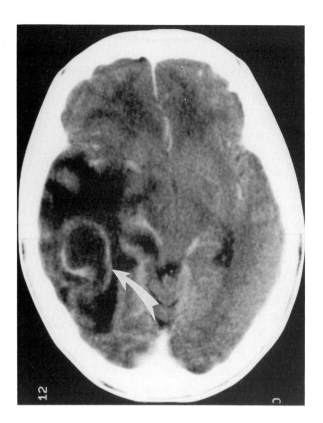

Fig. 8-21 **Toxoplasmosis, AIDS.** Biopsy-proven lesion demonstrating toxoplasmosis *(arrow)* has an irregular enhancing wall on post-contrast CT that resembles a primary brain glioma or pyogenic abscess. There is a large amount of edema surrounding the central focus of infection.

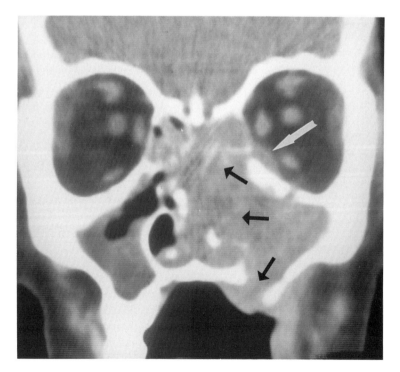

Fig. 8-22 **AIDS, sinonasal lymphoma.** Coronal CT scan of the paranasal sinuses shows a mass occupying the left nasal cavity, left maxillary sinus, and left ethmoid sinus. Bones are destroyed *(arrows)*. Tumor mass invades the orbit *(white arrow)*. Invasive fungal infection can produce this picture.

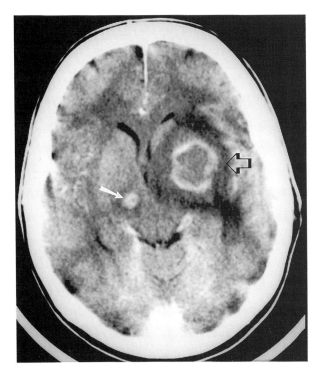

Fig. 8-23 Lymphoma, AIDS. Enhanced CT scan shows a large ring enhancing lesion within the left basal ganglia (*open arrow*), biopsy proven to represent lymphoma. Appearance of the lesion is nonspecific and could have been toxoplasmosis. Another lesion is present in the right thalamus (*white arrow*), but the diagnosis is unknown.

Kaposi sarcoma infrequently metastasizes to the brain. It has been shown on CT scan as a homogeneously enhancing mass. However, CT scanning has failed to diagnose some lesions found at autopsy. No data on MRI scanning are yet available.

AIDS in children, known as pediatric AIDS (PAIDS), differs from adults. Most infants acquire the disease through transmission from an infected mother. The most common abnormalities seen in the brain are atrophy, HIV encephalitis causing high T_2 signal within the white matter, and calcifications within the basal ganglia and the paraventricular region in 50 percent after age 1 year. Pyogenic infections in general are more common in children than in adults, particularly otitis media and meningitis and their complications. Toxoplasmosis and lymphoma are not common.

In summary, the lesions seen with CT scanning and MRI are often nonspecific in character. Therefore, no etiologic diagnosis can be given with certainty from the scan alone. Lesions associated with HIV, *T. gondii*, CMV, lymphoma, and other infections may appear alike; in fact, they frequently coexist, even in the same biopsy specimen. Before specific treatment is given, biopsy of one or more of the lesions is recommended. Biopsy may be done efficiently by the percutaneous needle aspiration technique, using CT or MRI guidance or stereotaxis (Table 8-1).

Herpes Simplex Virus Encephalitis

HSV encephalitis is the most common type of sporadic viral encephalitis. In adults, type 1 herpes simplex virus type 1 (HSV-1) is the infective agent. Most cases of encephalitis seem to result from primary infection, although some patients have had recurrent oral or cutaneous herpes. In neonates, herpes simplex virus type 2 (HSV-2) is the agent, transferred from a maternal genital infection. Symptoms begin at 5 to 15 days of age.

Adult cases are severe, associated with a 70 percent mortality rate, with great morbidity in those who survive. The brain shows necrotic lesions with microscopic hemorrhages that are characteristically bilateral, but often asymmetric. The limbic system is most often involved, especially the anterior medial temporal lobes, the insula, and the orbital surface of the frontal lobes. Those who survive are left with severe regional atrophy. The pattern is similar to limbic encephalitis (see Ch. 10).

In neonates, the infection may cause global devastation of the brain with cystic encephalomalacia similar to the pattern of severe hypoxic/ischemic encephalopathy

Table 8-1 MRI Patterns in AIDS

Pattern	Pathologic Correlation
Generalized atrophy	Subacute encephalitis
	AIDS–dementia complex
	Effects of chronic illness
Single or multiple discrete high-intensity foci on T_2-weighted images; enhancement	Opportunistic infection
	Toxoplasmosis
	Other infections
	Lymphoma
Large bilateral patchy white matter high-signal lesions on T_2-weighted images	HIV encephalitis
	Progressive multifocal leukoencephalopathy

(Fig. 8-24). Intense diffuse gyral enhancement may be seen with contrast CT scanning. Gross low-density abnormality in the white matter represents encephalomalacia. The ventricles are enlarged because of brain atrophy, and there may be late paraventricular calcification. Congenital intrauterine infection rarely occurs.

In adults, CT scans may be normal in mild cases. In severe forms, low-density abnormality is seen in the characteristically involved regions about 5 days after onset, and blotchy contrast enhancement occurs after 7 to 10 days (Fig. 8-25). The hemorrhages are usually not demonstrable. Late atrophy is seen with gross dilation of the temporal horns of the lateral ventricles. Radionuclide scanning may demonstrate abnormal accumulation of radionuclide in the temporal lobes 1 or 2 days sooner than with CT scan.

The most sensitive method for detecting infection appears to be MRI. High signal abnormality representing edema is seen on T_2-weighted images in the inferior frontal and medial temporal lobes and extending to the insula. Atypical distribution may occur (Fig. 8-26), and small hemorrhages may be seen. The coronal image is helpful in defining the characteristic distribution. The lesions are seen at least as early as with radionuclide scanning. Biopsy has been recommended for accurate diagnosis before initiation of antiviral treatment, although because of morbidity from the biopsy in herpes-involved tissue, many patients are treated on the basis of

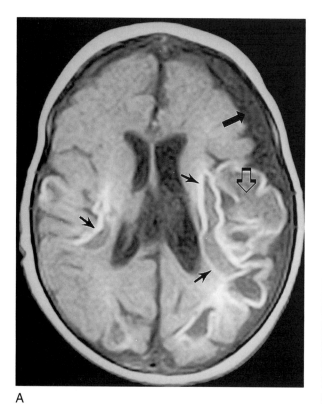

A

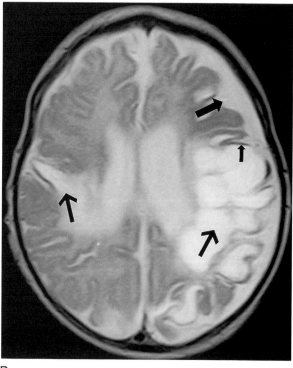

B

Fig. 8-24 Congenital herpes simplex encephalitis. (A) Nonenhanced sagittal T_1-weighted MRI shows a large destructive lesion in the left parietal lobe *(open arrow)* and a smaller lesion on the right. Lesions are outlined by high signal border peculiar to herpes simplex *(smaller black arrows)*. This does not represent hemorrhage or calcium, but a gliotic capsule. There is also a subdural effusion over the left frontal cortex *(larger black arrow)*. **(B)** Axial T_2-weighted MRI shows the central necrotic portions of the lesions as high signal intensity *(larger black arrows)*. Capsule is dark *(smallest black arrow)*. Left frontal subdural effusion has high signal *(medium black arrow)*.

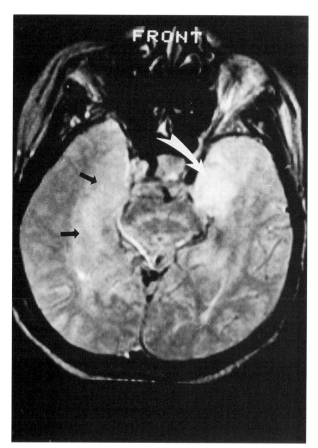

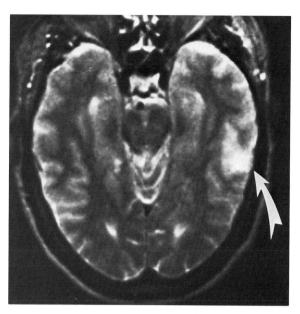

Fig. 8-26 Herpes encephalitis. T$_2$-weighted MRI shows high signal intensity in the left middle temporal gyrus *(arrow)*. This abnormality is nonspecific, and the diagnosis must be made by biopsy. This lateral temporal location is unusual for herpes encephalitis.

Fig. 8-25 Herpes simplex encephalitis, adult. Axial T$_2$ MRI identifies the region of infection as high signal intensity within the anterior medial temporal lobes, most prominently on the left *(large white arrow)*. There is lesser involvement on the right *(black arrows)*.

the MRI results and the clinical diagnosis. Limbic encephalitis, an idiopathic paraneoplastic inflammatory process, affects the medial temporal lobes, insula, and cingulate gyrus in a pattern that may be indistinguishable from herpes encephalitis.

Subacute Sclerosing Panencephalitis

Subacute sclerosing panencephalitis is a post-measles viral encephalitis of obscure etiology. It usually occurs in persons aged 4 to 20 years. Onset is insidious and may not occur until years after the measles infection. The disorder gradually produces relentless mental deterioration and eventual death. CT scanning and MRI show nonspecific multifocal regions of edema in the subcortical and paraventricular white matter associated with atrophy. Contrast enhancement does not occur. Both MRI and CT scans may show only progressive atrophy.

Progressive Multifocal Leukoencephalopathy

Progressive multifocal leukoencephalopathy is caused by a papovavirus infection, affecting only immunodeficient persons. It is most commonly associated with Hodgkin's disease, chronic lymphocytic leukemia, and AIDS. The infection causes demyelination primarily of the posterior parietal and occipital white matter; it may be bilateral by asymmetric. It begins as smaller lesions in the subcortical white matter; with progression, it enlarges and extends deep into the paraventricular region, and sometimes the cortex. CT shows the lesions as ill-defined low-density regions in the posterior white mat-

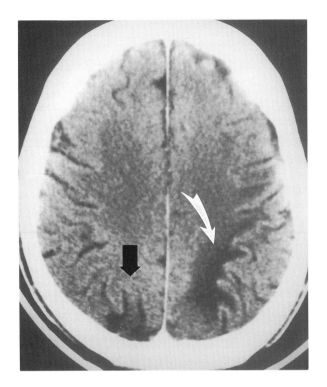

Fig. 8-27 **Progressive multifocal leukoencephalopathy, AIDS.** Axial contrast CT scan identifies a region of nonenhancing hypodensity within the left posterior parietal subcortical white matter *(white arrow)* representing progressive multifocal leukoencephalopathy. No mass effect or atrophy is associated with the lesion, helping differentiate it from subcortical infarction. The lesion in the posterior right parietal lobe involves the cortex and is thought to represent an old infarct *(black arrow)*.

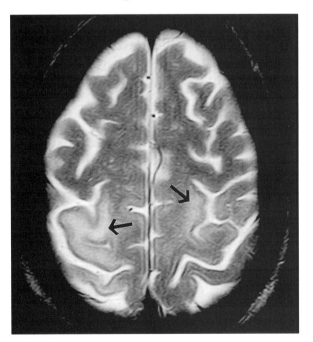

A

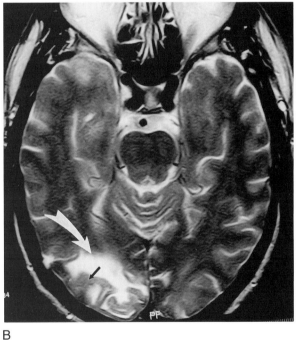

B

Fig. 8-28 **Progressive multifocal leukoencephalopathy. (A)** Axial fast spin echo T₂ MRI indicates the high signal intensity abnormality in the subcortical white matter of both parietal lobes *(black arrows)*. The process is more advanced on the right. **(B)** Lower slice shows very high signal within the subcortical white matter of the right occipital lobe *(large white arrow)*. With advanced progressive multifocal leukoencephalopathy, the cortex becomes involved as well, mimicking infarction *(small black arrow)*. Progressive multifocal leukoencephalopathy shows neither mass effect nor contrast enhancement.

ter, characteristically without contrast enhancement or mass effect (Fig. 8-27). T_2-weighted MRI shows high signal abnormality in the characteristic locations (Fig. 8-28). Occasionally, active demyelination will demonstrate slight contrast enhancement from the associated inflammation. The disease can be treated successfully with cytosine arabinoside, hence the importance of making the diagnosis. Biopsy is necessary for an accurate diagnosis.

Creutzfeldt-Jakob Disease

Creutzfeldt-Jakob disease is a moderately rapid progressive degenerative disease of the brain transmitted by an incompletely defined infectious agent, probably a prion. It usually infects middle-aged persons, causing a severe progressive dementia that results in death within 2 years. The cortex, basal ganglia, and white matter are involved in a vacuolar necrosis termed *status spongiosus*. MRI is the most sensitive imaging test; it shows progressive atrophy and high T_2 signal abnormality in the basal ganglia and white matter. Positron emission tomography (PET) scanning with ^{18}F-fluorodeoxyglucose may show metabolic abnormalities within the cortex, helping to make the correct diagnosis in the early stages and differentiating the disease from Alzheimer's dementia. The electroencephalogram (EEG) is characteristic, with periodic synchronous discharges (see also Ch. 10).

Chronic Mononucleosis-like Syndrome, Chronic Fatigue Syndrome

A syndrome of chronic fatigue, recurrent infections, headache, memory disturbance, and cognitive malfunction has been recognized. It is thought to be caused by infection with the Epstein-Barr virus, or possibly the human B-cell lymphotrophic virus (HBLV) or others. It most commonly affects late adolescents and young adults. MRI of the brain has shown small punctate regions of high T_2 signal intensity in the immediate subcortical frontal and parietal lobes. Occasionally, the lesions have been found to be larger and slightly deeper but are not distributed in the paraventricular region. This pattern is different from the usual distribution of lesions in multiple sclerosis. Single photon emission computed tomography (SPECT) imaging may show multiple small perfusion defects. No pathologic correlation is available at this time.

OTHER INFLAMMATORY DISEASES

Fungal Infections

Fungal infections of the CNS occur as part of disseminated systemic infection. The infection may affect normal persons in endemic regions (coccidiomycosis, histoplasmosis, blastomycosis, chromomycosis) or immunodeficient persons without geographic limitation (candidiasis, aspergillosis, cryptococcosis, nocardiosis, mucormycosis, and actinomycosis).

Basal meningitis is common and is demonstrated by contrast enhancement of the leptomeninges on both CT scan and MRI. The basal cisterns may show increased intensity on T_1-weighted MRI. Hydrocephalus is common (see also Ch. 11), and there may be focal ventricular entrapments from ependymitis. Focal fungal abscesses occur within the brain and may be single or multiple. On contrast-enhanced CT scans and MRI, the lesions appear as round homogeneous enhancing lesions when small, but when larger abscesses they appear ring-like with central necrosis. MRI shows the cerebral infections as nonspecific regions of high signal intensity on T_2-weighted imaging. More mature granulomatous lesions have lower central intensity. The lesions are indistinguishable from metastatic tumor, multiple pyogenic abscesses, or tuberculosis. Lytic bone lesions occur.

Most fungal infections appear similar on imaging, with a few exceptions. Large fungi (hyphal fungi), particularly *Aspergillus*, tend to cause occlusion of the major cerebral arteries, with secondary brain infarction. They may cause invasive sinusitis, particularly within the sphenoid sinus, with scattered high density on CT, and low intensity on T_2-weighted MRI. *Mucormycosis* (phycomycosis) occurs primarily in diabetic patients with ketoacidosis. It involves the nasal cavity and paranasal sinuses. The infection is highly invasive, entering the base of the frontal fossa through the cribriform plate (rhinocerebral phycomycosis) and extending into the orbits. Here a fungal inflammatory mass may mimic tumor (Fig. 8-29). *Cryptococcus* primarily causes a meningitis with occasional secondary involvement of the brain parenchyma.

Syphilis

Neurosyphilis may occur at any time after the primary stage of the infection. Most commonly, it is an asymptomatic aseptic meningitis. However, if it is more severe,

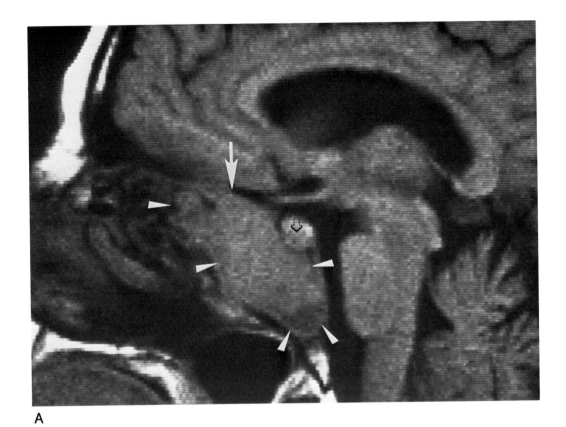

A

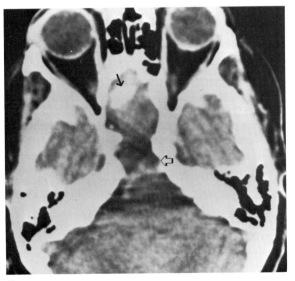

B

Fig. 8-29 Fungal infection of sphenoid sinus. (A) Parasagittal T₁-weighted MRI shows an isointense, nearly homogeneous mass within the sphenoid sinus *(arrowheads)*. It has destroyed the upper clivus and eroded through the planum sphenoidale into the inferior frontal fossa *(arrow)*. The pituitary is preserved *(open arrow)*. A carcinoma could have this appearance. **(B)** Transaxial CT scan shows an isodense mass within the sphenoid sinus, causing destruction of the upper clivus *(open arrow)*. Calcification is present within the lesion *(arrow)*, a relatively common finding with fungus infections, but not with tumor. *(Figure continues.)*

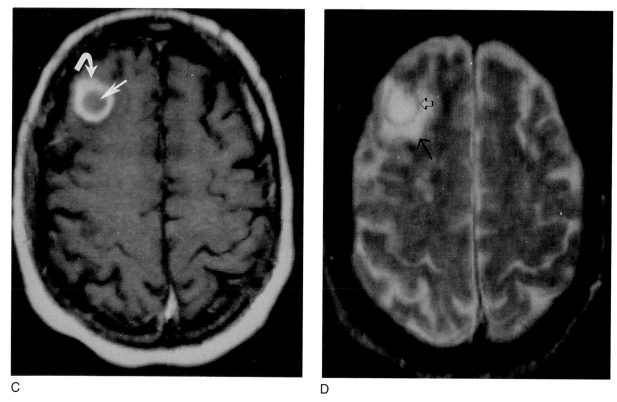

C   D

Fig. 8-29 *(Continued)*. **(C)** Aspergillus. Axial T$_1$ Gd-enhanced MRI illustrates a typical fungal abscess within the brain, *Aspergillus* in this case. The capsule intensely enhances with contrast *(curved arrow)* surrounding the hypointense abscess contents *(straight arrow)*. **(D)** Axial T$_2$ MRI of the same case shows the T$_2$ hypointense abscess capsule *(open arrow)*, a typical finding with abscess, but not with most neoplasia. Edema surrounds the capsule *(black arrow)*. Both pyogenic and fungal abscesses have the characteristics shown here.

meningovascular syphilis causes endarteritis and vascular thrombosis of the major arteries at the base of the brain; this can produce cerebral infarction indistinguishable from that due to other causes. The pattern of basal meningitis is similar to that found with tuberculosis and fungal infections; however, the CSF VDRL is positive with meningovascular syphilis. CT scanning and MRI show infarction and contrast enhancement at the base of the brain; angiography shows narrowing of the lumen or thrombosis of major vessels at the base of the brain. In the chronic stage, gummas may occur within the brain parenchyma and appear as nonspecific, focal, round, contrast-enhancing lesions. Severe atrophy is also present with tertiary syphilis. Lytic skull lesions occur.

Granulomatous Meningitis and Pachymeningitis

Granulomatous meningitis results primarily from fungal or tuberculous infections, but it may occur in association with other systemic inflammatory granulomatous processes, such as sarcoidosis. The meninges may become thick, particularly at the base of the brain. This change causes cisternal block to the flow of CSF, resulting in hydrocephalus, the primary cause of morbidity in this group of patients (see also Ch. 11). Cranial nerve palsies may occur as well.

MRI or CT scans demonstrate the hydrocephalus. The meninges show varying degrees of contrast enhancement, from none at all to pronounced enhancement, confined primarily to the basal cisterns (Fig. 8-30). Dense calcification of the basal meninges may occur in chronic inflammations, particularly tuberculosis. The fungal and tuberculous infections may be associated with cerebral granulomas, usually appearing as ring-enhancing focal lesions. The pattern may be indistinguishable from that of metastatic disease to the brain and meninges. The diagnosis is usually made by biopsy at ventricular shunting.

Sarcoidosis

Sarcoidosis is a relatively common granulomatous process of unknown etiology. The CNS is involved in 5 percent of cases. There may be meningitis, meningoencephalitis, or granulomas within the brain, spinal cord, or lytic round lesions in the skull.

Hydrocephalus is the most common finding on MRI and CT scans, usually occurring as a result of obstruction to the outlet of the fourth ventricle or basal CSF path-

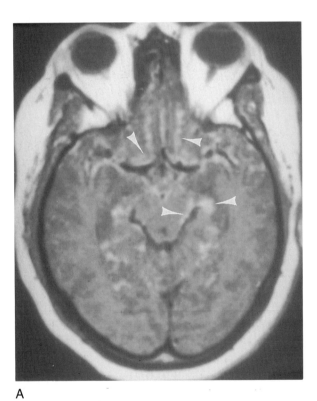

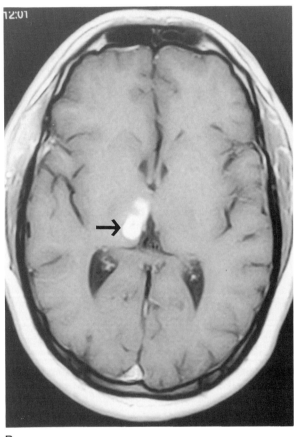

A B

Fig. 8-30 Sarcoidosis. (A) Axial Gd-enhanced T₁ MRI shows intense contrast enhancement of the meninges at the base of the brain, including the pia (*arrowheads*). This represents the diffuse granulomatous form of intracranial sarcoid meningitis. **(B)** Axial T₁ Gd-enhanced MRI identifies an intensely enhancing focal lesion within the right thalamus (*arrow*). Ring-like or solid appearing lesions are typical of parenchymal sarcoid granulomas. They are nonspecific and indistinguishable from lesions associated with metastasis or tuberculosis, among others.

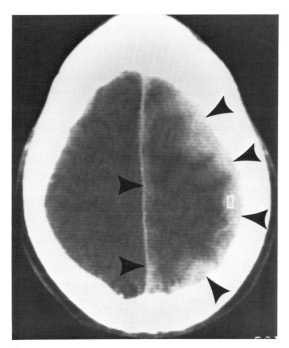

A

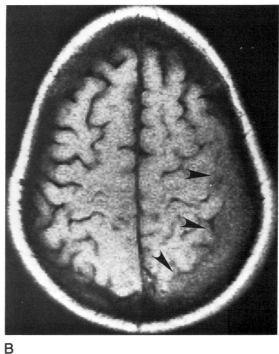

B

Fig. 8-31 Sarcoidosis, pachymeningitis. (A) Contrast-enhanced CT scan shows homogeneous enhancement of the grossly thickened meninges over the left hemisphere and in the left interhemispheric fissure *(arrowheads)*. **(B)** T$_1$-weighted MRI shows the thickened dura as being slightly hypointense to the brain *(arrowheads)*. **(C)** Slightly T$_2$-weighted image shows the extracerebral inflammatory tissue as very hypointense *(arrowheads)*. (MRI courtesy of Dr. Paul Markarian, Springfield, MA.)

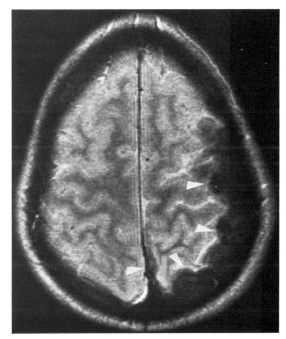

C

ways by granulomatous meningitis, as described in the preceding section. The dural surfaces, including the falx and tentorium, may show lesions that are indistinguishable from meningioma. These lesions may become thick and extensive (Figs. 8-30 and 8-31). Encephalitis may appear as a diffuse patchy enhancement of the brain.

Parenchymal sarcoid granulomas are slightly hyperdense on the nonenhanced CT scan and show homogeneous enhancement with contrast. Calcification sometimes occurs. The lesions are most common in the hypothalamus, pituitary stalk (Fig. 8-32), or pons, but they may occur anywhere, including the spinal cord. The clinical presentation may suggest tumor.

MRI demonstrates CNS granulomas as regions of nonspecific high signal intensity on T_2-weighted images, sometimes with slightly less intensity centrally. MRI may show lesions not seen with CT scans, but calcified lesions are poorly seen. It is important to recognize neurosarcoid, as the outcome is much improved with steroid therapy. The hydrocephalus is treated with shunting. Biopsy is usually necessary for diagnosis.

Tuberculosis

Tuberculosis remains a common and important intracranial infectious disease worldwide. The intracranial infection results from hematogenous spread of *Mycobacterium* tuberculosis, usually from a pulmonary site. Untreated miliary tuberculosis increases the likelihood of intracranial infection.

The intracranial infection may be focal in the brain parenchyma (tuberculoma) or may cause meningitis, when the infection spreads from a tuberculoma into the subarachnoid space. Both processes may occur simultaneously. It is usually a subacute illness, developing insidiously over 3 to 6 weeks.

The tuberculoma is equivalent to the pulmonary granuloma and forms a discrete mass within the brain parenchyma. On CT scan, it appears as a contrast-enhancing, round, nonspecific mass that is variable in size and may be multiple. Like all metastatic processes to the brain, it tends to occur at the corticomedullary junction (Fig. 8-33). Larger lesions appear ring-like (Fig. 8-34). On

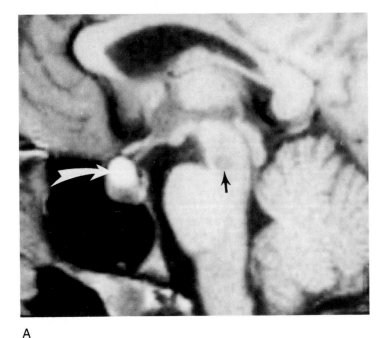

A

Fig. 8-32 Sarcoidosis, suprasellar. (A) Sagittal T_1 Gd-enhanced MRI defines a round intensely enhancing mass located at the upper border of the sella turcica *(arrow)*. This slice clearly demonstrates the red nucleus within the mesencephalon *(small black arrow) (Figure Continues.)*

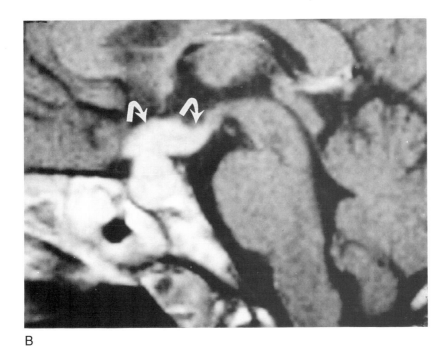

B

Fig. 8-32 *(Continued)* **(B)** Gd-enhanced sagittal T$_1$ MRI shows a more extensive intensely contrast-enhancing mass of tissue within the suprasellar cistern engulfing the optic chiasm and abutting the hypothalamus *(curved arrows)*. Both cases are biospy proven.

Fig. 8-33 Tuberculoma. Contrast-enhanced CT scan shows a ring-like enhancing lesion at the corticomedullary junction extending to the cortical surface of the left posterior frontal lobe. There is moderate surrounding edema. This lesion is indistinguishable from a metastasis or early phase glioma. (Courtesy of Dr. Sam Mayerfield, UMDNJ.)

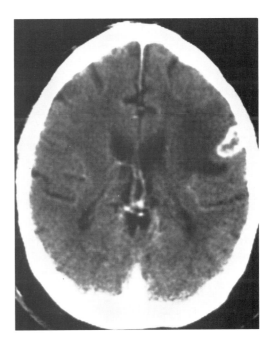

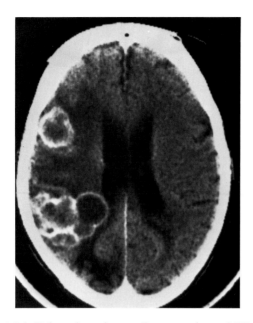

Fig. 8-34 Tuberculous abscess. Contrast-enhanced CT scan shows multiple large tuberculomas in the right frontal and posterior temporal lobes. This picture is indistinguishable from that of multicentric glioma.

MRI, the lesions have high signal on T_2-weighted images and may have a ring-like appearance. Because of frequent hemorrhage within the lesions, they have high signal intensity on T_1-weighted images and regions of low signal on T_2-weighted spin echo (SE) and gradient echo (GRE) images. Edema around the lesion is variable and may be minimal.

If the infection spills over into the subarachnoid space, meningitis occurs. Characteristically, a thick exudate is present at the base of the brain, often producing hydrocephalus from blockade of either the cisternal flow of CSF or the outlet of the fourth ventricle. CT scans and MRI show the process as intensely contrast-enhancing and thickened meninges, frequently with associated hydrocephalus and demonstrable tuberculoma. This constellation of findings suggests the diagnosis of tuberculosis. Heavy calcification of the meninges may be seen during the chronic phase, and cranial nerve compression and vascular occlusion of major arteries may occur (see Ch. 2).

PARASITIC INFESTATIONS

Parasites that infest the lungs, intestines, liver, and muscle may infest the brain as well. These organisms include cestodes (tapeworms: cysticercosis, hydatidosis), protozoans (unicellular organisms: toxoplasmosis), nematodes (roundworms: trichinosis), and trematodes (flukes: schistosomiasis, paragonimiasis). The cestodes and protozoans involve the CNS relatively commonly, but involvement by nematodes and trematodes is rare. Only the first two categories are discussed here.

Cysticercosis

Cysticercosis (cestode infestation) is cased by infection with the larval form of the pork tapeworm, *Taenia solium*. This infection is becoming more prevalent not only in nonindustrialized regions of the world but also in industrialized countries without endemic disease, principally because of the migration of infected persons. The disease is particularly common in Mexico, Central America, South America, Africa, India, China, eastern Europe, and Indonesia. CNS involvement occurs in up to 90 percent of those infected.

Knowledge of the life cycle of *T. solium* is essential for understanding the disease cysticercosis. The pig is the primary source of the infection. Humans acquire the adult tapeworm by ingesting undercooked pork. The tapeworm then attaches to the human intestine, producing a large number of eggs, which are discharged in the feces. This stage can result in the contamination of food or in autoinfection if the person harboring the tapeworm contaminates his or her hands and thereby ingests the eggs. The ingested eggs hatch in the small intestine, burrow into venules, and are carried in the blood to distant sites, most commonly muscle and brain. Here they develop into mature larvae, or cysticerci, within about 2 months.

The larvae remain for years, as the viable cysts do not cause any significant inflammatory reaction. After about 4 to 5 years, the larvae die, the cysts degenerate, and the inflammatory reaction begins, resulting in enlargement of the lesions and surrounding inflammatory edema. It is only at this time that the patient becomes symptomatic.

The cysticerci (5- to 10-mm spheres) are found in the brain parenchyma, the ventricular system, or the subarachnoid space. The location of the cysts determines the nature of the clinical presentation. The symptoms are produced by the mass effect, the surrounding inflammatory reaction, and the obstruction of the flow of CSF in the ventricles or the subarachnoid cisterns. Any combination of findings may occur, as multiple cysts usually exist in various locations. Seizures, focal deficits, and headache from hydrocephalus with increased intracranial pressure are the most common symptoms, mimicking tumor.

The diagnosis is based on clinical, CT, or MRI findings. Subcutaneous nodules are present in 50 percent of those afflicted, and biopsy of the nodules is diagnostic. A serologic antibody test (enzyme-linked immunosorbent assay [ELISA]) is usually positive. CSF examination shows the nonspecific findings of chronic meningitis, with a lymphocyte-predominant pleocytosis and elevated protein and pressure. Eosinophilia occurs in only 20 percent of patients. Plain film findings are helpful only when there is calcification of the cysts in the brain or muscle.

MRI is overall the most useful test for the diagnosis of cysticercosis. The parenchymal form is the most common. During the earliest stages, the initial infection of the brain with larva shows multifocal lesions of edema which later enhance. After about 6 to 12 months, the cisticercus becomes fully grown, with an enlarged bladder forming the visible cyst. This is the vescicular stage. The scolex (head) of the larva is identified at the margin of the cyst. These cysts within the brain parenchyma appear as spherical or nodular 1- to 2-cm lesions of hypointensity on T_1 and hyperintensity on T_2-weighted images, roughly equivalent to CSF. They are usually located at the gray–white matter junction, but they may be found in the depth of the cerebral sulci as well. The cysts may become large, measuring 5 cm or more (Fig. 8-35). CT will show the cystic with CSF equivalent fluid and may also demonstrate the scolex within the cyst. There is little surrounding inflammatory reaction, hence the lack of enhancement during the vescicular stage (Fig. 8-36).

After 3 to 5 years, the lesions degenerate, ensuing in an intense inflammatory reaction in the surrounding parenchyma. The cyst contents increase in intensity on the T_1-weighted MRI because of colloid accumulation. This is the colloid vescicular stage. During this stage, the lesions exhibit contrast enhancement. Smaller lesions homogeneously enhance, appearing as solid nodules, whereas large lesions tend to show central low density with peripheral enhancement of the cyst wall. Thin ring-like enhancements may occur (Fig. 8-37). The picture may mimic that of metastatic neoplastic disease. Edema frequently surrounds the lesion. At this stage, MRI shows the lesions as moderately high T_2 signal with higher-intensity perifocal edema mimicking a tumor. Infarctions may result from the intense perivascular inflammatory reaction, arteritis, and vascular occlusion. Angiography may show the typical changes of arteritis with narrowing, "beading," or occlusion.

With the death of the cysticercus, either naturally or after treatment, the cyst collapses and the scolex partially or totally calcifies (Fig. 8-36). The calcified lesions are easy to detect with CT scanning but are more difficult to detect with MRI. Enhancement occurs at this stage, showing small lesions that may have a target-like appearance.

Within the ventricular system or the subarachnoid space, the cysts have about the same CT absorption spectrum as that of the CSF, making them difficult to detect. The cysts are attached to the ependyma or are free floating. Ring-like abnormal contrast enhancement or the scolex may be seen, allowing one to discern the lesions, but it occurs less frequently than with cysts in the brain parenchyma. A racemose form exists without a scolex, making the lesion harder to identify. The cysts assume the shape of the ventricular cavity, so they may appear to be the ventricle itself, particularly within the fourth ventricle. Positive-contrast cisternography or ventriculography may be necessary to define the cyst, which is seen as a negative defect within the contrast-filled ventricle. MRI may be helpful here, as sometimes the rim of the cyst and adjacent ependyma may have very high T_2 signal intensity, or the cyst may contain proteinaceous fluid with higher signal intensity than that of CSF, allowing for its identification within the ventricle. Contrast enhancement may identify inflamed meninges. The intraventricular cysts may cause obstruction of the foramina, with the development of hydrocephalus. Surgical excision of critical lesions can be helpful in relieving symptoms.

Calcification does not occur in cysts within the ventricles or the subarachnoid space. Rarely, a cyst within the CSF shows homogeneous contrast enhancement, mimicking a solid tumor. Rupture of a cyst may lead to diffuse meningitis. The infection may occur along the spinal cord in the subarachnoid space and is seen with myelography as intradural extramedullary masses. Intramedullary spinal cord lesions are rare but may be seen as enhancing cystic lesions.

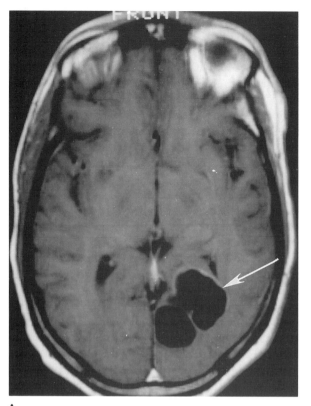

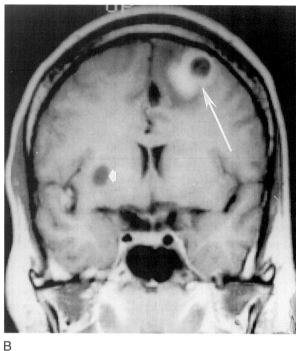

A   B

Fig. 8-35 Cysticercosis. (A) Axial T$_1$ Gd-enhanced MRI shows a large cyst in the occipital lobe with a contrast-enhancing cyst wall *(arrow)*, indicating early degeneration. **(B)** Coronal T$_1$ Gd-enhanced MRI shows considerable enhancement around a degenerating cyst near the cortex *(white arrow)*. Edema surrounds the cyst. A viable cyst is present
in the right lateral basal ganglia *(small white arrow)*; there is no contrast enhancement or edema.

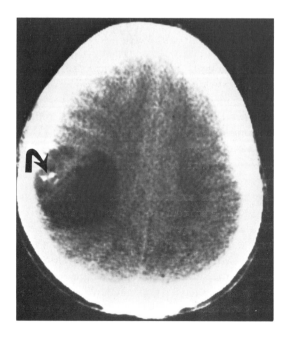

Fig. 8-36 Cysticercosis. A large cyst is present in the right parietal
lobe containing fluid of near-CSF density. The calcification represents the dead larva *(arrow)*.

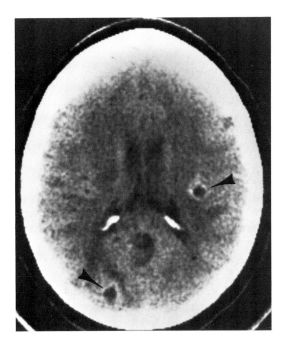

Fig. 8-37 Cysticercosis. Contrast-enhanced CT scan shows multiple ring-enhancing lesions with central low density (*arrowheads*). (Courtesy of Dr. Sam Mayerfield, UMDNJ, New Jersey Medical School, Newark, NJ.)

Treatment consists of shunting of the hydrocephalus and removal of the critical lesions in the ventricular system, as well as administration of steroids, antiseizure medication, indium-131-tagged specific immunoglobulins, and praziquantel (a broad-spectrum antitrematodal agent).

Hydatidosis

Hydatidosis (cestode infestation) is caused by the larval form of the tapeworms *Echinococcus granulosa* or *Echinococcus multilocularis*. It is endemic in South America, northern and eastern Africa, central Europe (particularly the Mediterranean countries), and the Middle East. The dog is the primary host; humans, sheep, cattle, and horses are the intermediate hosts. The life cycle is similar to that described for cysticercosis. Humans become infected when they come into contact with soil or with dogs contaminated with eggs.

The CT appearance is characteristic. Hydatid cysts in the brain are usually large and unilocular, containing fluid of nearly CSF density. There is little surrounding edema and only occasional calcification. The cyst is round (Fig. 8-38). This feature may help differentiate hydatid disease from a congenital arachnoidal cyst, which tends to have at least one straight border. The cyst itself produces a mass effect, leading to increased intracranial pressure; it may rupture, causing dissemination of the scolices throughout the subarachnoid space.

Paragonimiasis

Paragonimus westermani is the most common lung fluke infection in humans. The disease is endemic to Africa, southeast Asia, and South America. Larvae are ingested from crabs or crayfish. The major infection is in the lung and pleura. About 1 percent of infected persons will have CNS involvement.

The larvae penetrate the meninges to invade the temporal or parietal lobes. Multiple 1- to 3-cm cystic granulomatous lesions form with surrounding edema in the parenchyma. The central portion of the lesion contains necrotic debris, ova, inflammatory cells, and even-

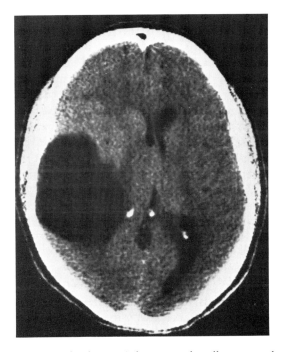

Fig. 8-38 Hydatid cyst. A large, round, well circumscribed cyst of near-CSF density is present in the right temporal lobe. There is little surrounding edema but considerable mass effect from the cyst itself.

tually the dead larva. The walls of the granuloma enhance with Gd. The central portion of the lesion is hypointense on T_1 and hyperintense on T_2-weighted MRI. Late in the disease, the granuloma calcifies and is easily seen with CT scanning. The lesions take many forms, but the characteristic pattern is a "soap-bubble" pattern of the calcified multiple cysts.

Sparganosis

Sparganosis is a rare infection caused by a tapeworm. It is worldwide, but most common in eastern Asia. The infection is acquired from drinking infected water or from eating undercooked frogs or snakes. CNS infection occurs in 1 percent of those infected. The lesion is granulomatous and occurs within a hemisphere. It causes a region of edema, multiple punctate calcifications, and adjacent ventricular dilation. Enhancement occurs in the granuloma and in the parenchyma from the inflammation in response to the granuloma. There is no known treatment.

Toxoplasmosis

Toxoplasmosis is a protozoan infection that occurs either as a severe congenital form manifested by destructive lesions of the CNS, the eyes, and viscera or as an acquired form with acute lymphadenopathy, ocular lesions, and an illness resembling infectious mononucleosis. The disease is caused by *Toxoplasma gondii*, which multiplies intracellularly and persists in the CNS and the eye. Elsewhere it subsides in response to the development of antibodies. The infectious agent resides in cattle, swine, sheep, dogs, and rabbits, as well as in several domestic birds. The prevalence of this disease is increasing, primarily because of the relatively high incidence in immunodeficient persons, particularly those with AIDS.

With congenital toxoplasmosis, maternal infection is transferred to the infant. The maternal infection may be chronic and occult. The brain and eyes are the primary target organs. A severe encephalitis causes extensive tissue destruction, resulting in microcephaly, severe psychomotor retardation, and chorioretinitis. Calcifications are deposited in granulomas in the ependymal walls of the lateral ventricles, basal ganglia, peripheral frontal and parietal lobes, and choroid plexi. The calcifications may take any form (i.e., curvilinear, nodular, or plaque-like), or they may be in clumps. The ventricular enlargement is most commonly the result of the severe brain

destruction, although rarely a granuloma will obstruct the aqueduct of Sylvius, causing hydrocephalus (Fig. 8-39). Congenital toxoplasmosis and cytomegalic inclusion disease are the two most common causes of intracranial calcifications in infants. Cortical and basal ganglionic calcifications are more common in congenital toxoplasmosis, although the two processes cannot be reliably differentiated by imaging along.

Acquired adult toxoplasmosis is now occurring with increasing frequency in immunosuppressed patients. It is seen in those with tumors of the hematopoietic system, those with AIDS, and those receiving chemotherapy for malignant disease. Because of the impaired immune reaction in these patients, the serologic test for toxoplasmosis may be negative. Most commonly, a multifocal encephalitis causes densely enhancing lesions seen on CT scans (Fig. 8-18). They may be up to 4 cm in diameter and are distributed throughout the brain, but preferentially in the paraventricular white matter or at the level of the corticomedullary junction. The appearance of the lesions on CT scans is nonspecific, ring-like with central low density due to necrosis and surrounded by a

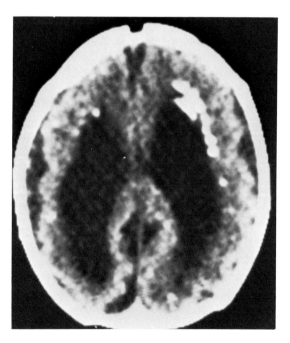

Fig. 8-39 Congenital toxoplasmosis. CT scan shows gross ventricular dilatation and periventricular calcifications. Small extracerebral fluid collections are present as well.

rim of edema. MRI shows the lesions as nonspecific round regions of high T_2 signal intensity (Fig. 8-19). The lesions themselves have the same appearance as metastatic tumor or pyogenic abscesses. Because the disease is relatively acute in onset, the lesions tend to be of the same size within a given patient. Metastatic tumors to the brain, when multiple, are frequently varied in size, a result of blood-borne embolization occurring over an extended period. In most cases, biopsy of a lesion is required for definitive diagnosis. Many clinicians believe that these multiple lesions, when seen in the appropriate clinical setting, are sufficiently diagnostic of toxoplasmosis to warrant immediate treatment.

Lyme Disease

Lyme disease is the result of infection with the spirochete *Borrelia burgdorferi*. The disease is widespread, occuring in the United States, Europe, Russia, and Asia. The infection is transmitted to humans through the bite of an infected tick of the ixodid group (*I. ricinus complex*). The ticks may be found on a wide variety of animals, deer being the most common. Humans pick up the tick when in infected regions.

The infection begins locally around the site of the tick bite, usually producing a characteristic skin lesion, erythema migrans. Within days to weeks, the infection becomes disseminated to a wide variety of organs, including the meninges and brain. Afflicted persons may suffer severe headache from meningitis and Bell's palsy and other cranial or peripheral neuropathies from radiculoneuritis. Imaging of the brain is usually normal during this phase. Malaise, migratory muscle and joint pains, lymphadenopathy, and symptoms of myopericarditis may be present.

Late neurologic abnormalities may occur from 6 months to as long as 6 years after infection. Most commonly, this is attributable to a progressive encephalomyelitis, which may produce paraparesis, bladder dysfunction, ataxia, cranial nerve palsies, and dementia. The symptoms may mimic a demyelinating disease, such as multiple sclerosis, and oligoclonal bands may be found in the CSF. During this stage, MRI has shown focal high-intensity lesions within the white matter in some patients, suggesting demyelination. Contrast enhancement may be demonstrated within the active lesions. The spine may be involved. The correct diagnosis is made by serologic testing with a specific ELISA.

SUGGESTED READING

Bazan C III, Rinaldi MG, Ranch RR et al: Fungal infections of the brain. Neuroimag Clin North Am 1:57, 1991

Chang KH, Cho SY, Hesselink JR et al: Parasitic diseases of the central nervous system. Neuroimag Clin North Am 1:159, 1991

Chang KH, Lee JH, Han MH et al: Role of contrast enhanced MR imaging in the diagnosis of neurocysticercosis. AJNR 12:509, 1991

Dalakas M, Wichman A, Sever J: AIDS and the nervous system. JAMA 261:2396, 1989

Dowe DA, Heitzman ER, Larkin JJ: Human immunodeficiency virus infection in children. Clin Imag 16:145, 1992

Enzmann DR, Britt RR, Obaua WG et al: Experimental *Staphylococcus aureus* brain abscess. AJNR 7:395, 1986

Enzmann DR, Britt RR, Placone RC Jr: Staging of human brain abscess by computed tomography. Radiology 146:703, 1983

Haimes AB, Zimmerman RD, Morgello S: MR imaging of brain abscesses. AJNR 10:279, 1989

Hayes WS, Sherman JL, Stern BJ et al: MR and CT evaluation of intracranial sarcoidosis. AJNR 8:841, 1987

Heck AW, Phillips LH II: Sarcoidosis and the nervous system. Neurol Clin 7:641, 1989

Kieburtz K, Schiffer RB: Neurologic manifestations of human immunodeficiency virus infections. Neurol Clin 7:447, 1989

Krüger H, Meesmann C, Rohrbach E et al: Panencephalopathic type of Creutzfeldt-Jakob disease with primary extensive involvement of the white matter. Eur Neurol 30:115, 1990

Post MJD, Tate LG, Quencer RM et al: CT, MR, and pathology in HIV encephalitis and meningitis. AJNR 9:469, 1988

Schwartz RB, Garada BM, Komaroff AL et al: Detection of intracranial abnormalities in patients with the chronic fatigue syndrome: comparison of MR imaging and SPECT. AJR 162:935, 1994

Sze G, Zimmerman RD: The magnetic resonance imaging of infectious and inflammatory diseases. Radiol Clin North Am 26:839, 1988

Teitelbaum GP, Otto RJ, Lin M et al: MR imaging of neurocysticercosis. AJNR 10:709, 1989

Weingarten K, Zimmerman RD, Becker RD et al: Subdural and epidural empyemas: MR imaging. AJNR 10:81, 1989

Zimmerman RD, Leeds NE, Danziger A: Subdural empyema: CT findings. Radiology 150:417, 1984

9

White Matter Disease; Demyelinating Disease

Most diseases of the white matter affect myelination. Myelin is the layered wrap around axons, consisting predominantly of lipids and of a much lesser amount of protein. The myelin axonal sheath is produced by oligodendrocytes. If myelin has formed normally, it may be broken down by a disease process, such as demyelination or a myelinoclastic process. There may also be a delay in normal myelination in response to neonatal or infantile insults. Alternatively, the myelin may not have formed properly in the first place (dysmyelination). The dysmyelinating diseases are discussed in Chapter 12.

Myelin is hydrophobic. Therefore, myelinated white matter normally has a low water content. If myelin is deficient, broken down, or abnormally formed, the water content of the tissue is much greater than that of normally myelinated tissue. Regions of abnormal or absent myelin are seen on CT scans as regions of hypodensity. On MRI, they are of low signal intensity on T_1-weighted images and of high signal intensity on T_2-weighted images. MRI is more sensitive than CT scanning for the detection of a demyelinating process.

A myelinoclastic process, the prime example of which is multiple sclerosis (MS), may have an associated inflammatory component that sometimes produces a mass effect or abnormal breakdown in the blood-brain barrier, which results in enhancement after contrast infusion. Dysmyelination does not produce these changes. Demyelination has many causes, including MS, ischemia, trauma, infection, radiation damage, and metabolic factors.

DEMYELINATING DISEASES

Multiple sclerosis
Vascular disease
    Infarct
    Subcortical arteriosclerotic encephalopathy (Binswanger's disease)
Trauma
    Postsurgical demyelination and gliosis
    Contusion
    Diffuse axonal injury
Radiation damage
Viral and postviral disease
    AIDS
    Herpes simplex encephalitis
    Progressive multifocal leukoencephalopathy (PML)
    Immune-related encephalopathy (acute disseminated leukoencephalitis [ADEM])
    Subacute sclerosing panencephalitis (SSPE)

Toxic and paraneoplastic syndromes
    Central pontine myelinolysis
    Marchiafava-Bignami disease
    Limbic encephalitis
    Carbon monoxide
    Diffuse hypoxia/ischemia
    Some genetic metabolic disorders
        Aminoacidoses
        Mitochondrial disorders
    Chemotherapy for neoplasm

NORMAL MYELINATION PATTERN

The CT and MRI appearance of the central white matter varies with age. There are infantile, transitional, adult, and aging patterns. Myelination is a dynamic process that progresses in an orderly fashion over the first 3 years of life. The most ancient phylogenetic tracts are myelinated first and the telencephalon last. This development proceeds in a more or less caudal to rostral direction (see Ch. 12). Before myelination, the white matter has relatively high water content, displayed as relative hypodensity with CT scanning and as relative high signal with T_2-weighted MRI. Transition in the cerebrum occurs at about 10 to 12 months, when the white matter of the centrum semiovale first becomes isointense, and then hypointense on T_2-weighted images. The white matter assumes the adult level of hypointensity at about 36 months. Delays in myelination can be determined if the infantile pattern persists longer than normal. After 3 years, the pattern of the white matter remains the same, until late adulthood. White matter maturation is discussed in greater detail in Chapter 12.

In normal persons of all ages, there is frequently a small focal region of high T_2 signal intensity within the white matter adjacent to the frontal and occipital horns of the lateral ventricles. It is thought to represent a concentration of water about to be absorbed into the lateral ventricles. The flow of CSF here is centripedal; it is opposite that seen with transependymal absorption associated with hydrocephalus. There may also be a thin "pencil line" of high intensity outlining the whole ventricular system. This pattern of high T_2 signal is accentuated in certain MRI sequences, such as inversion recovery (fluid attentuated inversion recovery [FLAIR]) and magnetization transfer contrast (MTC). These regions must not be mistaken for pathologic change. However, in some cases it is impossible to determine whether the amount of high signal in the paraventricular white matter is physiologic or pathologic.

UNIDENTIFIED BRIGHT OBJECTS

More frequently with increasing age, small focal or punctate regions of high T_2 signal intensity, or unidentified bright objects (UBOs), are seen within the white matter on MRI. Sometimes they are slightly larger or coalesce. They occur predominantly in the paraventricular white matter but may be in the subcortical region and brain stem as well. These high-signal foci have been found to represent a variety of lesions. Most commonly, they represent myelin pallor seen at autopsy, thought to be secondary to ischemia. Other causes include gliosis, focal demyelination (possible subclinical multiple sclerosis), brain cyst, ventricular diverticulum, or dilated perivascular spaces of Virchow-Robin. In some cases, autopsy fails to detect the lesion within the brain.

The significance of these lesions is not fully known, although there is a positive correlation with atherosclerosis of the perforating arteries, as well as dementia, hypertension, hyperlipidemia, heart disease, and diabetes. On MRI scan, these lesions cause diagnostic confusion, as they sometimes cannot be differentiated from small metastases or regions of multiple sclerosis. Clinical correlation, contrast enhancement, and follow-up scanning can usually determine whether they are or relevance. If a small high T_2 signal lesion does not enhance, especially after high-dose gadolinium (Gd) or the use of MTC T_1-weighted images, there is little likelihood of metastatic tumor.

Multiple Sclerosis

MS is an idiopathic inflammatory demyelinating disease of the white matter of the brain and spinal cord. It is by far the most common primary demyelinating disease in adolescents and young adults. It typically occurs in persons aged 20 to 50 years who live in the northern parts of the United States, Europe, Asia, and in southern Australia. There is a strong female predominance, especially among children. It causes dissolution of the myelin sheaths of the brain and spinal cord with preservation of the nerves. Proliferating astrocytes form the firm "plaques" characteristic of the lesions. Inflammatory changes are present with edema and eventual scar formation.

MS produces variable neurologic signs and symptoms characterized by exacerbation and remission. The diagnosis is made clinically by history, physical findings, CSF analysis, and evoked potential tests. Imaging is used for confirmation of the clinical diagnosis, estimation of the severity of the disease, and the response to therapy. The diagnosis of MS cannot be made on the basis of imaging alone.

The lesions, which vary from a few millimeters to centimeters in diameter, are characteristically located in the deep white matter adjacent to the ventricles of the cerebral hemispheres. Lesions may also be found in the peripheral subcortical white matter, but there they tend to be smaller and less numerous. Lesions within the corpus callosum are common and nearly specific for MS (Fig. 9-1). They are especially likely at the junction

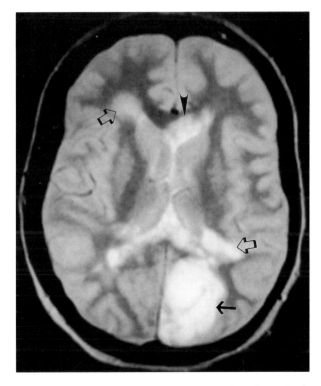

Fig. 9-1 Multiple sclerosis. Axial T_2 MRI identifies multiple lesions of MS. High signal lesions are found typically adjacent to the lateral ventricles *(open arrows)*. Lesions within the corpus callosum *(arrowhead)* occur with MS, and not with most other white matter diseases. Active lesions may be very large with mass effect *(black arrow)*, sometimes appearing as a tumor.

CHARACTERISTICS OF MULTIPLE SCLEROSIS LESIONS

CT
    Hypodense
    Contrast enhancement with active lesion
    Atrophy, with chronic disease

MRI
    Hyperintense on T_2-weighted images
    Hypointense on T_1-weighted images
    Contrast enhancement with active lesion
    Atrophy, with chronic disease

Location
    Deep, paraventricular (95%)
        Trigones
        Body
        Occipital
        Frontal
    Ovoid configuration (80%)
    Floor of fourth ventricle
    Brain stem (60%)
    Corpus callosum (75%)
    Later, subcortical white matter
    Spinal cord
    Any white matter tract
    Optic nerves

Female-to-male ratio
    Adults   (2:1)
    Children (10:1)

Contrast
    Enhancement represents active plaque

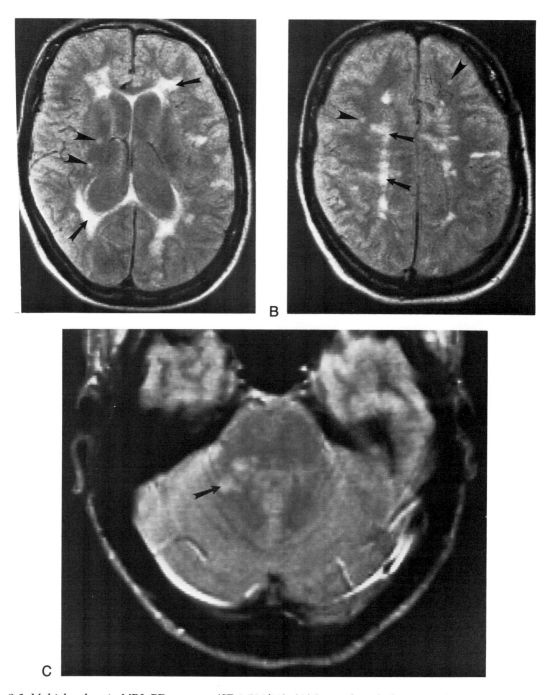

Fig. 9-2 Multiple sclerosis, MRI, PD sequence (SE 2,500/40). (A) Image through the ventricular system shows the CSF to be isointense with the white matter. This image is optimal for visualization of paraventricular white matter abnormality. Confluent high T_2 signal abnormality is seen in the paraventricular white matter *(arrows)*. Smaller discrete lesions can be seen in the internal capsule on the right *(arrowheads)*. **(B)** Scan through a higher level shows the confluent paraventricular high T_2 signal abnormality *(arrows)*. The more peripheral subcortical lesions are smaller and less intense *(arrowheads)*. This distribution is typical of MS lesions. **(C)** High T_2 signal abnormality is seen in the cerebellar peduncle *(arrow)*.

between the corpus callosum and the septum pellucidum, seen best on the parasagittal projection. As a rule, callosal lesions are not seen with vascular disease, but they have been seen with Lyme disease and hydrocephalus. Less commonly, lesions occur within the cerebellum, pons, and spinal cord. Children are more likely than adults to have posterior fossa lesions. Multiple lesions may become confluent. Often the paraventricular lesions assume an ovoid shape and are oriented perpendicular to the ependymal surface. These have been called "Dawson's fingers" and, while not specific for MS, they have a high association with the disease. Optic nerve involvement is frequent with MS and may be the presenting symptom in 20 percent of cases. Occasionally, a single large acute lesion has an appearance usually associated with glioblastoma (Fig. 9-1). Edema and contrast enhancement usually accompanies the larger brain lesions. Cerebral atrophy is seen after the disease has been present for some time, causing both the ventricles and the sulci to become enlarged.

Spinal cord lesions are common and may be seen in about 50 percent of patients. The cervical cord is the most likely location, but lesions may occur at any level. The lesions tend to be elongated and may show edema, enlargement of the cord, and contrast enhancement during the acute phase. Cord atrophy is a late result.

MRI is the most sensitive imaging modality for the detection of MS. Lesions are seen in 95 percent of those clinically diagnosed with MS. However, many more lesions are seen pathologically than are seen with MRI. The lesions are of high signal intensity on the T_2-weighted images. Those in the immediate paraventricular region are best seen with proton-density (PD) images (long time of repetition [TR] with relatively short time of echo [TE]; e.g., spin echo [SE] 2,500/30), or inversion recovery T_2(FLAIR); thus, the ventricular CSF appears isointense with the normal white matter, with the effect that the paraventricular lesions stand out. Most commonly, the lesions are discrete or confluent adjacent to the frontal and posterior horns. Those in the more peripheral white matter are best demonstrated with long echo T_2-weighted images (e.g., SE 2,500/120) (Fig. 9-2).

The lesions seen with MRI may be acute (inflammatory change) or chronic (scar). The activity of the lesions cannot be determined with nonenhanced MRI, although older lesions tend to display lower intensity on T_1-weighted images. Once established, most of the lesions on MRI remain indefinitely, although they sometimes change slightly in size. The size and number of the lesions imaged do not necessarily correlate with the clinical activity or clinical severity of the disease. On routine MRI, the lesions do not change with steroid or other therapy, so routine MRI is not used to follow treatment effectiveness.

Gd enhancement occurs only with inflammatory activity and may be used to define the progression or response to therapy (Fig. 9-3). However, unlike CT scanning, Gd generally does not demonstrate more lesions than were seen on the nonenhanced T_2-weighted scan. Initially, the lesions may be ring-like on immediate postinjection scans, but they fill in over 15 to 30 minutes. The degree of enhancement of specific lesions decreases with steroid therapy.

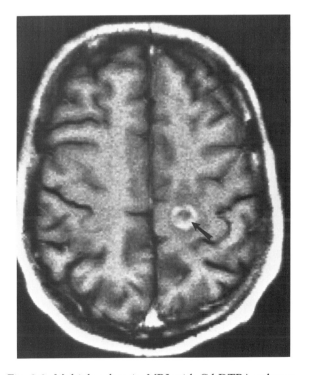

Fig. 9-3 Multiple sclerosis, MRI with Gd-DTPA enhancement. T_1-weighted MRI demonstrates a ring-like enhancing lesion of MS in the left hemispheric white matter (*arrow*). Such lesions are indistinguishable from some other lesions (e.g., metastasis or infection), except for their association with typical MS changes in the brain and the appropriate clinical history.

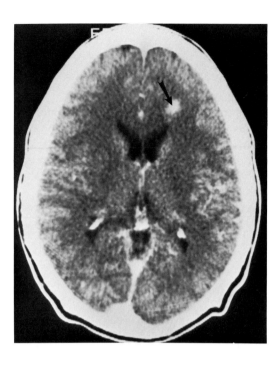

Fig. 9-4 Multiple sclerosis, CT scan with contrast enhancement. High-dose drip/delay contrast-enhanced CT scan shows a homogeneously enhancing lesion adjacent to the left frontal horn *(arrow)*. The lesion is nonspecific in appearance and can be diagnosed as MS only in the appropriate clinical setting.

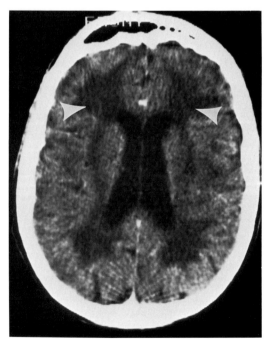

Fig. 9-5 Multiple sclerosis: severe and end stage. Nonenhanced CT scan shows generalized hypodensity of the hemispheric white matter. The ventricles are enlarged because of deep white matter atrophy.

The high T_2 signal lesions of MS must be differentiated from "normal" paraventricular high signal intensity, paraventricular high intensity associated with hydrocephalus, UBOs, multiple subcortical infarctions (e.g., Binswanger's disease), arteritis (e.g., lupus), axonal (shear) injury, migraine, Lyme disease, and postirradiation change.

The nonenhanced CT scan is relatively insensitive to the detection of MS lesions. High-dose contrast infusion (80 g of iodine) and 1-hour delayed scanning increases the sensitivity, so that about 50 percent of patients with clinical MS show focal lesions (Fig. 9-4). Only active lesions are detected with the enhancement. Low-density white matter lesions may be seen with the nonenhanced scan, but far fewer than are actually present (Fig. 9-5). Contrast-enhanced lesions have a positive correlation with the activity of the lesions and their response to treatment.

RADIATION DAMAGE

Radiation in therapeutic doses is potentially damaging to the brain, particularly in children. An acute encephalopathy may develop during the course of treatment. Imaging is usually normal. The encephalopathy is transient and responds to steroid therapy without residual effects (Table 9-1).

The early delayed injury is also a transient phenomenon seen 1 to 6 months after therapy. It may be mild or severe enough to cause death. The brain shows a diffuse inflammatory demyelination. The lesions are punctuate when the encephalopathy is mild, and confluent when severe. Contrast enhancement occurs within the lesions, which may mimic tumor growth. The lesions usually disappear after steroid treatment.

Late radiation injury causes white matter damage with nonspecific changes of demyelination, vacuolation,

edema, and gliosis. It is thought to be the result of small vessel hyalinization, with the most severe changes in the deep paraventricular region. Contrast enhancement does not occur. The damage resides only within irradiated tissue 1 to 10 years after the treatment. It is most common following a radiation dose of greater than 50 Gy, but it may be seen after lower levels. The process may be focal or diffuse. Atrophy of the brain is an almost constant feature after radiotherapy and may be seen with or without demonstrated diffuse late injury.

Focal radiation necrosis is seen more commonly now as a result of focal radiation boost, brachytherapy, and focused steroscopic radiation treatment (γ-knife). Such treatment delivers very high doses to the brain tumor, but also to the nearby surrounding tissue. Focal necrosis typically presents about 1 year after treatment as a progressive process within the white matter. The cortical gray matter is less affected. The necrotic tissue creates an intense inflammatory reaction with mass effect; on imaging, it mimics recurrent malignant tumor. CT and MRI show irregular contrast enhancement within and near the focus of necrosis, surrounded by edema and mass effect. Sometimes the tumor may be distinguished from necrosis by metabolic radionuclide imaging (see Ch. 5). Necrosis does not demonstrate metabolic activity, whereas active tumor demonstrates high metabolic activity. Focal necrotic tissue may be surgically removed, diminishing the inflammatory response.

Diffuse delayed injury results from whole brain radiation. This process, involving the white matter, may be mild and reversible or severe and progressive to death. CT scans most commonly show only atrophy with dilation of the ventricles and the sulci. Hypodensity of the white matter may occur. MRI shows regions of radiation effect as high signal intensity in the white matter on T_2 images (see Ch. 2). Mild lesions may be multifocal. With more severe injury, there is confluence of high T_2 signal lesions, beginning in the periventricular deep white mat-

Table 9-1 Radiation Injury

Type	Onset	Imaging
Acute injury	During treatment	Normal
Early delayed injury	1–6 months post-treatment; transient demyelination	Inflammatory demyelination
Delayed (late) injury	1–10 years post-treatment	
	Diffuse demyelination	Demyelination, without inflammation
	Focal necrosis	Inflammatory mass

ter and extending to the subcortical region. This finding represents edema, demyelination, and gliosis. There is no contrast enhancement with the delayed diffuse injury.

NECROTIZING LEUKOENCEPHALOPATHY

Necrotizing leukoencephalopathy refers to injury to the white matter as a result of tumor chemotherapy, often associated with radiotherapy. The disease may occur from chemotherapy alone. A wide spectrum of chemotherapeutic agents may cause the injury. The process appears similar to diffuse delayed radiation injury with abnormal T_2 signal abnormalities through-

out the white matter. It occurs 1 to 4 months after the onset of the chemotherapy. A mineralizing microangiopathy may occur in children in response to chemotherapy and radiotherapy, with basal ganglionic and paraventricular calcification a striking feature (see Ch. 6).

ACUTE DISSEMINATED ENCEPHALOMYELITIS (IMMUNE-RELATED ENCEPHALOPATHY)

Acute disseminated encephalomyelitis (ADEM) includes the syndromes of postinfectious encephalomyelitis, postvaccinal encephalomyelitis, and acute hemor-

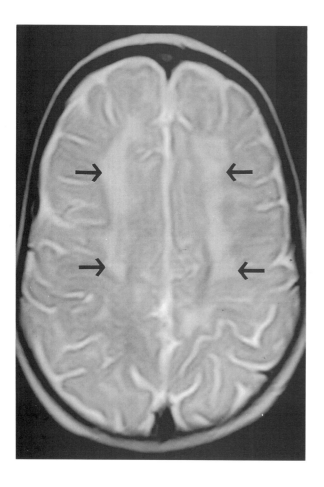

Fig. 9-6 Immune-mediated encephalitis, ADEM. Axial fast spin echo (4/4) 96/3,000 T_2 MRI shows nearly symmetric high signal within the centrum semiovale (*arrows*).

rhagic leukoencephalitis. It is now commonly referred to as immune-related encephalopathy. It consists of demyelination that follows 1 to 2 weeks after vaccination or of a viral illness that usually does not directly infect the nervous system, such as varicella or influenza. The lesions may be widespread or limited and discrete. The optic nerves and spinal cord may become affected, producing the clinical syndromes of optic neuritis and transverse myelitis.

CT scans and MRI show the typical pattern of demyelination with low density of the white matter on CT scans and high signal intensity of the white matter on T_2-weighted MRI (Fig. 9-6). The lesions are mostly asymmetric and involve both the cerebrum and the cerebellum. The optic nerves and spinal cord may be involved. The gray matter is infrequently affected. Hemorrhage and mass effect usually do not occur. Contrast enhancement is variable and is usually slight when present. The lesions can regress if the patient survives.

CENTRAL PONTINE MYELINOLYSIS

Central pontine myelinolysis is a relatively rare form of demyelination that occurs primarily within the pons but may extend upward into the mesencephalon, thalamus, basal ganglia, and internal capsules. Although first described in alcoholics, it also occurs with such other metabolic disorders as hyponatremia or poorly controlled diabetes. The course of the disease is variable and may be mild or severe. The demyelination can best be seen with MRI as diffuse symmetric high T_2 signal intensity within the brain stem and mesencephalon (Fig. 9-7). Peripheral sparing within the pons is an almost constant feature. MRI findings do not always correlate with the clinical severity. Extrapontine myelinolysis may be the only manifestation of the disease in about 10 percent of cases. Contrast enhancement is unusual but may occur around the periphery of the lesions within the pons. The process must be differentiated from infarction, ADEM, MS, lymphoma glioma, and bulbar encephalitis.

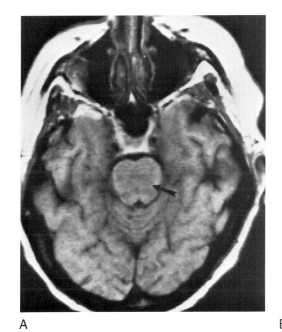

A

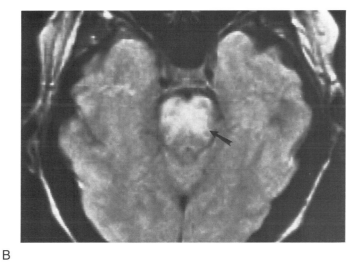

B

Fig. 9-7 Central pontine myelinolysis, MRI. (A) T_1-weighted MRI through the upper pons shows slight enlargement of the pons and central hypointensity *(arrow)*. **(B)** Lightly T_2-weighted MRI shows high signal intensity through most of the pons with minimal peripheral sparing *(arrow)*.

MARCHIAFAVA-BIGNAMI SYNDROME

Marchiafava-Bignami syndrome is a rare disorder that occurs in patients with severe malnutrition, often alcoholics. Middle-aged men are most often affected. The course is variable, but recovery is rare. The neurologic disturbance suggests bilateral frontal lobe disease. The lesion is demyelination of the corpus callosum, beginning anteriorly and extending contiguously posteriorly. The lesion can be seen as a high intensity signal in the corpus callosum with T_2-weighted MRI and sometimes as low density with CT scan.

MOTOR NEURON DISEASE

Amyotrophic Lateral Sclerosis

Amyotrophic lateral sclerosis (ALS) is the most common of the motor neuron diseases. It causes loss of white matter tracts. ALS is a specific system disease affecting only voluntary motor control. There is variable progression of bilateral symmetric degeneration of corticospinal tracts and the α-motor neurons within the anterior horns of the spinal gray matter. The etiology is unknown, although the disease is endemic in the southwestern Pacific islands. Demyelination and gliosis occur within the tracts without an inflammatory reaction. The disease usually begins after age 30, with gait disorder, muscle atrophy, fasciculations of muscle, dysarthria, and dysphagia. ALS must be distinguished from MS, cervical spondylosis, syringomyelia, Chiari I malformation, and other compressive lesions at the foramen magnum.

Different patterns of motor neuron disease exist. ALS refers to both upper and lower motor neuron dysfunction. Primary lateral sclerosis refers to upper motor neuron dysfunction. Progressive muscular atrophy refers to lower motor neuron dysfunction. Overlap of both syndromes occurs.

The primary goal of imaging patients with motor neuron disease is to detect the other disease in the differential diagnosis given above. In some patients, MRI has shown symmetric high T_2 signal in the corticospinal tracts of the brain and spinal cord and in the subcortical region of the motor cortex. A slight decrease of the T_2 signal in the cortical gray matter has also been noted, possibly representing the paramagnetic affect of increased iron or free radicals associated with the disease. Atrophy of the anterior and lateral spinal cord represents the loss of corticospinal tracts and anterior horn cells and can be detected on transaxial images of the spine. The posterior tracts are preserved. Gd enhancement is not observed. These MRI patterns strongly suggest the diagnosis of motor system disease.

SUGGESTED READING

Braffman BH, Zimmerman RA, Trojanowski JQ et al: Brain MR: pathologic correlation with gross and histopathology. 2. Hyperintense white matter foci in the elderly. AJNR 9:629, 1988

Friedman DP, Tartaglino LM: Amyotrophic lateral sclerosis: hyperintensity of the corticospinal tracts on MR images of the spinal cord. AJR 160:604, 1993

Grossman RI, Braffman BH, Brorson JR et al: Multiple sclerosis: serial study of gadolinium-enhanced MR imaging. Radiology 169:117, 1988

Maravilla K: Multiple sclerosis. p. 344. In Stark DD, Bradley WG Jr (eds): Magnetic Resonance Imaging. CV Mosby, St. Louis, 1988

Rowley HA, Dillon WP: Iatrogenic white matter disease. Neuroimag Clin North Am 3:379, 1993

Simon JH: Neuroimaging of multiple sclerosis. Neuroimag Clin North Am 3:229, 1993

Smith RR: Central pontine myelinolysis. Neuroimag Clin North Am 3:319, 1993

Uhlenbrock D, Seidel D, Gehlen W et al: MR imaging in multiple sclerosis: comparison with clinical, CSF and visual evoked potential findings. AJNR 9:59, 1988

10
Atrophy, Aging, Dementia, and Degenerations

Dementia is the deterioration of intellectual capacity. It is a clinical entity defined by behavior. Memory loss is a major feature. In addition, it is characterized by an inability to learn new material and to calculate, reason, and speak. Emotional disturbance, particularly depression, is common. Any pathologic process affecting the cerebral hemispheres may result in dementia. Although Alzheimer's disease is the most common specific cause of dementia, a small percentage of patients have other, correctable causes of dementia. The major reason for neuroradiologic examination of persons with dementia is to define the degree of atrophy and to identify any correctable pathologic process. This chapter discusses the causes of dementia.

Physiologic changes naturally occur with aging. These changes must be taken into account when evaluating the brain of patients with dementia and other degenerative syndromes.

CAUSES OF DEMENTIA IMPORTANT FOR IMAGING

Alzheimer's disease

Angiopathic dementia

(Continued)

Subcortical atherosclerotic encephalopathy (of Binswanger)

Multi-infarction dementia

Amyloid (congophilic) angiopathy

Low-pressure hydrocephalus

Creutzfeldt-Jakob disease

Pick's disease

Huntington's chorea

Wilson's disease

Parkinson's disease
Typical/atypical

Brain trauma
Diffuse axonal injury

Mass lesions
Tumor
Abscess
Large infarction
Chronic subdural hematoma

Pugilistic encephalopathy

AIDS

Toxic/metabolic causes
Alcohol
Lead

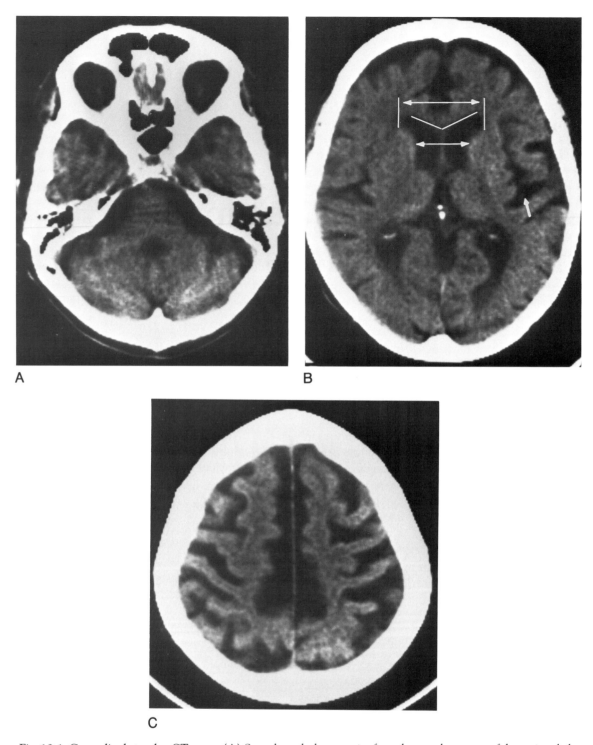

Fig. 10-1 **Generalized atrophy, CT scans. (A)** Scan through the posterior fossa shows enlargement of the pericerebellar and peripontine cisterns. **(B)** Note enlargement of the lateral ventricles, third ventricle, sylvian fissure *(arrow)*, and subarachnoid spaces over the frontal lobes. The long opposed arrows represent the bifrontal dimension. The short opposed arrows indicate the bicaudate dimension. The white lines outline an obtuse frontal horn (callosal) angle. **(C)** Gross widening of both the interhemispheric fissure and the cerebral sulci is seen.

ATROPHY AND AGING

Atrophy refers to the decrease in brain volume that occurs with the loss of brain tissue. This is an irreversible process, except perhaps in young infants, in whom there appears to be some regenerative capability. With CT scanning and MRI, atrophy is evidenced by enlargement of the CSF spaces within and surrounding the brain.

Atrophy is categorized as deep when it involves the white matter and basal ganglia and as cortical when it involves the gyral gray matter. Most commonly, both types are found together, and the atrophy is then said to be generalized or mixed (Fig. 10-1).

With deep atrophy, the frontal horns, bodies of the lateral ventricles, and the third ventricle are enlarged. The temporal horns enlarge relatively little, a finding that helps differentiate atrophy from hydrocephalus. A useful guide when determining the "normal" size of the lateral ventricles is to calculate the bicaudate or bifrontal indices. The *bicaudate index* (Fig. 10-1B) is the width of the ventricles at the most medial margins of the caudate nuclei, expressed as a percentage of the internal diameter of the skull at that level. The mean adult index is 0.15. The bifrontal index is the maximum width of both frontal horns expressed as the percentage of the internal diameter of the skull at the level of measurement. The mean bifrontal index is 0.30. The mean width of the third ventricle is 3.5 mm; greater values indicate ventricular enlargement. With deep atrophy, the callosal angle of the frontal horns is obtuse (more than 110 degrees). There is no periventricular edema with atrophy.

With cortical atrophy, the cerebral sulci enlarge as the gyri shrink. Fortunately, simple visual inspection and intuitive grading of the degree of atrophy into four categories—none, mild, moderate, severe—are, for practical purposes, adequate for defining the presence of atrophy.

A gradually accelerating loss of brain substance occurs with aging, so that the average adult brain diminishes from about 1,400 g at age 25, to 1,375 g by age 45, and to 1,200 g by age 80. Brain atrophy is considered an inevitable part of the aging process. Normally, little or no cortical atrophy is present up to age 50. From age 50 to 75, there is mild atrophy, and over age 75, moderate atrophy is present. Severe atrophy or premature atrophy suggests a degenerative process, possibly associated with dementia. There is about a 75 percent correlation of the degree of atrophy with the degree of mental impairment.

Additionally, a gradual development of paraventricular leukomalacia (softening of the white matter) is positively correlated with the blood pressure level and the degree of hyalinization of the small deep arterioles. This change has a slight positive correlation with intellectual function and gait disturbance. These changes are seen as hypodensity on CT scans in about 30 percent of those over 75 years and as high T_2 signal intensity foci with MRI in 90 percent of this population. The changes are not seen in normal persons under 45 years of age. They represent small regions of demyelination, necrosis, and gliosis, most likely from microinfarctions. When the changes are mild, they are considered part of the aging process. When more severe, they may be associated with Alzheimer's disease or subcortical arteriosclerotic encephalopathy (SAE), (Binswanger's disease). The line between normal aging and pathologic change is indistinct.

BRAIN CHANGES WITH AGING

Gradual cerebral volume loss (atrophy)

Multifocal high T_2 signal intensity lesions in white matter

Increased brain iron in the globus pallidus, substantia nigra, red nucleus, and dentate nucleus

---

CAUSES OF DIFFUSE CEREBRAL ATROPHY

Degenerative disease of the brain
  Alzheimer's disease
  Binswanger's disease
  Creutzfeldt-Jakob disease
  Huntington's disease
  Pick's disease
  Parkinson's disease
  Multiple sclerosis
  Leukodystrophies
Trauma
  Diffuse axonal injury
  Contusion

(Continued)

Vascular disease
    Migraine
    Vasculitis
    Anoxia/global ischemia
    Postmeningitis
Toxic causes
    Alcohol
    Marijuana and other substances
    Cancer chemotherapy
    Radiation therapy to brain
    Chronic illness

CAUSES OF FOCAL ATROPHY

Infarction
Contusion
Hemorrhage
Surgical removal of tissue
Arteriovenous malformation
Abscess

Focal atrophy results from a focal destructive process. It is identified by local enlargement of the CSF spaces surrounding the abnormality. Therefore, with CT scan or MRI, the lesion is indicated by focal enlargement of the cerebral sulci or adjacent ventricular cavity. Density changes may be seen within the brain on CT scan and MRI, representing encephalomalacia or gliosis. This finding is most commonly seen following head trauma or infarction. Rarely, dystrophic calcification occurs within focal atrophic regions, particularly around arteriovenous malformations (AVMs). Focal cerebellar atrophy may result from prolonged phenytoin (Dilantin) therapy. The causes of focal atrophy are given below.

BRAIN IRON

At birth the brain contains no ferric iron. At about 6 months of age, beginning traces of iron deposits can be found within the globus pallidus. Soon thereafter, iron also can be found in the zona reticulata (lateral portion) of the substantia nigra within the mesencephalon, and later in the red nucleus and the dentate nucleus of the cerebellum (Fig. 10-2). Ferric iron is paramagnetic and

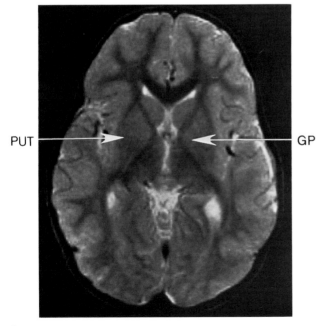

A

Fig. 10-2 Brain iron, MRI. (A) Normal 2-year-old brain. The globus pallidus (GP) is isointense with the putamen (PUT). The globus pallidus (medial) and putamen together constitute the lenticular nuclei (lentiform nucleus). (*Figure continues.*)

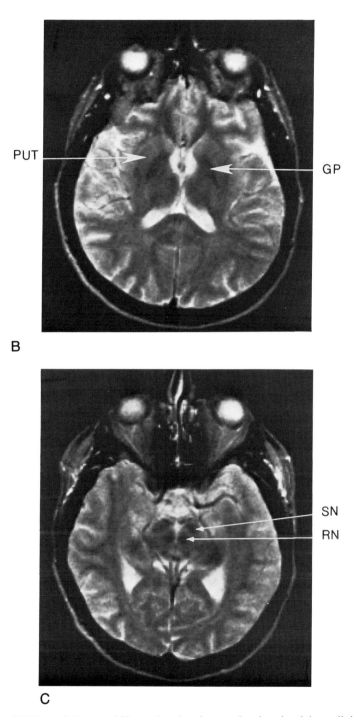

Fig. 10-2 *(Continued).* **(B)** Normal 40-year-old brain. Iron has deposited within the globus pallidus, which is now hypointense to the putamen on this T_2-weighted image. **(C)** Scan through the upper mesencephalon shows the hypointense T_2 signal from iron deposition in the red nucleus (RN) and the substantia nigra (SN). These regions become more hypointense with aging.

on MRI produces reduced signal intensity on the T_2-weighted spin echo (SE) and gradient echo (GRE) images. Iron deposition is less obvious on fast spin echo (FSE) T_2-weighted imaging. It cannot be imaged with CT scanning.

The amount of iron deposition increases with age, so the loss of signal intensity becomes prominent after age 60. Brain iron may deposit in unusual regions with some degenerative diseases and may be of some diagnostic use (see Multisystem Atrophy, below).

ALZHEIMER'S DISEASE

Alzheimer's disease is a devastating, progressive, terminal disease resulting from a degenerative process in the brain. It occurs in about 5 percent of the population over age 65. The onset is insidious, with memory loss of recent events the most common early manifestation. After a few years, this stage is followed by disturbance of other cognitive functions, such as judgment, comprehension, and abstract reasoning. Anxiety and depression are prominent features. Because of cortical involvement in the degenerative process, aphasia, apraxia (inability to perform tasks), and agnosia become evident. Focal or generalized seizures occur late in about 15 percent of persons affected. Myoclonus and extrapyramidal signs may occur in severe cases. Death usually occurs 6 to 10 years after onset. The diagnosis can be made with certainty only at autopsy.

Pathologically, the disease is characterized by generalized brain atrophy, with frontal and temporal lobe predominance. Atrophy is particularly striking in the region of the hippocampus, amygdala, and the inferior and middle temporal gyri. Microscopically, there are specific changes consisting of neurofibrillary tangles, "senile" plaques, and decreased numbers of neurons. In addition, there is a high incidence of paraventricular leukomalacia in the brain of the patient with the clinical diagnosis of Alzheimer dementia. There is no essential pathologic difference between presenile dementia (classic Alzheimer's disease), occurring before age 60, and senile dementia of the Alzheimer type (SDAT), occurring after age 60. The cause is unknown.

With CT scanning and MRI, atrophy is the major finding. It is both deep and superficial. Scans show generalized enlargement of the cerebral sulci, which may be more prominent in the parasylvian region resulting in enlargement of the sylvian fissure. The lateral and third ventricles dilate owing to deep atrophy, which may be pronounced. The degree of ventricular enlargement generally correlates with the deterioration of mental impairment. The moderate enlargement of the temporal horns, the choroidal–hippocampal fissure, and suprasellar cistern reflects the atrophy in the hippocampus and basal ganglia. Scans taken 2 years apart show progression of atrophy in those with Alzheimer's disease but little appreciable change in normal persons. The clinical diagnosis of Alzheimer's disease must be questioned if the ventricles are not enlarged, and especially if the temporal horns are normal. The major difficulty in differential diagnosis is Alzheimer's disease and normal-pressure hydrocephalus, which may also cause ventricular enlargement. The two processes may also coexist.

Paraventricular abnormalities of the white matter are demonstrated with both CT scans and MRI. On CT scan, these appear as low-density regions, predominantly adjacent to the frontal horns, but also extending more posteriorly with severe involvement. These lesions do not enhance with contrast. MRI shows the lesions best as high signal intensity with T_2-weighted images. CT scans demonstrate these lesions in about 30 percent of cases of SDAT, whereas MRI demonstrates their presence in about 90 percent of cases.

Positron emission tomography (PET) has shown changes in the brains of demented persons that appear specific for Alzheimer's disease. Using ^{18}F-2-fluoro-2-deoxy-d-glucose (^{18}FDG), PET demonstrates the rate of metabolism in the parietal cortex to be exceedingly low. Also, those who have specific cognitive defects (e.g., aphasia, apraxia, or spatial difficulties) have a corresponding regional decrease in measured cortical metabolism. These changes do not occur with other dementing diseases, such as Parkinson's or Huntington's dementia.

NORMAL-PRESSURE HYDROCEPHALUS

Normal-pressure hydrocephalus syndrome, also known as (NPH), low-pressure hydrocephalus and Hakim-Adams syndrome, consists of the triad of gait apraxia, dementia, and urinary incontinence associated with ventriculomegaly. The ventricular enlargement is from a communicating obstructive hydrocephalus that has minimal or intermittent increase in intraventricular pressure.

The gait disturbance manifests as unsteadiness, ataxia, shuffling, and rigidity, sometimes described as "magnetic." It precedes the onset of dementia and is initially more disabling. Typically, the dementia consists of memory impairment, with little disturbance in language function. Its onset is insidious over many years. Urinary incontinence occurs late in the course of the illness.

The disease must be considered in the differential diagnosis of dementia. It can occur without any known antecedent event, or it may result from meningitis, subarachnoid hemorrhage (SAH), meningeal carcinomatosis, or, rarely, head trauma.

The CT or MRI scans show the usual findings of hydrocephalus (see Ch. 11 for a detailed discussion). However, many of the persons afflicted also show atrophic changes, making interpretation of the scan difficult. In general, hydrocephalus is a possibility when there is considerable enlargement of the temporal horns, narrowing

of the callosal angle of the ventricles (less than 100 degrees), periventricular edema, and relatively small cerebral sulci for the person's age. When the bifrontal span is more than 50 mm, hydrocephalus is almost always present. Extensive paraventricular high T_2 signal abnormality on MRI also suggests hydrocephalus, rather than simple atrophy. The hippocampus is preserved with NPH.

Radionuclide cisternography may help with the diagnosis. It is performed by injecting indium-111-diethyl-enetriamine pentaacetic acid (DTPA) (or an equivalent tracer) into the lumbar subarachnoid space and imaging the migration of the tracer substance through the intracranial subarachnoid space and ventricles. Iodinated intrathecal contrast and CT scanning may also be used. Normally, the tracer reaches the basal cisterns within 1 hour; over the next 24 hours, it migrates over the cerebral convexities, concentrating in the region of the sagittal sinus within the interhemispheric fissure. The tracer

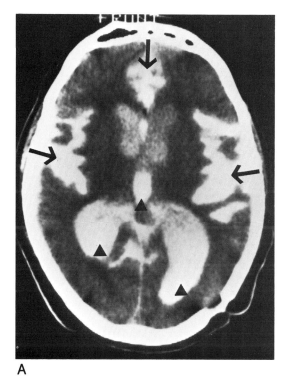
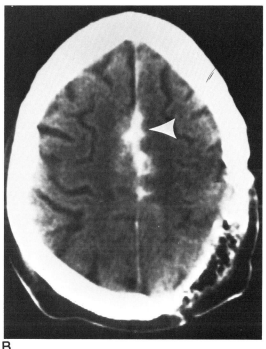

A

B

Fig. 10-3 Normal-pressure hydrocephalus, CT cisternography. Aqueous contrast is injected into the lumbar subarachnoid space. Serial CT scans are then obtained to follow the intracranial distribution of the contrast. **(A)** View through the ventricular system at 24 hours shows dense contrast within the sylvian and interhemispheric fissures *(arrows)*. Dense contrast is also seen within the visualized ventricular system *(arrowheads)*. **(B)** Scan through the vertex shows only a small amount of contrast collecting in the region of the sagittal sinus *(arrowhead)*. There is no contrast at other regions of the interhemispheric fissure or within the cerebral sulci.

does not normally enter the ventricular system. In persons with large ventricles due to atrophy, some tracer may reflux into the ventricular system, but the amount is generally small and clears completely by 48 hours.

In patients with NPH or any type of communicating hydrocephalus, the tracer shows continuous reflux into the ventricular system, where it remains for at least 72 hours (Fig. 10-3). In addition, the tracer does not reach the region of the sagittal sinus but is held up at some point, usually within the basal cisterns, at the incisura, or at the sylvian fissures. This pattern, thought to represent the "block" to the flow of CSF through the subarachnoid space, is diagnostic of communicating hydrocephalus. A "mixed" pattern may be seen that shows delay of the migration of tracer over the convexities (eventually concentrating in the parasagittal region) and transient reflux into the ventricular system that mostly clears by 48 hours. The positive predictive value of the mixed pattern for the diagnosis of hydrocephalus is low.

Treatment of NPH consists of shunting of CSF, usually by intraventricular catheter. About 50 percent of those shunted show improved gait and mental functioning. Selection of the appropriate persons for shunting remains inaccurate. There are no absolute criteria. In general, the closer the patient's symptoms are to the typical syndrome, including the order of onset of the symptoms, and the more firm the findings of hydrocephalus on the CT, MRI, and radionuclide scans, the more likely there is to be a favorable result from the shunting.

ANGIOPATHIC DEMENTIA

Multiple Infarction Dementia

Whenever 100 cc or more of brain volume has been destroyed by infarction, there is almost always some deficit in cognitive function. Smaller regions of infarction can produce cognitive dysfunction if they occur in the hippocampal gyri, the thalamic nuclei, or other regions of the limbic system. The mental decline may occur in a stepwise fashion. MRI and CT scans show the multifocal ischemic infarctions. However, it is often difficult to determine whether the infarcts seen on CT and MRI are sufficient cause for the mental impairment, and the diagnosis of multiple infarction dementia is most often a judgmental decision. Carotid or vertebral artery stenosis alone is not thought to produce dementia, even when the decrease in blood flow is significant.

Subcortical Arteriosclerotic Encephalopathy of Binswanger (SAE)

Although somewhat controversial, SAE is now thought to be a discrete disease entity. Most persons with the disease have had long-standing hypertension and a history of infarctions, particularly of the lacunar type. Most neuropathologists consider Binswanger's disease a form of multiple infarction dementia that results from diffuse small vessel arteriosclerotic disease found deep within the brain. The deep ischemic lesions are thought to be the cause for the dementia, gait apraxia, motor deficits, and urinary incontinence seen in affected patients. However, the disease is not associated with language disturbance, which is a distinguishing feature from Alzheimer's disease. SAE may be the cause of up to 55 percent of those with dementia over the age of 70.

The CT scan shows nonenhancing multifocal low density lesions within the paraventricular white matter (Fig. 10-4). The lesion extends more peripherally within the centrum semiovale when the involvement is severe. Ischemic infarctions, mostly lacunar, are usually present, and there is ventricular enlargement due to diffuse deep cerebral atrophy. Cortical atrophy is less severe. The temporal horns are usually near-normal in size, helping distinguish these patients from those with the clinically similar NPH syndrome.

MRI is the most sensitive modality for the detection of SAE. It shows the bilateral symmetric multifocal lesions as hypointense on T_1- and hyperintense on T_2-weighted images (Figs. 10-5 and 10-6). Confluence of the lesions into a diffuse region of high T_2 signal intensity usually occurs adjacent to the ventricles and when severe also involves the external and internal capsules. Smaller focal round lesions are generally seen in the more peripheral white matter and may reach the subcortical zones. In any given patient, the lesions appear more severe and widespread on MRI than on CT scans. SAE can probably be considered an advanced form of the lacunar state. The corpus callosum is spared.

Amyloid (Congophilic) Angiopathy

Amyloid (congophilic) angiopathy is a relatively rare form of angiopathy characterized by amyloid deposition in the walls of small intracranial vessels. It usually occurs only within the brain, although some cases show evidence of amyloid deposits in the vessels of other organs, particularly the gastrointestinal tract. Most often pa-

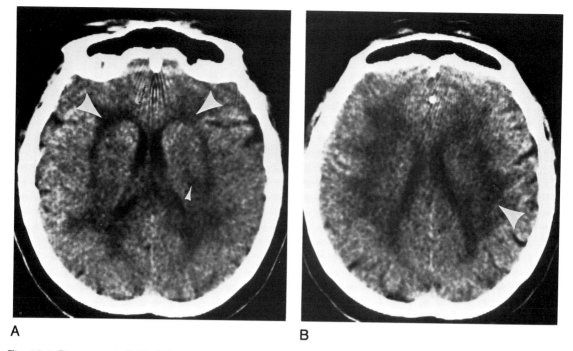

Fig. 10-4 Binswanger's SAE, CT scans. (A) Scan through the ventricular system shows hypodensity in the paraventricular white matter *(large arrowheads)*. A lacunar infarction is seen in the left posterior internal capsule *(small arrowhead)*. The ventricular system is dilated but less than average for a patient with this disease. **(B)** White matter changes are severe, extending into the subcortical regions *(arrowhead)*. The cerebral sulci are not dilated.

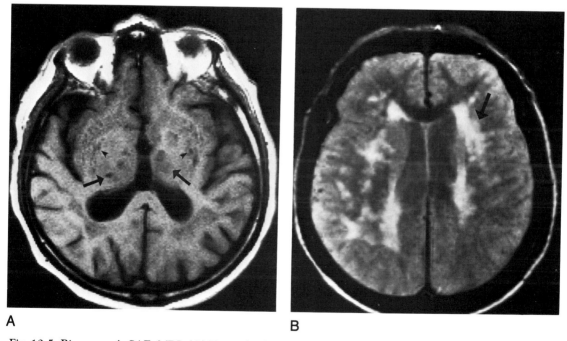

Fig. 10-5 Binswanger's SAE, MRI. (A) T$_1$-weighted MrI at the level of the third ventricle shows multiple lacunar-type infarctions in the thalamic nuclei bilaterally *(arrows)*. The Virchow-Robin spaces are enlarged *(arrowheads)*. **(B)** Proton density image through the ventricular system shows multifocal and confluent high T$_2$ signal intensity in the paraventricular white matter *(arrow)*. The ventricles are moderately enlarged.

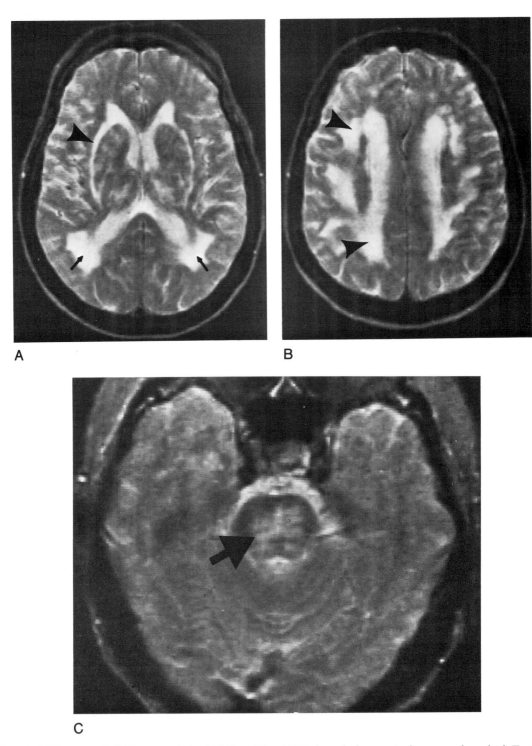

Fig. 10-6 **Binswanger's SAE, severe, MRI. (A)** T_2-weighted MRI through the ventricular system shows high T_2 signal in the white matter adjacent to the atria *(arrows)* and in the external capsule *(arrowhead)*. **(B)** Scan at a higher level shows an extreme confluent high T_2 signal white matter abnormality *(arrowheads)*. **(C)** Scan through the pons shows symmetric high T_2 signals throughout the white mater of the brain stem *(arrow)*. Brain stem involvement is present only in the most severe cases.

tients present with cortical or subcortical hemispheric hemorrhages that tend to be recurrent. The hemorrhages may be indistinguishable from the subcortical hemorrhages that occur secondary to hypertension and atherosclerosis. The diagnosis is made by biopsy of the meninges at surgical removal of the cerebral blood clot. Some patients present with a slowly progressing dementia that is clinically indistinguishable from SDAT. Amyloid angiopathy may be suggested if the CT or MR scans show evidence of old hemorrhage or infarction in the posterior regions of the brain.

OTHER CAUSES OF DEMENTIA

Creutzfeldt-Jakob Disease

Creutzfeldt-Jakob disease is a rare, severe, rapidly progressive, infectious degenerative disease of the brain resulting in presenile dementia. It is believed to be caused by a "slow virus" infection that is transmitted by contact with the diseased tissue, including organ translants of all types. The cerebral cortex is the region most involved, but there may be atrophy of the basal ganglia, thalamus, and cerebellum as well.

Onset generally consists of focal neurologic deficits, particularly visual, followed shortly thereafter by dementia, myoclonus, and gait disturbance. The dementia is severe, coming on much more rapidly than the dementia of Alzheimer's disease. Changes may even be observable on a day-to-day basis. Death usually occurs within 1 year from the onset of symptoms. The electroencephalogram shows characteristic periodic spike–wave complexes, often associated with viral infections of the brain. The usual onset is between the ages of 40 and 60, but the disease can be seen at any age, even rarely in teenagers.

The CT scans and MRI show nonspecific changes of cerebral atrophy that are rapidly progressive as compared with other dementing illnesses. There is demonstrable change from month to month. Sometimes there is evidence of demyelination and high T_2 signal in the basal ganglia and white matter.

Pick's Disease

Pick's disease is a relatively rare cause of dementia. Neuropathologically, there is degeneration of neurons with cytoplasmic swelling, called Pick's bodies. The dementia tends to involve language and behavior more

than memory. The atrophy is usually more focal than with Alzheimer's disease and may be limited to the anterior temporal and frontal lobes. The basal ganglia may be involved as well.

CT scans and MRI reflect these gross anatomic changes (Fig. 10-7). There is severe widening of the frontal horns of the lateral ventricles, third ventricle and temporal horns, sylvian fissures, and frontal and temporal gyri. The remainder of the brain exhibits much less atrophic change.

Huntington's Chorea

Huntington's chorea is a rare hereditary dementing disease that usually manifests during middle adulthood. Generalized atrophy is present, but the characteristic anatomic finding is shrinking of the caudate nuclei, which produces widening of the intercaudate region of the frontal horns of the lateral ventricles. The lateral walls of the frontal horns may become convex outward at the level of the caudate head. When the ratio of the maximum width of the frontal horns to the intercaudate dis-

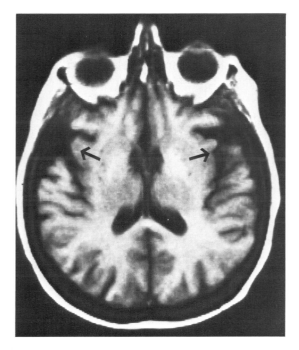

Fig. 10-7 Pick's disease, MRI. T_1-weighted MRI shows severe enlargement of the sylvian fissures (*arrows*). The posterior sulci show only slight enlargement.

tance is less than 1.6, the diagnosis of Huntington's disease can be made.

However, there is relatively poor anatomic correlation with the clinical symptomatology, and patients with severe chorea may have a nearly normal-appearing brain. No distinctive changes are present during the preclinical phase of the disease. The clinical diagnosis is usually made by the presence of a family history for the disease, decreased levels of α-aminobutyric acid in the CSF, and abnormality in the fourth chromosome.

Wilson's Disease

Hepatolenticular degeneration, or Wilson's disease, is an autosomal recessive inherited disease resulting from an abnormally low concentration of serum ceruloplasmin. This condition allows accumulation of abnormal amounts of copper within the lentiform nuclei (putamen and globus pallidus), caudate nucleus, thalamus, brain stem, dentate nuclei, and subcortical cerebral white matter, which leads to gliosis and cavitation.

The amount of copper is usually too low for detection with CT scanning. Atrophy is the major finding, most prominent in the basal ganglia and brain stem. Sometimes focal low-density nonenhancing lesions are seen involving the regions of copper deposition listed above. There is general correlation between the degree of atrophy and the severity of the dementia. The atrophy may progress even after treatment with penicillamine, a chelating agent. The CT scan of the liver is normal in persons with Wilson's disease.

MRI shows increased T_2 signal representing gliosis in the regions of copper deposition listed above, but most commonly within the lenticular nuclei and the brain stem. The lesions are most often symmetric and show little change with therapy. There is general correlation with the severity of the clinical disease and the amount of MRI signal abnormality found. Scattered hypointensity within the lenticular nuclei represents iron deposition seen with Wilson's disease.

Parkinson's Disease

Parkinson's disease is the most common disorder of the extrapyramidal system, affecting about 1 percent of the population of the United States over the age of 50. The disease has four cardinal signs: tremor at rest, bradykinesia (general slowing of movement, masked face, decreased frequency of blinking), rigidity, and the loss of postural reflexes. It is classified into three etiologic groups: the idiopathic primary form, referred to as typical Parkinson's disease; secondary, or acquired, parkinsonism; and parkinsonism-plus syndromes (atypical parkinsonism), in which nonparkinsonian neurologic findings are present in addition to those of parkinsonism. Atypical parkinsonism does not respond to L-dopa therapy. The final underlying pathologic process of all parkinsonism is loss of dopamine activity in the striatum (caudate nucleus and putamen). Degeneration of the cells of the substantia nigra (SN) leads to a loss of dopamine in the basal ganglia.

The findings of primary Parkinson's disease are depigmentation and loss of neurons in the substantia nigra and the locus ceruleus of the brain stem, the site of production of the neurotransmitters dopamine and norepinephrine. Focal gliosis may occur. Because of the striatonigral connections, this condition leads to a secondary loss of these transmitters in the striatum. Typical Parkinson's disease is also referred to as the nigral type of parkinsonism.

The etiology is unknown. Clinical improvement can occur after replacement therapy with L-dopa or dopamine agonist drugs. About 20 percent of persons with Parkinson's disease eventually become demented.

CT scans show moderate nonspecific generalized atrophy with no recognizable specific changes in the brain stem or basal ganglia. MRI may show changes not seen with CT scanning. Commonly, there is a decrease in the width of the pars compacta of the substantia nigra to about 2 mm (normal 2.5 mm). In practice, this change is difficult to detect. On heavily T_2-weighted images, some observers have described hypointensity of the putamen and slight increased intensity of the lateral substantia nigra (reticulata), so it becomes isointense with the pars compacta, but these findings have not been consistently confirmed. The moderate nonspecific atrophy and occasional focal high T_2 signal in the brain stem is most likely due to gliosis.

The findings of secondary parkinsonism on MRI or CT examination depend on the etiology of the process. Postinfarction parkinsonism often shows lucunar-type infarctions in the basal ganglia or thalamic nuclei. Carbon monoxide poisoning, trauma, and Wilson's disease result in bilateral basal ganglionic atrophy. Hypoparathyroidism may cause parkinsonism associated with calcium deposits in the basal ganglia, although idiopathic basal ganglionic calcification does not cause symptoms. Rarely, deep tumors cause parkinsonism.

Parkinson-plus refers to parkinsonism that occurs along with other degenerative diseases of the CNS. It is considered below.

Multisystem Atrophy

Multisystem atrophy syndrome, or atypical parkinsonism, refers to a group of similar idiopathic brain degenerations that have varied anatomic distributions and clinical manifestations. These disorders are uncommon. The specific types of multisystem atrophy are listed in the box. MRI shows increased iron deposition within the putamen associated with the atypical parkinsonism syndromes, which is demonstrated as low signal intensity on heavily T_2-weighted SE and GRE images. This is thought to represent neuronal degeneration within the putamen with loss of dopamine receptors along with the cells. This is referred to the striatal type of parkinonism. Atrophy in the distribution of the degeneration is also seen.

Hallervorden-Spatz Disease

Hallervorden-Spatz disease is a degenerative disorder that usually presents in the second decade of life, with motor disorders of rigidity, dystonia, pyramidal tract signs, and dementia. The globus pallidus and substantial

nigra show degeneration with accumulation of large amounts or iron. MRI shows the deposition as hypodensity within the involved regions on T_2 images.

Idiopathic Ferrocalcinosis

Idiopathic ferrocalcinosis is a degenerative disorder frequently referred to as Fahr's disease or striopallidodentate calcinosis. It is often familial. Pathologically, there is a deposit of a mixture of calcium, iron and additional metals within the basal ganglia, the dentate nuclei, the hippocampus, and subcortical regions. The mineralization is primarily within blood vessels and causes a noninflammatory neuronal necrosis. When the disorder is severe, the deposits are extraordinarily heavy. Patients suffer progressive psychosis, dementia, parkinsonism, and seizures. The deposits are heavy and white on CT (Fig. 10-8). MRI shows the deposits most often as high

---

MRI FINDINGS WITH ATYPICAL
PARKINSONISM AND
THE MULTISYSTEM
ATROPHY SYNDROMES

Olivopontocerebellar degeneration
    Atrophy of the medulla, pons, and cerebellum
    T_2 Hypointensity of the putamen
Striatonigral degeneration
    T_2 Hypointensity of the putamen
    Thin pars compacta of SN
Shy-Drager syndrome
    T_2 Hypointensity of the putamen
Progressive supranuclear palsy
    Atrophy of the midbrain and tectum
Hallervorden-Spatz disease
    Atrophy of the pars reticulata SN
    T_2 Hypointensity in the globus pallidus

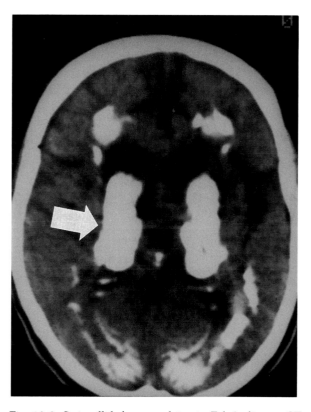

Fig. 10-8 Striopallidodentate calcinosis, Fahr's disease. CT scan shows the striking calcinosis that occurs throughout the basal ganglia (*arrow*) and hemispheric white matter. Severe atrophy and dementia also occur.

signal intensity on T_1 and hypointense on T_2-weighted imaging.

Basal Ganglia Calcification

Mild calcification often occurs within the globus pallidus, increasing with age. It is generally considered a normal physiologic finding, as it is not associated with clinical symptoms. Larger amounts of calcium deposition results from disorders of calcium metabolism, most commonly hypoparathyroidism, but also hyperparathyroidism, some congenital disorders, and prior infections. When severe, extrapyramidal movement disorders may occur.

Alcoholism

The effects of chronic excessive alcohol intake on the CNS are complex and serious. Nutritional deficiencies or trauma may produce direct toxicity or indirect effects.

CAUSES OF BASAL GANGLIA CALCIFICATION

Mild idiopathic

Idiopathic ferrocalcinosis (Fahr's disease)

Hyperparathyroidism

Hypoparathyroidism

Infection
  Tuberculosis
  Congenital toxoplasmosis
  Congenital cytomegalovirus
  Cisticercosis
  Pediatric AIDS

Tuberous sclerosis

Cockayne's disease

Mitochondrial myopathy, encephalopathy, lactic acidosis, and strokes/myoclonic epilepsy with ragged red fibers, mitochondrial disorders

Chemotherapy/radiotherapy

Post-severe acute hypoxia

Fabry's disease

Alcohol withdrawal may produce various CNS symptoms, from mild tremors to life-threatening delerium tremens. Withdrawal seizures are relatively common. These usually consist of one to six convulsions occurring within a 6-hour period and consisting of tonic-clonic movements. When there is no associated focal deficit or evidence of head trauma, CT or MR scanning is unnecessary.

Wernicke's encephalopathy is a relatively common disorder, caused by associated thiamine deficiency. It is probably more common than is usually recognized. Oculomotor abnormalities (e.g., nystagmus, palsy, gaze paresis), gait ataxia, and encephalopathy (indifference, disorientation, agitation, depressed consciousness or coma) are present to varying degrees. Treatment consists of prompt parenteral administration of thiamine. The lesions in the brain are characteristic, consisting of demyelination about the third and fourth ventricles and the intervening aqueduct. This may be imaged as hypodensity on CT and as hyperintensity on T_2-weighted MRI. In addition, the mamillary bodies are small, which can be diagnosed on the parasagittal T_1-weighted MR study.

True dementia may occur in alcoholics and is usually associated with one of two syndromes: (1) Marchiafava-Bignami disease, a rare disorder consisting of necrosis of the corpus callosum and adjacent white matter; or (2) Korsakoff syndrome, associated with necrosis in the dorsal medial thalamus. These syndromes may sometimes be diagnosed by finding the necrotic lesions in their characteristic locations on either CT or MRI.

Central pontine myelinolysis is a rare demyelinating disorder of the pons most commonly seen in alcoholics. It may be quite extensive, involving the thalami and corpera striatum as well. It has a variable association with clinical symptoms. It is best diagnosed with MRI as high T_2 signal abnormality in the characteristic distribution. The disease is most often found in those with severe hyponatremia sometimes occurring only after rapid over correction of the electrolyte abnormality (see Ch. 9).

Cerebral volume loss is common in alcoholics, but it is controversial as to whether true atrophy is present. Most alcoholics have enlarged ventricles and sulci, but this will usually revert to normal about 1 month after the cessation of drinking. The degree of cerebral volume loss does not correlate with either the duration of alcohol abuse or any associated mental impairment. Cerebellar loss is common, particularly of the anterior superior vermis, but it does not correlate with the degree of ataxia. Post-traumatic intracranial hemorrhage, particularly acute or chronic subdural hematoma, occur from trauma.

Cerebellar Degeneration

The cerebellum partakes in the atrophy of aging, although the atrophy seen with imaging is usually minor. Congenital degeneration or atrophy is associated with Friedreich's ataxia (atrophy of the cerebellum and the cervical spinal cord), ataxia-telangiectasia, the fragile-X syndrome, and various isolated partial dygenesis syndromes. Acquired cerebellar degeneration and atrophy occurs from long-term Dilantin therapy, chronic alcoholism, and non-CNS neoplasms (especially breast, ovary, the female genital tract, Hodgkin's disease, and small cell carcinomas).

PARANEOPLASTIC DEGENERATIVE SYNDROMES

Cerebellar degeneration

Encephalomyelitis
  Subacute spinocerebellar degeneration
  Limbic encephalitis
  Bulbar encephalitis
  Myelitis

Necrotizing myelopathy

Sensorimotor polyneuropathies
  Subacute progressive sensory neuronopathy

Myasthenia gravis
Retinitis
Eaton-Lambert syndrome
Opsoclonus-ataxia-myoclonus
Polymyopathy (polymyositis)

PARANEOPLASTIC SYNDROMES

Paraneoplastic disease refers to neural abnormality associated with tumors outside of the nervous system, usually malignancies. Most occur on an idiopathic basis, but some are related to antineuronal autoantibodies that cause degeneration or inflammation. Some are visible with imaging, such as necrotizing myelopathy, limbic encephalitis (Fig. 10-9), bulbar encephalitis (Fig. 10-10), cerebellar degeneration, and polymyopathy. As a group, these show high T_2 signal and swelling in the acute phase, and atrophy in the chronic phase. Other paraneoplastic effects target synaptic transmitters and cause cognitive or neurologic symptoms without visible change in structure. Small cell carcinoma of the lung is the most common tumor that makes a protein, triggering the manufacture of the antineural antibody by lymphocytes. Other tumors include carcinoma of the breast and prostate, carcinomas of other locations, Hodgkin's disease, and sarcoma.

Paraneoplastic syndrome may be the first indication that the patient harbors a cancer. The Eaton-Lambert syndrome, consisting of subacute progressive cerebellar dysfunction, indicates müllerian (ovary) or mammary tumors. The presence of opsoclonus-ataxia-myoclonus

in children indicates neuroblastoma. A specific polyclonal compliment fixing IgG antibody called anti-Hu has been identified with some of the conditions listed in the box.

Limbic Encephalitis

Limbic encephalitis can be identified with MRI. It is an idiopathic paraneoplastic inflammatory disorder most often associated with small cell lung carcinoma. However, it may occur without associated carcinoma. The inflammatory lymphocytic reaction produces a dementia-like clinical picture with severe progressive memory impairment. There is a lymphocytic pleocytosis in the CSF. MRI shows high T_2 signal intensity within one or both anterior medial temporal lobes, which may spread to the insula, and cingulate gyri. There is no mass effect, and the lesions do not enhance with Gd. The pattern often changes over time, sometimes with complete resolution of MRI and clinical abnormalities. The condition must be differentiated from herpes encephalitis and gliomatosis cerebri, which may have similar imaging findings. Biopsy may be necessary for accurate diagnosis.

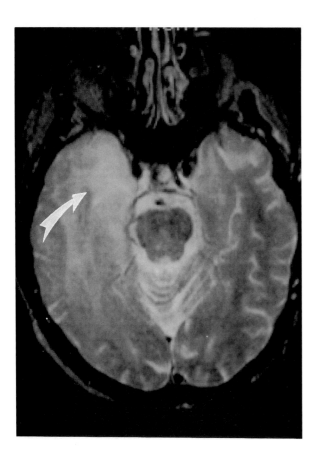

Fig. 10-9 **Limbic encephalitis.** Axial T$_2$ MRI identifies a region of high signal intensity within the medial right anterior temporal lobe *(arrow)*. This finding is nonspecific and could represent herpes encephalitis. The biopsy did not indicate herpes and given the clinical findings was most consistent with limbic encephalitis.

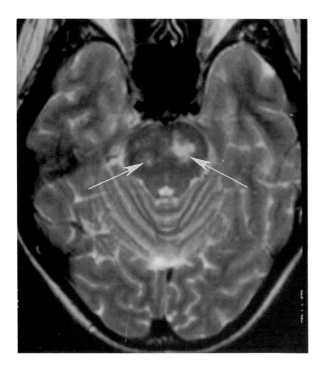

Fig. 10-10 **Bulbar encephalitis, paraneoplastic syndrome.** Axial T$_2$ MRI through the pons shows irregular multifocal regions of high signal *(arrows)* representing degeneration of brain tissue in a patient with small cell carcinoma of the lung.

SUGGESTED READING

Aisen AM, Martel W, Gabrielsen TO et al: Wilson disease of the brain: MR imaging. Radiology 157:137, 1985

Braffman BH, Grossman RI, Goldberg HI et al: MR imaging of Parkinson disease with spin echo and gradient echo sequences. AJNR 9:1093, 1988

Braffman BH, Zimmerman RA, Trojanowski JQ et al: Brain MR: pathologic correlation with gross and histopathology. 2. Hyperintense white matter foci in the elderly. AJNR 9:629, 1988

Drayer B, Burger P, Darwin R et al: Magnetic resonance imaging of brain iron. AJNR 7:373, 1986

Drayer BP: Imaging of the aging brain. I. Normal findings. Radiology 166:785, 1988

Drayer BP: Imaging of the aging brain. II. Pathologic conditions. Radiology 166:797, 1988

Dropcho EJ: The remote effects of cancer on the nervous system. Neurol Clin 7:579, 1989

Hendrie HC, Farlow MR, Austrom MG et al: Foci of increased T_2 signal intensity on brain MR scans of healthy elderly subjects. AJNR 10:703, 1989

Jack CR Jr, Mokri B, Laws ER Jr: MR findings in normal-pressure hydrocephalus: significance and comparison with other forms of dementia. J Comput Assist Tomogr 11:923, 1987

Kodama T, Numaguchi Y, Gellad FE et al: Magnetic resonance imaging of limbic encephalitis. Neuroradiology 33:520, 1991

Lemay M: CT changes in dementing diseases: a review. AJNR 7:841, 1986

Lotz PR, Ballinger WE Jr, Quisling RG: Subcortical arteriosclerotic encephalopathy: CT spectrum and pathologic correlation. AJNR 7:817, 1986

Olanow CW: Magnetic resonance imaging in Parkinsonism. Neurol Clin 10:405, 1992

Pastakis B, Polinsky R, DiChiro G et al: Multiple system atrophy (Shy-Drager syndrome): MRI imaging. Radiology 159:499, 1986

Rosenblum MK: Paraneoplasia and autoimmunologic injury of the nervous system: the anti-Hu syndrome. Brain Pathol 3:199, 1993

Rutledge JN, Hilal SK, Silver AJ: Study of movement disorders and brain iron by MR. AJNR 8:397, 1987

Simmons JT, Pastakis B, Chase TN: Magnetic resonance imaging in Huntington disease. AJNR 7:25, 1987

11
Hydrocephalus

Hydrocephalus is pathologic enlargement of the cerebral ventricles, accompanied by an increase in CSF volume. The term implies obstruction to the flow, or a decrease in absorption, of the CSF. About 70 percent of the CSF is produced by the choroid plexi within the cerebral ventricles at an average rate of 0.5 ml/min (600 to 800 ml/day), although there is considerable individual variation. The choroid is located within the atria, bodies, and temporal horns of the lateral ventricles, the roof of the third ventricle, and the roof of the fourth ventricle. There is active secretion of chloride into the CSF by choroid, so that the CSF becomes slightly hypertonic to blood plasma. Water is leached from the plasma to contribute to CSF volume.

Most of the CSF must flow through the ventricular system to reach the exit foramina of the fourth ventricle (Fig. 11-1). That which is formed within the lateral ventricle must egress through the foramina of Monro into the third ventricle, pass through the aqueduct of Sylvius into the fourth ventricle, and exit the lateral foramina of Luschka or the midline foramen of Magendie into the cisterna magna at the base of the brain. From this point, most of the CSF flows over the cerebral convexities to be absorbed through the arachnoid villi into the sagittal venous sinus. A much smaller amount is absorbed through the spinal villi at exiting root sleeves. If production of CSF is greater than absorption, there is an increase in the intracranial volume and pressure of CSF with secondary dilatation of the ventricular system.

CLINICAL SIGNS OF HYDROCEPHALUS AND INCREASED INTRACRANIAL PRESSURE

Infants
  Enlarging head and anterior fontanelle
  Distended scalp veins
Children/adults
  Headache
  Diplopia (CN VI compression)
  Vomiting
  Papilledema and visual blurring
  Gait apraxia

The usual cause of hydrocephalus is obstruction to the flow of CSF. Hydrocephalus from overproduction of CSF is rare, becoming a factor only with choroid plexus papilloma. Hydrocephalus is said to be "active" when increased intraventricular pressure causes progressive enlargement of the ventricles. It is "arrested" when compensatory mechanisms have caused the intraventricular pressure to return to normal with no further tendency for enlargement of the ventricular system. In infants and young children with open cranial sutures, hydrocephalus causes enlargement of the head. In older children and adults with closed cranial sutures,

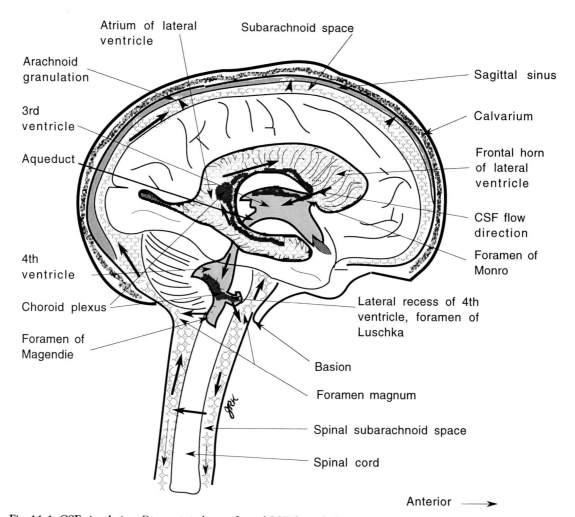

Fig. 11-1 CSF circulation. Diagnosis indicates flow of CSF through the ventricles and the subarachnoid space within the calvarium and spinal canal *(heavy lined arrows)*. Most of the CSF is produced from the choroid plexus shown as dark gray within the lateral, third, and fourth ventricles. The great majority of the CSF is resorbed into the sagittal dural venous sinus through the arachoid villi. There is small flow and resorption within the spinal subarachnoid space *(thin lined arrows)*. (Modified from Goss, 1963, with permission.)

CAUSES OF HYDROCEPHALUS

Noncommunicating type
  Infants
    Aqueduct obstruction
    "Congenital tumor," particularly of the fourth ventricle
    Hemorrhage, with adhesions within the ventricles or aqueduct
    Arachnoidal cyst
    Vein of Galen aneurysm
    Chiari II malformation
    Dandy-Walker malformation
  Children
    Posterior fossa tumor (medulloblastoma, astrocytoma, ependymoma, epidermoid
    Craniopharyngioma
    Germ cell tumors of the pineal region
    Arachnoidal cyst of the quadrigeminal cistern
    Hypothalamic or thalamic glioma
    Tuberous sclerosis
    Pontine glioma (rarely)
  Adults
    Tumors, particularly metastatic tumors, within the posterior fossa or near the aqueduct
    Cerebellar hemorrhage
    Cerebellar infarction
    Tentorial meningioma
    Large glioma
    Pituitary marcroadenoma
    Colloid cyst of the third ventricle
    Giant aneurysm, particularly of the basilar artery
    Transtentorial herniation
    Pachymeningitis of the basal cisterns
    Cysticercosis
Communicating type
  Infants/children
    Congenital "external" hydrocephalus
    Posthemorrhage
    Postmeningitis
    Meningeal tumor (leukemia, medulloblastoma)
  Adults
    Subarachnoid hemorrhage
    Meningitis, especially fungal or tuberculous
    Idiopathic low-pressure hydrocephalus syndrome
    Meningeal carcinomatosis (breast, small cell of lung)
    Very high CSF protein.

(Continued)

hydrocephalus causes increased intracranial pressure. The clinical signs and causes of hydrocephalus are given below. Obstructive hydrocephalus is further classified as noncommunicating or communicating. Ventriculomegaly from atrophy is not considered hydrocephalus.

Gradient echo (GRE) cardiac-gated MRI without flow suppression and phase-contrast imaging demonstrates the normal CSF flow through the ventricular system. With systole, flow is first seen as flow-void at the outlet of the fourth ventricle, and then the aqueduct. The flow is craniocaudal, out of the ventricular system. With diastole, the flow reverses. With obstruction at aqueduct, or outlet of the fourth ventricle, the flow-void is missing at the site of obstruction. With ventriculomagaly from any cause, there is increased flow within the ventricular system at many levels.

NONCOMMUNICATING HYDROCEPHALUS

Noncommunicating hydrocephalus results from an obstruction to the flow of CSF that occurs within the ventricular system or at the level of the outlet foramina of the fourth ventricle. The obstruction isolates the intraventricular CSF that is proximal to the obstruction. Noncommunicating hydrocephalus most commonly results from an intracerebral mass lesion or congenital narrowing of the aqueduct of Sylvius. Rarely, pachy-

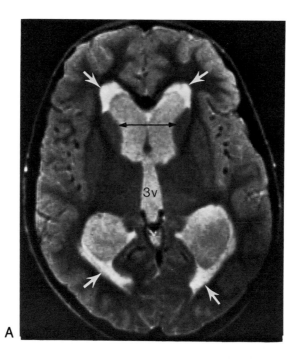

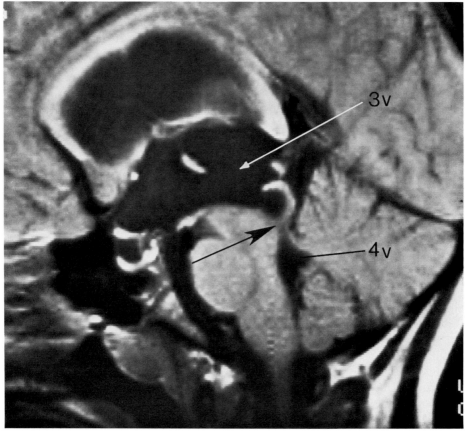

Fig. 11-2 **Hydrocephalus, aqueduct stenosis.** (A) Transaxial T$_2$-weighted MRI shows enlargement of the lateral (*opposed arrows*) and third ventricle (3V). High paraventricular signal intensity represents increased fluid content in the white matter from reversal of transependymal flow (*white arrows*). (B) Parasagittal T$_1$-weighted MRI shows enlargement of the third ventricle (3V) and a normal-size fourth ventricle (4V). The problem was caused by a short obstruction within the aqueduct (*black arrow*).

meningitis due to a chronic inflammatory process such as sarcoid obstructs the outlets of the fourth ventricle.

The ventricles enlarge proximal to the obstruction, and the cisterns and sulci are compressed by the expanded brain. There may be unilateral enlargement or enlargement of only a portion of the lateral ventricle by a mass within the ventricle. Occasionally, multiple lesions produce entrapment of portions of a ventricular cavity. Multiple shunts may be required for decompression.

Obstruction at the level of the foramina of Monro is most commonly due to a colloid cyst of the third ventricle (see Ch. 5). Other causes include a large pituitary macroadenoma, craniopharyngioma, meningioma (from the floor of the frontal fossa or rarely within the third ventricle), septal and hypothalamic gliomas, and tubers (tuberous sclerosis). The lateral ventricles enlarge, and the third and fourth ventricles are small. Bilateral intraventricular shunting may be necessary to relieve the hydrocephalus.

Obstruction at the level of the aqueduct of Sylvius or posterior third ventricle causes enlargement of both lateral ventricles and the third ventricle associated with a normal or small fourth ventricle. Congenital aqueduct stenosis is by far the most common cause during the first 15 years of life (Figs. 11-2 and 11-3). In young adults, pineal region tumors are most common (see Ch. 6). In later life, periaqueductal tumors, metastasis, and subependymoma are the most common. Other causes include compressive nearby lesions, such as arachnoid cyst of the quadrigeminal cistern, vein of Galen aneurysm, giant

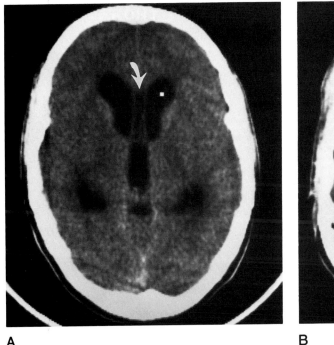

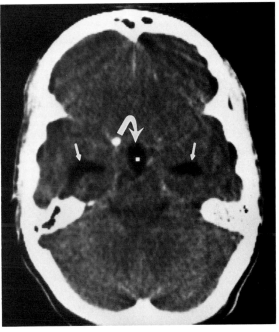

A
B

Fig. 11-3 Hydrocephalus, aqueduct stenosis. (A) Transaxial CT scan shows enlargement of the lateral and third ventricles and a cavum septum pellucidum *(arrow)*. There is only minimal paraventricular edema, as reflected by "smudginess" of the borders of the ventricles. **(B)** Transaxial CT scan at a lower level shows a dilated anterior third ventricle *(curved arrow)* and temporal horns *(small arrows)*. The fourth ventricle is not dilated and cannot be seen at this scan level. The basal cisterns are compressed from the hydrocephalus.

aneurysm of the basilar artery, and meningioma about the tentorial hiatus. It is possible for small tumors in this region to cause obstruction. Therefore a thorough investigation of the region is mandatory preferably with MRI in the parasagittal and transaxial projections with contrast enhancement. The characteristics of the various lesions are described under their specific headings elsewhere in the text. It may be only after a follow-up examination that a benign nonprogressive process can be diagnosed.

Tumor masses within the ventricular cavity or adjacent to the ventricle may cause obstruction to the fourth ventricle. The aqueduct and the third and lateral ventricles are dilated. The fourth ventricle may also be dilated, especially when the obstruction is at the foramina of Luschka (lateral) and Magendie (medial) from tumor or adhesive meningitis.

The diagnosis of basilar adhesions as the cause of fourth ventricular outlet obstruction is often difficult. The adhesions may occur, particularly after meningitis from tuberculosis, fungus infection, or sarcoidosis. The diagnosis may be suspected only by the generalized ventriculomegaly, including a large fourth ventricle and without a visible mass. Occasionally, there is contrast enhancement of the basal meninges on CT scan or MRI. Nevertheless, the exact etiology may be apparent only at surgery. An intraventricular cyst of cysticercosis may exactly mimic the ventricular cavity itself and may require CT ventriculography for the diagnosis. The clinical history of residence within an endemic region is suggestive of this diagnosis.

COMMUNICATING HYDROCEPHALUS

Communicating hydrocephalus is the most common type of hydrocephalus in adults. It is common with subarachnoid hemorrhage (see Ch. 3) meningeal carcinomatosis or meningitis. It is rare after trauma. With these conditions the hydrocephalus is usually mild and self-limited, the ventricles returning to near-normal with time and appropriate treatment. Sometimes the hydrocephalus is progressive and symptomatic, and it may be severe. Shunting is required in these instances (see also Normal-Pressure Hydrocephalus, in Ch. 10). Communicating hydrocephalus implies obstruction to the

flow of CSF outside the brain within the cisterns or at the level of the arachnoid villi. Sometimes, no antecedent cause can be determined for communicating hydrocephalus.

The ventricles enlarge, although the fourth ventricle may remain relatively small; and the cisterns show variable enlargement. There is free flow of CSF between the ventricular system and the subarachnoid space. Radionuclide or positive-contrast CT cisternography demonstrates "reflux" of the contrast agent into the ventricular system and block to the flow of radionuclide or contrast over the cerebral convexities.

Diagnosis of Hydrocephalus

The hallmark for the diagnosis of hydrocephalus is enlargement of the ventricular system, best demonstrated with MRI or CT scanning. However, there is no precise

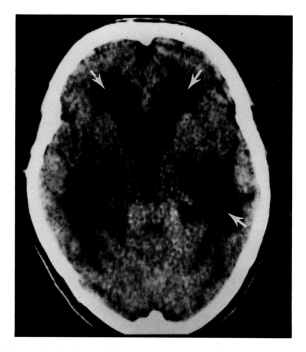

Fig. 11-4 Hydrocephalus. CT scan through the lateral ventricles shows striking periventricular edema (*arrows*) most commonly associated with acute hydrocephalus. It will disappear after adequate shunting of the hydrocephalus.

measurement for distinguishing normal from abnormal size. Some have defined ventricles as being abnormally enlarged when the frontal horn span measured at the widest point is more than 30 mm or the ventricular–intracranial index is more than 0.30. However, change from a baseline size is most helpful for defining ventricular enlargement.

With hydrocephalus, alternative pathways of CSF resorption are found. The most prominent is the transependymal absorption from the lateral ventricles. Fluid is forced through breaks in the ependyma into the paraventricular white matter; from there it is resorbed into the venous system. This process is seen as paraventricular edema which shows high paraventricular signal on T_2-weighted MRI and hypodensity on CT (Figs. 11-2 and 11-4).

The pattern of the ventricular enlargement is important for differentiating hydrocephalus from the ventricular enlargement of atrophy. The major anatomic features of hydrocephalus are the following (see also Ch. 10).

1. Enlarged lateral ventricles
2. Commensurate enlargement of the temporal horns, usually more than 7 mm in width (Fig. 11-3B)
3. An acute callosal angle, formed by the frontal horns (less than 95 degrees)
4. Bulbous enlargement of the frontal horns so there is a relatively wide frontal horn radius for the size of the ventricles
5. Periventricular white matter edema from compensatory transependymal absorption of CSF (Figs. 11-2 and 11-4)
6. Decreased size and number of visible cortical sulci
7. Decreased size of the subarachnoid cisterns

Hydrocephalus may also be diagnosed with angiography. The enlarged frontal horns can be outlined on the anteroposterior (AP) projection by the spread subependymal veins, particularly the thalamostriate vein, which is stretched and bowed laterally (Fig. 11-5A). Similarly, the lenticulostriate and the sylvian vessels are stretched and bowed laterally. The pericallosal artery is stretched in a smooth, curving arc around the stretched corpus callosum (Fig. 11-5B). On the lateral vertebral study, the posterior pericallosal arteries are displaced and stretched posteriorly, and the posterior lateral choroidal arteries

are pushed forward and downward by the enlarged atria of the lateral ventricles (see Ch. 1).

Nonspecific changes of the calvarium result from the increased intracranial pressure associated with hydrocephalus. They can be seen on plain film skull radiography (Fig. 11-6). Most of the skull changes listed occur after hydrocephalus has been present for some time.

1. Separation of the cranial sutures in infants and children
2. Elongated sutural interdigitations
3. Persistence or excessive prominence of the convolutional marking of the inner table of the calvarium (This sign is most useful in the 8- to 12-year age group. Before this age, prominent convolutional margins are normal. This sign infrequently occurs in adults.)

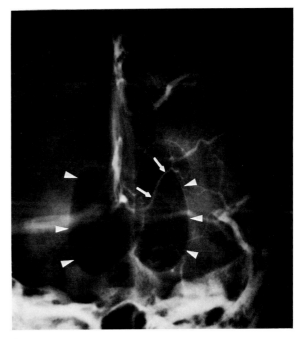

A

Fig. 11-5 Hydrocephalus, angiography. (A) AP venous view shows the thalamostriate vein (*arrows*) being stretched laterally by the hydrocephalus. The enlarged ventricles are outlined by air from prior pneumoencephalography (*arrowheads*). (*Figure continues*).

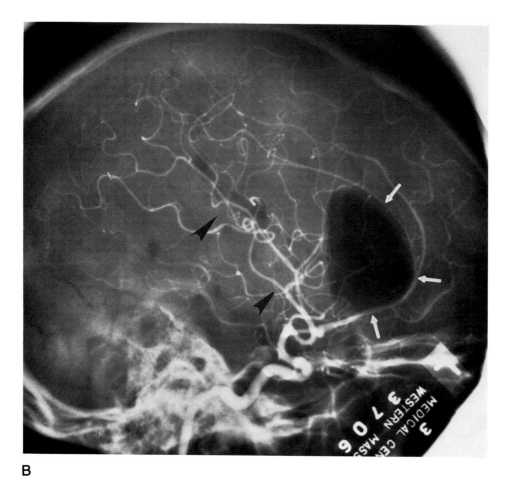

B

Fig. 11-5 *(Continued)*. **(B)** Lateral arterial angiography shows stretching of the pericallosal arteries around the enlarged frontal horns *(arrows)*. The sylvian vessels are also displaced upward by the enlarged temporal horn *(arrowheads)*.

4. Thinning and bone resorption of the floor of the sella, beginning at the posterior inferior portion (Thinning may also occur along the planum sphenoidale.)

5. Enlargement of the sella and truncation of the dorsum by downward expansion of an enlarged third ventricle

6. Enlargement of the emissary vein channels, particularly in the occipital region

7. Enlargement of the internal auditory canals

When severe hydrocephalus occurs acutely, transtentorial cranial-caudal herniation of the brain may occur, sometimes quickly, causing death. It is especially true when shunt occlusion occurs in "shunt-dependent" individuals. This potential may be recognized by obliteration of the basal cisterns caused by the compressed brain. Immediate action must be taken when this situation is observed. If the hydrocephalus is acute, there is usually complete recovery of the compressed brain after decompression. Prolonged hydrocephalus causes atrophy of brain tissue. Severe hydrocephalus may also cause occlusion of the posterior cerebral arteries (Fig. 11-7).

Whereas the diagnosis of hydrocephalus is easy in most circumstances, the differentiation from atrophy may be difficult if not impossible. In these situations, it is best to describe the ventricles as enlarged but indicate that an accurate etiology cannot be given at the time.

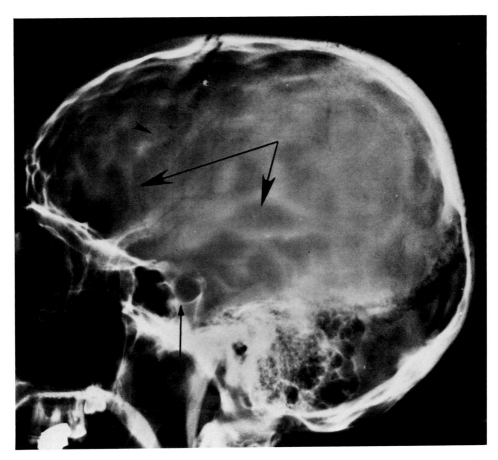

Fig. 11-6 Hydrocephalus, increased intracranial pressure. Lateral plain skull radiograph shows slightly enlarged sella turcica with loss of the cortical margin along the floor *(small arrow)* and at the chiasmatic sulcus. There are prominent convolutional markings *(large arrows)* and widening of the cranial sutures with elongation of the interdigitations *(arrowheads)*.

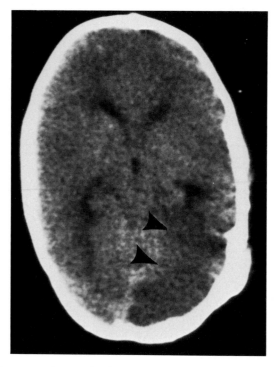

Fig. 11-7 Occipital infarction from hydrocephalus. Severe hydrocephalus has caused compression of the left posterior cerebral artery as it crosses over the tentorium. It has caused infarction in the left occipital and subtemporal regions (*arrowheads*). Hydrocephalus has been shunted.

Follow-up scans showing progression may be the only way to diagnose hydrocephalus if the clinical features are not obvious.

AQUEDUCT STENOSIS

The aqueduct of Sylvius is normally a small channel connecting the posterior third ventricle with the superior fourth ventricle. Congenital aqueductal stenosis refers to the maldevelopment of the aqueduct such that the normal single channel is subdivided, or "forked," into two or more small channels that have inadequate capacity to drain the CSF adequately. Atresia or focal stenosis is rare. The abnormality may occur as an isolated event or in association with the Chiari II malformation.

Aqueduct stenosis usually produces severe hydrocephalus, manifested shortly after birth as a rapidly enlarging head. The diagnosis is generally made with ultrasonography or CT scan. The scans show enlarged lateral and third ventricular cavities with no demonstrable mass in the periaqueductal region. The occipital horns are usually larger than the frontal horns. The left occipital horn is often asymmetrically enlarged. The fourth ventricle is small or normal in size (Figs. 11-2 and 11-3). Left untreated, the head becomes grotesquely enlarged.

The condition responds to ventricular shunting. It is remarkable how a think cortical mantle can reconstitute so that the brain returns to a relatively normal appearance. These infants are normally "shunt-dependent," meaning that the relief of hydrocephalus depends entirely on adequate shunt function. Irreversible damage in the brain tissue occurs if the hydrocephalus is active for more than about 3 months.

If the condition is mild and goes undetected, a few infants reach a state of compensation of CSF production and drainage. This state, often called "arrested hydrocephalus," usually occurs after a moderate degree of ventricular dilation has developed. A few of these persons function normally and go undetected into adulthood and are then discovered only by chance.

Occasionally, aqueductal stenosis develops during adulthood. Many are new cases that result from periaqueductal gliosis, which may be a response to some unknown inflammatory process. Rarely, prior intraventricular hemorrhage produces adhesions within the aqueduct. A careful search for small periaqueductal tumor must be made, with Gd-enhanced MRI the single best examination.

NORMAL-PRESSURE HYDROCEPHALUS

Normal-pressure hydrocephalus (NPH) is also known as low-pressure hydrocephalus, occult hydrocephalus, and Hakim-Adams syndrome. It is discussed in Chapter 10.

BENIGN ENLARGEMENT OF THE SUBARACHNOID SPACES

Benign enlargement of the subarachnoid spaces (external hydrocephalus, benign subdural effusions) is a condition in infants that consists of accelerated head growth with widened subarachnoid spaces and normal, or only slightly enlarged, ventricles. The subarachnoid

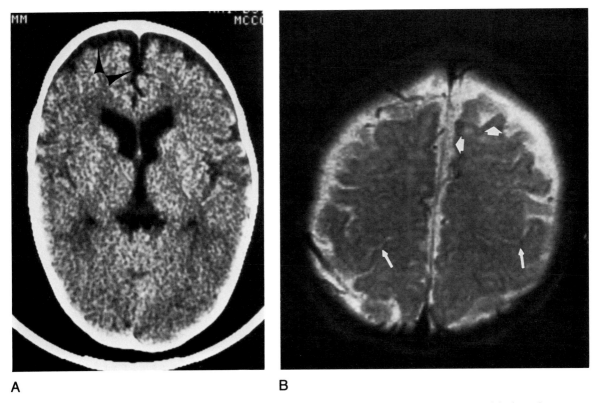

Fig. 11-8 "External" communicating hydrocephalus. (A) CT scan shows moderate enlargement of the lateral ventricles from hydrocephalus. There is disproportionate enlargement of the cerebral space over the frontal lobes and in the interhemispheric fissure *(arrowheads)*. **(B)** T_2-weighted MRI through the vertex shows the widened subarachnoid space over the frontal lobes and in the interhemispheric fissure *(fat arrows)*. The sulci and extracerebral spaces posteriorly are much smaller *(thin arrows)*.

spaces show disproportionate widening over the frontal lobes and in the frontal interhemispheric fissure. The cerebral sulci are prominent over the frontal lobes but are normal elsewhere (Fig. 11-8). The condition is thought to be a communicating hydrocephalus resulting from decreased absorption of CSF from immature arachnoid villi in the presence of open cranial sutures.

This condition occurs spontaneously or following subdural hematoma, meningitis, and prematurity with associated intracranial hemorrhage. The process is almost always self-limited and resolves without treatment in 2 to 3 years. Atrophy may be a delayed finding in those with the other associated conditions. The clinical course is determined more by the underlying damage than by the external hydrocephalus. Shunting is carried out only for neurologic symptoms or increased intracranial pressure.

MACROCRANIUM IN INFANTS AND CHILDREN

Hydrocephalus
    Noncommunicating
    Communicating
Subdural collections
Storage diseases
Leukodystrophy
Metabolic disorder (lead)
Congenital megalencephaly
Tuberous sclerosis

Conditions that can cause macrocranium in infants and children are given in the box.

Benign enlargement of the subarachnoid spaces must be differentiated from atrophy, and subdural collections, which may have a similar CT appearance. With atrophy, the head size is normal or small, the subarachnoid space is more uniformly enlarged, there is little widening of the interhemispheric fissure, and both the ventricles and the sulci are diffusely large. MRI is essential for the differentiation of subdural collections from enlarged subarachnoid spaces. With most subdural collections the fluid contains increased protein or blood elements. This produces a signal hyperintense to CSF on the T_1- and T_2-weighted sequences. The collection is peripheral against the inner table of the skull, with the normal CSF subarachnoid space seen as an inner compartment against the surface of the brain. The differential diagnosis is important as subdural collection may require external drainage and imply injury. They may also be seen with child abuse (see Ch. 7). The prognosis is good with benign enlargement of the subarachnoid spaces, whereas atrophy implies brain damage.

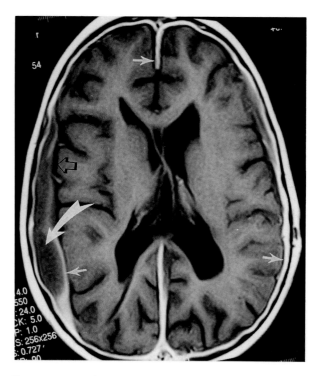

Fig. 11-9 Post-shunt, dural enhancement, subdural fluid collection. Gd-enhanced MRI shows the intense enhancement of the dural membranes that is common following ventricular shunting procedures (*small white arrows*). The pia is not enhancing (*open arrow*). A subdural fluid collection occurs with "overshunting" (*large white arrow*).

VENTRICULAR DECOMPRESSION: SHUNTS

The usual treatment for hydrocephalus is the surgical establishment of any one or more of a variety of shunts (see below). The object is to divert the intracranial CSF into the peritoneum, pleural space, or bloodstream, thereby reducing the intracranial volume and pressure of the CSF. The shunt may be temporary or permanent. Even when the hydrocephalus is severe, the brain may

---

TYPES OF SHUNT

Ventricular-peritoneal
Ventricular-pleural
Ventricular-right atrial
Ventricular-cisternal (Torkildsen)
Lumbar-peritoneal

---

may return to a normal appearance, especially in infants and young children. Acute hydrocephalus responds better to shunting than does chronic hydrocephalus. Long-standing hydrocephalus causes irreversible atrophic changes.

The ventriculoperitoneal shunt is the one most commonly employed. The intracranial portion of the tube is preferably passed through the nondominant hemispheric tissue into the ipsilateral ventricle, entering the skull in the posterior parietal region or through the coronal suture. Optimally, the tip of the catheter is within the frontal horn. It may cross through the septum pellucidum to the opposite side with no adverse consequence. A shunt that resides within the temporal horn or the atrium of the lateral ventricle can easily become plugged with the surrounding choroid. Care must be taken before diagnosing a catheter as being outside the ventricular cavity, as collapse of a decompressed ventricle or partial volume averaging may spuriously suggest a catheter within brain tissue.

The shunt function is usually evaluated by observing the change in the size of the ventricular system with CT. MRI may be used, but it has no particular advantage. If the ventricles become smaller, the shunt is considered to be working. With successful decompression, the ventricles begin to decrease in size within 24 to 48 hours. The lower the functioning pressure setting of the shunt, the faster the ventricles decompress. After acute hydrocephalus, the ventricles usually return to normal with fully functioning shunts. If atrophy is present, the ventricles become stabilized at some larger size. Higher-pressure shunts leave slightly larger ventricles. Following shunt placement, the dura, and sometimes the pia, enhances with Gd on MRI examination (Fig. 11-9). This enhancement does not imply meningitis

or other pathology. The phenomenon remains indefinitely.

Shunt malfunction can usually be diagnosed with CT scanning by observing enlargement of previously decompressed ventricles. Initially, the dilation may be subtle, but it can be recognized by careful comparison with prior examinations. The sulci and interhemispheric fissure become compressed (Fig. 11-10). The condition of the cisterns must be consciously noted, so that any potential for herniation can be predicted (Fig. 11-11). Particularly in children, fibrosis of the ventricular ependyma may prevent appreciable enlargement of the ventricles when the shunt has failed. This situation is especially apt to occur with the "slit ventricle syndrome," a condition in which the ventricles become extremely small because of shunting. In this condition, it is difficult to determine shunt malfunction by scanning, and the clinical symptoms alone must be used to make the correct diagnosis. Scans are repeated at frequent intervals

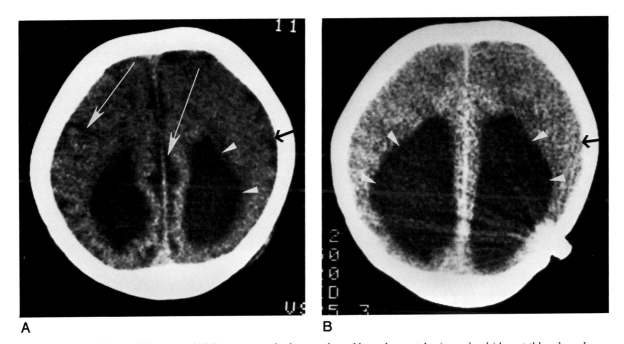

A **B**

Fig. 11-10 Shunt malfunction. (A) Scan superiorly shows enlarged lateral ventricles (*arrowheads*) but visible subarachnoid space in the interhemispheric fissure and cerebral sulci (*white arrows*). A small hygroma is present on the left (*black arrow*). The presence of these spaces indicates adequate shunt function of the preexisting hydrocephalus. **(B)** Shunt malfunction. A scan sometime later shows greater enlargement of the lateral ventricles (*arrowheads*). The cerebral sulci, interhemispheric fissure, and extracerebral fluid collection on the left have been obliterated. (*Figure continues*).

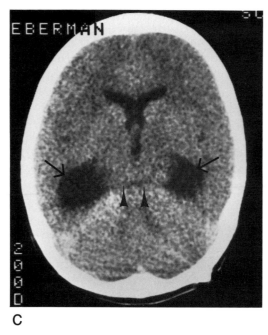

C

Fig. 11-10 *(Continued)*. **(C)** Scan at a lower level shows considerable enlargement of the temporal horns *(arrows)*, but the frontal horns are relatively small. The quadrigeminal plate cistern is compressed *(arrowheads)*.

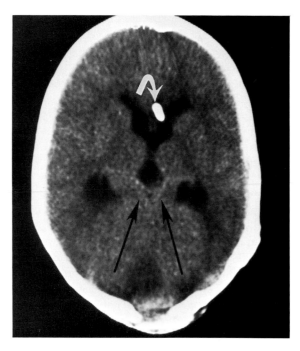

Fig. 11-11 **Shunt malfunction, imminent herniation.** The ventricles are mildly enlarged, and the shunt catheter is present in the left frontal horn *(curved arrow)*. The basal cisterns are totally obliterated *(arrows)*, indicating increased intracranial pressure in this child, who is shunt dependent. This finding indicates risk of imminent cranial caudal herniation, and the function of the shunt must be re-established immediately.

(hours to days) in questionable cases, to detect subtle changes.

Ventricular shunts may create extracerebral fluid collections, either subdural hygromas or hematomas (Fig. 11-12). This problem is especially likely to occur when the ventricles are rapidly decompressed, particularly in older persons and in those with underlying brain atrophy. In this situation, the brain parenchyma is relatively noncompliant and is unable to regain its original volume. The relative negative pressure within the collapsed ventricles causes the cerebral hemispheres to fall away from the inner table of the calvarium. The extracerebral space created is then filled with near-water-density fluid. The extracerebral collections are a sign of "overshunting." Subsequent hemorrhage into the hygroma occurs frequently, causing a mass effect. If the shunt is blocked, or a higher pressure valve is placed, the ventricles re-expand and the extracerebral fluid decreases or disappears. This complication can be prevented by initially using higher-pressure valves and antisiphon devices in the shunt. Other complications of shunts (Figs. 11-13 and 11-14) are given below.

With noncommunicating hydrocephalus, herniation of the brain can occur after shunting. In the presence of a large posterior fossa tumor, the decompression of the hydrocephalic lateral ventricles may permit upward transtentorial herniation of the cerebellum, resulting in severe hemorrhage or death. Similarly, when the obstruction is at the level of the foramina of Monro, decompression of only the lateral ventricle may cause subfalcial herniation of the opposite "trapped" ventricle. The trapped ventricle requires shunting with an additional catheter. These potential disasters must be anticipated from the findings on the preshunt CT or MRI scans.

Fig. 11-12 Subdural hygroma, overshunting. CT examination shows small, nearly slit-like ventricles *(white arrow)*. A low-density extracerebral collection has developed on the right *(black arrows)*, agenesis of corpus callosum.

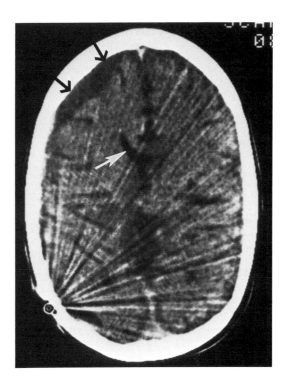

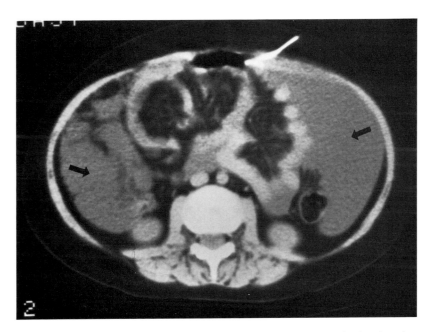

Fig. 11-13 CSF accumulation in the abdomen. CT scan shows a large amount of fluid within the peritonial cavity *(arrows)*, indicating failure of resorption of the shunted fluid. The collections may be focal or diffuse.

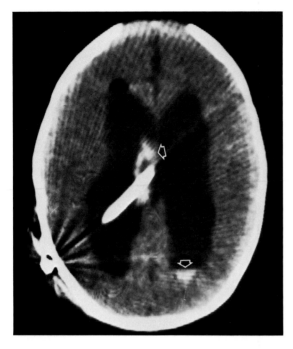

Fig. 11-14 Intraventricular hemorrhage from shunt placement. A small hemorrhage has occurred in the region of the choroid plexus from placement of the shunt. Blood is present in the dependent ventricle *(arrows)*.

COMPLICATIONS OF SHUNTS

Shunt malfunction (poor drainage)
    Poor placement
        Not within ventricle
        Plugged by choroid within temporal horn, atrium, or third ventricle
    Disconnection or kinking
    Loculated CSF fluid collection within peritoneum or pleural space (Fig. 11-13)
    Plugged by blood clot

Extracerebral fluid collection (overshunting) (Fig. 11-12)
Hemorrhage around shunt catheter (Fig. 11-14)
Local cerebral infarction
Infection with ventriculitis
Thrombosis and septic emboli from right atrial shunt
Intraperitoneal spread of primary CNS tumor or infection
Brain herniation from asymmetric decompression

SUGGESTED READING

Atlas SW, Mark AS, Fram EK: Aqueductal stenosis: evaluation with gradient-echo rapid MR imaging. Radiology 169:449, 1988

Bradley WG Jr: Hydrocephalus. p. 451. In Stark DD, Bradley WG Jr (eds): Magnetic Resonance Imaging. CV Mosby, St. Louis, 1988

Gammal TE, Allen MB Jr, Brooks BS, Mark EK: MR evaluation of hydrocephalus. AJNR 8:591 1987

Goss CM, Gray's Anatomy. p. 929. Lea & Febiger, Philadelphia, 1963

Maytal J, Alvarez LA, Elkin CM, Shinnar S: External hydrocephalus: radiologic spectrum and differentiation from cerebral atrophy. AJNR 8:271, 1987

Naidich TP, Schott LH, Baron RL: Computed tomography in evaluation of hydrocephalus. Radiol Clin North Am 20:143, 1982

Quencer RM: Intracranial CSF flow in pediatric hydrocephalus: evaluation with Cine-MR imaging. AJNR 13:601, 1992

Wilms G, Vanderschueren G, Demaerel PH et al: CT and MR in infants with pericerebral collections and macrocephaly: benign enlargement of the subarachnoid spaces versus subdural collections. AJNR 14:855, 1993

(Continued)

12
Congenital Abnormalities of the Brain

Congenital structural abnormalities result from insults to the developing brain. The timing of the insult is considered the most important factor. As most insults affect the entire brain, frequently several anomalies are seen. Malformations are classified on the basis of anatomic features.

Other congenital abnormalities include destructive disorders, dysmyelination, and the neurocutaneous syndromes. These conditions tend to exhibit inheritance, and some have a known metabolic basis.

MRI is the preferred imaging technique for identifying these structural malformations. T_1 imaging is usually better for anatomic definition, and the sagittal and coronal planes are often helpful. T_2-weighted MRI is better for dysmyelinating diseases. CT scanning may be used but is generally less precise.

EMBRYOLOGIC REVIEW

The development of the nervous system begins with the thickening of the dorsal neural ectodermal plate. At about the T12–L1 level, progressing rostrally, the ectodermal plate infolds to produce the neural tube, which then forms the brain and spinal cord. Below this level, a regressive type of differentiation process forms the lower spine and dural tube. Failures of the normal tube closure and differentiation process result in abnormalities such as anencephaly, cephalocele, myeloschisis, myelomeningocele, and the Chiari malformation. This process occurs at 4 to 7 weeks gestation.

Once the neural tube is formed, ventral induction occurs in the rostral end, forming the face and the brain. The prosencephalon, mesencephalon, and rhombencephalon are created. There is division into two hemispheres, which then roll over to form lobes and the major fissures. The connecting corpus callosum and the vermis of the cerebellum are formed in a rostral–caudal direction. Disorders during this phase result in various anomalies, including agenesis of the corpus callosum, holoprosencephaly, septo-optic dysplasia, cerebellar aplasia, and Dandy-Walker syndrome. These events occur at 5 to 10 weeks gestation.

Formation of the gross structure of the brain is followed by cellular migration and differentiation. Neurons are formed in the germinal matrix of the paraventricular region and then migrate outward to the cortex. Failure of this migration results in deformation of the gyral pattern—either too few and too thick (agyria, pachygyria) or too thin and too many (polymicrogyria)—and heterotopic gray matter, consisting of neurons that failed to reach the cortex. Abnormal differentiation or inclusion of foreign cells results in vascular malformations, teratomas, hamartomas, and other tumors from embryonal rests. These processes occur at 2 to 5 months gestation.

Maturation of Myelin

Myelination of the white matter tracts of the brain progresses in an orderly predictable sequential fashion (Table 12-1). It begins in utero after neuronal migration is complete and is essentially complete by 2 years of age. In general, sensory fibers are myelinated before motor fibers. The process begins in the phylogenetically oldest portion of the caudal brain and proceeds superiorly into the cerebrum. Within the cerebrum, the occipital and parietal lobes myelinate before the frontal and temporal lobes. Maturation of brain function correlates with myelination, so delay or interruption in the process is reflected in the clinical neurological development of the infant. At 2 years, the myelin, while complete in distribution, is immature in structure. Over the years to adulthood, this transitional myelin becomes more tightly wrapped in a lamellar fashion about the axons, changing its composition of lipids slightly to become mature myelin.

Table 12-1 Pattern of Myelination, SE T_1/T_2

Age	Structures Myelinated
Birth	Medulla
	Superior and inferior cerebellar peduncles
	Dorsal tracts of the pons
	Dorsal tracts of the midbrain
	Ventrolateral thalamus
	Posterior limbs of the internal capsule with or without postcentral white matter
3–6 mo	Central cerebellar white matter
	Middle cerebellar peduncle
	Complete pons
	Optic radiations and calcarine region
	Precentral white matter
	Anterior limb of the internal capsule
	Splenium of the corpus callosum
	Optic nerves and tracts
	Occipital U fibers
6–12 mo	Genu, then body of the corpus callosum
	Cerebral white matter, parietal to most of frontal
	Parietal to frontal U fibers
	Isointense gray/nonmyelinated white matter
12–18 mo	Completion to adult pattern

Table 12-2 MRI Signal Intensities of Myelin

	T_1	T_2
Immature	Hypointense	Hyperintense
Mature	Hyperintense	Hypointense

Spin echo (SE) MRI is the standard method for evaluation of myelination of the brain. Both T_1- and T_2-weighted sequences are used, but the T_2-weighted images are more closely aligned with the maturation of myelin seen on pathologic sections. The T_1-weighted sequence changes to a mature pattern earlier than actual maturation, although it does indicate the overall distribution of the myelin. The T_2-weighted MRI characteristics are based on the changes in water content of the white and gray matter of the brain with maturation. The inversion recovery T_1- and T_2-weighted sequences show slightly advanced maturation changes compared with SE sequences.

At birth, the brain is generally wet. The water content of the white matter is 87 percent, and the gray matter 89 percent. At 2 years of age, the myelinated white matter has a water content of 72 percent, and the gray matter 82 percent. The signal intensities of MRI with myelination are given in Table 12-2 (Figs. 12-1 to 12-3).

The congenital disorders discussed below are grouped according to their most likely etiology. When placed within this somewhat arbitrary framework, the disorders make some sense and are easier to remember.

DISORDERS OF CLOSURE OF THE NEURAL TUBE

Anencephaly

Anencephaly is the most severe of anomalies and causes death of the fetus or infant. It may be diagnosed with ultrasonography as early as 20 weeks gestation. The brain and calvarium may be totally deficient, or there may be rudimentary brain at the base and severe microcephaly. Polyhydramnios and increased levels of α-fetoprotein (AFP) in the amniotic fluid are usually associated with these defects of closure of the neural tube.

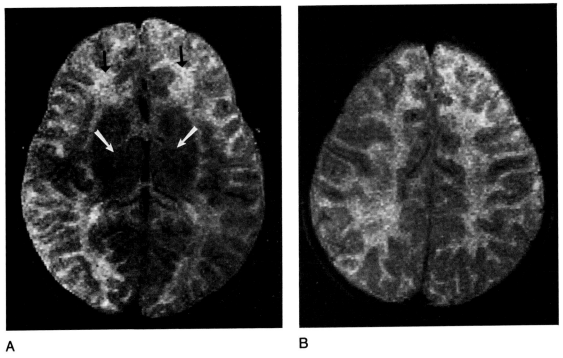

A
B

Fig. 12-1 Normal 3-month-old brain, immature white matter; T$_2$-weighted MRI SE2500/80. (A) Transaxial slice through the basal ganglia shows relatively high intensity of the hemispheric white matter, particularly the frontal lobes (*black arrows*). This pattern is opposite that of normal adult myelin. Myelination has occurred within the posterior limbs of the internal capsules and the thalamic nuclei (*white arrows*). These areas are the more "primitive" portions of the brain. The white matter of the cerebellum and pons also have low intensity by this age. **(B)** Slice through the centrum semiovale shows the typical high intensity signal of immature nonmyelinated white matter. It must not be interpreted as dysmyelination.

Cephalocele

Cephalocele is a congenital defect that results in herniation of intracranial contents outside the calvarium. Most occur in the midline. In Europe and the United States, the defect is most commonly occipital. In Asia, it is most commonly nasofrontal. It may also occur in parietal, frontal, and nasopharyngeal sites.

Meningocele refers to herniation of only the meninges. *Encephalomeningocele* indicates herniation of brain tissue along with the meninges. *Encephalocystomeningocele* includes the ventricle within the herniation (Fig. 12-4A). Microcephaly is present with large encephaloceles. The herniated brain may be normal or dysgenetic. It is impossible to predict this condition accurately with imaging or angiography. Low occipital encephaloceles containing the cerebellum are called Chiari III malformations (Fig.

12-1A&B). Low cephaloceles may include posterior dysraphism of the upper cervical spine. A midline dermoid may mimic the skull defect of cephalocele but does not have the herniated brain tissue. Frontal encephaloceles may be occult (Fig. 12-5A) but are important to recognize as a cause of meningitis and before sinus surgery is planned for infection. Sphenoethmoidal encephaloceles are frequently associated with facial midline defects, and always with hypertelorism, but these sacs usually do not contain herniated brain (Fig. 12-5A&B).

Chiari Malformations

The Chiari malformations are considered under a separate heading, as they are a diverse group with poorly understood pathogenesis. There is notorious confusion

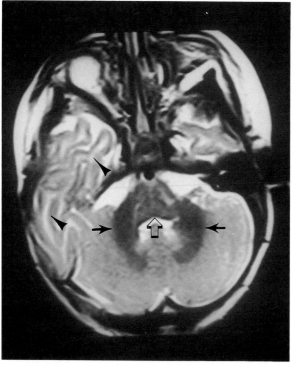

A

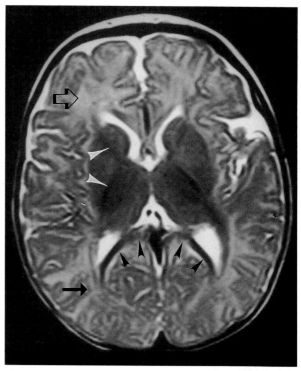

B

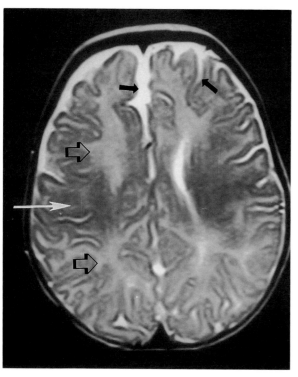

C

Fig. 12-2 Normal brain, 5-month-old. T_2 FSE (4'4) 96/3500 MRI. **(A)** There is conversion to low T_2 signal (myelination) of the deep cerebellar white matter and middle cerebellar peduncles *(black arrows)* and the white matter tracts through the brain stem *(open arrow)*. The temporal lobe white matter remains unmyelinated and has high T_2 signal *(arrowheads)*. **(B)** T_2 scan. There is conversion of the anterior and posterior limbs of the internal capsule *(white arrowheads)*, the thalamus, an the splenium of the corpus callosum *(black arrowheads)*. Partial conversion has occurred in the occipital lobe white matter *(black arrow)*. The frontal white matter is unmyelinated and bright *(open arrow)*. **(C)** T_2 scan through the centrum semiovale shows myelination in the posterior frontal and anterior parietal white matter *(white arrow)*, but little myelination within the frontal and occipital white matter *(open arrows)*. Widening of the sulci and subarachnoid space *(black arrows)* is normal up to age 2 years. *(Figure continues.)*

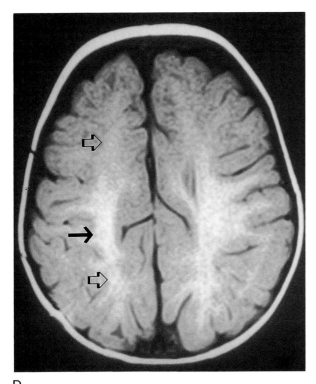

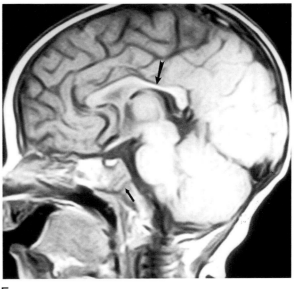

D   E

Fig. 12-2 *(Continued).* **(D)** T$_1$ MRI shows conversion to high signal in the posterior frontal and anterior parietal white matter representing myelination *(black arrow).* The frontal and occipital lobe white matter is just slightly brighter than the gray matter *(open arrows).* **(E)** Sagittal T$_1$ MRI shows the corpus callosum, which is normally thin in the posterior portion *(larger arrow).* It has converted to high T$_1$ signal. Note the basisphenoid synchondrosis *(smaller arrow).*

about them. Four entities were described by Chiari, which he labeled types I to IV. The four are unrelated pathologically.

CHIARI I

Chiari I refers to a malformation of both the cerebellum and the cervicomedullary junction of the spinal cord, sometimes associated with malformation of the cervical spin and craniovertebral junction. The cerebellar tonsils are small, pointed inferiorly, and situated through the foramen magnum into the posterior upper cervical spinal canal more than 5 mm below the posterior rim of the foramen magnum. The fourth ventricle is usually small and slightly low in position. The medulla may be elongated and thickened. The cisterna magna is small. Associated bone abnormalities include basilar impression, occipitalization of the atlas, incomplete ossification of the atlas, and Klippel-Feil anomaly (cervical spinal fusions). Hydromyelia is common. Chiari I is discussed further in Chapter 13.

CHIARI II

The Chiari II malformation is a complex abnormality almost always associated with a myelomeningocele. The malformation affects many diverse structures (Fig.

(Text continued on page 463)

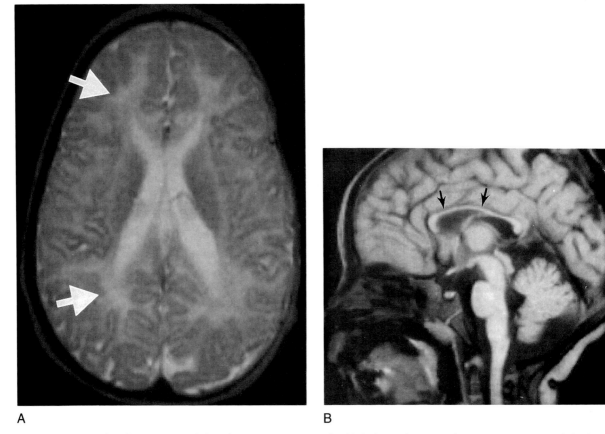

A   B

Fig. 12-3 Delayed maturation of the white matter, 16-month-old. (A) Axial T_2 MRI demonstrates persistently high T_2 signal within the frontal and occipital white matter *(arrows)*, which by this time should have converted to low signal. The white matter is small in volume. The etiology in this case is indeterminate. **(B)** Sagittal T_1 MRI shows the thin but complete corpus callosum *(arrows)*. The corpus callosum is thin if there is loss of white matter axons.

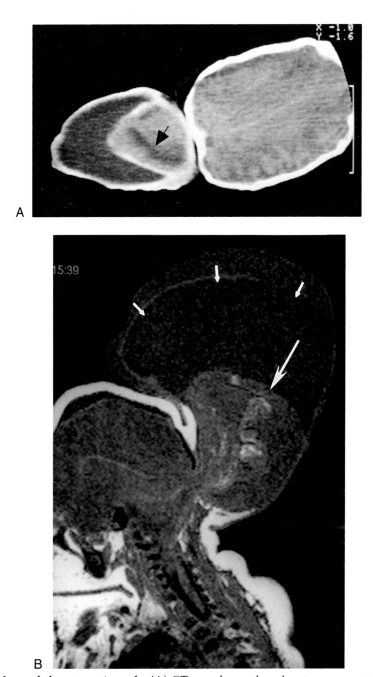

Fig. 12-4 Occipital encephalocystomeningocele. (A) CT scan shows a large hernia sac posteriorly. It contains CSF, brain, and what appears to be a rudimentary ventricular cavity *(arrow)*. **(B)** Sagittal T_1 MRI shows the herniation of brain *(large arrow)*, including cerebellum and ventricle, through an occipital defect into a large posterior sac *(small arrows)*. This is a Chiari III malformation.

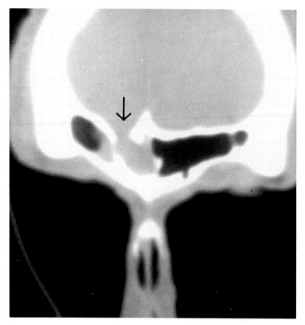

A

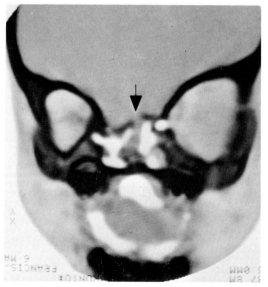

B

Fig. 12-5 (A) Encephalocele. Coronal CT scan shows a defect in the anterior floor of the right frontal fossa. The soft tissue density represents brain herniation into the frontal sinus *(arrow)*. **(B) Frontoethmoidal encephalocele.** Coronal view (reverse gray scale) of the brain of a 7-month-old infant. There is hypertelorism and hypoplasia of the right orbit. The encephalocele extends through the frontal bone and into the nasal ethmoid region through a small ethmoid defect *(arrow)*. (From Fitz, 1984, with permission.)

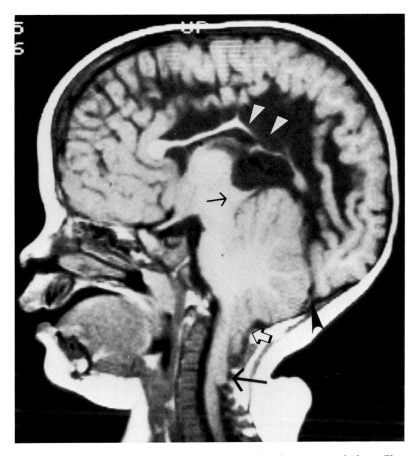

Fig. 12-6 Chiari II malformation. Sagittal T$_1$-weighted MRI shows the characteristic findings. Elongation and buckling of the medulla *(large arrow)*, a vermian peg *(open arrow)*, beaking of the tectum *(small arrow)*, agenesis of the posterior corpus callosum *(white arrowheads)*, and stenogyria (polymicrogyria). The foramen magnum is large, and there is a low position of the tentorium, allowing inferior extension of the occipital lobes *(black arrowhead)*. The fourth ventricle, not seen on this image, is small, elongated, and inferiorly positioned.

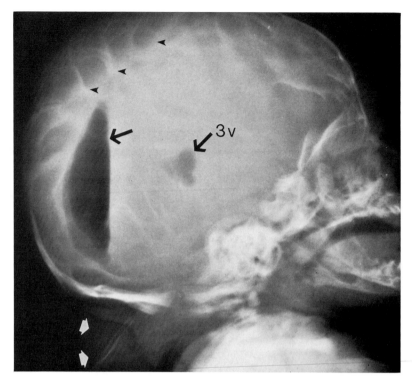

Fig. 12-7 Craniolacunia, Chiari III. Plain skull radiograph shows a scalloped appearance of the calvarium (*arrowheads*). This infant has undergone ventriculography, and air is seen in the dilated occipital horns in the posterior third ventricle (3V, *black arrows*). The posterior fossa is small, and there is a cervico-occipital encephalocele (*white arrows*). This picture represents a Chiari III malformation.

12-6). The tentorium is low, creating a small posterior fossa. The falx is fenestrated allowing interdigitation of the medial hemispheres across the interhemispheric fissure. The brain stem is positioned inferiorly within a large foramen magnum and upper cervical spinal canal. There may be actual buckling of the medulla behind the upper cervical cord, forming a characteristic kink. The small fourth ventricle is elongated inferiorly. The cerebellum is poorly differentiated, herniates inferiorly, and expands anteriorly around the pons. There is a postero-inferior vermian peg behind the medulla. Hydrocephalus is common (usually a result of aqueduct stenosis) and often requires shunting. There may also be a beaked collicular plate, agenesis of the corpus callosum, a large massa intermedia, inferior frontal horn beaking, anatomic polymicrogyria (stenogyria), anomalous cisterns, syringohydromyelia, clival scalloping, and carniolacunia (Fig. 12-7). Craniolacunia is a dysplastic scalloping of the calvarium that resolves after 6 months.

CHIARI III

The Chiari III malformation is a cervico-occipital encephalocele, with the sac containing nearly all of the cerebellum.

CHIARI IV

The Chiari IV malformation is severe cerebellar hypoplasia.

DISORDERS OF ORGANOGENESIS

Dysgenesis of the Corpus Callosum

The corpus callosum may be partially or completely absent. The corpus callosum begins development from the region of the anterior commissure, and the growth

Fig. 12-8 Agenesis of the corpus callosum. (A) Parasagittal T$_1$-weighted MRI shows complete agenesis of the corpus callosum *(open arrow)*. There is also cerebellar hypoplasia and an arachnoid cyst within the posterior fossa *(white arrow)*. **(B)** Agenesis of the corpus callosum, cerebellar hypoplasia, and posterior fossa arachnoid cyst. Parasagittal T$_1$-weighted MRI shows agenesis of the posterior portion of he corpus callosum *(open arrow)*. The medial sulci are oriented in a radial fashion *(arrowheads)*. There is a large collection of CSF-intensity fluid in the posterior fossa representing an arachnoid cyst. The cerebellum and vermis are hypoplastic. *(Figure continues.)*

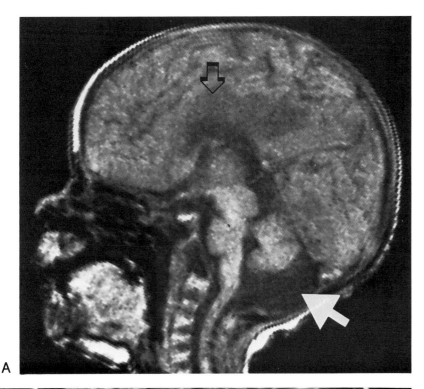

A

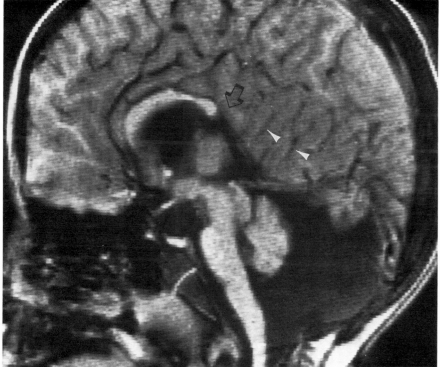

B

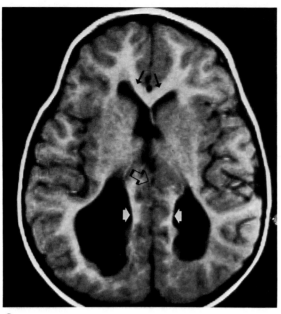

C

Fig. 12-8 *(Continued).* **(C)** Partial agenesis of the corpus callosum. Transaxial T$_1$-weighted MRI shows normal anterior corpus callosum and frontal lobe white matter *(black arrows).* The posterior corpus callosum and splenium are absent *(open arrow).* The occipital horns are dilated and extend more medially than normal because of the absence of the forceps major *(white arrows).* **(D)** Total agenesis of the corpus callosum, as seen by angiography. The pericallosal artery has a meandering posterior and superior course instead of its normal curving sweep around the genu of the corpus callosum *(arrows).*

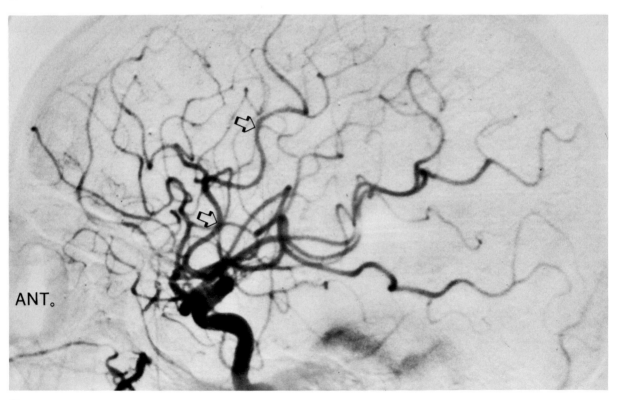

ANT.

D

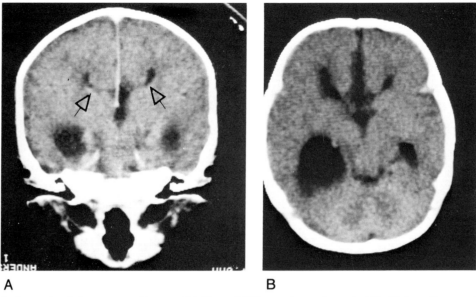

Fig. 12-9 Agenesis of the corpus callosum. (A) Coronal CT scan shows the typical "bull's horns" separation of the frontal horns *(arrows)*. The third ventricle rides high between the lateral ventricles. **(B)** Transaxial CT scan shows widening of the frontal horns and posterior extension of the interhemispheric fissure. (From Fitz, 1984, with permission.)

proceeds posteriorly. Therefore, with partial absence, the anterior genu is formed, but there is aplasia of the body or posterior splenium (Figs. 12-6 and 12-8B&C). The anomaly occurs most commonly in association with other defects of brain formation, particularly Chiari II, Dandy-Walker cyst, Aicardi syndrome (chorioretinopathy and mental retardation), and holoprosencephaly. Callosal dysgenesis may occur as an isolated defect, causing minor mental impairment.

Callosal dysgenesis is best diagnosed with sagittal MRI (Fig. 12-8) but may also be diagnosed with CT scanning (Fig. 12-9). The corpus callosum is absent or partially absent. Large longitudinal parasagittal white matter tracts, called the bundles of Probst, indent and splay the frontal horns, causing a "bull's horns" appearance on the coronal scan (Fig. 12-9). The frontal horns are small with sharp lateral angles. The third ventricle is usually large and superiorly positioned. Owing to the absence of the splenium and the forceps major, the atria and occipital horns may be large and extend medially to nearly the cortical interhemispheric surface of the occipital lobe (Figs. 12-8C and 12-10). This condition of posterior horn dilation is termed *colpocephaly*.

The interhemispheric sulci are radially oriented and reach the ventricular surface (Fig. 12-8B). There may be an associated interhemispheric arachnoid cyst or lipoma.

Lipoma

Intracranial lipomas almost arise near the midline, usually in or along the corpus callosum, at the collicular plate, or the tuber cinereum. They are thought to represent a form of dysraphism. Large interhemispheric lipomas are found in association with seizures and callosal dysgenesis (Fig. 12-10). Small lipomas are usually incidental findings (Fig. 12-11). Rarely, they are found within the cerebellopontine angle cistern or the internal auditory canal.

Holoprosencephaly

Holoprosencephaly refers to failure of the brain to separate into hemispheres. The spectrum of abnormality is classified into three forms.

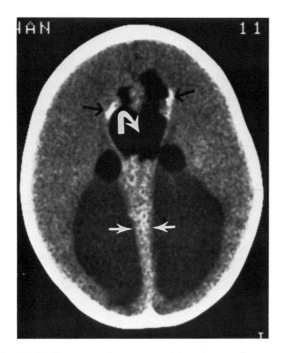

Fig. 12-10 Lipoma with agenesis of the corpus callosum. A large multilobulated interhemispheric lipoma is present (*white curved arrow*) associated with agenesis of the corpus callosum. There is calcification associated with the lipoma (*black arrows*). Note the medial enlargement of the atria of the lateral ventricles due to the absence of posterior white matter fibers (*white arrows*).

Alobar holoprosencephaly. There is complete failure to form separate hemispheres and lobes. The mass of the brain is fused, and there is a central monoventricle. A large posterior cyst is usually present, thought to arise from the roof of the third ventricle. It fills the posterior calvarium, compressing the brain anteriorly (Fig. 12-12). There is no falx. The thalami are completely fused.

Semilobar holoprosencephaly. This form is less severe. The posterior occipital and temporal lobes separate partially with partial differentiation of the ventricles into occipital and temporal horns. The frontal half of the brain is fused, but there is only partial fusion of the thalami (Fig. 12-13).

Lobar holoprosencephaly. This form is least severe. Only the anteroinferior part of the frontal lobes are fused.

The frontal horns of the lateral ventricles are "squared off," and the septum pellucidum is absent. The interhemispheric fissure and falx are nearly complete. The occipital and temporal horns are well formed; a third ventricle separates the thalami (Fig. 12-14).

Facial anomalies are common with holoprosencephaly. There is always hypotelorism. The other common facial abnormality is absence of the intermaxillary segment (the middle third of the upper lip, the palate, and the incisors). A spectrum of more severe abnormalities also occurs, including cyclopia.

Septo-optic Dysplasia

Septo-optic dysplasia (DeMosier syndrome) is a rare syndrome consisting of hypoplasia of the optic nerves and chiasm, absence of the septum pellucidum, and dilation of the anterior third ventricle and lateral ventricles. Occasionally, callosal dysgenesis is present. Clinically, this syndrome is characterized by short stature, visual loss, and nystagmus.

Dandy-Walker Syndrome and Cysts of the Posterior Fossa

The Dandy-Walker malformation is thought to be caused by defective development of the roof of the fourth ventricle and by failure of the foramen of Magendie (median aperture of the fourth ventricle) to open. This deformity causes variable expansion of the fourth ventricle. When mild dilation occurs, it is called the Dandy-Walker variant. When the "full-blown" malformation occurs from complete obstruction of the outlets, it is called the Dandy-Walker cyst. Large cystic dilation of the fourth ventricle causes inferior vermian aplasia, cerebellar hemispheric hypoplasia, and expansion of the posterior fossa.

With the classic Dandy-Walker cyst, the entire posterior fossa is filled with a cyst that balloons upward, often through the incisura of the tentorium. The torcula is high, and the falx cerebelli is absent. The foramen of Magendie is closed (Fig. 12-15).

The Dandy-Walker variant is seen as a less severe form of the malformation with a somewhat formed but enlarged fourth ventricle that communicates with a retro-

(*Text continued on page 472*)

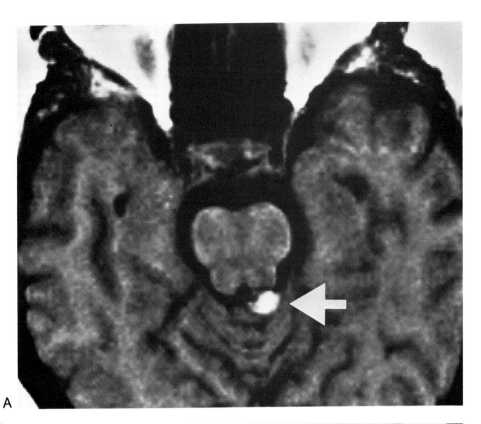

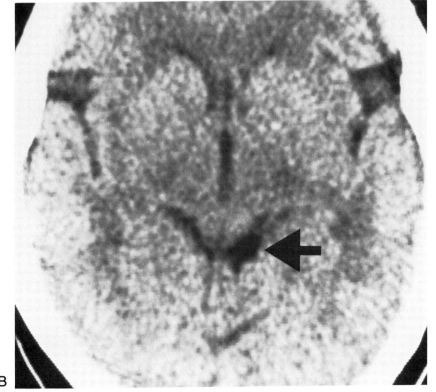

Fig. 12-11 Lipoma. (A) T$_1$-weighted MRI shows a small lipoma in the quadrigeminal plate cistern as being very hyperintense *(arrow)*. **(B)** CT scan shows the lipoma as a region of hypodensity *(arrow)*. This location is common.

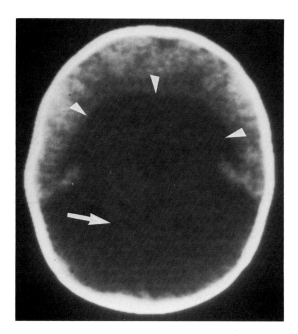

Fig. 12-12 Alobar holoprosencephaly. CT scan shows the anterior monoventricle *(arrowheads)*. The characteristic large cyst is seen posteriorly *(arrow)*.

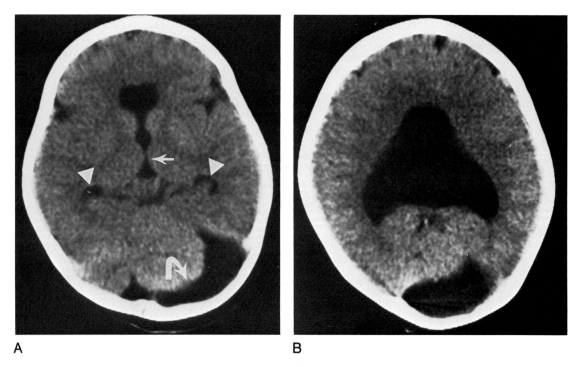

A   B

Fig. 12-13 Semilobar holoprosencephaly. (A) Transaxial CT scan shows anterior fusion of the frontal horns, partial fusion of the basal ganglia *(arrow)*, and differentiation into well formed temporal horns *(arrowheads)*. Note the arachnoid cyst of the posterior fossa *(curved arrows)*. **(B)** Scan at a slightly higher level shows the fusion of the body of the lateral ventricles and separational into atria. The falx is absent anteriorly.

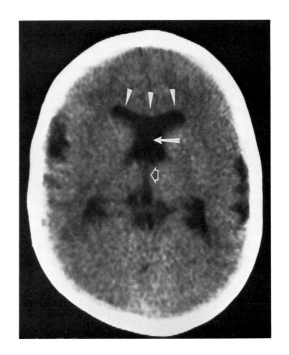

Fig. 12-14 **Lobar holoprosencephaly.** Transaxial CT scan shows enlargement of the fused frontal horns with a "squared off" anterior margin (*arrowheads*). The septum pellucidum is absent (*arrow*). There is a well-formed third ventricle and separation of the basal ganglia (*open arrow*).

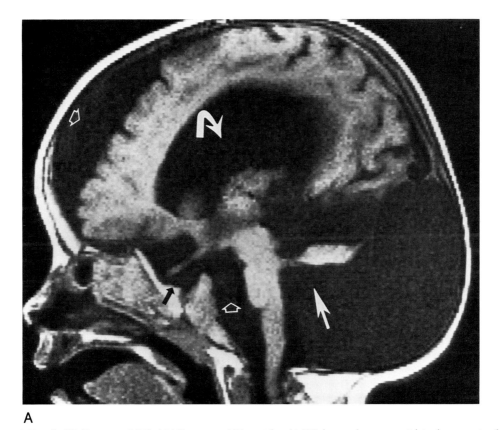

A

Fig. 12-15 **Dandy-Walker cyst, MRI. (A)** Parasagittal T$_1$-weighted MRI shows a large cyst within the posterior fossa in continuity with the fourth ventricle (*arrow*). There is hypoplasia of the cerebellum, and hydrocephalus is present (*curved arrow*). There are also arachnoid cysts over the frontal convexity and in the prepontine and suprasellar cisterns (*open arrows*). The sella turcica is grossly enlarged (*black arrow*). (*Figure continues.*)

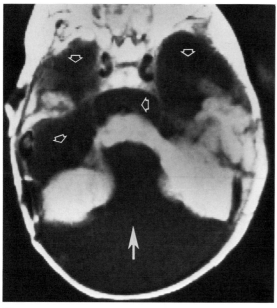

B

Fig. 12-15 *(Continued).* **(B)** Transaxial MRI shows the large cyst in the posterior fossa and its wide communication with the fourth ventricle *(white arrow)*. Arachnoid cysts are seen in the temporal fossae and the prepontine cisterns *(open arrows)*.

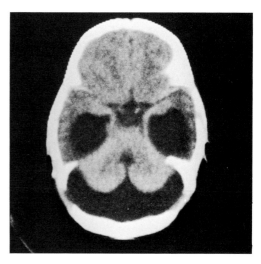

Fig. 12-16 Dandy-Walker variant. A moderately large posterior fossa cyst is connected to the relatively normal fourth ventricle by a small interhemispheric cleft. Hydrocephalus is present with large temporal horns. (From Fitz, 1984, with permission.)

cerebellar cyst. There is less severe vermian and cerebellar hypoplasia and a relatively normal-size posterior fossa (Fig. 12-16).

Hydrocephalus is commonly present but, surprisingly, not always. Dysgenesis of the corpus callosum, holoprosencephaly, and cortical dysgenesis are often associated anomalies.

The retrocerebellar pouch (of Blake) is an arachnoidal cyst that evaginates from the choroid of the fourth ventricle behind normally formed cerebellar hemispheres and vermis (Fig. 12-17). It communicates with the fourth ventricle and subarachnoid space. Although it usually does not cause problems, it is probably responsible for what is often called the "mega cisterna magna." Differentiation between a cyst and a large cisterna magna is not necessary unless hydrocephalus or a posterior fossa mass effect is present.

Cerebellar Hypoplasia

There are varying degrees of cerebellar and vermian hypoplasia. The vermis forms in a rostrocaudal direction. Therefore, there is either complete vermian agenesis or aplasia of the inferior portion (Fig. 12-18). It may form part of Down or Joubert syndrome. Severe cerebellar hypoplasia is Chiari IV malformation.

Craniosynostosis

Craniosynostosis is considered a result of deformity of the base of the calvarium, which secondarily causes failure of proper formation of the cranial sutures. The abnormality is most often present at birth. It is generally first recognized by skull deformity.

The abnormality consists of varying lengths of either fibrous or bone fusion of the cranial sutures. One or more of the sutures may be partially or completely fused. With time, a bone ridge forms at the suture line. Usually only some of the cranial sutures are involved, resulting in asymmetric deformity of the calvarium (plagiocephaly).

Cosmetic deformity of the skull is the major problem of craniosynostosis. Constriction of brain growth and increased intracranial pressure occur only with synostosis of all the cranial sutures. The diagnosis of craniosynostosis is made by plain skull radiography or by CT scanning

Fig. 12-17 Arachnoid cyst (retrocerebellar pouch of Blake). CT scan shows CSF-containing space behind normally formed cerebellar hemispheres and cerebellar vermis. There is no mass effect or hydrocephalus.

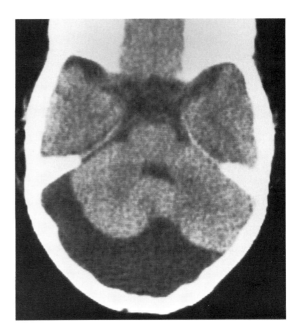

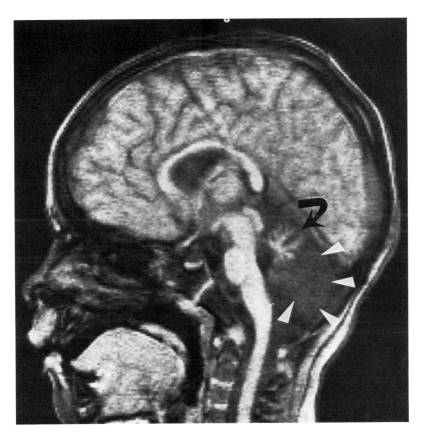

Fig. 12-18 Cerebellar hypoplasia. Severe hypoplasia of the vermis is present, although portions of the superior vermis remain *(curved arrow)*. There is an arachnoid cyst (pouch of Blake) in the posterior fossa *(arrowheads)*.

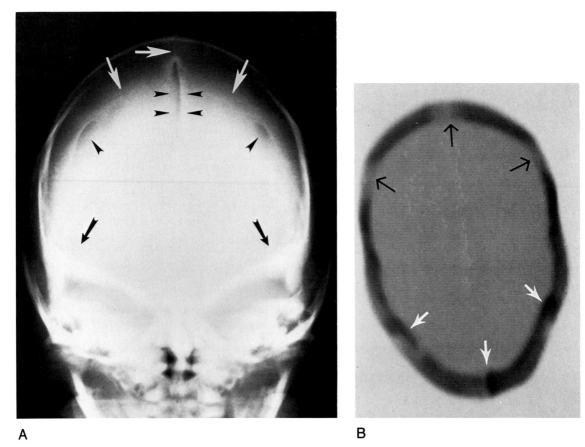

A   B

Fig. 12-19 Craniosynostosis. (A) Plain AP skull radiograph shows complete bone fusion of portions of the sagittal and coronal sutures (*white arrows*). Parts of the sutures show fibrous union with sclerotic margins (*arrowheads*). Note the deformity of the orbits from abnormal skull growth (*black arrows*). **(B)** Bone window CT images show the bone fusion and ridging of the posterior sagittal and lambdoid sutures (*white arrows*). The metopic and inferior coronal sutures appear open (*black arrows*).

with bone images. Multiple tangential views may be required if there is partial stenosis of only one suture (Fig. 12-19).

MIGRATION ANOMALIES

Cortical neurons originate in the paraventricular germinal matrix. From here they migrate peripherally in an orderly fashion along the glial fibers, to populate the cerebral cortex. Neuronal migration begins at about 4 weeks gestation following completion of tubulation of the brain. It reaches maximum activity at 8 to 16 weeks gestation. After the cortex is populated with neurons, organization into layers occurs to about 30 weeks gestation. Interrupted migration or cortical organization results in a thickened disorganized cortical layer.

There are multiple types of migration anomaly (Table 12-3). The pattern of cortical abnormality depends on the timing and severity of the interruption of the process and on whether the interruption is diffuse or focal. The abnormally formed cortex is always thicker than normal.

Table 12-3 Migrational Anomalies and the Cortical Dysplasias

Lissencephaly (agyria)	Diffuse, severe
Pachygyria	Diffuse, less severe
Schizencephaly	Focal, severe, early < 22 wk
Polymicrogyria (cortical dysplasia)	Focal, less severe, later (22–26 wk)
Polymicrogyria with white matter gliosis	Focal, > 26 wk
Gray matter heterotopia	Focal least severe

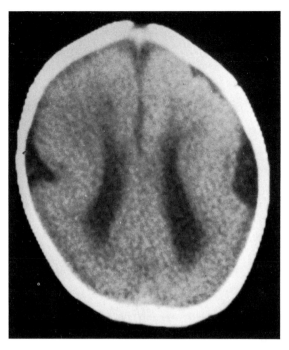

Fig. 12-20 Lissencephaly. CT scan shows a smooth cortical surface, thickened gray matter, wide sylvian fissures, and mild ventricular dilation. (From Fitz, 1984, with permission.)

The underlying white matter is small, as there are correspondingly fewer axons present. If the interruption of migration occurs after 26 weeks gestation, gliosis of the underlying white matter occurs also, and shows as high T_2 signal on MRI. There seems to be a strong genetic control of migration, as migrational abnormalities are found with chromosomal anomalies and autosomal recessive inheritance of unknown cause. Migration anomalies may also occur following fetal insult from interruption of oxygenation or of the vascular supply of the germinal martix or cortex (maternal asphyxia, fetal infection, cocaine exposure, and idiopathic causes).

Lissencephaly

Lissencephaly (agyria, pachygyria) is a rare, severe malformation consisting of a total loss of the gyral and sulcal organization of the cerebrum, so the brain appears smooth. This anomaly results from global insult early during gestation. The cortical gray matter is much thicker than normal and the white matter correspondingly smaller. The sylvian fissures have the appearance of shallow, open grooves, creating a figure-of-eight configuration of the brain as viewed in the transaxial plane. The ventricles are somewhat large and relatively formless (Fig. 12-20). It is sometimes familial. *Agyria* refers to absence of a gyral pattern and is the most severe form. *Pachygyria* refers to broadening of the gyri and is thought to be a milder form of the malformation. The white matter is smallest in volume with the more severe agyria pattern. These conditions are associated with microcephaly and severe retardation.

Schizencephaly

Although the precise etiology of schizencephaly is unknown, it probably results from an early focal vascular injury in the germinal matrix. This interrupts the neuronal migration process at its origin. The result is a full-thickness cleft in the cerebral cortex that may be unilateral or, more commonly, bilateral.

There is an infolding of the adjacent cortex along the margins of the cleft, with the intervening pia becoming fused with the ependyma of the subjacent ventricle. The ventricle communicates with the subarachnoid space through the cleft. Dysplastic gray matter lines the cleft from the ventricle to the cerebral cortex. The septum pellucidum is usually absent. Callosal dysgenesis occurs at the cleft, and cortical dysplasia is frequently found at distance sites, or in the opposite hemisphere.

There are two types of schizencephaly. Type I is mild with a small slit defect ("closed lips") and normal ven-

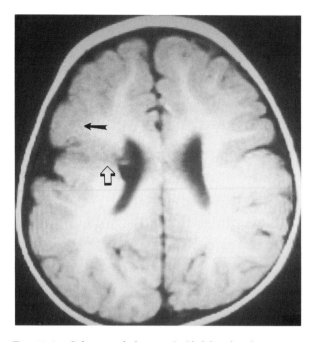

Fig. 12-21 Schizencephaly, type I. Cleft lined with gray matter extends from the cortex to the body of the right lateral ventricle *(open arrow)*. Note the adjacent pachygyria *(solid arrow)*.

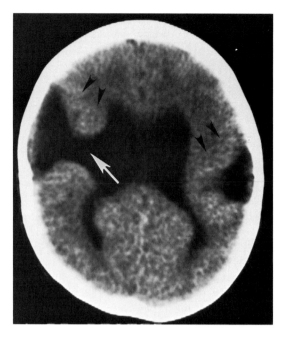

Fig. 12-22 Schizencephaly, type II. CT study shows the characteristic deep clefts extending to the ventricular system *(white arrow)*. The margins of the clefts are lined with cortical gray matter *(arrowheads)*.

tricles (Fig. 12-21). Type II is more severe, with larger fan-like cortical defects ("open lips") and ventriculomegaly (Fig. 12-22). The clinical dysfunction varies according to the severity of the defects.

Polymicrogyria

Polymicrogyria (cortical dysplasia) occurs relatively frequently and is less severe than the above migrational malformations. It is thought to result from vascular interruption of the cortex and white matter at 22 to 30 weeks gestation. It can be imaged only with MRI. It is characterized by a regional thickening of the cortex with excessive but shallow and poorly formed convolutions or surface pits (Fig. 12-23). The normal finger-like subcortical projections of the white matter are absent. High T_2 signal is present in the white matter, representing gliosis that occurs in response to

injuries after 26 weeks gestation (Fig. 12-24). The condition may be seen alone or in association with other anomalies, particularly schizencephaly. The opercular regions are characteristically involved. Polymicrogyria is to be distinguished from the condition of too many but normally formed gyri called stenogyria. Stenogyria is seen with the Chiari II malformation (Fig. 12-6).

Heterotopic Gray Matter

Heterotopic gray matter may occur anywhere in the white matter and is the least severe migrational abnormality. The abnormal nodules of neurons are thought to result from mild focal interruption of migration of neuroblasts. They are seen on MRI (and less obviously on CT scans) as foci of gray matter within the white matter (Fig. 12-25A&B). Most often there are small foci of gray

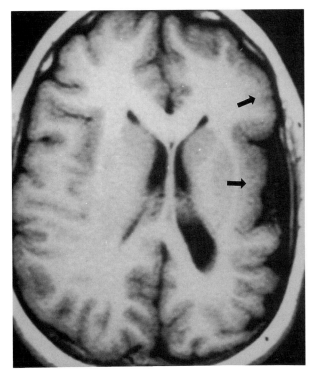

A

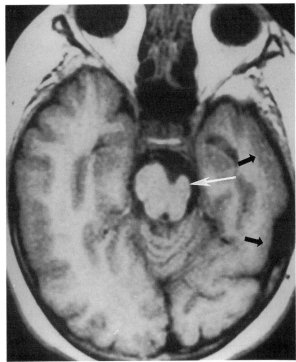

B

Fig. 12-23 Polymicrogyria, hypoplastic crus. (A) Axial T_1 MRI illustrates the nearly smooth cortex with surface crypts indicating polymicrogyria of the left frontal lobe and insula *(arrows)*. The left hemisphere is small, causing ipsilateral ventricular shift and enlargement of the left lateral ventricle. Compare this with the normal right cortex. **(B)** Axial T_1 MRI through the lower brain showing the severely hypoplastic left cerebral peduncle *(white arrow)* and the smooth cortical surface of the left temporal lobe *(black arrows)*. **(C)** Widespread pachygyria involves both frontal lobes, insulas, and posterior temporal lobes.

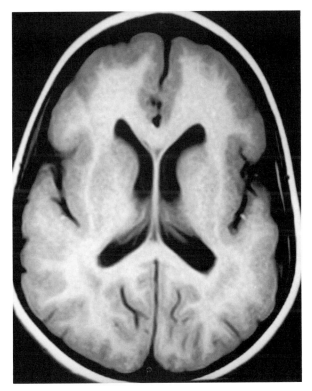

C

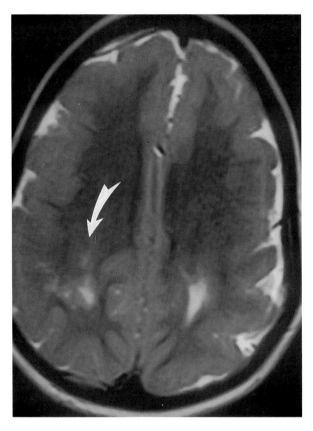

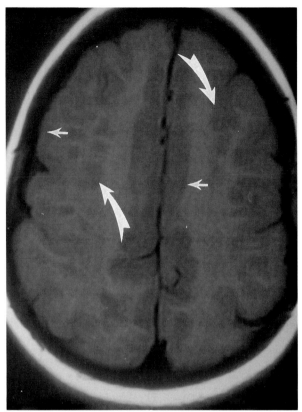

Fig. 12-24 Polymicrogyria, high T_2 signal. High T_2 signal may form within the white matter from gliosis in some with polymicrogyria that occurs later in utero (*arrow*, and see text).

Fig. 12-25 Neuronal heterotopia and polymicrogyria. Axial T_1 MRI shows abnormally positioned gray matter within the subcortical white matter bilaterally (*larger arrows*). When contiguous and extending along the whole subcortical region it is referred to as band type of hetertopia. The gyri are smooth (*smaller arrows*).

matter located near the ventricle. A larger contiguous mass of hetertopias seen in the white matter parallel to the cortex is called band heterotopia. There is no associated gliosis. Rarely, they may be found within the cerebellum. Heterotopia may occur as an isolated event or may be found as part of other migrational or structural malformations, such as dysgenesis of the corpus callosum or Chiari II. They may be associated with seizures.

Megalencephaly

Megalencephaly implies a larger-than-normal brain from any cause. Megalencephaly may be unilateral or bilateral, malformative, or associated with infiltrative or metabolic disorders. Unilateral megalencephaly may be

mild or severe, causing seizures, hemiplegia, and retardation. The MRI is characteristic, showing unilateral ventricular dilation and a thickened distorted cortex of lissencephaly or polymicrogyria. The hemisphere is overgrown, and there is contralateral shift of the falx. The underlying white matter has high T_2 signal intensity due to demyelination. Associate cortical migrational abnormalities are frequent. CT shows the anatomic abnormalities and low density in the white matter. Unilateral megalencephaly associated with somatic hemihypertrophy does not have cortical abnormalities. Bilateral megalencephaly may be isolate or found with many other conditions (see box). It is frequently associated with mental retardation.

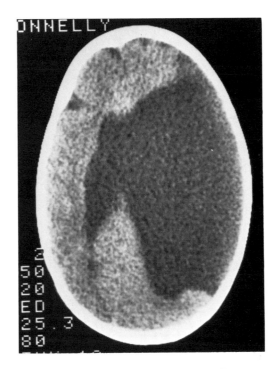

Fig. 12-26 **Porencephaly and hemiatrophy.** A large region of porencephaly is present on the left that has caused enlargement of the left side of the calvarium.

CAUSES OF BILATERAL MEGALENCEPHALY

Tay-sachs disease

Mucopolysaccharidoses

Alexander's disease

Canavan's disease

Tuberous sclerosis

Lead intoxication

DESTRUCTIVE LESIONS

Porencephaly and Atrophy

Porencephaly refers to a cyst in the brain parenchyma that usually communicates with the ventricular system, but not with the subarachnoid space. Most cysts are encephaloclastic; that is, they occur as a result of brain destruction, such as from infarction or severe hemorrhage. They are seen with CT or MRI as large CSF-containing structures within the brain (Fig. 12-26A). They normally do not cause any mass effect, except when large and partially obstructed or when hydrocephalus is also present. Porencephaly is the condition resulting from an encephaloclastic insult after neuronal migration is completed. Intrauterine events causing porencephaly occur after 30 weeks gestation.

Hemiatrophy of the brain may result from events in utero or during early infancy. One hemisphere becomes smaller with dilation of the ipsilateral ventricles and sulci. When it occurs early in life, the skull on the side of the side of the atrophy becomes thicker, and the sinuses enlarge to fill the void. This condition is called Dyke-Davidoff-Mason syndrome (Fig. 12-27A–C).

Hydranencephaly

Hydranencephaly is severe destruction of the cerebral hemispheres, thought to be a result of bilateral occlusion of the internal carotid arteries and massive infarction. It could be a result of abnormal histogenesis. CT scans and MRI show a fluid-filled cranial cavity with preservation of the occipital lobes, cerebellum, and basal ganglia (Fig. 12-28). If a small amount of brain tissue is present, this entity cannot easily be differentiated from severe hydrocephalus (Fig. 12-29). Angiography shows an atretic carotid vascular system with hydranencephaly and stretched but normal vessels with hydrocephalus.

METABOLIC AND NEURODEGENERATIVE DISORDERS

The metabolic encephalopathies generally affect both the gray and white matter, but it is the white matter change that is most easily recognized on imaging. Therefore, these diseases are often grouped as leukodystrophy. White matter abnormality with dysmyelina-

(Text continued on page 483)

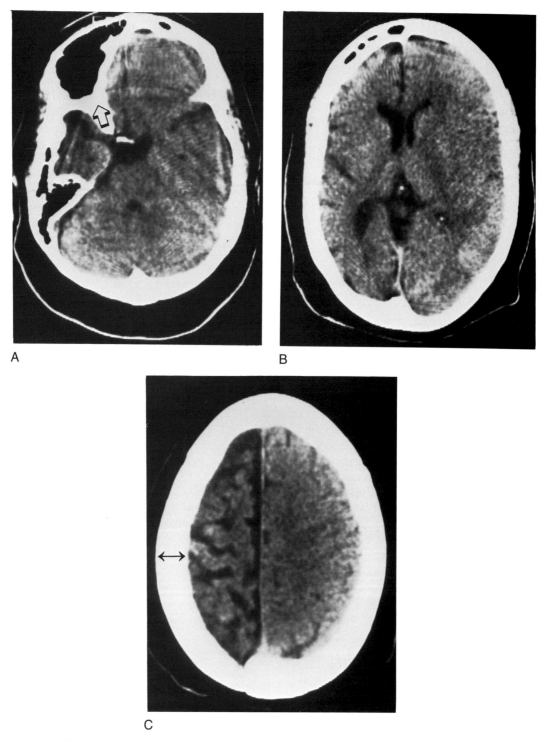

A

B

C

Fig. 12-27 Dyke-Davidoff-Mason syndrome. These three CT atrophic scans show the effects from hemiatrophy, which occurred in utero or during infancy. The right hemisphere is atrophic with dilation of the ventricular system and cerebral sulci. There is gross enlargement of the frontal *(open arrow)* and mastoid sinuses and elevation of the ipsilateral sphenoid ridge, anterior clinoid process, and petrous pyramid. The right hemicalvarium is thickened *(opposed arrows)*.

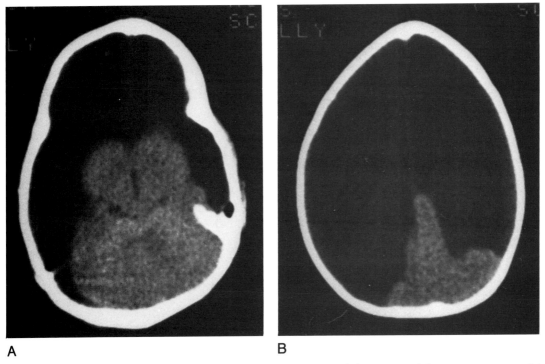

A   B

Fig. 12-28 Hydranencephaly. (A & B) Transaxial CT scan shows total destruction of the anterior portion of the cerebral hemispheres, with CSF filling the cranial cavity. The cerebellum, thalamic nuclei, and portions of the occipital lobes are preserved.

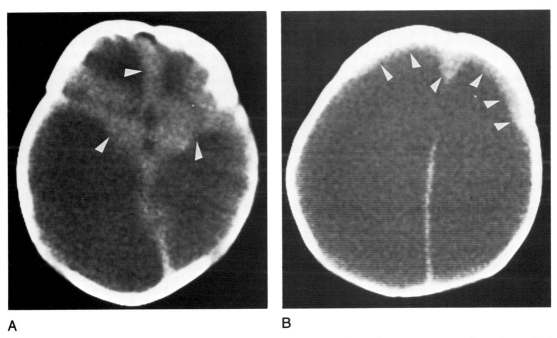

A   B

Fig. 12-29 Severe hydrocephalus. (A) CT slice through the base of the brain shows preservation of some hemispheric tissue *(arrowheads)*. **(B)** More superiorly, a small amount of cortical mantle can be seen *(arrowheads)*. Compare with Figure 12-28.

481

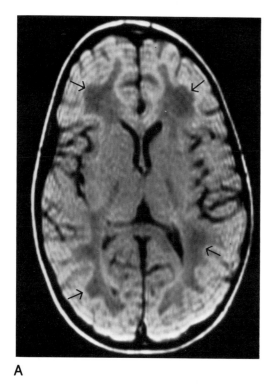

A

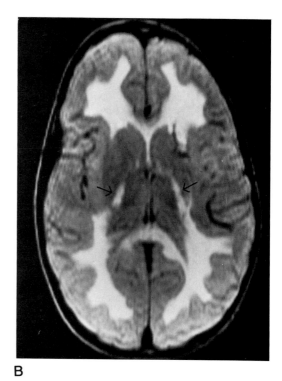

B

C

Fig. 12-30 Dysmyelinating disease. (A) T_1-weighted MRI shows the hypointense white matter in this 2-year-old with an undiagnosed leukodystrohy *(arrows)*. There is mild atrophy. (B) T_2-weighted MRI shows the extreme hyperintensity of the white matter, including the posterior limbs of the internal capsules *(arrows)*. (C) CT scan of the same patient shows the leukodystrophy as diffuse hypodensity of the centrum semiovale.

Table 12-4 Dysmyelinating Disease

Disease	Age of Onset	Distribution	Laboratory Findings	Specific Finding
Metachromatic leukodystrophy	<2 mo	Diffuse	Arylsulfatase A in urine	
Canavan's spongy degeneration	2–9 mo	Bilateral, variable	0	Large head
Krabbe's disease (globoid cell leukodystrophy)	4–6 mo	Severe atrophy (with dysmyelination)	β-Galactocidase deficiency	High CT density in thalami and cerebellum
Alexander's disease	<1 yr	Frontal white matter	Rosenthal bodies	Large head
Cockayne's disease	<2 yr	Patchy white matter; calcium in basal ganglia and cerebellum	0	Absent subcutaneous fat
Adrenoleukodystrophy	4–6 yr, males only	Occipital lobes		Primary adrenal failure
Pelizaeus-Merzbacher disease	Infancy, males	Diffuse	0	0
Aminoacidurias	Infancy	Diffuse	Specific deficit	0
Leigh's disease	Infancy, childhood	Basal ganglia, white matter	Mitochondrial disorder	0

tion is a prominent feature of most of the different entities.

Dysmyelination refers to the formation of abnormal myelin, most often a result of a deficient enzyme. Most are inherited as autosomal recessive. They share many features. MRI shows the regions of dysmyelination as high signal intensity on T_2-weighted imaging and as low intensity on T_1-weighted images (Fig. 12-30). CT scans demonstrate the abnormal myelin as regions of low density. Some specific types of dysmyelination have distribution signatures, but in general the specific diagnosis cannot be made accurately with imaging. Histologic or chemical confirmation is necessary. Atrophy usually accompanies the process. The characteristics of the dysmyelination diseases are given in Table 12-4. The leukodystrophies are now classified as disorders of lysosomes, peroximsomes, mitochondria, and amino acid metabolism. Some remain idiopathic.

LYSOSOMAL DISORDERS

Lysosomes are intracellular structures that package hydrolases and other enzymes within the endoplasmic reticulum of the cell. They are involved normally in the orderly breakdown of cytoplasmic constituents. When enzymic defects occur within the lysosome, an abnormal accumulation of the target material results in one of the storage diseases. The material may be lipid, ganglioside, lipofuscin, mucopolysaccharide, mucopolysaccharide and lipid, or glycogen. Both gray and white matter are most often affected with liposomal disease.

Metachromatic Leukodystrophy

Metachromatic leukodystrophy, with autosomal recessive inheritance, is the most common of the leukodystrophies. Its cause is a deficiency of arylsulfatase A, resulting in an accumulation of sulfatides in the brain. It produces a diffuse nonspecific pattern of dysmyelination and moderate atrophy on both MRI and CT scanning. Metachromatic material is found in the urine. Onset is usually during infancy, but delayed juvenile and adult forms exist. Progressive mental deterioration ending in death within 2 to 4 years is characteristic.

Krabbe's Disease

The enzyme deficit in Krabbe's disease causes an accumulation of galactocerebroside within the brain. Characteristic globoid cells are found within the white matter. It occurs in early infancy, with rapid progression to death by age 2 years. Macrocephaly is common. There is nonspecific diffuse dysmyelination with severe atrophy. Late-onset forms exist.

Fabry's Disease

Fabry's disease is rare with sex-linked inheritance. Glycosphingolipid accumulates within blood vessels

---

LYSOSOMAL DISORDERS

Storage diseases

Lipidoses

Metachromatic leukodystrophy

Krabbe's leukodystrophy

Fabry disease

GM_2 gangliosidosis (Tay-Sachs)

Neuronal ceroid lipofuscinosis

Mucopolysaccharidoses

I Hurler/Scheie (formerly V)

II Hunter

III Sanfilippo

IV Morquio

VI Maroteaux-Lamy

VII Sly

Mucolipidoses

I–IV Mucolipidosis

Mannosidosis

Fucosidosis

Glycogenosis

I1 Pompe's disease

---

throughout the body, including the brain. Infarcts are the most common imaging abnormality, particularly within the basal ganglia in young adults. Cutaneous angiectasis occurs over the lower abdomen and thighs in children, along with premature hypertension.

Tay-Sachs Disease

Tay-Sachs disease occurs in Ashkenazi Jews and is caused by the accumulation of a ganglioside. Neuronal death and white matter deterioration occur. Initially, there is enlargement of the caudate nuclei and high T_2 signal within the basal ganglia and thalamus, rapidly progressing to diffuse atrophy.

Neuronal Ceroid-Lipofuscinosis

Neuronal ceroid-lipofuscinosis is a rare disease of lipofuscin accumulation. It causes atrophy of the brain, with high T_2 signal on MRI within the white matter and low

signal within the basal ganglia and thalamus. Some forms demonstrate only atrophy on MRI.

Mucopolysaccharidoses

There are multiple forms of the mucopolysaccharidoses. Abnormal accumulation occurs within histiocytes in the Virchow-Robin spaces, causing striking enlargement of these structures in the basal ganglia and corpus callosum. Marked thickening of the meninges of the brain and spine are also characteristic. This causes hydrocephalus and cord compression in some cases.

PEROXISOMAL DISEASES

Adrenoleukodystrophy

Classic adrenoleukodystrophy is a sex-linked (X-linked) inherited disease in males associated with adrenal insufficiency. It is caused by a single peroxisomal enzyme deficit of acyl-CoA synthetase. There is visual and intellectual impairment. Abnormal fatty acids accumulate in the lipids of the cerebral white matter and adrenal glands. The occipital lobes are predominantly involved. The brain lesion consists of central necrosis, more peripheral inflammatory active demyelination and a still more peripheral penumbra of edema. Clinical onset is usually at age 4 to 6 years. Neurologic symptoms precede adrenal insufficiency in most cases.

The CT scan typically shows a bilateral, symmetric, low-density abnormality in the occipital lobe white matter (Fig. 12-31). There may be mass effect. Rim contrast enhancement with surrounding peripheral edema indi-

---

PEROXISOMAL DISEASES

Adrenoleukodystrophy

X-linked type

Classic adrenoleukodystrophy

Adrenomyeloneuropathy

Neonatal

Zellweger syndrome

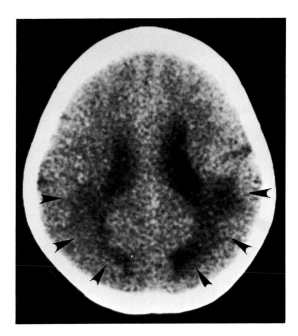

Fig. 12-31 Adrenoleukodystrophy. Transaxial CT scan shows bilateral symmetric hypodensity in the posterior temporal and occipital white matter (*arrowheads*).

cates the active phase of the disease. As the disease advances, the white matter changes extend anteriorly adjacent to the trigones of the lateral ventricles. The splenium of the corpus callosum is frequently affected. The entire hemispheres may be involved in advanced disease. Early in the disease, the changes may be unilateral, mimicking a tumor. Occasionally, calcification is seen within the central necrotic portion.

The white matter lesions are demonstrated on MRI as a high-intensity abnormality on T_2-weighted images. More extensive involvement can also be seen with MRI, including lesions in the brain stem and visual and auditory pathways.

An autosomal recessive form of the disease presents during infancy, causing early death. Polymicrogyria may also be present.

The less common adrenomyeloneuropathy has a later onset (20 to 30 years) and involves mostly the spinal cord and peripheral nerves. Thoracic spinal cord atrophy is the most common imaging finding. The brain is affected in 10 percent of cases.

Neonatal adrenoleukodystrophy results from lack of multiple peroxisomal enzymes. It presents at birth with a diffuse dysmyelination. Its inheritance is autosomal recessive.

Zellweger Syndrome

Zellweger syndrome is an autosomal recessive disorder caused by lack of many peroxizomal enzymes. A progressive psychomotor retardation presents during infancy. There are both dysmyelination and various migrational abnormalities (agyria/pachygyria).

MITOCHONDRIAL DISORDERS

The mitochondrial disorders are a spectrum of clinical entities caused by functional abnormalities of the mitochondria. This leads to various interferences in the oxygenation of the cells. Mitochondrial dysfunction causes abnormality in the gray matter, particularly the basal ganglia, but also the cerebral cortex and brain stem. The white matter is not primarily affected, unless by infarctions of larger regions of tissue, as with MELAS, discussed below.

Leigh's Disease

Leigh's disease has X-linked inheritance and a number of mitochondria deficiencies. It causes necrosis within the basal ganglia and tegmentum of the brain stem (Fig. 12-32). There are infantile, juvenile, and adult forms. It

---

MITOCHONDRIAL DISORDERS

Leigh's disease (subacute necrotizing encephalomyelopathy)
MELAS
MERRF
Kearns-Sayre syndrome
Menkes kinky hair syndrome

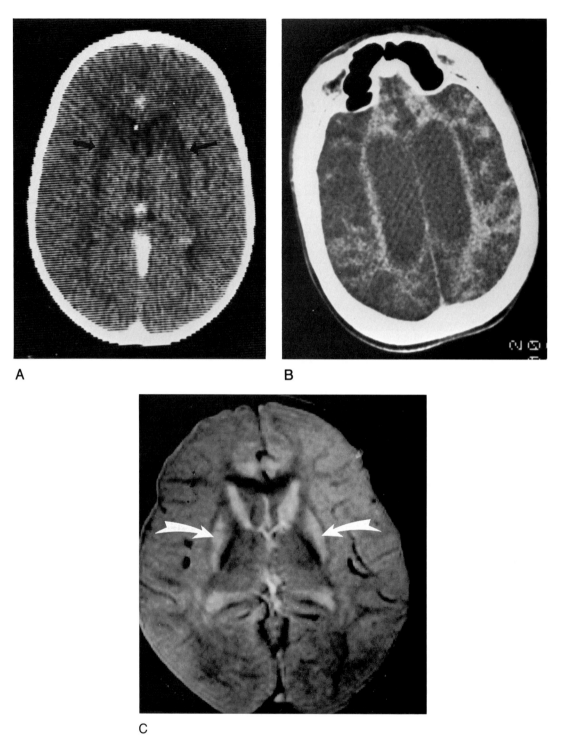

Fig. 12-32 Leigh's disease. (A) CT study shows hypodensity involving the basal ganglia and adjacent white matter. **(B)** Scan done many years later shows severe progression of the necrotizing process with development of global atrophy. The calvarium is thickened and the sinuses are enlarged because of the brain atrophy. **(C)** Axial T$_2$ MRI shows high signal bilaterally within the lenticular nuclei *(arrows)*.

is slowly progressive to death. Generalized atrophy occurs in the end stages of the disease. Kearns-Sayre is an autosomal dominant variant of Leigh's disease. CT and MRI show necrosis within the basal ganglia and brain stem, as well as the severe late atrophy.

Melas/Merrf/Menkes

Mitochondrial myopathy, encephalopathy, lactic acidosis, and strokes (MELAS) is associated with an inherited defect in transfer RNA (tRNA) within the mitochondria. Infarcts not corresponding with known vascular territories produce th characteristic CT and MRI findings. (Fig. 12-33).

Myoclonic epilepsy with ragged red fibers (MERRF) has brain findings similar to those of MELAS. However,

muscle disease is prominent and there are often endocrinopathies.

Menkes disease is a disorder of copper metabolism and mitochondria dysfunction. The vessels are abnormally tortuous. Infarctions occur within the brain. Hair is characteristically kinky.

AMINOPATHIES

The aminopathies (aminoacidurias) are autosomal inherited disorders of amino acid metabolism. Amino acids are essential for the formation of myelin, so the predominant MRI finding is dysmyelination in a nonspecific pattern. The changes associated with glutaric aciduria, type I are more similar to those of mitochondrial disorders, with predominant basal ganglia necrosis. Homocystinuria involves the blood vessels, primarily causing arterial and venous cerebral infarcts.

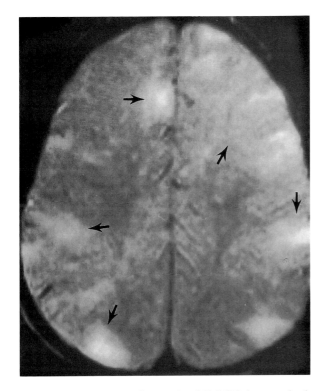

Fig. 12-33 MELAS syndrome. Axial T$_2$ MRI shows multiple regions of bright signal involving the cortex and white matter of both hemispheres (*arrows*). The lesions are widespread and variable. They suggest infarction but do not correspond with specific vascular territories.

AMINOPATHIES

Phenylketonuria
Homocystinuria
Glutaric acidemia, type I
Maple syrup urine disease
Tyrosinemia
Oculocerebral-renal syndrome
Nonketotic hyperglycinemia
Hartnup's disease

LEUKODYSTROPHY, PRIMARY

Spongy Degeneration (Canavan's Disease)

Many biochemical disorders cause spongy degeneration and dysmyelination, particularly the aminoacidurias. When no specific etiology can be determined, the process is referred to as Canavan's disease. It has autosomal inheritance and is particularly common among European Jews. Macrocephaly is common. Death usually occurs within 2 years.

Pelizaeus-Merzbacher Disease

Pelizaeus-Merzbacher disease is a rare diffuse nonspecific leukodystrophy that affects males starting in early infancy. Its cause is unknown. Cockayne's disease is similar, but with basal ganglia calcification.

Alexander's Disease

Alexander's disease presents during infancy and shows nonspecific diffuse dysmyelination associated with macrocephaly.

NEUROCUTANEOUS DISORDERS

Neurocutaneous disorders encompass a group of congenital disorders characterized by dysplasia or neoplasia of the skin, brain, peripheral nervous system, and eyes

(ectodermal structures). However, abnormalities of bone, blood vessels, and gut may also be included in the syndromes. These diseases are genetically determined. Equivalent terms for this group of disorders are *phakomatosis* (from Greek, meaning "birthmark") and *congenital neuroectodermal dysplasia*.

Neurofibromatosis

Neurofibromatosis is actually a group of disorders. The most common and well-known types are neurofibromatosis-1 (von Recklinghausen's neurofibromatosis, peripheral NF, NF-1) and neurofibromatosis 2 (bilateral acoustic neurofibromatosis, central NF, NF-2). Other types are, for the most part, variants of NF-1 and NF-2; overlap among groups occurs. These are autosomal dominant genetic disorders.

NEUROFIBROMATOSIS 1

NF-1 consists of multiple abnormalities of both ectoderm and mesoderm. The defining features of NF-1 are café-au-lait spots, peripheral neurofibromas, and iris hamarto-

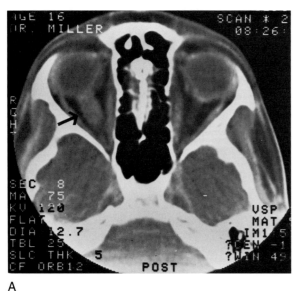

A

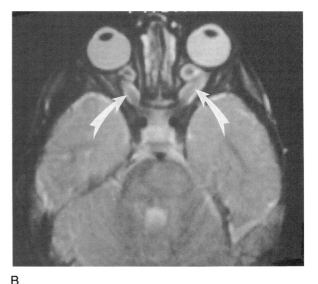

B

Fig. 12-34 NF-1, optic nerve glima. (A) CT scan of the orbits shows enlargement of the right optic nerve (*arrow*). **(B)** Axial T_2 MRI shows enlarged elongated and tortuous optic nerves bilaterally (*arrows*).

mas (Lisch nodules). All persons with NF-1 have these three features but only some of the other features listed below. The syndrome is highly variable.

A far-ranging variety of abnormalities may be seen in the brain. Optic pathway glioma is the most common tumor (Fig. 12-34A&B). This lesion may involve the optic nerves, chiasm, or tracts. It is a slow-growing tumor and is often asymptomatic. Astrocytomas occasionally occur elsewhere in the brain (Fig. 12-35). Malignant tumors are rare.

Ventricular dilation occurs most commonly because of central brain atrophy, but it may also indicate hydrocephalus due to aqueductal stenosis. Macrocephaly is also common. Vascular dysplasia, particularly stenosis of the arteries at the base of the brain, may lead to cerebral infarctions or even moyamoya (Fig. 12-36). Sphenoid wing dysplasia may allow herniation of temporal lobe tissue into the orbit, causing pulsating exophthalmos (Fig. 12-37). A round skull lucency near the lambdoid suture is characteristic. Large craniofacial plexiform neurofibromas may grow into the cranial cavity, par-

ticularly from the orbit (Fig. 12-37). Arachnoid cysts occur.

Regions of high T_2 signal intensity are seen with MRI anywhere in the white matter (Fig. 12-38A and B) and basal ganglia, most commonly within the internal capsules and globus pallidus. Their etiology and significance are presently unknown, but most likely they represent glial dysplasia.

The cranial nerves are only rarely involved with tumor in NF-1. Only the extracranial portion of cranial nerve V (CN V) is frequently involved with a large plexiform neurofibroma. Acoustic neurinoma is rare and occurs almost exclusively in NF-2.

The peripheral nerves, including the parasympathetic and visceral nerves, are extensively involved with neurofibromas (Fig. 12-39). Significant gastrointestinal hemorrhage occurs from tumors involving the bowel wall. Angiography demonstrates the vascular tumors, and often the bleeding site. Multiple small or large neurofibromas may be found anywhere. Large major nerve plexes are frequently involved. Spine erosion is caused by paraspinal tumors. Occasionally, sarcomatous degeneration occurs, which can cause increased pain, but there are no distinguishing imaging features. Pheochromocytoma occurs occasionally.

The spine and spinal cord are affected secondarily from adjacent neurofibromas and scoliosis. Severe spine dysplasia (Fig. 12-40) and intraspinal meningoceles (Fig. 12-41) can be found.

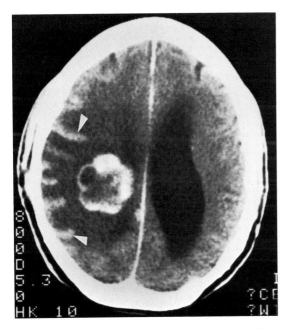

Fig. 12-35 NF-1, anaplastic astrocytoma. Contrast-enhanced CT scan shows an enhancing white matter tumor fund to represent a grade III astrocytoma. Note the pseudo-enhancement of compressed gyri *(arrowheads)*.

NEUROFIBROMATOSIS 2

Bilateral acoustic (vestibular nerve) neurinomas are characteristic of NF-2, occurring in 90 percent of patients with this form of the disease. These tumors may be small or large, slow- or fast-growing. Changes in the rate of growth occurs so the tumors need to be followed at frequent intervals, at least annually. They tend to present during the second or third decade, earlier than sporadic unilateral acoustic neurinoma. Acoustic neurinoma presenting in a person under age 40 suggests NF-2.

Other CNS tumors are common with NF-2. Schwann cell tumors may occur on the cranial and spinal nerves, and occasionally an intramedullary schwannoma is seen, indistinguishable from a spinal cord astrocytoma. Astrocytoma is seen within the cerebrum but almost never in the optic pathways. Meningiomas, often multi-

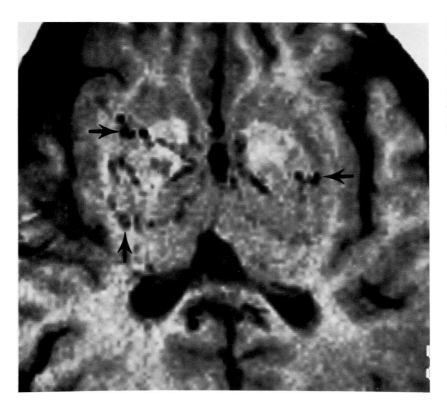

Fig. 12-36 NF-1, moyamoya. T₁-weighted MRI through the base of the brain shows the prominent collateral vascular channels of moyamoya (*arrows*). The collateral vessels have developed as a result of severe vascular dysplasia of the major arteries of the base of the brain.

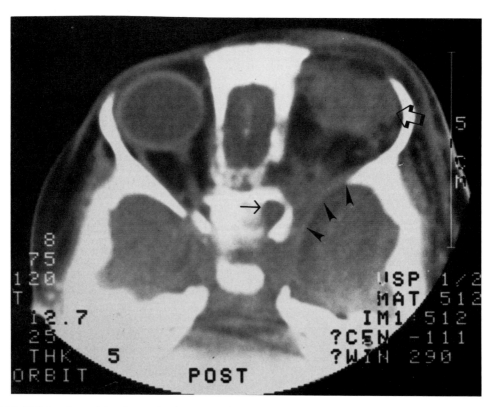

Fig. 12-37 NF-1, sphenoid dysplasia, plexiform neurofibroma. There is dysplasia of the left sphenoid bone, allowing expansion of the left temporal lobe into the posterior orbit (*arrowheads*). A plexiform neurofibroma is present in the anterior left orbit (*open arrow*). The left optic canal is enlarged because of an optic nerve glioma (*arrow*).

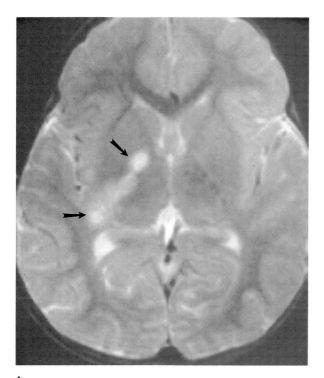

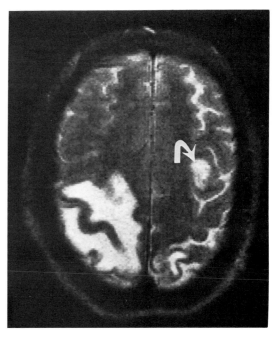

A

B

Fig. 12-38 NF-1, gliodysplasia. (**A**) Axial T₂ MRI (90/2500) illustrates multiple round and oval high signal lesions within the right internal capsule, globus pallidus, and deep posterior insula *(arrows)* — a typical appearance and location for gliodysplasia. These lesions should not be interpreted as tumor or infarction. (**B**) T₂ MRI shows a left hemispheric subcortical lesion thought to be gliodysplasia *(curved arrow)*. This and the cerebellum are less common locations for gliodysplasia. The edema within the right hemisphere represents a glioma not seen on this scan image (see Fig. 12-35).

Fig. 12-39 Plexiform neurofibroma. Axial FSE (4/4) 96/ 3555 T₂ MRI shows a plexiform neurofibroma in the right deep lateral face *(arrow)*. The bright signal within the mass is mostly from cystic change. The rim is dark.

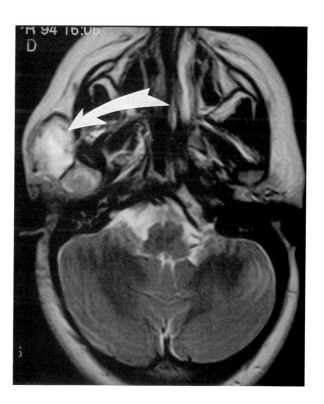

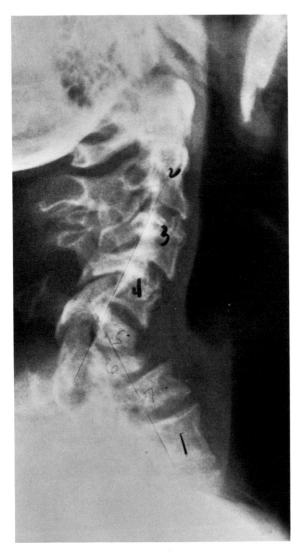

Fig. 12-40 NF-1, spinal dysplasia. Severe spinal dysplasias may occur, causing scoliosis and gibbus deformities.

Tuberous Sclerosis

Tuberous sclerosis is a genetic disorder, with one-half of cases inherited as autosomal dominant. It is widely distributed, with no sex predominance. The characteristic lesion is the hamartoma, which occurs in many organs. Reddish brown papules (adenoma sebaceum) characteristically occur on the face, particularly near the nose. Other skin lesions are shagreen patches, hypomelanotic spots, and subungual fibromas. The syndrome is associated with a high incidence of angiolipoma of the kidney, rhabdomyoma of the heart, and thickening of the calvarium. Periosteal reaction and bone cysts occur, especially in the digits. Rarely, there is lymphangiomatosis of the lung. The diagnosis may be made with MRI or CT scans, but MRI is more sensitive in younger patients before the lesions calcify.

The tuber, the characteristic lesion in the brain, consists of giant cells resembling astrocytes, focal gliosis, and abnormal myelination of regional axons. Tubers occur on the cortical surface, causing a widened gyrus. They may occur anywhere but are seen most commonly in the frontal lobes; they infrequently calcify. The lesions may be seen with MRI as thickened gyri that have central high T_2 signal intensity (Fig. 12-42). Radially oriented high T_2 signal intensity is also seen in the white matter, correlating with regions of abnormal collections of giant cells and gliosis.

Subependymal nodules occur along the ventricular surface, most commonly at the caudate nucleus and in the thalamostriate sulcus. Although these lesions are similar to the cortical tuber, some differences are detected with imaging. A high incidence of calcification is easily seen on CT scans (Fig. 12-43). These lesions tend to have low T_2 signal intensity with MRI. A subependy-

ple, are common. A meningioma in a child suggests NF-2.

NF-2 characteristics are described below. Many of the features of NF-1 are absent, particularly the cutaneous manifestation. However, the division between NF-1 and NF-2 is not clearly defined. NF-2 is much less common than NF-1. Other types of neurofibromatosis have been defined, but they are mixed variants.

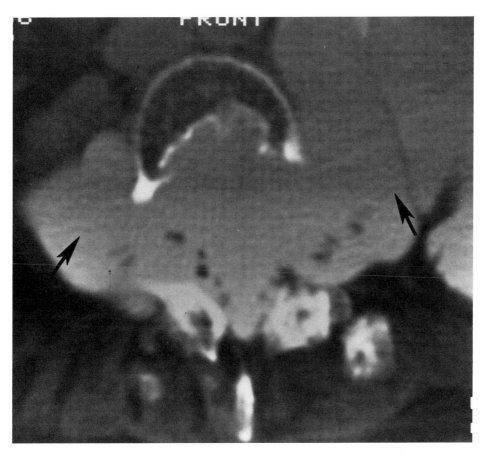

Fig. 12-41 NF-1, intraspinal meningocele. CT myelography shows a large contrast-filled intraspinal meningocele causing enlargement of the spinal canal, erosion of the posterior vertebral body, enlargement of the intervertebral foramina, and extension into the paraspinal space *(arrows)*. Dural dysplasia is common with NF-1.

mal nodule near the foramina of Monro, a characteristic location, may cause hydrocephalus.

Subependymal giant cell astrocytomas occur in about 10 percent of these patients. These lesions tend to be near the foramina of Monro and may become large; they are histologically benign and slow-growing. These tumors can be differentiated from subependymal nodules by their large size, high T_2 signal intensity on MRI, and contrast enhancement on CT scan (Fig. 12-44). Resection is done only for control of hydrocephalus. Other brain tumors are rare.

Von Hippel-Lindau Disease

Von Hippel-Lindau disease is an inherited autosomal dominant disorder that occurs equally in the sexes. The common abnormalities that occur are retinal hemangioblastoma (50 percent), cerebellar or spinal hemangioblastoma (50 percent), renal cell carcinoma (30 percent), and pheochromocytoma (10 percent). Other findings are polycythemia, cutaneous nevi, and pancreatic cysts. Cerebellar hemangioblastoma is described in the section on brain tumors.

Sturge-Weber Syndrome

Sturge-Weber syndrome is probably a noninherited genetic disorder consisting of a cutaneous vascular nevus of the face and of a leptomeningeal angiomatous malformation on the ipsilateral occipital cortical surface. Serpiginous cortical calcification is characteristic. The malformation is described in more detail in Chapter 4.

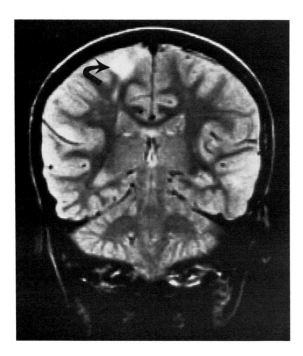

Fig. 12-42 Tuberous sclerosis. T₂-weighted MRI shows high intensity cortical and subcortical region (*arrow*), which represents a cortical tuber.

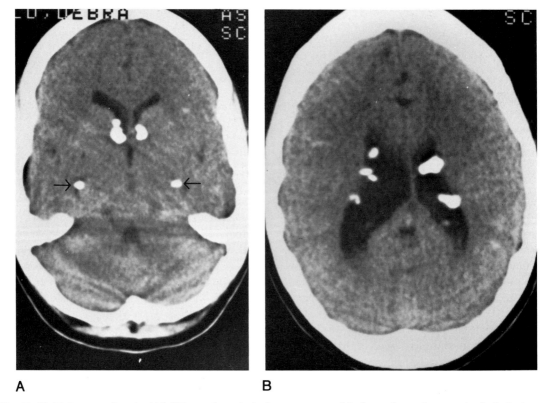

A   B

Fig. 12-43 Tuberous sclerosis. (A) CT scan through the lower portion of the brain shows the typical calcified tubers in the caudate nuclei adjacent to the foramina of Monro. Note also the tubers along the temporal horns (*arrows*). **(B)** Calcified subependymal nodules along the ventricular surfaces.

Fig. 12-44 **Tuberous sclerosis, subependymal giant cell astrocytoma.** Nonenhanced CT scan shows a large isodense tumor mass at the level of the foramina of Monro (*arrow*). It is causing hydrocephalus. Note the calcified cortical tubers (*arrowheads*).

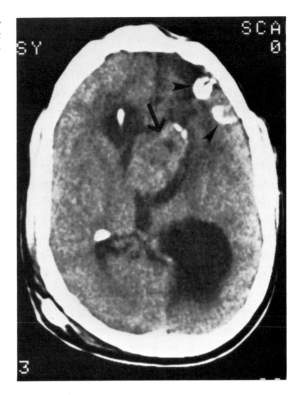

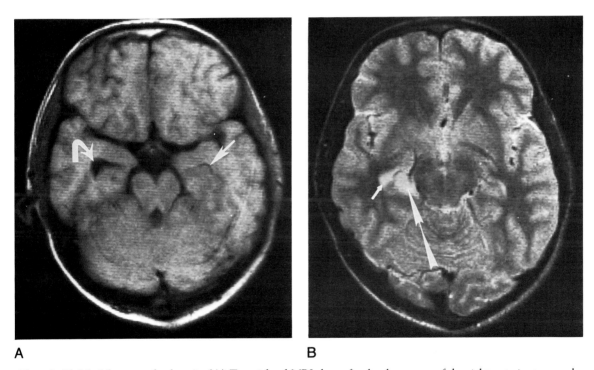

A   B

Fig. 12-45 **Mesial temporal sclerosis.** (A) T_1-weighted MRI shows focal enlargement of the right anterior temporal horn (*curved arrow*). The left temporal horn is normal (*arrow*). (B) T_2-weighted MRI shows high signal intensity in the right hippocampus (*long arrow*) medial to the temporal horn (*small arrow*).

MESIAL TEMPORAL SCLEROSIS

Mesial temporal lobe sclerosis is an uncommon abnormality consisting of gliosis of the anterior hippocampus of one or both temporal lobes. Its importance is related to its presumed role in the causation of temporal lobe type seizures. The etiology is not definitely known, although the lesion is thought to be a result of ischemia.

The abnormality is best imaged with MRI. The region of gliosis is seen as high signal on the T_2-weighted images. Local atrophic change causes slight regional enlargement of the temporal horn (Fig. 12-45). In some instances of intractable seizures, removal of the gliotic region has proved curative. Only about 50 percent of those with mesial temporal sclerosis show abnormalities on MRI; internally placed electroencephalography electrodes are generally a more sensitive means of defining abnormal seizure foci within the brain.

PITUITARY DWARFISM

Pituitary dwarfism results from growth hormone deficiency of less than 10 μ/L. The cause may be surgical removal or injury to the pituitary axis; trauma, tumor, radiation; or congenital or idiopathic. With MRI, about 45 percent of pituitary dwarfs will show a small anterior hypophyseal lobe. In a recent MRI series, 43 percent of pituitary dwarfs had an absent pituitary stalk and a posterior pituitary bright spot (neurohypophysis) displaced to the level of the median eminence. There is no bright spot in the region of the posterior lobe within the sella when there is a bright spot at the base of the hypothalamus. Severe panhypopituitarism accompanies this complex of findings, but diabetes insipidus is not present. The cause of the absent stalk and displaced posterior hypophyseal tissue is unknown.

SUGGESTED READINGS

Aoki S, Barkovich AJ, Nishimura K et al: Neurofibromatosis 1 and 2: cranial MR findings. Radiology 172:527, 1989

Barkovich AJ, Chuang SH, Norman D: MR of neuronal migration anomalies. AJNR 8:1009, 1987

Barkovich AJ, Kjos BO, Jackson DE Jr, Norman D: Normal maturation of the neonatal and infant brain: MR imaging at 1.5T. Radiology 166:173, 1988

Barkovich AJ, Lyon G, Evrard P: Formation, maturation and disorders of white matter. AJNR 13:447, 1992

Becker LE: Lysosomes, peroxisomes and mitochondria: function and disorder. AJNR 13:609, 1992

Bird CR, Hedberg M, Dreyer BP et al: MR assessment of myelination in infants and children: usefulness of marker sites. AJNR 10:731, 1989

Bognanno JR, Edwards MK, Lee TA et al: Cranial MR imaging in neurofibromatosis. AJNR 9:461, 1988

Brassman BH, Bilaniuk LT, Zimmerman RD: The central nervous system manifestations of the phakomatoses on MR. Radiol Clin North Am 26:773, 1988

Byrd SE, Naidich T: Common congenital brain anomalies. Radiol Clin North Am 26:755, 1988

Byrd SE, Darling CF, Wilczynski MA: White matter of the brain: maturation, and myelination on magnetic resonance in infants and children. Neuroimag Clin North Am 3:247, 1993

Curnes JT, Laster DW, Koubek TD et al: MRI of corpus callosal syndromes. AJNR 7:617, 1986

Fitz CR: Developmental abnormalities of the brain. p. 215. In Heinz ER (ed): Neuroradiology. Churchill Livingstone, New York, 1984

Jacobson RI: Abnormalities of the skull in children. Neurol Clin 3:117, 1985

Koenig SH, Brown RD III, Spiller M et al: Relaxometry of brain: why white matter appears bright on MRI. Magn Reson Med 14:482, 1990

Lee BCP: Magnetic resonance imaging of metabolic and primary white matter disorders in children. Neuroimag Clin North Am 3:267, 1993

Lee BCP, Engel M: MR of lissencephaly. AJNR 9:804, 1988

Medina L, Chi TL, DeVivo DC, Hilal SK: MR findings in patients with subacute necrotizing encephalomyelopathy (Leigh syndrome): correlation with biochemical defect. AJNR 11:379, 1990.

Mehta RC, Marks MP, Levin PS: Aicardi's syndrome: MR appearance of unusual orbital and ventricular cystic lesions. AJR 160:601, 1993

Nixon JR, Houser OW, Gomez MR et al: Cerebral tuberous sclerosis: MR imaging. Radiology 170:869, 1989

Nomura Y, Sakuma H, Takeda K et al: Diffusional anisotropy of the human brain assessed with diffusion-weighted MR: relation with normal brain development and aging. AJNR 15:231, 1994

Nowell MA, Grossman RI, Hackney DB: MR imaging of white matter disease in children. AJNR 9:503, 1988

Osborn RE, Byrd SE, Naidich TP et al: MR imaging of neuronal migrational disorders. AJNR 9:1101, 1988

Smith AS, Ross JS, Blaser SI et al: Magnetic resonance imaging of disturbances in neuronal migration: illustration of an embryologic process. Radiographics 9:509, 1989

Van der Knapp MS, Valk J: Classification of congenital abnormalities of the CNS. AJNR 9:315, 1988

Wolpert SM, Anderson M, Scott RM et al: Chiari II malformation: MR imaging evaluation. AJNR 8:783, 1987

13

The Cranial Nerves, Orbit, and Temporal Bone

OLFACTORY PATHWAY: CN I

The rhinencephalon accounts for the sense of smell. It includes the olfactory bulbs, tracts, the prepyriform cortex of the anterior temporal lobe and the amygdala. It is the most primitive of the senses. The olfactory tracts and bulbs are outgrowths of the brain, similar to the optic nerves (Fig. 13-1).

The bulbs receive impulses from the olfactory receptor cells of the nasal cavity. The olfactory receptors are specialized mucosa and cells in the posterior superior nasal cavity. Tiny hairs on the membrane surface react to aromatic chemicals circulating within the nasal cavity. The impulses from the hairs project onto the mucosal nerve cells. From there, the impulses are carried through the cribriform plate of the ethmoid by minute unmyelinated fibers, the olfactoria filia, which synapse within the olfactory bulb. As a group, the olfactoria filia constitute the olfactory nerve.

The paired olfactory bulbs lie paramedian on the cribriform plate of the ethmoid, immediately inferior to the olfactory sulci of the frontal lobes (Fig. 13-2). The sulci form the lateral margin of the gyrus rectus, the most medial inferior frontal gyrus. Impulses from the bulb travel backward through the olfactory tract, projecting laterally on the prepyriform cortex and the periamygdaloid area of the anteromedial temporal lobe (the uncus), known as the primary olfactory cortex. Additional medial connec-tions are made in the subcallosal and septal cortex of the medial frontal lobes. From these locations, multiple connections are made to the posterior inferior frontal lobe and to the limbic system. Disordered smell occurs from disease within the nasal cavity, the temporal lobes, and the posterior frontal lobes. Temporal lobe seizures often produce odor sensations.

Allergic rhinitis and inflammatory nasal polyposis are the most common causes of diminution in the sense of smell (Fig. 13-3). The precise reason is unknown, but the mechanism is thought to be mechanical blockage of aromatic compounds from reaching the olfactory membrane. Sinonasal inflammatory disease is discussed in the orbit section. Nasal cocaine use causes anosmia, sometimes with osteolysis of the cribriform plate, osteolytic sinusitis, and necrosis of the nasal septum. Human immunodeficiency virus (HIV) encephalitis, Wegener's granulomatosis, rhinocerebral fungal infection, and mucocele may also cause anosmia.

Tumors of the nasal cavity and adjacent paranasal sinuses account for the low percentage of all tumors, less than 0.5 percent in the United States. Most are of epithelial or salivary gland origin, but rare tumors arise from the neurosensory cells of the olfactory mucosa (see box) (Fig. 13-4). Squamous cell carcinoma is by far the most common tumor type; nasal origin of this tumor is second only to maxillary sinus origin. In the early stages, all these tumors cause obstruction to one

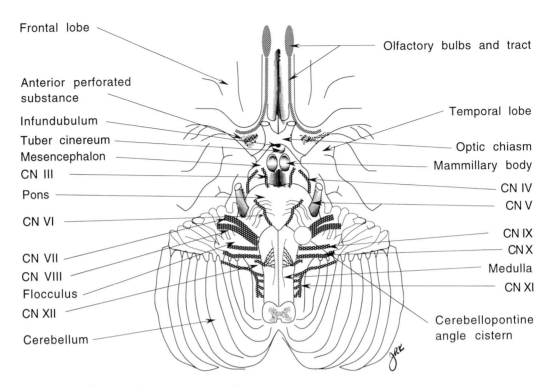

Fig. 13-1 Ventral brain and the cranial nerves. Illustration of the brain as seen from the ventral surface. (Adapted from Goss, 1963, with permission.)

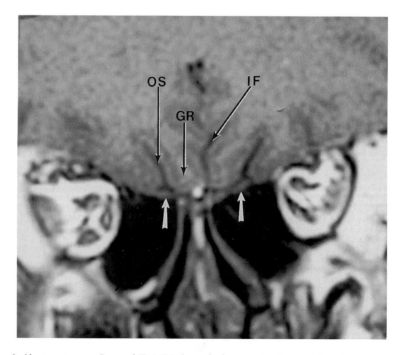

Fig. 13-2 Normal olfactory tracts. Coronal T_1 MRI through the anterior frontal lobes demonstrates the small normal olfactory tracts (*small white arrows*) located just inferior to the olfactory sulci (OS). The gyrus rectus (GR) of the inferior frontal lobe lies medial to the olfactory sulcus and is itself bounded medially by the interhemispheric fissure (IF).

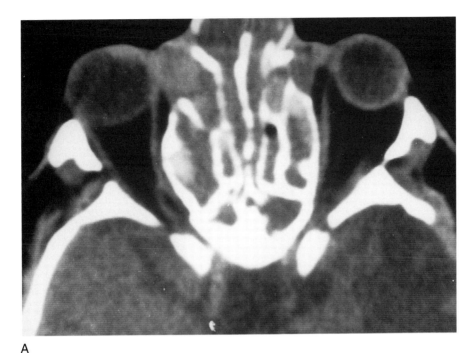

A

B

Fig. 13-3 Nasal polyposis. (A) Polypoid masses expand the nasal cavity producing hypertelorism and proptosis. **(B)** Coronal view shows soft tissue masses that expand the nasal cavity and destroy the nasal and ethmoidal septa.

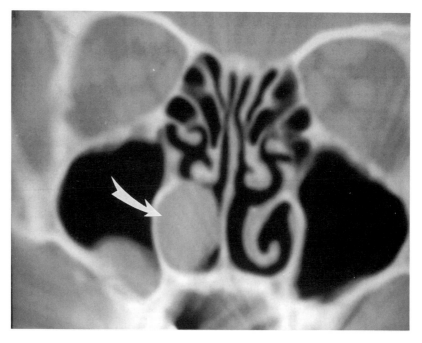

Fig. 13-4 Plasmacytoma. Well-circumscribed round tumor arises from the inferior turbinate and expands the nasal cavity (*arrow*). The turbinate is resorbed. Coronal CT scan.

side only of the nasal passageway, a differential point in suspecting tumor instead of inflammatory disease as the cause for mass density within the nasal cavity. In the later stages, tumors may spread to involve both passages, causing anosmia and region invasion of adjacent cavities, including the paranasal sinuses, the orbit, and the intracranial frontal fossa. Bone destruction generally indicates tumor, rather than infection. CT and MRI in the coronal plane are the best means

---

TUMORS OF THE NASAL CAVITY AND CRIBRIFORM PLATE

Epithelial
    Squamous carcinoma
    Adenocarcinoma
    Undifferentiated carcinoma
    Papilloma
        Exophytic
        Inverting
Salivary gland
    Adenoid cystic carcinoma
    Acinic cell carcinoma
    Pleomorphic adenoma
    Mucoepidermoid carcinoma
Neurosensory cell tumors
    Olfactory neuroblastoma
    Esthesioneuroblastoma
    Neuroendocrine carcinoma
Other
    Plasmacytoma
    Sarcoma
    Olfactory groove meningioma
    Lymphoma
    Metastasis

(Continued)

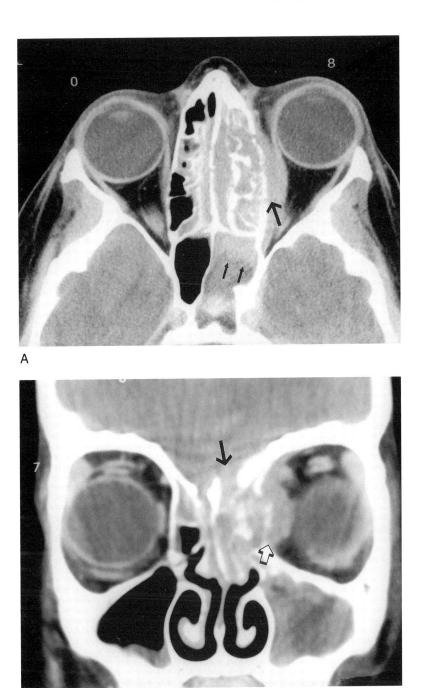

A

B

Fig. 13-5 Rhabdomyosarcoma, ethmoid. (A) Axial CT scan of this 9-year-old shows a mass arising within the left ethmoid, destroying bone as it extends into the medial extracoronal left orbit *(large arrow)* and the anterior sphenoid sinus *(small arrows)*. It crosses the midline, partially destroying the nasal septum. The differential diagnosis includes lymphoma and esthesioneuroblastoma. **(B)** Coronal CT shows extension of the mass into the frontal intracranial fossa *(black arrow)*. The tumor subsequently metastasized throughout the subarachnoid spaces of the head and spine. The tumor extends into the medial left orbit *(open arrow)*, eroding the lamina papyracea.

of detection and evaluation, but the tumor characteristics are nonspecific and cannot be differentiated by imaging alone (Fig. 13-5). Bone destruction and calcification may be seen with all types of tumors. Sometimes, more than one type of tumor is present at a time.

Papilloma

The fungiform exophytic type and the endophytic (inverted) type account for most of the papillomas. The exophytic type arises from the nasal septum and usually has a variegated surface. The endophytic (inverting) type arises from the lateral wall near the middle turbinate and, as its name implies, tends to grow inward from the base of origin (Fig. 13-6). It usually has a smooth medial surface. Both types begin as unilateral mucosal tumors that may be small and hard to detect in the early stage. With time, the tumor spreads into adjacent regions, and the destruction may be extensive. Bone is displaced or destroyed. They may be slow- or fast-growing, and malignant change sometimes occurs. Large amounts of bone destruction suggests malignant transformation or an associated squamous carcinoma. Cure is possible with complete removal of the tumor, but small residual tumor nests will result in recurrence.

Neurosensory Cell Tumor

Olfactory neuroblastoma, esthesioneuroblastoma, and neuroendocrine carcinoma are uncommon tumors that arise from the neurosensory cells of the olfactory mucosa high within the nasal cavity. Subtle histologic features distinguish between these similar olfactory neural tumors. They occur most commonly in adolescents or adults about age 60. The tumor begins on one side of the nasal cavity, but it is aggressive, and locally invades through the cribriform plate into the anterior frontal fossa and the orbit.

CT displays a homogeneous soft tissue mass with intense contrast enhancement. With MRI, it is isointense

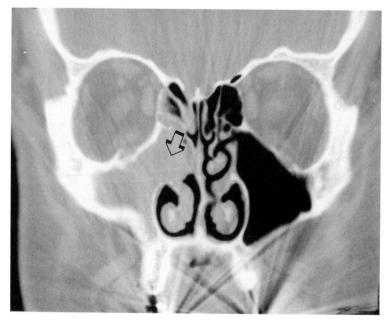

Fig. 13-6 Inverting papilloma. Coronal CT scan demonstrates a mass arising at the level of the middle turbinate and extending laterally into the maxillary sinus *(open arrow)*. It locally destroys the bone.

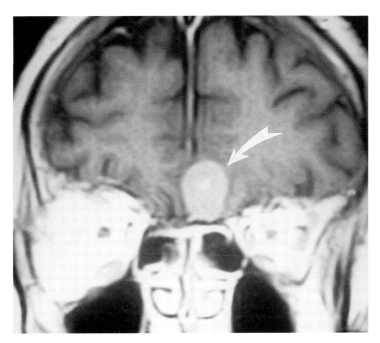

Fig. 13-7 Meningioma. Coronal T$_1$-weighted Gd-enhanced MRI shows an olfactory groove meningioma arising from the floor of the frontal fossa *(arrow)*.

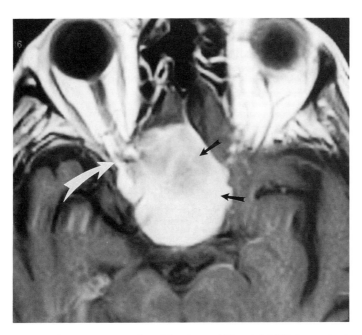

Fig. 13-8 Meningioma, invading posterior orbit. Gd-enhanced T$_1$ axial MRI shows the homogeneously enhancing meningioma arising from the planum sphenoidale *(small black arrows)*. The mass extends laterally, invading the posterior right orbit *(white arrow)*.

with the brain on T_1 and hyperintense on T_2-weighted images. Gadolinium (Gd) enhancement is intense. The intracranial subfrontal component of the tumor may become very large. It frequently invades the brain and may spread throughout the CSF spaces. Other tumors in this region include rhabdomyosarcoma (Fig. 13-5) and angiofibroma in children, and epidermoid carcinoma, metastasis, lymphoma (Fig. 8-22), and meningioma in adults (Figs. 13-7 and 13-8).

Congenital Anosmia

Kallmann syndrome refers to defective development of the rhinencephalon with aplasia or hypoplasia of the olfactory bulbs and tracts and with small or absent olfactory sulci. These patients have anosmia. Affected patients also have hypogonadotropic hypogonadism. The cells that produce gonadotropin-releasing hormone (GnRH) originate in the olfactory placode and are arrested from their normal migration to the hypothalamus. Calcification sometimes occurs within the lenticular nuclei, dentate nuclei, red nuclei, and other regions of the brain. There are multiple patterns of inheritance, and a variety of other associated anomalies outside of the CNS. Other congenital anomalies that may cause anosmia include septo-optic dysplasia, nasofrontal encephalocele, and the holoprosencephaly syndromes.

Anosmia may be seen in a variety of other CNS diseases including trauma in the region of the cribriform plate, multiple sclerosis, Parkinson's disease, Alzheimer's disease, and schizophrenia. Temporal lobe tumors may cause olfactory hallucination.

THE ORBIT AND CN II

The orbit contains and protects the ocular bulb and the ophthalmic nerve (CN II). It is a conical structure with its apex directed posteromedially, ending in the orbital orifice of the optic canal. The plane of the orbit is approximately parallel to a line drawn from the inferior orbital rim to the anterior superior attachment of the ear pinna with the scalp.

Seven bones form the orbit. The orbital plate of the frontal bone forms the anterior three-fourths of the roof. It is generally thin and convex upward into the frontal fossa. The inferior frontal lobe rests on the superior surface. The lesser wing of the sphenoid forms the flat and thicker posterior one-fourth of the roof. The greater wing of the sphenoid forms the posterolateral orbit. Between it and the lesser wing above there is a space, the superior orbital fissure, through which nerves and vessels enter the orbit (see below). The zygomatic process of the frontal bone (superior), the zygomatic bone (lateral and inferolateral), the anterior part of the maxilla (inferior), and the lacrimal bone (medial) form the anterior portion of the orbit. The orbital process of the palatine bone forms a small portion of the posterior medial apex. The thick rounded anterior margins of the orbit form the superior and inferior rims. The lamina papyracea of the ethmoid bone forms most of the medial wall and abuts the small orbital process of the palatine bone.

The thin orbital part of the maxillary bone forms the floor, which is also the roof of the maxillary sinus below. A central groove along the floor of the orbit carries the infraorbital nerve (maxillary division of cranial nerve V [CN V]) and vessel forward, exiting into the subcutaneous tissue of the face through the infraorbital foramen. The superior rim is pierced by the supraorbital foramen (sometimes only a notch), which carries the supraorbital nerve (ophthalmic division of CN V) and vessels to the tissues of the forehead. A shallow fossa along the anterosuperior lateral orbit holds the lacrimal gland. The lacrimal groove and canal for the lacrimal duct are present in the medial anterior inferior orbit.

The periosteal layer that lines the orbit is called the periorbita (Fig. 13-9). It is loosely attached to the subjacent bone, except at the trochlea, the optic foramen, and orbital fissures. Through the fissures and optic canal, it is continuous with the periosteal layer of the intracranial dura mater. Anteriorly, it blends with the firmly attached periosteum about the orbital rim of the skull. Here it contributes to the radially directed anterior orbital septum. This septum continues into the eyelids, forming a weak fascial barrier between the orbit and the anterior facial tissues. The periorbitum is pierced by vessels and nerves that enter the orbit.

Three layers (tunics) that form the ocular bulb are the sclera (and Tenon's capsule), choroid (uvea), and the retina. The sclera is the strong outer fascia that maintains the spherical shape of the eye. It forms the cornea anteriorly, and an opaque protective covering posteriorly, averaging about 1 mm thick. The middle choroid

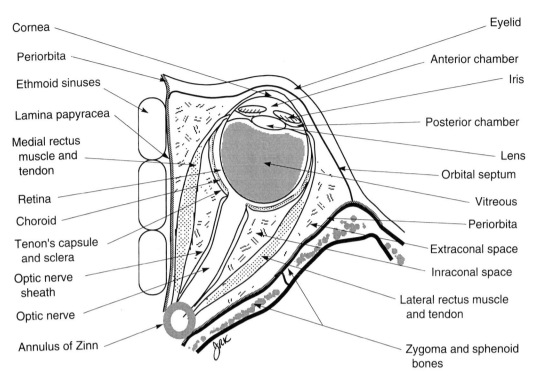

Fig. 13-9 The orbit. Axial illustration through the middle of the orbit illustrates the major orbital structures, and spaces, including the layers of the globe and optic nerve. This is a drawing of the left orbit using CT/MRI convention with lateral toward the right of this drawing.

layer, the uvea, carries the arborization of the ciliary arteries that supply the retina. It also forms the iris. The inner retina contains the specialized neural light receptors that output to the brain through the optic nerve. Anterior to the retina within the globe is the large gelatinous vitreous humor. The lens of the eye separates the vitreous from the anterior aqueous humor, which itself is separated by the iris into anterior and posterior chambers.

The thin membranous capsule of Tenon surrounds the optic nerve and ocular bulb. Tenon's membrane is the intraorbital continuation of the meningeal layer of the dura. It is separated from the optic nerve and sclera by a small space filled with fluid, called Tenon's space, continuous with the intracranial subarachnoid space. The capsule is firmly attached with the sclera at the site of entrance of the optic nerve into the bulb, separating the periscleral space from the more proximal perineural space. The perineural space dilates to accommodate an increased amount of CSF with the presence of increased intracranial pressure producing pseudoenlargement of

the optic nerve. Diseases of the meninges affect Tenon's space.

Six extraocular muscles operate the globe. They generally have opposing actions. The four rectus muscles—superior, inferior, medial, and lateral—arise from a fibrous ring surrounding the optic canal called the annulus of Zinn. The muscles continue anteriorly and perforate the capsule of Tenon to insert into the sclera just anterior to the equator of the globe. The rectus muscles, except the lateral rectus, are innervated by the oculomotor nerve (CN III). The lateral rectus muscle is innervated by the abducens nerve (CN VI). A thin membrane surrounds and connects the muscles to form a muscle cone. This confines an inner region of the orbit called the intraconal space. The space outside of the muscle cone is called the extraconal space. These spaces are filled with fatty reticulum containing nerves and vessels (Fig. 13-10).

The superior oblique muscle arises from the sphenoid bone adjacent to the optic canal and extends anteriorly,

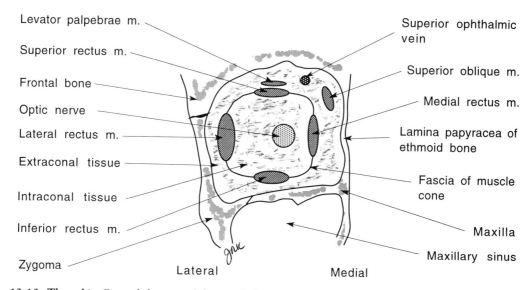

Fig. 13-10 The orbit. Coronal drawing of the retrobulbar orbit shows the extraocular muscles, fascia forming the muscle cone, and the optic nerve.

looping through the trochlea at the superomedial orbital rim. From here it swings backward to attach on the opposite side of the globe behind the superior equator. It rotates the globe downward and outward. It is innervated by the trochlear nerve (CN IV). Because the fibrous trochlear band lies on the frontal bone adjacent to the paranasal sinuses, inflammatory sinus disease and frontal bone fracture may result in constriction of the movement of the superior oblique tendon and simulate a CN IV palsy.

The inferior oblique muscle arises from the anteroinferior medial orbital wall of the maxilla lateral to the lacrimal groove and extends posterolaterally to insert behind the equator on the posterolateral surface of the globe. It is innervated by the occulomotor nerve (CN III). The levator palpebrae muscle arises from the lesser wing of the sphenoid just above the optic canal and runs forward just superior to the superior rectus to insert into the upper eyelid.

Imaging Studies

CT, MRI, and ultrasonography are the major imaging techniques used for evaluation of the orbit and ocular bulb. CT is the easiest and most versatile. Studies are done in the transaxial, coronal, or modified parasagittal plane (longitudinal) through the orbit, although this is often difficult or impossible for the patient. Scan slice width is varied from 1.5 to 5 mm, depending on the detail required. Overlapping scans are done if multiplanar reconstruction is contemplated. However, radiation dose to the lens and cornea is significant (up to 8 cGy), so it is imperative to keep the number of scans to as few as pos-

---

THE FOUR ORBITAL SPACES

Intraconal space
    Tenon's space
    Central space
Extraconal space
    Peripheral space
    Subperiosteal space

sible for diagnosis. Bone and soft tissue windowing should be used. Contrast enhancement is used for diagnosis of vascular and some neoplastic lesions. CT is of unique advantage for the evaluation of fractures, and calcification within tumors. CT and MRI are equal for the diagnosis of the presence of mass lesions, most inflammatory lesions, and the enlargement of extraocular muscles (EOMs).

MRI is performed using dedicated orbit coils to obtain detail. Multiplanar imaging is easily done, and there is no ionizing radiation. Thin sections (3-mm) with T_1- and T_2-weighted spin echo (SE) sequences are routinely used. Mass lesions are well seen, and the signal characteristics may help with diagnosis, particularly with hematoma. Gd contrast enhancement improves evaluation of the optic nerve and tumors. Fat-suppression techniques improve the conspicuity of abnormal contrast enhancement by decreasing the bright signal of the abundant fat within the orbit. MRI has advantages over other techniques in evaluating the optic nerve, the ocular bulb, the orbital apex, and the superior orbital fissure and in detecting intracranial extension of intraorbital disease.

Ultrasound is most useful for imaging the ocular bulb. The examination employs high frequency, which may reach 50 MHz if the water immersion method of contact is used, permitting high spatial resolution. B mode is most commonly used. The technique has high sensitivity and tissue specificity for the diagnosis of intraocular tumors. Detection is excellent for vitreous hemorrhage, retinal detachment, and cystic lesions, particularly of the anterior orbit (lacrimal sac hydrops in infants, lacrimal cysts). Ultrasound is not as useful for the diagnosis of

---

LESIONS AFFECTING THE VISUAL PATHWAY

Optic nerve
    Glioma (< 10 years of age) (50% associated with NF-1)
    Meningioma
    Optic neuritis
    Metastasis
    Sarcoidosis
    Berry aneurysm, ophthalmic art
    Any large intraorbital mass
Optic chiasm
    Pituitary adenoma (suprasellar extension)
    Craniopharyngioma
    Meningioma (parasellar)
    Sarcoidosis
    Metastasis
    Aneurysm (giant)
    Glioma (of nerve, hypothalamus)
    Rathke pouch cyst
Optic tract and radiations
    Tumors, infarctions, and abscesses in the temporal, parietal, and occipital lobes and in the distribution of the striate arteries
    Herniation of the medial temporal lobe

---

retrobulbar lesions, as the high echogenicity of the retrobulbar fat obscures lesions. If lesions are seen here, they are relatively echo poor.

The Optic Pathways

MRI examination provides the best overall evaluation of the optic pathway (Table 13-1). The segment of the optic pathway clinically involved determines the focus of the scan (optic nerve, chiasm, tracts). T_1 and T_2 se-

---

OCULAR TUMORS

Adults
    Melanoma
    Metastasis
    Hemangioma
    Hematoma
    Lymphoma
Children
    Retinoblastoma

Table 13-1 Clinical Syndromes of the Optic Pathway

Lesion	Finding
Optic nerve	Monocular vision loss
	Decreased visual acuity
Optic chiasm	Complex visual-field cuts
Optic tracts and radiations	Hemianopsia

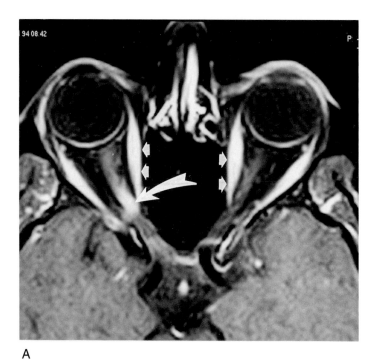

A

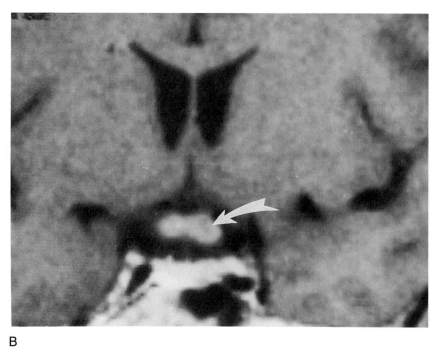

B

Fig. 13-11 Optic neuritis. (A) Axial fat-saturated (fat-sat) Gd-enhanced T₁ MRI through the orbits shows a slightly enlarged enhancing right optic nerve representing optic neuritis (*large white arrow*). The left optic nerve is normal. Note also the normally bright extraocular muscles with fat-saturation technique (*small white arrows*). **(B)** Coronal contrast-enhanced CT scan through the suprasellar cistern illustrates an enhancing enlarged optic chiasm (*white arrow*) in a patient diagnosed as having optic neuritis. In the correct clinical setting, this could represent optic nerve glioma.

quences, including fast spin echo (FSE) T_2, are done without and with Gd contrast enhancement in the axial or coronal planes, depending on the region of interest. Optic glioma, meningioma, and metastasis enlarge the nerve and show contrast enhancement. They may involve any segment of the optic nerve (orbital, canalicular prechiasmal). Calcification is common in meningioma and is best seen with thin-section CT. Regions of active demyelination from optic neuritis show bright signal following enhancement (Fig. 13-11). The nerve enlarges slightly. Optic neuritis is associated with multiple sclerosis in about 20 percent of cases, and may be the presenting manifestation of the disease. The other lesions affecting the optic pathways are discussed in detail in other chapters (see box).

ORBITAL LESIONS

Intraconal
  Tenon's space
    Inflammation (pseudotumor)
    Ocular melanoma, metastasis, and retinoblastoma
  Central space
    Optic nerve tumors
      Glioma
      Meningioma
    Vasculogenic lesions
      Cavernous hemangioma, lymphangioma
      Hemangiopericytoma
    Schwannoma
    Metastasis, including EOMs
    Lymphoma
Extraconal
  Peripheral space
    Lacrimal gland lesions
      Tumors, sarcoid, peudotumor, Wegener's
    Vasculogenic lesions
      Cavernous hemangioma, capillary hemangioma
      Hemangiopericytoma, lymphangioma, varix

(Continued)

Lymphoma, leukemia (especially children)
Rhabdomyosarcoma (infants)
Myeloma
Schwannoma
Teratoma
Metastasis, or direct invasion
Cellulitis (infection)
Rare, one-of-a-kind tumors (e.g., fibroma, sarcoma, neuroma)
  Subperiosteal space
    Subperiosteal hematoma
    Hematic cyst, granulomatous cyst
    Infection: effusion, phlegmon, abscess
    Sinonasal inflammatory polyposis
    Lacrimal sac mucocele (hydrops)
    Dermoid cyst
    Bone tumors: osteoblastoma, metastasis
    Sinus tumors
    Mucocele of ethmoid and frontal sinus

Masses and Infiltrations of the Orbit

OPTIC NERVE TUMORS

Optic nerve glioma and meningioma account for almost all tumors that originate in the optic nerve. They are the most common cause of enlargement of the nerve. Most people with optic nerve tumor present with monocular decreasing visual acuity, commonly a peripheral field constriction. There is edema of the disc in the affected eye. The degree of proptosis is generally less than the severity of the visual disturbance. Rare tumors of the optic nerve include metastasis, schwannoma, hemangiopericytoma, choristoma (fat and muscle tumor), and spread of melanoma into the nerve from the globe. Percutaneous fine-needle aspiration biopsy is used successfully to diagnose the tissue type of optic nerve tumors when the diagnosis cannot be made from clinical facts.

Glioma
Glioma of the optic nerve is most common in children under 10 years of age. Most are very slow-growing pilocytic astrocytomas. Fifty percent are associated with

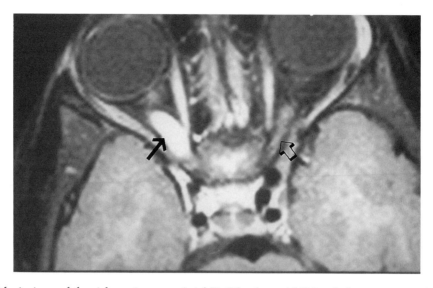

Fig. 13-12 Meningioma of the right optic nerve. Axial T_1 Gd-enhanced MRI with fat saturation technique clearly shows the enhancing meningioma surrounding the posterior right optic nerve (*black arrow*). Fat saturation (suppression) is needed to see the enhancing tumor within the normally high signal orbital fat. The normal optic nerve is seen on the left (*open arrow*). The extraocular muscles appear bright with fat-saturation technique.

neurofibromatosis type 1 (NF-1) (see Ch. 12). Bilateral optic nerve glioma is diagnostic of NF-1. The tumor is confined by Tenon's capsule and assumes a fusiform shape, buckling the nerve. It rarely extends posteriorly to invade the chiasm or brain and does not invade the globe. Optic nerve glioma is rare in adults, but when it occurs the tumor is likely to be a rapid-growing malignant glioblastoma with diffuse invasion.

CT scan shows the enlarged buckled nerve contour, with intense contrast enhancement and sometimes a few small calcifications. The optic canal will enlarge with intracanalicular tumor, but surrounding sclerosis does not occur. Cysts may be present within the tumor. MRI shows the tumor as slightly hypointense on T_1-weighted images and as slightly hyperintense on T_2-weighted images. Gd enhancement may be striking.

Meningioma

Meningioma occurs most often in women aged 35 to 60 years. They are uncommon in children, but are seen with NF-2 where they may be bilateral. Initially, the tumor is confined within Tenon's capsule, so that it takes on a tubular or fusiform shape and straightens the normally curved course of the nerve. With growth, the tumor breaks out of the capsule and assumes a more rounded shape (Fig. 13-12). The nerve is not buckled. Tumors arising from the intracanalicular portion of the nerve enlarge the optic canal, sometimes causing visible sclerosis of the lesser sphenoid wing. CT of the orbit, without and with contrast enhancement, is the best imaging modality, as calcifications may be detected within the tumor confirming the diagnosis of meningioma. The tumor enhances intensely. Indistinct margins indicate early extension outside Tenon's capsule into the central space. MRI is superior for the detection of tumor within the optic canal. The MRI signal of the tumor is variable but is usually slightly less intense than brain on T_1-weighted images, and slightly greater than brain on T_2-weighted images. Optic neuritis may enlarge the nerve and be indistinguishable from a small meningioma. Search for MRI findings of multiple sclerosis within the brain may help make the correct diagnosis.

CAUSES OF ENLARGEMENT OF THE OPTIC NERVE

Common
    Glioma (mostly in children)
    Meningioma (mostly in women, aged 35 to 60 years)
    Optic neuritis (associated with multiple sclerosis, in 20%)
Uncommon
    Inflammatory pseudotumor (painful proptosis)
    Graves ophthalmopathy (has EOM changes)
    Increased intracranial pressure
    Central retinal vein occlusion (fundal changes)
    Inflammation (sarcoid, Tuberculosis, toxoplasmosis)
    Meningeal carcinomatosis
    Other tumors (lymphoma, schwannoma, metastasis, spread of tumor from globe)

LYMPHOMA, LYMPHOID HYPERPLASIA, PSEUDOTUMOR, AND GRAVES DISEASE

These four entities consist of neoplasia or lymphoid reaction, although they vary considerably in pleomorphism and malignancy. The categories overlap, and there is considerable confusion in the terminology in the literature. It is reasonable to use these four categories of lymphoid orbit disease.

CHARACTERISTICS OF LYMPHOID DISEASES OF THE ORBIT

Lymphoma
    Monomorphic, monoclonal, usually B cell
    Variable malignancy
    Association with systemic lymphoma up to 75% with the most malignant tumors
    Painless proptosis

(Continued)

    Little visual loss or restriction of eye movement
    Tends to be localized within the extraconal orbit (conjunctiva, lacrimal gland, superior orbit)
Reactive lymphoid hyperplasia (pseudolymphoma)
    Monomorphic small normal lymphocytes
    Association with systemic lymphoma (25%)
    Pattern within orbit similar to pseudotumor
    Little pain and few inflammatory signs
Inflammatory pseudotumor
    Polymorphic inflammatory cells, with eosinophils
    Little or no malignant potential
    No association with systemic lymphoma or other disease
    Painful proptosis with inflammatory signs
    Multifocal involvement of the orbit
    Rapid regression with steroid therapy
    Most often unilateral disease
    EOM tendons involved with inflammation
Graves disease
    Most often associated with toxic goiter or Hashimoto's thyroiditis
    Painless proptosis
    Slight restriction eye movement
    Bilateral disease
    Involvement of multiple muscles, but not tendons
    Retrobulbar fat not involved with inflammation
    Increased volume of retrobulbar fat

LYMPHOMA AND REACTIVE LYMPHOID HYPERPLASIA

Lymphoma and reactive lymphoid hyperplasia are considered neoplastic processes, but with different malignant potential. They have the same pattern of involvement within the orbit and present as a slowly progressive painless mass with proptosis. The most common sites of involvement are the conjunctiva, the lacrimal gland, and the superior peripheral retrobulbar orbit. Less commonly, they involve the central space and the EOMs. The lesion tends to be localized to only one of these sites in an individual. No inflammatory reaction occurs within the retrobulbar fat. The masses of cells are soft, so there is little distortion of intraorbital structures and no bone erosion, differentiating lacrimal lymphoma from other lacrimal gland tumors. On CT, the masses are slightly hyperdense and homogeneous showing variable contrast enhancement, but the appearance is nonspecific. On MRI, the mass is slightly hypointense on T_1-weighted imaging, and slightly hyperintense to fat on T_2-weighted images. Contrast enhancement is variable and slight and does not help much with differential diagnosis.

INFLAMMATORY PSEUDOTUMOR

Inflammatory pseudotumor, or idiopathic orbital inflammatory syndrome (IOIS), is an inflammatory infiltrative process within the orbit, of unknown cause. It occurs at any age and is usually unilateral. Painful proptosis is characteristic, with restriction of eye movement and conjunctival inflammation.

The process is multifocal involving one or more of the following orbital structures: EOMs (myositis form), lacrimal gland, sclera, Tenon's capsule, optic nerve sheath (perineuritis form), and the retrobulbar fat. The EOMs may be enlarged, including their tendon insertions into the sclera, a differentiating point from Graves ophthalmopathy (Fig. 13-13). The sclera and lacrimal glands may be thickened with fuzzy margins. With CT, the retrobulbar fat may show streaks, diffuse increased density in more severe cases, or focal mass lesions. MRI shows the inflammatory tissue as isointense with muscle on T_1-weighted images and as isointense or hyperintense with fat on T_2-weighted images. The intraorbital fat loses its normally high T_1 signal intensity when it is invaded by the inflammatory process. There may be slight contrast enhancement of inflammatory masses. Similar inflammatory changes may occur with Wegener's granulomato-

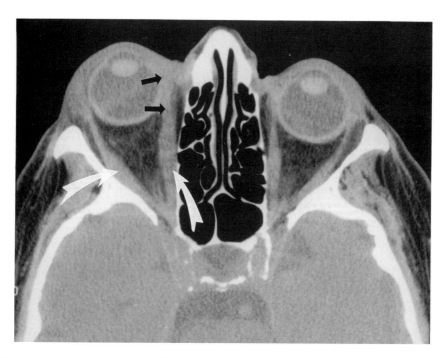

Fig. 13-13 Pseudotumor of the orbit. Axial CT scan through the orbits shows enlargement of the medial and lateral rectus muscles on the right (*large white arrows*). The muscles on the left are normal. There is inflammatory thickening of the tendonous insertions into the globe and the adjacent capsule (*black arrows*). Compare with Graves disease in Fig. 13-15B. The right globe is proptotic.

sis, systemic vasculitis, sarcoid, rheumatoid disease, and retroperitoneal or mediastinal fibrosis. Paranasal sinuses are not involved with IOIS.

TOLOSA-HUNT SYNDROME

Tolosa-Hunt syndrome is a painful ophthalmoplegia. The pathology is similar to that of idiopathic orbital pseudotumor, but occurring in the posterior orbital apex, and in the cavernous sinus, mostly the lateral portion. The tissue infiltration causes enlargement of the affected cavernous sinus, and sometimes narrowing of the cavernous portion of the internal carotid artery. Occlusion of the cavernous sinus and the superior orbital vein frequently accompanies the syndrome. CN III to VI and the sympathetic nerves to the orbit are variably affected. The inflammatory tissue rapidly regresses in response to corticosteroid therapy.

CT scan may show abnormal tissue within the cavernous sinus and orbital apex. MRI shows an enlarged cavernous sinus, with a convex lateral margin in most cases. The abnormal inflammatory tissue within the sinus and at the orbital apex is isointense with muscle on T_1-weighted images and isointense with fat on T_2-weighted images. Tissue may extend some distance away from the cavernous sinus along the intracranial dura. The tissue enhances with Gd and iodinated contrast.

GRAVES DISEASE

Graves disease (endocrine ophthalmopathy) is associated with abnormal thyroid function, usually toxic goiter and Hashimoto's thyroiditis. However, the onset may occur prior to clinically recognized thyroid dysfunction. It is the most common cause of proptosis in adults, which is usually painless and not associated with limitation of eye movement. Lymphocytes, plasma cells, and mucopolysaccharide accumulate in the orbital tissues. Fibrosis occurs later.

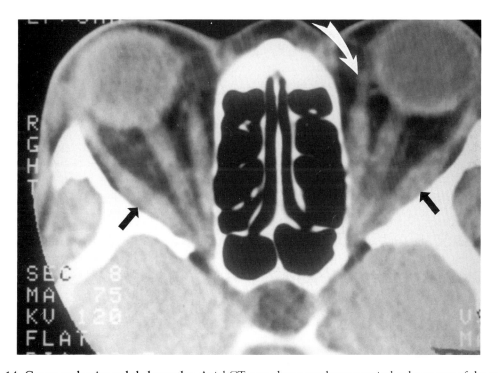

Fig. 13-14 Graves endocrine ophthalmopathy. Axial CT scan shows nearly symmetrical enlargement of the medial and lateral rectus muscles of the orbits *(black arrows)*. Note the lack of enlargement of the muscle tendons in the region of the globe *(white arrow)*.

The disease is almost always bilateral and nearly symmetric. The characteristic finding is infiltration and enlargement of multiple EOMs, particularly the inferior and medial rectus muscles. The tendons of the EOMs and the retrobulbar fat are not infiltrated, helping differentiate Graves disease from inflammatory pseudotumor (Figs. 13-14 and 13-15). The retrobulbar fat may increase in volume, contributing to the exophthalmus, but generally remains normal in texture. The lacrimal glands may be slightly enlarged, but this feature is not as prominent compared with orbital pseudotumor. CT or MRI may be used for diagnosis, with the coronal plane images most helpful for evaluation of the muscle enlargement. Contrast enhancement is not helpful.

OTHER INTRAORBITAL MASSES

Metastasis occurs to any intraorbital space, including the EOMs (Figs. 13-16 and 13-17). The globe is the most commonly involved structure (Fig. 13-18). Most metastases come from breast and lung primary tumors. Proptosis is common, although some cicatrizing tumors from breast carcinoma will cause globe retraction and enophthalmos. The tumor density is nonspecific on CT and MRI, with most tumors enhancing in a nonspecific fashion. Fat-suppression techniques must be used to visualize the tumors optimally on T_2-weighted MRI or contrast-enhanced T_1-weighted MRI.

Cavernous hemangioma is a common benign tumor that occurs within the intraconal central space in adults. It causes slowly progressive painless proptosis, with decreased visual acuity. It is most common from age 20 to 50 years. The lesion consists of dilated vascular channels with slow blood flow. Pheboliths may occur. CT shows the mass as generally ovoid, well circumscribed, and homogeneous with slight hyperdensity. Contrast enhancement is variable. MRI shows the lesion as hyperintense on T_2-weighted images.

Hemangiopericytoma arises from vascular pericytes and may be found within any region of the fatty orbital space. This lobular tumor is composed of vessels and septa. It may be found at any age. Fifty percent are malignant. The lesion is generally well circumscribed and shows intense contrast enhancement.

ORBITAL VARIX

Abnormally large veins (or varices) within the orbit occur from congenital malformation (primary varix) or from increased flow from an acquired abnormality (secondary varix). One dilated vein or many may be noted. The primary varix may be present at birth. Secondary varix results from increased carotid-cavernous fistula, occlusion of other draining veins of the orbit, or the very rare arteriovenous malformation. The varix may cause intermittent proptosis. The varices are most often found in the superior orbit. The Valsalva maneuver during imaging accentuates the veins, helping in the differential diagnosis. Orbital venography, performed by injecting a frontal forehead vein with aqueous contrast, fills the enlarged vascular channels and confirms the diagnosis when CT or MRI is equivocal.

LESIONS IN CHILDREN

Capillary hemangiomas and lymphangiomas occur in newborns and infants, mostly within the peripheral space. They are poorly marginated and show contrast enhancement (Fig. 13-19). Rhabdomyosarcoma is an extremely malignant and destructive tumor of infancy and childhood. It is the most common orbital malignancy of this age group, and it may be found anywhere within the orbit. The lesion is aggressive and invades regional bone and tissues, particularly the adjacent sinuses and nasal cavity, and can be multifocal within the orbit (Fig. 13-20). The tumor may begin in the retropharyngeal space, the pterygoid fossa, or the nasal cavity, invading the orbit and base of the skull. CT shows the lesion as isodense. Contrast enhancement is of little diagnostic help, but defines subtle invasion outside of the orbit. MRI shows the tumor as hyperintense on T_2-weighted images. There is Gd enhancement.

Orbital teratoma is a benign tumor consisting of all three layers of tissue. It occurs within any segment of the orbit, and tumor may be present at birth. It produces extreme exophthalmos and orbital expansion. It may grow more rapidly after birth, with regional extradural growth into the temporal fossa. MRI and CT show the tumor as multicystic with variable solid components, often with rim enhancement with contrast. Choristomas are similar tumors but consist of tissue not normally found within the orbit.

(*Text continued on page 519*)

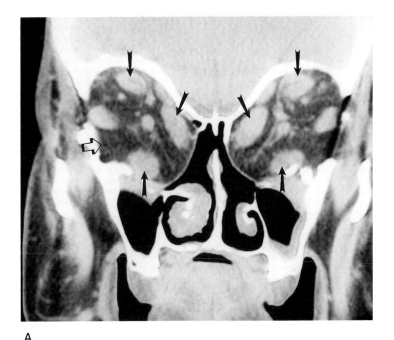

A

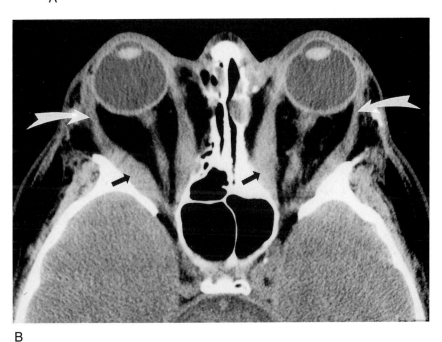

B

Fig. 13-15 Thyroid exophthalmopathy, Graves disease. (A) Coronal CT scan through the orbits shows the nearly symmetric enlargement of the extraocular muscles *(black arrows)*. There is also a slight increase in the density of the retrobulbar fat tissue *(open arrow)*. **(B)** Axial CT scan shows the marked bilateral proptosis and the enlargement of the medial and lateral rectus muscles *(black arrows)*. Note the lack of enlargement of the tendons at the insertion into the ocular bulbs *(white arrows)*.

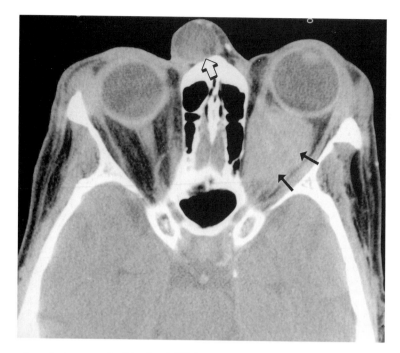

Fig. 13-16 Metastasis to the intraconal orbit. Axial CT shows a mass within the muscle cone of the left orbit *(arrows)*. This was a metastasis from a primary carcinoma of the esophagus. Any tumor may metastasize to the intraconal orbital tissue. It cannot be differentiated from lymphoma or hemangioma. There is a sebaceous cyst at the base of the nose *(open arrow)*.

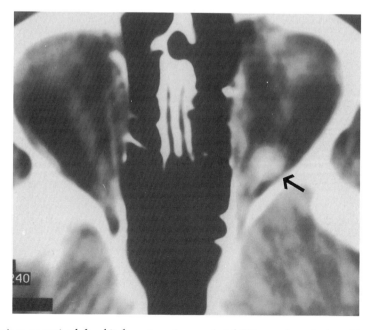

Fig. 13-17 Metastasis to posterior left orbit, breast carcinoma. Axial CT scan through the orbits shows a round mass in the posterior extraconal orbit *(arrow)*. The mass is nonspecific, and no tissue diagnosis can be given from the scan alone.

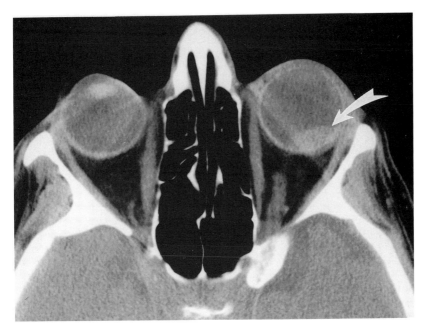

Fig. 13-18 Metastasis to choroid. Axial CT scan shows an oval mass in the posterior left globe arising from the choroid (*white arrow*). It proved to be metastasis from a carcinoma of the ovary, a very rare metastasis to the eye. More common tumors here are melanoma and metastasis from carcinoma of the breast. The CT characteristics are nonspecific.

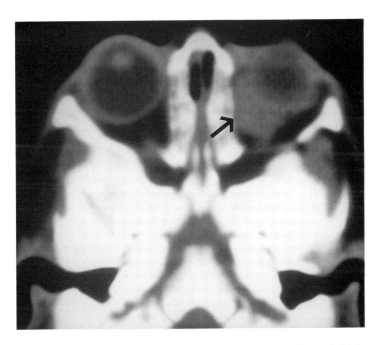

Fig. 13-19 Capillary hemangioma. Axial CT scan shows a nonspecific mass in the medial left retrobulbar orbit of an infant (*arrow*).

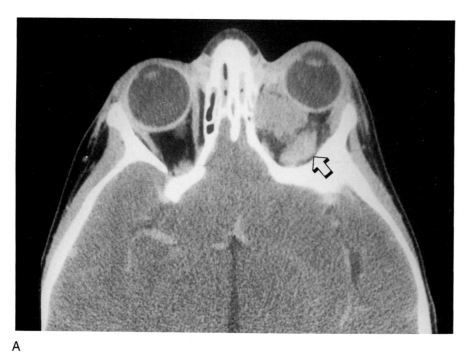

A

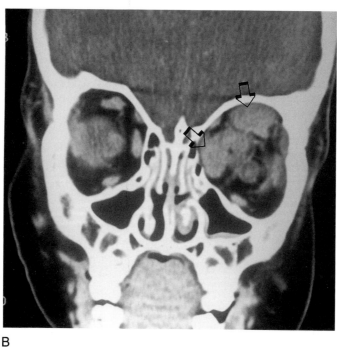

B

Fig. 13-20 Rhabdomyosarcoma. (A) Axial CT scan shows multiple nonspecific mass lesions within the left orbit causing proptosis *(arrow)*. **(B)** Coronal CT shows the masses to be predominantly extraconal *(arrows)*.

PROPTOSIS IN INFANTS

Rhabdomyosarcoma

Capillary hemangioma

Optic glioma

Lacrimal sac mucocele

Teratoma

Neuroblastoma

Leukemia

Choristoma

Large globe (pseudoproptosis)

Congenital glaucoma

Lymphangioma

LACRIMAL GLAND LESIONS

The lacrimal gland lies within a fossa in the anterolateral and superior orbit. There is an anterior palpebral lobe and a posterior orbital lobe. Solid tumors arise within the posterior lobe almost exclusively, whereas inflammatory processes and lymphoma involve both the anterior and posterior lobes. The lacrimal gland has a histology similar to that of other salivary glands.

Epithelial tumors account for 50 percent of mass lesions in the gland. Of these, 50 percent are benign mixed adenomas. The other 50 percent are malignant tumors, including adenocystic carcinoma (most common malignant tumor), pleomorphic malignant tumor, mucoepidermoid carcinoma, adenocarcinoma, and squamous carcinoma. Malignant tumors are difficult to control. Benign tumors may be successfully removed en bloc, although they have a tendency for local recurrence. The tumors occur within the posterior orbital lobe. On CT and MRI, the orbital lobe tends to be enlarged in a rounded shape. Benign tumors are well circumscribed, whereas malignant tumors often have irregular indistinct margination. The tissue characteristics are nonspecific with imaging.

Lymphomatous lesions account for most of the nonepithelial mass lesions. There is a broad spectrum of histology, similar to lymphomatous involvement of the retrobulbar orbit. These lesions involve both lobes of the gland and often extend forward of the orbital rim. Other regions of the orbit may be involved at the same time. The lesions are soft and tend to have a tapered posterior orbital mass, rather than the rounded configuration of the epithelial tumors.

Inflammation of the gland (dacryadenitis) is common. It may be acute or chronic, bacterial, viral, or idiopathic. Acute dacryadenitis is unilateral and responds to antibiotic therapy. Chronic dacryadenitis most often results from sarcoidosis, inflammatory pseudotumor, Graves disease, Mikulicz syndrome (diffuse salivary gland inflammation associated with marrow tumors, chronic infection, or sarcoid), Sjögren syndrome, and Wegener's granulomatosis. The glands may become grossly enlarged. Hydrops of the lacrimal duct is cystic dilation of the duct from obstruction. It appears as a cystic mass along the course of the duct (Fig. 13-21) (see below).

SUBPERIOSTEAL LESIONS

Infection, hemorrhage, and tumor involve the subperiosteal space. Many lesions involve this space by spread from the subjacent structures.

Tumor

In infants, dermoid is the most common lesion found in this region. It is attached to the periosteum or within the subperiosteal space. These lesions are most common in the superior anterolateral orbit near the frontozygomatic suture (Fig. 13-22). They are frequently cystic, and often have high T2 signal intensity with MRI because of contained fat. The common malignant tumors of childhood often metastasize to the periorbital bone. Neuroblastoma, Ewing sarcoma, Wilms tumor and leukemia can be found here.

In adults, primary tumors arising from the paranasal sinuses and nasal cavity and metastasis from distant tumor sites are most likely to invade the subperiosteal space of the orbit. They are usually contained by the periorbita. Squamous carcinoma is the most common primary tumor of the paranasal sinuses to invade the orbit. Eighty percent arise within the maxillary sinus, about 15 percent within the ethmoid sinuses, and the rest from the frontal and sphenoid sinuses. Bone destruction is almost always visible with CT. Large tumors invade the orbit by contiguous growth.

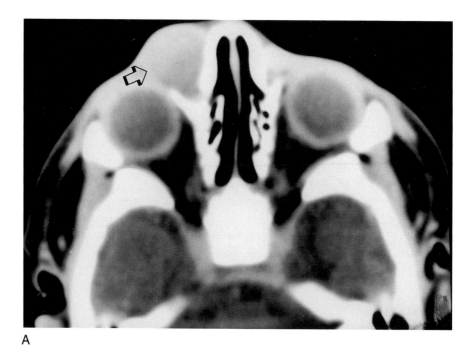

A

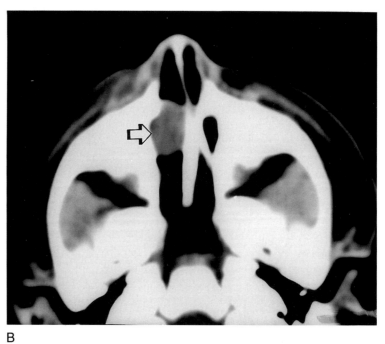

B

Fig. 13-21 Lacrimal sac mucocele (hydrops). (A) Axial CT scan through the orbits demonstrates a cystic mass in the anteromedial right orbit representing the cystic lacrimal sac *(open arrow)*. **(B)** Cystic dilation extends through the course of the lacrimal duct *(open arrow)*.

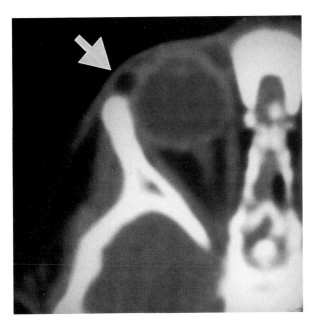

Fig. 13-22 Dermoid. Typical hypodense dermoid cyst occupies the region of the right lacrimal gland, CT scan (*arrow*).

Other primary tumors are adenoid cystic carcinoma, adenocarcinoma, lymphoma, plasmacytoma, melanoma, and inverting papilloma. Lymphoma is particularly likely within the nasal cavity and ethmoid sinuses of patients with acquired immunodeficiency syndrome (AIDS) (see Ch. 8). Metastasis to the bone surrounding the orbit may spread into the orbit. The most common metastases are from primary tumors of the breast, prostate, kidney, and lung. The tumors cause bone destruction or hyperostosis.

Hemorrhage

Subperiosteal hemorrhages develop most often secondary to trauma. Occasionally, they may be spontaneous secondary to hematologic disorder. Large hemorrhages may cause proptosis. Most resorb, but some become chronic collections that may lead to granulomatous masses called *chronic hematic granulomatous cysts*. These lead to local erosion of bone. The cysts may cause painless proptosis or globe displacement and constriction of ocular movement. They rarely involve the floor of the orbit. CT shows the lesions as mixed-density protein

fluid collections, sometimes with calcification. MRI will show high variable signal intensity because of proteinaceous fluid balanced by the low signal of hemosiderin from chronic blood.

Lacrimal Sac

Lacrimal sac mucocele (hydrops) appears as a cystic mass in the region of the lacrimal groove and canal in infants. It results from obstruction of the lacrimal duct. The cyst may become quite large. It shows low density on CT (Fig. 13-21) and water equivalent signal on MRI. Primary tumors of the sac are rare, but they tend to be malignant. Transitional cell carcinoma, lymphoma, and melanoma are the most common.

Infection

Infection enters the orbit most commonly from the adjacent paranasal sinuses. The ethmoid sinuses are the most important, as the lamina payracea separating the sinus from the orbit is thin, and often incomplete. This places the subperiosteal space in direct contact with the mucous membrane of the sinus. Infection within the sinus easily extends into the subperiosteal space. The collection elevates the periorbita, displacing the medial rectus muscle into the retrobulbar space (Fig. 13-23). It may extend onto the globe, causing effusion within Tenon's capsule over the sclera, or anteriorly into the preseptal space. The subperiosteal space may contain fluid (effusion), inflammatory tissue (phlegmon), or pus. The process produces pain, proptosis, lid edema, and limited eye movement (Fig. 13-24). Rarely, the infection breaks through the periorbita, to involve the peripheral or central orbital spaces (true orbital cellulitis). Abscess produces a central liquification seen with CT and MRI. Posterior extension of the infectious process produces the superior orbital fissure syndrome, cavernous sinus thrombosis, or intracranial epidural or subdural empyema. Brain abscess is possible by transmission of infection through valveless draining veins into the brain parenchyma.

Cavernous sinus thrombosis results from acute thrombophlebitis from infection in the region of draining orbital veins. It causes pain, ophthalmoplegia, and papilledema. With contrast-enhanced CT, the sinus fails to enhance normally, although enhancement of the lumen of the carotid artery occurs within it. Only subsegments

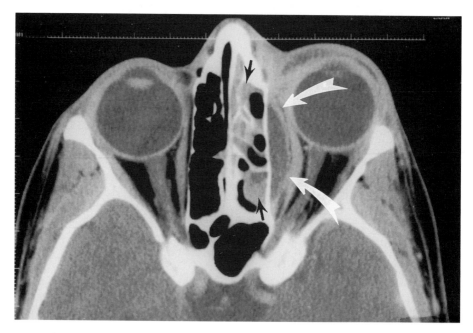

Fig. 13-23 Periorbital abscess, ethmoid sinusitis. Axial CT through the orbits demonstrates inflammatory fluid and edema within the ethmoid sinuses *(small black arrows)*. The infection has broken through the lamina papyracea into the medial left orbit but is contained by the periorbita *(large white arrows)*.

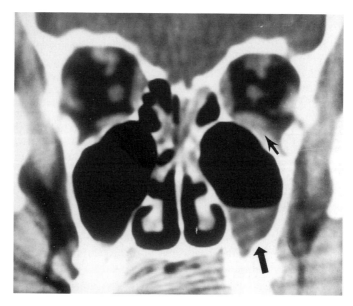

Fig. 13-24 Sinusitis. Fluid collects in the dependent left maxillary sinus from acute suppurative sinusitis *(thick black arrow)*. The inferior rectus muscle is swollen from the sinus inflammation causing restriction of upward gaze *(thin black arrow)*.

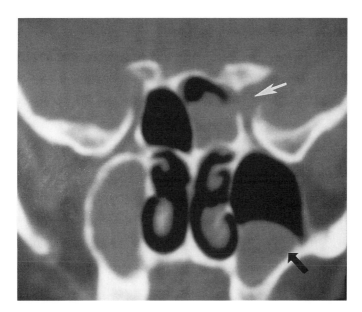

Fig. 13-25 *Aspergillus* **sinusitis.** Fluid and more solid-appearing material occupies the left sphenoid sinus. The infection extends through the lateral sinus wall into the posterior orbit and optic canal (*white arrow*). The right maxillary sinus is completely opacified. There is a seromucinous cyst in the inferior left maxillary sinus (*black arrow*).

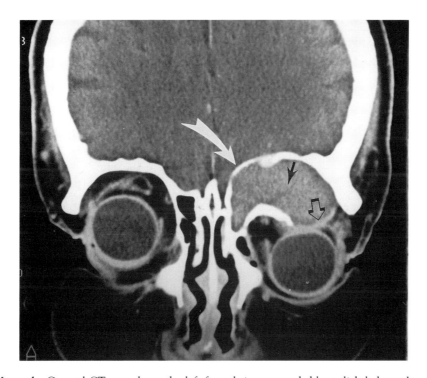

Fig. 13-26 **Mucocele.** Coronal CT scan shows the left frontal sinus expanded by a slightly hyperdense mass within (*black arrow*). The orbital roof is expanded downward causing proptosis of the globe. The bone around the mucocele is thinned (*white arrow*) and resorbed (*open arrow*).

of the cavernous sinus may be affected. With MRI, the normal flow void within the sinus is usually replaced by high signal on T_1- and T_2-weighted images. This represents blood clot, but the signal intensities vary with the age of the clot. Contrast enhancement may be difficult to interpret because of the high signal of the clot. CT shows nonenhancement within the sinus.

The most common orbital infection is bacterial, but fungal disease and tuberculosis can produce a similar picture. Radiographic and CT findings are nonspecific. Rhinocerebral mucomycosis and *aspergillus* infections (Fig. 13-25) are particularly aggressive in immuno compromised persons, causing bone destruction, with invasion of the orbit and brain. *Aspergillus* may cause calcification within the inflammatory tissue and loss of signal on MRI. Granulomatous diseases, such as Wegener's granulomatosis, sarcoidosis, and tuberculosis, produce sinus disease that may extend into the subperiosteal space that is indistinguishable from purulent processes.

Mucocele

Mucoceles are fluid-filled slowly enlarging expansions of the paranasal sinuses. The frontal and ethmoid sinuses are by far the most commonly affected, followed by the sphenoid and maxillary sinuses (Fig. 13-26). The etiology of a mucocele is not completely understood but is thought to result primarily from chronic obstruction of the ostium of the sinus and retention of secretions. The sinus expands and is surrounded by a thin rim of remodeled bone. Rarely, calcification rims the periphery. CT shows the thin rim of bone surrounding a generally homogeneous inner density equivalent to brain tissue. MRI shows variable signal within the mucocele. With primarily inflammatory fluid accumulation, the content of the mucocele is low on T_1- and high on T_2-weighted images. When inspissated, it is high on T_1- and low on T_2-weighted images. It may sometimes be high on both T_1- and T_2-weighted images or completely black from complete inspisation (extremely high protein content) appearing as air on MRIs. Peripheral Gd enhancement sometimes occurs if the mucocele is infected. Enhancement does not occur within the contents of a mucocele, differentiating the entity from tumor. The mucocele causes proptosis or ocular displacement by its expansion into the orbit.

THE GLOBE

Lesions of the Globe in Children

RETINOBLASTOMA

Retinoblastoma is a rare tumor, but it is the most common malignant intraocular tumor of infancy and childhood. It must be differentiated from other lesions of the globe that cause leukokoria (near white pupillary light reflex by clinical examination). Survival depends on the cellular behavior of the tumor and on the amount of spread of the tumor at diagnosis. Survival may be as high as 92 percent at 5 years for localized tumors, but mortality is 100 percent if the tumor has spread outside the globe. The tumors may be inherited, noninherited, associated with a deletion of the q14 band of chromosome 13 (13q14), unilateral (65 percent), bilateral (35 percent), or bilateral and associated with an intracranial tumor of the same cell type in the parasellar region or pineal, called trilateral retinoblastoma (rare). Those with inherited retinoblastoma (autosomal dominant with variable penetrance) are at increased risk of the development of other malignancies, especially osteogenic sarcoma. The tumor is congenital, but it may not be diagnosed until a later age. Calcification is characteristic of intraocular retinoblastoma; it occurs as a result of rapid cell turnover and the formation of a DNA–calcium complex. A calcified lesion within the globe in children under 3 years of age is most likely a retinoblastoma.

CT scan shows the tumor as variably hyperdense, with calcification in 90 percent of cases. Contrast enhancement is slight to moderate. The tumor may grow as a flat plaque along the retina, elevate the retina (detachment), grow into the vitreous, or extend outside of the globe, most commonly along the optic nerve sheath. MRI shows the tumor as slightly hyperintense to the vitreous on T_1-weighted images and as slightly hypointense to the vitreous on T_2-weighted images. MRI is best for separating the actual tumor mass from the commonly associated retinal detachment.

PERSISTENT HYPERPLASTIC PRIMARY VITREOUS (PHPV)

Persistent hyperplastic primary vitreous (PHPV) is a congenital entity, consisting of persistence of a hypervascular embryonic vitreous, the second most common

CAUSES OF LEUKOKORIA IN INFANTS

Retinoblastoma

Persistent hyperplastic primary vitreous (PHPV)

Coat's disease

Retrolental fibroplasia

Endophthalmitis (*Toxocara canis*)

Retinal astrocytoma, glial dysplasia

cause of leukocoria in infants. In early gestation, the posterior system of the globe consists of a primary vitreous supplied richly by vascular branching from the central hyaloid artery. Normally, the hyaloid vascular primary vitreous regresses by 8 months of gestation and is replaced by the avascular secondary vitreous gel of hyaluronic acid and ground substance. The primary vitreous becomes a remnant in the form of a small linear space in the central vitreous, called the canal of Cloquet, connecting the lens with the optic nerve. With PHPV, the primary vitreous fails to regress, leaving the central posterior space of the globe filled with highly vascular opaque primitive tissue. The condition may be unilateral or bilateral and is sometimes associated with cataract, seizures, hearing loss, and mental retardation. It is found in association with fetal alcohol syndrome, fetal hydantoin syndrome, and midline cranial defects. Bilateral PHPV is almost always associated with one of the chromosomal trisomy syndromes.

A retrolental mass of vascular connective tissue and dysplastic retina displaces the lens anteriorly into the anterior chamber. This causes glaucoma, which initially produces buphthalmos (large globe from increased intraocular pressure) and then phthisis as an end-stage finding. Hemorrhage within the globe and retinal detachment are frequent findings. CT scan shows a diffusely hyperdense vitreous with variable contrast enhancement, particularly in the posterior vitreous. The canal of Cloquet containing the persistent hyaloid artery and surrounding vascular tissue mass may enhance in a triangular (cone-shaped) structure, extending between the lens and optic nerve. The primitive vitreous does not enhance, but the vascular mass of tissue does. Layering of

hyperdense fluid, with a visible fluid level changing with patient position, indicates vitreous hemorrhage. The globe may be normal, slightly increased in size, or microphthalmic, depending on the stage of the process at imaging. Calcification is not present, a helpful sign in distinguishing the process from retinoblastoma. MRI shows the vitreous as hyperintense on both T_1- and T_2-weighted images. The connective tissue mass may be isointense or hyperintense to the EOMs. The vascular mass enhances with Gd, while the primitive vitreous does not.

COAT'S DISEASE

Coat's disease is an uncommon unilateral congenital lesion, consisting of a telangiectatic vascular malformation of the retina. The malformation exudes a lipoproteinaceous material into the retinal and subretinal space, causing retinal detachment, usually after 4 years of age. The globe is usually normal, but it may be slightly microphthalmic. CT scan shows the retinal detachment with slightly hyperdense material within the vitreous that cannot be differentiated from a noncalcified retinoblastoma. MRI show the exudate and detachment as high signal intensity on T_1- and T_2-weighted images.

RETROLENTAL FIBROPLASIA

Retrolental fibroplasia is a retinopathy that occurs in premature infants exposed to high oxygen tension over a prolonged period for the treatment of respiratory distress syndrome. It is uncommon today because of improved respiratory therapy technique. The condition produces retinal detachment and a small anterior chamber of the eye, which can be imaged with CT and MRI.

SCLEROSING ENDOPHTHALMITIS

Sclerosing endophthalmitis is a severe intraocular granulomatous inflammatory reaction to infestation of the retina by the larva of the nematode, *Toxocara canis*. It is an affliction of infants and children who are in close contact with the host dogs and who ingest the eggs of the nematode. It is bilateral in 85 percent of cases. The inflammatory reaction causes retinal detachment and extends into the vitreous. It is seen on CT as slightly high density and on MRI as high signal intensity on T_1- and T_2-weighted images. Calcified lesions often involve the cerebrum and are best visualized with CT. An enzyme-linked immunosorbent assay (ELISA) for *Toxocara* is highly sensitive and will make the differentiation from other retinal lesions.

RETINAL ASTROCYTOMA

Retinal astrocytoma (glial dysplasia, astrocytic hamartoma) of the retina occurs most commonly as part of tuberous sclerosis, but the other stigmata of the phakomatosis may not be present at birth. The lesion involves the posterior retina. With imaging, it is indistinguishable from a small retinoblastoma.

DRUSEN

Drusen are small hamartomas of the optic nerve head. They form small round calcifications at the junction of the optic nerve and the globe that are easily seen with CT.

COLOBOMA

Coloboma (from the Greek *koloboun*, "to multilate") is a rare congenital, often inherited, defect of the globe, consisting of a hole or fissure in the head of the optic nerve, sclera, iris, or ciliary body. Defects within the sclera allow bulging of the vitreous into the defect. Sometimes, there is associated glial hyperplasia at the site of the defect (morning glory deformity). The lesion is the result of failure of complete closure of the embryonic vascular fissure in the inferonasal portion of the optic stalk (nerve). The usual defect seen with imaging is a bulge of the globe at the head of the optic nerve (the isolated coloboma), but clefts may be seen along any portion of the inferior globe. Atypical coloboma refers to any defect that occurs at a site other than the inferior portion. The lesion is bilateral in 60 percent of cases. It may also occur in combination with microphthalmos and a separate cyst attached along the inferior globe. The cyst may be huge, dwarfing the globe and expanding the orbit. The cyst forms from inner layer retinal tissue that herniates through an exceptionally large fissure.

CONGENITAL GLAUCOMA

Glaucoma consists of increased intraocular pressure from abnormally increased resistance in the outflow pathways of the aqueous humor. The increased pressure causes macrophthalmos (pseudoproptosis), especially enlargement of the anterior chamber of the eye. It is bilateral in 80 percent of cases. It occurs sporadically or with the phacomatoses.

HYPOPLASIA OF THE OPTIC NERVE

A hypolastic small optic nerve is a congenital defect consisting of optic nerves with fewer than normal axons. This is commonly an isolated finding, but it may be associated with other abnormalities of the face or CNS.

Septo-optic dysplasia is a rare midline developmental anomaly, consisting of hypoplasia of the optic nerves and chiasm, slight enlargement of the frontal horns of the lateral ventricles, absence of the septum pellucidum, poor eyesight, developmental delay, and endocrine abnormality, particularly hypothyroidism. Coronal T_1-weighted MRI or direct coronal CT are the best approaches to the diagnosis.

ANOPHTHALMOS

Primary anophthalmos is congenital complete absence of the ocular bulb from failure of the formation of the optic vescicle. It occurs rarely, most often with trisomy 13 to 15, Klinefelter syndrome, or severe craniofacial malformations. The orbit is small and shallow and filled with rudimentary tissue. Secondary anophthalmia results from the cystic degeneration of an optic vesicle that forms but fails to mature. A variable-sized cyst replaces the globe within the orbit.

Lesions of the Globe in Adults

MELANOMA

Melanoma is the most common neoplasm of the globe in adults. The tumor is rare in children. It arises within the uveal tract, mostly from a pre-existing small nevus. Blacks are rarely found to have this tumor. Melanoma causes an associated hemorrhaghic or serous retinal detachment, making the diagnosis difficult with ophthalmoscopy and imaging. CT shows the tumor to be a slightly hyperdense mass with intermediate intensity contrast enhancement. The retinal detachment appears as a slightly hyperdense collection of fluid along the scleral margin. Melanoma cannot be differentiated accurately from choroid metastasis. With MRI, the melanoma is hyperintense on both T_1- and T_2-weighted images. The shortening of the T_1 relaxation time results from the paramagnetic effect of the stable free radicals in melanin. This signal pattern differentiates melanoma from other tumors and metastases. Hemangioma of the retina exhibits more rapid and more intense contrast enhancement than occurs with melanoma or metastasis.

RETINAL AND CHOROIDAL DETACHMENT

Retinal detachment occurs spontaneously or from injury, tumor, and inflammation. The tissue layers may separate between the hyaloid membrane and retina (subhyaloid space), between the two layers of the retina (true retinal detachment), or between the choroid and the sclera (suprachoroidal space). With detachment hemorrhage, exudate or serous effusion fills the space between the layers. The diagnosis of detachment is made indirectly by visualizing the fluid with ultrasound, CT, or MRI. CT shows the fluid to be of varying density, with pure effusions difficult to see against the fluid density of the vitreous. Exudate and blood have higher density. MRI shows the serous as a slightly high protein collection with a small increase in T_1 signal. Exudate is bright on both T_1- and T_2-weighted images. Hemorrhage has the expected variable signal intensities depending on whether it is acute, subacute, or chronic. MRI with contrast enhancement is best for visualizing an associated tumor.

Optic nerve and chiasm
    Subarachnoid hemorrhage within sheath
    Contusion of the nerve
    Compression by post-traumatic aneurysm
Occulomotor nerves
    Intraneural hemorrhage
    Compression by post-traumatic
Orbit
    Edema
    Fibrosis
    Subperiorbita hemorrhage
    Fracture
        Pure blowout
        Orbitozygomatic
        Nasoethmoid-orbital
        Le Fort II and III
    Injury to nasolacrimal duct
    Injury to the suspensory ligaments of the globe
Foreign bodies
    Glass
    Metal
    Bullet

TRAUMATIC LESIONS OF THE ORBIT AND THE VISUAL APPARATUS

Globe
    Retinal detachment
    Lens dislocation
    Hemorrhage (scleral, retinal)
    Scleral laceration

(Continued)

KEY OBSERVATIONS IN ORBITAL TRAUMA

Intraorbital hemorrhage

Decreased volume of the orbit with inwardly displaced fracture fragments

Increased volume of the orbits with outwardly displaced fracture fragments

Floor fracture with herniation of orbital tissue

Enophthalmus

Proptosis

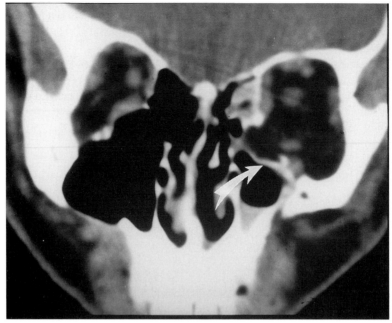

A

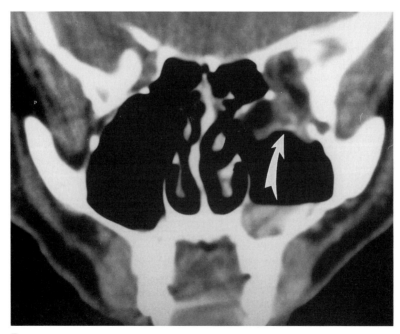

B

Fig. 13-27 Orbital blow-out fracture. (A) Coronal CT shows a left orbital floor fracture with downward displacement of bone fragments into the upper maxillary sinus. There is downward herniation of the orbital fat and inferior rectus muscle. **(B)** The fracture extends to involve the posterior orbital floor. Blood layers in the dependent maxillary sinus.

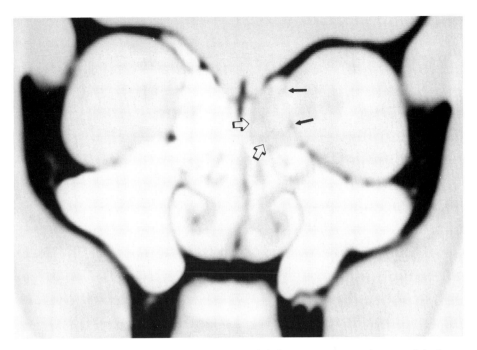

Fig. 13-28 Blow-out fracture through lamina papyracea. Coronal CT scan shows a fracture of the lamina papyracea (*small black arrows*). Orbital contents herniate into the left ethmoid sinus (*open arrows*). Blow-out fractures occur through the floor, roof, and lamina papyracea.

ORBITAL TRAUMA

Orbital trauma is commonly associated with head and facial injury, particularly facial bone fractures. The injury causes loss of vision as well as diplopia and epiphora. CT scanning is generally most useful, but MRI has the advantage in defining contusion of the optic nerve, and hemorrhage within the EOMs. Movement of the globe is impaired by hemorrhage and edema within the orbit or muscles, enophthalmos, herniation of orbital fat or muscles through a fracture (Figs. 13-27 and 13-28), and injuries to CN III, IV, and VI. Change in the intraorbital volume—both an increase and a decrease—impairs motion of the globe by affecting the function of the EOMs.

Injuries to the globe include foreign bodies, scleral laceration, hemorrhage (Fig. 13-29), retinal detachment, and lens dislocation (Fig. 13-30). Laceration of the globe produces a hypotonous eye (low intraocular pressure) from extravasation of fluid. The globe is small and has a crumpled appearance along the posterior margin from uveoscleral infolding. These may be demonstrated by CT

scan, although ultrasonography may have the advantage for evaluation of the anterior chamber and retinal detachment. Severe injury ultimately causes severe atrophy of the globe, called *phthisis bulbi*.

CT scanning is best for the evaluation of orbital fractures. Coronal and direct sagittal planes are preferred for the evaluation of the floor and roof, while transaxial planes are excellent for the medial and lateral walls. Nondisplaced fracture of the orbital floor is most common. It occurs with zygomatic fractures (tripod), more complex zygomatic fractures, and Le Fort II and III fractures. In addition, Le Fort II and III fractures may cause disruption of the medial canthal ligament and injury to the lacrimal canal, with interruption of the lacrimal drainage system. Diastatic fracture through the zygomaticofrontal suture may disrupt the lateral suspensory and canthal ligaments, resulting in distortion of the cantal fold and enophthalmus.

Blow-out-type orbital fractures occur with or without associated orbital rim fracture (Figs. 13-27 and 13-28). The fracture is usually in the weakest central portion of

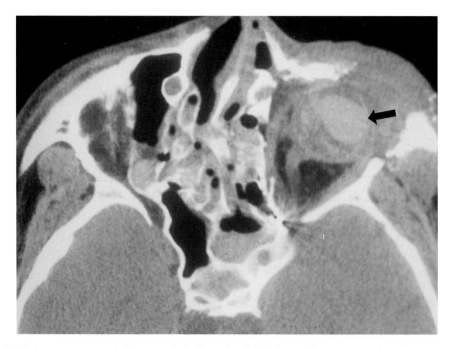

Fig. 13-29 Globe hematoma, smash fracture of the face. Axial CT shows a hematoma within the anterior left globe (vitreous) *(arrow)*. There are comminuted fractures of the left ethmoid, maxilla, and zygoma.

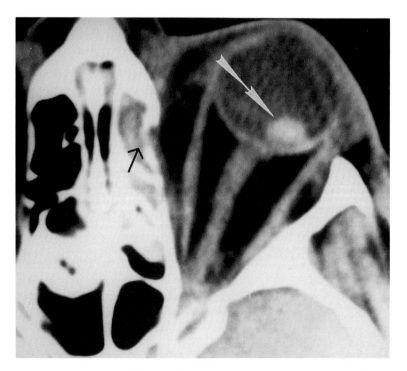

Fig. 13-30 Lens dislocation. The lens of the left ocular bulb is dislocated and rests against the retinal *(white arrow)*. The lamina papyracea is also fractured and displaced medially into the ethmoid sinus *(black arrow)*.

the floor near the infraorbital canal. The fracture results from a blow to the inferior orbit rim, transmitting force to the floor, augmented by an increase in intraorbital pressure. Most commonly, only the inferior orbital connective tissue herniates into the upper portion of the maxillary sinus. The inferior rectus muscle herniates through the largest defects in the floor. The herniation of the connective tissue alone causes restriction of eye movement because of its fibrous septa, which attach to the globe. Most floor fractures do not extend posteriorly to the pterygoid strut of the apex. Controversy surrounds the treatment of floor fractures, but authorities generally agree that fractures that cause restriction of eye movement require repair. Blow-out-type orbital fractures may also affect the lamina papyracea and the orbital roof (Fig. 13-28).

The volume of the orbit is affected by some orbital fractures. Subperiorbital hematoma may compress the muscle cone and other orbital structures (Fig. 13-31). Inward rotation of a zygoma fracture about a vertical axis

---

CAUSES OF IMPAIRED EYE MOVEMENT AFTER TRAUMA

Edema
Herniation of orbital tissue
Scar tissue between muscle and bone
Palsy of CN III, IV, and VI
Severe enophthalmos (lax muscles)

---

causes a decrease in orbital volume, while outward rotation causes an increase in orbital volume. A blow-out-type floor fracture with herniation of orbital tissue increases the orbital volume. Nasoethmoidal fractures usually displace the medial wall inward, thereby decreasing orbital volume. Decreased orbital volume constricts the retrobulbar tissue, causing proptosis and constriction

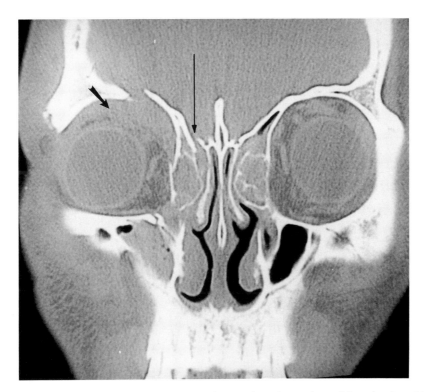

Fig. 13-31 Subperiorbital hematoma. A large hematoma elevates the subperiorbita of the orbital roof *(small arrow)*. There are fractures of the orbital roof and cribriform plate *(long arrow)*.

of ocular movement. An increase in the orbital volume causes enophthalmos, removing the tension on the extraocular muscles so that their action is no longer effective. It is important to recognize alterations in orbital volume.

The cranial nerves supplying the orbit are frequently injured, producing varying patterns of ophthalmoplegia. CN IV is most commonly affected. The nerve is thought to be injured by compression as it courses near the free edge of the tentorium. The nerves are injured by hemorrhage, avulsion, or tears. Occasionally, other lesions, such as rare meningeal vascular malformations associated with the nerves, have been implicated as a cause of hemorrhage following injury. Increased intracranial pressure from traumatic brain edema may compress CN VI against the clivus of the skull base.

CRANIAL NERVES (CN) III, IV, AND VI

CN III, IV, and VI control the six EOMs, which control the movements of the eye. The complete system of nerves and muscles works with precise coordination to accomplish visual tracking of objects, and conjugate gaze. Disturbance of this system results in binocular diplopia (double vision when both eyes are uncovered).

These cranial nerves have a brain stem segment (Figs. 13-32, and 13-33) (nucleus and fasciculus), a cisternal segment, a cavernous segment (Fig. 13-34), and an orbital segment. Lesions affecting one or more of the nerves at any segment may result in oculomotor palsy. The clinical pattern of abnormality varies with the segment involved and may consist of an isolated neuropathy

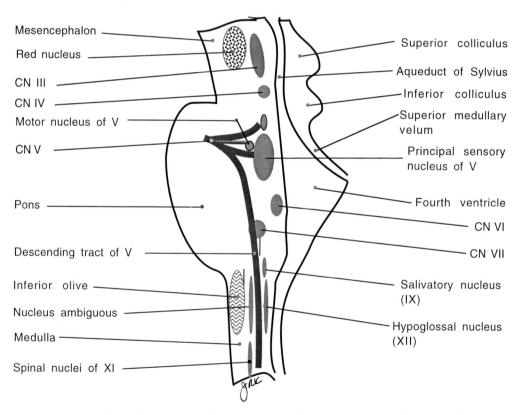

Fig. 13-32 Brain stem and cranial nerve nuclei. Sagittal drawing of the brain stem shows the levels of the cranial nerve nuclei and the major tracts of CN V.

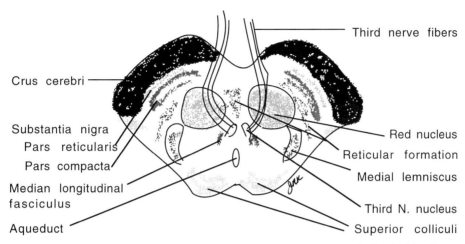

Fig. 13-33 Mesencephalon. Axial drawing of the mesencephalon at the level of CN III.

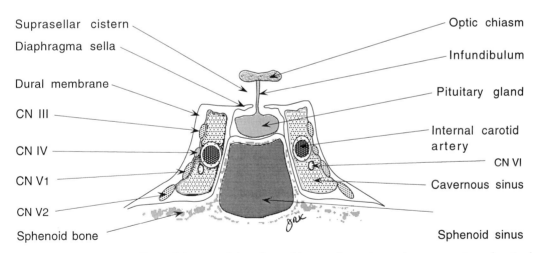

Fig. 13-34 Cavernous sinus. Coronal drawing of the sphenoid bone, sella turcica, and cavernous sinus, showing how the dura forms the sella and the cavernous sinus and contains CN III to VI. CN III to $V_{1,2}$ are incorporated into the inner layer of the dura that forms the lateral wall of the cavernous sinus. CN VI lies within the cavernous sinus. The diaphragm sellae is a dural roof over the sella turcica pierced by the pituitary stalk (infundibulum).

or a complex pattern of multiple nerve dysfunction. Only CN III is consistently imaged with CT or MRI.

Oculomotor Nerve: CN III

CN III originates in a longitudinal group of nuclei in the dorsal mesencephalic tegmentum at the level of the superior colliculus, immediately anterior to the aqueduct, and posterior to the red nuclei. The Edinger-Westphal subnucleus is along its dorsal margin, providing the parasympathetic (vasoconstrictor) neurons of CN III for the pupil and ciliary nerves. The fasciculus (axon tract) of CN III runs anteriorly through the red nuclei, to exit from the mesencephalon between the peduncles (crura cerebri) into the interpeduncular cistern. The nerve runs anteriorly between the posterior cerebral artery (PCA) and the superior cerebellar artery (SCA), parallels the undersurface of the posterior communicating artery (PCoA), and enters the posterior superior cavernous sinus. The parasympathetic axons are in the peripheral superior portion of the nerve. The sympathetic axons from the cervical sympathetic ganglia course along the

internal carotid artery (ICA) and join the periphery of CN III in the anterior cavernous sinus, to enter the orbit. CN III is carried forward within the dural layers of the superior lateral wall of the cavernous sinus, entering the intraconal orbit through the superior orbital fissure and the annulus of Zinn.

ISOLATED CN III PALSY

An isolated CN III palsy is an important clinical sign. The palsy may be complete (exotropia and diplopia, ptosis) or partial, and pupil-involving (dilated pupil) or pupil-sparing. In general, pupil-sparing neuropathy results from intrinsic disease of the nerve, most commonly ischemia associated with diabetes mellitus or hypertension. Pupil-involving neuropathy implies a compressive lesion affecting the nerve, most commonly an ICA aneurysm at the PCoA, but sometimes at the PCA or SCA or basilar ectasia (Fig. 13-35). Pain is associated with both types of lesions, although with compressive lesions, the pain is usually more severe and longer-lasting. An acute pupil-involving CN III palsy constitutes an emer-

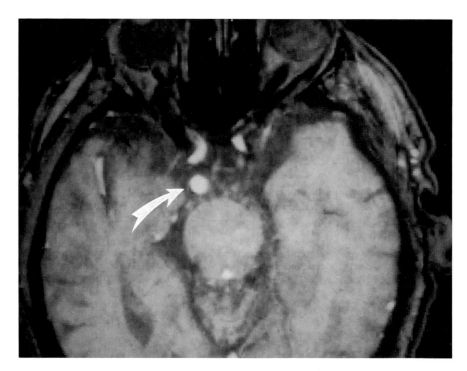

Fig. 13-35 Basilar artery ectasia. An ectatic basilar artery extends superiorly and laterally in the prepontine cistern, compressing CN III as it crosses the cistern. GRE MRI, Flash 70 degrees, 20/10, axial scan done through the upper pons.

gency and requires cerebral angiography or high-detail magnetic resonance angiography (MRA) for the diagnosis of a possible berry aneurysm.

Trochlear Nerve: CN IV

CN IV arises from its nucleus in the dorsal tegmentum of the mesencaphalon, just caudal to CN III at the level of the inferior colliculus. Its fasciculus courses posteriorly around the aqueduct, decussates within the superior medullary velum of the fourth ventricle, and exits on the backside of the brain stem just below the inferior colliculus. The nerve then swings anteriorly about the mesencephalon, within the ambient cistern at the level of the medial free edge of the tentorium. It then runs between the PCA and the SCA and enters the cavernous sinus just inferior to CN III. It courses anteriorly within the lateral wall of the sinus, just inferior to CN III.

An isolated CN IV palsy is relatively uncommon. It occurs most commonly from head trauma and injury of the nerve against the close free edge of the tentorium.

> **CAUSES OF ISOLATED THIRD NERVE PALSY**
>
> Adults
> > Ischemia of the nerve or brain stem
> > Aneurysm of the PCoA
> > Vascular ectasia
> > Post-trauma
> > Tumor (extrinsic)
> > Multiple sclerosis
> > Herpes zoster
> > Meningitis
> > Sinusitis
>
> Children
> > Trauma
> > Meningitis
> > Sinusitis
> > Tumor

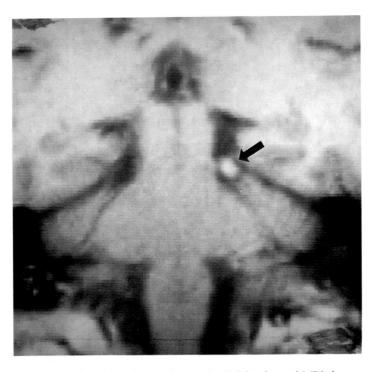

Fig. 13-36 Cavernous angioma of CN IV. Coronal T_1-weighted Gd-enhanced MRI shows a region of high signal intensity along the course of the left CN IV *(arrow)*. Neuritis, neuroma, and metastasis might also present this appearance.

Brain stem lesions, nerve tumors, viral infections, (including HIV), or rare vascular malformations of the nerve may also produce the palsy (Fig. 13-36). In children, CN IV palsy is most common after trauma and viral infections. Any of these lesions may result in Gd enhancement of the nerve on MRI.

Abducens Nerve: CN VI

CN VI arises from its nucleus within the tegmentum of the pons, just anterior to the middle portion of the floor of the fourth ventricle (Fig. 13-37). CN VII loops posteriorly around the nucleus, creating a small bulge on the floor of the fourth ventricle, the facial colliculus, a recognizable landmark on axial MRI (Fig. 13-38). The CN VI fasciculus courses anteriorly through the pons, exits medially at the pontomedullary junction, and takes a long path superiorly within the prepontine cistern along the clivus. It passes through Dorello's canal beneath the petroclinoid ligament, to enter the inferior cavernous sinus medial to CN V. The nerve lies within the cavernous sinus, medial to CN III and CN IV, and inferior to the ICA (Fig. 13-34).

Isolated CN VI palsy is relatively common. Often it is transient and idiopathic. In adults, the most common identifiable causes are ischemia, increased intracranial pressure, multiple sclerosis, middle ear or mastoid infection with medial petrositis (Gradenigo syndrome), and tumor, particularly nasopharyngeal tumor invading the base of the skull. Otitis media with CN VI palsy is an uncommon combination requiring urgent treatment.

Complex Neuropathies

Complex neuropathies and syndromes consist of multiple cranial nerve involvement or of a cranial nerve with brain stem tract involvement. Brain stem lesions may affect one or more cranial nerves, as well as adjacent structures or tracts. These lesions produce a variety of alternating hemiplegias, consisting of ipsilateral cranial nerve dysfunction and contralateral motor or sensory dysfunction. The lesions may be infarction, hemorrhage,

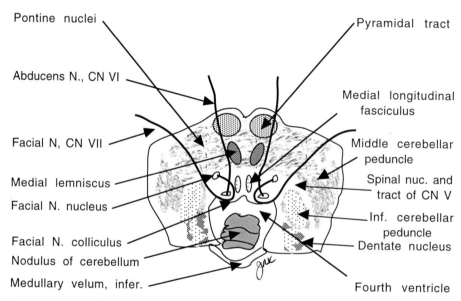

Fig. 13-37 Lower pons. Axial drawing of the lower pons at its junction with the medulla and at the level of the exiting CN VI. It shows the posterior loop of CN VII around CN VI nucleus, producing a small bump called the facial nerve colliculus. Anterior is toward the top.

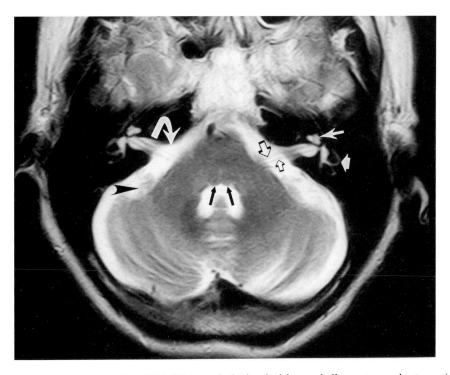

Fig. 13-38 Normal posterior fossa. Axial T$_2$ MRI through the level of the cerebellopontine angle cistern (*curved white arrow*). It shows CN VIII with its anterior cochlear division (*large open arrow*) and posterior vestibular division (*small open arrow*). The facial nerve (not seen) lies just above the cochlear division. The facial colliculi protrude into the anterior fourth ventricle (*small black arrows*). The flocculus protrudes into the left cerebellopontine angle cistern (*black arrowhead*). The cochlea (*white arrow*) and semicircular canals (*fat white arrow*) can be seen well on T$_2$ MRI.

trauma, tumor, or demyelination. The more common combinations are presented in Table 13-2.

Trigeminal Nerve: CN V

CN V has primarily sensory function, with the exception of the small motor root that innervates the masticatory muscles, the myohyoid muscle, and the digastric muscle. The sensory neurons reside in the gasserian ganglion. These neurons receive axons from three major peripheral branches: the ophthalmic (V$_1$), the maxillary (V$_2$), and the mandibular (V$_3$) nerves. The ganglion projects posteriorly through the large fifth nerve trunk across the prepontine cistern to the pons (Fig. 13-39). Here, the heavily myelinated A fiber tracts carry tactile sensation and end within the primary sensory nucleus of the pontine tegmentum at the level of the root

(Fig. 13-32). The lightly myelinated C-fiber tracts carry pain and temperature and descend to end with topographical organization along the spinal trigeminal tract. This longitudinal tract extends from the lower pons to the upper cervical spinal cord. Tracts carrying proprioception from the mandibular region ascend within the pons to end within the mesencephalic nucleus of CN V. The motor nucleus lies adjacent to the primary sensory nucleus. Its efferent fibers course with the main trunk of CN V but pass through the gasserian ganglion without synapse, to enter the mandibular nerve. The chorda tympani branch of the sensory nervus intermedius of CN VII crosses the tympanic cavity and joins the lingual nerve of V$_3$, providing taste sensation from the anterior two-thirds of the tongue (see Fig. 13-45). The CN V is categorized into four anatomic regions: brain stem, cisternal, ganglionic, and peripheral.

Table 13-2 Complex Brain Stem CN Neuropathies

Syndrome	Location	Nerve and Tract	Symptoms
Weber	Mesencephalon	III Corticospinal	Ophthalmoplegia Contralateral hemiplegia
Claude	Mesencephalon	III Red nucleus	Ophthalmoplegia Contralateral ataxia/tremor
Benedikt	Mesencephalon	III Red nucleus Corticospinal	Ophthalmoplegia Contralateral ataxia/tremor Contralateral hemiplegia
Nothnagel	Mesencephalon (tectum)	III Superior cerebellar peduncle	Ophthalmoplegia Gaze palsy Ataxia
Millard-Gubler	Pons	VI, VII Corticospinal	VI palsy VII palsy Contralateral hemiplegia
Foville	Pons	V, VI, VII, VIII Corticospinal	Gaze palsy Facial palsy and sensory loss Hearing loss Contralateral hemiplegia
Raymond	Pons	VI Corticospinal	Abducens palsy Contralateral hemiplegia

(Modified from Laine and Smoker, 1993, with permission.)

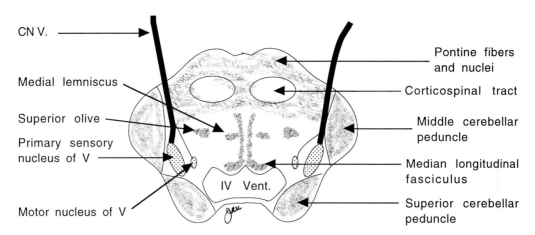

Fig. 13-39 Middle pons. Axial drawing of the pons at the level of the exiting CN V. The tegmentum is that portion posterior to the corticospinal tracts.

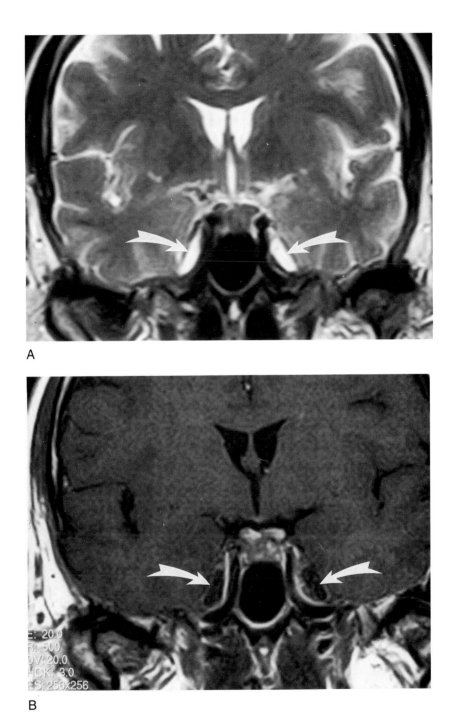

Fig. 13-40 Meckel's cave. (A) Coronal FSE T$_2$ MRI shows the CSF within Meckel's cave as high signal (*arrows*). The trigeminal ganglion cannot be imaged normally, as it is made up of a myriad of rootlets. **(B)** T$_1$ Gd-enhanced MRI shows the cave as regions of low signal intensity adjacent to the posterior cavernous sinus (*arrows*).

The gasserian ganglion lies within Meckel's cave, an outpouching of the prepontine cistern adjacent to the posterolateral inferior cavernous sinus (Figs. 13-40 and 13-41) and sellar turcica. The ganglion is too small and gossimer to be resolved with MR or CT imaging, but the cave shows as a collection of CSF at the posteroinferior cavernous sinus. The absence of CSF signal or density characteristics in this region implies a mass lesion within the cave.

PATHOLOGY

Clinical abnormalities cause loss of sensation, pain, or motor function in the distribution of the nerve. The location of a mass lesion determines the topographic distribution of the sensory or motor changes. Lesions of the upper basisphenoid bone, the gasserian ganglion, the posterior sella turcica, and the posterior cavernous sinus may involve all three divisions, in-cluding the motor nerve. Lesions of the anterior inferior cavernous sinus involve V_1 and V_2 (Fig. 13-34). Tumors of the nasal cavity, the floor of the mouth, and the pterygoid fossa may infiltrate along any of the nerve branches, causing enlargement of the nerves and their foramina. Precise clinical location of the peripheral nerve involved is not always possible (Table 13-3).

The Pontine Segment

Lesions within the pons may produce trigeminal neuropathy, often in combination with other clinical signs of alternating hemiplegia or ataxia. Infarction, multiple sclerosis, and metastasis are the most common lesions in adults. About 5 percent of patients with multiple sclerosis will have trigeminal neuropathy as their presenting symptom. Glioma is the most common pontine lesion to affect the trigeminal nerve in children.

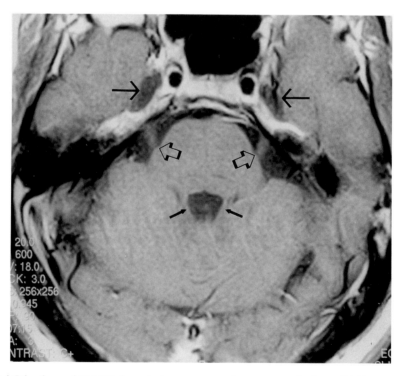

Fig. 13-41 Axial Gd-enhanced T_1 MRI through the upper pons demonstrates the large CN V coursing anteriorly from the pons through the pontine cistern (*open arrows*). The nerves will enter Meckel's cave (*large black arrows*), filled with hypodense CSF. The superior cerebellar peduncles run along the posterolateral upper fourth ventricle (*small black arrows*).

Table 13-3 Lesion Location versus Cranial Nerve Dysfunction[a]

Location	CN Dysfunction
Posterior cavernous sinus Upper basisphenoid bone	All divisions of V, including motor function, III, IV, VI
Anterior cavernous sinus	V_1 and $\pm V_2$, III, IV, $\pm VI$
Superior orbital fissure	V_1, III, IV, VI
Inferior orbital fissure	V_2
Cisternal	All divisions of V, VI, III, IV
Pons	Loss of function determined by the level of pontine lesion

[a] Not all nerves may be affected at each anatomic site.

LESIONS WITHIN THE PONS AFFECTING CRANIAL NERVES

Adults
  Infarction
  Multiple sclerosis
  Metastasis
  Hematoma; angioma
Children
  Glioma

The Cisternal Segment

The large trigeminal root crosses the prepontine cistern, to reach Meckel's cave. Mass lesions that involve the central base of the skull or meninges in this region may cause trigeminal neuropathy, generally associated with both pain and neurologic deficit. The most common mass lesions are metastasis to the basisphemoid, direct extension of nasopharyngial carcinoma through the skull base (see Ch. 5), myeloma (Fig. 13-42), and chordoma. Metastasis to the meninges, or inflammatory pachymeningitis, may also cause neuropathy. Larger tumors common within the cerebellopontine angle cistern may extend superiorly to affect CN V. These are most commonly acoustic neurinoma, meningioma, and epi-

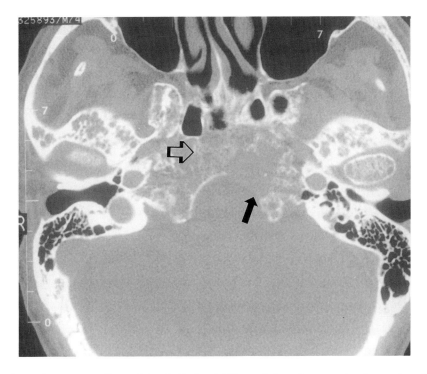

Fig. 13-42 Myeloma, destruction of base of the skull. Axial CT scan defines the extensive destruction of the skull base from myeloma *(open arrow)*. It has broken through the surface of the clivus on the left *(black arrow)*. High-resolution CT reconstructed with bone algorithm is excellent for definition of bone destruction at the skull base. MRI defines better tumor extending beyond the bone margins.

dermoid. Epidermoid tumors have signal characteristics close to those exhibited by CSF and are difficult to detect.

Tic Douloureux

Tic douloureux (trigeminal neuralgia) is the most common abnormality of the trigeminal nerve in adults. The true tic syndrome has severe pain without neurologic deficit. The pain is paroxysmal and is frequently triggered by mininal neural tactile or motion stimulation within the distribution of the nerve. The cause of tic is thought by some to be compression of CN V root entry zone (REZ) at the surface of the pons, and many patients have demonstrated vascular loops of the SCA, anterior inferior cerebellar artery (AICA), or basilar artery compressing the REZ. Surgical decompression of the REZ has been curative in a number of cases. However, many patients do not have vascular compression, and the syndrome is thought to be caused by an imbalance of the nuclear interactions within the pons. Mass lesions or infarctions affecting the nerve do not produce trigeminal neuralgia.

CISTERNAL CAUSES OF TRIGEMINAL NEUROPATHY

Tumors of the basisphenoid bone
   Metastasis
   Direct extension of nasopharyngeal carcinoma
   Chordoma
   Chondrosarcoma
   Meningioma
Large tumors of the cerebellopontine angle cistern
   Acoustic neurinoma (schwannoma)
   CN V neurinoma (schwannoma)
   Metastasis (especially melanoma)
   Epidermoid
Lesions of the leptomeninges
   Metastasis (neoplastic meningitis)
   Meningioma
   Pachymeningitis (sarcoid, tuberculosis, idiopathic)

(Continued)

Vascular compression
   Loops of the SCA, AICA, basilar artery
   Aneurysm
Infection (rhombencephalitis, if pons affected)
   HSV-I
   Varicella-zoster
   HIV

Meckel's Cave and Ganglion

Primary lesions within Meckel's cave are rare. The most common are trigeminal schwannoma, meningioma, and epidermoid. Metastatic tumors of the meninges may spread into the cave, which is an extension of the subarachnoid space. The trigeminal schwannoma arises most commonly from the ganglion, causing neuropathy in all three divisions of the nerve. It has the typical characteristics of schwannoma, being isointense with brain of T_1-weighted images, and isointense to hyperintense on T_2-weighted images. It shows intense enhancement with MRI and CT. Most commonly, this is homogeneous but sometimes it is heterogeneous. The larger tumors extend in dumbbell fashion into the prepontine cistern. These tumors may arise along any part of the nerve between the pons and cave. Tumors within the cave displace the CSF, so that the normally high T_2 signal within the cave on MRI is lost.

Viral neuritis causes acute CN V neuropathy. Herpes simplex virus (HSV) and varicella-zoster virus (VZV) are the most common causes, but any virus may be the culprit. These latent viruses reside within the gasserian ganglion. With reactivation, the virus spreads anteriorly through the peripheral branches, causing aphthous ulcers within the oral cavity. Less commonly, retrograde spread along the cisternal segment of the nerve carries the virus to the brain stem, producing rhombencephalitis. The few cases that have been reported demonstrate striking Gd enhancement of the nerve, and less intense enhancement within the pons, reverting to normal after about 6 weeks.

The Cavernous Sinus

Giant carotid artery aneurysm is the most common mass lesion within the cavernous sinus. The aneurysm is best demonstrated with MRI, although CT angiography of

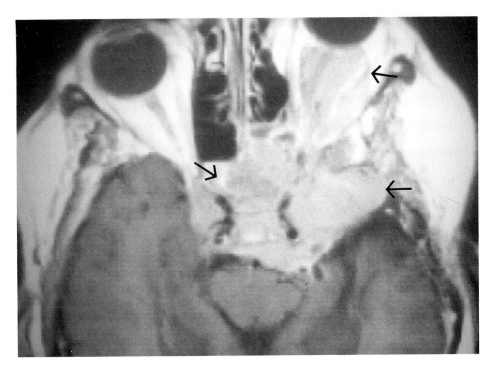

Fig. 13-43 Meningioma. A large meningioma of the sphenoid and ethmoid sinuses, the anterior left temporal fossa, and the posterior left orbit. The tumor enhances homogeneously with Gd *(arrows)*. Axial T$_1$-weighted MRI.

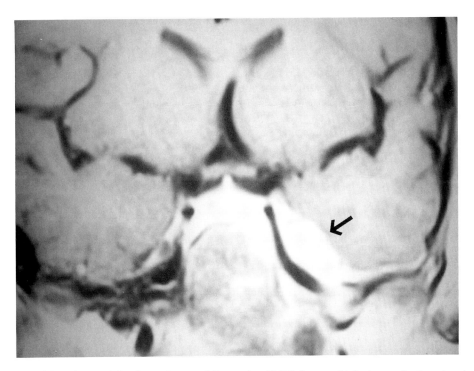

Fig. 13-44 Pseudolymphoma. Gd-enhanced coronal T$_1$-weighted MRI shows a thick plaque of enhancing tissue along the left side of the cavernous sinus *(arrow)*. It extends inferiorly into the inferior temporal fossa. This biopsy-proven lesion regressed after steroid therapy.

**LESIONS OF THE
CAVERNOUS SINUS**

Giant aneurysm of the ICA

Pituitary adenoma, direct extension

Metastasis to the upper skull base

Meningioma, within the sinus

Direct extension of tumor from nasopharynx

Cavernous sinus thrombosis

     Pyogenic infection of the sinuses and face

     Tolosa-Hunt syndrome, pseudolymphoma

Carotid-cavernous fistula

---

the region is also excellent—and diagnostic in most cases. CT shows the frequent calcification within the wall of the aneurysm, as well as the characteristic erosion of the sella turcica from laterally inward. Angiography confirms the diagnosis in equivocal cases. The ophthalmic division of CN V (V_1) is the most commonly affected, along with CN VI. Other lesions within the cavernous sinus (Figs. 13-43 and 13-44) are listed in the box.

The peripheral segments of CN V are most commonly affected by inflammatory disease in the sinuses or by tumors of the face, sinuses, oropharyngeal region, or parotid gland. Here, fat-suppressed Gd-enhanced MRI is the best imaging modality. Tumor masses are seen within the region; perineural spread is recognized by enhancement of enlarged peripheral nerves. CT shows the enlarged foramina or bone destruction at the base of the skull.

Facial Nerve: CN VII

The voluntary motor neurons of facial movement are located in the precentral and postcentral gyri of the frontal and parietal cerebral cortex. The axons descend through the genu of the internal capsule to the corticobulbar tract and decussate within the pons at the level of the facial nerve motor nucleus. Many of the fibers controlling the upper one-half of the face do not decussate, providing bilateral cortical representation for this upper region. Lesions affecting the cortex or corticobulbar tracts produce contralateral facial paresis for the lower face only. Lower motor neuron or facial nerve lesions

produce ipsilateral paresis of the entire face, although the topographic organization within the facial nerve permits upper face sparing with incomplete peripheral lesions.

The facial nerve consists of two nerves: a larger trunk of heavily myelinated motor fibers, and a smaller trunk of lightly myelinated sensory fibers called the nervus intermedius. The motor nucleus of CN VII lies within the inferior pons anterolateral to the abducens nucleus. The motor fasciculus loops posteriorly around the nucleus of CN VI, creating a bump on the floor of the fourth ventricle called the facial colliculus (Fig. 13-37), before passing anteriolaterally through the pons, exiting at the pontomedullary junction. The sensory roots join the nerve within the pons. The nerve traverses the cerebellopontine angle cistern, lying close, but anterosuperior, to CN VIII. It enters the internal auditory canal (IAC) in its anterosuperior quandrant and goes deep into the fundus, to enter the facial nerve canal (fallopian canal). A short narrow horizontal segment (the labyrinthine segment) runs to the anterior geniculate ganglion of the facial nerve. At the ganglion, the greater and lesser superficial petrosal nerves originate and course anteriorly through the skull (vidian canal) into the pterygopalatine fossa, to synapse in the pterygopalatine and otic ganglia. These parasympathetic efferent nerves supply the lacrimal glands and salivary glands of the mouth.

The nerve continues curving sharply posteriorly into the horizontal typanic segment beneath the horizontal semicircular canal and just above the oval window. The canal then curves inferiorly around the posterior tympanic cavity to course through the mastoid to exit the temporal bone at the stylomastoid foramen (Fig. 13-45). In the descending segment, the sensory chorda tympani nerve comes off, coursing anteriorly between the malleus and incus, and then through the skull, to join the lingual nerve of CN V. It supplies taste sensation to the anterior two-thirds of the tongue. Theoretically, lesions can be located by observing the pattern of loss of motor and sensory functions of the nerve, but in practice this is difficult and imprecise. MRI, with and without contrast enhancement, is best for evaluation of most pathology of the nerve within the brain stem, cistern, and IAC. CT scan is best for evaluation of trauma of the temporal bone and for demonstrating the anatomy and destructive lesions of the temporal bone and facial canal.

Within the fallopian and vidian canals, the facial and greater superficial petrosal nerves are surrounded by a rich vascular network. The network is less prominent within the labyrinthine segment of the canal. This net-

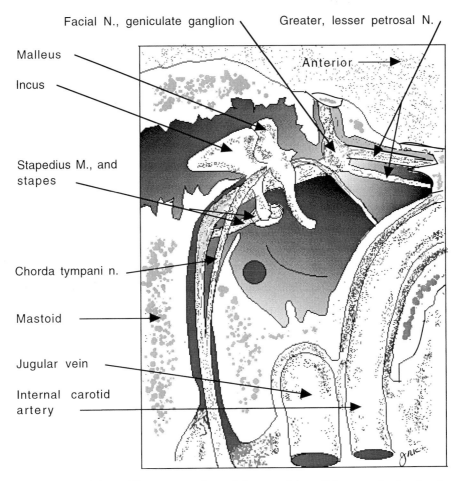

Facial N., geniculate ganglion

Greater, lesser petrosal N.

Malleus

Incus

Anterior

Stapedius M., and stapes

Chorda tympani n.

Mastoid

Jugular vein

Internal carotid artery

Fig. 13-45 Facial nerve and the middle ear cavity. Sagittal drawing of the middle ear cavity showing the oscicles, the facial nerve, and the tympanic cavity. The facial nerve exits the skull inferiorly through the styloid foramen. (Adapted from Shambaugh, 1967, with permission.)

work normally becomes moderately bright with contrast-enhanced MRI and should not be misinterpreted as pathology of the nerve (Fig. 13-46).

PATHOLOGY

The facial nerve is generally resistant to extrinsic pressure, consisting of predominantly heavily myelinated motor fibers. Therefore, the most mass lesions do not produce facial nerve palsy, until there is extreme pressure on the nerve. For example, mass lesions within the IAC and cerebellopontine angle cistern produce sensorineural hearing loss (CN VIII neuropathy) long before facial nerve palsy. Intrinsic lesions of the nerve cause facial nerve palsy much sooner in the course of the disease.

Bell's palsy is the most common affliction of the facial nerve, representing 50 to 80 percent of all cases of facial nerve palsy. Onset is acute, with recovery in 4 to 6 months. Only atypical cases are imaged, those with progressive palsy over weeks, delayed recovery beyond 6 months, and recurrence of palsy. Reports indicate contrast enhancement of the facial nerve with Bell's palsy, but enhancement is seen normally as well, often on the side opposite the palsy; in such cases, the finding has no diagnostic or prognostic value.

Progressive or insidious onset of facial palsy and post-traumatic facial palsy require an imaging study appropriate for the clinical problem suspected. Disease processes affecting the facial nerve and the anatomic location are listed in the box.

LESIONS AFFECTING THE FACIAL NERVE (CN VII)

Brain stem
    Infarction
    Metastatic tumor
    Pontine glioma (children)
    Multiple sclerosis
    Hematoma; angioma
Cistern
    Acoustic neurinoma
    Meningioma
    Epidermoid
    Metastasis
    Facial N. neuroma
    Vascular compression (hemifacial spasm)
Nerve within canal
    Bell's palsy
    Trauma (transverse fracture primarily)
    Cholesteatoma, middle ear
    Facial N. neuroma
    Metastasis to the temporal bone
        Lung, breast, kidney, prostate, histiocytosis X
    Lyme disease
    Herpes zoster oticus (Ramsey-Hunt syndrome)
    Cavernous hemangioma of the geniculate ganglion
    Malignant external otitis (*Pseudomonas* aeruginosa)
Extracranial segment
    Parotid gland tumors (mostly malignant)
    Sarcoid of the parotid gland
    Facial N. neuroma
    Toxoplasmosis of parotid gland
    Lymphoepithelial cysts (HIV)

Cavernous Hemangioma
Cavernous hemangioma arises from the vascular plexus surrounding the facial nerve. It occurs most commonly at the geniculate ganglion or the fundus of the IAC, but it can be found at any segment of the nerve. It causes a facial nerve palsy despite its small size of less than 2 cm. The lesion is most often isointense with brain on T_1-weighted images, and hyperintense on T_2-weighted images. Some are hyperintense on T_1-weighted MRI, and this is diagnostic of the lesion. The lesion enhances with Gd. The hemangioma produces a focus of bone destruction with the peripheral infiltration, creating a serrated margin seen with bone window CT scans.

Hemifacial Spasm
Hemifacial spasm is a condition of involuntary twitching of facial muscles that occurs mostly in women. It is usually unilateral. The syndrome is analogous to trigeminal neuralgia and is often caused by vascular compression of the REZ of the facial nerve. MRI demonstrates the vascular compression in a high percentage of cases.

Vestibular Cochlear Nerve: CN VIII

The vestibulocochlear nerve has two components: the vestibular nerve originating from the utricle and saccule within the vestibule of the temporal bone, and the cochlear nerve originating from the organ of Corti within the cochlea of the temporal bone. The vestibular nerve courses medially across the cribriform plate of the fundus (depth) of the IAC, adjacent to the vestibule, to enter the IAC and the vestibular ganglion (Scarpa's ganglion). Meninges extend into the depth of the canal, allowing CSF to surround the nerves. The IAC is divided into the superior and inferior channels by the thin plate of bone called the crista falciformis. The vestibular nerve is split into superior and inferior bundles that pass through the IAC in the posterior half. The nerve exits the canal at its medial opening, the porus acousticus, continuing medially through the cerebellopontine angle cistern, to enter the dorsolateral margin of the upper medulla. The fibers synapse within the paired vestibular nuclei in the lateral floor of the inferior fourth ventricle. The cochlear nerve also enters the fundus of the IAC and courses through the anterior inferior quadrant, through the cerebellopontine angle cistern, enters the upper medulla adjacent to the vestibular nerve, and synapses within the paired cochlear nuclei along the lateral portion of the medulla. The facial nerve occupies the anterior superior quadrant of the IAC.

Disorders of the vestibular nerve and nuclei produce vertigo and nystagmus. Disorders of the cochlear nerve produce tinnitus and sensorineural hearing loss. Conductive hearing loss results from interruption of the transmission of sound from the tympanic membrane to the perilymph within the cochlea produced by middle

COMMON CAUSES OF VERTIGO

Benign positional vertigo
Labyrinthitis, vestibular neuritis
Migraine syndrome
Menière's disease
Multiple sclerosis
Vertebrobasilar ischemia

LESIONS PRODUCING ENHANCEMENT OF CN VIII

Neurinoma (schwannoma)
Meningioma
Neuritis (viral)
Meningitis

ear disease. The two types of hearing loss are differentiated by audiometric testing. Lesions within the cerebellopontine angle cistern almost always first produce tinnitus and insidious hearing loss. Vertigo and facial nerve symptoms are late findings with tumors in this region.

The bony labyrinth refers to the complex system of canals within the temporal bone that constitutes the inner ear (the cochlea, vestibule, and semicircular canals. These canals contain specialized sensory receptor tissues: the membranous labyrinth. Perilymph is a CSF-like fluid found within the bony labyrinth surrounding the membranous labyrinth. The stapes transmits sound waves to the perilymph through the oval window. Endolymph is CSF-like fluid contained in sacs within the membranous labyrinth. Movement of this fluid within the vestibular apparatus provides information for equilibrium.

Thin-section (3-mm) MRI is the most useful imaging technique for evaluation of CN VIII and membranous labyrinthine disorders. T_1- and T_2-weighted images are used, followed by contrast enhancement in most cases. A precontrast T_1-weighted image is needed to identify high signal tissue from hemorrhage or fat, to avoid confusion with true tissue contrast enhancement. Small lipomas occur within the IAC, and hemorrhage occurs within the labyrinth. High-resolution thin-section (less than

COMMON LESIONS OF THE CEREBELLOPONTINE ANGLE CISTERN

Acoustic neurinoma ($\approx$ 80%)
Meningioma
Epidermoid
Metastasis
Aneurysm
Chronic meningitis
Facial nerve neuroma
Cholesterol granuloma

MORE COMMON CAUSES OF HEARING LOSS

Sensorineural hearing loss
    Acoustic neurinoma
    Meningioma
    Meningitis
    Viral infections of CN VIII
    Brain stem lesions
    Perilymph fistula
    Congenital malformation
Conductive hearing loss
    Otitis, acute, chronic
    Cholesteatoma
    Otosclerosis
    Oscicular fracture or dislocation
    Congenital malformation
Both sensorineural and conductive hearing loss
    Tumors of the temporal bone, with bone destruction
    Otitis, with bone destruction
    Paget's disease
    Fibrous dysplasia

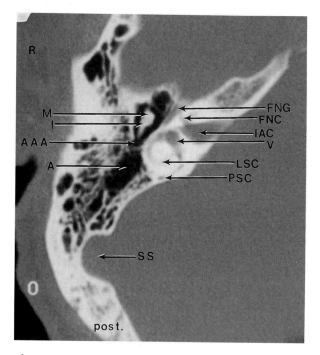

A

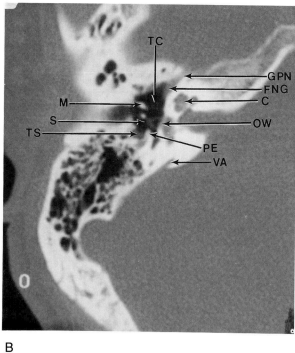

B

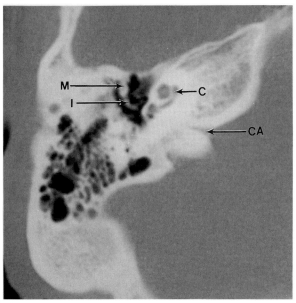

C

Fig. 13-46 Normal temporal bone anatomy, CT. (A–C)
Axial high-resolution CT scans of the right temporal bone
from superior to inferior. *(Figure continues.)*

Fig. 13-46 *(Continued).* **(D)** Coronal CT scan through the middle ear cavity at the level of the vestibule. Epitympanum (attic) *(curved white arrow).* A, antrum; AAA, aditus ad antrum; C, cochlea; CA, cochlear aqueduct; CP, cochlear promontory; FNC, facial nerve canal; FNG, facial nerve ganglion; GPN, greater petrosal nerve; I, incus; IAC, internal auditory canal; LSC, lateral semicircular canal; M, malleus; OW, oval window; PE, pyramidal eminence; PSC, posterior semicircular canal; R, right; S, stapes; Sc, scutum (bony spur); SS, sigmoid dural venous sinus; TC, tympanic cavity; TS, tympanic sinus; V, vestibule; VA, vestibular aqueduct.

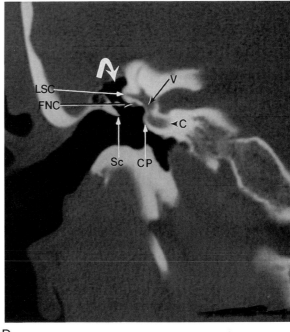

D

2-mm) CT is most useful for evaluation of bone anatomy, conductive hearing loss, and congenital abnormalities of the temporal bone and for most infections involving the middle ear (Fig. 13-46).

SENSORINEURAL HEARING LOSS

Sensorineural hearing loss is produced by lesions of CN VIII or cochlear labyrinth. Most of the afflictions produce contrast enhancement within CN VIII, IAC, or membranous labyrinth. Viral or bacterial infections frequently involve CN VIII or membranous labyrinth and cause a mild to moderate amount of high T_1 signal in the membranes on Gd-enhanced MRI (Fig. 13-47). Reports indicate a high correlation of enhancement with clinical deficit, but the sensitivity of the technique is unknown. Like meningeal enhancement with tumor or infection, there seems to be a threshold level of involvement, so that early or mild infection may not show enhancement.

The enhancement is nonspecific and may occur with small tumors, ischemia, perilymphatic fistula (leak into the middle ear), syphilis, and viral and bacterial infections. Enhancement of a thickened meningeal lining of the IAC from meningitis, neoplastic meningitis, or sar-

coid may fill the IAC completely and mimic intrinsic disease of the nerve. There is no good way to differentiate between viral infection of CN VIII and a small CN VIII tumor when first seen, if the enhanced nerve has a thin and linear appearance. Follow-up scans may be necessary to show disappearance of the enhancement with infection, and growth with tumor. Larger tumors of the nerve have a more rounded appearance.

Acoustic Neurinoma (Schwannoma)
Acoustic neurinoma (schwannoma) is the most common mass lesion of CN VIII. The tumor usually arises from the sheath of the vestibular nerve at the level of the vestibular ganglion in the depth of the IAC. This is the level of the pial–dural junction. When examined early, the tumor may be small and linear, so the enhancement seen on T_1-weighted MRI is indistinguishable from neuritis. With growth, the tumor assumes a more rounded appearance, enlarging the IAC. Most tumors show homogeneous intense enhancement with Gd, but some are inhomogeneous and contain cysts (see Ch. 5). Meningioma may arise from within the IAC and may be indistinguishable from an acoustic neurinoma, but most meningiomas in this region originate from the dura adjacent to

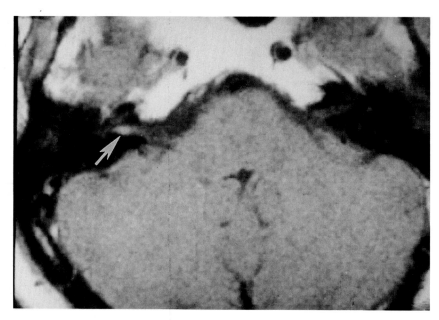

Fig. 13-47 CN VIII neuritis. Contrast-enhanced T_1-weighted MRI demonstrates enhancement within the depth of the right IAC *(arrow)*. This disappeared on a later scan, indicating a high probability of viral neuritis.

the canal. Although the vestibular division of CN VIII and the facial nerves are both compressed by the tumor, the predominant deficit is hearing loss (see the discussion of CN VII).

Cholesterol granuloma is a growing reactive type of mass lesion that is probably a response to chronic inflammation, and local bleeding. It is most common at the petrous apex and cerebellopontine angle cistern, where it compresses regional nerves, particularly the facial nerve. The mass has high signal intensity on both T_1- and T_2-weighted MRI. When seen, this pattern is almost always a cholesterol granuloma, although some epidermoid tumors may have high T_1 signal.

Hemorrhage within the membranous labyrinth occurs with trauma, infection, and tumor invasion through the bony labyrinth. Some occur spontaneously. Most patients are studied during the subacute period of the hemorrhage, so the blood shows as a small region of bright signal within the labyrinth on the nonenhanced T_1-weighted MRIs.

Labyrinthine schwannoma is a rare lesion, occurring mostly in patients with NF-2. The tumor arises within the cochlea or vestibule. The tumor shows as a very bright small round lesion after Gd-enhanced MRI.

CAUSES OF THE Gd-MRI ENHANCEMENT OF CN VIII, IAC, AND LABYRINTH

IAC
> Acoustic neurinoma
> Neuritis
> Meningioma
> Neoplastic meningitis
> Meningitis (acute and chronic, sarcoid)
> Hemangioma, arteriovenous malformation of IAC

Labyrinth
> Labyrinthitis (viral, bacterial, idiopathic)
> Recent trauma (especially with fracture)
> Labyrintine schwannoma (most with NF-2)
> Perilymphatic fistula
> Postoperative (cerebellopontine angle, inner ear)
> Menière syndrome (endolymphatic sac)
> Otosclerosis (within the abnormal bone)

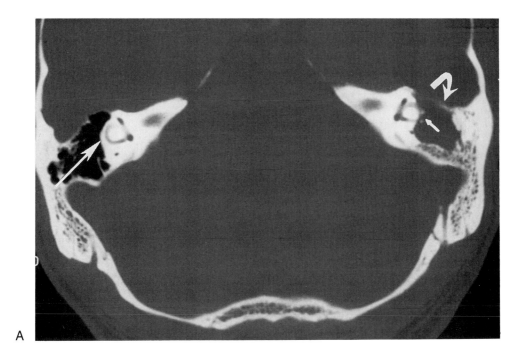

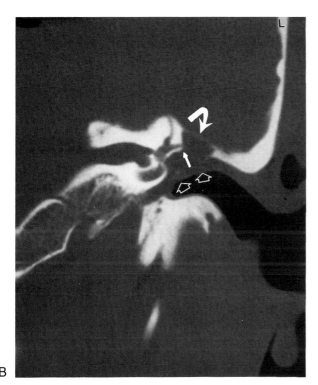

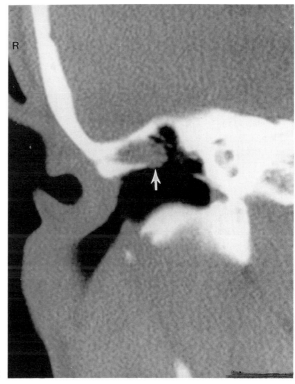

Fig. 13-48 Cholesteatoma. Endolymphatic fistula; eroded scutum. (A) Axial CT scan through the middle ear cavities shows a soft tissue density within the left epitympanum and mastoid antrum causing destruction of the temporal bone *(curved white arrow)* and the lateral aspect of the horizontal semicircular canal *(small white arrow)*. Note the normal semicircular canal on the right *(large fat white arrow)*. **(B)** Coronal view of the same patient shows the soft tissue mass *(small fat white arrows)*, the eroded lateral margin of the semicircular canal *(small white arrow)*, creating an endolymphatic fistula, and the eroded tegmentum of the temporal bone *(curved white arrow)*. **(C)** Coronal CT scan of a different patient shows a typical cholesteatoma arising from the region of the pars flaccida of the tympanic membrane and eroding the scutum *(arrow)*.

Perilymphatic fistula refers to a communication between the middle ear and the perilymphatic space of the inner ear labyrinth. Perilymph leaks into the middle ear, usually in small amounts, and intermittently. The diagnosis is difficult, even with direct visualization at surgery. The communication is most often through a damaged oval or round window resulting from trauma, barotrauma, infection, cholesteatoma, or tumor. Gd-enhanced MRI frequently shows enhancement of the membranous labyrinth. High-resolution CT will show fracture or bone destruction (Fig. 13-48). Direct visualization of the fistula may not be possible.

Lesions along the auditory pathway within the brain stem and cerebrum may cause sensorineural hearing loss. This includes tumor, infarction, hemorrhage, multiple sclerosis, and trauma. With sensorineural hearing loss, the imaging must cover the complete auditory system, including the labyrinth, the pons, and the cerebrum (see box).

AUDITORY PATHWAY

Cochlea
Cochlear division of CN VIII
Cochlear nuclei in medulla
Lateral lemniscus in the brain stem
Inferior colliculus
Medial geniculate body
Superior temporal gyrus

Congenital Malformation
Malformations of the otic capsule cause congenital deafness. The most common is the Mondini malformation, consisting of varying aplasia of the cochlea, beginning at its apex. The cochlea appears deformed on CT, with a

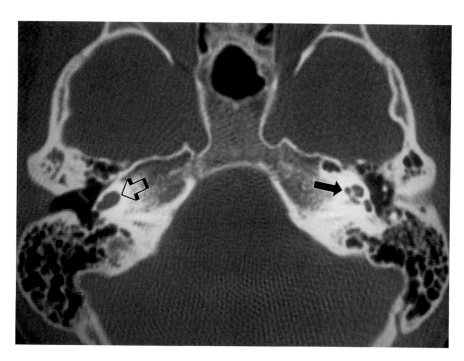

Fig. 13-49 Mondini anomaly. Axial CT scan through the temporal bones shows the large single cavity of the deformed right cochlea *(open arrow)*. The left cochlea is normal, with two visible turns *(black arrow)*.

single cavity representing the remaining basal turn (Fig. 13-49). Complete aplasia of the cochlea leaves sclerotic bone in its place.

Abnormality of the semicircular canals frequently accompanies malformation of the cochlea. Severe narrowing of the IAC to less than 3 mm has a high association with congenital deafness. The cochlear and vestibular aqueducts are often widened to greater than 5 mm in diameter in association with other inner ear malformations.

CONDUCTIVE HEARING LOSS

Conductive hearing loss refers to hearing deficit resulting from failure of mechanical transfer of sound waves in air to the perilymph within the bony labyrinth. The sound transfer normally occurs through the external auditory canal, the tympanic membrane, the middle ear ossicles, and the oval window. The final transfer is from the piston-like motion of the footplate of the stapes on the movable membrane across the oval window. The major causes of conductive hearing loss are infection of the middle ear, otosclerosis, trauma, and congenital malformation.

Acute Otitis

Acute otitis is an inflammation of the mucous membranes of the middle ear cavity, caused by bacterial or viral infection. It begins with a serous transudate, which converts to a purulent exudate. The eustachian tube may or may not be obstructed. Mastoid reaction refers to extension of the exudate into the mastoid air cells (Fig. 13-50). Mastoiditis refers to osseous involvement with septolysis, forming large cavities within the mastoid. Severe progression results in osteitis with bone destruction and in fistulization, particularly into the temporal and posterior intracranial spaces, forming epidural, subdural, or intracerebral abscesses (otitic abscess).

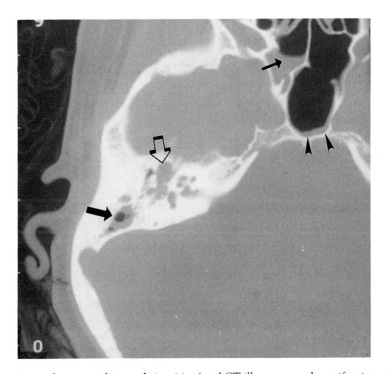

Fig. 13-50 Acute otitis media, mastoiditis, and sinusitis. Axial CT illustrates total opacification of the epitympanum of the right middle ear cavity representing otitis media *(open arrow)*. Inflammation also causes fluid to accumulate within the mastoid air cells, with septolysis indicating true mastoiditis *(fat black arrow)*. There is also fluid within the right sphenoid sinus *(small black arrow)* and membrane thickening of the left sphenoid sinus *(arrowheads)*.

Chronic otitis results from obstruction of the eustachian tube and the accumulation of seromucinous fluid, which becomes "glue-ear" with time. Chronic inflammation may lead to organization and fibroadhesive otitis. This causes fibrous stranding within the tympanic cavity and retraction of the drum. The incorporation of the ossicles within the fibrosing process causes conductive hearing loss. Osteitis is recognized by sclerosis or diffuse surface erosion of the walls surrounding the tympanic cavity. Cholesteatoma will form in some individuals.

Cholesteatoma (Keratoma)

A cholesteatoma (keratoma) is a sac-like mass bounded by an outer inflamed epithelial membrane, filled with an inner collection of keritinized necrotic debris. The lesion is almost always associated with chronic middle ear inflammation, although the precise mechanism of formation is unknown. Rarely, congenital cholesteatoma occurs. It originates from the pars flaccida (superior part), or the posterior pars tensa (inferior part) of the tympanic membrane, most often in association with chronic perforation of the membrane.

A cholesteatoma arising from the pars flaccida occupies Prussak's space and spreads superiorly into the attic of the middle ear cavity, the aditus ad antrum, and the antrum of the mastoid. Here, it initially erodes the lateral wall of the attic and the inferior lateral wall (the scutum, or bony spur), medially dislacing the malleus and incus. When large, it expands the attic, erodes the tegmen tympani, the lateral semicircular canal, the facial canal, and the mastoid antrum (Fig. 13-48). The long process of the incus may be destroyed. Cholesteatoma from the pars tensa initially fills the posterior tympanic cavity in the region of the tympanic sinus and facial recess. It then spreads superiorly into the attic and mastoid, causing erosion of the medial wall of the attic, and displacing the ossicles laterally. Complications include temporal lobe abscess, perilymphatic fistula, and facial nerve involvement.

The diagnosis is best made with thin-section (less than 2-mm) high-resolution CT scanning, using bone and soft tissue windows. Small cholesteatomas are suggested when nondependent tissue density is seen in the appropriate location within the tympanic cavity. The tissue density is nonspecific and may represent fibroadhesive otitis, or granuloma. The lesion is more obvious when bone erosion is present. Glomus tumor and epidermoid carcinoma of the middle ear may produce identical CT findings. The inflamed membranes enhances with Gd on MRI but cholesteatoma does not.

Malignant Otitis Externa

Malignant otitis externa is an extremely aggressive infection of the external canal that is most often caused by *Pseudomonas aeruginosa*. It may be so aggressive that it causes death in a short time. Elderly diabetics and those with immune compromise are most susceptible. Initially, the infectious process is seen with CT scan as a soft tissue density within the external canal, sometimes associated with fluid in the mastoid air cells. The tympanic membrane protects the middle ear cavity, which often remains clear. Bone destruction occurs early as the infection extends into the mastoid and in the region of the facial nerve. Deep extension medially causes involvement of CN IX–XI. Erosion of the infection into the labyrinth may cause meningitis through the labyrinthine communication with the subarachnoid space. Epidermoid carcinoma, metastasis, glomus tumor, sarcoma, and Wegener's granulomatosis are included in the differential diagnosis.

Otosclerosis

Otosclerosis is a hereditary dystrophy of the otic capsule, occurring most commonly in middle-aged women. The lesion is an otospongious hypervascular ossifying focus that arises from a cartilaginous inclusion within the otic capsule. It begins most commonly near the oval window. It first incorporates the footplate of the stapes in its ossifying process, causing a conductive hearing loss. The dystrophy later spreads to involve the cochlea, producing a mixed hearing deficit.

The very early stages cannot be visualized by imaging, and the CT scan will be normal when hearing loss is already evident. Later, the CT scan in the axial plane shows narrowing of the oval window and thickening of the footplate of the stapes. With advancement of the lesion, there is first demineralization and then sclerosis of part or all of the cochlea. Gd-enhanced MRI may show enhancement of the hypervascular dystrophic bone of the otic capsule.

Trauma

With trauma, the ossicles may be dislocated or fractured. Fracture of the temporal bone is associated in only 50 percent of cases. Thin-section CT scan is the best mode of diagnosis. The axial plane shows the incudo mallear articulation, and the coronal plane shows the incudosta-pedial articulation.

Incudostapedial luxation is most common, followed by incudo-mallear luxation. The displacement may be small or large. The oscicles may be rotated or angled from their normal position. Displacement takes place in any direction. When the incus and malleus are luxed together, their position within the middle ear is altered, although their relationship to each other may remain the same. This is difficult to recognize, although there is often associated incudostapedial dislocation. Fractures occur through the crura of the stapes, the long process of the incus, the neck (base of the handle) of the malleus, and the manubrium (midshaft) of the handle of the malleus. Blast injuries and barotrauma may rupture the tympanic membrane and cause ossicular fracture.

Congenital Malformation

Congenital malformations of the external ear lead to a conductive hearing deficit of 40 to 60 decibels. Many of the malformations are associated with minor deformities of the auricle. Microtia, an abnormally small and often severely deformed auricle, usually indicates a more severe external ear malformation. The oscicles, especially the incus, may be deformed and thickened. The most severe defect replaces the oscicular chain with a plate of bone protruding from the attic into the middle ear cavity.

With stenosis of the external auditory canal, the canal may be replaced with fibrous tissue, or there may be narrowing of the bony canal itself. Eburnated bone may fill the canal. The malleus and incus are usually fused with the lateral wall of the attic; the middle ear and inner ear structures are otherwise normal. With complete atresia, the walls of the canal do not form, and the head of the mandible rests on the anterior surface of the mastoid. Frequently, the head of the mandible is small. A bony atresia plate of varying thickness replaces the tympanic membrane. The middle ear cavity may be small or slit-like, and the ossicles are often deformed or fused to the atresia plate.

TUMORS OF THE TEMPORAL BONE

Benign and malignant tumors involve the temporal bone, but they are relatively unusual. Many cause bone destruction that is best diagnosed with CT scanning. The soft tissue mass of the tumor often enhances with contrast. When small, the tumor type can be suggested by the region of origin. Ceruminous tumors and epithelial carcinomas begin as soft tissue masses within the external canal, but they may extend through the tympanic membrane into the middle ear cavity. When more advanced, bone destruction occurs. Choristomas arise from ectopic salivary gland tissue within the external canal or middle ear. All these tumors are indistinguishable from granulomatous processes, such as Wegener's granulomatosis. Neurinomas are most common along the course of the facial nerve or within the IAC. Osteoma is a dense benign bone tumor that occurs on the external surface of the mastoid or temporal squamosa. Osteochondroma, chondroma, and osteosarcoma are rare and cause expansile osseous lesions from any portion of the temporal bone. The osteoblastoma is an expansile lesion arising from the bone, and usually surrounded by a thin calcium rim. Metastasis occurs anywhere. Nasopharyngeal tumors erode the base of the temporal bone by direct posterior extension.

Glomus Tumor (Paraganglioma)

The glomus tumor (paraganglioma) is a highly vascular slow-growing neuroendocrine tumor that may be small and insignificant, or it may be large, causing extensive bone destruction. Ninety-five percent have benign histology. They are related to other neural crest tumors, such a pheochromocytoma, and may secrete catecholamines. Most present in women age 40 to 50; they are very rare before age 20 years. In the region of the temporal bone, they arise from the chemoreceptor tissue within the middle ear (glomus tympanicum), the jugular fossa (glomus jugulare), and the ganglia of the vagus nerve within the jugular fossa (glomus vagale). Tumors arising within the carotid body (glomus caroticum) in the crotch of the carotid bifurcation may grow superiorly, to invade secondarily the base of the skull and the temporal bone. These tumors produce conductive or sensorineural hearing loss, depending on their pattern of destruction. Growth in the region of the jugular foramen damages CN IX to XI, causing loss of taste in the posterior third of the tongue, hoarseness, and dysphagia, as well as shoulder weakness.

The tumor shows on CT and MRI as an intensely contrast-enhancing mass with internal inhomogeneity. Flow-voids are frequent on MRI, representing the internal vessels of the tumor. Early invasion of bone is best seen with MRI, while extensive destruction is best seen with CT bone imaging. When small, the glomus tympanicum is a soft tissue mass within the middle ear cavity that cannot be distinguished from inflammatory tissue. Clinically, the lesion presents as a vascular mass behind the tympanic membrane that may not be differentiated from a displaced carotid artery within the middle ear. Here, observation on CT of an absent posterior wall of the carotid canal or a displaced carotid canal makes the diagnosis of a displaced carotid artery.

Dynamic high-dose (0.3-mmol/kg) Gd-enhanced scans (rapid-acquisition turbo GRE, short TE, low flip angle) show a decrease (dip) in the signal intensity of the glomus tumor at 20 to 50 seconds. The extremely high concentration of Gd within this hypervascular bed of the tumor produces a superparamagnetic effect, shortening of T_2 relaxation time, and a transient drop in signal prior to tissue equilibration of contrast. As the contrast accumulates within the tumor interstitium, the overall concentration of Gd decreases, and the signal intensity increases again. This transient dip phenomenon does not occur in other less vascular, but intensely enhancing, tumors seen in this region, such as meningioma.

Angiography shows the tumor with striking hypervascularity. The tumor may extend into the lumen of the jugular vein. The tumors feed from the ICA, the VA, and branches of the ECA (the internal maxillary artery, the ascending pharyngeal artery, the occipital artery, and the posterior auricular artery). This provides a map for catheter embolization of the tumor with coils, Gelfoam, or ethanol for control of bleeding at surgery.

Tumors in Children

Rhabdomyosarcoma is a relatively common tumor of children. In the region of the temporal bone, it arises within the middle ear cavity. Most often, these tumors are rapidly growing destructive mass lesions. Langerhans

---

TUMORS OF THE EAR AND TEMPORAL BONE, BASE OF SKULL

Children
    Dermoid (usually within or near a bone suture)
    Rhabdomyosarcoma (orbit, pharynx, and middle ear are the most common sites)
    Histiocytosis (Langerhans cell histiocytosis)

(Continued)

Adults
    Epidermoid carcinoma, primary
    Epidermoid carcinoma, direct spread from pharynx
    Metastasis
    Cerebellopontine angle tumors (neurinoma, meningioma, epidermoid)
    Cholesteatoma
    Glomus tumors (paraganglioma, glomus tympanicum, jugulare, vagale)
    Hemangioma (facial canal)
    Neurinoma (facial, vestibular nerves, CN IX–XI)
    Ceruminous tumors (adenoid cystic carcinoma)
    Choristoma (ectopic salivary tissue)
    Osseous and fibrous tumors and sarcomas

---

cell histiocytosis (histiocytosis X) is also a tumor of childhood that causes lytic destruction of the temporal bone. Lesions in other locations may be present to help make the correct diagnosis of this tumor. Congenital epidermoid and dermoid tumors are nonenhancing, often of near CSF equivalency, that occur near the petrous apex or within a cranial suture.

THE VESTIBULAR AND COCHLEAR AQUEDUCTS

Two small channels connect the internal cavity of the bony labyrinth with the posterior intracranial fossa. The cochlear aqueduct arises at the basal turn of the cochlea near the round window and extends inferomedially to open into the posterior fossa near the jugular foramen. It is narrowest near the cochlea, widening gradually like a funnel to its intracranial opening. The opening lies just posterior and inferior to the IAC on the coronal CT. It is large, in association with many developmental abnormalities of the inner ear. Through this aqueduct, the perilymph communicates with the subarachnoid space, providing a conduit for internal ear infection to the meninges.

The vestibular aqueduct arises at the medial aspect of the vestibule, curving backward and downward in the parasagittal plane, to open into the posterior fossa on the backside of the pyramid 10 mm lateral to the IAC. It can

THE CRANIAL NERVES, ORBIT, AND TEMPORAL BONE **557**

usually be seen on axial CT scans. With Menière's disease, the aqueduct is small (less than 3-mm diameter) in about 50 percent of persons, and the endolymphatic sac may become bright with Gd-enhanced MRI. The aqueduct may be large, with congenital malformations of the inner ear.

CN IX–XII

CN IX–XII are considered together, as they have their origin in the medulla, course through the medullary cistern, exit the skull through adjacent foramina, and run close together in the immediate suboccipital region. Many disease processes at the base of the skull or within the medulla produce a complex cranial neuropathy involving all or parts of the whole group. Isolated (single) cranial neuropathy is more likely to occur from lesions involving the supranuclear tracts of the nerves or the more peripheral portions of the nerves after the roots diverge.

The glossopharyngeal (CN IX), the vagus (CN X), and the spinal accessory (CN XI) arise from a paired common nuclear complex, each made up of three parallel columns within the dorsal medulla (Fig. 13-51). The most dorsomedial is the motor nucleus of the vagus nerve, which lies in the floor of the inferior fourth ventricle, just off the midline. Lateral to this is the nucleus solitarius, the sensory nucleus of the glossopharyngeal and vagus nerves. It receives visceral sensation and special sensory inputs from taste (posterior third of the tongue) and the carotid body. The ganglia for these sensory nerves are in the carotid body, along the carotid sheath and the vagus nerve, and in the jugular foramen, the sites of origin of the glomus tumors. Ventrolateral to this group is the nucleus ambiguus, the motor nucleus of the pharynx and larynx, contributing primarily to the vagus nerve, and the vagus portion of the spinal accessory nerve. The visceral part of the spinal accessory nerve also arises from the nucleus ambiguus and travels within the vagus nerve. The nerves exit the medulla at the dorsolateral sulcus, cross the medullary cistern, and enter the jugular foramen.

The spinal component of the spinal accessory nerve arises from the anterior horn cells of C1–C5 and ascends through the foramen magnum, to exit as CN XI through the pars vascularis of the jugular foramen. It innervates the sternocleidomastoid and trapezius muscles.

The hypoglossal nucleus lies paramedian in the floor of the fourth ventricle immediately medial to the vagus nucleus. Together they form small bumps that protrude into the ventricle, which can be seen with axial MRI: the hypoglossal and vagal eminences. CN XII exits the medulla in the pre-olivary sulcus, crosses the medullary cistern, and enters the hypoglossal canal through the rim of the foramen magnum. This nerve provides motor innervation of the tongue and pharynx. Lesions within the medulla or along the peripheral nerve produce lower motor neuron findings with atrophy and fasciculations of the ipsilateral tongue. Upper motor lesions produce paresis without fasciculations.

Motor regions of the cerebral cortex near the sylvian fissure project to the motor nuclei of these nerves

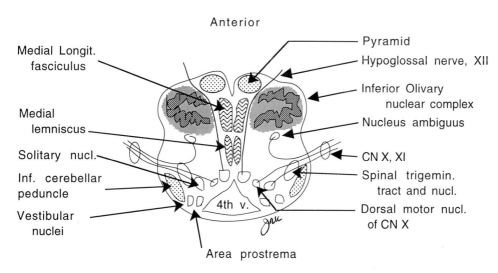

Fig. 13-51 The medulla. Axial drawing through the medulla at the level of the inferior olivary nuclear complex.

Table 13-4 Primary Findings from Lesions of CN IX–XII

Nerve	Primary Function	Sign
Glossopharyngeal	Sensory	Loss of gag
		Ear pain (otalgia)
Vagus	Motor	Hoarseness
		Dysphagia
Spinal accessory	Motor	Weakness of muscles (sternocleidomastoid and trapezius)
Hypoglossal	Motor	Paresis of tongue (denervation atrophy)

through the internal capsule and the corticobulbar tracts and provide voluntary control. Lesions along this system produce supranuclear nerve palsy (pseudobulbar palsy). Lesions of the cranial nerves and their clinical findings are given in Table 13-4.

Lesions involving the periphery of the nerves, after they have diverged from each other a short distance below the base of the skull, produce an isolated cranial nerve dysfunction. The vagus nerve has the longest course and passes inferiorly into the mediastinum on the left, where the recurrent laryngeal nerve loops around the aorta in the aortopulmonary window. The nerve travels along the carotid artery in the neck and is intimately related to the posterior aspect of the thyroid and parathyroid glands. Tumors within the left mediastinum, the thyroid, and along the carotid artery may cause laryngeal palsy. Paralysis of a vocal cord causes it to lie flaccid in the midline of the larynx, which can be seen with CT scan, but the position varies with the location of the lesion.

Lesions within the medulla and the medullary cistern are those that are discussed in more detail in other sections of the text (Fig. 13-52). These lesions are listed in Table 13-5.

THE JUGULAR FORAMEN

The jugular foramen lies at the petro-occipital fissure at the base of the skull lateral to the foramen magnum. It is divided into the larger pars vascularis and the smaller

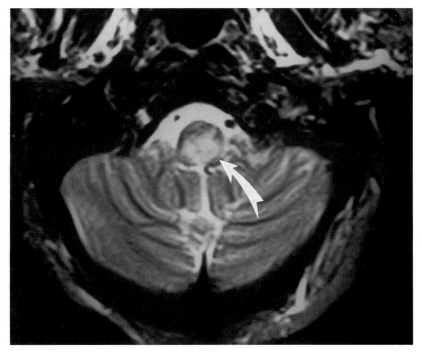

Fig. 13-52 Lymphoma. Axial T$_2$-weighted FSE MRI shows homogeneous high signal intensity within the medulla (*arrow*). Multiple sclerosis and other demyelinating processes may also have the same appearance. Infarction is almost always involves only one side of the medulla.

Table 13-5 Lesions Causing Dysfunction of CN IX–XIII

Location	Lesion	Location	Lesion
Supranuclear	Infarction Tumor Multiple sclerosis Trauma (axonal injury, contusion)	Jugular foramen	Nasopharyngeal carcinoma Metastasis (to temporal bone) Glomus tumor Malignant infection (otitis externa, *Aspergillus*) Tumors of bone (osteoma, chordoma)
Medulla	Infarction Multiple sclerosis Tumor, primary and metastasis astrocytoma, lymphoma Infection (HIV, mononucleosis, viral, sarcoid) Amyotrophic lateral sclerosis	Peripheral nerves	Cervical soft tissue tumor, adenopathy Cervical abscess Thyroid mass Mediastinal mass, lung carcinoma, mediastinitis Schwannoma, neuroma Surgical dissection
Cistern	Infection, inflammation, tuberculosis, sarcoid, meningitis Neoplastic meningitis Schwannoma Meningioma Epidermoid Vascular extasia, aneurysm		

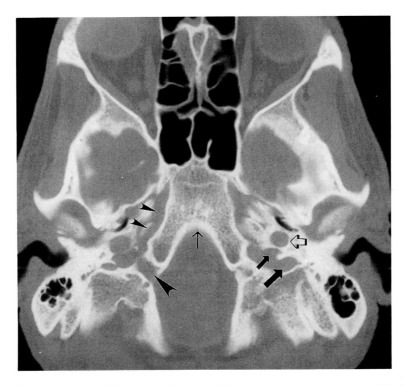

Fig. 13-53 Normal anatomy. Axial CT scan through the skull base shows the pars nervosa (*small thick black arrow*) and pars venosa (*large thick black arrow*) of the jugular foramen. CN IX courses through the pars nervosa. CN X and XI and the jugular vein course through the pars venosa. The carotid canal is the round foramen anterior to the jugular foramen (*open arrow*). The hypoglossal canal is seen on the right rim of the foramen magnum (*large arrowhead*). The foramen lacerum lies lateral to the basiocciput (*small arrowheads*). The clivus is the posterior surface of the basiocciput (*thin black arrow*).

Table 13-6 Medullary Infarction Syndromes

Location	Syndrome	Symptoms
Dorsolateral	Wallenberg's	Ipsilateral Horner's
		Ipsilateral loss of pain and temperature of the face
		Contralateral loss of pain and temperature
		Hiccups and dysphagia
		Ipsilateral ataxia
		Vertigo, nystagmus
Paramedian		Ipsilateral paresis of the tongue
		Contralateral hemiplegia and loss of accurate touch and position

anteromedial pars nervosa by a variable bony spur. The pars vascularis contains the jugular bulb of the dural venous system and CN X and XI. The pars nervosa contains CN IX and the inferior petrosal sinus as it crosses, to reach the jugular bulb (Fig. 13-53).

Mass lesions within the jugular foramen cause CN IX to XI dysfunction. Direct invasion of the foramen from a primary nasopharyngeal carcinoma is the most common lesion in adults. Other lesions are glomus jugulare tumor, lymphoma, metastasis to the temporal bone, schwannoma, and meningioma. Rarely, destructive invasive osteomyelitis of the temporal bone, such as malignant otitis externa, or invasive aspergillus of the mastoid or middle ear, will invade the jugular foramen. MRI is best for evaluation of most of the lesions, using Gd enhancement with fat-suppression sequences. Also, T_1-weighted GRE imaging (short TE, nonspoiled, large flip angle) enhances flow within vessels and is excellent for the diagnosis of small lesions within the foramen that compress the vein. High-resolution CT with bone windows may show subtle erosion of the spur of the jugular foramen or marginal erosion with smaller lesions, as well as the gross bone destruction with advanced tumor and infection. CT may be necessary for diagnosis of fungal infections, as the inflammatory tissue is dark on T_1- and T_2-weighted MRI and may not be apparent.

VASCULAR LESIONS OF THE MEDULLA

The vertebral posterior inferior cerebellar artery (PICA), the basal, and the anterior spinal arteries supply the medulla with blood. Two major territories are defined: the dorsolateral (PICA and vertebral arteries), and the paramedian. Dorsolateral infarction affects the descending sympathetic pathway, the descending tract and nucleus

of CN V, the lateral spinal thalamic tract, nuclei of CN IX and X, the inferior cerebellar peduncle, and the lower vestibular nuclei. The paramedian infarction affects the nucleus of CN XII, the medial lemniscus, and the pyramidal tract. The infarction patterns are listed in Table 13-6.

SUGGESTED READINGS

Abdul-Rahim AS, Savino PJ, Zimmerman RA et al: Cryptogenic oculomotor nerve palsy. The need for repeated neuroimaging studies. Arch Ophthalmol 107:387, 1989

Atlas SW, Grossman RI, Savino PJ: Surface coil MR of orbital pseudotumor. AJNR 8:141, 1986

Bilaniuk LT, Farber M: Imaging of developmental anomalies of the eye and orbit. AJNR 13:793, 1992

Brogan M, Chakeres DW: Gd-DTPA-enhanced MR imaging of choclear schwannoma. AJNR 11:407, 1990

Brogan M, Chakeres DW: Gd-DTPA-enhanced MR imaging of cochlear neuroma. AJNR 11:407, 1990

Dillon WP, Som PM, Fullerton GD: Hypointense MR signal in chronically inspisated sinonasal secretions. Radiology 174:73, 1990

Gentry LR, Jacoby CG, Turski PA et al: Cerebellopontine angle-petromastoid mass lesions: comparative study of diagnosis with MR imaging and CT. Radiology 162:513, 1987

Gentry LR, Mehta RC, Appen RE, Weinstein JM: MR imaging of primary trochlear neoplasms. AJNR 12:707, 1991

Goss CM: Gray's Anatomy. p. 853. Lea & Febiger, Philadelphia, 1963

Hammerschlag SB, Hesselink JR, Weber AL: Computer Tomography of the Eye and Orbit. Appleton & Lange, E Norwalk, CT, 1983

Hopper KD, Sherman JL, Boal DK, Eggli KD: CT and MRI Imaging of the pediatric orbit. Radiographics 12:485, 1992

Jungreis CA: Skull-base tumors: ethanol embolization of the cavernous carotid artery. Radiology 181:741, 1991

Kaste SC, Jenkins JJ III, Meyer D, et al: Pictorial essay. Persistent hyperplastic vitreous of the eye: imaging findings with pathologic correlation. AJR 162:437, 1994

Knorr JR, Ragland RL, Brown RS, Gelber N: Kallmann syndrome: MR findings. AJNR 14:845, 1993

Kumar A, Maudelonde C, Mafee M: Unilateral sensorineural hearing loss: analysis of 200 consecutive cases. Laryngoscope 96:14, 1986

Lane JI: Facial nerve disorders. Neuroimag Clin North Am 3:129, 1993

Laine, Smoker: Neuroimag Clin North Am 3:85, 1993

Lanzieri CF, Shah M, Krauss D et al: Use of gadolinium-enhanced MR imaging for differentiating mucoceles from neoplasms in the paranasal sinuses. Radiology 178:425, 1991

Li C, Yousem DM, Doty RL, Kennedy DW: Neuroimaging in patients with olfactory dysfunction. AJR 162:411, 1994

Li C, Yousem DM, Doty RL, Kennedy DW: Review: neuroimaging in patients with olfactory dysfunction. AJR 162:411, 1994

Lo WWN: Tumors of the temporal bone and cerebellopontine angle. p. 1046. In Som PM, Bergeron RT (eds): Head and Neck Imaging. 2nd Ed. Mosby-Year Book, St. Louis, 1991

Mafee MF, Jampol LM, Langer BG, Tso M: Computed tomography of optic nerve colobomas, morning glory anomaly, and colobomatous cyst. Radiol Clin North Am 25:693, 1987

Mark AS, Seltzer S, Harnsberger HR: Sensorineural hearing loss: more than meets the eye? AJNR 14:37, 1993

Murphy BL, Griffin JF: Optic nerve coloboma (morning glory syndrome): CT findings. Radiology 191:59, 1994

Nadol JB: Hearing loss. N Engl J Med 329:1092, 1993

Parker GD, Harnsberger HR: Clinical-radiological issues in perineural tumor spread of malignant diseases of the extracranial head and neck. Radio-Graphics 11:883, 1991

Rubin J, Curtin HD, Yu VL, Kamerer DB: Malignant external otitis: utility of CT in diagnosis and follow-up. Radiology 150:729, 1990

Shaumbaugh GE: Surgery of the Ear. p. 55. WB Saunders, Philadelphia, 1967

Som PM, Bergeron RT (eds): Head and Neck Imaging. 2nd Ed. Mosby-Year Book, St. Louis, 1991

Tien RD, Dillon WP: Herpes trigeminal neuritis and rhombencephalitis on Gd-DTPA-enhanced MR imaging. AJNR 11:413, 1990

Truwit CL, Kelly WM: The olfactory system. Neuroimag Clin North Am 3:47, 1993

Vignaud J, Jardin C, Rosen L: The Ear, Diagnostic Imaging. Masson Publishing USA, New York, 1986

Vogl T, Dresel S, Bilaniuk LT et al: Tumors of the nasopharynx and adjacent areas: MR imaging with Gd-DTPA. AJNR 11:187, 1990

Vogl TJ, Mack MG, Juergens M et al: Skull base tumors: gadodiamide enhanced MR imaging-drop-out effect in the early enhancement pattern of paragangliomas versus different tumors. Radiology 188:339, 1993

Woodruff WW, Vrabec DP: Pictorial essay. Inverted papilloma of the nasal vault and paranasal sinuses: spectrum of CT finding. AJR 162:419, 1994

Yousem DM: Imaging of sinonasal inflammatory disease. Radiology 188:303, 1993

Yousem DM, Atlas SW, Grossman RI et al: MR imaging of the Tolosa-Hunt syndrome. AJNR 10:1181, 1989

Yuh WTC, Wright DC, Barloon TJ et al: MR imaging of primary tumors of the trigeminal nerve and Meckel's cave. AJNR 9:665, 1988

14
Spine

ANATOMY

The spine consists of a series of segmental bones called vertebrae. There are 33: 7 cervical, 12 thoracic, 5 lumbar, 5 sacral, and 4 coccygeal. Transitional conditions are relatively frequent, so that an additional segment may appear in one region at the expense of an adjacent region. The total number is almost always 33. The cervical, thoracic, and lumbar regions have unfused segments that allow motion through intervertebral discs and uncovertebral joints. The sacrum and coccyx fuse in adults and are rigid.

Each vertebral segment consists of the body, which forms the anterior segment, and of the neural arch, which forms the posterior segment. The body of each vertebral segment is a cylinder of central trabecular bone filled with marrow and encased in a thin layer of compact bone. The marrow contains a moderate amount of fat, which produces the characteristic intermediate signal intensity (greater than muscle, less than epidural fat) within the vertebral body on T1-weighted MR imaging. The percentage of fat within the marrow increases with age from about 20 percent at age 20, to 60 percent at age 75. The endplates of the vertebral bodies consist of compacted trabecular bone covered by a cartilaginous lining, which is hyaline cartilage in the central region and fibrocartilage circumferentially. The endplates are relatively weak and allow herniation of disc into the central body, forming Schmorl's nodes (see p. 572).

The neural arch consists of paired thick pedicles arising from the posterolateral and superior aspect of the vertebral body and projecting backward a short distance. Two flat broad lamina extend posteromedial from the pedicles, to meet in the midline. At the junction of the

pedicles and lamina, the articular processes give rise to the superior and inferior articular surfaces (facet joint surfaces). A transverse process projects laterally from this site on each side, to form the attachment for spinal muscles. A large spinous process projects posteriorly from the midline junction of the lamina. There is considerable variation in the size and shape of the spinous processes, some being bifid. The ring of the vertebral body, pedicles, and lamina forms the central spinal canal. Incomplete formation of the ring is called spina bifida and, when associated with a dural or neural abnormality, it is called *dysraphism* (see p. 666).

The first cervical (C1) segment has only anterior and posterior arch segment without a vertebral body. C2 has a superiorly oriented bone process arising from the body, called the odontoid. This process protrudes into the arch of C1 and articulates with its anterior arch.

The transverse processes of C1–C6 are perforated by a canal for the bilateral ascending vertebral arteries (Fig. 14-1). Superior ridges along the posterolateral superior borders of vertebral bodies C3–C7 are called the uncinate processes. These form synovial joints with the adjacent superior vertebral body, called the uncinate (uncovertebral) joints. Degenerative change of these joints often narrows the adjacent intervertebral foramen or central spinal canal. The articular pillar is the mass of bone that supports the superior and inferior articular processes. The articular facet is the flat surface of the articular process. The superior facet forms an incline plane facing posterosuperior that articulates with the inferior facet of the vertebral segment above. The facet joints are also synovial joints subject to degenerative arthritis.

The first cervical segment (C1) is called the atlas be-

ANATOMY OF
VERTEBRAL SEGMENT

Anterior segment
    Vertebral body (except C1)
Posterior segment (neural arch)
    Pedicles
    Laminae
Other structures
    Articular processes
    Transverse processes
    Spinous process

cause it supports the head. It consists of a ring of bone with large lateral articulating processes (the lateral masses) onto which rest the occipital condyles of the skull. The spinous process is absent or rudimentary. What would have been the body of C1 is replaced by the odontoid process (dens) of C2, protruding upward into the anterior portion of the ring. The odontoid is held close to the anterior arch by the strong thick transverse ligament that attaches to the anterior inner margin of the ring of C1. This ligament gives off a superior process that attaches to the anterior rim of the foramen magnum and an inferior process that attaches to the posterior body of C2. The entire ligament complex is called the cruciate ligament. Additional paired ligaments, the strong alar ligaments, extend from the upper odontoid to the inner margin of the ring. A small apical odontoid ligament goes from the tip of the odontoid to the anterior

rim of the foramen magnum. Synovial joints surround the dens. The atlas is held to the skull by the strong circumferential atlanto-occipital membranes. The cervical central spinal canal is generally triangular in shape.

The thoracic spine has a generally similar structure to the lower cervical and lumbar spine, with the addition of facets to receive the medial ends of the ribs at the superior posterior margin of the vertebral bodies and the lateral transverse processes. The muscles and ligaments associated with the ribs add considerable support to the thoracic spine. The vertebral bodies are intermediate in size compared with the smaller cervical and larger lumbar segments and increase in size from above downward. The central spinal canal is generally round or oval.

The lumbar segments are the largest, and the vertebral bodies become quite massive. The anterior aspect of L5 is usually slightly taller than the posterior aspect, to accommodate the lumbar lordotic curve. The facet joints tend to have a sagittal orientation in the more superior segments and a coronal orientation in the lower segments. The L5 segment is susceptible to sacralization with large transverse processes that may articulate with the sacrum. This creates what is called a transitional lumbosacral spine that may cause confusion in counting and labeling the vertebral segments (see section Lumbar Disc Disease). The central spinal canal is generally triangular in shape.

Each vertebral segment is separated by a fibrocartilage intervertebral disc. The intervertebral disc is a fibrocartilaginous cushion occupying the interval between two vertebral bodies. It has three distinct anatomic units:

1. A *cartilaginous lining* of the endplate of the adjacent vertebral body: This consists of fibrocartilage around the periphery near the ring apophysis, and the thinner

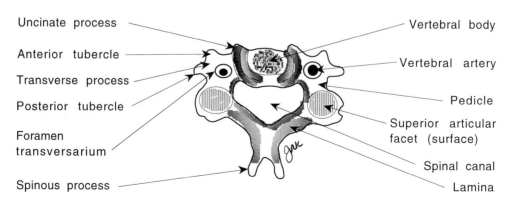

Fig. 14-1 Cervical vertebral body. Axial drawing through a mid-cervical vertebral body, showing the major spinal structures.

central hyaline cartilage covering the central portion of the vertebral body.

2. An *annulus fibrosis*, which forms a thick ring of concentric fibers that is the largest component of the intervertebral disc: The peripheral portion is collagenous and forms a strong ring. It is attached to the vertebral rim through Sharpey's fibers, and to the anterior and posterior longitudinal ligaments. The attachment is strongest anteriorly.

3. The *nucleus pulposus*, a soft ovoid gelatinous core within the central disc space: It is contained by the annulus fibrosis and the vertebral endplates. A horizontal plate of fibrous tissue often transects the nucleus and this is imaged with T2-weighted MRI as a low-intensity band within the normally bright nucleus (the intranuclear cleft).

The disc interspaces are smaller in the cervical and thoracic regions and larger in the lumbar region. In the lumbar spine, the interspaces gradually widen through the lower levels. The disc spaces are normally atretic in the sacrum and coccyx, as these segments are fused.

The strong ligaments provide much of the support the spine. The anterior longitudinal ligament is strong and extends along the anterior surface of the vertebral bodies from the axis (C2) to the upper sacrum. At each level, it is attached to the anterior annulus and the rims of the vertebral bodies, but it is not attached to the central portions of the bodies. It thickens to fill in the anterior concavities of the vertebral bodies. Above, it is continuous with the atlantoaxial ligament. The posterior longitudinal ligament runs along the posterior surface of the vertebral bodies from the axis to the sacrum. Above, it is continuous with the occipitoaxial ligament (membrana tectoria) extending from the axis to the base of the skull. It is attached to the annulus and vertebral margins of the disc space, but not as strongly as the anterior longitudinal ligament. Behind the vertebral bodies, it forms a narrow thick band separated from the bone by the basivertebral venous plexus.

The ligamenta flava are paired posterior elastic ligaments that connect the laminae of adjacent segments. The anterior margin is at the intervertebral foramen; they meet at the midline posteriorly, forming the posterior margin of the central spinal canal between the interlaminar segments. The ligaments are thin in the cervical region and thick in the lumbar region. The interspinal and supraspinal ligaments connect the spinous processes. The ligamentum nuchae is the name given the supraspinous ligament in the cervical segment. It extends from the spinous process of C7 to the external occipital protuberance of the occipital bone of the skull. Articular capsules surround and connect the articular processes of adjacent vertebral segments. The ribs are held to the transverse processes by the anterior costrotransverse ligaments.

The dural tube runs within the central spinal canal. It is formed by the downward continuation of the meningeal layer of the intracranial dura (Fig. 14-2). It normally ends inferiorly at S2, but it may end as high as L4.

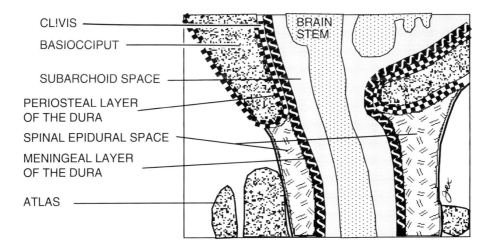

CLIVIS
BASIOCCIPUT
SUBARCHOID SPACE
PERIOSTEAL LAYER OF THE DURA
SPINAL EPIDURAL SPACE
MENINGEAL LAYER OF THE DURA
ATLAS
BRAIN STEM

Fig. 14-2 Spinal dural tube. Sagittal drawing of the base of the skull and the upper cervical spine illustrates the formation of the dural tube by the inferior continuation of the meningeal layer of the dura within the spinal canal. The periosteal layer of the dura continues around the basion, to become the periosteum of the anterior basiocciput.

Within the dural tube there is arachnoid, CSF, and the spinal cord. The cord is held within the central portion of the tube by the lateral denticulate dural ligaments extending from the lateral spinal cord to the inner margin of the dura. The spinal cord normally ends in adults at the T12–L1 level (see p. 666). The cord is covered by pia-arachnoid. The inferior extension of the pia below the lowest level of the cord (the conus medullaris) forms the filum terminale internum, which insets into the posterior fundus of the dural at about S2. The filum penetrates the dural tube at its lower tip, becomes invested with a dural continuation, and continues beyond the dural tube as the filum terminale externum (coccygeal ligament), to insert into the upper posterior coccyx.

The epidural space surrounds the dural tube and occupies the space within the central spinal canal and foramina between the dura and the vertebral structures. It contains predominantly fat, with epidural veins, arteries, and radicular nerve roots coursing through it. The volume of epidural fat generally increases caudally and is variable. It may become quite large. Sometimes the fat contained within the canal becomes so great that it compresses the dural tube and nerve tissue (see epidural lipomatosis). The fat is bright on T_1-weighted MRI, forming a natural contrast against herniated disc tissue or tumor within the spinal canal.

The epidural venous plexus may be quite prominent. The veins usually show as signal void with MRI and enhance with contrast because of the relatively slow flow within. The basivertebral vein of the vertebral body communicates with the plexus of veins between the posterior longitudinal ligament and the posterior vertebral body. The veins within the vertebral body form a Y-shaped channel. Sometimes the pattern of the venous channels is irregular, suggesting vertebral body fracture. A thin strip of sclerotic bone lines the vascular channels; this is not the case with fracture.

The spinal cord is a cylindrical structure that is the downward continuation of the medulla of the brain stem. It lies within the subarachnoid space of the spinal dural tube and is covered by the pia-arachnoid that is continuous with the pia of the brain. There are 31 segments: 8 cervical, 12 thoracic, 5 lumbar, 5 sacral, and 1 coccygeal. A spinal segment is that portion of the cord that gives rise to one segment of spinal rootlets that combine to form a segmental pair of spinal nerves. The cord is enlarged in the lower cervical (C4–T1) and lumbar (L1–S2) segments, where the nerves arise that supply the upper (brachial plexus) and lower extremities (lumbar and sacral plexus). The lowest portion of the cord tapers to the end and is called the conus medullaris. It contains the sacral spinal segments that provide cutaneous innervation of the perineum and motor control of the bladder and bowel sphincters, and the autonomic control of the bladder. At birth, the inferior tip of the conus medullaris is at L3; in adults, it is at T12–L2. From the lower cord the lumbar and sacral segmental roots course through the dural tube as a conglomeration called the cauda equina.

The spinal cord consists of a butterfly-shaped central region of gray matter containing cell bodies. It is divided into the posterior (sensory) and anterior (motor) horns. Surrounding the central gray matter is white matter consisting of ascending and descending axon tracts (fasciculi). The white matter is subdivided into the posterior, lateral, and anterior funiculi (a group of tracts). The posterior funiculus consists of the dorsal columns (touch and proprioception). The lateral funiculus is made up of important ascending (lateral spinal thalamic) and descending (corticospinal) tracts. The anterior funiculus consists mostly of descending reticulospinal and vestibulospinal tracts.

The segmental ventral motor root filaments have their cell bodies in the anterior horn gray matter; the axons exit the cord at the anterolateral sulcus. From here they course to their respective intervertebral canal (foramen),

---

MAJOR TRACTS OF THE SPINAL CORD

Posterior (dorsal column)
  Light touch
  Two point discrimination
  Joint position
  Vibration
Lateral (ascending spinothalamic)
  Pain, slow and fast
  Temperature
Lateral (descending corticospinal)
  Pyramidal pathway
Anterior (ascending ventral spinothalamic)
  Light touch, poorly localized
Anterior (descending reticulospinal, vestibulospinal)
  Coordination

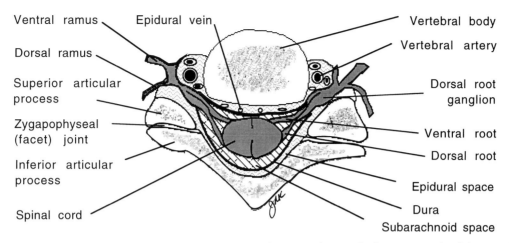

Fig. 14-3 Relationship of the cervical spinal cord and spinal roots to the vertebral structures. Axial drawing shows the spinal cord, the spinal roots, dorsal root ganglion, and their position within the spinal canal and intervertebral formen. The intervertebral foramen is the space shown here between the posterolateral vertebral body and the superior articular process from the segment below. The dorsal root ganglion lies within the foramen.

bypass the dorsal root ganglion, and join with the segmental spinal sensory fibers, to form the segmental nerve distal to the ganglion. The posterior sensory root filaments have their cell bodies in the dorsal root ganglion, which resides within the intervertebral foramen. The filaments then course medially, to enter the spinal cord at the posterolateral sulcus. The nerve root filaments are covered by segmental nerve root sleeves of the dural tube, which act as conduits for the roots between the formaina from the cord (Fig. 14-3).

The anterior spinal artery is the main conduit that supplies the central portion of the spinal cord. This vessel arises from two branches, one from each vertebral artery, that immediately combine behind the odontoid to form one vessel which descends in the anterior median fissure of the cord. It receives major contributions from radicular arteries, usually at two sites. The upper contribution enters at C3 or C4 from the costocervical or thyrocervical trunk, arising from the subclavian artery. This contributes the major supply of the cervical and upper thoracic region. The lower contribution is from the large artery of Adamkiewicz, which arises normally from an intercostal artery at T10–T12. This supplies the bulk of the thoracic cord. The conus medullaris and roots of the cauda equina are supplied by variable branches that enter along lumbar nerve roots. Small posterior arterial branches supply the superficial posterior portion of the cord, but they are not often significant in ischemic cord disease.

CAUSES OF ISCHEMIA OF THE SPINAL CORD

Atherosclerosis
    Diabetes mellitus
    Hypercholesterolemia
    Myocardial infarction
    Embolization

Aortic dissection

Vasculitis

Sickle cell disease

Surgical interruption of radicular arteries
    Aortic aneurism repair
    Scoliosis surgery
    Neck dissection
    Intercostal nerve blocks
    Sympathectomy
    Retoperitoneal exploration

Angiography/interventional catheter procedure

Spine trauma/fracture-dislocation of the spine

Spondylosis with central canal stenosis

Large disc herniation

Decompression sickness (barotrauma)

VASCULAR DISEASE OF THE SPINAL CORD

Most vascular lesions of the cord are the result of atherosclerosis or surgical damage to the feeding radicular vessels or vertebral arteries, or from direct compression of the anterior spinal artery. When blood flow is diminished through a segment of the anterior spinal artery, the anterior half of the cord becomes ischemic. This produces local flaccid paraplegia, loss of pain and temperature from the level and below (sensory level), and urinary retention, but preservation of position sense from the joints and touch (posterior column function). Cord infarction shows best with MRI as high T_2 signal in the anterior one-half of the cord (Fig. 14-4).

IMAGING

Spine imaging has changed markedly with the introduction of CT scanning, MRI, and the water-soluble intrathecal contrast agents. Examination of the spine

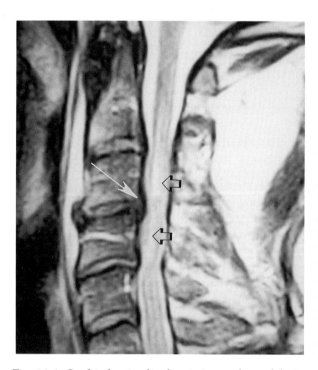

Fig. 14-4 Cord ischemia, disc herniation, and spondylosis. Sagittal FSE T_2 MRI of the cervical spine demonstrates the high signal within the anterior portion of the spinal cord (*open arrows*) representing ischemia in this person with acute myelopathy. This is thought to be secondary to compression of the anterior spinal artery by disc herniation associated with spondylosis and a narrowed spinal canal (*white arrow*).

and its contents constitutes a major part of neurodiagnosis.

MYELOGRAPHY

New nonionic, hypo-osmolar, iodine-containing contrast agents are now in general use for myelography. There are a number of similar compounds, including iohexol, iopamidol, iotrol, and ioglunide. They have many advantages over Pantopaque, the previously used oily agent. Because the materials are less viscous and less radiodense, they produce much improved definition of nerve root sleeves and small structures within the subarachnoid space. Moreover, the compounds are completely resorbed from the subarachnoid space within hours.

The aqueous agents are absorbed unaltered from the CSF into the blood through the arachnoid villi located along the sagittal sinus and in the thecal sac at the nerve root sleeves. The normal flow of the CSF always carries some of the contrast over the convexities. Once in the bloodstream, the agent is handled like any ionic water-soluble contrast material. About 90 to 95 percent is excreted through the kidneys, with a small amount exiting in the bile and feces. The biologic half-life of the contrast in the blood is approximately 4 hours. Only 5 percent of the contrast remains in the CSF after 24 hours.

The toxicity of contrast agents is based primarily on the ionic properties, osmolality, and the intrinsic chemotoxicity of the particular molecule. With the nonionic hypo-osmolar agents, the intrinsic toxicity of the compound becomes the most significant factor, and it is low. Seizures are rare with the newer agents but were relatively common with the earlier ionic material metrizamide. Electroencephalographic (EEG) changes occur infrequently (10 percent) with the newer agents and are usually the less significant slow-wave type. The more ominous spike activity is uncommon. Nevertheless, use of the newer water-soluble agents is contraindicated in persons using drugs that lower the seizure threshold (monoamine inhibitors, tricyclic antidepressants, phenothiazines, and CNS stimulants). Myelography is safe if these medications are withheld for 48 hours prior to the examination, and premedication with diazepam or phenobarbital is given. Alcoholics and epileptics are also given anticonvulsive prophylaxis.

The major risks from myelography are postmyelographic headache, nausea, muscular twitching, sciatica, seizures, mental dysfunction (psychic symptoms), and al-

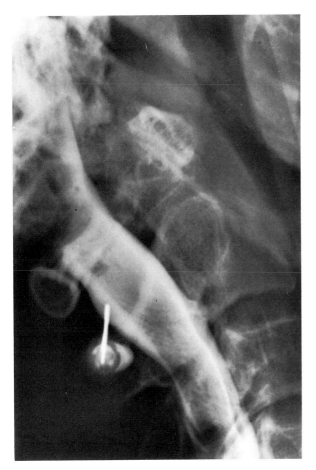

raphy helps decrease the incidence and severity of the headache. Postmyelographic arachnoiditis is not a problem with the water-soluble agents. A history of adverse allergic-type reaction to iodinated contrast agents is associated with an increased incidence of similar reaction from myelography.

Ordinarily, a lumbar puncture is performed with a 22- to 23-gauge spinal needle at L3–L4. The puncture is not done above L2–L3. For lumbar examinations, 170 to 200 mg I/ml is used. Contrast is carried superiorly to outline the conus medullaris. The cervical examination can usually be accomplished by lumbar instillation of the contrast, particularly if the agent is initially pooled in the upper lumbar lordotic curve with the patient in the slightly head-down position. If the patient is large or difficult to position, the lateral C1–C2 puncture is used for instillation (Fig. 14-5), which allows better control of the contrast and less chance of having the contrast enter the head in high concentrations. It may be necessary to perform CT scanning (CT myelography) for adequate evaluation of regions that are difficult to examine. Contrast is carried only to the region of interest to keep the intracranial concentration to a minimum. Excessive hyperextension of the neck should be avoided. The indications for myelography are discussed under the various subsections of this chapter.

Fig. 14-5 C1–C2 needle puncture for cervical myelography. Excellent opacification of the cervical region can be obtained by administration of contrast at C1–C2. The needle is always placed in the posterior 20 percent of the C1–C2 space. The cervical examination is done promptly without lowering the head. Then, as soon as the fluoroscopic pictures are completed, the head is raised and the contrast medium is dumped into the lumbar space to decrease the cranial resorption. (From Sage, 1984, with permission.)

lergic reactions. These symptoms may be delayed, occurring as late as 24 hours after the injection, with the average delay of onset about 8 hours. The overall incidence of postmyelographic reaction is about 33 percent. The incidence of adverse reaction has a positive correlation with the total amount of contrast instilled and the amount reaching the brain.

Headache is the most common problem. It may be severe and last up to 10 days. Once it has occurred, little can be given for treatment except analgesics. Hydration and a slightly upright position immediately after myelog-

MAGNETIC RESONANCE IMAGING

The MRI technique has broad applications for imaging the spine. The examination is done in multiple planes, depending on the problem to be studied. T_1- and T_2-weighted images are basic, and an interslice gap (25 percent) is usually necessary to clean images. Surface coils and phased-array configuration of the coils greatly improve the quality of the image and the ease and speed of the procedure and are necessary for detailed examinations. Fast spin-echo (FSE) techniques are substituted routinely when available to save time.

Gradient echo (GRE) imaging is useful for examining the spine and has the advantage of short imaging times. T_1 and T_2^* weighting is accomplished by manipulating the flip angle. Flip angles of less than 20 degrees with nonspoiled GRE provide T_2 weighting, which is more useful. It produces the myelographic appearance of the image and good demonstration of the gray and white matter of the spinal cord. As with T_2-weighted spin echo (SE) technique, edema in the cord is seen as high signal intensity, although it is generally less sensitive to cord disease than routine SE T_2-weighted imaging. The bones and ligaments are especially well defined as very low signal. Short echo times (time of echo [TE] less than 15 msec) are used to reduce the error produced by the magnetic susceptibility effect of bone. With a longer TE, protuberances such as osteophytes appear spuriously large.

Gadolinium (Gd) enhancement is useful for examining the spine in specific situations. The contrast defines abnormalities of the blood-brain barrier. As in the brain, the blood-brain is normally present within the spinal cord and the intradural segments of the spinal nerves. As the dorsal root ganglia do not have a blood-brain barrier, they enhance consistently under normal conditions. Lesions that destroy the blood-brain barrier show enhancement, but the intensity of the enhancement, are not a predictable as in the brain. Strong enhancement is most common with primary cord tumors and intradural–extramedullary metastasis. Other metastatic tumors, inflammatory lesions, and multiple sclerosis show less consistent enhancement. Enhancement of the nerve roots is always abnormal. Nerve roots may enhance with radiculitis from any cause, but most commonly with disc herniation and nerve root compression, after spinal surgery, or with arachnoiditis. There is moderate correlation with clinical symptoms of radiculopathy. Osseous metastases within vertebral bodies enhance slightly, so they may become isointense with surrounding fatty marrow and be obscured. Therefore, Gd is not used routinely before a nonenhanced T_1-weighted MRI of the spine if vertebral metastasis is sought. Epidural postoperative scar (fibrosis) shows consistent contrast enhancement.

The epidural venous plexus strongly enhances because of its slow blood flow. Small intradural veins may also enhance, mimicking intradural nerve root enhancement.

As a rule, Gd enhancement is used for specific problems. It is best for evaluation of intramedullary tumors, intradural–extramedullary tumors (particularly meningeal metastasis and meningioma), spinal infections, and differentiation of scar from disc herniation in the postoperative spine. Other potential uses include multiple sclerosis and transverse myelitis, although the value here is not as clear. The major uses are given in the box.

Artifacts are a significant problem with MRI of the spine. Motion of the CSF and vascular system can produce periodic linear artifacts. Multiple strategies have been devised to control these degradations, the most useful of which are the presaturation and motion rephasing techniques. Truncation (Gibbs) artifact also produces linear alternating bands of high and low signal parallel with sharp borders of strongly differing signal

---

CAUSES OF Gd ENHANCEMENT OF NERVE ROOTS ON SPINAL MRI

Disc herniation with root decompression

Spinal stenosis with root constriction
    Central
    Lateral

Arachnoiditis

Ischemia

Trauma

Postoperative

Neuritis

Idiopathic

USES FOR Gd IN MRI OF THE SPINE

Common uses
    Differentiation of postoperative scar from disc herniation
    Intradural–extramedullary tumors, particularly meningeal metastasis
    Cord tumor
    Osteomyelitis and disc space infection

Uncommon
    Epidural phlegmon/abscess
    Distinguishing large extradural tumor from cord
    Defining viable tumor for biopsy
    Arachnoiditis
    Determination of tumoricidal effect of radiotherapy
    Transverse myelitis
    Multiple sclerosis

intensity. It is a function of the Fourier transform that is inherent in the image reconstruction. Truncation artifact can be reduced with larger matrix size (256×256 or larger). Chemical shift artifact occurs along the frequency-encoded direction and produces changes in the appearance of the cortical bone, which lies in the orthogonal plane (Fig. 14-6). Patient motion must be kept to a minimum.

It is better to use MRI, rather than CT scanning, for the evaluation of intrinsic spinal cord pathology, spinal cord compression, intradural–extramedullary lesions, osteomyelitis, and vertebral metastasis. It is the preferred examination for the evaluation of myelopathy. Its accuracy is about equal to that of CT scans for the diagnosis of herniated disc disease. MRI is not generally as accurate for defining most other bone pathology or for some small disc herniations, particularly in the cervical region.

COMPUTED TOMOGRAPHY

CT remains an excellent examination for the spine, especially for evaluation of lumbar disc disease, spinal stenosis, and fracture. It is fast, available, and relatively inexpensive. The examination is done in the high-resolution mode, using thin contiguous slices. Disc spaces are usually examined individually with a scan plane parallel with the endplates, unless reformating the images into other planes is planned. The thin overlapping contiguous scans are best. Intravenous contrast is generally of little use, except for differentiation of a postoperative scar from disc herniation. Cord tumors show inconsistent enhancement, which is also difficult to detect because of the artifact from surrounding bone. CT shows better than MRI nerve root compression within the intervertebral foramen and lesions that are calcified.

CT myelography is valuable in defined situations. It can be performed up to a few hours after the injection of a small amount of contrast in the subarachnoid space. It is used to outline the cord and to define intradural–extramedullary lesions, arachnoiditis, perineural cysts, meningoceles, extradural lesions compressing the theca (metastasis), indistinct disc herniations, spinal stenosis, and empyema.

The major disadvantages of CT scanning for examination of the spine are the relatively high radiation dose given and the lack of facilitation of defining the internal structure of the spinal cord and intrathecal contents. It is accurate, however, for the diagnosis of degenerative disc disease and spinal stenosis, the most common problems involving the spine. It remains essential in cases of trauma.

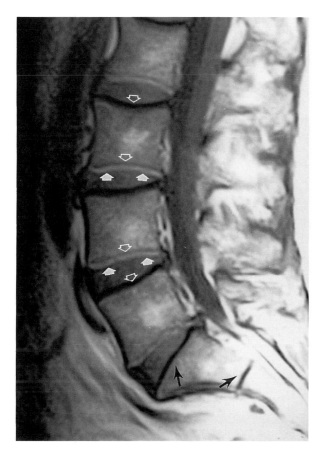

Fig. 14-6 Chemical shift artifact. Sagittal T_1 MRI done on a 1.5-T magnet demonstrates a downward shift of fat signal along the frequency encoded direction (*small solid white arrowheads*) so that it overlaps the inferior margin of the vertebral bodies. The signal from bone is not shifted and no longer aligns with the shifted fat signal (*small open white arrows*). The shift may be up or down, depending on the direction of the spatial encoding gradient of the magnet. In this case, there is no shift within the first sacral segment as it is oriented in the phase encoded direction (*small black arrows*).

RADIONUCLIDE SCANNING

Radionuclide scanning remains important for spinal imaging. Bone scanning is still effective for the demonstration of metastatic carcinoma, although MRI may have slightly greater sensitivity for the detection of all sites of tumor. The techniques are useful in demonstrating osteomyelitis and disc space infection, and occult lesions in persons with unexplained back pain. The specific techniques and uses are discussed in the appropriate sections that follow.

DEGENERATIVE DISC DISEASE AND SPONDYLOSIS

The water content of the nucleus decreases with age from 88 percent at birth to 66 percent at age 70 and accounts for much of the loss in height of the spine that occurs with aging. On MRI, a decreased T_2 signal intensity of the disc reflects this process (Fig. 14-7) and correlates with the degree of degenerative change demonstrated by discography or postmortem examination.

Biochemical explanation for the pathogenesis of disc degeneration is incomplete. It is known that the end result of the degeneration of the nucleus is loss of water, protein content, and turgor, which reduces the functional efficiency necessary for normal movement and stress resistance of the spine. The disc becomes brittle, fibrotic, and inelastic. Ultimately, the nucleus disappears. The degenerative process begins during the late teens or early twenties and probably starts with an initial increase in water content. This initial expansion is thought to contribute to the development of Schmorl's nodes, which are herniations of the nucleus through weak points in the degenerating cartilaginous plate. This stage is followed by dehydration and the progressive collapse and loss in height of the disc space.

Spondylosis is vertebral osteophytosis that occurs secondary to disc degeneration. Degeneration of the disc causes the disc space to narrow, which in turn causes the annulus to bulge in a circumferential manner. The bulging lifts the periosteum at the sites of attachment to the rim of the vertebral body, causing the subperiosteal osteogenesis that forms the osteophytes. This change is most prominent about the anterior two-thirds of the vertebral body, where the annulus and ligament attachment

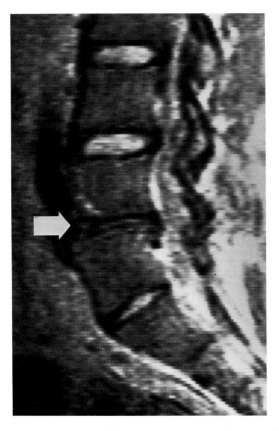

Fig. 14-7 Degenerated disc. Parasagittal proton-density MRI (SE 2500/40) shows loss of T_2 signal within the L4–L5 disc interspace (*arrow*). The disc interspace is narrowed.

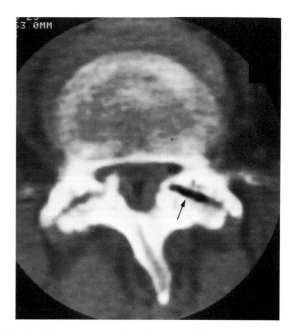

Fig. 14-8 Facet arthropathy. Transaxial CT scan shows severe bilateral degenerative arthropathy of the facet joints. The joint spaces are narrow, and the left space contains gas (*arrow*). Periarticular osteophytes may protrude into the the spinal canal, a lateral foramen, or a lateral recess. A perarticular cyst may develop and protrude into the canal. The cyst sometimes contains gas. (From Haughton, 1984, with permission.)

is most easily elevated. Posterior osteophytes are unusual in the lumbar and thoracic regions. However, because uncinate joints are present in the cervical spine and are subject to degenerative arthritis, posterior osteophytes are common here (see Fig. 14-38). Osteophytes always indicate degenerative disc disease. Those arising from the vertebral rim may cause nerve compression by narrowing the foramina and the lateral recesses.

Posteriorly in the spinal ring, there are zygoapophyseal joints between the opposing superior and inferior facets of the articular processes. They are the only true diarthrodial articulations in the spine. These joints are surrounded by a capsule and are lined by a synovium. With degeneration of the intervertebral disc, abnormal stress is transferred to these facet joints, producing painful degenerative arthritis with secondary hypertrophic changes (Fig. 14-8). Degenerative cysts form about these joints, which may protrude into the spinal canal or lateral intervertebral foramen. They have variable intensity on T_1-weighted MRI but are high intensity on T_2-weighted MRI (Fig. 14-9).

Disc Herniation

As a result of the degenerative process, the annulus develops fissures, first in a circumferential orientation, but later becoming radial. The annulus becomes weak, so it may bulge or rupture, allowing herniation of the nu-

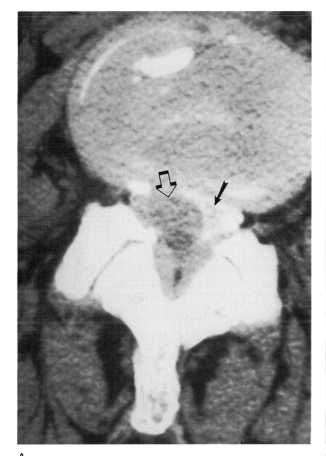

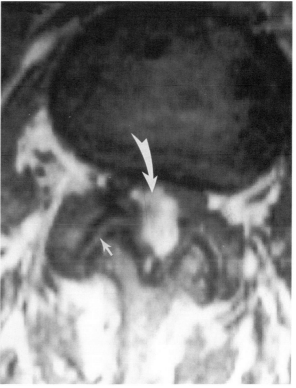

A   B

Fig. 14-9 Degenerative synovial cyst from facet joint. (A) CT myelogram of lumbar region shows a large slightly hypodense sharply circumscribed mass within the right epidural spinal space *(open arrow)*. It is contiguous with the degenerated right intervertebral facet joint. **(B)** T_2 MRI shows the epidural mass to have high T_2 signal indicating fluid within the mass, consistent with a cyst *(large white arrow)*. The T_2 signal within a degenerative synovial cyst may vary and be dark with very high protein viscous fluid or gas within it. Note the high signal fluid within the joint space *(small arrow)*.

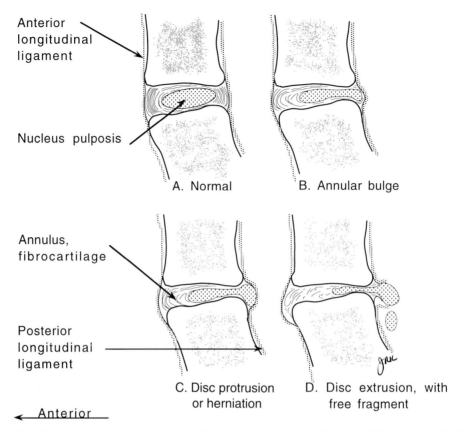

Fig. 14-10 Classification of stages of intervertebral disc herniation. Sagittal drawings of the intervertebral disc in the four stages from normal to disc extrusion. **(A)** The normal disc has a thick annulus posteriorly and just a slight convex border. **(B)** Disc bulge. The posterior annulus is thinned and bulges along with the posteriorly migrated nucleus pulposis. The progression from bulge to protrusion is a continuum, and the distinction of moderate bulge versus disc protrusion is arbitrary and imprecise. **(C)** Disc herniation. The annulus is completely effaced, and the nucleus protrudes posteriorly into the spinal canal. The posterior longitudinal ligament is intact, containing the herniated disc material. **(D)** Disc extrusion. The herniated nucleus pulposis breaks through the posterior longitudinal ligament into the epidural space. Fragments may separate and migrate in any direction.

cleus pulposus to a position underneath the posterior longitudinal ligament. Most often, this change occurs in the posterior lateral portions of the annulus where the posterior longitudinal ligament is thinnest. Consequently, central disc herniation is less common.

The herniated disc material may perforate the posterior longitudinal ligament into the epidural space. Rarely, with chronic extrusion, the dura may also be perforated, with the result that some of the disc material becomes intradural. The strength of the annulus determines whether there is simple bulging of the annulus or herniation of disc material. In order to herniate, the nucleus must be mobile and not completely fibrotic. Therefore, herniated disc disease is generally an affliction of

younger adults (under 45 years of age), although herniated disc material may be seen at any age.

The terminology used to describe degenerative disc disease and disc herniation is important but confusing. Different textbooks use different terms. The following discussion is a somewhat arbitrary compilation intended to describe the stages of disc herniation (Fig. 14-10).

BULGING OF THE ANNULUS

Bulging of the annulus refers to a circumferential annular bulge, so that the annulus extends beyond the margins of the rim of the adjacent vertebral bodies. The annulus is thinned but visibly intact. It usually does not cause nerve root compression, except when associated with spinal stenosis.

PROTRUSION OF THE NUCLEUS

Partial herniation of the nucleus pulposus occurs through some of the layers of the annulus, but the outer annular layers remain intact. This is sometimes called inner annular disruption or subtotal annular disruption. For the most part, this specific type of partial herniation can be described accurately only by direct vision at surgery. The partial herniation can sometimes be diagnosed with MRI, but it cannot be differentiated from complete herniation on CT scans. It has the same clinical significance as complete herniation, the size of the focal protrusion being the critical factor. This form is often called "prolapsed disc."

HERNIATION OF THE NUCLEUS PULPOSUS

Herniation of the nucleus pulposus is complete herniation of the nucleus through all the layers of the annulus fibrosis. The herniated material is still contained by the posterior longitudinal ligament (subligamentous herniation). The herniated material may migrate up or down behind a vertebral body but still be contained by the posterior longitudinal ligament. Some refer to this category as disc extrusion and as disc protrusion.

EXTRUSION OF THE NUCLEUS

Complete herniation of the nucleus pulposus occurs through all layers of the annulus and the posterior longitudinal ligament. If a fragment separates from the disc level, it is called a free or sequestered fragment. This fragment may migrate up or down the epidural space or into a lateral neural canal or recess. Rarely, the herniated disc fragment erodes into the dural tube. Most large herniations are extrusions.

Herniated disc disease is most common in the lumbar and cervical segments of the spine. Thoracic disc hernia-

tion is uncommon. The diagnostic problems are different for each region.

Lumbar Disc Disease

Most commonly, posterolateral disc herniation in the lumbar region causes nerve root compression, resulting in radiculopathy (pain, sensory loss, muscular weakness, and hyporeflexia). The nerve root syndromes are listed in Table 14-1. Small central disc herniations generally cause deep poorly localized scleratogenous-type pain. Large central herniations, particularly in association with spinal stenosis, may cause cauda equina syndrome, which if severe constitutes a surgical emergency. Deep pain that is poorly localized may also be caused by annular bulge and facet disease.

Plain film examination of the lumbar spine is recommended as the initial examination. Anteroposterior (AP) and lateral projections are all that are routinely required. Narrowing of the disc spaces and osteophytes indicate degenerative disc disease, but not necessarily disc herniation. The lumbar disc spaces normally increase slightly in width through successively lower levels, with the exception of L5–S1. If the L5 vertebral body is sacralized, it is referred to as a "transitional vertebra." Here the width of the L5–S1 disc space correlates inversely with the degree of sacralization of the L5 vertebral segment. Plain films can also be used to diagnose other processes, such as spondylolisthesis, osteomyelitis, discitis, and osseous tumors. Degenerative facet disease and narrowing of the neural foramina and spinal canal can be seen. Plain film examination provides the most accurate means of numbering the vertebral body levels and can be correlated with intraoperative films.

Lumbar myelography includes the AP, both obliques, and the upright lateral projections. The contrast is carried superiorly, outlining the conus medullaris. Normal lumbar myelography shows the lumbar roots as they exit the dural sac in their sleeves. The root exits the spine underneath the pedicle, corresponding at the root segment. For example, the L5 root courses inferolaterally

Table 14-1 Synopsis of Lumbar Root Syndromes

Root	Usual HNP	Sensory	Muscle	Reflex
L4	L3–L4	Posterolateral thigh, anteromedial leg	Quadriceps	Knee jerk
L5	L4–L5	Anteromedial leg and foot	Anterior tibial, ext. hallucis long	None
S1	L5–S1	Posterolateral leg	Calf	Ankle jerk
Cauda equina L2-sacrum	Any level large	Buttocks, back of legs	Legs and feet, bladder, rectum	Knee jerk

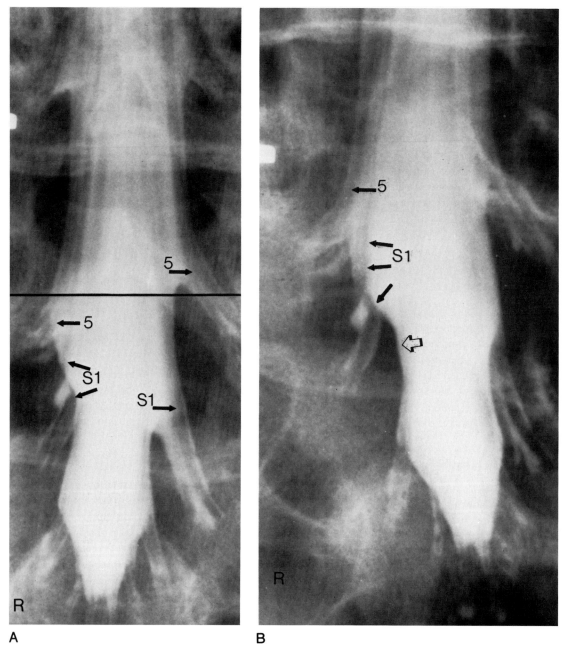

Fig. 14-11 Normal myelography, conjoined right L5–S1 root sleeves. (A) AP myelography shows the conjoined right L5 and S1 root sleeves. The normal arrangement of the root sleeves is seen on the left. Note the lower level of exit of the right L5 root compared with the left. Two preganglionic nerve roots can be seen within each sleeve. **(B)** Oblique view shows the conjoined roots on the right, which changes the contour of the dural tube. It may be misinterpreted as showing disc herniation *(open arrow)*. *(Figure continues.)*

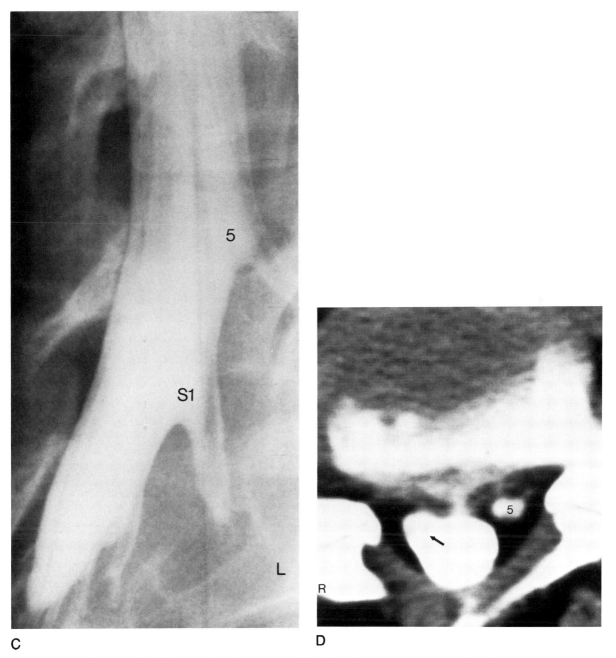

C

D

Fig. 14-11 *(Continued)*. **(C)** Normal left oblique view. **(D)** CT myelography shows the conjoined root sleeves on the right *(arrow)*. The normal left L5 sleeve is separated from the dural tube (5). The horizontal line in Fig. A shows the level of the CT scan. Conjoined sleeves are reported, as they may cause difficulties during surgical exploration.

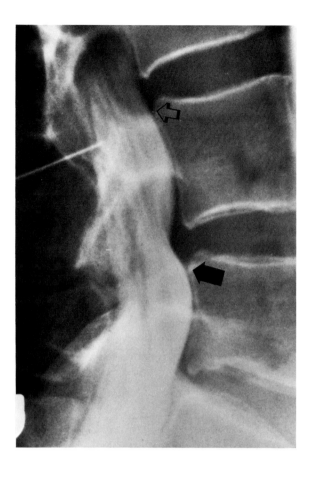

Fig. 14-12 Annular bulge. Lateral myelography shows a large annular bulge at the L4–L5 level. Note that the displacement of the dural tube does not extend significantly beyond the disc margins *(black arrow)*. A small central disc herniation could have the same appearance. A normal posterior disc margin is present at the level above *(open arrow)*.

underneath the pedicle of L5 in the superior portion of the intervertebral foramen. Conjoined root sleeves must be recognized, so this anomaly is not misinterpreted as an extradural defect (Fig. 14-11).

In the upper lumber region, the anterior margin of the subarachnoid space is closely applied to the posterior margins of the vertebral bodies and disc spaces. Normally, there are slight indentations in the constant column at the level of the disc spaces, representing the annulus and the thick posterior longitudinal ligament (Fig. 14-12). These indentations are accentuated with extension. The thickness of the anterior epidural space at the lower levels, especially L5–S1, is variable and may be large. The lower lumbar thecal sac may taper and be small. This pattern renders myelography less accurate for the detection of epidural lesions (Fig. 14-13).

The "extradural defect" is the hallmark of either bulging of the annulus (Fig. 14-12) or herniation of the nucleus pulposus (HNP). The most common defect caused by an HNP has some specific features. The dural sac is

deviated away from the disc space by the herniation. The nerve root crossing the disc space is compressed and may appear widened (Fig. 14-14). For lateral herniations the displacement is best seen on the oblique projections. For central herniations, the lateral projection is best.

The nerve root sleeve exiting the intervertebral foramen at the level of disc herniation is compressed only by lateral herniations, by superior migration or by herniations that have migrated into the lateral neural canal (see Fig. 14-21). Lateral disc herniations into the neural canal may show no abnormality on the myelogram.

Centrally occurring annular bulge or central disc herniation produces similar defects on the anterior portion of the dural sac, as seen on the lateral film (Fig. 14-12). Disc herniation tends to produce a larger, more "angulated" defect that extends beyond the margins of the disc space. It is difficult to differentiate bulge from HNP reliably with myelography alone. Myelography has a sensitivity of about 80 percent and a specificity of about 90 percent for the diagnosis of HNP.

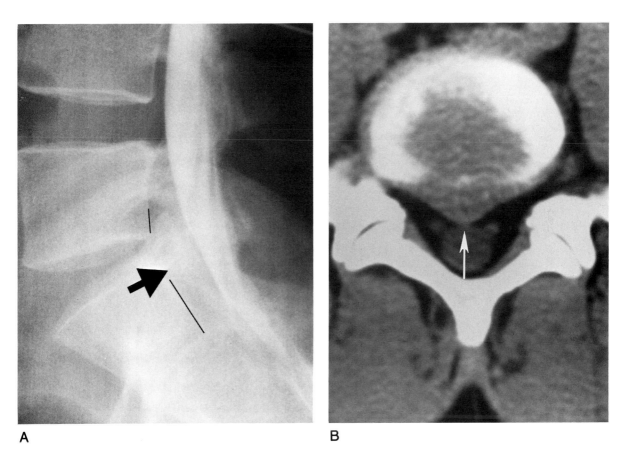

A   B

Fig. 14-13 Large lumbar epidural space, with herniated disc. (A) Lateral lumbar myelography shows a large epidural space at L5 and S1 with a tapered dural tube. It is a normal variation. When this anatomy is present, lower lumbar myelography has low sensitivity for detection of disc herniation. **(B)** CT scan at L5–S1 in the same patient. The large central disc herniation was not seen with myelography. It just barely abuts the dural tube *(arrow)*.

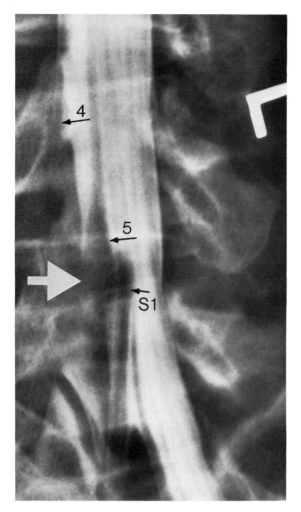

Fig. 14-14 Posterolateral L4–L5 disc herniation. An oblique myelographic view shows compression and flattening of the right L5 root as it crosses the disc interspace *(arrow)*. The right S1 nerve root is also medially displaced.

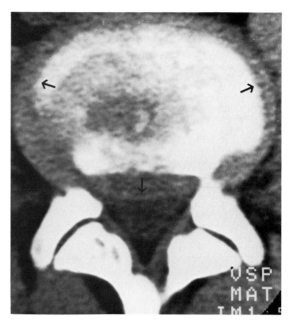

Fig. 14-15 Annular bulge. CT scan shows the circumferential bulging of the annulus about the entire disc margin *(arrows)*. The bulge is generally uniform with slight posterior convexity. The dural tube is indented.

On CT scans, the diagnosis of both bulging of the annulus and herniation of the nucleus is based on detection of hyperdense disc material displacing the hypodense epidural fat of the spinal canal. The shape of the posterior disc margin is also important. The normal disc margins conform to the rims of the vertebral bodies. Bulging of the annulus causes the entire margin of the disc to extend beyond the margins of the vertebral body (Fig. 14-15). The posterior margin of the bulged annulus usually becomes convex posteriorly, although it some-

times retains its normally central concavity, held by the strong central portion of the posterior longitudinal ligament. The bulge may indent the anterior margin of the dural sac. In general, the bulged annulus does not cause nerve root compression, and surgery is not recommended. However, annular bulge may contribute to cauda equina compression when there is associated spinal stenosis. It is impossible to differentiate a small bulge of the annulus from a small, broad, central herniation of the nucleus (Fig. 14-16).

The CT findings for the diagnosis of a herniation of the nucleus pulposus are as follows (Figs. 14-12 and 14-17): (1) a focal hyperdensity in the epidural space arising from the posterior, posterolateral, or lateral portion of the disc margin; and (2) displacement of a nerve root or the thecal sac. Occasionally, there is calcification or gas in the herniated nuclear material (Fig. 14-18). Most commonly, the abnormal epidural density of the herniated disc is at the level of the disc space, but it is sometimes separated and has migrated away from the disc margin, which may appear normal (Fig. 14-19). A large central herniation may have an appearance similar to the

Fig. 14-16 Annular bulge. The disc annulus is bulged about the posterior two-thirds of the interspace. There is slightly greater bulge centrally along the posterior margin *(arrow)*. In this instance, it is impossible to differentiate between annular bulge and small broad-based central herniation.

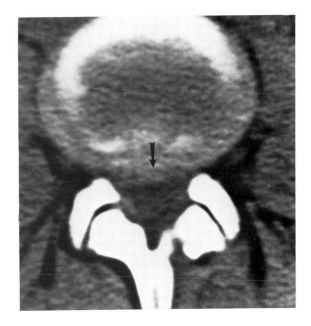

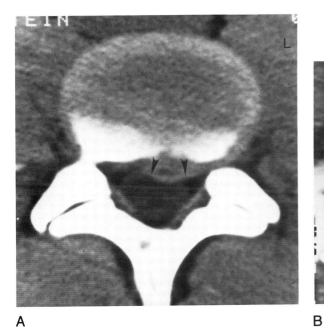

A

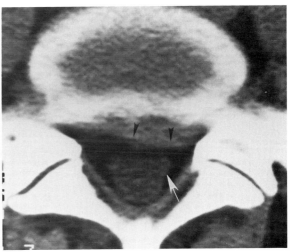

B

Fig. 14-17 Common posterolateral disc herniation. (A) CT scan shows the high-density disc herniation extending posteriorly into the epidural space on the left *(arrowheads)*. The dural sac is indented. **(B)** A more broad-based herniation extends into the epidural space on the left *(arrowheads)*. There is just subtle posterior displacement of the left S1 nerve root *(white arrow)* because the epidural space is large and can accommodate the herniation.

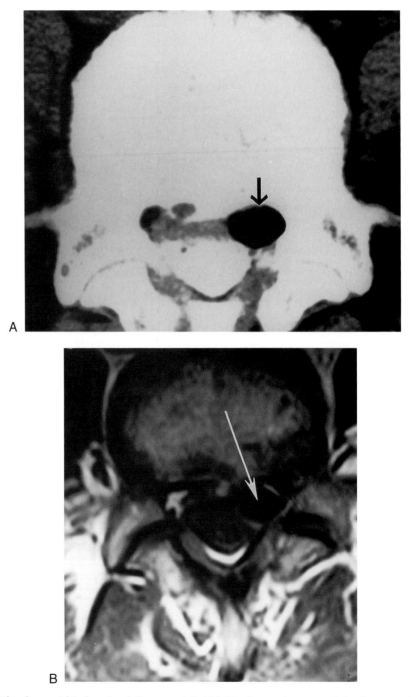

Fig. 14-18 Epidural gas within herniated disc material. (A) Gas from a degenerated disc has extruded into the left anterior epidural space *(arrow)*. This indicates perforation of the posterior longitudinal ligament. There may or may not be associated herniated disc material. **(B)** Axial T$_1$ SE 20/823 MRI shows the gas as a region of signal void *(arrow)*.

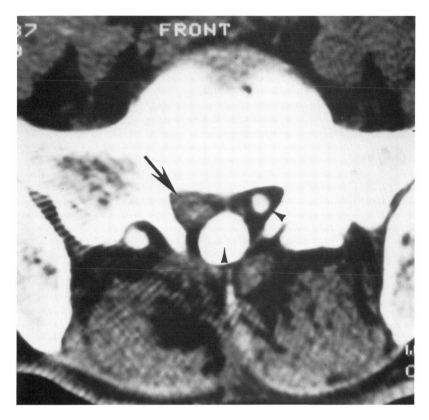

Fig. 14-19 Migrated free fragment. A fragment from disc herniation has migrated inferiorly behind the S1 vertebral body (*arrow*). It obliterates and displaces the right S1 root. A normal thecal sac and left S1 root are filled with contrast (*arrowheads*). The left S1 root is seen within a normal-size lateral recess.

thecal sac and go unrecognized (Fig. 14-20). In this situation, the epidural fat is usually totally obliterated, and the region of the thecal sac is of higher density than expected. If there is any doubt, myelography or MRI can establish the diagnosis. The disc may herniate from the lateral portion of the disc into the intervertebral foramen (lateral neural canal) (Fig. 14-21). The rare neurofibroma may have a similar appearance but almost always enlarges the neural canal.

The herniated disc must be differentiated from dilated epidural veins (which appear round in the epidural space and may enhance with intravenous contrast infusion), conjoined spinal root sleeves, spondylolisthesis, metastatic or primary tumors, infection, and hemorrhage. Both the sensitivity and specificity of CT scanning for the evaluation of HNP are about 90 to 95 percent. Routine CT scans are relatively insensitive to intrathecal pathology.

MRI has replaced much of CT scanning for the lumbar spine. It is generally preferred, as it avoids irradiation of the pelvic region that is inherent with CT scanning. Its sensitivity and specificity are approximately the same. Sagittal and transaxial planes are routine. T_1-weighted SE and T_2-weighted GRE images are the most useful for disc disease. T_2-weighted SE or FSE images show degenerative discs as having low signal intensity compared with the normal disc interspaces. Because GRE T_2-weighted imaging shows all disc interspaces as higher signal intensity, correlation with degenerative change cannot be made with this sequence.

The posterior margin of the disc space is well seen on the sagittal and transaxial projection. The annulus and the posterior longitudinal ligament are seen as a single linear structure of very low signal intensity behind the vertebral bodies. The ligaments cannot be separated from the cortical bone.

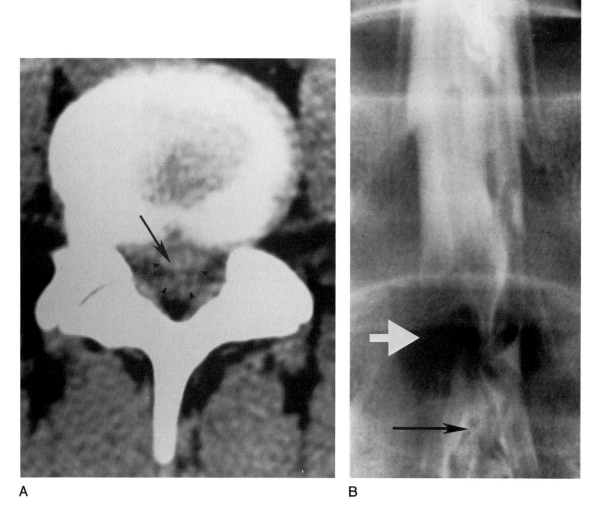

A   B

Fig. 14-20 Huge central disc herniation. (A) CT examination shows huge disc herniation occupying the central spinal canal *(arrow and arrowheads)*. The density of the herniated material is slightly greater than the density of the thecal sac, which would normally occupy this space. This type of herniation is difficult to recognize. **(B)** Myelography performed on the same patient shows the impression of the dural tube and high grade partial block from the large disc herniation *(white arrow)*. MRI or myelography is performed when huge disc herniation is suspected. Note the buckled roots below the dural constriction *(black arrow)*.

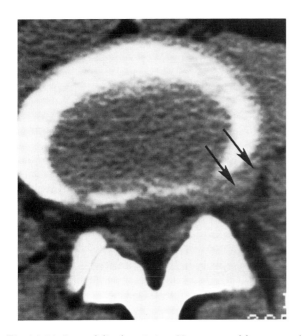

Fig. 14-21 Lateral disc herniation. Herniation of disc material may not occur laterally into the intervertebral foramen *(arrows)*. Extreme lateral herniations compress the nerve root exiting at the corresponding level. This type of herniation cannot be detected with myelography but is seen with CT as high-density material within the intervertebral canal.

Herniation of disc material is seen as disc-equivalent intensity tissue in the epidural space. On the T_1-weighted images, it is slightly hyperintense to the CSF and hypointense to the epidural fat (Fig. 14-22). On T_2-weighted images, including GRE images, the disc material may be isointense with the CSF, and the disc herniation is primarily diagnosed by defining the elevated posterior ligament and annulus. Large herniations are obvious (Fig. 14-23). Small herniations may be difficult to differentiate from annular bulge, except by the contour on the transaxial images, as discussed in the section on CT scanning (Fig. 14-24). Lateral herniations into the intervertebral foramen are best seen on the transaxial images. Tears of the annulus may be seen as interruptions of the annular-ligament line (Fig. 14-25). Small acute annular tears show bright signal in the posterior annulus on T_2-weighted (Fig. 14-26) or post-Gd T_1-weighted sequences.

Degenerative changes of the bone adjacent to disc space produce characteristic signal abnormalities on the T_1-weighted images. In the early stages, the signal intensity is decreased adjacent to the endplates, which is thought to represent edema infiltrating the normal marrow. Fibrofatty infiltration occurs in the later stages, producing high signal intensity (Fig. 14-27). The endplates remain sharply defined, differentiating this process from the early changes of osteomyelitis and discitis. Large regional osteophytes may result in very low signal along the vertebral margins. Discs with normal T_2 signal intensity are almost never associated with disc herniation, except in cases of acute trauma.

Postoperative Lumbar Spine

A significant number of persons who have had prior lumbar surgery develop recurrent symptoms, sometimes termed the "failed back surgery syndrome." These patients are often referred to the imaging specialist to differentiate between the possible causes.

The most common causes of the "failed back" are inappropriate patient selection for surgery, a lateral recess or canal stenosis that was not treated, arachnoiditis, and

CAUSES OF THE FAILED-BACK SURGERY SYNDROME

Inappropriate patient selection for surgery
Lateral stenosis
Postoperative epidural scarring
Recurrent disc herniation or lateral disc
Central spinal stenosis
Infection
Operation at the wrong level
Arachnoiditis
Facet subluxation or fracture
Bone overgrowth and spinal stenosis
Failure of bone graft fusion
Foreign body
Pseudomeningocele
Hemorrhage

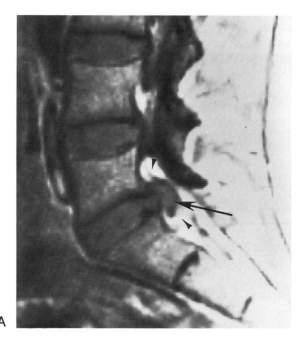

A

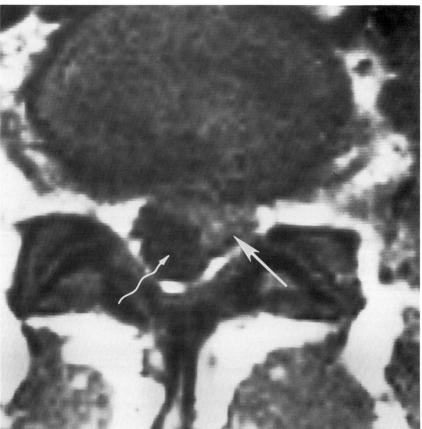

B

Fig. 14-22 Posterolateral disc herniation, MRI. (A) Parasagittal T_1-weighted image (SE 900/20). A large disc extrusion is present in the epidural space *(arrow)* and is clearly outlined by the high signal intensity epidural fat *(arrowheads)*. **(B)** Para-axial view shows the herniated disc material in the left epidural space *(arrow)*. The disc material is intermediate density between high intensity epidural fat and low intensity CSF within the dural tube *(wavy arrow)*. Aside from intensity factors, the principles for diagnosis of herniated disc with MRI are the same as with CT scanning.

space. Herniated discs produce rounded defects in the contrast column, but it is difficult to differentiate the defect produced by disc from that of postoperative scar and arachnoiditis.

CT scanning is useful for postoperative imaging. The level of the prior surgery can be defined reliably by the partial absence of the ligamentum flavum and a defect in the lamina at the site of the disc exploration (Fig. 14-28). A partial facetectomy is identified by the absence of the medial part of the superior articular process with a

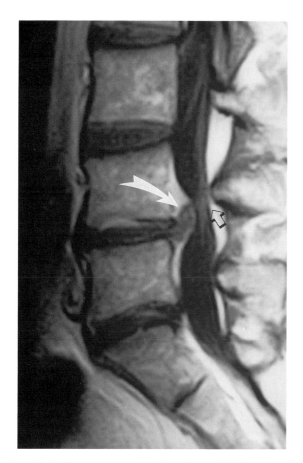

Fig. 14-23 Disc extrusion, lumbar. Sagittal T$_1$ MRI through the lumbar spine demonstrates a disc extrusion at L4–5 (*white arrow*). Large herniations such as this rupture through the posterior longitudinal ligament and are classified as extrusions. The extrusion compresses the thecal sac nearly completely obliterating the subarachnoid space (*open arrow*). This patient is at risk of the development of an acute cauda equina syndrome, and this should be reported promply to the clinician.

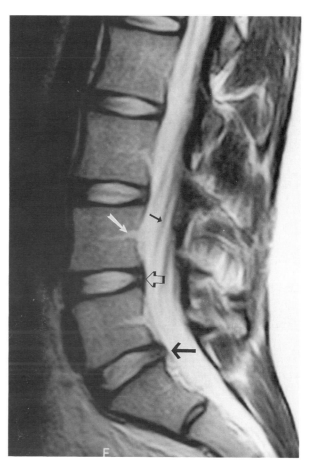

Fig. 14-24 Early protrusion into inner annulus with slight bulge. Sagittal FSE (8) 80/4000 T$_2$ MRI shows the normally high signal within the disc interspaces. At L5–S1, there is early protrusion of disc into the annulus with a slight annular bulge (*large black arrow*). The L4–L5 annulus and disc space is completely normal (*open arrow*). Note the high signal of the basivertebral vein (*white arrow*) and the nerve roots of the cauda equina within the dural tube (*small black arrow*).

postoperative scarring. Recurrent disc herniation is relatively uncommon. The most difficult diagnostic problem is the differentiation of recurrent disc from scar (fibrosis) at the level of the prior surgery. Recurrent disc herniation requires surgical removal, whereas scarring cannot be treated successfully with excision.

The myelogram has little use in the evaluation of the postoperative patient. Most postoperative myelograms are abnormal to some degree, showing failure of contrast filling of the nerve root sleeves at the level of surgery and minor irregularities of the margin of the subarachnoid

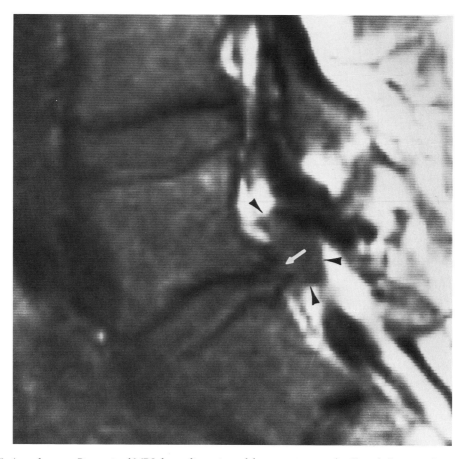

Fig. 14-25 Annular tear. Parasagittal MRI shows disruption of the posterior annular fibers *(white arrow)* associated with a large disc extrusion *(arrowheads)*. The disc interspace is narrowed (SE 900/20).

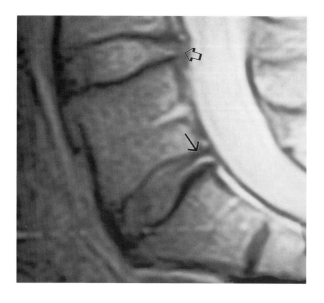

Fig. 14-26 Acute annular tear. Sagittal FSE T_2 MRI indicates the tear in the posterior annulus fibrosis of L5–S1 as high signal *(black arrow)*. There is no disc herniation here. At the L4–L5 level, there is a small disc protrusion *(open arrow)*.

sharp margin remaining. Facet fractures, spinal stenosis, and infections can all be diagnosed. Bone windows are essential.

The most difficult differential is between postoperative scar and recurrent disc herniation. The diagnosis of recurrent disc herniation is based on the same criteria as in the nonoperated spine. There must be a well-circumscribed rounded density greater than that of the dural sac causing a mass effect. Displacement of the theca and nerve roots is critical for making the diagnosis. Displacement away from the disc space is unusual with scarring alone. If present, calcification is diagnostic of herniation. Fibrosis tends to be a relatively homogeneous cuff of tissue density in the epidural space confined to the side of surgery. Most often, the theca is retracted toward the scar density. Nerve roots may be surrounded and difficult to identify. High-dose intravenous contrast enhancement can sometimes distinguish scar from disc herniation. The scar generally enhances, whereas the disc material does not (Fig. 14-29). Chronic recurrent herniation may be outlined by an intensely enhancing rim of dense scar.

The MRI technique is useful for recognition of recurrent disc herniation and its differentiation from scar. Both T_1- and T_2-weighted images are obtained. The transaxial plane is most useful. Scar is relatively hyperintense with the CSF of the dural sac on T_1-weighted sequences, isointense on proton-density images, and mod-

(*Text continued on page 593*)

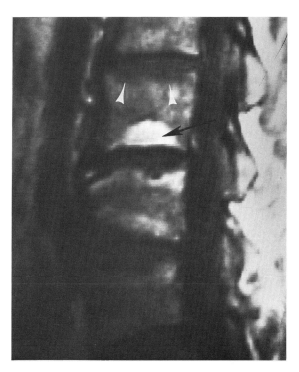

Fig. 14-27 Bone changes of degenerative disc disease. Parasagittal MRI (SE 900/20) shows early-stage changes of hypointensity adjacent to the disc space (*white arrowheads*) and high intensity signal representing the late stage of fatty infiltration (*black arrow*).

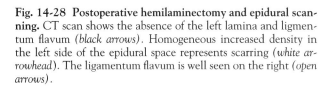

Fig. 14-28 Postoperative hemilaminectomy and epidural scanning. CT scan shows the absence of the left lamina and ligmentum flavum (*black arrows*). Homogeneous increased density in the left side of the epidural space represents scarring (*white arrowhead*). The ligamentum flavum is well seen on the right (*open arrows*).

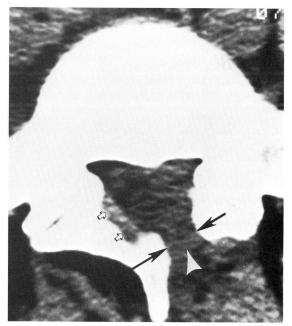

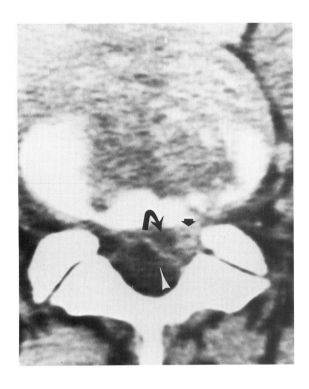

Fig. 14-29 **Postoperative recurrent disc herniation.** Contrast-enhanced CT scan shows homogeneous enhancement of epidural scar on the left *(black arrow)*. The nonenhancing region more centrally represents recurrent disc herniation *(curved arrow)*. The herniated disc material is causing compression of the dural tube *(white arrowhead)*.

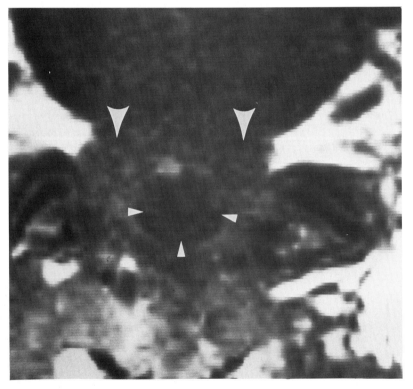

A

Fig. 14-30 **Postoperative epidural scar, bilateral laminectomy, MRI.** **(A)** T_1-weighted study shows epidural scar as homogeneous signal intensity *(large arrowheads)* that is relatively hyperintense to the CSF within the dural tube *(small arrowheads)*. The scar is well outlined by the high-intensity epidural fat. *(Figure continues.)*

Fig. 14-30 *(Continued)*. **(B)** Proton-density images (SE 2500/40) show the epidural scar *(white arrowheads)* as nearly isointense with the CSF within the dural tube *(black arrowheads)*. **(C)** T$_2$-weighted image (SE 2500/90) shows the epidural scarring as homogeneous hypointensity *(white arrowheads)* compared with the high signal intensity of the CSF *(black arrowhead)*. The epidural fat and epidural scar are nearly isointense.

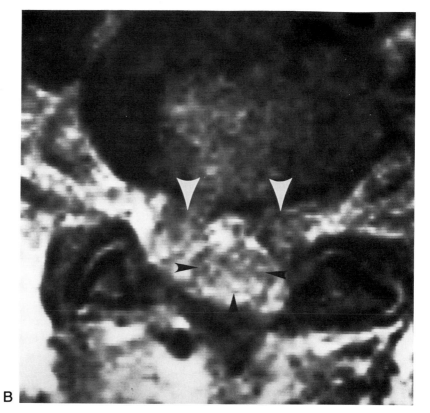

B

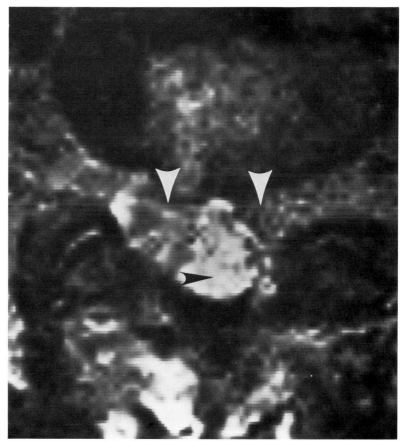

C

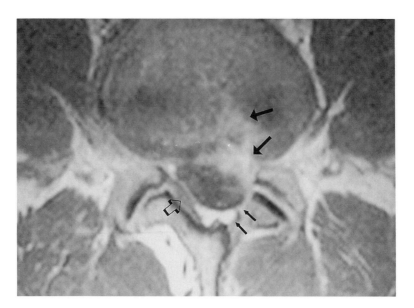

Fig. 14-31 Postoperative disc excision, epidural fibrosis. Axial T_1 Gd-enhanced MRI shows intense enhancement of the left-sided epidural fibrosis *(large black arrows)*. The enhancement extends into the left side of the disc space. The left ligamentum flavum is removed *(small black arrows)*. The normal ligament is present on the right *(open arrow)*.

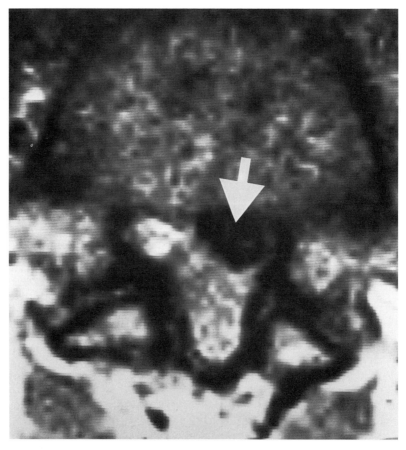

Fig. 14-32 Postoperative recurrent disc herniation, MRI. T_2-weighted MRI (SE 2500/90) shows the focal disc herniation as extreme hypointensity *(arrow)*.

erately hypointense on T_2-weighted images (Figs. 14-30 and 14-31). Herniated disc material is usually of low signal intensity on heavily T_2-weighted images (Fig. 14-32). Gd can be used as with MRI scanning, producing enhancement of the epidural scar on T_1-weighted images. The herniated disc material does not enhance (Fig. 14-33). There is often disc enhancement of the annulus and disc space in the region of disc excision (Fig. 14-34). This defines the extent of the surgery. Enhancement within the disc space cannot be taken as evidence of postoperative discitis unless there is edema (high T_2 signal) and enhancement within the adjacent vertebral bodies. No changes are detected within the vertebral bodies seen with MRI following discectomy. The imaging characteristics of scar and recurrent disc herniation are summarized in the box.

ARACHNOIDITIS

Arachnoiditis is most common after lumbar surgery but may occur after trauma or infection. Previously, Pantopaque myelography was the most common cause. Intra-arachnoid adhesions develop, producing characteristic

---

CT AND MRI CHARACTERISTICS OF EPIDURAL SCAR

Scar
　　CT density slightly lower than HNP
　　MRI T_2 intensity slightly higher than HNP
　　Relatively homogeneous
　　Occurs only on the side of the surgery
　　Conforms to the contour of the dural sac
　　May retract the sac (rarely displaces the sac)
　　May be away from the disc space
　　Show enhancement with intravenous contrast
Disc herniation
　　CT density greater than scar
　　MRI T_2 intensity less than scar
　　Occurs at disc space level
　　Causes displacement of thecal sac and nerve roots
　　Does not contrast-enhance

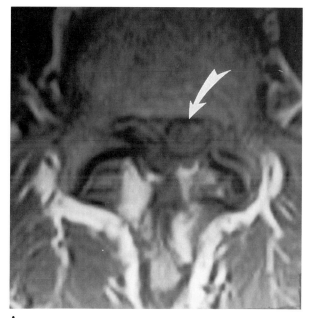

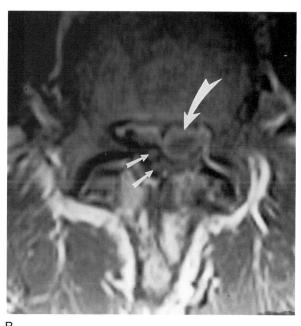

A   B

Fig. 14-33 Postoperative recurrent disc herniation; enhancing nerve roots. (A) Nonenhanced T_1 MRI shows a mass of tissue in the left anterolateral epidural space (*arrow*). There are faint concentric rings within the lesion. **(B)** Gd-enhanced T_1 MRI through the same level shows a ring of enhancing fibrosis surrounding an nonenhancing recurrent disc fragment (*large white arrow*). This displaces the dural tube, an uncommon finding with fibrosis alone. The nerve roots of the cauda equina enhance slightly (*small white arrows*), which probably means nerve root irritation or arachnoiditis.

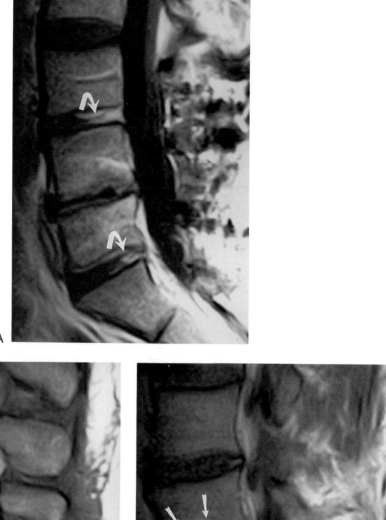

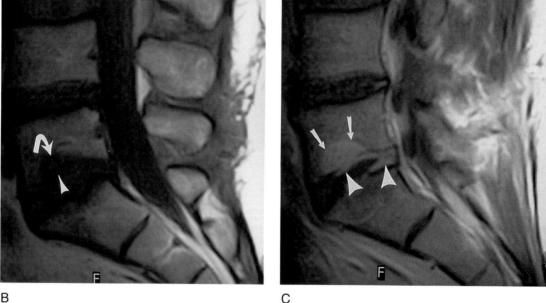

B   C

Fig. 14-34 Postoperative disc excision. Discitis. (A) Normal. Sagittal T$_1$ Gd-enhanced MRI shows enhancement within the disc space along the site of disc excision at L3–L4 and L5–S1 *(curved arrows)*. The enhancement pattern varies from peripheral along the endplate to central within the disc space. This is a common finding after disc excision **(B)** Discitis. Nonenhanced T$_1$ MRI shows a focal region of loss of the normal fat signal within the marrow adjacent to the inferior endplate of L5 *(curved white arrow)*. The line of the endplate is lost *(white arrowhead)* **(C)** Discitis, Gd, enhancement. Gd-enhanced T1 sagittal MRI shows enhancement of the fibrocartilage and disc space after discectomy *(white arrowheads)*. This pattern alone is not diagnostic of infection. The enhancement of the marrow within L5 indicates that this person has discitis and local osteomyelitis *(white arrows)*.

findings with imaging. The nerve roots within the dural tube show clumping (Figs. 14-35 and 14-36), and the roots may become adherent to the margins of the dura. If all the nerve roots are adherent laterally, the thecal sac may appear empty. Bizarre patterns can develop on the myelogram (Fig. 14-37). Focal clumps may produce myelographic defects that suggest disc herniation. If the fibrosis is particularly severe, the dural tube can be blocked by adhesions. This situation is seen as high signal intensity within the sac on T_1-weighted MRI. The pain of arachnoiditis may be severe and unrelenting.

Cervical Spondylosis and Disc Herniation

Cervical spondylosis results from disc degeneration, which creates a pathology similar to that described for the lumbar spine, except for some variation in the distribution of the lesions. The uncovertebral joints (of Luschka) are affected, creating osteophytes that protrude from the posterolateral uncinate processes into the intervertebral foramina (Fig. 14-38). Posterior osteophyte formation from the disc margins may be large, creating transverse spondylytic bars. Partial fusion of the segments may result from the hypertrophic bridges. The C5–C6 and C6–C7 levels are those most frequently diseased (80 percent). It is less common at the C4–C5 and C3–C4 levels. The osteophytes produce radiculopathy when they compress the nerve root in the lateral part of the spinal canal or intervertebral foramen. The exiting nerve root normally occupies no more than 25 percent of the neural foramen, so severe narrowing must occur before radicular symptoms are produced. Posterior growth of the osteophytic bars narrows the spinal canal and, if severe, produces a myelopathy from cord compression.

Significant calcification or ossification of the posterior

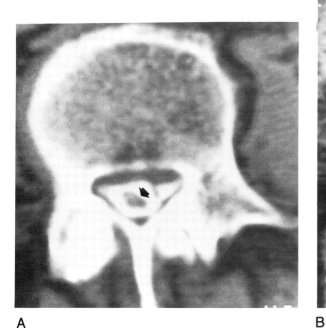

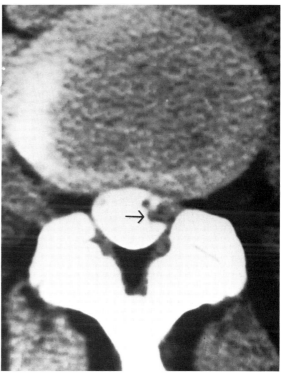

A
B

Fig. 14-35 Arachnoiditis. (A) CT myelography shows central clumping of the roots of the cauda equina from adhesive arachnoiditis *(arrow)*. **(B)** Here the nerve roots are adherent with the lateral margin *(arrow)*. This abnormality may simulate disc herniation on myelography.

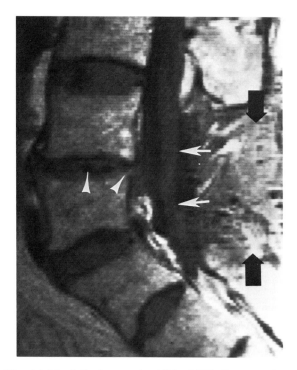

Fig. 14-36 Adhesive arachnoiditis, MRI. Parasagittal T_1-weighted image shows adhesion of the roots of the cauda equina into a central cord (*white arrows*). This patient has had laminectomy and removal of the spinous processes from L3–L5 (*black arrows*). Degenerative disc disease with gas disc phenomenon and a slight retrolisthesis is present at L3–L4 (*arrowheads*).

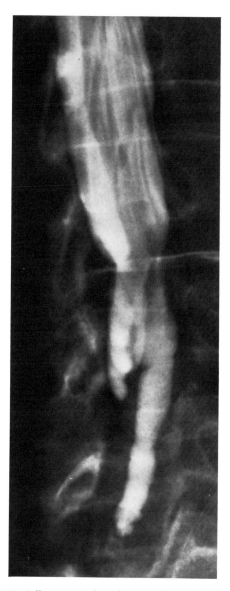

Fig. 14-37 Adhesive arachnoiditis, myelography. The roots of the cauda equina are adherent, and the lower lumbar subarachnoid space is obliterated. The myelographic patterns of arachnoiditis are infinite.

longitudinal ligament may occur and is most frequent in the cervical region. The ossified ligaments may be so thick that they severely narrow the spinal canal, causing a myelopathy. The thick calcific density can be seen on plain films as well as CT scans (Fig. 14-39) and MRI.

Herniation of the nucleus occurs spontaneously or with trauma. It may be central or posterolateral and is often referred to as a "soft disc" herniation. "Hard disc" (a misnomer) refers to osteophytosis. Disc herniations and spondylytic bars are both capable of producing radiculopathy or myelopathy.

Plain film radiography is basic to the evaluation of cervical spondylosis. Oblique views are necessary to evaluate the intervertebral foramina. In the early stages, there is loss of the normal lordosis, indicating some restriction of motion. The disc spaces become narrowed. Large osteophytes are seen on both the anterior and posterior margins of the vertebral bodies. The parasagittal dimension of the spinal canal can be estimated by measuring from the posterior margin of an osteophyte to the anterior margin of the spinous process. From C4 to C7, this dimension is normally more than 12 mm. A dimension of less than 10 mm indicates a high probability of spinal cord compression and myelopathy. Myelography, CT myelography, or MRI is necessary for the anatomic diagnosis of cord compression.

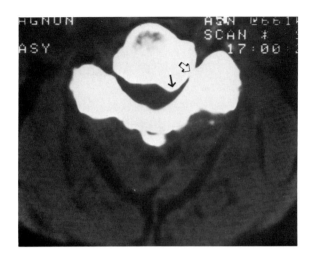

Fig. 14-38 Cervical osteophytes. Transaxial CT scan shows a large osteophyte from the left uncinate process *(black arrow)*. It protrudes into the spinal canal and into the intervertebral foramen *(open arrow)*, where it may cause constriction of the exiting nerve root.

Myelography is still an important test for evaluating cervical spondylosis. The contrast should fill the entire cervical dural sac. Osteophytes from uncinate process or a small (2- to 4-mm) posterolateral disc herniation produce a small defect in the contrast-filled nerve root sleeves. These defects occur commonly with aging, and most are asymptomatic. Correlation of myelographic defects with the precise level of the radiculopathy is mandatory. Suspected symptomatic lesions can be confirmed with CT myelography or MRI.

Anterior transverse defects are produced by both osteophytes and central disc herniations. Central herniations are more common in the cervical region than in the lumbar region. "Ridges" may be present at many levels; their significance is determined primarily by correlation with the level of any radiculopathy or myelopathy. If the spinal cord is compressed, the subarachnoid space is attenuated, and the cord appears widened on the frontal projection (Fig. 14-40A). It may mimic an intramedullary tumor. The lateral projection shows the hypertrophied bone ridges or disc herniation responsible for cord widening (Fig. 14-40B&C). Lateral disc herniations produce lateral defects on the contrast column at the disc level (Fig. 14-41).

With CT, cervical disc herniation may be seen as a high-density epidural lesion on routine examination (Fig. 14-40C), but the lack of much epidural fat makes

most herniations difficult to detect. CT myelography has much greater sensitivity, which shows the disc herniation displacing the contrast-filled thecal sac (Fig. 14-42). Thin sections (2 mm) are required. Disc herniations can be differentiated from osteophytes. Bone window images allow evaluation of the intervertebral foramina for encroachment by osteophytes.

The diagnosis of cord compression by disc, osteophyte, or a narrow canal requires CT myelography. To diagnose cord compression, the subarachnoid space must be nearly obliterated and the spinal cord flattened or distorted (Fig. 14-39).

The preferred initial method for evaluating cervical spondylosis disc herniation and myelopathy is MRI. Thin-section parasagittal and transaxial images are obtained (Figs. 14-43 and 14-44). T_2-weighted FSE and T_2*-weighted GRE sequences are the most useful. The degree of cord compression can be determined. GRE imaging with T_2* weighting produces a myelographic effect to better define the posterior margin of disc herniation and the degree of spinal stenosis. The transaxial plane is used to detect small herniations that cause radiculopathy. GRE imaging is excellent for defining osteophytes, but very short TE sequences (less than 15

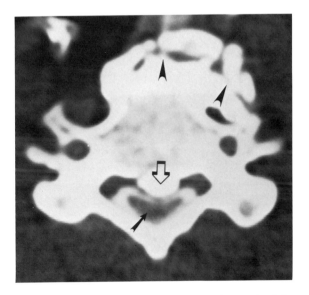

Fig. 14-39 Ossification of the posterior longitudinal ligament. CT myelography shows thickening and ossification of the posterior longitudinal ligament *(open arrow)*, which causes significant compression of the dural tube and spinal cord *(black arrow)*. The cord compression produces myelopathy. Large anterior osteophytes are present *(arrowheads)*.

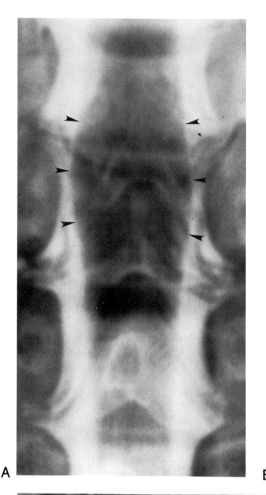

A

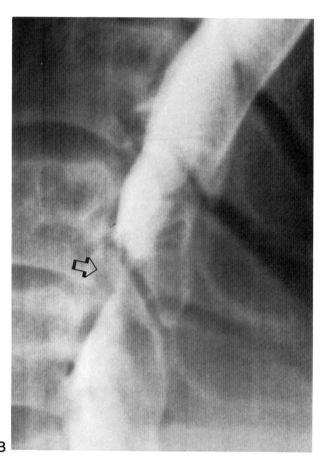

B

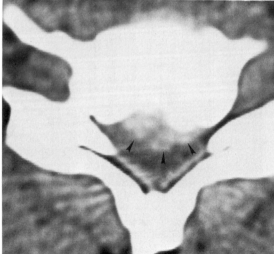

C

Fig. 14-40 **Large cervical disc herniation. (A)** AP myelography shows apparent widening (pseudoenlargement) of the cord due to compression by the large disc herniation *(arrowheads)*. **(B)** Lateral view shows the posterior displacement of the contrast-outlined dural tube by the large disc herniation *(arrow)*. Herniated disc material has migrated superiorly behind the vertebral body. **(C)** CT scan shows the high-density herniated disc material within the spinal canal *(arrowheads)*.

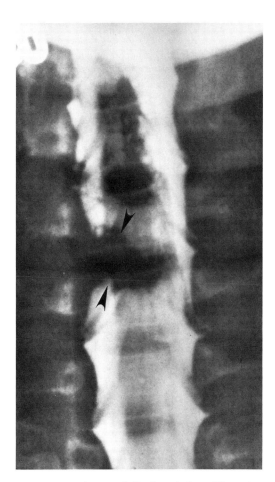

Fig. 14-41 Lateral cervical disc herniation. AP myelography shows a large, lateral, extradural-type defect at the level of the disc space representing the disc herniation (*arrowheads*).

msec) are needed to reduce magnetic susceptibility artifacts, which accentuate the size of osteophytes. In the cervical region, the intensity of the T_2 signal in the disc spaces (on SE images) is a less reliable indicator of degenerative disc disease because of the small size of the cervical nucleus pulposus. Myelography followed by CT myelography remains the most sensitive technique for evaluating cervical radiculopathy caused by small disc herniations not seen with MRI.

Thoracic Disc Herniation

Thoracic disc herniation is uncommon. It accounts for about 0.5 percent of all disc herniations; 75 percent of the herniations occur below T3, with most being at T11–T12. Multiple disc herniations may occur. Plain film myelography is reportedly diagnostic in only 56 percent of cases. CT scanning, usually as CT myelography, has improved the diagnostic results. The CT examination is performed in a manner similar to the lumbar study. The same criteria are used for the diagnosis of disc herniation, although many disc herniations are calcified and may be confused with meningioma. MRI is generally not as sensitive as CT myelography for the detection of thoracic disc herniation, but it has the advantage of facilitating examination of a greater length of the spine. The clinical diagnosis is also difficult, as there is no distinct symptom complex.

Discography

Discography is the direct injection of a water-soluble contrast material into the disc space and radiographically observing its distribution. The contrast injected into the nucleus pulposus pushes aside the disc matrix, forming pools of contrast. The pattern of the pools depends on the degree of degeneration of the disc nucleus and the surrounding annulus.

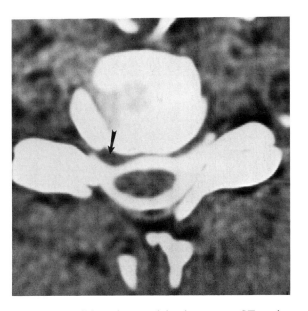

Fig. 14-42 Small lateral cervical disc herniation. CT myelography shows the disc herniation as a slight displacement of the contrast-outlined dural tube and nerve root sleeve (*arrow*). CT myelography is the most sensitive examination for detecting small cervical disc herniations.

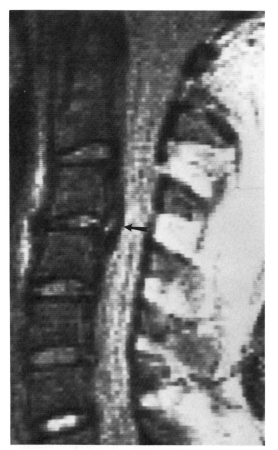

A

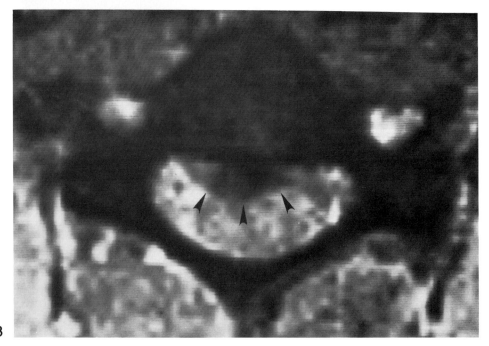

B

Fig. 14-43 Cervical disc herniation, MRI. (A) Proton density image shows disc herniation with displacement of the dural tube *(arrow)*. **(B)** Transaxial GRE T$_2$*-weighted image shows the hypointense disc herniation displacing the hyperintense dual tube *(arrowheads)*.

Fig. 14-44 Cervical disc herniation. Cord compression, spinal stenosis. (A) Sagittal FSE T$_2$ MRI shows the large C3–C4 disc herniation as dark material protruding into the spinal canal and compressing the spinal cord *(thin arrow)*. There is also anterior osteophyte and disc herniation *(thick arrow)*. A previous disc herniation and osteophyte have been removed and resorbed following anterior disc excision at C4–C5 *(arrowhead)*. **(B)** Axial GRE through C3–C4 shows the disc herniation *(arrowheads)*. The herniation extends laterally, narrowing the intervertebral foramina *(open arrows)*. The spinal canal is narrowed, and the spinal cord is significantly compressed *(arrow)*. This type of herniation may cause both radiculopathy and myelopathy. *(Figure continues.)*

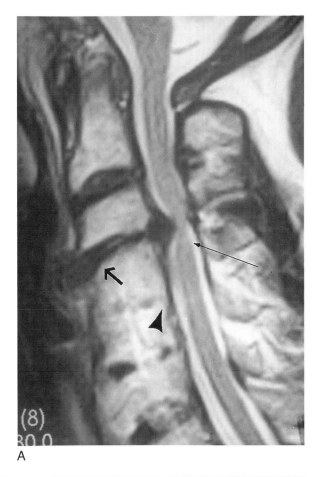

A

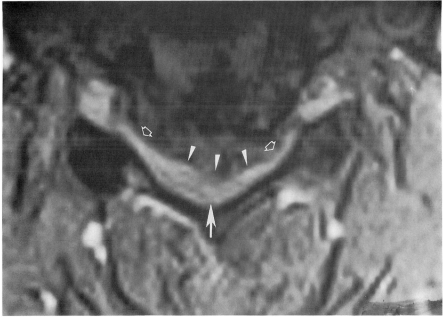

B

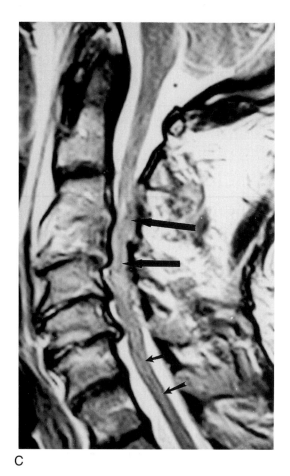

C

Fig. 14-44 *(Continued).* **(C)** Cervical spondylosis with spinal stenosis *(arrows).* Osteophytes compress the spinal cord, obliterating the subarachnoid space *(larger arrows).* The cord is atrophic below *(smaller arrows).*

The needle is placed under fluoroscopic control, using an oblique extradural approach. The nature of any pain produced is noted and correlated with the patient's symptoms.

In the completely normal disc, the contrast forms a coalescent collection of pools around the uniformly compressed nucleus. It appears as a more or less elliptical dense blob of contrast within the center of the disc space. Only 0.5 ml of contrast is accepted, and there is no pain.

When early degenerative changes have occurred, varying degrees of fibrosis of the nucleus and separation of the nucleus from the vertebral endplates result. The separation causes the injected contrast to pool adjacent to the endplates, which appear bilobular with the central portion free of contrast.

With more severe degrees of degeneration, irregularity of the increasingly fibrous nucleus is greater. Clefts form in the annulus. The annular clefts are demonstrated by contrast extending peripherally from the nucleus and, in the more degenerated conditions, completely through the annulus to the posterior or anterior longitudinal ligaments. The severely degenerated disc accepts large amounts of contrast, up to 1.2 ml, and the patient's pain may be reproduced by the injection.

With annular rupture, the injected contrast extends through the annulus and escapes the disc space, often deflected upward or downward by a still intact posterior longitudinal ligament. Annular rupture may occur at any stage of disc degeneration. There is positive correlation of the degree of hydration of the discs as demonstrated by T_2-weighted MRI and the degenerative changes as defined by discography.

The value of discography has not been clearly established and remains controversial. Because degenerative patterns have been seen in studies in asymptomatic persons, many radiologists believe that with the availability of CT scanning and MRI there is no place for discography. However, some clinicians and radiologists believe that discography can provide specific information regarding patients who have persistent long-term back pain without radiculopathy and no evidence of disc herniation on myelography, CT scans, or MRI. In this situation, some believe that demonstration of a posterior annular tear, associated with reproduction of the patient's pain, is significant, and that anterior interbody fusion at that level can relieve the pain.

Lumbar Facet Joint Arthropathy

Lumbar facet arthropathy is almost always caused by increased stress placed on the joints as a result of degenerative disc disease. The lumbar facets receive sensory innervation from overlapping segments of the posterior rami of the spinal nerves, so the pain produced by the arthropathy is deep and poorly localized. Pain of the "facet syndrome" is difficult to distinguish from the pain produced by degenerated disc disease.

The most frequent CT finding of facet arthropathy is osteophyte formation at the margins of the joint. The superior facet is the one most frequently affected. These osteophytes may narrow the spinal canal, the lateral recesses, and the lateral neural canal (intervertebral foramen) (Fig. 14-45). The facet joint space narrows to less than 2 mm, and a gas disc phenomenon may occur. Hy-

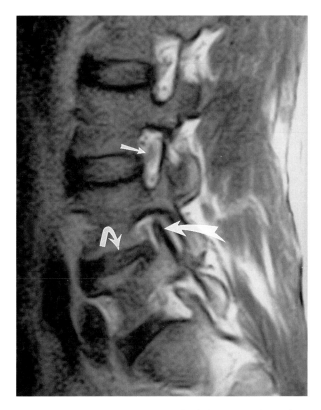

Fig. 14-45 Narrowed intervertebral foramen. Sagittal T$_1$ MRI of the lumbar spine shows a severely narrowed L4–L5 intervertebral foramen (*large white arrow*). The L4 vertebral body settled downward because of a narrowed degenerative disc (*curved white arrow*). The superior articular process of L5 protrudes into the foramen, compressing the exiting L4 nerve root. The L3–L4 foramen above is normal (*small white arrow*).

permobility results from the laxity of the periarticular ligaments, producing spondylolisthesis or retrolisthesis. Ossification may extend into the ligamentum flavum. There is subchondral sclerosis and cystic degeneration, and a synovial cyst sometimes containing gas may enlarge into the epidural space (see Fig. 14-9). It may produce symptomatic root compression. Some cysts have been successfully aspirated with percutaneous technique. CT scanning is the most accurate technique for defining facet joint arthropathy and its consequences (see Fig. 14-8).

Lumbar Spinal Stenosis

Spinal stenosis is characterized as either lateral or central and as developmental or degenerative (acquired). Degenerative lateral stenosis is the most common form of spinal stenosis. With this disorder, osteophytes narrow the lateral recesses or the lateral neural canals, causing radiculopathy. Central stenosis, narrowing of the spinal canal, is less common and when severe causes the cauda equina syndrome. All forms of degenerative spinal stenosis occur during the later phases of the spondylytic process, and usually in persons over the age of 50. Other causes of acquired stenosis include the degenerative type of spondylolisthesis (L4–L5), ossification of the posterior longitudinal ligament or ligamentum flavum, Paget's disease, trauma, and spinal fusion.

Developmental narrowing of the spinal canal rarely causes symptoms unless there is superimposed degenerative change. Small lesions become symptomatic in this group. Congenital stenosis may occur as an isolated spinal deformity, with achondroplasia, and with Morquio's disease.

Lateral spinal stenosis refers to narrowing of the lateral recesses or lateral neural canals. Axial CT sections through the level of the pedicles demonstrate the lateral recesses. They are bounded anteriorly by the posterolateral vertebral body, posteriorly by the base of the superior facet, and laterally by the pedicle. The lateral recess contains the segmental nerve root as it exits the dura but before curving, to exit below the pedicle. Osteophytes from the superior facet encroach on the recess, causing nerve root entrapment. Bulging of the annulus can further compromise the recess. The nerve root is flattened, and the epidural fat surrounding the root in the canal is obliterated. This type of stenosis is seen most frequently at the L5–S1 level.

There are no absolute criteria for the diagnosis of lateral recess stenosis. The recess is said to be narrowed if it measures 3 mm or less in height, but the diagnosis depends more heavily on visualization of the compressed nerve root within the recess (Figs. 14-46 and 14-19).

The lateral neural canal (intervertebral foramen) may also become narrowed by a similar process. Degeneration of the disc and narrowing of the disc space cause upward and forward movement of the superior facet into the foramen resulting in nerve root compression. Bulging of the annulus and osteophyte formation from the posterolateral rim of the adjacent vertebral body may further compromise the canal. Thin-section scanning or reformating of the images in the lateral or oblique planes may be necessary to make a definitive diagnosis. MRI gives excellent views of the foramen and segmental root (Fig. 14-47). Disc material can also herniate into the lateral canal.

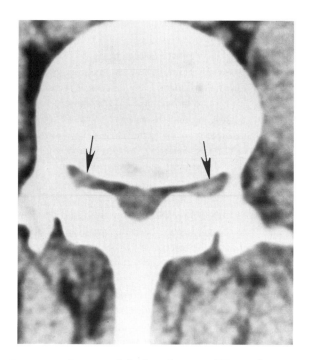

Fig. 14-46 Stenosis of the lateral recess. CT scan shows significant narrowing of both lateral recesses with compression of the exiting nerve roots *(arrows)*. The epidural fat within the recesses is nearly totally obliterated.

The myelographic findings for recess and canal stenosis are nonspecific. Shortening, widening, and poor opacification of the nerve root sheaths may be seen, but these changes can also be seen with disc herniation, arachnoiditis, and anatomic variation. CT and MRI scans are far superior for evaluating lateral stenosis.

Central spinal stenosis is narrowing of the spinal canal. Acquired stenosis from degenerative change is the most common cause of central spinal stenosis; it occurs most frequently at the L4–L5 level. The canal is narrowed from the posterolateral aspect by the enlarged osteophytic and encrusted inferior facets. Osteophytes and a bulging annulus also narrow the canal from the anterior aspect. In addition, the lamina and the ligamentum flavum may be thickened (more than 5 mm). Because of hypertrophy of the inferior facets, the interfacet distance is narrowed. It usually measures more than 16 mm but may become less than 10 mm with severe stenosis. A parasagittal dimension of less than 12 mm suggests stenosis. This dimension is the least important and may remain normal, even with severe stenosis of the canal.

The CT scan, particularly as CT myelography, is the most accurate technique for diagnosing central spinal stenosis (Fig. 14-48). Bone hypertrophy is easily seen. The epidural fat is obliterated as the dural sac is constricted. Contrast within the thecal sac demonstrates compromise of the subarachnoid space and displacement of the CSF with cauda equina constriction (see Fig. 14-51B).

With myelography, the dural sac usually shows undulating constrictions that are narrowest at the level of the disc spaces. Severe stenosis causes a complete block. The nerve roots are displaced centrally in the canal; buckling of the roots may occur above or below the level of the constriction (Fig. 14-49). Although many of the anterior defects may mimic herniation of the disc, these defects are almost always caused by annular bulging and osteophytes. Herniation of disc material is generally not a part of this late phase of spondylosis.

The MRI scan is also capable of defining spinal stenosis, but the degree of cauda equina constriction is generally more difficult to evaluate than with CT myelography. Transaxial T_2-weighted SE or GRE images to produce a myelographic effect help to diagnose nerve root constriction (Fig. 14-50). The sagittal plane may understate the degree of canal narrowing as the canal is most severely narrowed from side to side. On the T_1-weighted images are dural tube is more intense than average at the site of constriction because nerve roots rather than CSF fill the intradural space. Buckling of the nerve roots may be seen. Spinal stenosis may be acquired from other disease processes, particularly Paget's disease (Fig. 14-51) and fracture.

Spondylolisthesis

Spondylolisthesis is the anterior displacement of a superior vertebral body over the body below. It occurs in about 5 percent of the adult population and is most frequent at the L5–S1 level. There are two types: lytic and degenerative.

Lytic spondylolisthesis, by far the most common type, is the result of a fracture or fibrous cleft in the pars interarticularis of L5. The cleft is called spondylolysis. It divides the posterior elements into two portions: (1) an anterosuperior portion consisting of the pedicle, superior facet, and transverse process; and (2) a posteroinferior portion consisting of the inferior facet, lamina, and spinous process. The disconnection of the support of the inferior facet allows the forward slippage of L5 over S1. The clefts

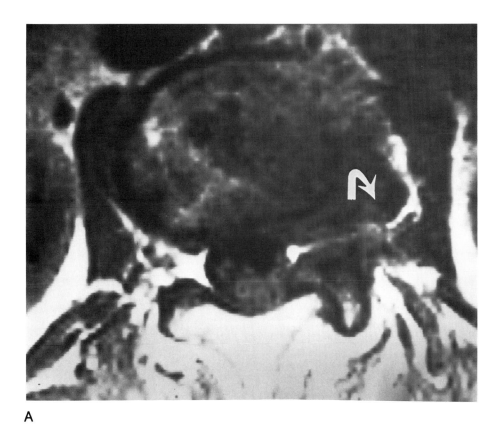

A

Fig. 14-47 Lateral canal stenosis, MRI. (A) T_1-weighted image in the transaxial view shows the low signal intensity of a large osteophyte obliterating the left lateral neural canal *(arrow)*. It may be difficult to differentiate disc herniation from an osteophyte. **(B)** Parasagittal view shows obliteration of the lateral neural canal, with displacement of epidural fat and compression of the nerve root *(black arrow)*. The normal intervertebral foramina are seen above and below *(open arrows)*.

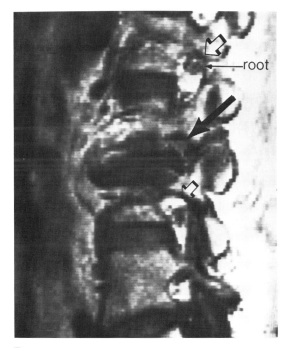

B

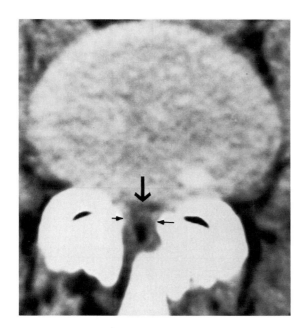

Fig. 14-48 Central spinal stenosis, CT scan. Transaxial view shows extreme narrowing of the central spinal canal *(large arrow)* due to inferior facet hypertrophy *(small arrows)*. With this degree of narrowing there is a high probability of cauda equina compression. Spinal stenosis may be congenital or acquired.

characteristic double appearance somewhat resembling a figure-of-eight. Sagittal reformation images give an accurate estimate of the size of the canal. Disc herniation at the level of slippage is unusual (less than 25 percent).

(Text continued on page 610)

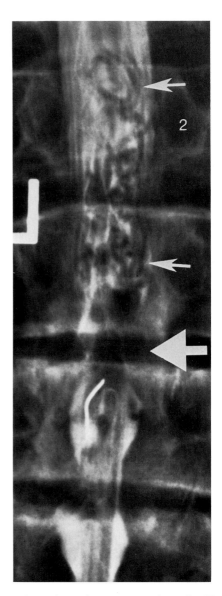

Fig. 14-49 Central spinal stenosis, myelography. High-grade constriction is present at the L3–L4 level *(large white arrow)*. Less severe constriction is present at L4–L5, but it is still causing displacement of the nerve roots to the central portion of the canal. Note the buckling of the nerve roots above the severe constriction *(small arrows)*.

are bilateral 85 percent of the time. The etiology is controversial but is generally considered post-traumatic.

Plain films demonstrate the forward displacement of L5 with relation to S1 (Fig. 14-52). It is classified according to the degree of slippage, from grade I (less than 25 percent anterior displacement) by quarters through grade IV (75 percent to total displacement). Generally, oblique films are best for showing the cleft in the pars interarticularis between the two facets.

The CT scan shows the pars defect as a horizontal cleft, often with surrounding sclerosis, just anterior to the facets at the level of the pedicles (Fig. 14-52B). It must be differentiated from the clefts of the facet joints themselves, which occur at the level of the disc space. The jagged defect can be seen on sagittal reformatted images. L5 slips forward on S1, and the annulus is heaped up between the displaced posterior margins of the vertebral bodies. It appears as a broad band of soft tissue density just above the posterior superior rim of S1 and must not be misinterpreted as herniated disc material (Fig. 14-52C). The spinal canal appears enlarged, often with a

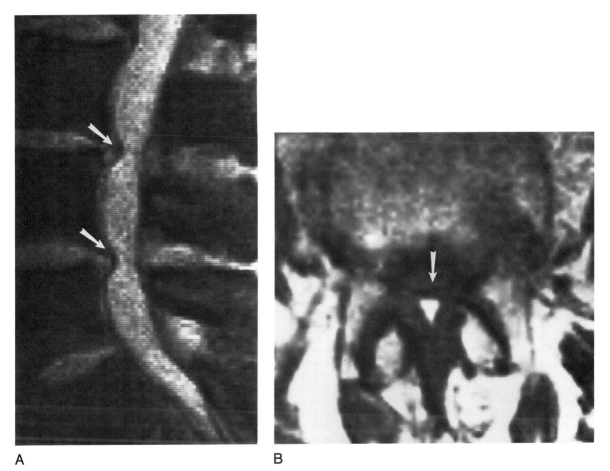

Fig. 14-50 Central spinal stenosis, MRI. (A) FISP 20 GRE 300/10 image shows good myelographic effect within the dural tube. Two disc herniations are present, causing spinal stenosis *(arrows)*. **(B)** There is hypertrophy of the facets and ligamentum flavum. Only a small amount of high intensity epidural fat remains in the posterior angle. The cauda equina fills the constricted dural sac *(arrow)*.

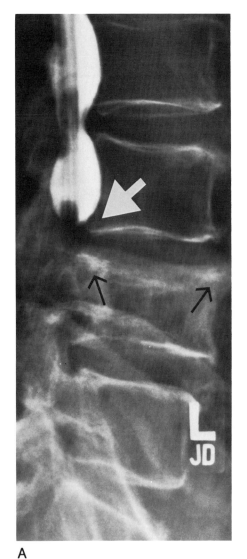

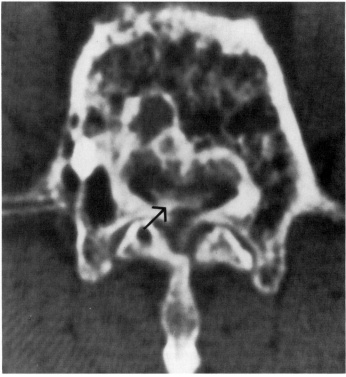

A   B

Fig. 14-51 Spinal stenosis, Paget's disease. (A) Lateral myelographic study shows a complete block to the inferior flow of contrast at the L3–L4 level *(white arrow)* secondary to enlargement of L4 by Paget's disease *(black arrows)*. A moderate annular bulge is present at L2–L3. **(B)** Transaxial CT scan shows the typical vertebral changes of Paget's disease with enlargement of the vertebral body, irregular ossification, and thickening of the cortex, which constricts the spinal canal. Minimal contrast is seen in a constricted dural tube *(arrow)*.

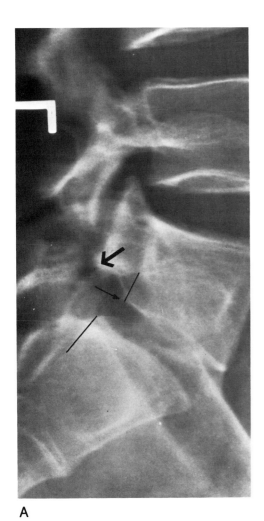

A

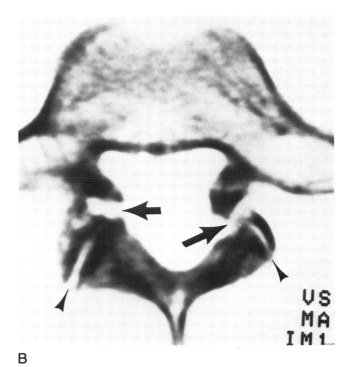

B

Fig. 14-52 Spondylolisthesis, lytic type. (A) Grade I spondylolisthesis is present at L5–S1 (thin arrow). Note the cleft in
the pars interarticularis (large arrow). (B) Transaxial CT scan
shows the bilateral clefts through the pars interarticularis
(arrows). These clefts are distinct from the facet joints (arrowheads). Note the enlargement of the spinal canal with a slight
figure-of-eight configuration. (C) CT scan through the level of
the disc space shows the forward slightly rotated slippage of L5
over S1 (opposed black arrows). The annulus and disc bridge the
increased gap (curved black arrow) between the posterior margins of L5 and S1 (arrowheads). This density must not be misinterpreted as disc herniation. Note that the spinal canal is not
narrowed with lytic spondylolisthesis.

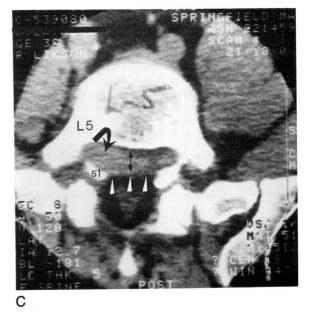

C

However, there is about a 30 percent incidence of disc herniation at the level above.

With the forward displacement, there is vertical narrowing of the intervertebral foramina at L5 (lateral canals) as the annulus and osteophytes protrude into the canal. The degree of foraminal narrowing increases with the degree of forward slippage. Significant compromise of the lateral canals may occur in 30 to 40 percent of patients.

Degenerative spondylolisthesis is secondary to destructive degenerative disease of the facet joints. Facets oriented in the sagittal plane augment the process. The key factor is hypermobility of the facet joints almost always at L4–L5. The inferior facet of L4 rides forward on the superior facet of L5, compressing the lateral recesses. Still attached to the pedicles, the laminae move forward, causing spinal stenosis and compression of the thecal sac from behind. There is a characteristic S-shaped defor-

mity of the dural tube, when seen from the side by MRI or myelography, that may cause significant cauda equina compression (Fig. 14-53).

Other types of spondylolisthesis are traumatic, iatrogenic (removal of the articular process), and pathologic (Paget's disease and rarely tumors).

Retrolisthesis

Retrolisthesis commonly occurs in both the certical and lumbar spine secondary to disc degeneration with disc space narrowing. The upper vertebral body is drawn backward as the superior facets descend the posteriorly inclined inferior facets (Fig. 14-36). Retrolisthesis may cause compromise of the lateral canals because of the relative upward displacement of the inferior facets into the foramen.

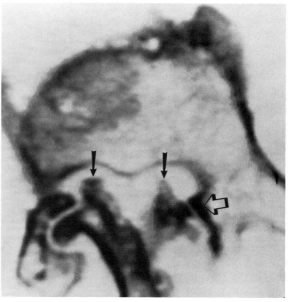

A
B

Fig. 14-53 Spondylolisthesis, degenerative type. (A) Parasagittal T$_1$-weighted MRI shows grade I spondylolisthesis at L4–L5 *(short arrow)*. Note the significant spinal stenosis caused by the anterior slippage of L4 *(long arrow)*. The increased signal intensity within the dural tube at the level of stenosis represents bunched nerve roots. **(B)** Transaxial CT scan shows the hypertrophic change and forward overriding of the superior facets of L4 into the spinal canal and lateral recesses *(black arrows)*. The lamina are also moved forward and cause posterior constriction of the central spinal canal. The left facet is oriented in a near-sagittal plane *(open arrow)*.

TUMORS AND TUMOR-LIKE LESIONS OF THE SPINE AND SPINAL CORD

Tumors and tumor-like conditions of the spine can be classified into four topographic categories. Such classification is useful, as the presentation and differential diagnosis are unique to each category. In order of frequency of occurrence, these groups are (1) tumors of the vertebral columns, (2) tumors of the epidural space, (3) tumors of the intradural–extramedullary space, and (4) tumors of the spinal cord (intramedullary). Listings of the frequency of tumors according to various categories are given in Table 14-2 (see also boxes).

Clinical Presentation

Extramedullary tumors generally impinge on the cord as they enlarge. First, there is compression of the nerve roots, causing localized pain at the level of the tumor or radicular pain in the distribution of the dermatome involved. Paresthesia may occur and is an important sign. Subsequently, the cord is involved. The fiber tracts and cord neurons that are closest to the expanding lesion are first compressed, resulting in a Brown-Séquard syndrome or some modification of it. This syndrome consists of ipsilateral spastic paralysis below the level of the lesion, ipsilateral loss of position and vibratory sensation, and contralateral loss of pain and temperature sensation. Because of irritation or destruction of both the posterior and anterior horn cells, there is pain, paresthesia, or an-

Table 14-2 Type of Tumor and Tumor-like Lesion by Tissue Space

Mass Location	Adults	Children
Extradural		
Common	Herniated disc	Neuroblastoma
	Metastasis	Ganglioneuroma
	Osseous with epidural spread	Leukemia
	Spread from retroperitoneum	Lymphoma
	Direct to epidural space	Lipoma
Uncommon	Myeloma	
	Lymphoma	
	Hematoma	
	Abscess (epidural empyema)	
Intradural–extramedullary		
Common	Neurofibroma (45%)	"Drop" metastasis
	Meningioma	Medulloblastoma
Rare	"Drop" metastasis	Ependymoma
	Melanoma	Dermoid
	Carcinomatosis	
	Arachnoiditis	
	Subdural empyema	
Intramedullary		
Common	Glioma (95%)	Glioma
	Ependymoma (65%)	Ependymoma
	Astrocytoma (35%)	Astrocytoma
Rare	Hemangioblastoma	Lipoma
	Neuronal tumors	
	Melanoma	
	Neurofibroma	
	Metastasis	
	Granuloma (sarcoid)	
	Myelitis and multiple sclerosis	
	Syringohydromyelia	

MOST COMMON TUMOR TYPES ACCORDING TO AGE

Children
　　Astrocytoma
　　Ependymoma
　　Congenital (teratoma, dermoid, lipoma)
　　Ewing sarcoma
　　Eosinophilic granuloma
　　Neuroblastoma
　　Ganglioneuroma
Adolescents/young adults
　　Osteoid osteoma
　　Osteoblastoma
　　Giant cell tumor
　　Aneurysmal bone cyst
　　Eosinophilic granuloma
　　Ewing sarcoma
　　Osteochondroma
　　Syringohydromyelia
　　Hemangioma (AVM)
　　Hemangioblastoma
Adults
　　Metastatic tumors (over 50 years), breast, prostate, kidney, lymphoma, lung
　　Neurofibroma (average age 39 years)
　　Meningioma (average age 45 years; 85% occur in females)
　　Myeloma (50–70 years)
　　Astrocytoma
　　Ependymoma
　　Lymphoma

MOST COMMON TUMORS BY LEVEL

Any level
　　Metastasis
　　Myeloma
Craniocervical junction
　　Meningioma
Cervical level
　　Neurofibroma (especially C2)
　　Astrocytoma
Thoracic level
　　Meningioma
　　Neurofibroma
Thoracolumbar level
　　Ependymoma of the conus medullaris
　　Hemangioma
　　Lymphoma
Lumbar level
　　Ependymoma of the filum terminale
　　Lipoma
　　Dermoid
　　Lymphoma
Sacral level
　　Chordoma
　　Giant cell tumor

esthesia in dermatome distribution, as well as flaccid paralysis of muscles innervated by the motor neurons of the level. Complete compression leads to the development of paraparesis and total loss of sensory function below the lesion.

The development of any of these symptoms constitutes an emergency. The potential for reversibility critically depends on early diagnosis and treatment. Once paraparesis develops, the prognosis for recovery is poor. This pattern of neurologic involvement of the spinal cord is most often caused by metastatic tumor or mye-

loma of the spinal column. Lymphoma, epidural abscess, epidural hematoma, and trauma with disc herniation or bone displacement are the other relatively common entities encountered.

Intramedullary lesions produce a different set of findings. Classically, these lesions first compress the lateral spinal thalamic sensory tracts as they cross ventral to the central canal, producing bilateral loss of pain and temperature sensation from dermatomes at the level of involvement. Because other sensory functions remain intact, the syndrome is referred to as sensory dissociation. As the lesion within the cord expands, it compresses the anterior horn cells, causing segmental flaccid paralysis with subsequent muscular atrophy. With compression of the lateral pyramidal tracts, there is spastic paresis of lower segmental muscle groups. This type of presentation occurs most commonly with syringomyelia. Spinal cord tumors usually present with motor findings and pain as the prominent features.

Tumors of the Vertebral Column

As a group, tumors of the vertebral column are most commonly encountered. Most are metastatic. The tumors are listed below.

Metastatic Tumors

Metastases account for most of the tumors of the spinal column. They usually result from hematogenous spread, often via Batson's valveless venous plexus. The tumors most commonly involve the vertebral bodies, and there are often multiple lesions at diagnosis. The less common solitary lesion is most likely to be from a carcinoma of the lung, kidney, or thyroid. The thoracic segments are those most commonly involved, followed by the lumbar, sacral, and cervical segments. Hematogenous metastatic tumor of the spinal cord and epidural space is uncommon.

T_1-weighted MRI and technetium-99m phosphate and diphosphonate radionuclide bone scans (see below) are the most sensitive diagnostic tests for the presence of metastatic tumor in the spine. The radionuclide accumulates in the region of increased bone formation, produced by most metastatic tumors (Fig. 14-54A). The sensitivity of the test is about 95 percent—double the sensitivity of plain film radiography. The specificity (true negative) of the radionuclide scan is also about 95 percent. However, the predictive value of a positive test is not as high, as a number of disease processes can produce nonspecific radionuclide uptake, particularly degenerative change. Lesions diffusely distributed throughout the axial skeleton may produce a uniformly dense scan, sometimes referred to as the "super scan." False-negative scans result from lesions that are purely lytic or are growing rapidly.

Plain film radiography is basic to the evaluation of metastatic tumors of the spine. The fundamental changes are alterations in the bone density and architecture. Metastatic tumors may have an osteolytic, osteoblastic, or mixed pattern. Fifty percent of the bone mineral content must be lost for the osteolytic lesions to be detectable with plain film techniques. The metastases have a strong predilection for the vertebral bodies but frequently spread, to involve the base of a pedicle, producing the "missing pedicle sign" (Fig. 14-54B). Although the entire vertebral body may be destroyed or

demineralized, the endplates and adjacent disc spaces are usually preserved. Pathologic compression fracture is common. Loss of posterior height of the vertebral body usually occurs, differentiating a pathologic fracture from simple post-traumatic wedging (Fig. 14-55). Osteoblastic metastases are most commonly from metastatic carcinoma of the prostate, mucin-secreting carcinoma of the gastrointestinal tract (colon or pancreas), and breast carcinoma. Mast cell tumors cause sclerosis within bone but are rare. As a rule, the cortex is not thickened, and the bone is not enlarged with blastic metastatic disease, differentiating it from Paget's disease (Fig. 14-51). Mixed osteoblastic and osteolytic metastases represent a combination of the resorptive and reparative functions and

---

TUMORS OF THE VERTEBRAL COLUMN

Adults
- Metastatic tumors
  - Breast
  - Prostate
  - Lung
  - Kidney
  - Gastrointestinal tract
  - Lymphoma
  - Thyroid
- Myeloma
- Hemangioma
- Leukemia
- Chordoma
- Giant cell
- Osteoid osteoma
- Osteoblastoma
- Aneurysmal bone cyst
- Chondrosarcoma
- Osteosarcoma

Children/adolescents
- Ewing sarcoma (primitive tumor)
- Langerhans cell histiocytosis
- Neuroblastoma
- Sarcoma
- Leukemia

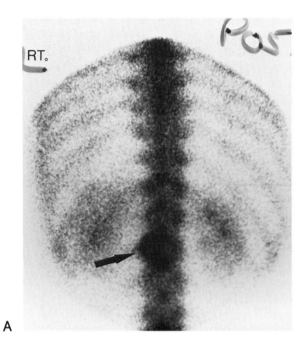

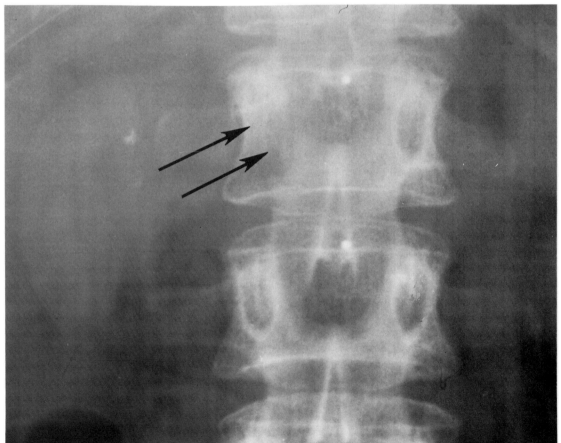

Fig. 14-54 Metastatic tumor to the spine. (A) Radionuclide bone scan shows uptake of label in the right L2 vertebral body, representing bone reaction to a metastatic deposit *(arrow)*. **(B)** Lytic metastasis. The "absent pedicle sign" indicates lytic involvement of the right pedicle and proximal lamina of L1 *(arrows)*.

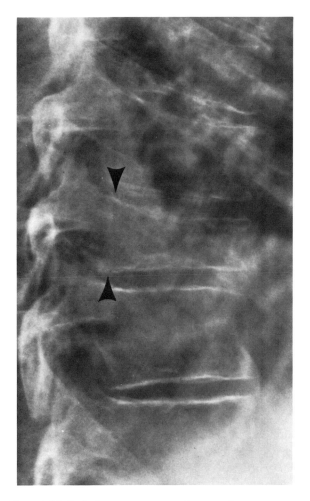

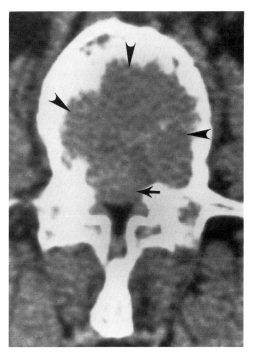

Fig. 14-56 **Lytic metastasis, CT scan.** Transaxial CT scan shows a large destructive tumor within the vertebral body *(arrowheads)*. It has broken through the cortex posteriorly into the epidural space *(arrow)*. Large, purely lytic lesions are most commonly metastases from lung or kidney, or they are myeloma. The radionuclide scan and plain film radiography could be normal with a tumor such as this one.

Fig. 14-55 **Pathologic compression fracture.** Lytic metastasis involves a thoracic vertebral body, resulting in a compression fracture. Note the loss of the height of the posterior portion *(arrowheads)*. This point differentiates the pathologic fracture from a post-traumatic wedge compression fracture (see Fig. 15-23).

produce a complex inhomogeneous pattern. These lesions most commonly occur with metastasis from breast carcinoma.

CT scans with bone windows are far more sensitive for detecting the presence of tumor than plain films and may demonstrate metastatic destruction when the radionuclide bone scan is negative (Fig. 14-56). The principles are the same as for plain film radiography. Epidural extension of tumor may be seen, especially with myelography or CT myelography (Fig. 14-57). Percutaneous biopsy can be done with CT guidance.

The MRI technique is sensitive to the presence of metastatic tumor and defines epidural growth and cord compression. Metastatic tumor causes prolongation of both T_1 and T_2 relaxation times. Tumor deposits appear relatively hypointense on T_1-weighted images (Figs. 14-58 to 14-60) and hyperintense on T_2-weighted images. Replacement of the high-intensity marrow fat contributes to the visibility of the deposits on T_1-weighted images, making this sequence the most useful (Fig. 14-61). GRE T_2*-weighted images may show the tumors as high signal intensity (Fig. 14-59C). The use of fat suppression sequences shows the tumors within the marrow as high signal within a dark background. With SE T_1-weighted imaging, Gd enhancement is generally not useful for defining bone metastasis; in fact, it may enhance tumor deposits slightly, so they become isointense with the marrow and therefore undetectable. With fat-suppression techniques, Gd enhancement makes the tumor mass more visible. MRI has the advantage of permitting

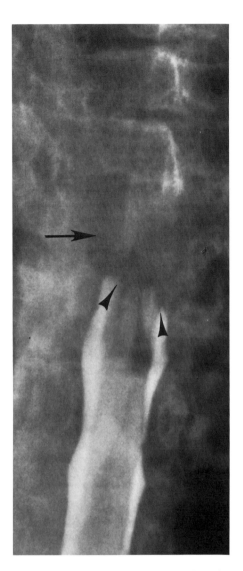

Fig. 14-57 Metastatic tumor to vertebral body with epidural extension. Myelography shows a complete block to the cephalad flow of contrast (*arrowheads*) by an epidural mass. This appearance is typical of an epidural block seen en face. Note the bone destruction of the vertebral body and posterior neural arch (*arrow*).

examination of the entire spine and epidural space at one time, without manipulation of the patient.

Acute post-traumatic compression fracture produces acute MRI signal changes within the marrow, simulating marrow tumor. Edema within the marrow from the injury lowers the normally high T_1 signal within the marrow cavity (Fig. 14-62A) and also creates increased T_2 signal.

The pattern of the compression and the lack of bone destruction on CT may make the correct diagnosis. However, it may be impossible to reliably exclude tumor during the acute phase. Stress fractures of the spine may create unusual patterns on plain film radiographs and bone scanning (Fig. 14-62B & C).

Postirradiation Change in the Spine and Cord

Radiation kills marrow cells, but not fat. The extent of the cellular depletion is dependent on the amount of radiation exposure. Low doses demonstrate little change. With higher doses, of greater than 30 Gy, there is no visible change in the vertebral column within the first 2 weeks. During the 3- to 6-week period, a heterogeneous mottled appearance develops with scattered regions of high T_1 signal intensity. This results from the loss of hematopoietic marrow cells, leaving the high T_1 fat signal unopposed. After 6 weeks, there is uniform high T_1 signal within the marrow cavities (Fig. 14-62D & E). If the amount of radiation delivered to the spine is greater than 50 Gy, the conversion to high T_1 signal intensity is permanent. Otherwise, there is a gradual reversion to normal over the next year or two. In children, radiation will cause growth retardation in the radiated field.

Radiation damage to the spinal cord occurs, especially if the dose to the cord exceeds 70 Gy. This may happen with treatment of head and neck carcinoma. During the 18 months after therapy MRI may show an enlarged cord with high T_2 signal within the treatment field. Contrast enhancement is variable. After 18 months, the cord shows atrophy and no signal abnormalities.

Bone Marrow Disease

Bone marrow is the source of blood production by birth. The composition of the marrow depends on the hematopoietic activity and the age of the person. In infants, the marrow is very cellular with a low fat content. By adolescence, the marrow contains 40 percent fat and 40 percent water. At age 75, the marrow contains about 60 percent fat and 30 percent water. In the newborn, the marrow exhibits a T_1 signal that is isointense with muscle. The fat content of the marrow gradually increases so that at middle age, the marrow is hyperintense to muscle. Mottling occurs with islands of very cellular marrow or islands of fat; this may become quite prominent in those over 75 years of age, making it difficult to define metas-

(*Text continued on page 620*)

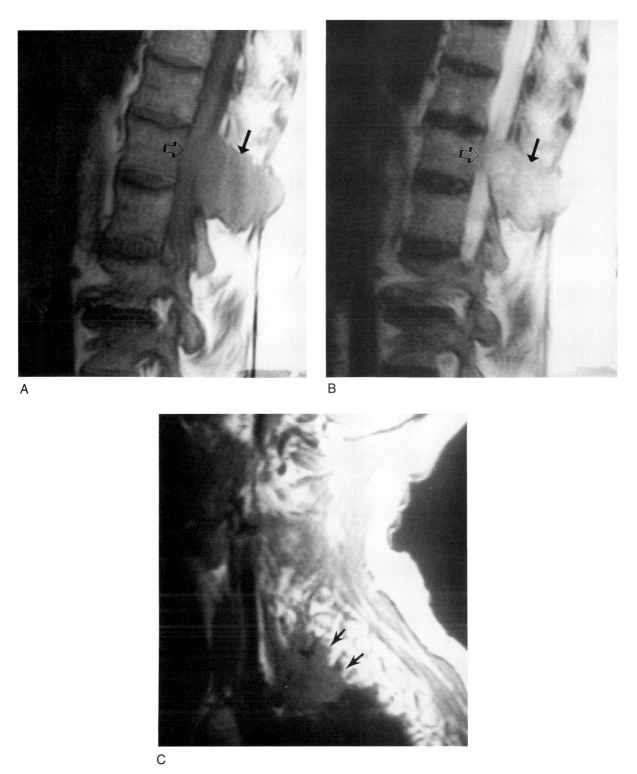

Fig. 14-58 Metastatic spine and paraspinal tumor. (A) Sagittal T$_1$ MRI shows a metastatic breast tumor deposit expanding out of the spinous process *(black arrow)*, to cause spinal cord compression *(open arrow)*. The tumor is slightly hypodense compared with normal marrow. **(B)** Sagittal T$_2$ MRI at the same level shows the tumor tissue as a bright signal mass *(black arrow)*. The tumor extends into the spinal canal *(open arrow)*. **(C)** Pancoast-type primary lung tumor extends from the right apex into the lower right paraspinal cervical tissue *(arrows)*. The tumor replaces the fat tissue normally seen in this region. Clinically, this tumor may mimic cervical spondylosis with radiculopathy.

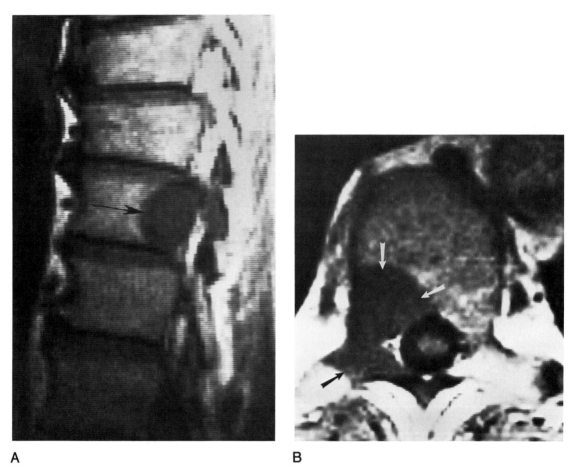

A

B

Fig. 14-59 Metastatic tumor, lung carcinoma, MRI. (A) T_1-weighted parasagittal MRI shows the tumor mass as a relative hypointensity replacing the relatively hyperintense fatty marrow *(arrow)*. **(B)** Transaxial T_1-weighted study shows the tumor in the posterior vertebral body *(white arrows)* extending into the right pedicle *(black arrow)*. *(Figure continues.)*

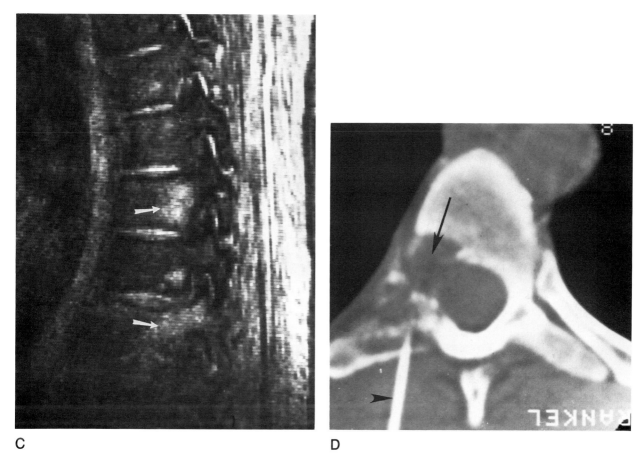

C

D

Fig. 14-59 *(Continued).* **(C)** Parasagittal GRE flash 20 450/10 shows metastatic deposit as high signal intensity *(arrows).* **(D)** CT scan shows the lytic destruction of the right pedicle and transverse process *(arrow).* A percutaneous biopsy needle is seen entering posteriorly *(arrowhead).*

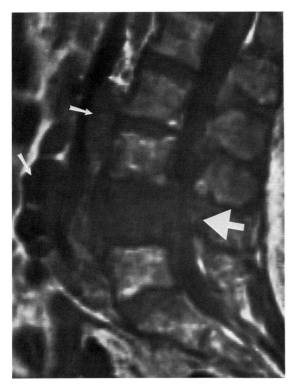

Fig. 14-60 Lymphoma, MRI. There is loss of the usual high signal intensity within the L4 vertebral body, representing involvement with lymphoma. Lymphoma is within the spinal canal as well *(large arrow)*. Note the lymphadenopathy in the prespinal region *(small arrows)*. Patchy loss of signal in other vertebral levels indicates diffuse involvement.

MARROW DISEASE

Decreased T_1 MRI marrow signal
    Metastasis (variable increase in T_2 signal)
    Myeloma
    Lymphoma (tumor has increased T_2 signal)
    Polycythemia vera
    Myelofibrosis
        Metastatic tumor
        Gaucher's disease
        Preleukemia
        Infection
        Idiopathic
    Leukemic infiltration (has increased T_2 signal in addition)
    Sickle cell anemia (marrow reconversion)
Increased marrow T_1 MRI signal
    Aplastic anemia
    Radiation effect

tasis or marrow infiltration. The signal intensity of the marrow with marrow disease processes is given in the box.

Myeloma

Myeloma, a malignant tumor of marrow plasma cells, most often presents at multiple sites, but may occur as a solitary lesion. It is most common in the 50- to 70-year-age range and is almost never seen before the age of 40. It most frequently begins in the axial skeleton. Because it is a diffuse disease, marrow aspiration is often positive for myeloma cells even at sites at which the bone appears normal. A serum M-component immunoglobin is present in more than 90 percent of patients.

Myeloma produces purely osteolytic lesions within the vertebral bodies. It is infrequent in the posterior neural arch of the spine, a helpful differential point for distinguishing myeloma from osteolytic metastatic carcinoma. The early bone changes mimic osteoporosis or hemangioma (Fig. 14-63). The tumor may break out of the confines of the vertebral body, producing a paraspinal or epidural mass. Because bone destruction is complete, vertebral collapse is common.

Hemangioma

Hemangioma is a common benign tumor of capillaries or other blood vessels. It occurs within the spine in up to 10 percent of persons, most frequently at the thoracolumbar junction. Almost all are within the vertebral body, where the collection of abnormal vessels replace the marrow and cancellous bone. The remaining trabeculae thicken and are imaged as thick striations running through the lesion. With CT scanning, the trabeculae are seen on end as multiple rounded bone densities within the marrow cavity. This pattern occurs rarely with myeloma and metastasis. Most hemangiomas have high signal intensity on both T_1- and T_2-weighted MRI (Fig. 14-64). Those with recent fracture of the vertebral body may have low signal on T_1-weighted images. A large soft

(Text continued on page 624)

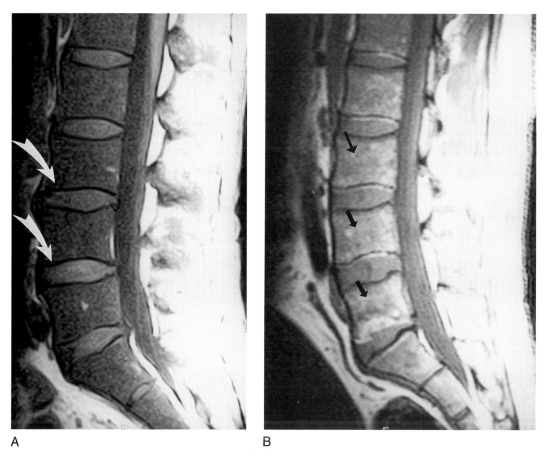

Fig. 14-61 Diffuse marrow infiltration with tumor. (A) Sagittal T$_1$ MRI shows a general decrease in the normally high signal intensity within the vertebral marrow *(arrows)*. The intensity of the marrow is less than that of the disc interspaces. This is leukemia. **(B)** Sagittal T$_1$ MRI shows an earlier stage of marrow infiltration, producing a mottled pattern within the vertebral bodies *(arrows)*. The darker regions represent the tumor within the marrow cavity. This is lymphoma.

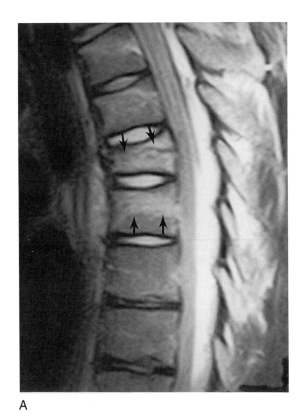

A

B

Fig. 14-62 Fractures, compression, and stress types. Post-irradiation effects. (A) Acute compression fracture, no tumor. Sagittal T$_2$ MRI shows bright signal within the marrow representing edema following compression fractures (*arrows*). (B) Stress fracture. Axial CT scan through the sacrum shows hypertrophic bone reaction around a stress fracture through the left sacrum (*arrows*). (*Figure continues.*)

Fig. 14-62 *(Continued)*. **(C)** Radionuclide bone scan shows high radioactivity in a characteristic butterfly pattern within the upper sacrum *(arrows)*. **(D)** Postirradiation effect. Sagittal T₁ MRI shows increased homogeneous signal intensity within the vertebral marrow at the segments exposed to radiotherapy *(white arrows)*. **(E)** T₁ sagittal lumbar spine examination in another patient shows more subtle increase in the marrow signal intensity of L5 and the sacrum *(black arrows)*. L3 and L4 vertebral bodies were outside of the field and remain normal *(white arrows)*. There is a small central disc herniation (protrusion) at L5–S1 *(curved white arrow)*.

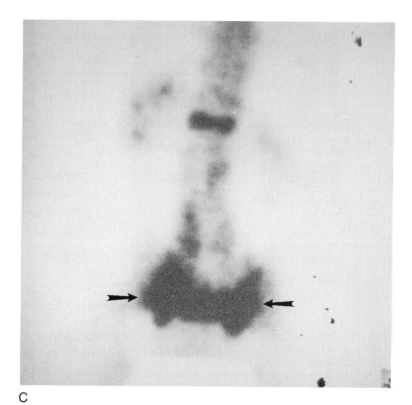

C

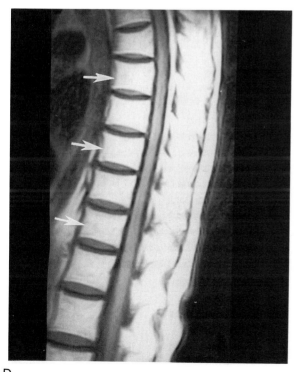

D

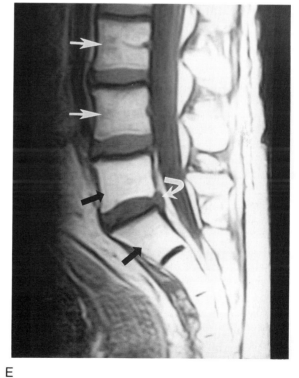

E

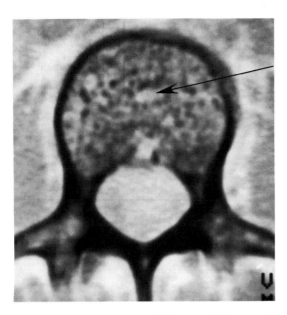

Fig. 14-63 Myeloma. Transaxial CT scan with bone windows show enlargement of the marrow spaces of the vertebral body without focal bone destruction (*arrow*). This change is typical of early myeloma and is difficult to differentiate from osteoporosis. The pedicles and transverse processes are not involved.

tissue component may grow out of the confines of the vertebral body, invading the epidural space and causing neural compression similar to metastatic tumor. Vertebral collapse may occur.

Hemangiopericytoma

Hemangiopericytoma is a malignant vascular tumor originating from the pericytes. Lytic destruction is the most common finding with the soft tissue signal characteristics indistinguishable from metastatic carcinoma. The tumor occurs in middle aged and older adults.

Langerhans Cell Histiocytosis

Langerhans cell histiocytosis (formerly called histiocytosis X) takes many forms. Eosinophilic granuloma is the most benign form, consisting of a neoplastic granulomatous proliferation of reticulum cells. It is a lesion of childhood and young adulthood. The lesions are lytic and may be solitary or multiple, sometimes superimposed within the same vertebral body, producing a "bone-within-a-bone" appearance on radiographs. Calve's vertebra plana in a child represents an isolated complete collapse of a vertebral body, resembling a dense disc of

bone that is almost always caused by eosinophilic granuloma. Letterer-Siwe disease is the fulminant fatal form of the disease, disseminated through the bones, liver, spleen, and lymph nodes, in infants under 3 years of age. Hand-Schüller-Christian disease is a chronic dissemi-

CAUSES OF VERTEBRA PLANA

Langerhans cell histiocytosis
Osteosarcoma
Giant cell tumor
Ewing sarcoma
Metastasis
Hemangioma
Osteomyelitis
Neuroblastoma
Idiopathic osteonecrosis
Leukemia

TUMORS OF THE VERTEBRAL COLUMN BY LOCATION

Vertebral body
  Metastasis
  Myeloma
  Hemangioma
  Giant cell tumor
  Chordoma
  Osteosarcoma
  Ewing sarcoma
  Chondrosarcoma
  Hemangiopericytoma
  Bone marrow diseases
  Aneurysmal bone cyst (rare)
  Langerhans cell histiocytosis
Posterior neural arch
  Metastasis
  Osteoid osteoma
  Osteoblastoma
  Osteochondroma
  Aneurysmal bone cyst

Fig. 14-64 Hemangioma, MRI. (A) Parasagittal T_1-weighted MRI shows a hemangioma of the vertebral body as hyperintense *(arrow)*. Hemangiomas are also hyperintense on T_2-weighted images. **(B)** A hemangioma may act as an aggressive tumor, expanding outward from the vertebral body to produce cord compression *(arrows)*. This hemangioma also involves the posterior neural arch *(open arrow)*. The signal intensity of an aggressive hemangioma is low on T_1, similar to metastatic carcinoma, and is usually only slightly bright on T_2-weighted sequences. **(C)** Axial T_1 MRI shows trabeculae remaining within the vertebral mass, a feature that distinguishes hemangioma from metastatic tumor *(arrowheads)*. Note the constricted spinal canal *(open arrow)* and the paraspinal mass *(black arrows)*.

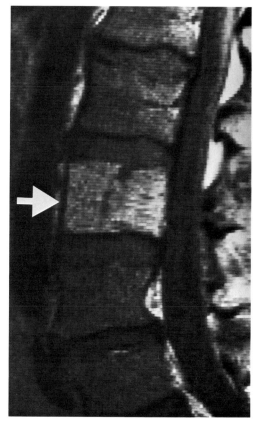

A

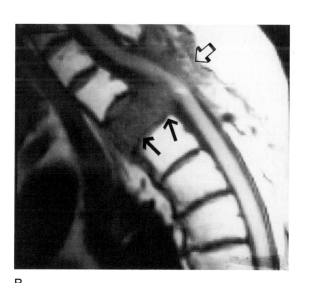

B

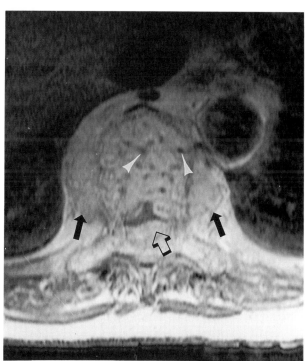

C

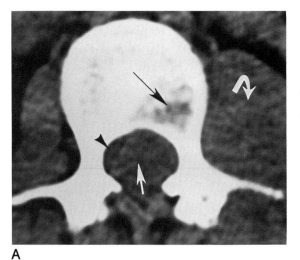

A

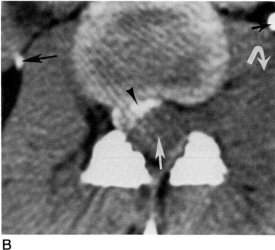

B

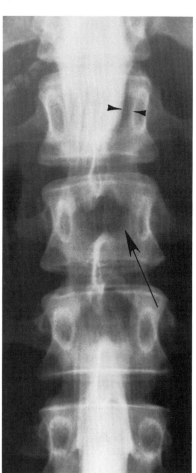

C

Fig. 14-65 Ewing sarcoma. (A) Routine transaxial CT scan shows a small destructive lesion in the left side of the vertebral body *(black arrow)*. A large paraspinal mass associated with it *(curved arrow)* extends into the epidural space *(white arrow)*, compressing the hypodense dural tube *(arrowhead)*. **(B)** CT myelography proves the epidural extension of tumor *(white arrow)* severely compressing the contrast-filled dural tube *(arrowhead)*. The large left parasagittal mass is seen *(curved arrow)*. Note the excretion of the myelographic contrast within the ureters *(black arrows)*. **(C)** Myelography shows a typical large epidural mass. The dural tube is severely compressed over two segments *(arrow)*. The dural tube is displaced away from the pedicle at the superior end *(arrowheads)*, proving the epidural location of the tumor mass.

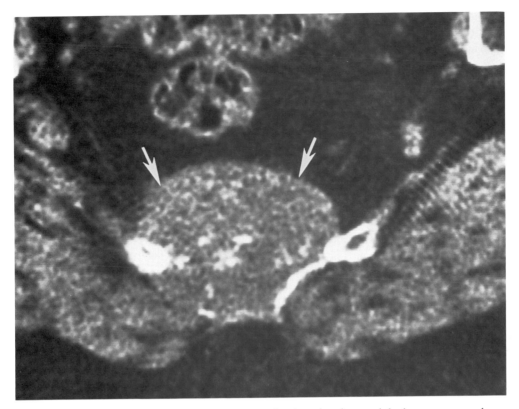

Fig. 14-66 Chordoma, CT scan. A large mass representing chordoma has destroyed the lower sacrum and coccyx and extends anteriorly into the pelvis (*arrows*).

nated form that causes lytic lesions of the skull and other bones, as well as involvement of the liver, skin, and lymph nodes.

Ewing Sarcoma

Ewing sarcoma is a highly malignant undifferentiated small cell tumor, most likely of neuroectodermal origin (primitive neuroectodermal tumor), that occurs rarely in children and young adults. Large soft tissue masses develop, often associated with only a small lytic lesion within the bone (Fig. 14-65). Mixed lytic and blastic lesions may also be found. The tumor does not calcify and is low signal on T_1 and high signal on T_2-weighted images. It enhances with Gd. The bone scan is positive, and the tumor accumulates ^{67}Ga-citrate radionuclide.

Chordoma

Chordoma is a malignant tumor arising from the fetal notochord. It is found most commonly in men aged 50 to 70 years. Eighty-five percent occur within the sacrum.

The spheno-occipital bone at the upper clivus, the cervical spine, and the petrous bone are other sites. The tumor most commonly is midline, but it may be eccentric or may arise lateral to the axis in rare cases.

These tumors are slow growing and may be large at diagnosis. They cause lytic destruction with a large exophytic tumor growth (Fig. 14-66). Calcification is common within the tumor, particularly those from the clivus, and there may be a thin peripheral rim of calcium. Biopsy may be the only means of differentiating the tumor from metastasis or giant cell tumor. Sacral tumors cause pain and bladder dysfunction from sacral neuropathy.

Giant Cell Tumor

Giant cell tumors arise from the stromal elements of the marrow cavity. Ninety percent are benign when first diagnosed, but up to 30 percent will become malignant, especially if radiotherapy is given for control of residual tumor. They occur overwhelmingly within the vertebral bodies of the sacrum in young female adults, and characteristically become large. In older persons, they may occur along with polyostotic Paget's disease. They are rare in other segments of the spine. The bone is destroyed from the large soft tissue mass, which most commonly has a thin peripheral rim of calcium. The tumor has low to intermediate signal on T_1 and intermediate signal on T_2-weighted MRI. Inhomogeneity within the tumor is common.

Enostosis

Enostosis, or bone island, is a focus of compact bone found deep to the cortex within cancellous bone. It is most commonly found within the spine and the skull. The lesion is variable and, although usually small and about 1 cm, it may become quite large. The lesion may be difficult to differentiate from a solitary blastic metastasis, as the margins of the bone island may be indistinct, and there may be slight growth over time. However, the lesion does not accumulate radionuclide tracer and does not cause symptoms.

Osteoid Osteoma

Osteoid osteoma is a small benign lesion of bone consisting of a 1- to 5-mm nidus of vascularized osteoid that may or not be calcified. The nidus is surrounded by a dense reaction of bone sclerosis. Deformity of the spine may result with local remodeling and scoliosis. A soft tissue mass does not occur. It is seen almost exclusively within the posterior neural arch, most often in the cortex of a facet joint, with sclerosis spreading into the pars interarticularis, lamina, and pedicle. The peak incidence is at 10 to 25 years of age, with a 2 : 1 male predominance.

Characteristically, the lesion is painful, although the pain may be well relieved by aspirin. The nidus needs to be removed for control of pain. Plain films, tomography, CT scan, and radionuclide scan may be used to locate and identify the lesion. Radiographic verification of removal of the nonossified nidus is desirable.

Osteoblastoma

An osteoblastoma resembles the tissue of the nidus of an osteoid osteoma, yet the tumor is much more aggressive and progressively enlarges. It is found almost exclusively in the posterior neural arch of persons aged 10 to 20 years, with a 2 : 1 male predominance. It is an expansile lesion, with the soft tissue component often breaking

through the cortical bone, producing a paraspinal soft tissue mass. Speckled calcification within the osteoid matrix is the rule, often associated with a thin rim of peripheral calcification. The lesion may resemble osteosarcoma, chondrosarcoma, or aneurysmal bone cyst. CT scan is the best imaging modality. On T_2-weighted MRI, considerable edema is seen extending into adjacent soft tissue and bone. The soft tissue mass may extend into the epidural space and cause neural compression. Scoliosis is common.

Aneurysmal Bone Cyst

An aneurysmal bone cyst is a benign expansile lesion most commonly found in the posterior neural arch consisting of blood-filled spaces and osteoclasts. It occurs in children and young adults 5 to 25 years of age. It is possibly a reaction to trauma or hemorrhage, often associated with other tumors of the spine, such as giant cell tumor, osteoblastoma, and osteosarcoma. This may create a confusing imaging picture. The lesion may grow rapidly, causing neural compression. Trabeculae may cross the central portion of the cyst. They have a margin of a thin shell of bone or unossified periosteum. Fresh or old blood products may be present within the lesion, creating inhomogeneity on MRI. Fluid levels are common, but they are not diagnostic, and levels may also be sometimes seen with osteosarcoma and giant cell tumors.

Chondrosarcoma

Chondrosarcoma is a malignant tumor that occurs within a vertebral body that produces cartilage. It occurs in the 30- to 50-year age range, growing slowly with the potential to metastasize. They cause osteolysis with punctate calcification within the tumor matrix. The lesion usually does not cross a disc space. When it occurs in the sacrum, it may be impossible to differentiate from chordoma. The tumor is radioresistant.

Osteochondroma

An osteochondroma represents local dysplasia of the cartilage at an epiphyseal growth plate, usually within a pedicle. It generally becomes symptomatic during adolescence. There may be spinal cord compression, deformity of the spine, and, rarely, malignant degeneration. The cortex of the lesion is continuous with the cortex of the neural arch. CT scanning is the best method to identify this rare lesion.

Lipoma

Lipomas arise frequently within the marrow spaces of the vertebral bodies. They are not often visible with CT scan but are obvious on MRI. Most commonly, they appear as round regions of high intensity within a vertebral body on T_1-weighted images. The signal drops off on the SE T_2-weighted images but remains high on FSE T_2-weighted images (Fig. 14-67). The lesions are inconsequential.

EXTRADURAL TUMORS AND MASS LESIONS

Extradural tumors are the most common type of tumor found within the spinal canal. The extradural nature of the mass can be defined by the displacement of the dural tube away from the margins of the spinal canal. This characteristic displacement is seen with myelography (Fig. 14-65C), CT myelography, and MRI. When a lesion is large, or when there is complete "block," it may be impossible to categorize its precise origin. Most neoplastic masses in the extradural space are malignant.

Metastasis

Metastasis is by far the most common epidural tumor encountered. It almost always results from exophytic growth of vertebral metastasis into the epidural space. Hematogenous epidural metastasis is rare. This subject was discussed in a preceding section about tumors of the vertebral column.

Lymphoma

Lymphoma is a relatively common extradural tumor most commonly found in the lumbar region. It may be primary within the epidural space in the spinal canal (Fig. 14-68), or it may spread into the epidural space from a retroperitoneal focus (Figs. 14-60 and 14-69) or from an adjacent vertebral body. The adjacent bone is sometimes secondarily involved as a result of spread from a primary epidural focus. By the time of recognition, the tumor is usually widespread in the retroperitoneum, intravertebral foramina, and the spinal canal; it may extend over several levels. On CT and MRI, the tumor is seen as a homogeneous mass displacing the epidural and paraspinal fat. It is slightly hypointense on T_1 and slightly hyperintense of T_2-weighted images. Percutaneous biopsy may make the diagnosis. Core biopsy rather than cytology may be necessary for complete diagnosis, especially because some lymphomas are fibrotic and yield few cells with simple aspiration.

Neuroblastoma and Ganglioneuroma

Neuroblastoma and ganglioneuroma are generally tumors of children and young adults. These tumors arise from either the adrenal gland or the sympathetic paraspinal neural chain that extends from the base of the skull to the pelvis. They may present in later adulthood. Direct intraspinal spread of tumor, which occurs in about 20 percent of cases, is best recognized with CT myelography or MRI (Fig. 14-70). Widening of the intervertebral foramen is common, and there may be local bone destruction. Calcification within the tumor is common as

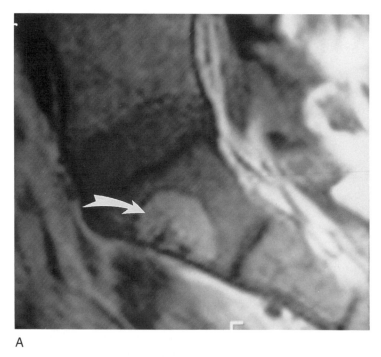

A

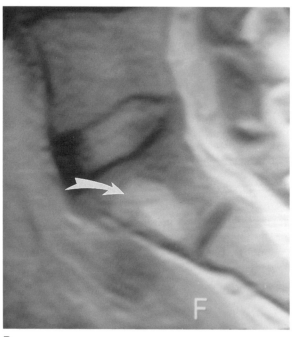

B

Fig. 14-67 Lipoma. (A) Sagittal T_1 MRI shows the lipoma as a well-delineated high signal region within the marrow cavity of the S1 vertebral body *(arrow)*. **(B)** Sagittal FSE T_2 MRI shows the lipoma with bright signal *(arrow)*. With SE T_2 MRI, the lipoma is only slightly hyperintense to the marrow space. The lack of internal trabecular pattern differentiates this from an intraosseous hemangioma.

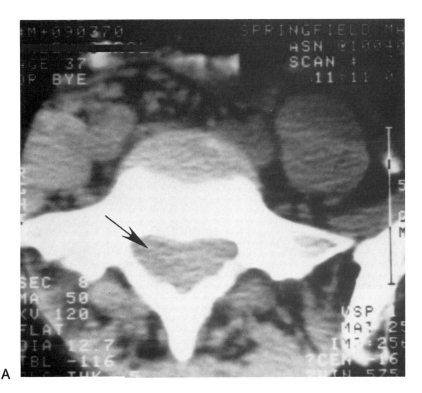

A

Fig. 14-68 Lymphoma. (A) Transaxial CT scan
shows soft tissue density obliterating the epidural fat
of the spinal canal *(arrow)*. It represents primary epidural lymphoma. **(B)** Lumbar myelography shows the
epidural mass *(arrow)*. It could be misinterpreted as a
large disc herniation. However, note the normal disc
interspace *(arrowheads)*, which would be unusual in
association with such a large herniation.

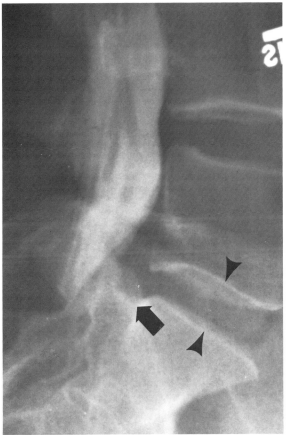

B

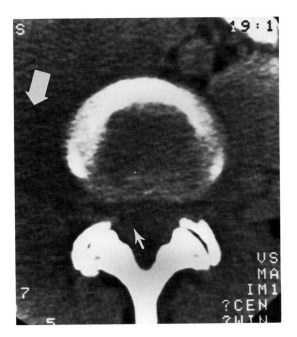

Fig. 14-69 Burkitt's lymphoma, AIDS. Transaxial CT scan shows a large extranodal mass of lymphoma in the paraspinal region on the right (*large arrow*). It extends through the intervertebral foramen into the epidural space of the spinal canal (*small arrow*).

Fat is normally present within the epidural space, and it is not always clear when the amount of fat is too great and pathologic.

Abnormal compression of the dural tube is diagnosed when the subarachnoid space is compressed to obliteration, as seen with T_2-weighted MRI or CT myelography. The fat may compress the dural tube enough to create a complete block with myelography. Epidural lipomatosis occurs most commonly following steroid therapy or in association with Cushing's disease or syndrome, but it also may occur on an idiopathic basis. Surgical removal of the tissue relieves the compression. The fat regresses with removal of steroid excess.

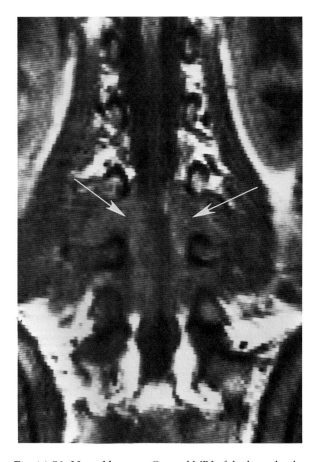

Fig. 14-70 Neuroblastoma. Coronal MRI of the lower lumbar region shows retroperitoneal neuroblastoma extending through the intervertebral foramina into the epidural space bilaterally (*arrows*). Normal epidural space and fat within the intervertebral foramina are seen at superior levels.

well (Fig. 14-71). The epidural spread of tumor may extend a few segments away from the primary paravertebral mass. It is important to recognize intraspinal spread before attempting surgical excision of these tumors. Often it is not possible to differentiate benign from malignant tumors reliably without biopsy or excision. Core biopsy is usually necessary to diagnose these tumors by percutaneous technique.

Epidural Lipomatosis

Benign epidural lipomatosis is a condition of an increased amount of fat tissue within the epidural space. It is most often localized to the mid-thoracic region, but may occur in the lumbar region as well (Figs. 14-72 and 14-73). The accumulation of fat is mostly in the posterior spinal canal. It may be great enough to cause compression of the dural tube and neural tissue, producing symptoms of cord compression or the cauda equina syndrome.

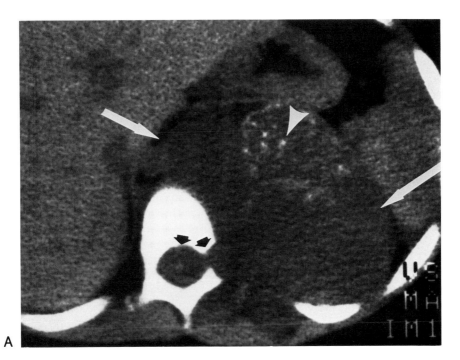

A

Fig. 14-71 Ganglioneuroma. (A) Transaxial CT scan shows a large relatively hypodense mass in a left paraspinal region *(white arrows)*. The mass has enlarged the left intervertebral canal and extends into the epidural space *(black arrows)*. Calcification is present within the tumor *(arrowhead)*. The tumor has locally invaded through the posterior chest wall. (B) AP myelography shows characteristic findings of an epidural mass. The dura is displaced away from the pedicle *(curved arrow)*. The subarachnoid space tapers to obliteration at the maximal point of the mass *(black arrow)*. A left intervertebral foramen is widened *(opposed arrows)*, and there is inferior erosion of a pedicle *(white arrow)*.

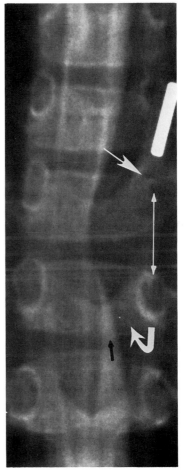

B

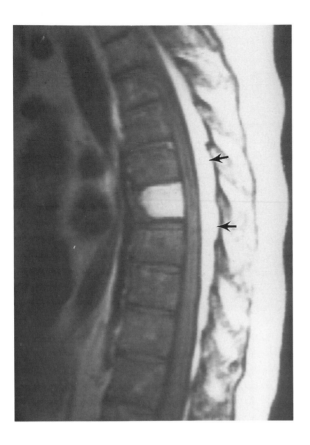

Fig. 14-72 Epidural lipomatosis, thoracic region. Sagittal T_1 MRI shows a large deposit of high-signal fat within the posterior spinal epidural space (*arrows*). This deposit is large enough to compress the cord anteriorly against the vertebral bodies, producing a thoracic myelopathy. The bright lesion in the thoracic vertebral body is thought to be a hemangioma.

Infection

Most commonly, an epidural inflammatory mass (epidural empyema or cellulitis) occurs spontaneously from hematogenous spread, but it may be from direct extension from vertebral osteomyelitis or retropharyngeal abscess. Acute inflammatory abscess has signal and density characteristics of a high protein fluid collection. In this situation it is difficult to differentiate the mass from a hyperacute hematoma. When the inflammation is more phlegmonous (indurated tissue) or granulomatous, the inflammatory tissue may mimic tumor, particularly lymphoma. Gallium or tagged white cell radionuclide studies may help make the correct diagnosis. MRI without and with Gd contrast enhancement is the best initial examination. Epidural infection is discussed in greater detail in the subsequent section on inflammation of the spine.

Hematoma

Spontaneous hematoma can occur in the epidural space, particularly in elderly persons who are taking anticoagulants of any type, including aspirin. Occasionally, an epidural hemorrhage occurs in younger persons who have no apparent predisposition. The mass of blood may be large, causing cord compression, but more often it simply causes pain in a dermatome distribution. The blood changes on MRI generally follow the same pattern described for intracerebral hematoma, although the changes may be slightly delayed (see Ch. 2).

The typical findings of an epidural mass are seen with myelography. CT scan usually demonstrates a nonspecific mass with density slightly greater than that of the dural sac (Fig. 14-74). In the acute phase, MRI shows the signal intensity of a high protein fluid but high signal intensity blood on T_1- and T_2-weighted images in later subacute and chronic phases (Fig. 15-36).

INTRADURAL EXTRAMEDULLARY LESIONS

Intradural extramedullary lesions are mostly benign and constitute the second most common location for tumors within the spinal canal. In adults the tumor is

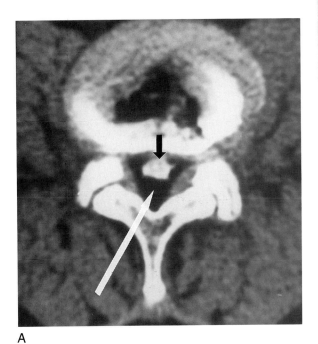 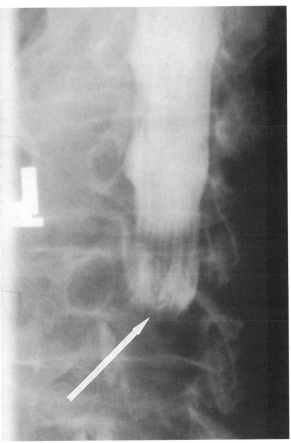

A   B

Fig. 14-73 Epidural lipomatosis, epidural compression of the cauda equina. (A) Axial CT myelogram identifies the excessive amount of fat within the lumbar epidural space *(white arrow)*. This constricts the dural tube and compresses the cauda equina *(black arrow)*. **(B)** Myelogram indicates a complete block to the caudal flow of intrathecal contrast *(arrow)*. The inferior edge of the contrast column has a "feather" appearance, a finding typically produced by the roots of the cauda equina with compression from an epidural mass.

Fig. 14-74 Hematoma, CT myelography. An epidural hematoma occurred spontaneously in this patient. A nonspecific epidural mass is seen *(long arrow)* compressing the dural tube, which is filled with contrast *(short arrow)*.

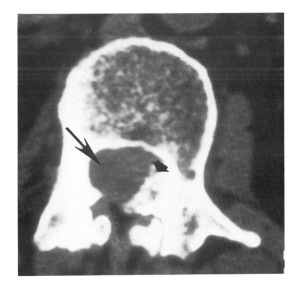

IMAGING SIGNS OF INTRADURAL EXTRAMEDULLARY LESIONS

"Capping" defect with sharp margin to lesion

Displacement of the cord away from the dural margin

Widening of the subarachnoid space on the side of the lesion

No bone abnormality (except with "dumbbell" lesions)

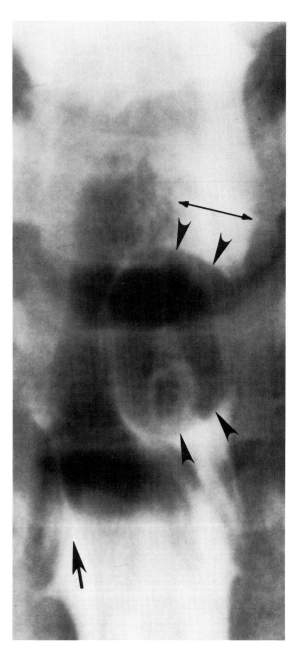

Fig. 14-75 **Neurinoma, myelography.** A large intradural neurinoma is outlined by the myelographic contrast material. A cap defect is seen *(arrowheads)*. The subarachnoid space is widened on the side of the tumor *(opposed arrows)*, and the cord is displaced away from the lesion, narrowing the contralateral subarachnoid space *(arrow)*. These abnormalities are the "classic" findings of an intradural extramedullary lesion.

almost always either a meningioma or a neurofibroma. Intradural metastases are uncommon. In children, intradural lesions are usually congenital, such as a lipoma or epidermoid inclusion, but intradural "drop metastases" occur with specific intracranial tumors, particularly medulloblastoma, ependymoma, and germ cell tumors.

The intradural extramedullary location of the lesion is determined by myelographic or MRI-aided detection of displacement of the spinal cord away from the margins of the dural tube. The lateral margins of the dura remain in close relation with the spinal canal. Often "capping"-type defect is seen with myelography, still the most accurate method of diagnosing these lesions. Some tumors have an extradural component as well. Gd shows enhancement of many of the lesions, increasing their visibility with MRI.

Neurinoma (Schwannoma, "Neurofibroma")

Neurinoma may also be correctly termed a schwannoma, as it arises from the sheath cells of the posterior nerve roots. It is seen most commonly in the 20- to 40-year age group. There is no sex predilection, and the tumor may occur at any site along the neuraxis, except the C1 level. It may be familial. Multiple neurinomas occur with neurofibromatosis.

The lesions produce the characteristic intradural extramedullary type of appearance on plain film myelography. A "cap" defect is almost always seen within the subarachnoid space, and the cord is usually displaced (Fig. 14-75). The lesions tend to grow out of the dura through the nerve root sleeves, entering the lateral

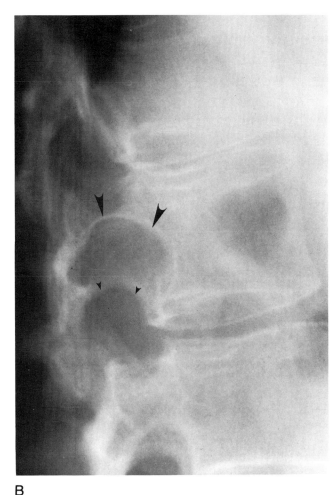

A   B

Fig. 14-76 "Dumbbell" neurinoma. (A) Contrast-enhanced CT scan shows a homogeneously enhancing dumbbell neurinoma extending from the intradural space through the intervertebral foramen into the right paraspinal space *(long arrow)*. The enhanced tumor is hyperdense to the dural tube and spinal cord *(open arrow)*. The intervertebral foramen is widened *(opposed arrows)*. (B) Lateral spine radiograph shows enlargement of the intervertebral foramen on the side of the tumor *(large arrowheads)*. A normal intervertebral foramen is seen on the opposite side *(small arrowheads)*.

neural foramen and enlarging it ("dumbbell tumors") (Figs. 14-76 and 14-77). The resulting paraspinal mass can become enlarged. Calcification does not occur. The tumors are vascular and show hypervascular "blushing" with angiography and contrast enhancement with MRI and CT scanning (Fig. 14-76A). MRI easily identifies these tumors and differentiates them from cystic lesions. They are generally isointense or hypointense to the spinal cord on the T_1-weighted images and moderately hyperintense on T_2-weighted images (Fig. 14-78). Rarely, neurofibromas may be predominantly cystic.

When the tumors are large or multiple within the spi-

nal canal, they may cause scalloping of the posterior margins of the vertebral bodies and separation or thinning of the pedicles. Most commonly, neurinomas cause a radicular syndrome, but if large enough, the spinal cord can become compressed with the production of a Brown-Sequard syndrome.

Meningioma

Meningioma is a benign tumor that arises from the meninges of the dura. It occurs most commonly after the age of 40 in the thoracic region and almost always from

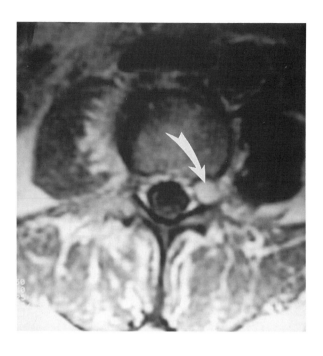

Fig. 14-77 Neurofibroma. Axial T_1 Gd-enhanced MRI through the lumbar spine shows a round enhancing mass within the left intervertebral foramen (*arrow*). This represented a neurofibroma arising near the dorsal root ganglion. The contrast enhancement differentiates neurofibroma from a perineural cyst.

the posteriolateral dura. Other common sites are the foramen magnum and the cervical region. There is a 4 : 1 female predominance. The lesion may calcify, differentiating it from neurofibroma. A rare calcified anterior intraspinal meningioma may be difficult or impossible to differentiate from the much more common calcified herniated thoracic disc. Meningiomas rarely grow out of the dural sac and do not produce significant paraspinal mass. Except perhaps at the foramen magnum, associated bone sclerosis does not occur with spinal meningioma.

With myelography, the typical intradural extramedullary type of defect is seen. The characteristic location, age, and sex of the patient suggest the correct tissue diagnosis. A meningioma is difficult to demonstrate with plain CT scanning, but it is easily seen with CT myelography. The lesion enhances with intravenous contrast, differentiating it from a fluid-filled cyst or meningocele.

The lesions are easily seen by MRI, which is the preferred method of diagnosis. The tumor is isointense to the cord on T_1-weighted images (Fig. 14-79) and slightly hyperintense on T_2-weighted images. It is seen as a negative defect within the CSF on the T_2-weighted images. Meningioma is hyperintense after Gd (Fig. 14-80).

Metastasis

Metastasis may occur within the subarachnoid space. It is most likely to result from lymphoma, leukemia, breast carcinoma, or melanoma. Drop metastases from primary intracranial tumors are more common in children, especially from a medulloblastoma, ependymoma, neuroblastoma, or germ cell tumor of the pineal region. On myelography and CT myelography, the tumor deposits appear as rounded filling defects within the contrast column (Fig. 14-81) and may be matted, mimicking the changes of arachnoiditis. Complete block to the flow of intrathecal contrast is possible. MRI has high sensitivity and shows these lesions (Fig. 14-82), particularly on T_1-weighted Gd-enhanced images, was hyperintense masses within the CSF (Fig. 14-83). Triple-dose contrast may show lesions not seen with the conventional dose. The intradural metastasis cannot be seen on plain CT scans.

INTRAMEDULLARY TUMORS AND MASS LESIONS

The hallmark of intramedullary tumor is enlargement of the spinal cord, usually over several vertebral levels (Fig. 14-84). With small lesions, the widening may be subtle or undetectable. Myelography and CT myelography outline the dimensions and configuration of the cord but cannot identify the inner anatomy.

The MRI method has significant advantage over other techniques for defining intramedullary pathology. Not only can the outline of the cord be seen, but signal abnormalities within the cord identify focal lesions, differ-

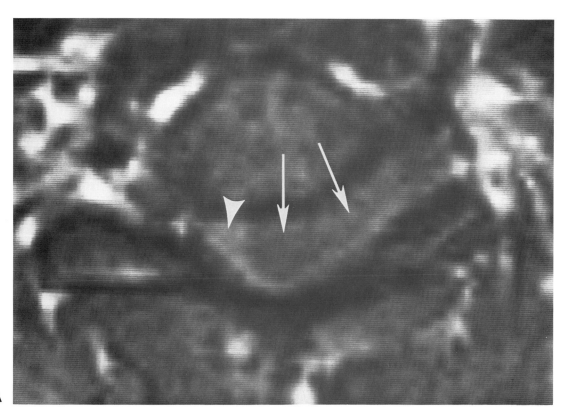

A

Fig. 14-78 Neurinoma, MRI. (A) Transaxial T₁-weighted MRI through the cervical spine shows a slightly hypointense dumbbell neurinoma within the spinal canal and left intervertebral foramen *(arrows)*. The more intense-appearing spinal cord *(arrowhead)* is compressed to the right. **(B)** Parasagittal T₂-weighted MRI shows the neurinoma as a hyperintense mass *(arrow)*.

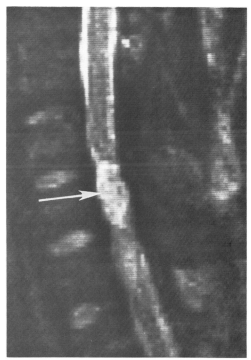

B

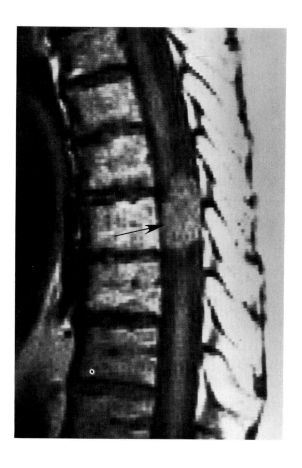

Fig. 14-79 Thoracic meningioma. Parasagittal T_1-weighted MRI shows an intradural extramedullary meningioma *(arrow)*. It is nearly isointense with the spinal cord. The cord is displaced anteriorly and laterally with posterior widening of the subarachnoid space, findings typical of an intradural extramedullary mass.

IMAGING SIGNS OF INTRAMEDULLARY LESIONS

Cord enlargement, usually over several segments

Circumferential narrowing of the subarachnoid space at the site of cord enlargement

Tapering of margins with normal cord

No displacement of the dura

---

entiating between cyst and solid tumor. Gd can further define focal lesions and increases the sensitivity of the examination, although the enhancement of cord tumors is not as consistent as with intracranial tumors. MRI cannot diagnose reliably the specific tumor type signal characteristics.

Glioma

Ependymoma of the spinal cord and filum terminale accounts for about 60 percent of all intramedullary neoplasms. This tumor almost always presents in the lumbar or sacral region and rarely extends as high as the thoracic cord (see Fig. 14-85). As with all intramedullary tumors, the cord is widened and fills the subarachnoid space at the level of the tumor mass. On plain film myelography, the cord must be seen to be widened in two projections. This criterion excludes the spurious apparent enlargement seen when the cord is compressed by an extradural lesion, such as a herniated disc or spondylytic ridge (pseudoenlargement) (Fig. 14-40). In the absence of

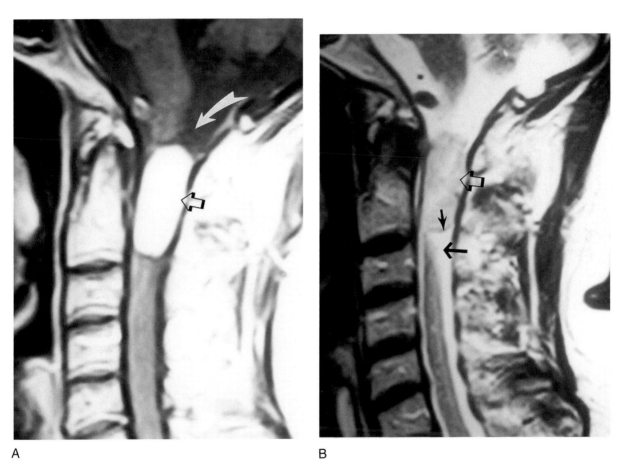

A   B

Fig. 14-80 Meningioma. (A) Sagittal Gd-enhanced MRI displays an intensely enhancing (bright) meningioma within the upper cervical dural tube *(open arrow)*. The cord is displaced anterolaterally by the extramedullary mass, widening the space behind the lower medulla *(white arrow)*. **(B)** On T₂ MRI, the mass is slightly hyperdense compared with the spinal cord *(open arrow)*. At this level, the subarachnoid space widens in the fashion typical of extramedullary intradural mass lesions *(large black arrow)*. The inferior edge of the meningioma is sharply outlined by the hyperintense CSF *(small black arrow)*.

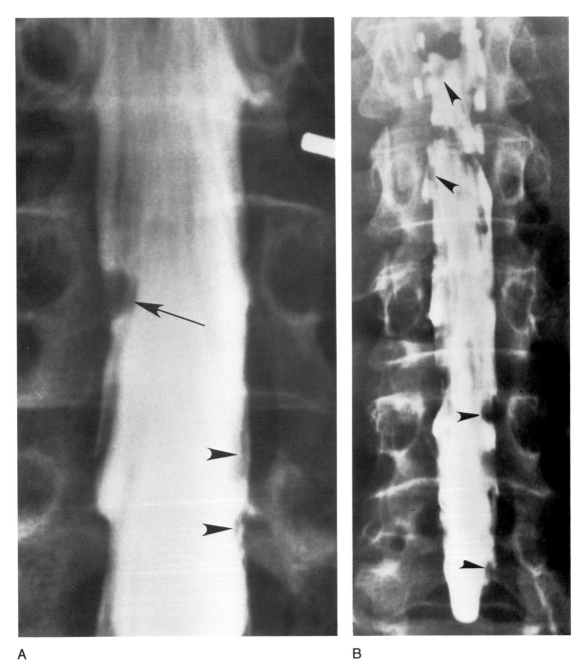

A   B

Fig. 14-81 Intradural metastases, myelography. (A) Lumbar myelography shows a rounded intradural mass (*arrow*) and irregular, flat masses along the lateral dural surface (*arrowheads*). They represent metastatic tumor deposits. **(B)** Multiple round masses are seen in the lumbar region, representing metastases (*arrowheads*).

Fig. 14-82 Intradural metastasis. T_1-weighted parasagittal MRI shows an intradural thoracic mass representing metastatic deposit from breast carcinoma. When seen as a single lesion, it cannot be differentiated reliably from a meningioma.

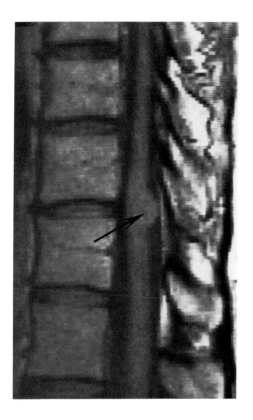

dysraphism, a mass within the filum terminale is almost always an ependymoma.

Astrocytoma is the next most common type of intramedullary tumor, accounting for about 33 percent of tumors of the spinal cord. It occurs predominantly in the cervical cord. It is the most frequently encountered cord tumor in children. The tumor may become large, extending through the entire cord. Cystic change is common. The imaging findings are identical to those of ependymoma. Oligodendroglioma and glioblastoma multiforme are rare.

The tumors may become large, extending over many segments; rarely, nearly the whole spinal cord is involved. The spinal canal may become enlarged to accommodate the tumor with the pedicles separated and thinned along their medial margins. This situation occurs most commonly at the L1–L3 level (Fig. 14-85). Because the discs are resistant to remodeling from tumor, the posterior margins of the vertebral bodies may become scalloped and can be pronounced.

MRI more clearly demonstrates the tumor within the cord, as well as any cystic component. The solid tumor is seen as high intensity on heavily T_2-weighted images (Fig. 14-86). Compared with syringomyelia, tumor cysts show relatively high signal intensity on both T_1- and T_2-weighted images because of protein content and the lack of fluid motion (Fig. 14-87). Gd contrast enhancement occurs in the solid components (Fig. 14-88) and clearly differentiates solid tumor from cysts and syringomyelia. The full extent of the tumor is most easily evaluated with MRI.

Hemangioblastoma

Hemangioblastoma is a malignant vascular tumor that may present within the cord as a single mass or as multiple tumors. It occurs as an isolated tumor or in association with the Von Hippel-Lindau syndrome. The lesion enlarges the cord and may be associated with enlarged vascular channels on the cord surface, which can be

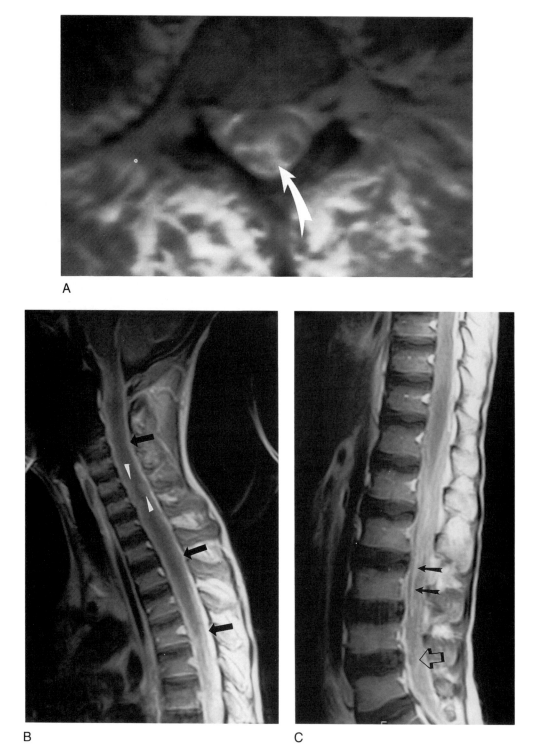

Fig. 14-83 Intrathecal metastasis (neoplastic meningitis). (A) Melanoma. Axial T_1 Gd-enhanced MRI of the lumbar spine shows the bright signal contrast enhancement within the dural tube in a region normally occupied by darker CSF (*white arrow*). The enhancement may represent solid tumor or contrast accumulation within the CSF from breakdown of the blood-brain barrier of the spinal meninges due to neoplastic meningitis. **(B–D)** Rhabdomyosarcoma. **(B)** Sagittal T_1 Gd-enhanced MRI shows high signal within the dural space and over the spinal cord (*black arrows*). Some of the enhancement may be contrast that diffuses into the CSF. Small lesions are on the cord (*white arrowheads*). **(C)** The intradural space is bright, as a result of CSF enhancement or tumor bulk within the thecal sac (*open arrow*). The nerve roots enhance from tumor sheets (*black arrows*). (*Figure continues.*)

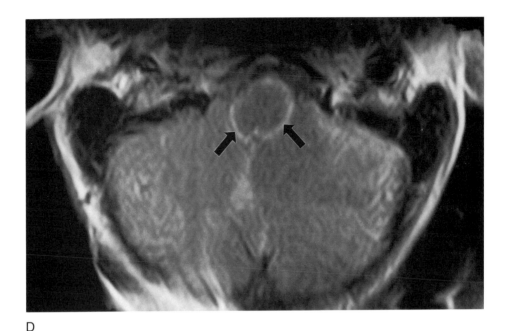

Fig. 14-83 *(Continued).* **(D)** Enhancing meningeal tumor surrounds the brain stem near the foramen magnum *(arrows).*

demonstrated with plain film myelography, CT myelography, and MRI as flow-void. Selective spinal cord angiography shows a highly vascular lesion with small intrinsic vessels, creating an intense blush. Arteriovenous shunting may be present to some extent but not nearly so predominantly as with an arteriovenous malformation. The blush lasts into the venous phase, reminiscent of the blush seen with meningioma. Cysts may be present. The solid tumor shows contrast enhancement and high T_2 signal on MRI.

Lipoma

Lipoma is a rare tumor that occurs in adults, even without associated spinal dysraphism. The lesion may be in any intradural space, although it tends to arise in the posterior part of the thecal sac or cord, most commonly in the lumbosacral region (Fig. 14-89). It is frequently related to the filum terminale. Widening of the spinal canal can occur with large lesions. CT scans and MRI can specifically identify these lesions on the basis of their characteristic signal signatures. Occasionally, these tumors extend outside the dura and may reach the skin.

Dermoid

Although usually associated with the dysraphic syndrome, the epithelial cystic lesions known as dermoids may occur as an isolated lesion. They have no specific CT characteristics and may be of high or low density, depending on the nature of the debris within the cyst. On MRI they tend to be inhomogeneous with variable signal intensities. They are usually intradural–extramedullary in location but may be within the cord (Fig. 14-90). They are known to occur following lumbar puncture or surgery, probably as a result of introducing viable epithelial cells deep into the spinal canal.

INFLAMMATION

Pyogenic Infections

Pyogenic infections of the spine are becoming increasingly common. The process may be difficult to diagnose; frequently, patients suffer for months without appropriate diagnosis or treatment.

(Text continued on page 650)

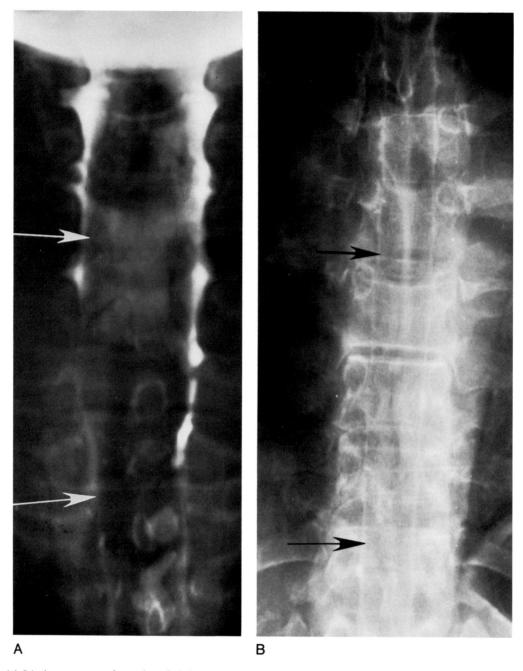

A   B

Fig. 14-84 **Astrocytoma of spinal cord. (A)** Cervical myelography shows gross enlargement of the cervical spinal cord over many segments (*arrows*). The subarachnoid space is circumferentially narrowed. **(B)** The tumor extends throughout the the entire cord (*arrows*). The compressed subarachnoid space is represented by a thin line of contrast.

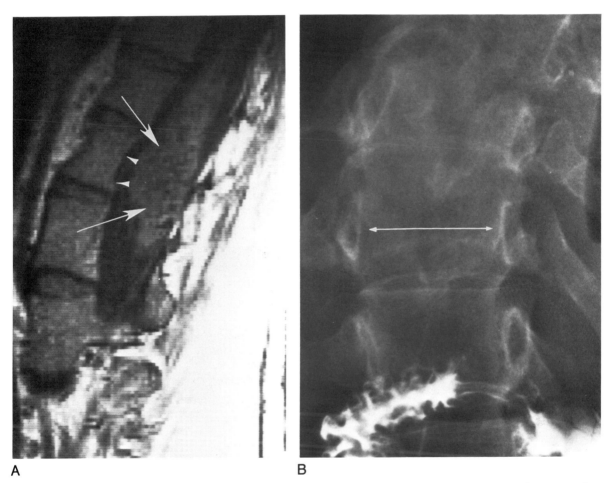

A   B

Fig. 14-85 Ependymoma of the conus medullaris. (**A**) Parasagittal T$_1$-weighted MRI shows isointense enlargement of the conus medullaris (*arrows*). Note the scalloping of the posterior margins of the vertebral bodies (*arrowheads*). (**B**) AP spine view shows scoliosis, widening of the spinal canal (*opposed arrows*), and flattening of the medial margins of the pedicles.

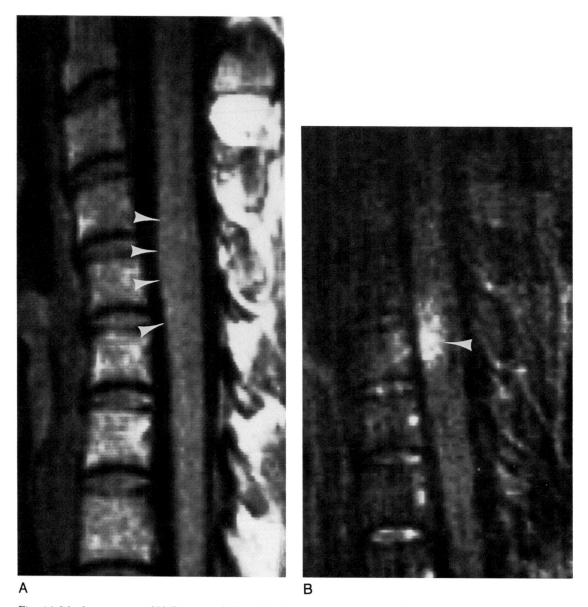

A   B

Fig. 14-86 Astrocytoma. (A) Parasagittal T_1-weighted image shows subtle enlargement of the mid-cervical cord (*arrowheads*). **(B)** T_2-weighted study shows high signal intensity representing the tumor mass (*arrowheads*). This appearance is nonspecific, and multiple sclerosis, cord metastasis, and inflammatory lesions such as sarcoid could produce these findings.

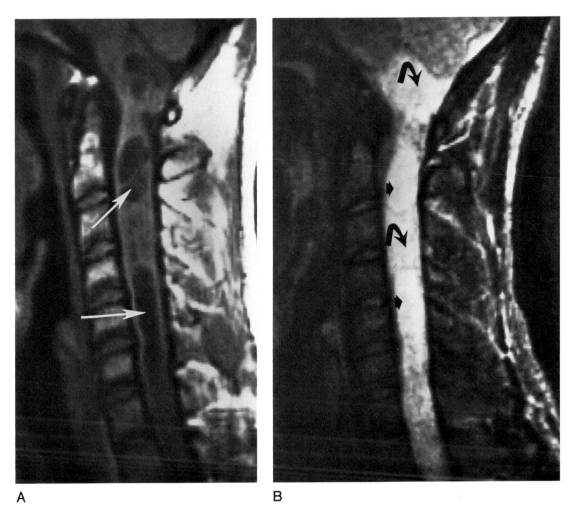

Fig. 14-87 Cystic astrocytoma. (A) Transaxial T$_1$-weighted MRI shows diffuse enlargement of the medulla and cervical cord. Multiple cysts are seen containing fluid slightly hyperintense to CSF *(white arrows)*. **(B)** T$_2$-weighted study (SE 2500/80) shows that the cyst contains very hyperintense fluid *(small arrows)*. The intervening solid tumor shows high T$_2$ signal intensity *(curved arrows)* but slightly less intense than the cysts.

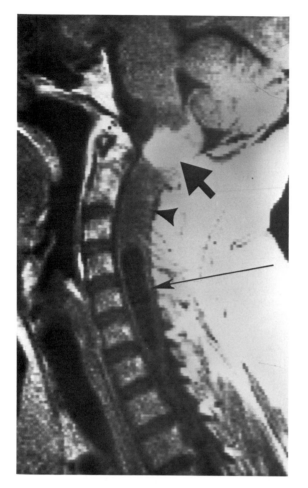

Fig. 14-88 Astrocytoma, Gd-DTPA, enhancement. Parasagittal T_1-weighted MRI shows intense enhancement of the medullary portion of the tumor (*fat arrow*). The upper cervical solid tumor does not enhance (*arrowhead*). A cyst is present at the inferior portion of the tumor (*long arrow*).

Pain is the most common symptom, and it is often relentless. Fever is inconsistent and low grade. The process is relatively slow, and the problem may go on for months or even years. The erythrocyte sedimentation rate (ESR) is almost always elevated. Commonly, patients have had superficial skin infections or have undergone a recent urologic or gynecologic procedure. Intravenous drug abusers are at risk. Most often no risk factor is present.

These infections may occur at any age but are more common in persons over age 60. Batson's plexus probably plays an important role in the spread of infection to the spine, and because of its proximity to the pelvic organs the lumbar spine is the area most frequently involved. Rarely, the infectious process is multicentric. *Staphylococcus aureus* is by far the most common pyogenic infectious agent affecting the spine, but any organism may cause the infection.

The hallmark radiographic findings are vertebral endplate erosion and disc space narrowing (Fig. 14-91). Pyogenic infection of the spine is most commonly an osteomyelitis of the vertebral body with secondary involvement of the disc space (discitis) and the opposite vertebral body. It may take 2 to 3 months from the onset of pain before the spine changes become detectable. Endplate erosion may be subtle, and plain film tomography may help in better definition.

Characteristically, the anterior two-thirds of the vertebral bodies are involved, and the posterior spinal arch is not affected. Eventually, the two adjacent vertebral bodies show symmetric erosion with associated obliteration of the disc space (Fig. 14-92A & B). Only late in the untreated process does vertebral body collapse occur. These changes are distinct from tumor of the spine. Tumor causes destruction or sclerosis of the vertebral body and may involve multiple levels, but as a rule there is preservation of the endplates and the disc space.

After 2 to 3 months, hypertrophic bone change

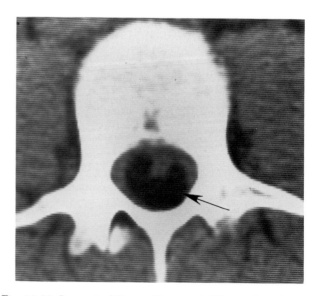

Fig. 14-89 Intraspinal lipoma. Transaxial CT scan shows hypodense tissue within the spinal canal, representing a lipoma (*arrow*). Most lipomas are related to the filum terminale.

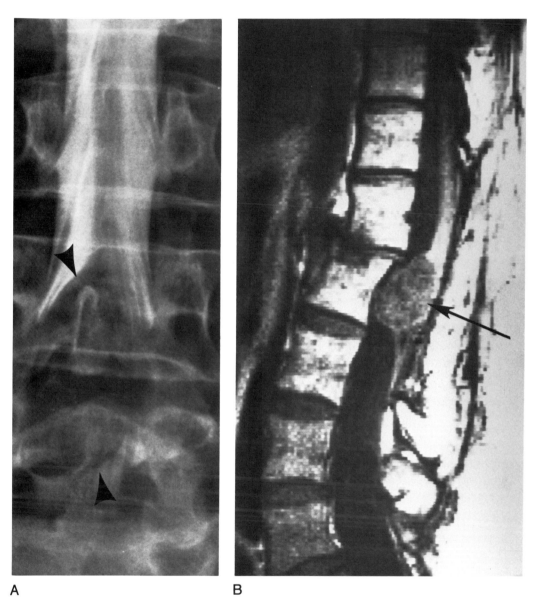

A   B

Fig. 14-90 Dermoid. (A) Myelography shows a large intradural mass in the lower lumbar region *(arrowheads)*. The differential diagnosis includes lipoma and ependymoma. **(B)** Parasagittal T_1 MRI in another patient shows an intradural slightly hypointense mass attached to the conus medullaris *(arrow)*. This dermoid was thought to result from prior trauma and surgery.

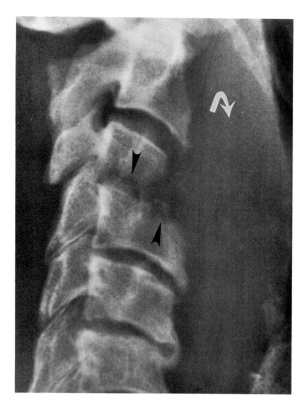

Fig. 14-91 Osteomyelitis and disc space infection. Lateral cervical spine in a patient with extensive laminectomy shows vertebral endplate erosion on both sides of a narrow disc interspace (*arrowheads*). Considerable destruction has occurred anteriorly at C3. A large prevertebral abscess is present (*curved arrow*).

occurs, causing osteophytes and sclerosis about the vertebral body. Eventual fusion between the adjacent vertebral bodies may result. A small paravertebral abscess mass is common, but it is not seen on the plain films, unless the mass is unusually large and displaces or obliterates the psoas muscle. The abscess within the disc space may bulge into the anterior portion of the spinal canal, but cord or nerve root compression and empyema are unusual.

Sometime before the plain film changes, the radionuclide technetium bone scan becomes positive, the process being represented as a hot spot in the vertebral body or two adjacent vertebral bodies. The three- or four-phase ^{99m}Tc phosphonate bone scan is used and is designed to help distinguish between paraspinal cellulitis and osteomyelitis. With cellulitis only, series of scans over 24 hours show initial high uptake of radionuclide

which decreases over time with clearing of the radionuclide from the body. With osteomyelitis, the radionuclide also accumulates within the bone; thus, that bone shows increasing activity over background. Sensitivity is greater than 90 percent in adults, but less sensitive in children and may be negative with infection. Even with four-phase scanning, the test remains somewhat nonspecific, and the diagnosis of osteomyelitis must be surmised because of the elevated ESR and any suggestive history.

Indium-111 leukocyte is useful in the acute stage of the disease and is positive in more than 90 percent of cases. With chronic infection, the sensitivity drops to 50 percent for vertebral osteomyelitis. ^{67}Ga citrate scanning has a good sensitivity of greater than 80 percent for chronic infection, and is diagnostic of infection when the gallium uptake exceeds the bone scan uptake. Gallium has a relatively poor sensitivity of less than 40 percent for acute-stage infection. With children, gallium scanning is more sensitive in all stages of infection, and may be positive when the bone scan is negative. Indium scanning is avoided in children because of its higher radiation exposure.

MRI is the most sensitive imaging examination for the detection of vertebral osteomyelitis. The findings are characteristic. Loss of height of the disc space and blurring of the endplates are seen. On T_1-weighted images, decreased signal intensity occurs within the portions of the vertebral bodies adjacent to the disc space and usually within the disc space itself. High signal intensity is seen in these regions on the T_2-weighted images, representing edema in the bone marrow and disc space. High T_2 signal in a narrowed degenerative appearing disc is characteristic of disc space infection. These changes are not seen following uncomplicated discectomy. The disc space and adjacent vertebral bodies may show enhancement with Gd. These disc space changes are almost never seen with tumor. Effective antibiotic therapy brings about resolution (Figs. 14-92F & G and 14-93).

Also useful for diagnosing and managing spinal infections is CT scanning. Regions of bone destruction adjacent to the endplates can be seen with appropriate windows. There is a small paravertebral swelling or mass visible in virtually all cases. If the infection is chronic, reactive bone formation is detected surrounding the disc space, usually anteriorly (Fig. 14-92C).

There are no specific findings to differentiate among the various pyogenic organisms. The etiology of infection of the spine can be determined only by tissue biopsy and culture. The CT scan is especially useful for selecting

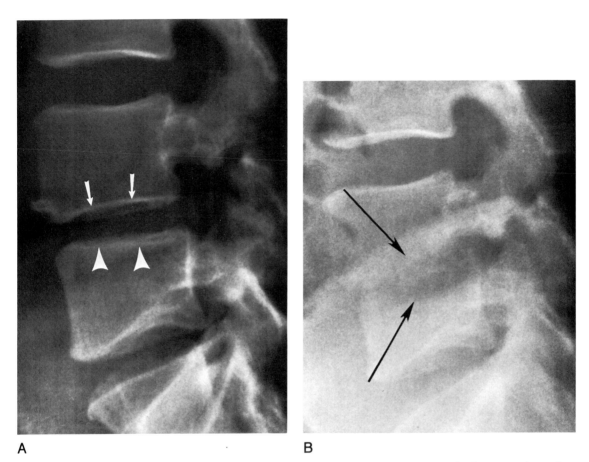

A   B

Fig. 14-92 Osteomyelitis and disc space infection. (A) Initial radiograph of the lower lumbar spine shows disc interspace narrowing at the L4–L5 level. The superior endplate of L5 is indistinct *(arrowheads)*, whereas the inferior endplate of L4 is near-normal *(arrows)*. These plain film changes are typical of early vertebral ostemyelitis and discitis. **(B)** Radiograph obtained 6 weeks later shows marked progression of the findings with severe erosion of both endplates *(arrows)* surrounding the obliterated disc space. There is partial collapse of the L5 vertebral body with spondylolisthesis. *(Figure continues.)*

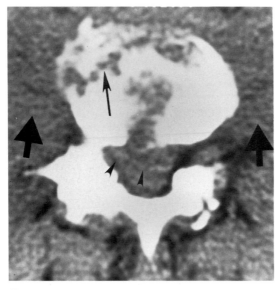

C

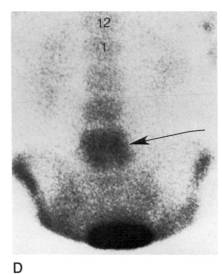

D

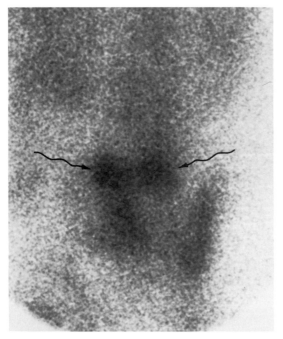

E

Fig. 14-92 *(Continued)*. **(C)** CT scan through the inferior L4 level shows the combination of lytic destruction with reactive new bone formation *(thin arrow)*. A large paraspinal inflammatory mass is present *(large arrows)*. The inflammation has extended posteriorly, causing obliteration of the epidural fat on the right and compression of the dural tube *(arrowheads)*. **(D)** Technetium radionuclide bone scan shows increased uptake at both L4 and L5 vertebral bodies with no visible intervening disc interspace *(arrow)*. **(E)** Gallium scan shows increased uptake at the L4–L5 level. Note that the uptake extends beyond the margins of the vertebral body, representing the paraspinal spread of infection *(arrows)*. *(Figure continues.)*

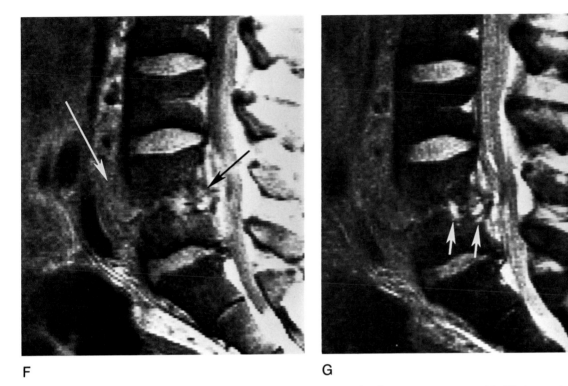

F G

Fig. 14-92 *(Continued).* **(F)** SE 1000/22 MRI shows erosion of the disc interspace at L4–L5 *(black arrow).* The inflammatory reaction extends into the prevertebral space *(white arrow).* **(G)** Proton-density MRI (SE 2500/40) shows some high signal within the posterior portion of the disc interspace and epidural extension of the infection *(arrows).*

the region of the largest inflammatory mass and greatest bone destruction, to guide placement of a percutaneous biopsy needle in obtaining material for microbiologic analysis prior to antibiotic therapy.

Nonpyogenic Infections

Coccidioides immitis, Cryptococcus neoformans, Aspergillus fumigatus, Brucella, and *Echinococcus* all cause spinal infection, especially in endemic regions. The lesions are generally osteolytic with no surrounding sclerosis. There may be disc space narrowing and a small paravertebral mass, but these findings are not as constant as with pyogenic infections. *Actinomyces israelii,* found in the mouth and often associated with mandibular infection, may cause infection of the spine, especially the cervical segments. This infection spreads to involve adjacent structures. The process is mixed lytic and sclerotic, and there is a large paravertebral abscess with sinus tracts. Biopsy and serologic findings may make the specific diagnosis.

Tuberculosis

Spinal tuberculosis results from hematogenous spread from pulmonary tuberculosis. Some evidence of the lung infection is almost always present. The infection is an osteomyelitis usually at the thoracolumbar junction. It begins adjacent to the vertebral endplate, spreading quickly into the adjacent disc space. The bone lesion is purely osteolytic with poor margination and regional osteoporosis. Cartilage is destroyed, resulting in a narrowed or obliterated disc space. The infection spreads to the opposite vertebral body. Vertebral collapse is common, resulting in a gibbus deformity. Uncommonly, the posterior arch becomes involved; but almost never with pyogenic infection.

Characteristically, there is a large paravertebral abscess, often large enough to be seen on plain films. The size of the mass may be disproportionately larger than the degree of vertebral body destruction. Multiple contiguous vertebral bodies may become involved with the infection as the infection spreads superiorly and inferiorly

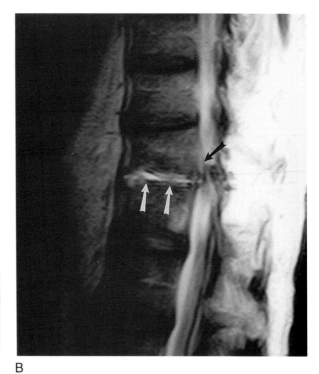

A   B

Fig. 14-93 Osteomyelitis and discitis. (A) Sagittal T_1 MRI of the thoracic spine shows obliteration of an intervertebral disc space *(small white arrows)*. There is loss of the normally high fat signal within the adjacent vertebral bodies *(large white arrows)* from infiltration of the marrow with inflammatory cells and edema. There is a small epidural extension of the inflammatory process *(small black arrows)*. **(B)** Sagittal T_2 MRI through the same level shows high signal intensity within the narrowed intervertebral disc interspace from inflammation *(white arrows)*. The epidural extension of high signal material is barely seen *(black arrow)*.

underneath the poorly attached anterior longitudinal ligament. The infection erodes the anterior borders of the more distant vertebral bodies, causing a "gouge" defect. In the chonric phase, extensive mottled calcification is deposited within the paravertebral abscess.

Epidural Abscess

An epidural abscess (empyema) usually occurs as a metastatic infection from a distant site. Less commonly, it may occur secondary to an adjacent osteomyelitis of the spinal column (Fig. 14-91) or with iatrogenic infection from surgery or lumbar puncture. *Straphylococcus* is the most common organism. It is a serious infection that may spread rapidly throughout the loose fatty tissue of the spinal canal. Patients are acutely ill with fever, progressing motor and sensory neurologic deficits, and spine tenderness. The infection may cause phlegmon (tissue induration) or abscess formation. Cord compression and

death may result. The abscess must be drained, which often requires extensive laminectomy. The infection occurs with equal frequency at all levels of the spine.

The best initial examination is MRI. The inflammatory change is seen on T_1-weighted images as obliteration of the normal epidural fat density and displacement of the dural tube. On the T_2-weighted images the infection is seen as high signal intensity epidural tissue (Fig. 14-94). Gd enhancement occurs within the indurated tissue. Paraspinal inflammatory reaction may also be present. CT myelography can be used to make the diagnosis (Fig. 14-95). The spinal puncture is made at an asymptomatic level to avoid contamination of the subarachnoid space.

Subdural Empyema

The spinal subdural space is a rare location for infection within the spinal canal. Such infection usually results from the spread of adjacent epidural infection

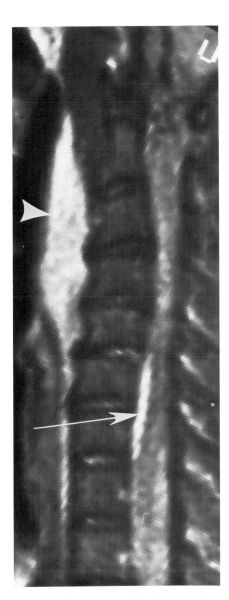

Fig. 14-94 Epidural empyeme, MRI. Parasagittal proton-density (SE 2500/40) image shows high signal intensity in the anterior epidural space causing dural compression (*arrow*). Also note the prevertebral retropharyngeal abscess (*arrowhead*).

Hypertrophic Pachymeningitis

Hypertrophic pachymeningitis may occur as an isolated finding or in association with the mucopolysaccharidoses. The meninges and dura are markedly thickened, causing compression of nerve roots and, in severe cases, the spinal cord. The thickened tissue enhances with Gd. Plain films are normal or show the bone changes associated with Morquio's disease.

Acute Transverse Myelitis

Acute transverse myelitis occurs primarily in young adults, producing rapid onset of symptoms mimicking cord compression. Although the etiology is elusive, the syndrome is associated with viral infections, postvaccinal myelitis, autoimmune disease, paraneoplastic syndromes, multiple sclerosis, and radiotherapy. Some investigators believe that most cases are a form of multiple sclerosis.

During the acute phases, the myelographic findings are usually normal, and the diagnosis is one of the exclusion of other disease processes. However, the cord may show intrinsic enlargement that is indistinguishable from a small cord tumor. On MRI, the signal on T_2-weighted images is variable (Fig. 14-96) but may be increased. Patchy Gd enhancement sometimes occurs. The CSF shows a low-grade pleocytosis. The cord disease causes both motor and sensory loss with local radicular changes. Long tract signs appear later.

Guillain-Barré Syndrome

Guillain-Barré syndrome is another form of myelitis and radiculitis. It is strictly limited to the anterior horn cells or peripheral nerves, causing a flaccid, usually ascending paralysis. Sensory changes do not occur. Myelography is normal, and the cord shows no enlargement.

Infections of the Spinal Cord

Infections of the spinal cord are rare. Pyogenic infection, especially with *Straphylococcus*, is the most common and usually occurs in association with known infection elsewhere. Granulomas, particularly of fungal and sarcoid origin, may also occur. The cord shows nonspecific enlargement and high T_2 signal with MRI. The differentiation from tumor may be possible only at surgical exploration.

through the dura into the subdural extraarachnoid space, although it may occur in isolation. When seen in isolation, the presentation is similar to that of epidural infection, except that there is no associated localized pain or tenderness in the back. Myelography is the best for diagnosis and the findings are the same as with epidural infection. MRI may show increased intensity within the dural tube on T_1-weighted images.

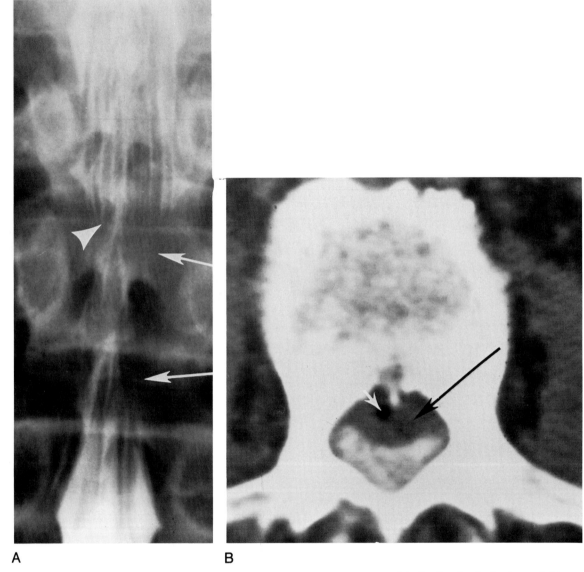

A   B

Fig. 14-95 Epidural empyema. (A) Myelography shows epidural compression of the lumbar dural tube *(arrows)*. Note the "feathering" at the superior margin of the block produced by the nerve roots *(arrowhead)*. This myelographic finding is nonspecific and can be produced by any epidural mass, particularly lymphoma. Acute fever suggests the diagnosis of epidural infection. **(B)** CT scan in the same patient shows the anterior epidural inflammatory mass *(black arrow)*. The gas bubble *(white arrow)* makes the diagnosis of infection likely. Note that there are no bone or paravertebral abnormalities. Contrast fills the compressed dural sac.

Fig. 14-96 Transverse myelitis. T_1-weighted MRI shows slight expansion of the upper cervical cord *(arrow)*, a nonspecific finding. (The T_2-weighted image showed no high signal intensity.) It was thought to represent transverse myelitis. The same change might be found with a small tumor, multiple sclerosis, or focal cord infection, although the latter processes usually show high signal intensity on the T_2-weighted images. (Courtesy of Dr. Paul Markarian, Mercy Hospital, Springfield, MA.)

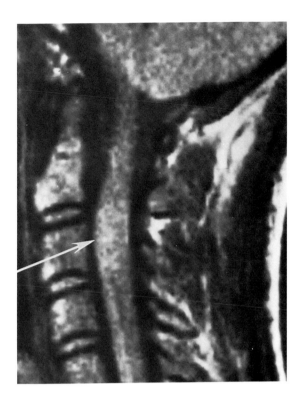

Multiple Sclerosis

Multiple sclerosis involves the spinal cord, but usually to a lesser degree compared with cerebral or cerebellar involvement. The lesions cause demyelination with MRI as high T_2 signal intensity. A slight inflammatory component produces contrast enhancement of the active lesions, along with slight regional swelling of the spinal cord (Fig. 14-97).

CRANIOVERTEBRAL JUNCTION

The craniovertebral junction is considered separately, as unique features of pathology affect this region. Lesions at this level of the spine are notoriously difficult to diagnose clinically. Symptoms of ataxia, weakness of the lower extremities, and tingling of the fingers are nonspecific and often suggest disease at other levels.

The superior imaging technique for evaluating this region is MRI. If MRI is unavailable, plain film radiography and CT myelography/cisternography may be used. Developmental abnormalities are common, producing symptoms of compression of the upper spinal cord and

medulla that do not become a clinical problem until adulthood. The average age for the onset of symptoms is 28 years.

Basilar Impression

Basilar impression (invagination) can be a developmental abnormality or may occur as a result of softening of the bone at the base of the skull, as in Paget's disease or osteogenesis imperfecta. The abnormality consists in superior displacement of the upper cervical spine into the base of the skull secondary to upward curving of the margins of the forament magnum. The diagnosis is generally made by plain skull film analysis, using any one of a number of reference lines drawn at the base (Fig. 14-98). Using the lateral projection, Chamberlain's line is drawn from the posterior margin of the hard palate back to the anterosuperior margin of the posterior rim of the foramen magnum. If more than 8 mm of the odontoid process projects above this line, basilar impression is a strong possibility.

The abnormality may cause direct compression of the lower brain stem because of the small size of the foramen

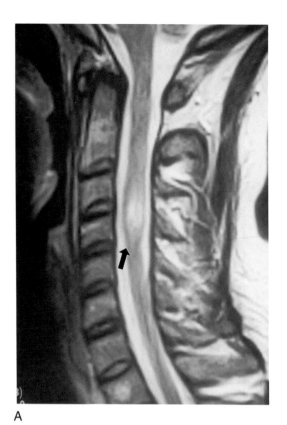

A

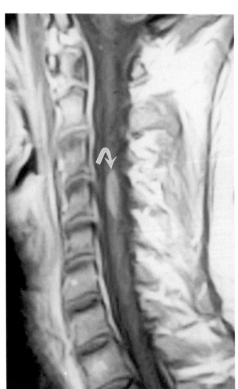

B

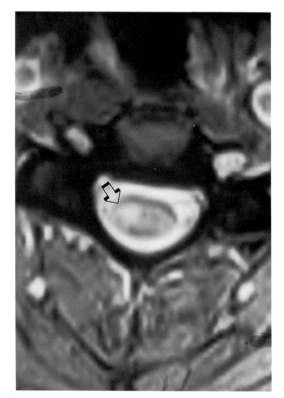

C

Fig. 14-97 Multiple sclerosis. (A) Sagittal FSE (8) 80/3500 T$_2$ MRI identifies an MS plaque in the cervical spinal as a region of high signal *(arrow)*. The lesion is active and swells slightly the spinal cord. **(B)** Sagittal T$_1$ MRI after Gd shows an intense region of enhancement within the active plaque *(arrow)*. **(C)** Axial T$_1$ MRI after Gd shows the lesion predominantly within the right side of the spinal cord *(open arrow)*. Note the slight swelling of the right side of the spinal cord.

Fig. 14-98 Basilar impression and platybasia. Lateral tomography demonstrates upward invagination of the cervical spine, with 50 percent of the ondontoid projecting above Chamberlain's line. This finding represents basilar impression. Note the flattening of the basal angle formed between the planum sphenoidale and the posterior margin of the clivus, called platybasia. (From Heinz, 1984, with permission.)

magnum. Importantly, basilar impression is frequently associated with other abnormalities, particularly the Chiari type I malformation and syringohydromyelia. MRI of the region is performed in patients with congenital basilar invagination.

Platybasia

Although often misunderstood, the term platybasia refers only to a relatively obtuse angle formed by the planum sphenoidale and the dorsal margin of the plane of the clivus (Fig. 14-98). If this angle is more than 143 degrees, the skull has a flattened appearance and platybasia is determined to exist. It has no significance, except as an alert to the presence of other possible anomalies. particularly at the craniovertebral junction.

Occipitalization of the Atlas

Occipitalization of the atlas results from failure of separation of the anterior arch of C1 from the inferior tip of the clivus. There may also be fusion of the transverse processes and the posterior arch into the occipital bone. The atlanto-occipital joints may be normal or fused. The odontoid process is often deformed and is high in the anterior portion of the forament magnum (Fig. 14-99).

The major importance of this malformation is its association with compressive syndromes of the medulla, es-

pecially if the AP dimension of the foramen magnum is reduced to less than 2 cm. Thickening of the adjacent ligaments, dural bands, and the Chiari type I malformation (in 15 percent) are the usual causes of the neural compression. Weakness and ataxia in the lower extremities are the most frequent symptoms, followed by upper extremity findings and neck pain.

Other segmental fusions (Klippel-Feil) of the cervical spine are frequently present, especially at the C2–C3 level (Fig. 14-100). The observation of fusion anomalies at any level of the cervical spine calls for a careful search for occipitalization of the atlas. CT myelography and cisternography as well as MRI scanning can demonstrate the associated anomalies (Chiari I, syrinx), cord compression, or angulation that may be present.

Odontoid Dysplasia and Atlantoaxial Hypermobility

Odontoid dysplasia, including aplasia and hypoplasia, is most commonly seen in association with dwarf syndromes but may occur in normal children or adults. The odontoid may not completely fuse with the axis, leaving a persistent synchondrosis (fibrous union). At best, non-fusion is difficult to differentiate from a fracture. The absence of prior neck trauma is important in making the diagnosis, as well as the performance of both flexion and extension views in the lateral projection to determine whether instability is present.

Os odontoideum (ossiculum terminale) refers to complete separation from the axis of a malformed small odontoid (Fig. 14-101). This situation permits hypermobility at the C1–C2 level, which can be seen with flexion and extension lateral plain films. The odontoid remnant may be fused with the clivus.

Hypoplasia of the dens may be associated with instability (Fig. 14-99). The odontoid may be absent (aplasia), and this is always associated with instability.

Atlantoaxial subluxation may occur in the absence of bone anomalies at the craniovertebral junction. Most commonly, it results from abnormality of the transverse ligament, usually from erosive change produced by rheumatoid disease (Fig. 14-102). It also occurs in persons with ligamentous laxity, such as in Down or Ehlers-Danlos syndrome. Rarely, it occurs in isolation.

The diagnosis is made when the lateral flexion view shows more than 5 mm of separation of the odontoid from the anterior arch of C1 in children under 12 years of

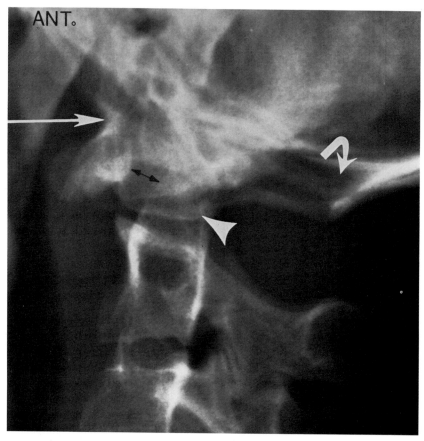

Fig. 14-99 Occipitalization of the atlas. Lateral plain radiograph shows fusion of the anterior arch of C1 with the base of the clivus *(arrow)*. The posterior arch is also fused to the occipital bone *(curved arrow)*. The odontoid is hypoplastic *(arrowhead)*, and there is widening of the dentoatlantal interval *(opposed arrows)*.

age or more than 3 mm separation in adults. Chronic instability is almost always associated with cord compression syndromes. Ligamentous instability may also result in hypermobility at the craniovertebral junction, especially in Down syndrome (Fig. 14-100). The instability may be at the C1–C2 level or at the craniovertebral junction (atlanto-occipital joints).

Chiari I Malformation

Chiari I malformation is probably an acquired abnormality distinct from the Chiari II type malformation. It consists of a downward position of the cerebellar tonsils, usually to the level of the posterior arch of C1. To make the diagnosis, the tonsils must be at least 5 mm below the posterior lip of the foramen magnum. The tonsils are often deformed into a pyramidal shape and are closely applied to the posterolateral portion of the medulla. The medulla may appear normal, elongated or slightly deformed. To be certain that a true Chiari I malformation exists, the medulla should be abnormal along with the low position of the tonsils. Ligamentous thickening at the foramen magnum may also be present. The anomaly occurs with or without other craniovertebral bone abnormalities and is frequently associated with syringomyelia. The abnormality most often becomes symptomatic during early adolescence or young adulthood, causing pain in the posterior neck with a cord compressive syndrome or ataxia. Chiari I is not associated with other cerebral abnormalities.

The diagnosis is best made with MRI (Fig. 14-103) or CT myelography. Plain CT scans of the upper cervical

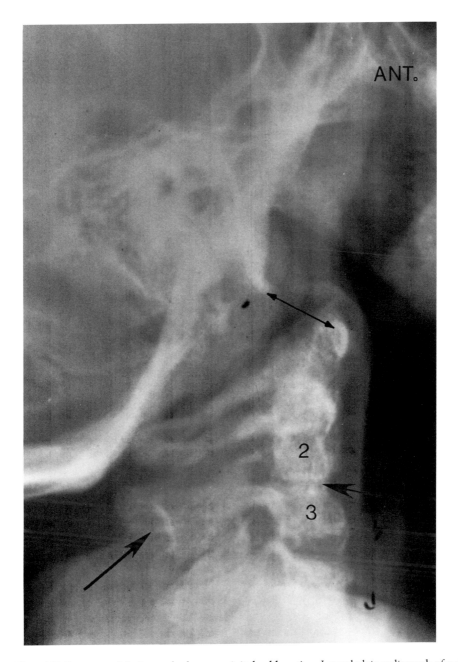

Fig. 14-100 Kippel-Feil segmental fusion and atlanto-occipital subluxation. Lateral plain radiograph of a child shows congenital fusion of the C2–C3 level involving the vertebral bodies and the posterior arches *(arrows)*. In addition, this patient has atlanto-occipital luxation *(opposed arrows)*, which may be due to laxity of ligaments in patients with Down and Ehlers-Danlos syndromes. (Courtesy of Dr. Leon Kruger, The Shriner's Hospital for Crippled Children, Springfield, MA.)

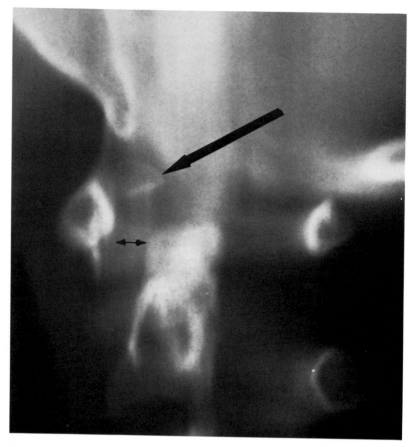

Fig. 14-101 Os odontoideum. A small malformed odontoid (*large arrow*) is widely separated from the axis. There is instability of the atlantoaxial articulation with anterior displacement of the atlas (*opposed arrows*).

region cannot reliably define the position of the tonsils. Relief of the compression usually results in diappearance of the symptoms. The anomaly may be seen with Paget's disease of the skull (Fig. 14-104).

Tumors

Meningioma is by far the most common tumor encountered at the craniovertebral junction. Patients most often present in the 35- to 55-year range with local pain, corticospinal tract symptoms, or high cervical radicular pain. The tumors almost always occur anteriorly or anterolaterally in the intradural extramedullary space.

MRI shows the mass to be anterior to the cord with the typical signal characteristics of meningioma. With CT scanning, the tumor is seen as a homogeneously enhanc-

ing lesion. Bone sclerosis is rare. Vertebral or ipsilateral external carotid angiography shows the typical hypervascular tumor with a prolonged homogeneous blush. Myelography demonstrates the mass in the upper subarachnoid space.

Neurinoma occurs rarely at the craniovertebral level. It may involve the lower cranial nerves and usually enlarges the neuroforamina. Neurinoma is almost never found on the C1 root. Because the tumors involve the nerve roots, they are laterally positioned in the spinal canal.

Metastatic tumor can be found in the craniovertebral region. Destructive or blastic lesions occur in the atlas, odontoid, and skull base. Large carcinomas of the posterior nasopharynx and hypopharynx may directly invade the anterior margin of the foramen magnum or upper cervical spine.

Fig. 14-102 **Rheumatoid arthritis, atlantoaxial hypermobility; invagination of C2.** (A) Increased space is present between the anterior arch of the atlas and the odontoid *(long arrow)*. The odontoid is eroded by the inflammatory process *(small arrow)*. Rheumatoid changes are also present in the facets *(arrowhead)*, which allows anterior listhesis at multiple levels. (B) Transaxial view shows severe odontoid erosion *(arrow)* as well as erosion of the anterior arch of C1 *(arrowheads)*. The atlantoaxial displacement is reduced when the patient is supine. *(Figure continues.)*

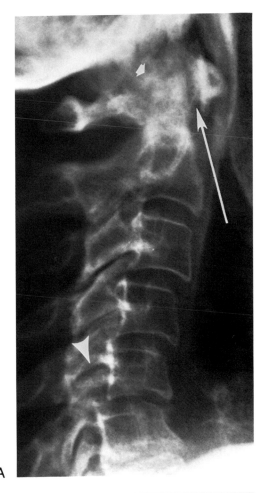

A

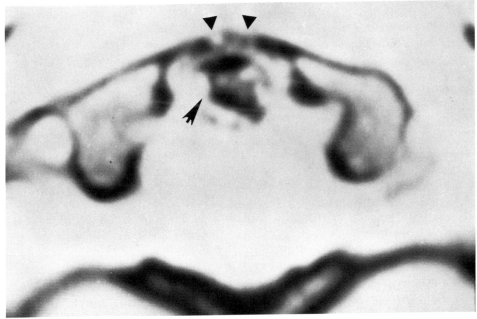

B

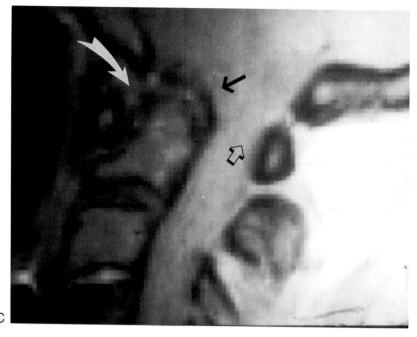

C

Fig. 14-102 *(Continued)*. **(C)** Sagittal T$_1$ MRI demonstrates the erosion of the bone surrounding the atlantodental articulation *(white arrow)*. The destruction of the ligamentous support allows the upward migration of eroded dens of C2 into the anterior foramen magnum *(black arrow)*. The result is medullary compression against the posterior arch of C1 *(open arrow)*.

CONGENITAL ANOMALIES

At birth, there are usually one or two ossification centers in the vertebral bodies. If two are present, they are divided by either a sagittal or a coronal cleft. The neural arch (pedicles, lamina, transverse processes, spinous process) has a single ossification center on each side. The odontoid process contains two ossification centers divided by a sagittal cleft, and the tip has one or two separate centers. The odontoid forms from the same somite as the distal clivus.

The neural arches fuse posteriorly beginning in the first year. The lumbar segments fuse first, followed by the thoracic segments and sacral segments. The cervical arches fuse last at the end of the third year. The neurocentral synchondrosis, the junction between the posterior neural arch (pedicles) and the vertebral bodies, begins to fuse at age 2 in the cervical region; the process is completed in the lumbar region by age 5. The sacrum does not become fully ossified until about age 5.

The dural sac and spinal cord have a rounded configuration during the first year of life but become somewhat more oval thereafter with the long dimension in the transverse plane. The spinal canal is originally larger than the vertebral bodies but becomes equal to the bodies at about 6 to 8 years. Thereafter the canals are smaller than the transverse dimension of the adjacent vertebral body. The cord normally enlarges at the C3–C7 level and again at the T10–T11 level. Because of the normal thoracic kyphosis the thoracic cord is anterior in the spinal canal, whereas at other levels it lies posteriorly. The conus medullaris comes to lie at T12–L1, its usual adult level, by age 3 months. There is some variation in the level, but a conus below the middle of L2 is considered abnormally low. The filum terminale, normally measuring 1.0 to 1.5 mm in diameter, extends from the lower point of the tapered conus medullaris and courses inferiorly to attach to the sacrum in the posterior aspect of the canal. The descending nerve roots form the cauda equina and are usually oriented in a V configuration within the sac, as viewed on transaxial CT myelography or MRI.

Dysraphism

The term spinal dysraphism encompasses a large group of disorders that originate from defects in the formation of the neural tube. These fusion abnormalities may in-

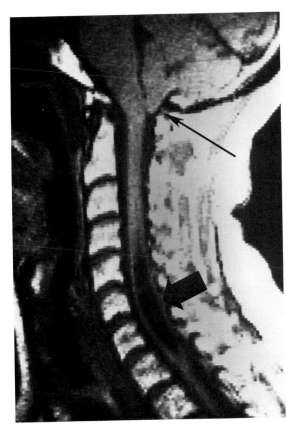

Fig. 14-103 Chiari I, syringomyelia. T₁-weighted MRI shows the downward displacement of pointed cerebellar tonsils into the upper cervical spinal canal (*thin arrow*). Syringomyelia is present in the lower cervical spinal cord (*fat arrow*). The medulla is slightly thickened.

volve any part of the spinal axis. The skeletal deformities include hemivertebrae, cleft vertebrae, segmental fusions of vertebral bodies or lamina, spinal bifida occulta, or widely spread neural arches. Neural tissue involvement with the dysraphic abnormality may be manifested by a thickened filum terminale with tethering of the cord, dural ectasia, myelomeningocele, myelocele, lipomeningocele, myelocystocele, simple meningocele, diastematomyelia, diplomyelia, neurenteric cyst, or the Chiari type II malformation. Overgrowth of sequestered tissue may lead to abnormal tissue collections, such as lipoma, epidermoid, dermoid, and teratoma. In 50 percent of cases, the dysraphic abnormalities are associated with an overlying cutaneous lesion, such as a hairy patch, pigmentation, vascular malformation, dermal sinus, subcutaneous lipoma, or obvious meningocele sac (spina bifida cystica). When there is a back mass it may or may not

be covered with a layer of skin. Dyraphism without a back mass or skin abnormality is called occult dysraphism. Spina bifida occulta is the most innocuous of the dysraphic lesions. Combinations of these lesions are the rule. Almost always, the dysraphic syndromes present during infancy or childhood. The more severe syndromes cause lower extremity paresis and bladder and bowel dysfunction. It is important to realize that the degree of neural derangement does not always correlate with the degree of the vertebral segmental abnormality.

Spina Bifida Occulta

Spina bifida occulta is a posterior defect in the fusion of the lamina, but without cutaneous abnormality or meningeal protrusion through the defect. Some refer to any dysraphic abnormality without posterior skin abnormality as spina bifida occulta. It occurs most commonly at the L5 or S1 level. With simple spinal bifida occulta, the posterior arches are closely approximated with cortical margins on the adjacent lamina. There may be overlap and peculiar configurations. Simple spina bifida occulta is usually of no clinical significance.

Tethered Cord

Tether refers to abnormal fixation of the cord within the dural sac. Tethering of the conus medullaris is the most common of the dysraphic syndromes. It is almost always associated with some form of spinal dysraphism. The spinal cord does not disjoin from the surrounding tissue; it becomes relatively fixed and stretched, resulting in functional derangements within the conus and neuronal damage, which may become irreversible.

With tethering, the conus is deformed and almost always below the middle of L2. The causes of tethering include spinal lipoma (Fig. 14-105A), thickened filum terminale (more than 2 mm), diastematomyelia, lipomeningocele (Fig. 14-106), and myelomeningocele. Syringomyelia may be present (Fig. 14-105B). These abnormalities can all be readily recognized with MRI or CT myelography. The conus may not form as a recognizable structure.

Symptoms usually begin during infancy or early childhood, but occasionally as late as adulthood. Foot deformities (especially pes cavus), scoliosis, leg weakness or numbness, muscular atrophy, and bladder dysfunction are the common problems. Surgery is directed at reliev-

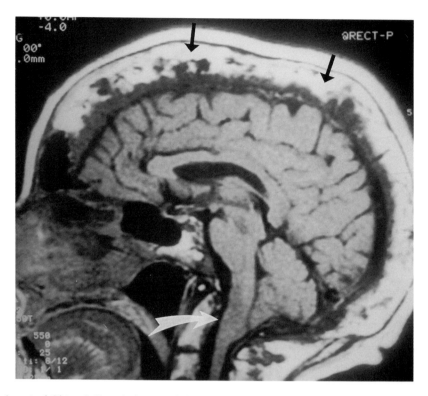

Fig. 14-104 Acquired Chiari I, Paget's disease of the skull. Sagittal T₁ MRI shows the downward positioning of the medulla and cerebellar tonsils into the upper cervical spinal canal *(white arrow)*. The calvarium shows thickened bone and a widened marrow cavity *(black arrows)*.

ing the tethering and often prevents further progression, although the neurologic deficits rarely improve.

Myelomeningocele, Myelocele, Lipomenigocele, and Myelocystocele

MYELOMENINGOCELE AND MYELOCELE

Myelomeningocele and myelocele is the herniation of exposed primitive neural tissue and meninges posteriorly through a large dysraphic defect in the posterior spine (Fig. 14-107). This severe defect consisting of a flattened low-lying tethered primitive cord forms a placode posteriorly over the large herniated CSF-filled sac lined by meninges. The nerve roots exit forward through the CSF-filled sac. The posterior neural elements may be split and splayed open, a condition called *myeloschisis*. There is no skin covering the posterior placode, exposing the neural tissue that protrudes posteriorly beyond the plane of the normal back above. These lesions are most common at the lumbosacral level, but they may occur anywhere up to the mid-thoracic spin. A large dysraphic

defect over several segments of the posterior neural arch leads to herniation of the meninges and neural tissue. The cord is always tethered in this condition. *Myelocele* refers to a smaller identical lesion. These malformations result from failure of separation of cutaneous ectoderm from neural ectoderm.

LIPOMENINGOCELE

Lipomeningoceles are rare (Fig. 14-106). The disorder consists of a large dysraphic defect of the posterior spine with a subcutaneous mass in the back consisting of fat (lipoma), neural tissue, and meninges. The lipoma tissue, easily recognized with CT scans and MRI, is located underneath a covering of skin. A placode of neural tissue, often splayed open, lies over the anterior surface of the fat tissue. Fat intertwines with the neural tissue causing tether. Fat may also grow upward into the spinal canal. Chiari I malformation and syringomyelia occur in 10 percent. Partial sacral agenesis may be present. Lipomenigocele results from premature separation of the neuroectoderm from the cutaneous ectoderm which permits interposed fat tissue.

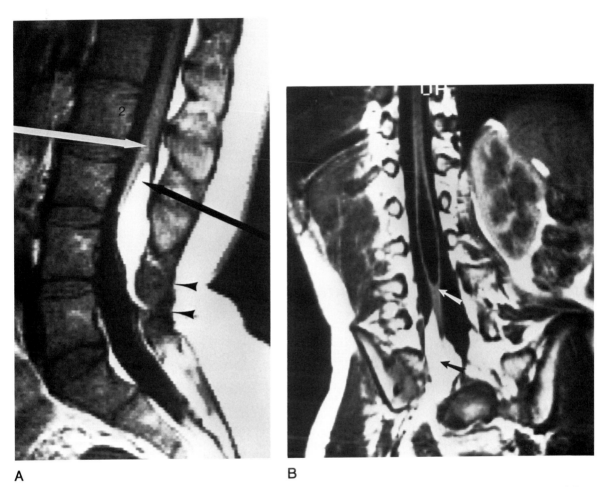

A   B

Fig. 14-105 Spinal dysraphism: tethered cord, lipoma. (A) Parasagittal T_1-weighted MRI shows a low conus medullaris due to tethering *(white arrow)*. A lipoma is present and extends into the conus medullaris *(black arrow)*. Dysraphism is present is the lower lumbar region *(arrowheads)*, and the lipoma tracks posteriorly through the defect. The spinal canal is widened. **(B)** Coronal MRI in another patient shows the low conus medullaris and tethering with syringomyelia of the lower spinal cord *(white arrow)*. A large lipoma fills the lower sacral canal *(black arrow)*. The pedicles are widely separated from the dysraphism.

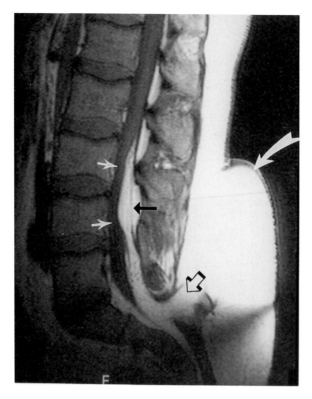

Fig. 14-106 Lipomeningocele. Sagittal T_1 MRI shows the large bright posterior lipoma beneath the skin, which covers the posterior surface (*large white arrow*). The lipoma extends through a dysraphic spinal defect into the spinal canal (*open arrow*). The lipoma entwines nerve roots (*black arrow*). The low-lying tethered neural placode lies on the anterior surface of the lipoma (*white arrows*).

MYELOCYSTOCELE

Myelocystocele is a cystic CSF containing sac that herniates posterior through a large dysraphic defect into the subcutaneous tissues of the lower back. The sac represents the dilated terminal end of the central neural canal. Neural tissue splits in a funnel fashion over the superior aspect of the cyst. The subcutaneous fat is increased posterior to the cyst, but is always separated from the neural tissue by a membrane, a condition that is different from lipomeningocele.

Meningocele

Simple meningocele is herniation of the meninges through a defect of the spinal canal (Fig. 14-107). The herniated meninges form a sac of CSF that does not con-

tain neural elements. The conus medullaris is normally formed and in the normal location without tether. The posterior meningocele, herniation through a small posterior dysraphic defect over one or two segments, is the most common. It may occur at any level of the spine, including the cervical region. Associated severe abnormalities are rare, although some meningoceles are associated with neurofibromatosis type 1 (NF-1). The her-

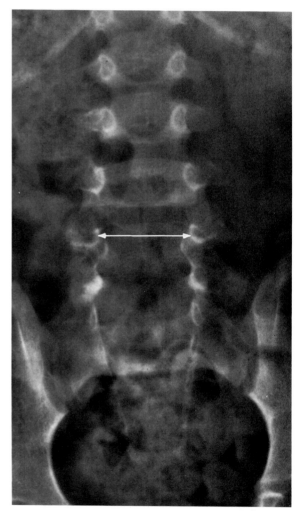

Fig. 14-107 Dysraphism. AP view of the lumbar spine shows gross enlargement of the lower lumbar spinal canal with separation of the pedicles (*opposed arrows*). The lamina and spinous processes are absent at the L4 and L5 levels. This abnormality was associated with a meningocele.

niated sac may be anterior, lateral or within the spinal canal (intraspinal, or intrasacral meningocele). Herniations other than posterior do not produce a back mass and are therefore considered occult dysraphic lesions (Fig. 14-108).

Diastematomyelia

Diastematomyelia is a congenital splitting of the spinal cord into halves. Each half contains its corresponding anterior and posterior horn cells and the ipsilateral nerve roots. Most commonly, the separated cord is contained within one dural sac (type II); but the sac may be split by a bone or cartilaginous spicule dividing both the dura and the cord (type I)(Fig. 14-109). If a spicule is present, it arises from the posterior of the vertebral body and crosses posteriorly through the spinal canal to a lamina. When a spicule is present the cord is always tethered. True dipolmyelia, duplication of the cord into two complete but separate structures, is rare.

Plain radiographs almost always show a spinal bifida occulta at the level. The bone of the spicule may be seen (Fig. 14-109A). The vertebral body and the local disc space are narrowed or fused and often deformed (Fig.

14-110B). Butterfly vertebra, hemivertebra, and block vertebra are common at the level of the lesion. The pedicles are separated but retain their normal shape. Most lesions occur at the upper lumbar level, but rarely they have been reported as high as C3 and in the sacrum. Scoliosis and other spinal deformities are usually present. Some report that the incidence of diastematomyelia may be as high as 5 percent in persons with congenital scoliosis.

The diagnosis is best made with MRI (Fig. 14-110) or CT myelography (Fig. 14-109B). The spur and the division of the cord and dural sac are easily identified. The ipsilateral nerve roots can be seen exiting from the cord. The split in the cord usually extends superiorly from the level of the spur. Without the spur, the split may extend inferiorly completely through to the conus. With a spur, there is tethering of the bifida conus. Rarely, multiple spurs are present. Often there is also an intradural lipoma of the filum terminale, or hydrosyringomyelia.

Clinically, the lesion may have any of the cutaneous markers associated with the dysraphic syndromes. About 50 percent have a hairy patch, hemangioma or pionidal cyst on the back. Pes cavus and other orthopaedic deformities are common. The clinical symptoms are less se-

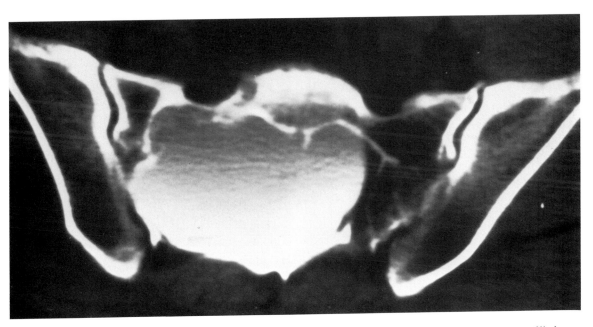

Fig. 14-108 Intraspinal meningocele. Transaxial CT myelography through the sacrum shows a large contrast-filled sac within a grossly dysraphic and expanded sacral spinal canal. The meningocele is confined to the region of the canal. There are no nerve roots within this sac.

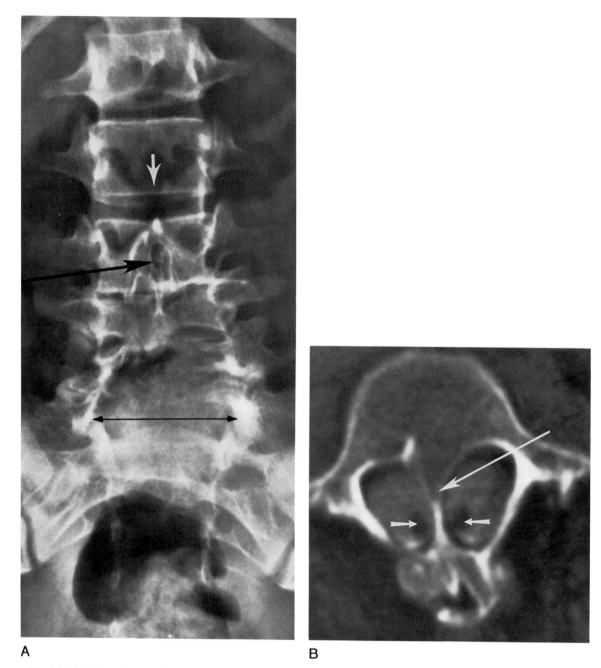

A   B

Fig. 14-109 Dysraphism, diastematomyelia. (A) A meningocele is present with a grossly widened lower lumbar and sacral canal *(opposed arrows)*. Above it, at L3, a diastem (bone spicule) is seen coursing between the lamina and the vertebral body *(black arrow)*. Spina bifida is present at L2 *(white arrow)*. **(B)** Transaxial CT myelography shows the diastem dividing the equal spinal canal *(long arrow)*, a tethered and divided spinal cord *(small arrows)*, and a divided dural tube (type I).

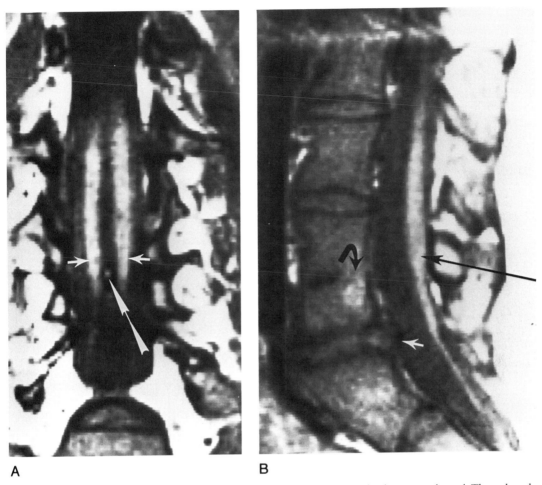

Fig. 14-110 Dysraphism, diastematomyelia. (A) Coronal MRI shows a dysraphic lower spinal canal. The tethered cord is divided in half *(small arrows)* by a diastem, which is difficult to recognize *(long arrow)*. Note the small amount of marrow fat within the center of the bone spicule. **(B)** Parasagittal MRI just off the midline shows one-half of the tethered cord *(long arrow)*. Vertebral body fusion is present at the level of the diastem *(curved arrow)*. A moderate-size disc herniation is seen at L5–S1 *(white arrow)*. The filum is thickened. (SE 600/15).

vere in those without the spur. Surgery is directed at relieving the mechanical tension on the cord caused by the tether.

Dermal Sinus

Congenital sinus tracts present frequently at the lower end of the spine. Those that occur above the gluteal crease have a potential for entrance into the dural sac. Most true dermal sinus tracts enter the dura. They may connect with intradural dermoid cysts or extend into the conus. Those that extend inward are usually associated with demonstrable spinal dysraphism, although the defect may be subtle. The tracts need to be removed as they have the potential to cause bacterial meningitis.

Syringohydromyelia

Practically, it is impossible to distinguish syringomyelia (cystic formation within the cord not lined by ependyma and separate from the central canal) from hydromyelia (cystic enlargement of the central canal, which is lined by ependyma). Gardner and colleagues (1957) proposed the term syringohydromyelia, which combines the two processes into one category. Post-traumatic cysts and cysts associated with tumors are still commonly termed *syringomyelia* in the literature.

The cystic change within the cord is most common in the cervical region but often extends inferiorly into the thoracic, and rarely the lumbar, region. It is unusual for the cyst to extend above C2, although *syringobulbia* (involvement of the pons and upper medulla) is found infrequently. In children, the cyst almost always causes cord enlargement, which may be associated with widening of the spinal canal. Syringohydromyelia is most often associated with one of the dysraphic anomalies, the Chiari I malformation, craniovertebral anomalies, and congenital scoliosis. It is seen uncommonly as an isolated finding.

The superior method for diagnosis is MRI (Fig. 14-111A). The cystic cavity is seen as CSF-equivalent signal intensity within the spinal cord. The cord may be wi-

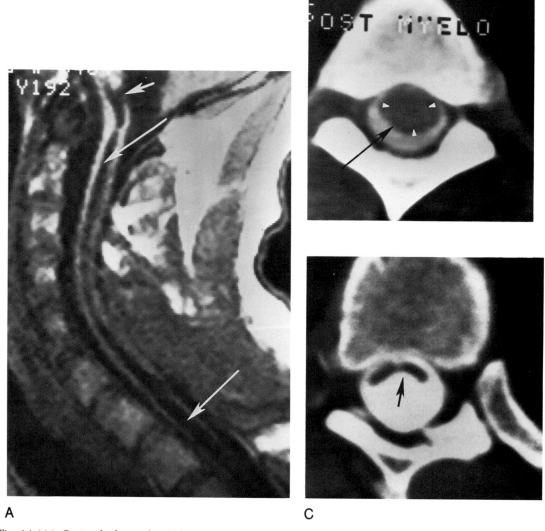

A   C

Fig. 14-111 Syringohydromyelia. (A) Parasagittal T_1-weighted MRI shows a long central cyst within the cervical spinal cord *(long arrows)*. There appears to be an open obex *(small arrow)*. **(B)** Transaxial CT myelography shows an enlarged cord *(arrow)*. Faint contrast filling of the large central cyst is barely seen *(arrowheads)*. **(C)** At a lower level, the atrophic cord is collapsed when the patient is supine *(arrow)*.

dened or collapsed (Fig. 14-111B & C), depending on patient position. Atrophy is common; when collapsed, the cord may appear small.

With CT myelography, the outline of the cord is obvious. The cyst may be seen on the initial postcontrast examination as a low-density abnormality within the cord. More likely, it is found with a 6-hour delayed examination. At this time some of the contrast agent enters the cyst cavity, either through an open obex at the inferior floor of the fourth ventricle or by diffusion through the cord parenchyma (Fig. 14-111). The region of the foramen magnum is studied to detect a Chiari malformation.

Neurenteric Cyst

The neurenteric cyst is a rare lesion that results from persistence of the embryonic canal (of Kovalevsky) between the yolk sac and the notochordal canal (split notochord syndrome). It is usually found in the thoracic region. It is always associated with some form of anterior dysraphism of the vertebral body, and the sclerotic margins of the persistent canal can be seen in the frontal projection on plain films and on CT examination. There are associated duplications of the bowel. The cyst of the lesion may be anterior to the spine or within the spinal canal (Fig. 14-112) or cord. The cyst may fill with contrast from the subarachnoid space.

Caudal Regression Syndrome

Caudal regression syndrome refers to agenesis of a portion of the lower spine. The mildest form is agenesis of the coccyx, which is usually asymptomatic. More severe forms consist of absence of the sacrum and below, the lumbar spine and below, and rarely the lower thoracic spine and all below. Urinary tract anomalies and other dysraphic abnormalities are frequently associated lesions. The conus medullaris is deformed in patients who are symptomatic.

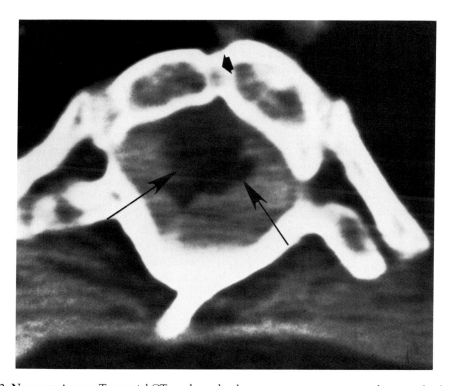

Fig. 14-112 Neurenteric cyst. Transaxial CT myelography shows a neurenteric cyst extending into the dural tube and spinal cord (*arrows*). The anterior canal of Kovalevsky can be seen coursing through the deformed vertebral body (*small arrow*).

Chiari II Malformation

Chiari II malformation is discussed in Chapter 12 as a congenital anomaly of the brain. It occurs with the severe dysraphic lesions, especially myelomeningocele.

DEVELOPMENTAL MASS LESIONS

Arachnoidal Cyst

Arachnoidal cysts occur rarely with the dysraphic syndrome, with neurofibromatosis, or from a previous bout of meningitis. They may also occur as isolated idiopathic lesions. They are usually intradural but may extend outside the dura within the spinal canal or paravertebral tissues. They may be either communicating or noncommunicating with the subarachnoid spaces. Arachnoid cysts characteristically occur in the posterior portion of the spinal canal, often causing a complete block at myelography. The lesion may be large, extending over many segments and producing widening of the canal and posterior scalloping of the vertebral bodies. Both CT scanning and MRI usually diagnose the cystic nature of the lesion. At the caudal end of the dural sac, it may be impossible to differentiate an arachnoid cyst from an intraspinal meningocele.

Perineural enlargements of the arachnoid occur frequently but are usually asymptomatic. The main problem is to differentiate a cyst from disc herniation on CT scans (Fig. 14-113). Myelography and MRI easily identify the cysts.

Almost always midline, developmental mass lesions are the result of sequestration or overgrowth of normal tissues. They are most often encountered in association with the dysraphic syndrome, although isolated tumor masses do occur.

Lipoma

Lipoma is the most prevalent form of the midline development mass; it is usually associated with a spinal defect or myelomeningocele. It is most commonly located in the subcutaneous tissue over the spinal defect but is often seen anteriorly within the dural tube and within the filum terminale. Here it may be small and insignificant. However, the caudal nerves are usually simply displaced and not incorporated within the fatty mass. Occasionally, it extends superiorly within the filum into the conus medullaris (Fig. 14-105). The extent of the mass can be evaluated easily with either CT scanning or MRI. Lipoma is bright on T_1 and FSE T_2-weighted MRI.

Teratoma

A teratoma is a rare lesion that may occur within the dura or spinal cord. It is usually associated with regional dysraphism. The lesion may be large, extending over many levels and widening the canal. It most commonly occurs in the thoracic region. A sacrococcygeal teratoma presents as a pelvic mass and is usually not associated with the dysraphic syndrome.

SCOLIOSIS

Seventy percent of scoliosis is idiopathic, occurring primarily in adolescent girls. Most frequently, the primary thoracic curve is convex to the right extending from T4 to L1 with a secondary curve convex to the left from L1 to L5. In most cases, only plain film radiographs are required for evaluation, obtained with the patient in the erect position. The angle of the scoliosis is determined by drawing lines parallel to the endplates of the most superior and inferior vertebral bodies of the primary curve, constructing perpendicular lines from them, and then measuring the angle at the point of intersection. Bending films may be requested to determine the flexibility of the spine before surgical intervention.

Scoliosis can occur because of tumor, infection, hemivertebrae (Fig. 14-114), dysraphism, muscular imbalance, and postirradiation damage to the spine. Congenital spine anomalies and early onset of scoliosis are often associated with syringohydromyelia, diastematomyelia, and Chiari I malformation. Scoliosis may also occur in association with neurofibromatosis. Myelography, CT myelography, or MRI scanning may be indicated for congenital scoliosis patients to detect associated lesions before corrective surgical manipulation.

ATROPHY

Spinal cord atrophy occurs after trauma (see Ch. 15), with syringomyelia (Fig. 14-111), or as a consequence of progressive spondylosis. The cord is decreased in size, and the visualized subarachnoid space is necessarily enlarged, except in those with spinal stenosis. The cord tends to be

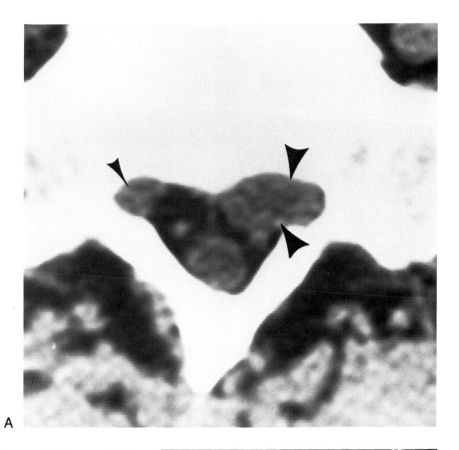

A

Fig. 14-113 Perineural cyst. (A) Transaxial CT scan shows a CSF-equivalent-density lesion in the left lateral recess of S1 *(large arrowheads)*. The lateral recess is enlarged, and the left S2 nerve root is displaced posteriorly. A small cyst is present on the right *(small arrowhead)*. **(B)** T$_2$-weighted MRI shows the high CSF-intensity signal of perineural cyst at S1 and S2 *(arrows)*.

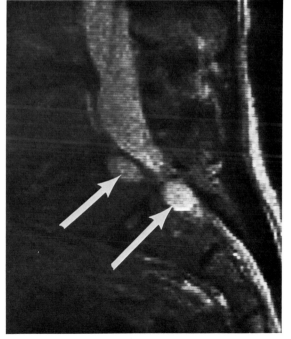

B

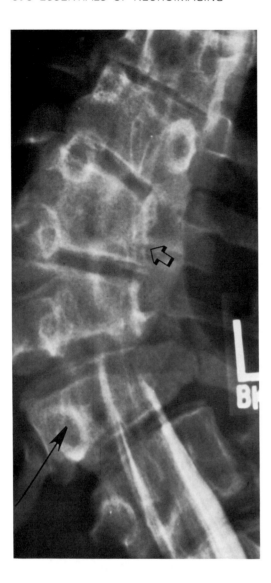

Fig. 14-114 Congenital scoliosis. Congenital scoliosis is often caused by vertebral anomalies such as hemivertebrae *(black arrow)* and segmental fusion *(open arrow)*. Congenital scoliosis is often associated with other spinal anomalies affecting the cord.

smaller in the AP dimension than in the transverse dimension, as the cord is tethered bilaterally by the nonelastic dentate ligaments. The anterior median-fissure is widened. The atrophy may be focal or may extend over many segments, particularly when associated with syringomyelia.

ARTERIOVENOUS MALFORMATIONS OF THE SPINAL CORD

Arteriovenous malformations (AVMs) of the spinal cord are rare lesions, most commonly encountered as unexpected findings on myelography. Rarely, they cause spinal subarachnoid hemorrhage or cord hematoma. A cutaneous hemangioma is present in 25 percent of these cases. Most occur in the thoracolumbar region.

Three types of malformation are encountered: (1) "adult type," consisting of a small, compact, serpiginous malformation often on the posterior surface of the cord and supplied by a single feeding artery; (2) a tangle of small vessels, often fed by a single artery; and (3) the "juvenile type," with a large malformation fed by many arteries and found mostly in children.

The diagnosis is made with selective spinal angiography (Figs. 14-115 and 14-116). A compulsive selective study of all the radicular arteries must be done to define all the feeders of the malformation. It is important to define the AP position of the AVM within the cord. An

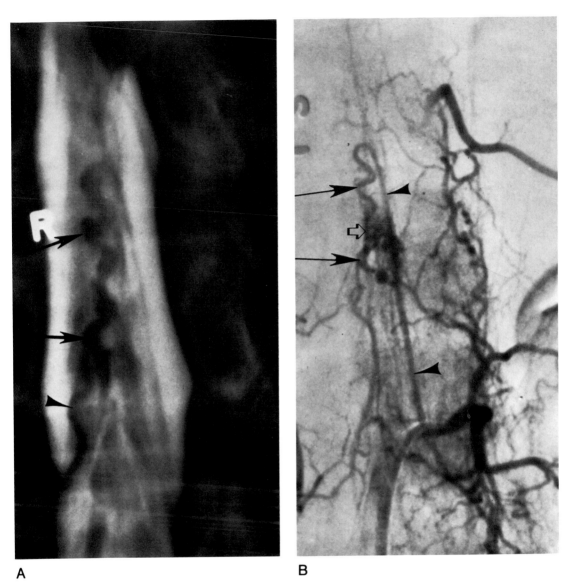

A   **B**

Fig. 14-115 Spinal cord AVM. (A) Myelography shows a serpiginous filling defect on the anterior surface of the conus medullaris, representing the enlarged feeding artery *(arrows)*. A radicular feeder is seen entering from below *(arrowhead)*. **(B)** Spinal angiography with the catheter in a left lumbar artery opacifies the large artery of Adamkiewicz *(arrowheads)*, which fills the enlarged, serpiginous anterior spinal artery *(arrows)*. A small tangle of abnormal vessels represents the malformation *(open arrow)*. A smaller segmental artery enters from the right inferiorly as seen on myelography. This malformation is type 2.

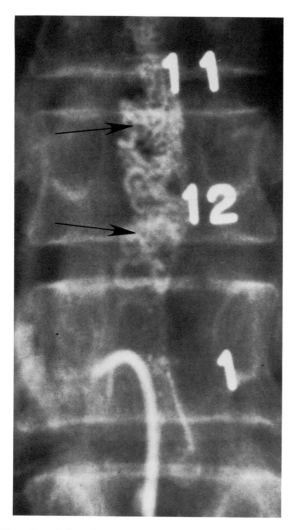

Fig. 14-116 Spinal AVM. A large vascular malformation is opacified in the lower spinal cord *(arrows)*.

Ring-like Gd enhancement may be seen. The cord changes resemble tumor and the correct diagnosis of dural AVM ismade when enlarged intradural veins are visualized. This is best done with CT myelography, although flow-voids within intradural veins are sometimes seen with MRI. Spinal angiography is the best means of diagnosis. A complete study is necessary, as feeders may arise from the vertebral, thyrocervical, or costocervical vessels above; the internal iliac vessels below; and any radicular artery in between.

SUGGESTED READING

Alazraki NP: Radionuclide imaging in the evaluation of infections and inflammatory disease. Radiol Clin North Am 31:783, 1993

Austin SG, Zee CS, Waters C: The role of magnetic resonance imaging in acute transverse myelitis. Can J Neurol Sci 19:508, 1992

Bassett LW, Gold RH, Webber MM: Radionuclide bone imaging. Radiol Clin North Am 19:675, 1981

Beres J, Pech P, Berns TF et al: Spinal epidural lymphomas: CT features in seven patients. AJNR 7:327, 1986

Breger RK, Williams AL, Daniels DL et al: Contrast enhancement in spinal MR imaging. AJNR 10:633, 1989

Brown BM, Schwartz RH, Frank E, Blank NK: Preoperative evaluation of cervical radiculopathy and myelopathy by surface-coil MR imaging. AJNR 9:859, 1988

Bundschuk CV: Imaging of the postoperative lumbosacral spine. Neuroimag Clin North Am 3:499, 1993

Bundschuk CV, Modic MT, Ross JS et al: Epidural fibrosis and recurrent disk herniation in the lumbar spine. AJNR 9:169, 1988

Byrd SE, Darling CF, McLone DG: Developmental disorders of the pediatric spine. Radiol Clin North Am 29:711, 1991

Christenson PC: The radiologic study of the normal spine: cervical, thoracic, lumbar and sacral. Radiol Clin North Am 15:133, 1977

Czervionke LF, Daniels DL, Ho PSP et al: Cervical neural foramina: correlative anatomic and MR imaging study. Radiology 169:753, 1988

Daniels DL, Hyde JS, Kneeland JB et al: The cervical nerves and foramina: local-coil MR imaging. AJNR 7:129, 1986

Donovan-Post MJ (ed): Computed Tomography of the Spine. Williams & Wilkins, Baltimore, 1984

Dorwart RH, Genant HK: Anatomy of the lumbar spine. Radiol Clin North Am 21:201, 1983

Dorwart RH, Vogler JB III, Helms CA: Spinal stenosis. Radiol Clin North Am 21:301, 1983

Epstein BS, Epstein JA, Jones MD: Lumbar spinal stenosis. Radiol Clin North Am 15:227, 1977

AVM must be differentiated from a hemangioblastoma of the cord, which appears similar but tends to have tightly packed small vessels and produces a more homogeneous tumor "blush." Some spinal AVMs are successfully treated by embolization through microcatheters.

Dural AVMs (fistulae) occur in the spine, particularly the lower thoracic and lumbar region. They are similar to the dural AVMs found intracranially (see Ch. 4). The fistula is fed by an intradural branch of a radicular artery, and large intradural veins may be present. They may produce subarachnoid hemorrhage and cord ischemia at a level some distance from the malformation. The spinal cord condition is called *subacute necrotizing myelopathy.*

Epstein BS, Epstein JA, Jones MD: Cervical spinal stenosis. Radiol Clin North Am 15:215, 1977

Flannigan BD, Lufkin RB, McGlade et al: MR imaging of the cervical spine: neurovascular anatomy. AJNR 8:27, 1987

Gardner WJ, Abdullah AF, McCormack LJ: The varying expressions of embryonal atresia of the fourth ventricle in adults. J Neurosurg 14:591, 1957

Haughton VM: Disc disease, degenerative spine disease, and tight spinal canal. p. 865. In Heinz ER (ed): Neuroradiology. Churchill Livingstone, New York, 1984

Haughton VM, Williams AL: Computed Tomography of the Spine. CV Mosby, St. Louis, 1982

Heinz ER: Craniovertebral abnormalities. p. 849. In Neuroradiology. Churchill Livingstone, New York, 1984

Hinks RS, Quencer RM: Motion artifacts in brain and spine MR. Radiol Clin North Am 26:737, 1988

Horton JA, Latchaw RE, Gold LHA, Pang D: Embolization of intramedullary arteriovenous malformations of the spinal cord. AJNR 7:113, 1986

Hyman RA, Gorey MT: Imaging strategies for MR of the spine. Radiol Clin North Am 26:505, 1988

Jackson DE Jr, Atlas SW, Mani JR et al: Intraspinal synovial cysts: MR imaging. Radiology 170:527, 1989

Karnaze MG, Gado MH, Sartor KJ, Hodges FJ: Comparison of MR and CT myelography in imaging the cervical and thoracic spine. AJNR 8:983, 1987

Lee SH, Coleman PE, Hahn FJ: Magnetic resonance imaging of degenerative disk disease of the spine. Radiol Clin North Am 26:949, 1988

Minami S, Sagoh T, Nishimura K et al: Spinal arteriovenous malformation: MR imaging. Radiology 169:109, 1988

Mirich DR, Kucharczyk W, Keller MA, Deck J: Subacute necrotizing myelopathy: MR imaging in four pathologically proven cases. AJNR 12:1077, 1991

Modic MT, Masaryk T, Boumphrey F et al: Lumbar herniated disk disease and canal stenosis; prospective evaluation by surface boil MR, CT and myelography. AJNR 7:709, 1986

Newton TH, Potts DG (eds): Computed Tomography of the Spine and Spinal Cord. Vol. 1: Spinal Dysraphism. Clavadel, San Anselmo, CA, 1983

Orrison WW, Eldevik OP, Sackett JF: Lateral C1–2 puncture for cervical myelography. III. Historical, anatomic and technical considerations. Radiology 146:401, 1983

Pagani JJ, Libshitz HI: Imaging bone metastasis. Radiol Clin North Am 20:545, 1982

Parizel PM, Baleriaux D, Rodesch G et al: Gd-DTPA-enhanced MR imaging of spinal tumors. AJNR 10:249, 1989

Penning L, Wilmink JT, van Woerden HH, Knol E: CT myelographic findings in degenerative disorders of the cervical spine: clinical significance. AJNR 7:119, 1986

Pettersson H, Harwood-Nash DCF: CT and Myelography of the Spine and Cord. Techniques, Anatomy and Pathology in Children. Springer-Verlag, Berlin, 1982

Porter BA, Shields AF, Olson DO: Magnetic resonance imaging of bone marrow disorders. Radiol Clin North Am 24:269, 1986

Raghavan N, Barkovich AJ, Edwards M, Norman D: MR imaging in the tethered spinal cord syndrome. AJNR 10:27, 1989

Robertson HJ, Smith RD. Cervical myelography: survey of modes of practice and major complications. Radiology 174:79, 1990

Ross JS, Masaryk TJ, Modic MT: MR imaging of lumbar arachnoiditis. AJNR 8:885, 1987

Ross JS, Delamarter R, Huestle MG et al: Gadolinium-DTPA-enhanced MR imaging of the postoperative lumbar spine: time course and mechanism of enhancement. AJNR 10:37, 1989

Ross JS, Modic MT, Masaryk TJ: Tears of the annulus fibrosis: assessment with Gd-DTPA-enhanced MR imaging. AJNR 10:12, 1989

Sage MR: Techniques in imaging of the spine. Part I. Plain-film radiology. p. 777. In Heinz ER (ed): Neuroradiology. Churchill Livingstone, New York, 1984

Sandhu FS, Dillon WP: Spinal epidural abscess: evaluation with contrast-enhanced MR imaging. AJR 158:405, 1992

Smoker WRK, Godersky JC, Knutzon RK et al: The role of MR imaging in evaluating metastatic spinal disease. AJNR 8:901, 1987

Teplick JG, Lassey PA, Berman A, Haskin ME: Diagnosis and evaluation of spondylolisthesis and/or spondylolysis on axial CT. AJNR 7:479, 1986

Tien RD: Fat-suppression MR imaging in neuroradiology: techniques and clinical application. AJR 158:369, 1992

Wang PY, Shen WC, Jan JS: MR imaging in radiation myelopathy. AJNR 13:1049, 1992

Warnock NG, Yuh WTC: Magnetic resonance imaging in the discrimination of benign from malignant disease of the lumbosacral vertebral column. Neuroimag Clin North Am 3:609, 1993

Williams AL: CT diagnosis of degenerative disc disease. Radiol Clin North Am 21:289, 1983

Yang PJ, Seeger JF, Dzioba RB et al: High-dose IV contrast in CT scanning of the postoperative lumbar spine. AJNR 7:703, 1986

Yu S, Haughton VM, Rosenbaum AE: Magnetic resonance imaging and the anatomy of the spine. Radiol Clin North Am 29:697, 1991

Zimmerman RA, Bilaniuk LT: Imaging of tumors of the spinal canal and cord. Radiol Clin North Am 26:965, 1988

15
Spinal Trauma

The diagnosis of spinal injury is one of the most significant tasks of the radiologist. An understanding of both the forces of injury and the structural and functional components of the spine makes it possible to analyze spinal derangements. Almost all fractures and dislocations occur in specific patterns. Multiple fracture patterns often occur from the multiple forces that apply in severe injuries.

MECHANISM OF INJURY

According to the two-column theory of the spine, proposed by Holdsworth (1963) and colleagues, there are two gross segments of the spine: the anterior and the posterior. These segments can be thought of as two parallel supporting columns. The anterior component consists of the vertebral body, intervertebral disc, annulus fibrosis, and anterior and posterior longitudinal ligaments. The posterior component consists of the pedicles, articular masses (pillars), lamina, spinous process, and posterior ligaments (ligamentum flavum, interspinous ligament, joint capsules). The three-column theory proposed by Denis categorizes the posterior vertebral body, annulus, and posterior longitudinal ligament as being a third significant component.

Basic principles relate vector force to the type of injury produced. The fundamental two-column theory indicates that a flexion injury causes compression of the anterior column (vertebral body) and distraction of the posterior column (spinous process, facet joints). An extension injury causes distraction of the anterior column and compression of the posterior column. Axial compression causes fractures of the anterior and the posterior columns. Shear leads to craniovertebral separation or fracture-dislocations at the disc space, particularly at the thoracolumbar junction. Lateral flexion causes fracture of the articular masses, and rotation results in unilateral facet joint dislocation. Because of anatomic variations, certain injuries are common to specific levels. The details of specific injuries are discussed below and are organized according to the spinal level and mechanism of the injury.

The motion of the spine occurs at the intervertebral disc and the synovial diarthrodial (free motion) facet joints. At all levels, the major stability of the spine is provided by the ligaments, particularly the ligamentum flavum and the posterior longitudinal ligament. When these ligaments are disrupted, the spine becomes unstable.

Spinal Stability

The goal of treatment for spinal injuries is to provide complete healing with preservation of normal, painless function of the spine and protection from future neurologic damage. When planning treatment, the concept of spinal stability is employed. Injuries that are unstable need fixation with prolonged traction, casting, or surgical fusion.

Stability has a complex definition that implies (1) maintenance of normal physiologic motion and limits of motion under normal conditions and during conditions of minor trauma; (2) stability of structure, to prevent further deformation or displacement of the spine; and (3) no progressive compression of neural tissue. There is no completely satisfactory system for classification of acute injuries as stable or unstable, although

683

White and Panjabi have offered a useful measurement. From cadaver studies, they found that the spine becomes unstable if there is more than a 3.5-mm anterior displacement of a vertebral body or an intervertebral angulation of more than 11 degrees. When this amount of displacement occurs, there is always disruption of the major posterior ligaments. In general, if only the anterior column (vertebral body) or the posterior column (articular process, lamina, and spinous process) is fractured and there is no displacement, the spine is stable. If both are fractured and there is displacement, the spine is unstable. For many injuries, however, it is difficult or impossible to determine stability.

IMAGING

Cervical Spine

Plain film radiography demonstrates most injuries and therefore remains the primary and essential imaging study. CT scanning is of use for certain fractures and is used to image levels that are not well examined with plain films. It is especially good for defining retropulsion of fragments into the spinal canal and for detecting fractures of the posterior arch and atlas. MRI is often useful for examining the acute injury with incomplete spinal cord injury, some subtle fractures (Fig. 15-1), and late complications resulting from the injury. With the availability of CT scanning, plain tomography has little advantage, except for horizontally oriented fractures, particularly of the odontoid process.

The lateral projection plain film radiograph is crucial in evaluating cervical spine injury. About 70 percent of fractures are recognized on this view and, except for the Jefferson fracture of the atlas, it demonstrates most important unstable fractures. When severe trauma has occurred to the cervical spine, this initial lateral view is obtained with a horizontal beam and the patient supine. Most significant injuries show some displacement on this view. However, it is a relocating position for cervical displacements, and the alignment of the spine may appear normal, even in the presence of significant unstable spinal injury. If the supine study is normal, additional

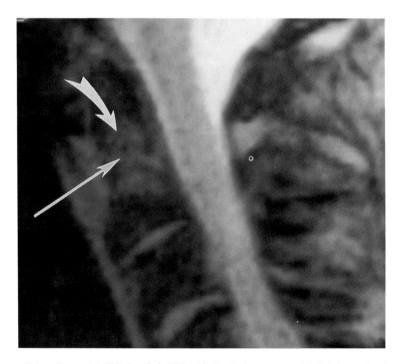

Fig. 15-1 Fracture of the odontoid, MRI. Spoiled GRE 10/260 10-degree sagittal MRI shows the fracture line through the base of the odontoid as high signal intensity *(thin white arrow)*. High signal represents edema within the marrow adjacent to the fracture *(large arrow)*.

lateral radiographs are obtained with controlled flexion-extension views to detect previously occult displacements.

Because a large number of injuries occur at the C7 level, it is essential that this level be examined. An oblique "swimmer's" projection or CT scans of this level can be obtained if the lateral film is inadequate.

In the neutral position, the anterior and posterior borders of the vertebral bodies, as well as the anterior margin of the posterior vertebral arch, form a gentle C curve, with slight lordosis (Fig. 15-2). Except at the C1–C2 level, there is equal space between the posterior spinous processes. The margins of the articular processes are aligned (Fig. 15-3). Any sudden change in angulation, displacement, or apposition suggests an injury. An exception is at the C7 level, where the superior facet is long and the posterior margin of the inferior facet surface of C6 does not align with the posterior margin of the superior facet of C7. A groove is frequently present on the superior facet of C7 (Fig. 15-2A).

Predictable normal displacements (less than complete dislocations) occur with motion of the cervical spine. With flexion, the spinous processes separate nearly equally, with somewhat larger separation of the C1–C2

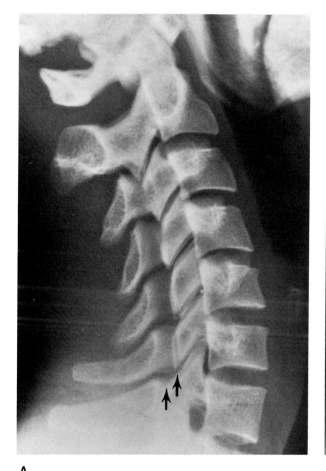 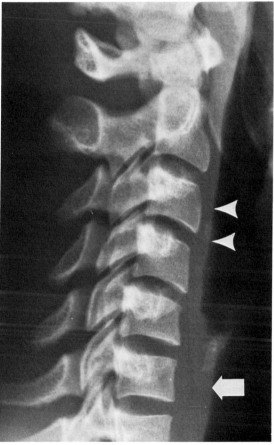

A   B

Fig. 15-2 Normal flexion-extension of the cervical spine. (A) Extension view. There is a slight posterior overlap of the facet joints with extension. Note the normal anterior offset of the C6 facet with relation to C7 *(black arrows).* **(B)** Normal neutral position. There is slight anterior displacement of the middle vertebral bodies secondary to the effects of gravity. Note the normal retropharyngeal space *(white arrowhoeads)* and retrotracheal space *(white arrow). (Figure continues.)*

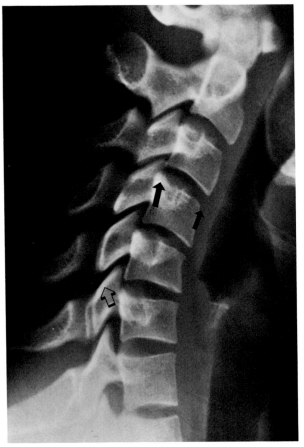

C

Fig. 15-2 *(Continued).* **(C)** Normal flexion view. There is stepwise anterior displacement of the vertebral bodies *(black arrows).* Separation may also occur at the posterior facet joints *(open arrow).*

interspinous space. The posterior borders of the vertebral bodies define a smooth anterior curve, about uniform in children but maximal in the lower regions in adults. As a consequence of the forward sliding of the vertebral bodies, small steps are formed at each level (Fig. 15-2C).

Pseudosubluxation is the term generally applied to the hypermobility of the C2–C3 level and, to a somewhat lesser extent, to the C3–C4 level in infants and children. The anterior displacement is attributed to laxity of the ligaments. The physiologic displacement may be difficult to differentiate from pathologic displacement associated with flexion sprain. The posterior cervical line is helpful. This line is drawn from the anterior margin of the posterior arch of C1 to the anterior margin of the

posterior arch of C3. The anterior margin of the posterior arch of C2 does not normally extend more than 2 mm in front of this line. If the posterior arch moves forward more than this, flexion sprain may be diagnosed.

Extension produces reverse motion. The spinous processes are closely apposed, and the cervical lordosis is accentuated. Slight posterior steps result, again defining a gentle curve (Fig. 15-2A).

Cervical spondylosis affects the alignment of the spine. With degenerative narrowing of the disc space, a slight retrolisthesis may result. With arthrosis of the intervertebral facets joints, a slight anterolisthesis may result from laxity of the articular ligaments.

The relationship of the anterior arch of C1 and the

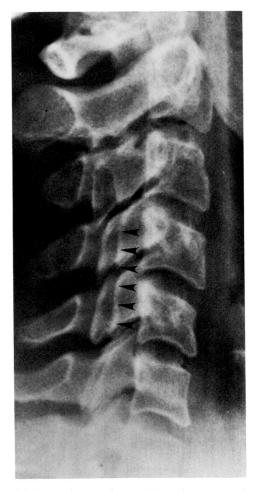

Fig. 15-3 Normal cervical spine. Note alignment of the posterior margins of both columns of pillars *(arrowheads)*, which is best seen on this slightly oblique view.

dens (the atlantodental interval) is evaluated on the lateral projection. In the neutral and extension positions, the interval is small, on the order of 1 mm. With flexion, the interval may widen, but it should not exceed 3 mm in the adult or 5 mm in young children. If these dimensions are exceeded, C1–C2 instability may be diagnosed.

The lateral projection also provides a good view of the anterior cervical soft tissues (Fig. 15-2B). With many types of cervical injury, there is an associated prevertebral hematoma. This causes a mass effect that can be seen as increased thickness to the prevertebral retropharyngeal or retrotracheal spaces. There may be anterior displacement of the prevertebral fat stripe. The hematoma may be large or small. It may be the only finding with occult fractures of the upper cervical spine, especially hyperextension injuries, but significant cervical spine injury frequently occurs without soft tissue hematoma or edema. The normal prevertebral soft tissue measurements are given in Table 15-1.

On the anteroposterior (AP) projection, the posterior spinous processes are normally midline. Rotational injury results in displacement of the spinous processes away from the direction of the rotation. Other findings may include widening of a vertebral body, which indicates burst fracture, displacements from shear, and fractures of the pillars. The space of the uncovertebral joints are normally small and uniform throughout. Rotational injuries may produce separation of the joints.

The pillar projection better defines the articular pillars. The central ray is angled about 45 degrees caudad with the head turned away from the side of interest. This view demonstrates compression fractures of the pillar, which are most common at the C6 level.

The AP open-mouth view of the C1 level is part of the complete cervical spine examination and shows the axis and atlas to advantage. It is used to detect fractures of the atlas (Jefferson type) and dens. The evaluation of this view is often difficult, mainly because of the numerous variations in the relation of the atlas and axis with head position. Fracture of the ring of the atlas results in lateral offset of the lateral masses of the atlas on both sides. Whenever there is doubt about a fracture of the atlas, a CT scan is performed.

There are normally two ossification centers for the odontoid and occasionally an accessory ossification center at the tip. Fusion of the dens at the subdental synchondrosis occurs at about 5 to 7 years of age. This fused synchondrosis may be visualized for many years as a condensation of bone along the fusion line at the base of the dens. It must not be confused with a compression fracture. Frequently the dens does not completely fuse with the axis, leaving an open horizontal synchondrosis at the base. This lucent line has slightly sclerotic borders. The dens may have a split in its superior portion from nonfused ossification centers. A vertical nonfused synchondrosis may also occur in the anterior arch of the atlas.

SIGNIFICANT SIGNS OF CERVICAL TRAUMA

Retropharyngeal and retrotracheal hematoma
Abnormal vertebral alignment
    Acute kyphotic angulation
    Widened interspinous space
    Rotation of a segment
    Displacement of a vertebral body
Abnormal joints
    Widened atlantodental interval
    Widened or narrowed intervertebral disc
    Widened facet joint
    Widened uncovertebral joint
    Widened or displaced atlantoaxial joint
Fractures
    Vertebral body
        Avulsions from anterior vertebral body
        Compression of the vertebral body
        Linear fracture, vertical or horizontal
    Vertebral arch
        Pillar
        Spinous process
        Lamina, horizontal or vertical
        Pedicle
        Transverse process

Table 15-1 Cervical Prevertebral Soft Tissue Measurements

Site	Average (mm)	Range (mm)
Retropharyngeal space	3.5	2–7
Retrotracheal space		
Children	7.9	5–14
Adults	14	5–22

(Modified from Gehweiler et al., 1980, with permission.)

The right and left oblique views are performed with the patient's body turned 45 degrees, visualizing the intervertebral foramina on the side opposite to the direction of turning; this approach is best for visualizing rotational injuries. In the event of a rotational injury, the superior pillar at the level of the injury is displaced anteriorly into the upper portion of the intervertebral foramen, and the facets do not align. The offset of the uncovertebral joints can also be seen in this projection. A list of the significant signs of cervical spinal trauma is given in the box.

Thoracic and Lumbar Spine

Injuries in the thoracic and lumbar spine are less common, and there are fewer injury patterns. Lateral and AP views suffice for detection in most cases. The same principles discussed above apply to these regions. CT scanning is useful for defining the full extent of the injury and the presence of fragments within the spinal canal.

HYPERFLEXION INJURIES

Complete Bilateral Facet Joint Dislocation

Complete bilateral facet joint dislocation is the most common type of hyperflexion injury and accounts for about 13 percent of all cervical spine injuries. It occurs most often at the C5–C6 and C6–C7 levels and results from severe hyperflexion. The injury is due to falls (particularly down stairs), automobile accidents, and contact sports.

The force is directed from below at the occipital region of the skull. This strong upward and forward force causes complete disruption of all the posterior ligaments, the annulus of the disc space, and usually the anterior longitudinal ligament. Such injury permits the facet joints to dislocate completely, with the pillars above becoming locked anteriorly to the pillars below. Occasionally, the dislocation of the facet joints is partial, with the posterior tip of the superior pillar just in contact with the anterosuperior aspect of the inferior pillar.

The dislocation is best seen on the lateral radiograph (Fig. 15-4). There is anterior displacement of the vertebral body by 50 percent or more, overriding and locking of the pillars into the intervertebral foramina, and widening of the interspinous distance. Other minor fractures

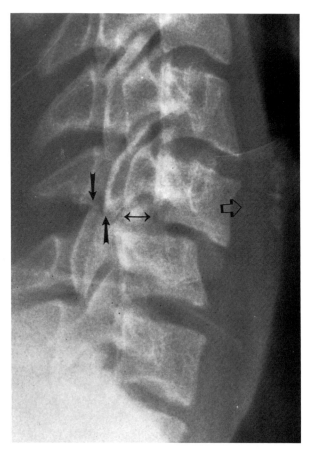

Fig. 15-4 Hyperflexion dislocation (bilateral "locked facets"). There is 50 percent anterior displacement of C4 on C5 (*horizontal connected arrows*). The articulating pillars are totally displaced and locked (*opposing black arrows*). Note the prevertebral hematoma (*open arrow*).

may occur of the vertebral body or pillars. Significant neurologic injury is common. The injury is unstable and requires reduction and fixation.

Hyperflexion Sprain

Hyperflexion sprain (subluxation) is a much less severe form of the hyperflexion dislocation described above. The force of injury is much less, resulting in partial injury to the posterior ligament complex. The anterior displacement stops short of complete dislocation, and the spine reduces to nearly its normal alignment.

The horizontal beam lateral radiograph may be normal, and the injury may become apparent only on the flexion or upright lateral view. Extension realigns the spine. Usually, there is only slight separation of the posterior facet joints, the posterior intervertebral disc, and the interspinous distance at the level of injury. As associated avulsion fracture of the posterior arch or compression of a vertebral body sometimes occurs (Fig. 15-5). Neurologic damage is rare.

The posterior ligament complex is always weakened or disrupted with this injury. However, the severity of the injury can be estimated only by the displacement with flexion. The posterior ligaments heal poorly, and up to 50 percent show delayed evidence of instability with increased anterior displacement at the level of injury. Follow-up radiographs are performed 3 months after the injury to detect progressive instability.

It may be difficult, if not impossible, to distinguish hypermobility from slight sprain. Subtle compression of a vertebral body or avulsion of the posterior arch is sought, as it may be the only firm indication of injury. Clinical correlation and follow-up films are necessary in this situation.

Spinous Process Fracture

About 15 percent of persons with cervical spine injury have a fracture of the spinous process (Fig. 15-5). Most of these fractures occur as part of a fracture or sprain complex. Only 2 percent have an isolated spinous process fracture (clay-shoveler's fracture).

It occurs predominantly at C6–T1 near the base of the spinous process and results from hyperflexion and the pull of the trapezius and rhomboid muscles and the nonelastic interspinous ligament. Small avulsion fractures may occur at the posterior tip of the process, resulting from the attachment of the ligamentum nuchae being torn out. There is a high association with occult hyperflexion sprain. An isolated spinous process fracture is stable.

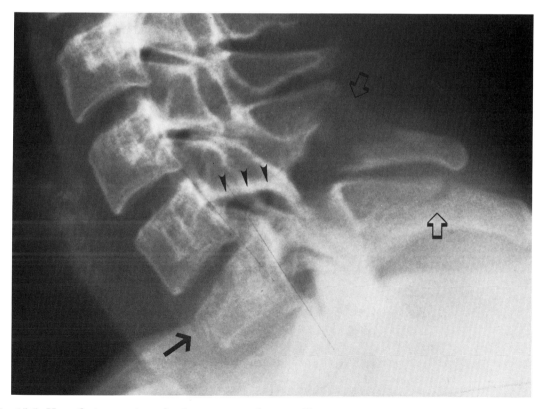

Fig. 15-5 Hyperflexion sprain and spinous process fracture. There is anterior displacement of C6 on C7 with separation of the facet joints *(arrowheads)* indicating sprain. Fractures are seen through the spinous processes *(open arrows)*. There is also compression of the C7 vertebral body *(black arrow)*.

Hyperflexion-Rotation Injuries (Unilateral Facet Joint Dislocation)

Unilateral dislocation of a facet joint is the characteristic injury of simultaneously occurring hyperflexion and rotation forces. It often results in the unilateral "locked facet." This injury accounts for about 4 percent of all cervical spine injuries and is most common at the C4–C7 levels (Fig. 15-6).

The superior vertebral body at the level of injury rotates around a pillar, causing the opposite superior pillar to dislocate forward over the inferior pillar, finally coming to lie in the intervertebral foramen. Usually the pillar

becomes locked in this forward position. The superior tip of the inferior pillar may be fractured. The capsules of the facet joints on both sides are torn, and the interspinous ligament is always ruptured. However, the posterior longitudinal ligament and the annulus of the disc space are variably injured and may remain intact. Laminar fracture is rare.

Multiple radiographic views are necessary to define the injury. The oblique projection is the best for showing the dislocated facet, and the AP projection is the best for showing the rotational component of the injury.

When the dislocated pillar is in the locked position, the injury is stable, and there may be difficulty reducing

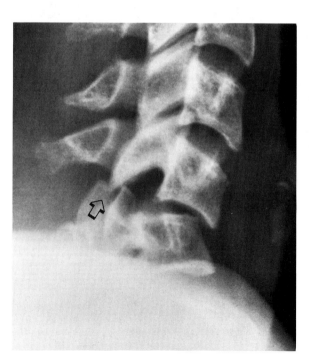

A

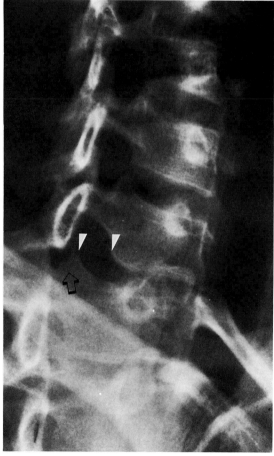

B

Fig. 15-6 Hyperflexion rotation (unilateral "locked facets"). (A) The lateral projection shows the unilateral facet locking *(arrow)* with less than 50 percent anterior offset of the displaced segment. There is widening of the interspinous space posteriorly. **(B)** Oblique view shows rotational displacement of the superior articular pillar into the intervertebral foramen *(open arrow)*. Note the wide separation of the uncovertebral joint *(arrowheads)*.

the spine to its normal alignment. Once the dislocation
has been reduced, the spine is usually unstable and requires fixation for proper healing. Dislocations that are
more than 4 months old are considered stable. No attempt at reduction is made at this point, as it might result
in new neurologic injury. Almost all neurologic injuries
involve a nerve root or the brachial plexus.

On the lateral radiograph, the hyperflexion unilateral
locked facet injury often cannot be differentiated from
the hyperextension-compression fracture-dislocation.
The two injuries can be differentiated by demonstration
of more severe pillar and posterior arch fractures that
occur with the hyperextension compression injury. CT
scanning is useful for identifying these fractures.

HYPERFLEXION-
COMPRESSION INJURIES

"Teardrop" Hyperflexion Fracture-
Dislocation

The teardrop hyperflexion fracture dislocation injury
occurs less often than hyperflexion dislocation, accounting for about 5 percent of cervical injuries. Almost all
fractures occur at the C5 level. The injury is the worst
result of a severe compressive hyperflexion force. Lesser
force results in only wedge compression of the vertebral
body or wedge fracture with hyperflexion sprain (Fig.
15-5).

As a result of the combination of hyperflexion and
compressive forces, the compressive vector is moved in
an arc. The compression force begins with a straight

downward vector but converts to a backward-directed
vector, causing the fractured vertebral body to be rotated
downward and backward. The resistance of the vertebral
body below produces the avulsion "teardrop" fracture.

The avulsed fragment is displaced and rotated anteriorly (Fig. 15-7). The fractured vertebral body is shoved
backward into the spinal canal, producing a kyphotic
angulation. The interspinous distance is widened because of the disruptive forces applied to the posterior
column. The disc space, the posterior longitudinal ligament, and the posterior ligament complex are disrupted,
rendering the spine unstable. Frequently there are fractures of the pedicles or lamina. Additional posterior injuries may occur at other levels.

This fracture complex is particularly common after a
diving injury, but it can also result from falls and automobile accidents. The injury is best demonstrated on the
lateral radiograph. Neurologic damage is frequently seen,

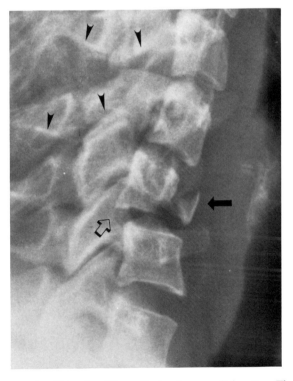

Fig. 15-7 "Teardrop" hyperflexion fracture-dislocation. The
anterior avulsed fracture can be seen with its typical rotation
(*black arrow*). There is retropulsion of the vertebral segment
into the spinal canal (*open arrow*). Injuries have also occurred
at more superior levels with widened interspinous spaces and
facet joint displacement (*arrowheads*).

correlating with the degree of retropulsion of the vertebral body into the spinal canal. This hyperflexion-compression injury is distinguished from the hyperextension fracture-dislocation and the compression "burst" fractures discussed below.

Simple Wedge Compression Fracture

The isolated simple wedge compression fracture is the simplest type of hyperflexion-compression injury. It results from the effect of a relatively mild compressive force acting along the anterior aspect of the vertebral column. The fracture almost always occurs at the C5 or C6 level and may be seen at both levels simultaneously. Although there is a hyperflexion component to the forces applied, usually the force is slight, so the posterior ligament complex stretches but does not rupture. However, with stronger force, the fracture may be associated with hyperflexion sprain. Occasionally, there is an associated arch fracture (lamina, spinous process, pedicle, or transverse process) from the pull of the intact ligaments. When it occurs as an isolated injury, it is stable and is almost never associated with significant neurologic damage.

The injury is best demonstrated on the lateral radiograph and is identified as a loss of the height of the anterior portion of the vertebral body. Care must be taken to differentiate the injury from the normally occurring slight decrease in vertical height of the anterior portions of C5 and C6. Such differentiation may be impossible in the case of minor fractures, and the clinical setting determines the likelihood of fracture. MRI may show edema with the vertebral marrow during the acute phase (Fig. 15-1), and the radionuclide (RN) bone scan will become positive after a few days.

The compression wedge fracture is most frequently associated with other injuries of the cervical spine, particularly hyperflexion sprain and unilateral or bilateral locking facet (Fig. 15-5). Therefore, whenever a wedge fracture is identified, a flexion view and a CT scan are obtained to determine the presence of occult posterior injury. Follow-up radiographs are repeated in 3 months to detect delayed instability from further collapse or associated sprain.

Avulsion Fracture of the Vertebral Body

With slightly greater compressive force than with simple wedge fracture, an avulsion of the anterosuperior aspect of a vertebral body may occur (Fig. 15-8). It must

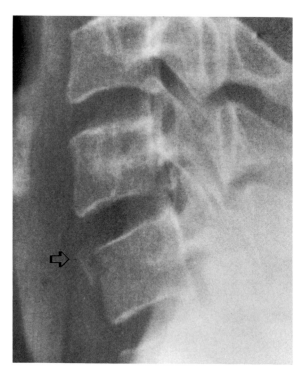

Fig. 15-8 Avulsion fracture of the vertebral body. This fracture of the anterosuperior aspect of the vertebral body is due to hyperflexion with mild compression *(arrow).*

be differentiated from the avulsion from the anteroinferior aspect of the vertebral body that generally occurs with hyperextension and may indicate severe occult disruption (see next section).

HYPEREXTENSION-DISRUPTION INJURIES

Accounting for about 2 percent of all spine injuries, the hyperextension sprain is infamous for its lack of radiologic findings in association with severe spinal cord injury. The injury, which may occur at any level, is the result of a hyperextension-disruptive force applied to the forehead, face, jaw, or throat. The upward and backward force compresses the spinous processes and the articular pillars, so they act as a fulcrum around which the anterior portion of the vertebral bodies rotate. This condition causes rupture of the anterior longitudinal ligament and the intervertebral disc, which is either bisected or torn from the inferior endplate of the vertebral body above. From the anteroinferior aspect of this vertebral body, a small bone fragment may be avulsed by the annulus. It is important to note that, aside from prevertebral soft tissue swelling, this fragment may provide the only clue of the significant occult injury.

The high incidence of severe neurologic damage results from the nature of the spinal disruption. The vertebral bodies above the level of the disruption dislocate posteriorly, pinching the spinal cord on the fixed upper margin of the lamina below. Minor injuries cause posterior cord syndromes, with loss of position sense and motor weakness, but complete cord injury is more common. This injury is a momentary dislocation, and the cervical spine reduces to its normal anatomic position, which accounts for the normal radiographic findings in the presence of severe neurologic dysfunction. The injury is unstable to extension motion.

The diagnosis may have to be made by assumption, in the characteristic clinical setting. The injury almost always occurs in persons over 50 years of age. It is considered a result of the greater susceptibility of degenerated discs to the forces of disruption.

When presented with a patient who has spinal cord injury and a normal spine radiograph, it is imperative that there be no extension of the neck. Because of the "pincer effect," small posterior movements of the neck may cause significant neurologic damage. It is especially true if the patient is somnolent or if tracheal intubation is attempted. It is crucial never to attempt a deliberate extension or hanging head radiograph in an effort to establish the diagnosis. MRI may diagnose this condition by the demonstration of cord injury, prevertebral hematoma and disruption of the anterior longitudinal ligament (Fig. 15-9). The ligament disruption is best demonstrated with gradient echo recalled (GRE) imaging sequences.

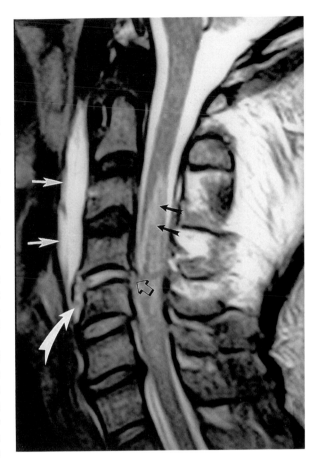

Fig. 15-9 Hypertension-disruption injury. Sagittal FSE T_2 MRI demonstrates the rupture of the anterior longitudinal ligament *(large white arrow)* and the posterior longitudinal ligament *(open arrow)*. The high signal within the C4–5 disc interspace is edema and hemorrhage. Prevertebral hemorrhage and edema is almost always present *(small white arrows)*. High signal within the cord represents the edema from cord injury that is common with this injury *(small black arrows)*.

HYPEREXTENSION-DISRUPTION INJURY

Severe cord injury with little radiographic abnormality

Prominent prevertebral swelling

Small avulsion "chips" from the anteroinferior margin of vertebral body

Gas phenomenon of a disc space

Widening of the injured disc space (may be subtle)

HYPEREXTENSION-COMPRESSION INJURY

Hyperextension-Compression Fracture-Dislocation

Hyperextension-compression fracture-dislocation is a common injury complex in the cervical spine, accounting for about 10 percent of cervical spine injuries. It occurs at the C4 through C7 level. The injury is thought to be a result of both hyperextension and compression forces moving the head in a posterior arc. At first, the downward compression force is applied to the posterior arch, causing variable fractures of the pillars, spinous process, and pedicles. Once they have fractured, the vertebral body is free to move anteriorly because of the combined forces acting through the arc, causing rupture of the disc space below.

In most cases, only one articular pillar is fractured, owing to a rotational component to the force. The pillar is comminuted and is characteristically rotated into the horizontal plane (Figs. 15-10 and 15-11), permitting visualization of the cleft of the facet joint on the AP projection. The horizontal position can also be recognized on the lateral and oblique projections. Varying degrees of comminution of the arch occur. Bone fragments may be displaced into an intervertebral foramen. The injuries

---

HYPEREXTENSION-COMPRESSION INJURY

Anterior displacement of the vertebral body with rotation

Horizontal rotation of the fractured articular pillar

Anterior displacement of a pillar into the intervertebral foramen

Posterior arch fractures

Separation of uncovertebral joint

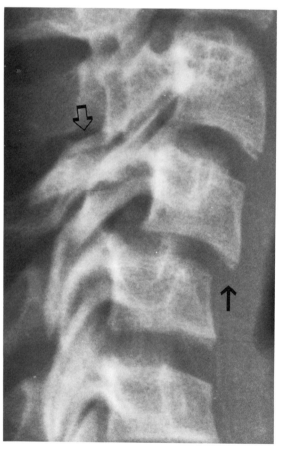

A

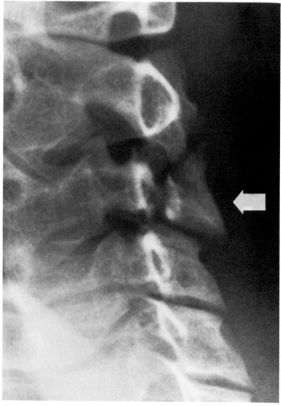

B

Fig. 15-10 **Hyperextension compression fracture-dislocation.** (A) Lateral radiograph shows anterior displacement of a vertebral segment (*black arrow*), with a fractured and horizontally rotated pillar (*open arrow*) (B) Oblique view shows fractured pillar, which is horizontally rotated and laterally displaced (*arrow*).

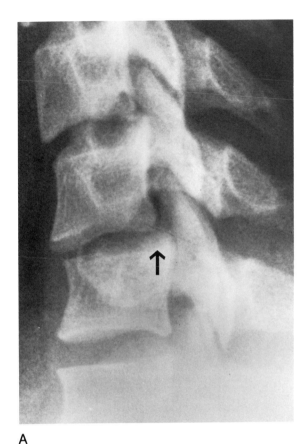
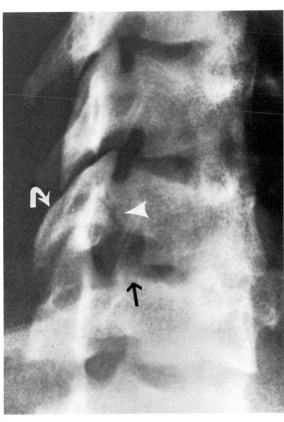

A

B

Fig. 15-11 Hyperextension compression fracture-dislocation. (A) Lateral radiograph shows subtle anterior offset of a vertebral segment *(arrow)*. **(B)** Oblique view shows fracture through the pedicle *(white arrowhead)*, horizontal rotation of the pillar *(curved arrow)*, and widening of the uncovertebral joint *(black arrow)*.

may occur at two contiguous levels (Fig. 15-12). Multiple projections are needed for an accurate diagnosis. Most of the injuries occur as a result of automobile accidents.

The CT scan shows the injury to good advantage (Fig. 15-13). Rotation of the vertebral body is well demonstrated with separation of the uncovertebral joints. The fractures of the pillar and posterior arch are easily seen, and any displaced bone fragments can be identified.

This fracture-dislocation is generally considered unstable because of the posterior arch fractures and the disconnection of the arch from the vertebral body. There is usually little or no neurologic injury. Treatment consists in restoration of anatomic alignment and stabilization by prolonged immobilization. Early posterior fusion is not satisfactory because of the disconnection of the vertebral body from the posterior arch.

The anterior displacement of the vertebral body suggests a flexion-rotation type of injury, with unilateral locking facets. The two injuries may appear identical on the lateral radiograph. The differentiation of the two injuries is based primarily on the presence of a pillar fracture other than simple avulsion. Pure hyperflexion injuries do not produce compression fractures of the pillars.

Pillar Fracture

The isolated pillar fracture results from a lesser force, with more rotational component than occurs the hyperextension fracture-dislocation, discussed in the preceding section. A vertical fracture of the pillar causes retropulsion and sometimes lateral displacement of a fracture fragment, which is seen on the lateral, oblique, and pillar

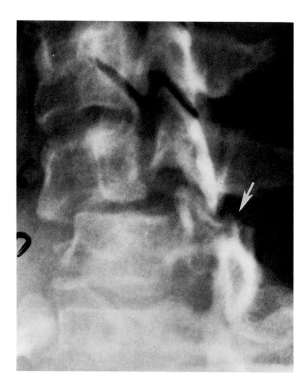

Fig. 15-12 Hyperextension compression fracture-dislocation. Sometimes it is the pillar from the nondisplaced segment that is fractured *(arrow)*.

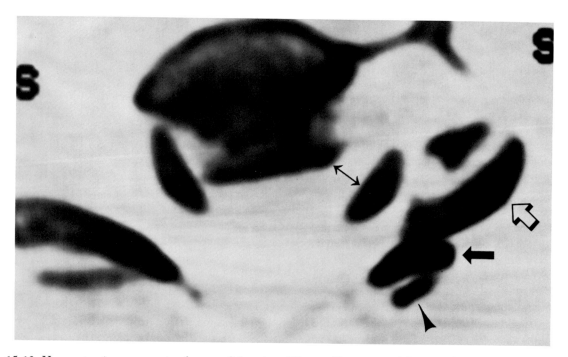

Fig. 15-13 Hyperextension compression fracture-dislocation, CT scan. The transaxial CT scan shows rotation of a vertebral body segment with widening of the uncovertebral joint *(opposed arrows)*. The left superior pillar is fractured and rotated horizontally and anteriorly into the intervertebral foramen *(open arrow)*. It is displaced anterior to the intact inferior pillar *(solid black arrow)*. A fragment of the superior pillar remains posteriorly *(arrowhead)*.

views. The CT scan shows the fracture well. The injury may result in radiculopathy.

VASCULAR INJURY

Injury to the vertebral artery following fracture dislocation of the cervical spine or pillar fracture is probably more common than is generally thought. The artery may be occluded, or it may be injured with dissection or pseudoaneurysm formation. Most injuries to the vertebral arteries remain asymptomatic, but infarction of the brain may occur if there is congenital abnormality of the vertebral supply to the brain or if emboli are produced by a dissection or intimal tear. X-ray angiography (XRA) is the best test for definitive diagnosis of vascular injury, although MRA may be adequate for screening of possible injury.

VERTICAL COMPRESSION INJURY

The "burst" fracture of the vertebral body results from a pure compressive force that is applied at precisely the instant the spine is held straight. It occurs almost exclusively in the cervical and lumbar spinal segments. The normal kyphosis of the thoracic spine converts an axially directed force into a forward rotational-compressive force, causing a wedge-type fracture. Vertical compression may cause fracture of the C1 ring (Jefferson fracture), which is discussed in a following section.

The burst-type injury is the result of the nucleus pulposus breaking through the thin inferior endplate of a vertebral body. This upward herniation of the nucleus into the centrum of the vertebral body causes the comminuted fracture and outward displacement ("bursting") of the fracture fragments. There is always a vertical fracture of the vertebral body (Fig. 15-14), which differentiates the burst fracture from the simple wedge fracture. The "burst" fracture may occur alone or as part of the hyperflexion-compression teardrop fracture dislocation. Major neurologic injury often results from retropulsion of a fracture fragment into the spinal canal. The CT scan is useful for defining all the components of the injury.

The burst compression fracture may be severe and associated with considerable disruption of the posterior arch, pillars, pedicles, and disc space. In this situation, the spine is unstable and requires internal fixation. Sometimes axial traction relocates retropulsed fragments, but often surgical removal of the fragment is necessary.

BURST FRACTURE

Vertical fracture through the vertebral body

Widening of the margins of the vertebral body

Loss of height of vertebral body

Retropulsion of a fracture fragment

Fractures of the posterior arch

FRACTURES AND DISLOCATIONS OF THE CRANIOVERTEBRAL JUNCTION

Odontoid (dens)
  Superior type I (rare)
  Transverse, type II (most common)
  Body of axis, type III
Pure atlantoaxial dislocation
  Transverse ligament rupture
Atlantoaxial rotatory dislocation
Vertebral arch fractures
  Bilateral pedicles (hangman's)
  Unilateral pedicle (unusual)
  Spinous process and lamina
Vertebral body avulsion
Fractures of the atlas
  Jefferson fracture
  Posterior arch fracture
  Anterior arch fracture
  Avulsion, anteroinferior margin
  Horizontal fracture of the arch
Atlanto-occipital junction
  Craniovertebral separation
Associated injuries at lower levels

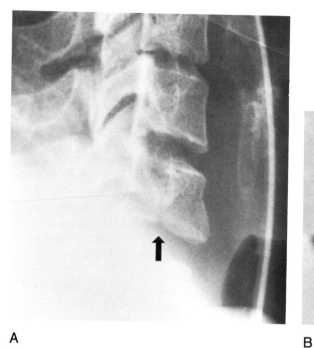

A

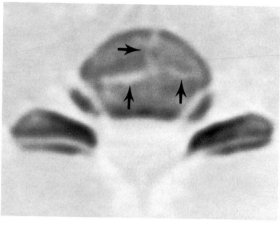

B

Fig. 15-14 Burst fracture. (A) Lateral cervical spine view shows vertically oriented fracture through the mid-vertebral body with subtle anterior displacement of the fracture fragment (*arrow*). (B) Transaxial CT scan shows the vertically oriented fracture lines of the burst injury (*arrows*).

CRANIOVERTEBRAL JUNCTION

The occiput, atlas, and axis together constitute the region of the craniovertebral junction. The anatomy of this region differs from the remainder of the spine. The same forces that affect the lower cervical spine cause the injury to the craniovertebral junction, but unique fractures occur as a result of the impact of the base of the skull on C1 and C2. Neurologic injury is uncommon, probably because of the wide spinal canal at this level.

Fractures and dislocations of the craniovertebral region account for 35 percent of all cervical injuries. The injuries may result from hyperextension, hyperflexion, and vertical compression. Axis (C2) injuries occur about four times as frequently as atlas (C1) injuries. The two most common injuries of the axis are fracture of the dens, accounting for about 13 percent of all cervical injuries, and bilateral pedicle fracture with dislocation, the hangman's fracture, accounting for about 7 percent of all cervical injuries. Because of the often subtle findings on radiographs and the infrequent association of neurologic injury, fractures and dislocation may be difficult to diag-

nose. Controlled flexion-extension lateral filming, plain film tomography, and CT scanning may be necessary. There is a high association of injuries at lower levels of the cervical spine.

Fracture of the Dens (Atlantoaxial Fracture-Dislocation)

The most common injury at the atlantoaxial level is fracture of the dens. The mechanism of the injury determines the direction of the displacement of the dens. The most common anterior displacement results from hyperflexion, with the strong transverse component of the cruciate ligament holding fast and shearing off the dens. With posterior dislocation, hyperextension causes the anterior arch of the atlas to move posteriorly, shearing the dens. Lateral displacement is rare. The displacements may be subtle, and there may simply be slight angulation with respect to the posterior margin of the body of the axis.

The fracture may be at any level in the dens. The rare type I fracture occurs near the superior tip of the dens.

Type II is a transverse fracture near the base of the dens but not within the body of the axis; it heals poorly. Type III is a fracture of the foundation of the dens in the anterosuperior portion of the body of the axis (Fig. 15-15); it is more stable than the type II fracture and heals much more readily.

In the absence of any displacement, radiographic diagnosis of a fracture of the dens is often difficult. Subtle changes in alignment or minute breaks in the cortical margin of the dens or anterior axis may be all that is apparent on the plain radiographs. Tomography may be necessary to demonstrate the fracture, particularly if it is oriented in an oblique plane. Delayed tomography may demonstrate a fracture not seen initially. MRI may demonstrate the fracture when other imaging is normal (Fig. 15-1). A fracture must be differentiated from (1) failure of fusion of the dens with the body of the axis at the subdental synchondrosis, and (2) Mach bands that result from the overlap of the margins of the anterior and posterior arches of C1 and the teeth. With persistent synchondrosis, there are sclerotic margins to the linear defect and no instability. CT may demonstrate a type III fracture that is occult on plain films. Controlled flexion-extension views may be necessary to make the diagnosis but are avoided if possible.

Neurologic injury occurs in only a few of the patients with odontoid fracture-dislocation. Many persons with this fracture complain only of neck stiffness or pain, sometimes with an associated torticollis. Persons with odontoid fracture often seek medical attention weeks after sustaining the injury.

Pure Atlantoaxial Dislocation

Pure atlantoaxial dislocation occurs infrequently. It results from a severe hyperflexion injury that causes disruption of the transverse ligament of the atlas, which normally holds the odontoid close against the anterior

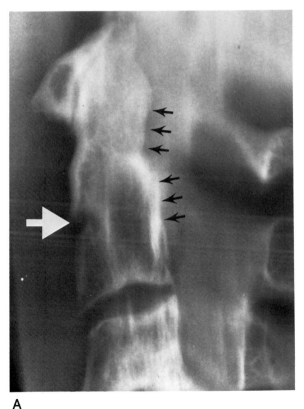

A

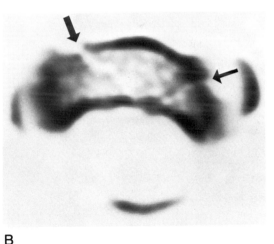

B

Fig. 15-15 Odontoid fracture, type III. (A) Lateral tomography shows a cortical fracture of the anterior axis (*white arrow*). There is a subtle change in the alignment of the odontoid with the inferior body of the axis (*black arrows*). **(B)** Transaxial CT scan shows a fracture through the body of the axis (*arrows*).

arch of the atlas. The factors that determine whether the ligament ruptures or the dens fractures are unknown, except in the obvious cases of rheumatoid arthritis, ankylosing spondylitis, Down syndrome, infection, metastatic tumor, or other processes that weaken the ligaments.

The diagnosis is made with the lateral projection. The predental space is widened to more than 3 mm in adults and 5 mm in children under age 12. Controlled flexion views may be necessary to make the diagnosis.

The injury is unstable and requires fixation. Patients present with either upper neck pain or cord dysfunction. Cord injury is more common with this injury than with the fracture-dislocation of the dens.

Atlantoaxial Rotatory Dislocation (Fixation)

Atlantoaxial rotatory dislocation is uncommon. It results from excessive rotation of the atlas, although the degree of rotation necessary to cause fixed dislocation is variable. It is usually at least 45 degrees with respect to the transverse plane of the axis. The mechanism of the injury is not well understood.

The dislocation may occur spontaneously or after trauma, respiratory infection, or surgical procedures, particularly those of the head and neck region. The fixed rotation causes torticollis. However, most people with

torticollis do not have demonstrable rotatory dislocation of the atlas.

The diagnosis depends on the demonstration of abnormal motion of the atlantoaxial junction with attempted rotation, which is best seen with fluoroscopy or cineradiography. Normally, the atlas moves at least the first 20 degrees without motion of the axis or the lower cervical spine. With rotatory fixation, the abnormal relationship of the atlas and the axis persists in all positions. When the transverse ligament is disrupted as well, the atlas may displace forward as well as rotate, and the predental space is abnormally wide. The spinal canal may become narrowed, causing cord compression. A CT scan can also demonstrate the fixed rotatory dislocation (Fig. 15-16).

Hangman's Fracture

Sometimes referred to as traumatic spondylolisthesis of C2, the fracture-dislocation known as hangman's fracture is complex. It may result from variable forces, but hyperextension is the most common cause. The pedicles are fractured bilaterally, and there is variable anterior displacement of the vertebral body, depending on the degree of disruption of the anterior ligaments and disc space. There may be dislocation of one or both of the

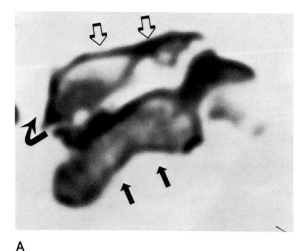
A

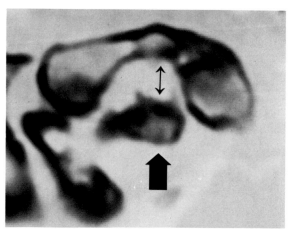
B

Fig. 15-16 Atlantoaxial rotatory dislocation, CT scans. (A) The rotated atlas (open arrows) is locked in front of the axis (black arrows). (B) The atlas is also displaced anteriorly with widening of the predental space (opposed arrows). The large black arrow points to the dens.

HANGMAN'S FRACTURE (C2)

Bilateral fractures of the pedicles

Anterior dislocation of the body of the axis

Anterior rotation of C2

Wedge compression of C3 (occasional)

Injuries at lower levels

facet joints, in which case the posterior ligaments are also disrupted and the posterior arch moves forward as well. A prevertebral hematoma is generally seen. Simultaneous hyperflexion injuries are frequent in the lower cervical spine. The injury is best diagnosed with the lateral plain radiograph (Figs. 15-17 and 15-18).

The bilateral pedicle fracture disconnects the posterior arch from the vertebral body of the axis, making the spine unstable at this level. Neurologic injury is uncommon except with facet dislocation. Most hangman's fractures can be reduced with traction, and they heal after immobilization in a cast or brace. With facet dislocation, anterior fusion may be necessary because of the instability of the posterior arch.

Vertebral Body Fractures

An avulsion fracture from the anteroinferior margin is the most common fracture of the body of the axis (Fig. 15-17). It is thought to be the result of a hyperextension. Its main importance is its high association with other injuries of the cervical spine.

Jefferson Fracture

The classic Jefferson fracture complex consists of four fractures: two fractures on each side involving the anterior and posterior arches of C1. It is a burst fracture of the

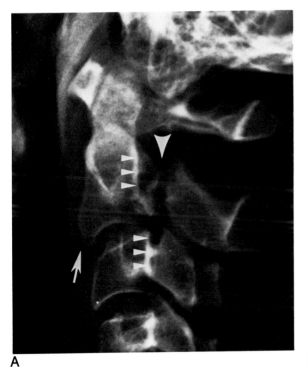

A

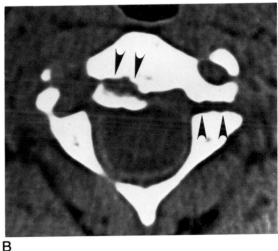

B

Fig. 15-17 Hangman's fracture. (A) Lateral radiograph shows fractures through the pedicles (*large white arrowhead*) and anterior displacement of the C2 vertebral body (*small arrowheads*). A small avulsion fracture is also present (*arrow*). **(B)** Transaxial CT scan shows fracture through the left pedicle and the base of the right pedicle within the vertebral body (*arrowheads*).

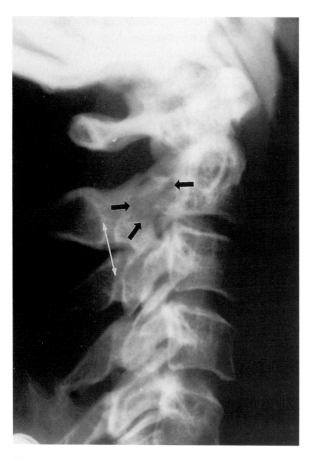

Fig. 15-18 Hangman's fracture, subtle. Crossfire lateral C-spine radiograph shows only subtle irregular lucency of the lamina of C2 (*black arrows*) and slight posterior displacement of the spinolaminar line of C2 (*white opposing arrows*). There is no offset of the vertebral bodies.

atlas that results from an axial compressive force transmitted through the occipital condyles to the lateral masses. A uniformly distributed compressive force produces symmetric fractures of the ring, allowing the ring to expand equally in all directions. Both lateral masses are displaced laterally about the same distance. If the force is strong enough, the transverse ligament is ruptured. A compressive force distributed more to one side fractures the anterior and posterior ring on that side only, producing more displacement of one lateral mass. This injury is a common variation of the Jefferson-type fracture, sometimes referred to as a lateral mass fracture.

The through-the-mouth AP projection is the most useful plain film examination. With the true Jefferson fracture the lateral margins of both lateral masses overhang the lateral margins of both superior articular surfaces of the axis (Fig. 15-19). When the total lateral displacement—the sum of the overhang of each lateral mass—is 9 mm or more, the transverse ligament is also ruptured. A controlled flexion view in the lateral projection may be necessary to define ligament stability. Unilateral displacement from unilateral fracture of a lateral mass may be difficult to diagnose by plain film. CT scans are diagnostic of fractures of the atlas (Fig. 15-19) and are obtained in most cases, particularly when the plain film diagnosis is uncertain. A comminuted fracture of a lateral mass may cause disruption of the transverse ligament, as well as instability.

When associated with disruption of the transverse ligament, the Jefferson fracture is unstable; otherwise it is a stable injury. Neurologic injury is infrequent. Vertebral artery damage may result from the lateral displacement. Traction may reduce the separation of the fragments. Immobilization usually results in healing.

Posterior Arch Fracture

Isolated posterior arch fracture of C1 usually occurs just posterior to the transverse process. These fractures result from hyperextension, crushing the thin posterior ring between the occiput and the large spinous process of the axis. The fracture may be unilateral or bilateral. The diagnosis can be made with the lateral radiograph or CT scanning (Fig. 15-20).

The fracture is stable and heals well with collar immobilization. There is a frequent association with hyperextension fractures at lower levels, particularly type II fracture of the dens.

Anterior Arch Fracture

An anterior arch fracture of C1 may occur with hyperextension. The dens remains intact, but the anterior arch fractures in two places. The fracture fragment remains attached to the dens, but the stability of C1–C2 is destroyed in extension. The fracture may be difficult to recognize on supine plain films as the spine may be in alignment anteriorly. The only finding may be posterior position of the laminar line of the posterior arch with relation to the laminar of C2 below. CT scan demonstrates the fracture of the anterior arch.

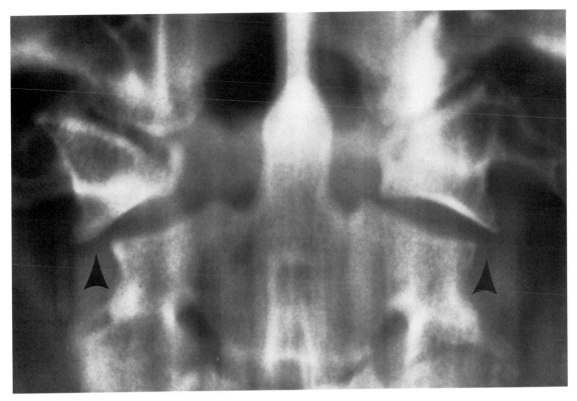

A

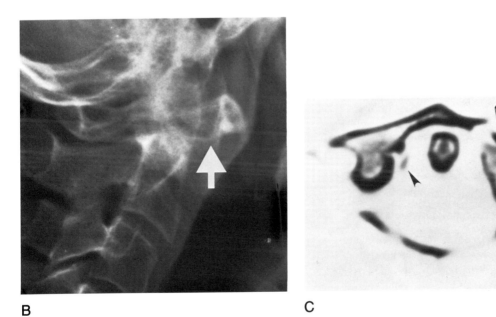

B   C

Fig. 15-19 Jefferson fracture of C1. (A) AP view through the mouth shows bilateral lateral offset of the lateral masses of C1 *(arrowheads)*. **(B)** Lateral view shows widening of the predental space, indicating rupture of the transverse ligament *(arrow)*. **(C)** Transaxial CT scan shows fracture through the left anterior ring *(arrow)*. It is a unilateral Jefferson fracture. Note avulsion of the transverse ligament *(arrowhead)*.

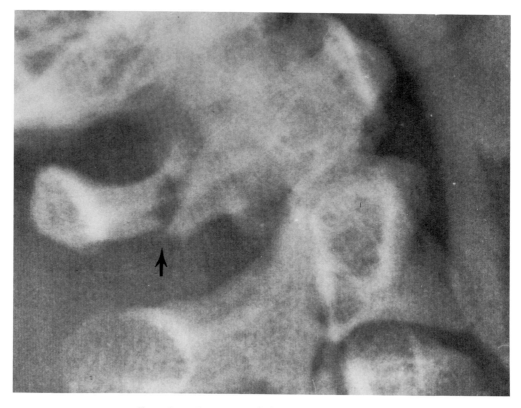

Fig. 15-20 Posterior arch fracture of C1 *(arrow)*.

Horizontal Fracture of the Atlas

The horizontal fracture of the atlas is also produced by hyperextension and can be considered a variant of an avulsion-type fracture. The avulsion probably results from the pull of the anterior longitudinal ligament and the longus colli muscles. The fracture itself has little importance but is highly associated with other hyperextension injuries of the axis or lower cervical spine.

Atlanto-occipital (Craniovertebral) Dissociation

Separation of the cranium from the spine is a result of severe shear and is almost always immediately fatal. The medulla is usually transected. All surviving persons have neurologic damage. Almost all cases are of the anterior-type dislocation; that is, the skull is anteriorly displaced off the spine (Fig. 15-21). A large local hematoma is a constant finding.

The most accurate measurement for determination of this injury is the basion-axial interval (BAI) and the basion-dental interval (BDI), as determined on supine lateral cervical spine radiographs. The BAI is determined by drawing a line extending the posterior cortical margin of the axis superiorly, and measuring the distance of the basion from this line. The limits are 12 mm anterior, to 4 mm posterior to the line. The BDI is the straight line distance from the tip of the odontoid to the basion. This distance should not exceed 12 mm (Fig. 15-22). Measurements that are greater than normal indicate atlanto-occipital dissociation (Harris et al., 1994a). This method is superior to the Powers method of measurement that was previously used.

THORACIC AND LUMBAR SPINE INJURIES

Compression Fractures

Anterior compression fracture is the most common fracture of the thoracic and lumbar spine in all age ranges. As with all injuries of the lower spine, it occurs most frequently at the thoracolumbar junction.

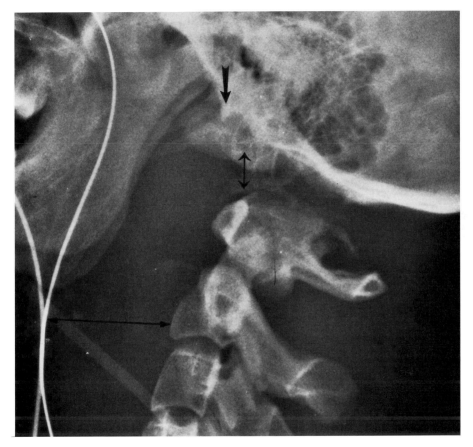

Fig. 15-21 Atlanto-occipital dislocation. Lateral radiograph shows anterior displacement of the skull with relation to the cervical spine. There is wide separation between the base of the clivus *(black arrow)* and the anterior arch of C1 *(vertical opposed arrows)*. A large prevertebral hematoma is a constant finding with this injury *(horizontal opposed arrows)*.

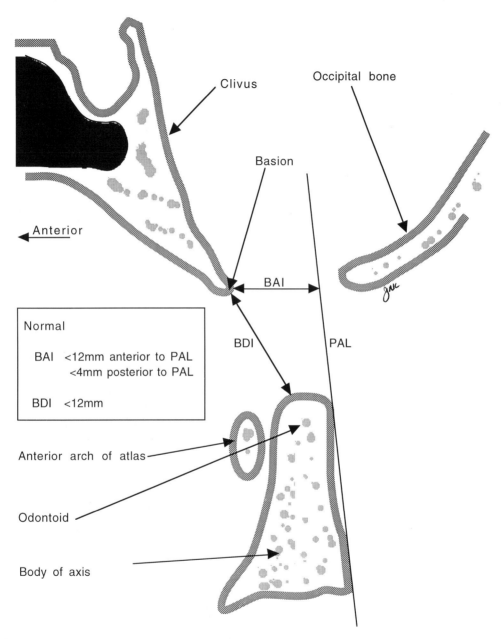

Fig. 15-22 Craniovertebral dissociation measurements. The basion–axial interval (BAI) and the basion–dental interval (BDI). These measurements determine abnormal relationship of the calvarium with the spine. The basion may be forward or back of the posterior axial line (PAL). Normal measurements are given in the box. (Adapted from Harris et al., 1994a, with permission.)

The injury results from a compressive force applied with the spine held in flexion. Because the flexion motion pivots about a fulcrum centered at the nucleus pulposus, it is the anterior portion of the vertebral body and intervertebral disc that suffers the greatest force. The anterosuperior portion of the vertebral body is the first to fracture, usually seen as a buckling of the cortex. With increasing force, there is increasing compression of the vertebral body into a wedge shape. With severe injury, the picture is one of near-total flattening of the anterior body and anterior angulation of the spine. Any herniation of the nucleus is almost always anterior. In the lumbar spine, the large disc space allows the upper vertebral segment to slide forward, producing a shearing force on the vertebral disc. This sequence may cause avulsion of the superior endplate of the vertebral body below (Fig. 15-23).

It may be difficult to determine the age of a vertebral compression fracture. It is generally not possible to be more specific than classifying the fracture as either "recent" or "remote." Healing takes about 3 months and results in loss of the zone of bone condensation and smoothing of the margins of avulsions or cortical breaks. MRI shows loss or focal decrease of the normally high fat signal of the marrow on T_1-weighted images.

Uncomplicated compression fractures are stable. However, hyperflexion with a disruptive force applied posteriorly may rupture the ligamentous complex, recognized by widening of the interspinous distance. Occasionally, a fractured vertebral body will sustain a delayed collapse, resulting in a significant kyphotic gibbus. Therefore, follow-up radiographs are necessary 3 months after the injury.

A vertebral compression fracture must be differentiated from a pathologic fracture (see section on metastatic disease to the spine, Ch. 14). A compression fracture in persons under 50 years of age without a clear history of appropriate trauma is likely to be a pathologic

Fig. 15-23 Compression fracture. Lateral radiograph showing preservation of the posterior height of the vertebral body (*opposed arrows*) and anterior avulsion (*arrow*). The anterior vertebral body is compressed from above.

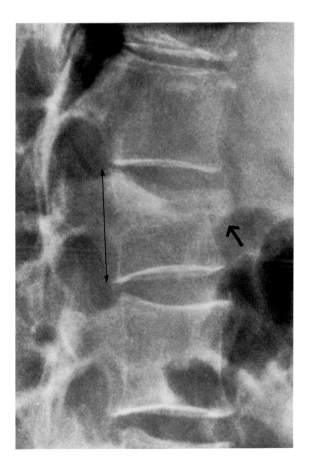

fracture, and tumor may be found in as many as 50 percent of these cases. In older persons with osteoporosis, compression fractures are much more commonly the result of minor trauma; in these patients the incidence of pathologic fracture decreases to about 20 percent.

Fracture-Dislocation

Almost all fracture-dislocations of the thoracic and lumbar spine are produced by hyperflexion force, alone or with rotation, compression, or shear. Because of the inherent strength of the thoracic and lumbar regions of the spine, dislocations are the result of severe forces. The fractures tend to involve multiple levels and have a high association with severe neurologic damage (Fig. 15-24).

The same principles of hyperflexion injury discussed for injuries of the cervical spine apply to injuries of the lumbar spine. There is anterior compression and posterior distraction. Stability of the spine is determined by the integrity of the posterior ligament complex and the posterior longitudinal ligament (the third column of Denis). This complex is disrupted when there is widening of the interspinous distance, displacement of the vertebral body, or widening of the posterior disc space. If a rotational force is involved, there may be unilateral offset of the facets and a characteristic "slice"-type avulsion fracture of the superior endplate of the subjacent vertebral body. A shearing force may result in fracture of the pars interarticularis.

The injury may be difficult to detect radiographically, as a supine position of the patient tends to relocate the spine into a near-normal alignment. Subtle displacements may be the only plain film clue to the presence of a severe fracture. CT scanning defines the injuries.

Seatbelt-Type Injury

The seatbelt-type injury is similar to the hyperflexion injury, but in this case the fulcrum of the flexion motion is more anteriorly positioned in the body. This situation

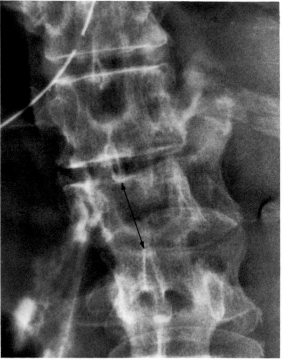

A

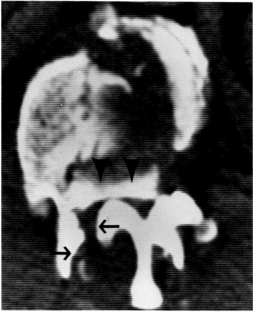

B

Fig. 15-24 Fracture-dislocation of T12–L1. (A) AP radiograph showing displacement with compression. There is wide separation between the spinous processes posteriorly *(opposed arrows)*. **(B)** Transaxial CT scan shows dislocation of the facets *(arrows)*, avulsion of the left anterolateral vertebral body, burst fracture, and retropulsion into the spinal canal *(black arrowheads)*.

produces more of a distractive force posteriorly and little compression of the vertebral bodies. The injury occurs not only with lap-type seatbelts, but with any restraint that allows the upper body to be thrown into acute hyperflexion over the restraint. For example, the injury occurs when a person riding a snowmobile or trail bike encounters a waist-high fence wire or rail.

The spine is ripped apart by the bending disruptive force. The nature of the fracture depends on the outcome of the tension between ligaments and bone. There may be only ligamentous rupture with intact bone, or there may be varied horizontal fractures through the posterior arch and vertebral body. Two types of fracture are common.

1. *Smith fracture*. This horizontal fracture goes through the transverse process, extends into the pedicle and pars interarticularis, and goes forward into the posterosuperior vertebral body, where a small fragment is avulsed from the posterosuperior endplate.
2. *Slice ("Chance") fracture*. There is complete horizontal splitting of the posterior arch, extending into the superior portion of the vertebral body, often curving upward to reach the endplate. The fracture goes posteriorly through the spinous process (Fig. 15-25).

Burst Fracture

The burst fracture is a result of axial compression. The mechanism is the same as that described for fracture of the cervical spine. Upward forcing of disc material into the vertebral body causes it to split. Retropulsion of the posterior fragment of the vertebral body is common and causes neurologic compression (Figs. 15-26 and 15-27). More than 25 percent compromise of the spinal canal is

Fig. 15-25 Slice fracture. A horizontal fracture extends through the posterior arch into the spinous process *(arrow)*.

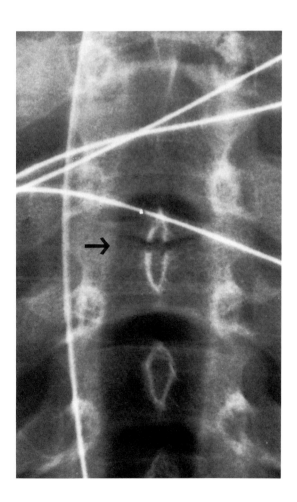

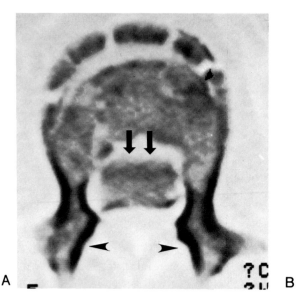

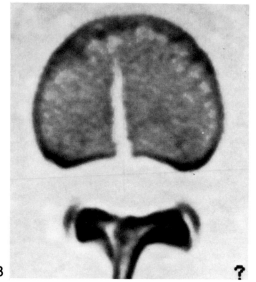

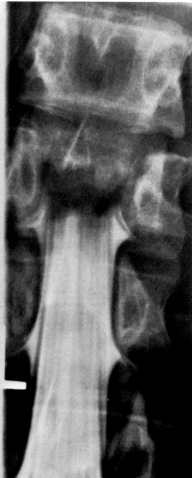

Fig. 15-26 Burst fracture, lumbar spine. (A) Trans-axial CT scan shows typical retropulsion of a fragment into the spinal canal *(arrows)*. Note the non-visualization of the inferior facets from the level above, indicating facet joint dislocation *(arrowheads)*. **(B)** Lower view of vertebral body shows the vertical burst fracture. **(C)** Myelography showing total extradural type block by the retropulsed fragment.

Fig. 15-27 Burst fracture of L1. AP radiograph shows loss of height and widening of the L1 vertebral body *(white arrows)*. Fracture can be seen extending through the lamina posteriorly on the right *(black arrow)*. These subtle changes must be recognized, as this view is frequently the only one available of the acutely traumatized patient.

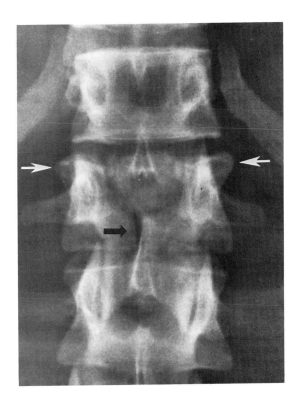

considered significant. Frequently, there is fracture of the posterior arch. Once thought to be a stable injury, the fracture is now considered unstable. Treatment is directed at removing fragments within the spinal canal and stabilizing the spine to avoid later collapse.

HYPEREXTENSION FRACTURE-DISLOCATION

Hyperextension fracture-dislocation is a rare injury that consists of fracture of the pars interarticularis or lamina. It leads to spondylolisthesis.

INJURIES WITHOUT FRACTURE OR DISLOCATION

Injury of the spinal cord may occur without disruption of the spine. These injuries cannot be diagnosed with plain film radiography or routine CT scanning but require MRI, myelography, or CT myelography.

Nerve Root Avulsion

With injuries of the upper extremities or pelvis, avulsion of nerve roots from the cervical or sacral region may occur as a result of distraction of the nerve roots. The roots are pulled from their origins in the cord in the cervical region or are torn from the cauda equina in the lumbosacral region. Usually there is tearing of the nerve root sleeve from the dural sac so a pseudomeningocele is formed.

Myelography, CT myelography, or MRI will make the diagnosis (Figs. 15-28 and 15-29). The myelographic contrast outlines the disruption of the nerve root sleeve. Sometimes the avulsion of the nerve root from the cord can be demonstrated.

Spinal Cord Contusion

Contusion or hemorrhage may occur within the spinal cord, producing a central cord syndrome (Figs. 15-30 and 15-31). It may occur without fracture or dislocation of the spine and is seen primarily in patients with spondylosis. The hypermobility of the spine and osteophytes

MAJOR DELAYED POST-TRAUMATIC COMPLICATIONS

Delayed malalignment of an injury thought to be stable

Failure of healing with malalignment

Unrecognized displaced fracture fragment that later results in neurologic dysfunction (Fig. 15-26)

Post-traumatic syringomyelia (Fig. 15-27)

Post-traumatic arachnoiditis (Fig. 15-28)

Occlusion or pseudoaneurysm of the vertebral artery

Post-traumatic atrophy of the spinal cord (Fig. 15-29)

compress the cord, causing the hemorrhage or contusion. This problem is best demonstrated with MRI, which shows the signal intensity of edema or hemorrhage within a swollen cord. Myelography or CT myelography shows only the cord swelling. If spinal stenosis is present, decompression sometimes produces improvement.

Acute Post-Traumatic Disc Herniation

Acute disc herniation may result from trauma. It causes pain or neurologic dysfunction, with little or no radiographic change seen. MRI is the best means of demonstrating the herniation (Fig. 15-30).

Delayed Complications of Fracture and Dislocations

Most delayed problems due to injury present with persistent or new symptoms following the injury. Focal radiculopathy suggests a change in position or a displaced bone fragment (Fig. 15-32), progressive myelopathy suggests syringomyelia (Fig. 15-33), and arachnoiditis causes pain (Fig. 15-34). Cord atrophy may be seen (Fig. 15-35). Plain film radiography, CT scanning, and MRI can

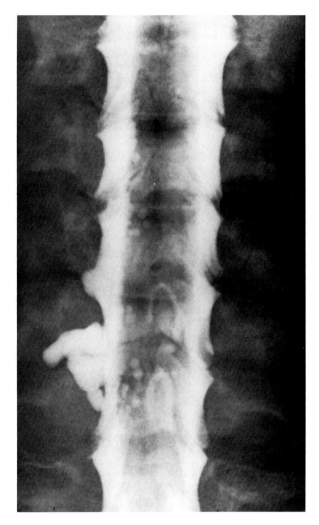

Fig. 15-28 Torn dural sac. This study followed an avulsion injury of the cervical nerve roots. A pseudomeningocele is demonstrated with a Pantopaque myelogram. (From Osborne, 1984, with permission.)

be used appropriately to make the diagnosis. Angiography of the vertebral artery may be necessary if vascular injury is suspected, although MRI may determine vessel occlusion or dissection.

MRI and Myelography in Acute Spinal Injury

A complete spinal cord injury is present when the patient has no neurologic function below the level of injury. Little can be done to alleviate this situation except
(Text continued on page 719)

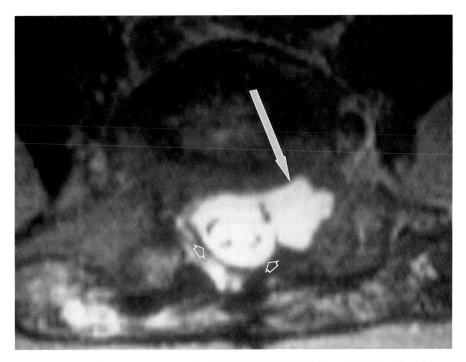

Fig. 15-29 Pseudomeningocele, post-traumatic. Axial SE 90/2500 T2 MRI through L5 identifies a large pseudomeningocele containing CSF extending laterally into the left intervertebral foramen (*arrow*). This was caused by root avulsion from an injury that also severely fractured the pelvis. The normal dural outline is seen away from the meningocele (*arrowheads*).

Fig. 15-30 Acute post-traumatic disc herniation with cord hemorrhage. MRI 1 week after injury shows disc herniation with increased intensity within the cord (*arrow*) on T_1-weighted image. The cord is swollen. (SE 600/20)

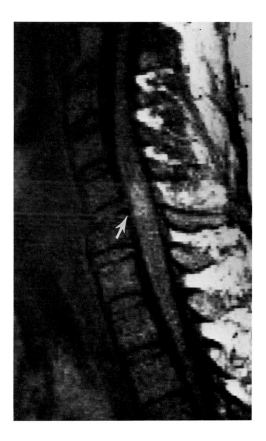

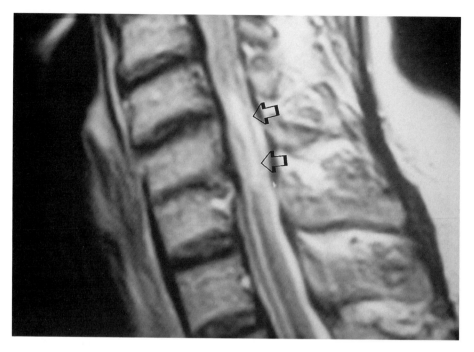

Fig. 15-31 Spinal cord contusion. Sagittal T$_2$ MRI identifies central high signal intensity within the spinal cord, representing acute edema *(open arrows)*. This occurred from injury without fracture in a patient with spondylosis.

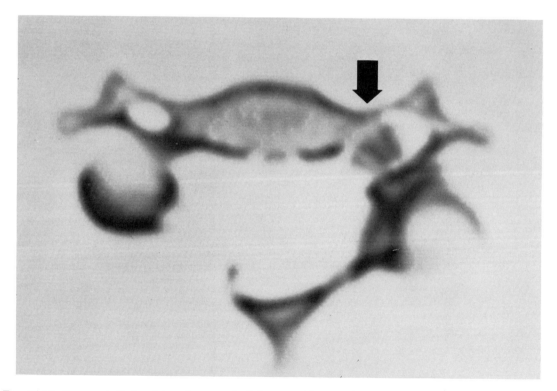

Fig. 15-32 Fragment displaced into intervertebral foramen. Transaxial CT scan easily shows a fragment in the foramen, causing radiculopathy *(arrow)*. Such fragments can also cause vertebral artery injury.

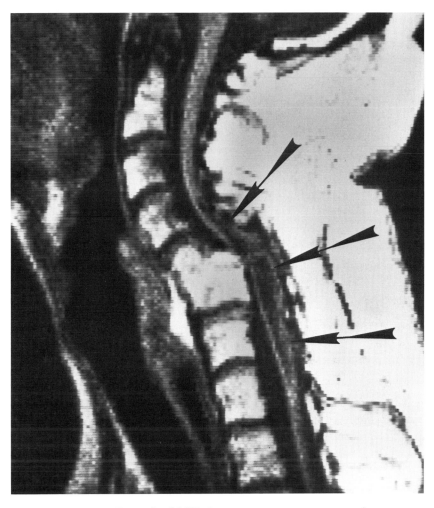

Fig. 15-33 Post-traumatic syrinx. T_1-weighted MRI shows post-traumatic syringomyelia as a region of low signal within the cord (*arrows*). Expansion of the cyst can lead to progressive neurologic dysfunction.

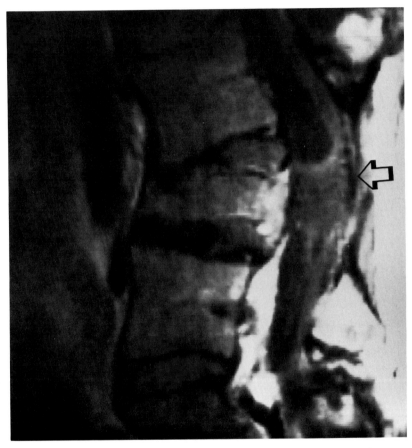

Fig. 15-34 Post-traumatic arachnoiditis. MRI T_1-weighted image shows increased signal intensity in the region normally occupied by the CSF within the dura *(arrow)*.

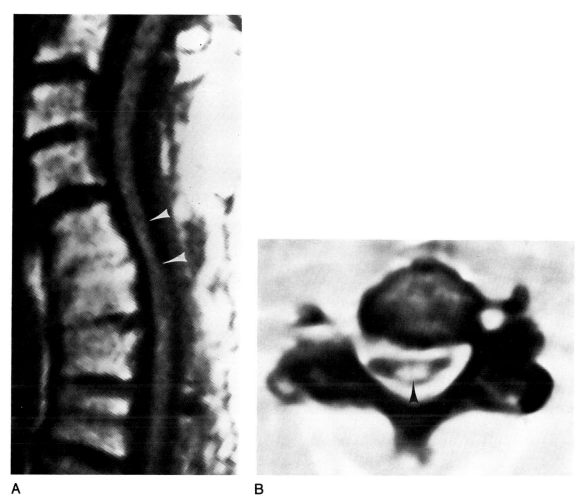

A   B

Fig. 15-35 Post-traumatic cord atrophy. (A) MRI examination shows a small cord through the region of the fracture site *(arrowheads)*. **(B)** Transaxial CT myelography shows the small size of the cervical cord *(arrowhead)*.

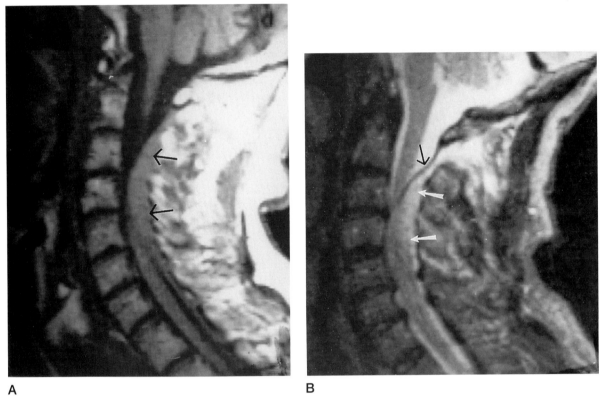

A   B

Fig. 15-36 Spinal epidural hematoma. (A) Sagittal T_1 MRI of the cervical spine identifies an early subacute hematoma as a mass in the posterior spinal canal of slight hyperintensity *(black arrows)*. This occurred spontaneously after strenuous dancing. **(B)** The sagittal T_2 MRI shows the hematoma as slightly hyperintense *(white arrows)*. The dura is displaced anteriorly *(black arrow)*. In the hyperacute phase, the signal characteristics are those of a high protein fluid and cannot be differentiated from an epidural abscess.

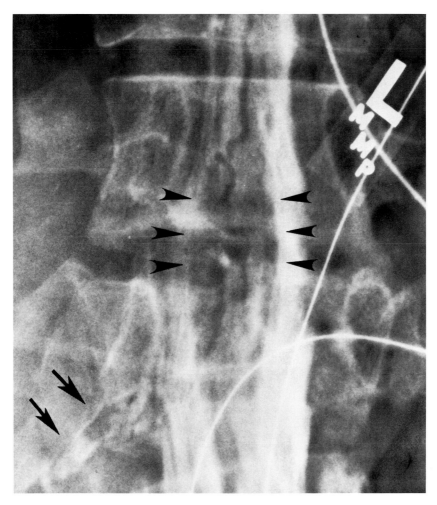

Fig. 15-37 Hematomyelia and dural tear. Myelography shows widening of the conus medullaris *(arrowheads)*. In addition, there is extravasation of contrast from the dural tear *(arrows)*.

to prevent ascending neurologic decline by stabilizing the bone injury and providing good physiologic supportive measures. MRI and myelography are unnecessary.

With incomplete spinal cord injury, decompression and stabilization can result in neurologic improvement. MRI and myelography may be of some benefit in this group. The primary aim is to identify mechanical cord compression caused by spinal stenosis, displaced fragments of bone, disc herniation, or epidural hematoma (Fig. 15-36), that is, extramedullary compressive lesions. Relief of the cord compression may lead to partial re-

covery of function and may prevent additional neurologic decline. If an intramedullary hematoma is found, no benefit is accrued by its decompression (Fig. 15-37).

MRI scans may provide prognostic information after injury, as the finding of cord edema, rather than hemorrhage, implies a greater chance for neurologic recovery. During the acute phase (less than 7 days), MRI shows both cord edema and hemorrhage as low intensity on the T_1-weighted images, but only edema produces high signal on the T_2-weighted images. GRE images also show acute hemorrhage as low signal. After a few days to a week, hemorrhage within the cord is demonstrated as high signal intensity on both T_1- and T_2-weighted images (Fig. 15-30).

SUGGESTED READING

Atlas SW, Regenbogen V, Rogers LF, Kim KS: The radiographic characterization of burst fractures of the spine. AJNR 7:675, 1986

Cervical Spine Research Society: The Cervical Spine. 2nd Ed. JB Lippincott, Philadelphia, 1989

Denis F: Updated classification of thoracolumbar fractures. Orthop Trans 6:8, 1982

Donovan-Post MJ, Green BA: The use of computed tomography in spinal trauma. Radiol Clin North Am 21:327, 1983

Donovan-Post MJ: Computed tomography of spinal trauma. In: Computed Tomography of the Spine. Williams & Wilkins, Baltimore, 1984

Edeiken-Monroe B, Wagner LK, Harris JH Jr: Hyperextension dislocation as the cervical spine. AJNR 7:135, 1986

Gehweiler JA Jr, Osborne RL, Becker RF: The Radiology of Vertebral Trauma. WB Saunders, Philadelphia, 1980

Harris JH Jr, Edeiken-Monroe B: Radiology of Acute Cervical Spine Trauma. 2nd Ed. Williams & Wilkins, Baltimore, 1987

Harris JH, Carson GC, Wagner LK: Radiologic diagnosis of traumatic occipitovertebral dissociation. 1. Normal occipitovertebral relationships on lateral radiographs of supine subjects. AJR 162:881, 1994a

Harris JH, Carson GC, Wagner LK, Kerr N: Radiologic diagnosis of traumatic occipitovertebral dissociation. 2. Comparison of three methods of detecting occipitovertebral relationships on lateral radiographs of supine subjects. AJR 162:887, 1994b

Holdsworth FW: Fractures, dislocations and fracture-dislocations of the spine. J Bone Joint Surg Am 52:1534, 1970

Holdsworth FW: Fractures, dislocations and fracture-dislocations of the spine. J Bone Joint Surg Br 45:6, 1963

Kim KS, Chen HH, Russell EJ, Rogers LF: Flexion teardrop fracture of the cervical spine: radiographic characteristics. AJNR 9:1221, 1988

Kowalski HM, Cohen WA, Cooper P, Wisoff JH: Pitfalls in the CT diagnosis of atlantoaxial rotatory subluxation. AJNR 8:697, 1987

Kulkarni MV, Bondurant FJ, Rose SL et al: 1.5 Tesla magnetic resonance imaging of acute spinal trauma. Radiographics 8:1059, 1988

Osborne D: Spinal trauma. p. 887. In Heinz ER (ed): Neuroradiology. Churchill Livingstone, New York, 1984

Quencer RM: The injured spinal cord. Radiol Clin North Am 26:1025, 1988

White AA, Panjabi MM: Clinical Biomechanics of the Spine. JB Lippincott, Philadelphia, 1978

Willis BK, Greiner F, Orrison WW et al: The incidence of vertebral artery injury after midcervical spine fracture or subluxation. Neurosurgery 34:435, 1994

Yamashita Y, Takahashi M, Matsuno Y et al: Acute spinal cord injury: magnetic resonance imaging correlated with myelopathy. Br J Radiol 64:201, 1991

Appendix

Abbreviations

ACA	Anterior cerebral artery		PCoA	Posterior communicating artery
ACoA	Anterior communicating artery		PCr	Phosphocreatine
ADC	Apparent diffusion coefficient		PD	Proton density
ADL	Adrenoleukodystrophy		PET	Positron emission tomography
AFP	α-fetal protein		PHPV	Persistent hyperplastic primary vitreous
AIDS	Acquired immunodeficiency syndrome		PML	Progressive multifocal leukoencephalopathy
AVM	Arteriovenous malformation		PNET	Primitive neuroectodermal tumor
BA	Basilar artery		PNS	Peripheral nervous system
CJD	Creutzfeldt-Jakob disease		PVL	Periventricular leukomalacia
CMV	Cytomegalovirus		rem	Roentgen equivalent in man (radiation dosage)
DAVF	Dural arteriovenous fistula		REZ	Root entry zone
DSA	Digital subtraction angiography		RIND	Reversible ischemic neurologic deficit
ECA	External carotid artery		RN	Radionuclide
ELISA	Enzyme-linked immunosorbent assay		SAE	Subcortical arteriosclerotic encephalopathy
EOMs	Extraocular muscles		SAH	Subarachoid hemorrhage
FSE	Fast spin echo		SBE	Subacute bacterial endocarditis
GCT	Germ cell tumor		SDH	Subdural hematoma
GRE	Gradient echo recalled MRI		SDAT	Senile dementia of the Alzheimer type
hCG	Human chorionic gonadotropin		SE	Spin echo MRI
HIE	Hypoxic/ischemic encephalopathy		SIE	Stroke in evolution
HNP	Herniated nucleus pulposis		SPECT	Single photon emission computed tomography
IAC	Internal auditory canal		SN	Substantia nigra
ICA	Internal carotid artery		S/N	Signal-to-noise ratio
IR	Inversion recovery		TCD	Transcranial Doppler
IOIS	Idiopathic orbital inflammatory syndrome		TE	Time of echo
IVH	Intraventricular hemorrhage		TIA	Transient ischemic attack
MCA	Middle cerebral artery		TOF	Time-of-flight
MIP	Maximum intensity pixel projection		tPA	Tissue plasminogen activator
MOTSA	Multiple overlapping thin slab acquisition		TR	Time of repetition of MRI sequence
MRA	Magnetic resonance angiography		TSC	Tuberous sclerosis complex
MTC	Magnetization transfer contrast (technique)		UBO	Unidentified bright object (T_2 MRI)
NF-1	Neurofibromatosis, type I		VB	Vertebral basilar
NF-2	Neurofibromatosis, type II		XRA	X-ray angiography
NGGCT	Non-germinoma germ cell tumor			
NPH	Normal-pressure hydrocephalus			
PAIDS	Pediatric AIDS			
PCA	Posterior cerebral artery			

Index

Note: Page numbers followed by *f* indicate figures; page numbers followed by *t* indicate tables.

723